Adhesives and Finishes for Wood

Adhesives and Finishes for Wood

For Practitioners and Students

Moon G. Kim

WILEY-VCH

Author

Prof. Moon G. Kim
5517 Paseo del Lago, East
Laguna Woods
CA
United States of America

Cover Image: © optimarc/Shutterstock

Library of Congress Card No.: applied for

British Library Cataloguing-in-Publication Data
A catalogue record for this book is available from the British Library.

Bibliographic information published by the Deutsche Nationalbibliothek
The Deutsche Nationalbibliothek lists this publication in the Deutsche Nationalbibliografie; detailed bibliographic data are available on the Internet at <http://dnb.d-nb.de>.

Print ISBN: 978-3-527-35236-4
ePDF ISBN: 978-3-527-84276-6
ePub ISBN: 978-3-527-84277-3
oBook ISBN: 978-3-527-84278-0

Typesetting Straive, Chennai, India
Printing and Binding CPI Group (UK) Ltd, Croydon, CR0 4YY

C9783527352364_310723

Contents

Preface *xix*
Author Biography *xxi*

1 **Introduction to Polymers, Wood Adhesives, and Wood Finishes** *1*
1.1 Good Wood Adhesives Must Be Optimum Polymers with Optimum Secondary Forces *1*
1.2 Polymeric Materials *2*
1.3 Synthetic Polymer Preparation Methods *2*
1.4 Typical Synthetic Polymer Materials *3*
1.5 Typical Natural Polymers *8*
1.6 Summary *10*
References *11*

2 **Principles of Polymer Chemistry for Wood Adhesives and Finishes** *12*
2.1 Degree of Polymerization and Molecular Weight *12*
2.1.1 Determination of Molecular Weight or Degree of Polymerization *12*
2.2 Properties of Polymer Solutions or Suspensions *13*
2.2.1 Viscosity of Polymer Solutions *13*
2.2.2 Dependence of Viscosity on the Shear Rate (Stirring Rate) *14*
2.2.3 Temperature Dependence of Viscosity of Polymer Solutions *15*
2.2.4 Dependence of Viscosity on the Concentration and Molecular Weight of Dissolved Polymers *15*
2.2.5 Methods of Viscosity Measurement of Liquid Materials *17*
2.3 Polymer Solids Level and Specific Gravity of Polymer Solutions *18*
2.4 pH of Polymer Solutions and Buffers *18*
2.4.1 pH *18*
2.4.2 Acidic Buffers and Buffer Capacity *19*
2.5 Solid Properties of Polymer Materials *20*
2.5.1 Adhesive Property of Polymer Solids *20*
2.5.2 Cohesive Property of Adhesive Polymer Solids *20*
2.5.3 Cohesive Strength Measurement and Viscous and Elastic Responses *20*
2.5.4 Measurement of the Elasticity of Solid Polymer Materials *21*
References *22*

3 Thermosetting and Thermoplastic Wood Adhesives and Practices *23*
3.1 Selection Criteria of Wood Adhesives *23*
3.2 High Temperature-Curing Thermosetting Adhesives and Curing Conditions *24*
3.3 Room Temperature-curing Thermosetting Wood Adhesives and Processes *25*
3.4 Room Temperature-curing Thermoplastic Wood Adhesives *25*
3.5 Adhesive Application Methods and Loading Rates and Costs *26*
3.6 Adhesive Curing by Hot Presses and Other Heating Methods *26*
3.7 Evaluation of Adhesive Bonds *28*
3.8 Summary *28*
References *28*

4 Principles of the Curing of Thermosetting and Thermoplastic Wood Adhesives *29*
4.1 Principles of Curing of Thermosetting Adhesive Resins at Elevated Temperatures *29*
4.1.1 Viscosity Changes of Adhesive Layer Before Hot-Pressing *30*
4.1.2 Spreading and Penetration of Adhesive Layers in Hot-Pressing *30*
4.1.3 Viscosity Decreases by Moisture Content Changes *31*
4.1.4 Viscosity Increases and Curing of Adhesive Layer by Polymerization Reaction *31*
4.1.5 Flow and Wetting of Adhesive Layers and Molecular Adsorption *31*
4.1.6 Gelation of Adhesive Layer *32*
4.1.7 Vitrification of Adhesive Layer *32*
4.1.8 Post-Curing of Adhesive Layer *33*
4.2 Relationship of Temperature and Moisture Content in Hot-Pressing of Wood Composite Mats *33*
4.2.1 Platen Pressure, Mat Thickness, and Density Profiles *33*
4.2.2 Temperature Changes and Moisture Movements in Board Mat *34*
4.2.3 Extents of Resin Cure in Hot-Pressing *35*
4.3 Curing of Thermosetting Adhesives at Room Temperature *35*
4.4 Curing of Thermoplastic Emulsion Adhesives *36*
4.5 Volume Contractions of Adhesive Layers upon Curing *37*
4.6 Thermal and Moisture Expansion/Contraction of Adhesive Layers *38*
4.7 Summary *38*
References *39*

5 UF and MUF Wood Adhesive Resins (Manufacturing and Resin Chemistry) *40*
5.1 Raw Materials of UF Resins *40*
5.2 Urea-Formaldehyde (UF) Resins *42*
5.3 Chemistry Occurring in UF Resin Synthesis *46*
5.3.1 Second Step of Resin Synthesis *47*
5.4 Polymer Chain Branching in UF Resins *50*
5.5 Other Reactions Occurring in UF Resin Synthesis *52*
5.6 F/U_1 Mole Ratio Effects in the Polymerization Step *53*
5.7 Final F/U Mole Ratios and Formaldehyde Emission Problem *54*

5.8 Physical and Chemical Tests and Properties of Industrial UF Resins *55*
5.9 Resin Changes Occurring After Manufacture of UF Resins *58*
5.10 Bond Performances and Durability Tests of UF Resins in Industry *60*
5.11 UF Resins vs. Various Operating Parameters in PB, MDF, and Hardwood Plywood Industry *61*
5.12 Other UF Resin Synthesis Procedures *62*
5.13 Polymeric Chemical Structures of UF Resins *63*
5.14 Melamine-Urea-Formaldehyde (MUF) Resins *63*
5.15 Urea-Melamine-Formaldehyde (UMF) Resins *66*
5.16 Summary *69*
References *69*

6 Urea-Formaldehyde and MUF/UMF Wood Adhesive Resins (Curing) *72*
6.1 Typical Latent/External Catalysts Based on Ammonium Salts for UF Resins *72*
6.2 Organic Tertiary Amine Salts of Strong Acids and Other Catalysts *73*
6.3 Catalyzation of UF Resins by Wood Acids *74*
6.4 Buffering of Acidic Catalysts *77*
6.5 Thermosetting Curing Reactions, Cured Resin Structures, and Formaldehyde Emission Problem *78*
6.6 Control of the Curing Speeds of UF Resins *81*
6.7 Side-Reactions Occurring in the Curing of UF Resins *81*
6.8 Phenomenological Changes of UF Resins in Curing *82*
6.9 Chemical Curing Mechanisms, F/U Mole Ratio, and Cured Resin Structure *84*
6.10 Composition of Cured UF Resins *85*
6.11 Uses of UF Resins *86*
6.12 Properties of UF Resin-Bonded Wood Composite Boards *86*
6.13 Formaldehyde Emission Mechanisms, Mole Ratios, and Board Strength Properties *87*
6.14 Various Methods Proposed/Practiced for Reducing the Formaldehyde Emissions of Boards *88*
6.15 Curing of Melamine-Urea-Formaldehyde (MUF) Resins *89*
6.16 Curing of Urea-Melamine-Formaldehyde (UMF) Resins *90*
6.17 Cost Increases Estimated for Boards with Very Low Formaldehyde Emission Values *91*
6.18 Summary *92*
References *92*

7 Particleboard, MDF, and Hardwood Plywood Bonding with UF Resin Binders *94*
7.1 Particleboard *94*
7.1.1 Wood Furnishes in Particleboard Manufacture *94*
7.1.2 Blending of Resin, Catalyst, and Wax Emulsion in Particleboard Manufacture *95*
7.1.3 Mat Formation and Transport *97*
7.1.4 Hot-pressing *98*
7.1.5 Equilibration and Finishing *100*

7.1.6 Bison–Mende process *100*
7.1.7 Properties of Commercial Particleboards *100*
7.1.8 US Particleboard (PB) Industry *100*
7.1.9 Summary *101*
7.2 Medium Density Fiberboard (MDF) Bonded with UF Resins *102*
7.2.1 Fiber Manufacture and Blowline Resin Blending *102*
7.2.2 Paddle Blending System of Resins *104*
7.2.3 Comparison of MDF Resin Blenders *105*
7.2.4 Mat-Forming and Hot-Pressing Parameters of PB and MDF in Industry *107*
7.2.5 Physical Properties of Commercial MDF *108*
7.2.6 Medium Density Fiberboard Industry *109*
7.2.7 Summary *109*
7.3 Hardwood Plywood Bonding with UF Resin Adhesives *109*
7.4 Paraffin Wax and Uses in Wood Composites *112*
7.4.1 Use of Paraffin Wax *112*
7.4.2 Properties of Paraffin Wax *112*
7.4.3 Analyses of Wax *113*
7.4.4 Wax Application Methods *114*
7.4.5 Wax Use Levels *114*
7.4.6 Summary *114*
7.5 Effects of UF Resins' Curing Catalysts on Tool Wear in Machining of Boards – An Example of Laboratory PB Manufacturing and Testing *115*
7.6 Summary *120*
References *120*

8 PF Novolac Wood Adhesive Resins (Manufacturing and Chemistry) *122*
8.1 Raw Materials of PF Resins *122*
8.2 Reaction Chemistry of Phenol *123*
8.3 Synthesis and Chemistry of Novolac Phenol-Formaldehyde (PF) Resins *124*
8.4 Compounding and Curing of Novolac PF Resins *129*
8.5 Curing Speed and Testing Methods of Molding Compounds of PF Novolac Resins *130*
8.6 Wood Adhesive Uses of Novolac PF Resins *132*
8.7 Summary *133*
References *134*

9 PF Resole Wood Adhesive Resins (Manufacturing and Chemistry) *136*
9.1 Alkaline PF Resole Wood Adhesive Resins *136*
9.2 Synthesis Chemistry and Manufacturing Practices of Alkaline PF Resole Resins *136*
9.3 Typical Synthesis Procedures of Alkaline PF Resole Resins for Various Uses *140*
9.4 Properties and Polymer Structures of Alkaline PF Resole Resins and Measurements *143*

9.5 Resin Characteristics and Uses of Various PF Resole Resins *147*
9.6 Molecular Weights (Sizes) of Alkaline PF Resole Resins and Wood Cell Wall Penetration *148*
9.7 Powder PF Resole Resin Manufacturing by Spray-drying *149*
9.8 Curing of Alkaline PF Resole Resins *150*
9.8.1 Pre-dry and Pre-cure, Resin Spreading, and Flow *150*
9.8.2 Final Curing of Alkaline PF Resole Resins in Wood Bonding *151*
9.8.3 Gel Times and Stroke Cure Times of Liquid PF Resins *153*
9.8.4 Dynamic Mechanical Analysis (DMA) Method of Curing Tests for Thermosetting Resins *153*
9.8.5 Differential Scanning Calorimetric (DSC) Analysis of PF Resole Resins *157*
9.9 Good Wood Adhesion and Exterior Durability of PF Resin Adhesives *158*
9.10 Formaldehyde Emission Problem of PF Resin-bonded Products *159*
9.11 Various Binder Uses of PF Resins in the US Wood Products Industry *159*
9.12 Acid-curing PF and PMF Resole Resin and Dispersion Wood Adhesives *162*
9.13 Summary *163*
References *163*

10 PRF Novolac Wood Adhesive Resins and Lumber Lamination *166*
10.1 Manufacturing and Chemistry of PRF Resins *167*
10.1.1 Chemical Structures of PRF Resins *169*
10.1.2 Relationship Between Target Viscosity, Resin-Solids, and Storage Stability of PRF Resins *171*
10.1.3 P/R Ratio Measurements from ^{13}C NMR Spectra *172*
10.2 PRF and RF Resin Adhesives Currently Available in Industry *172*
10.2.1 Some Nonideal Aspects of PRF Resins *173*
10.3 Hardeners for PRF Resin Adhesives *174*
10.4 Mixing of a PRF Resin and Hardener, Pot-Lives, Working Life, and Assembly Time *175*
10.5 Applying and Curing of PRF Adhesives in Wood Lamination *176*
10.6 Curing Chemistry of PRF Resin Adhesives and Industry Practices *179*
10.7 Wood Lamination Industry *179*
10.8 Recent Developments in PRF Resin Adhesives *182*
10.9 Laminating Wood Adhesives Based on Other Resorcinolic Materials *184*
10.10 Advanced Reading Materials on Resorcinol–Formaldehyde (RF) Resins *185*
10.10.1 Synthesis and Fractionation of an RF Resin and GPC Analysis *185*
10.10.2 Assignments of ^{13}C NMR Chemical Shift Values of RF Resin Fractions *187*
10.10.3 Polymer Structures of RF Resins Derived from ^{13}C NMR Spectra *188*
10.10.4 Relationship Between Intrinsic Viscosity and Molecular Weight Values from Equation n_3 *190*
10.11 Summary *191*
References *191*

11 Softwood Plywood Adhesives and Manufacturing Technology *194*
11.1 Softwood Plywood Manufacturing Technology *194*
11.2 PF Resole Resins for Bonding of Softwood Plywood *195*

11.3 Veneer, Veneer Drying, and Adhesion Problems *196*
11.4 Bond Performance Characteristics of Softwood Plywood Adhesives *197*
11.5 Softwood Plywood Adhesive Mixing Procedure *197*
11.6 Softwood Plywood Adhesive Formulation and Characteristics *198*
11.6.1 Characteristics of PF Resin Adhesive Mixes *198*
11.6.2 Gap-Filling Capability of Softwood Plywood Adhesives *199*
11.6.3 Viscosity Targets of Softwood Plywood Adhesive Mixes *200*
11.6.4 Fillers in Softwood Plywood Adhesives *200*
11.6.5 Extenders in Softwood Plywood Adhesives *201*
11.6.6 PF Resin Adhesives, Adhesive Spread Rates, and Application Methods *202*
11.7 Open and Closed Assembly Times *204*
11.8 Pre-pressing of Softwood Veneer Assembly *204*
11.9 Hot-Pressing Parameters of Softwood Plywood *204*
11.10 Laminated Veneer Lumber (LVL) *207*
11.11 Parallel (Veneer) Strand Lumber (PSL) *207*
11.12 New Developments in the Softwood Plywood Adhesive Technology *208*
11.13 Phenolic Components Present in PF Resin-Bonded Softwood Plywood *209*
11.14 Softwood Plywood Manufacturing Industry *209*
11.15 Plywood Adhesive Fillers Made from Hydrolysis Residues of Municipal Newsprint Wastes *210*
11.15.1 TVA Fillers and a Control Filler *210*
11.15.2 Plywood Adhesive Mix *211*
11.15.3 Plywood Manufacture *211*
11.15.4 Plywood Test Results *212*
11.16 Conclusion *214*
11.17 Summary *214*
References *214*

12 Isocyanate Wood Adhesive Resins *217*
12.1 Chemical Compositions of Isocyanate Wood Adhesive Resins *217*
12.2 Reactivity, Polymerization, and Curing Reactions of Isocyanate Resins *218*
12.3 Mat Moisture Levels in Using pMDI Resins for OSB Bonding *223*
12.4 Bond Properties and Uses of pMDI Resins as OSB Binders *224*
12.5 pMDI Resins Used in Bonding of Other Wood Composite Products *226*
12.6 Summary *227*
References *227*

13 OSB Manufacturing with PF and Isocyanate Wood Adhesive Resins *229*
13.1 Oriented Strand Board (OSB) Manufacturing Processes *229*
13.1.1 Wood Species Used in OSB Manufacture *229*
13.1.2 Stranding (Flaking) *230*
13.1.3 Drying and Screening of Strands *231*
13.1.4 Binder Resin Types *231*
13.1.5 Blending of Binder Resin and Slack Wax *232*
13.1.6 Mat Forming with Strand Orienters *234*

13.1.7 Hot-Pressing of Mat *235*
13.1.8 Oriented Strand Lumber (OSL) *238*
13.1.9 Adhesive Improvements Needed in the OSB Industry *239*
13.1.10 History of Wafer Board and Flakeboard *239*
13.1.11 Durability and Span Ratings of Oriented Strandboard in Use *239*
13.2 OSB Bonded with PF Resins with Various Levels of Urea Added at the End of Resin Synthesis *240*
13.2.1 Syntheses of PF Resole Resins with Urea Added at the End of Synthesis *241*
13.2.2 Testing of Synthesized Resins *242*
13.2.3 Resin Curing Rates Determined by DMA *242*
13.2.4 Laboratory OSB Manufacturing and Testing *243*
13.2.5 Test Results of Manufactured OSB Panels *244*
13.2.6 Conclusion *246*
13.3 Summary *246*
References *247*

14 Polyvinyl Acetate (PVAc) Emulsion Wood Adhesives *249*
14.1 Polyvinyl Acetate (PVAc) Emulsion Wood Adhesive Resins *249*
14.1.1 Manufacturing of PVAc Emulsion Resins *249*
14.1.2 Properties of PVAc Emulsions and Polymers *250*
14.1.3 Commercial PVAc Emulsions *250*
14.1.4 Additives Used in Formulating PVAc Emulsion Wood Adhesives *251*
14.1.5 Properties of Formulated PVAc Emulsion Wood Adhesives *252*
14.1.6 Various Commercial Formulated PVAc Emulsion Wood Adhesives *252*
14.1.7 Application Methods of PVAc Emulsion Wood Adhesives *253*
14.1.8 Curing Mechanism of PVAc Emulsion Wood Adhesives *253*
14.2 Summary *253*
References *254*

15 Troubleshooting in Wood Bonding with PVAc Resin Adhesives *255*
15.1 Effects of Moisture Content of Wood *255*
15.2 Troubleshooting Methods in Various Gluing Operations *259*
15.2.1 Troubleshooting in Cold-Press Lamination *260*
15.2.2 Troubleshooting in Hot-Press Laminating Process *260*
15.2.3 Troubleshooting in Radiofrequency Edge-Gluing *261*
15.2.4 Troubleshooting in Edge and Face Gluing by Clamping *261*
15.2.5 Troubleshooting in Assembly Gluing by Clamping *262*
Reference *262*

16 Hot-melt and Other Specialty Wood Adhesives *263*
16.1 Introduction to Hot-melt Adhesives *263*
16.2 Requirements For Hot-melt Adhesives *263*
16.3 Materials Used for Hot-melt Wood Adhesive Formulations *265*
16.4 Advantages and Disadvantages of Hot-melt Adhesives *266*
16.4.1 Advantages *266*

16.4.2 Disadvantages *266*
16.4.3 Uses of Hot-melts *267*
16.5 Thermosetting Hot-melt Adhesives – Recent Development *267*
16.5.1 A Typical Thermosetting Hot-melt Formulation *267*
16.5.2 A Comparison of a Thermosetting Hot-melt Against Other Types *267*
16.6 Key Variables in Hot-melt Edge-banding Operation *268*
16.7 Other Specialty Wood Adhesives *269*
16.7.1 Mastic Adhesives *269*
16.7.2 Contact Adhesives *270*
16.7.3 Epoxy Resin Adhesives *271*
References *273*

17 Casein, Soybean Flour, Animal Blood, and Lignin Wood Adhesives *275*
17.1 Casein Wood Adhesives *275*
17.2 Animal Protein-Based Wood Adhesive *276*
17.3 Soybean Meal and Soybean Protein Wood Adhesives *276*
17.4 Animal Blood-based Wood Adhesives *278*
17.5 Various Lignins and Uses in Wood Adhesives *278*
References *281*

18 Theory and Practices of Adhesive Bonding for Wood *283*
18.1 Formation of Interphase in Wood Adhesive Bonds and Failure Modes *283*
18.1.1 Interphase Present Between Wood and Adhesive Layer in Adhesive Bond *283*
18.1.2 Failure Modes in Mechanical Testing of Wood Bonds *284*
18.1.3 Causes of Adhesive-layer or Interphase Failures in Wood Adhesive Bonds *284*
18.1.4 Possible Further Divisions of Interphase in Wood Adhesive Bonds *285*
18.1.5 Adhesive Failure vs. Wood Failure in Testing of Adhesive Bonds *285*
18.1.6 Monitoring is a Must: Adhesive Quality, Adhesion Operation, and Bonded Products *286*
18.2 Wettability of Solid Surface, Contact Angle, and Surface/Interface Tensions *287*
18.3 Work of Adhesion *291*
References *291*

19 Physical and Chemical Mechanisms of Adhesive Bonding for Wood *292*
19.1 Adsorption (Secondary Bond Forces) Theory of Adhesion *292*
19.2 Mechanical Interlocking Theory of Adhesion *296*
19.3 Diffusion Theory of Adhesion *297*
19.4 Primary Chemical Bond (Covalent Bond) Theory *298*
19.5 Summary of Adhesion Mechanisms *299*
19.6 Glueline Layer Thickness and Bond Strengths *299*
19.7 Summary *301*
References *301*

20 Evaluation of Wood Adhesive Bonds, Quality Control, and Bond Durability *302*
20.1 Mechanical Testing Modes and Methods for Measuring the Adhesive Bond Strength *302*
20.1.1 Shear Strength Test *302*
20.1.2 Tensile Strength Tests *303*
20.1.3 Cleavage and Peel Strength Tests *306*
20.1.4 Bending Strength Tests *307*
20.2 Quality Control, Certification Tests, and Adhesive Bond Durability *309*
20.2.1 Objectives and Scope of Quality Control Processes *309*
20.2.2 Durability Evaluation of Wood Composite Boards and Certification *309*
20.2.3 Certification of Exterior-Use Wood Composite Products *310*
20.2.4 Certification of Exterior-Use Wood-Based Structural Panels (PS2-92) *311*
20.2.5 Cyclic Delamination Test Procedure (AITC Test 110) *312*
20.2.6 Other Common Laboratory Aging Test Methods Used in the Industry *312*
20.2.7 Testing and Certification Organizations Involved *313*
20.3 Comparison of Various Wood Adhesive Bonds by Accelerated Aging or Exterior Exposure Tests *313*
20.4 Nondestructive Testing of Wood and Wood Adhesive Bonds *315*
20.5 *In-situ* Adhesion Testing *315*
References *316*

21 Introduction to Coatings Technology for Wood *317*
21.1 Three Components of Coatings *317*
21.2 Pigment Volume Concentration (PVC) *317*
21.3 Various Kinds of Vehicle Polymers *319*
21.3.1 Fatty Oils and Modified Fatty Oils *319*
21.3.2 Alkyd Resins *320*
21.3.3 Polyester – Unsaturated and Saturated Polyesters *321*
21.3.4 Cellulosics *322*
21.3.5 Acrylic Resins *323*
21.3.6 Vinyl Resins *324*
21.3.7 Epoxy Resins *325*
21.3.8 Isocyanate Resins (urethane resins) *326*
21.3.9 Silicone Resins *327*
21.3.10 PF, MF, UF Resin Derivatives *327*
References *328*

22 Introduction to Coatings Technology for Wood. II *329*
22.1 Pigments and Fillers *329*
22.2 Manufacturing and Kinds of Pigments and Fillers *329*
22.3 Color Control Methods *329*
22.4 Color Scales – Hunter L, a, b, and CIE 1976 $L^*a^*b^*$ (CIELAB) Color Scales *331*
22.5 Carriers of Coatings *332*
22.6 Additives to Coatings *332*
22.7 Manufacturing Procedures of Coatings *332*

22.8 Film Formation Mechanisms *333*
22.9 Water-Borne Coatings and Coatings for Less VOC Emissions *334*
22.10 Exterior Coatings for Wood *336*
22.10.1 Stains *336*
22.10.2 Varnishes *336*
22.10.3 Paints *337*
22.11 Summary *337*
References *338*

23 Industrial Coating Application Processes *339*
23.1 Application Methods of Coatings *339*
23.2 Sanding Processes *342*
23.3 Sanding Abrasives, Construction, and Flexing *344*
23.4 Typical Furniture Finishing Procedures and Materials *344*
23.5 Flat Line Finishing Procedures of Wood Composite Boards for Furniture Production *346*
23.6 Kitchen Cabinet Finishing Procedures *347*
23.7 Hardwood Plywood Panel Prefinishing *348*
23.8 Hardboard Panel Finishing Procedures *349*
23.9 Summary *350*
References *350*

24 Advanced Reading Materials on UF Wood Adhesive Resins *351*
24.1 Introduction to the ^{13}C NMR Spectroscopic Analysis Method *351*
24.2 Introduction to ^{13}C NMR Analysis Methods of UF and UMF Wood Adhesive Resins *356*
24.3 ^{13}C NMR Analysis Results of Reaction Intermediates Taken in UF Resin Syntheses *360*
24.4 ^{13}C NMR Analysis Results of Reaction Intermediates Taken in UF Resin Syntheses with a Higher Power Instrument *366*
24.4.1 Effects of the Addition of Second Urea and Mild Heat Treatments *369*
24.4.2 Polymerization in the Acidic Polymerization Step Observed with Second Urea Addition *370*
24.4.3 Polymerization In the Acidic Reaction Step Observed Without the Second Urea Addition *372*
24.4.4 Summary *373*
24.5 Chemical Changes Occurring in UF Resins on heat/stirring and Room-temperature Storage Treatments by ^{13}C NMR and Formaldehyde Emission Tests of Particleboard *374*
24.5.1 Effects of Mild and Intermediate Level Heat/Stirring Treatments (Samples A_2–A_6) *374*
24.5.2 Effects of Room-temperature Storage Treatments of UF Resins *378*
24.5.3 Formaldehyde Emission Test Results of Particleboard (PB) *381*
24.5.4 Summary *382*
24.6 Effects of Mild Heating/Stirring Treatments on UF Resins Synthesized with Different F/U_1 Mole Ratios by^{13}C NMR *382*

24.6.1 Syntheses of UF Resins with Varying F/U_1 Mole Ratios *382*
24.6.2 Heat/Stirring Treatments of Synthesized UF Resins *383*
24.6.3 Viscosity and Turbidity Changes in Heat Treatments *383*
24.6.4 F/U_1 Mole Ratio Effects on UF Resin Structures in Heat Treatments Observed by ^{13}C NMR *384*
24.6.5 Summary *388*
24.7 Effects of Room-temperature Storage Treatments on UF Resins Synthesized with Various F/U_1 Mole Ratios by ^{13}C NMR and Formaldehyde Emission Tests of Particleboard *388*
24.7.1 F/U_1 Mole Ratio Effects on Viscosity/Turbidity Changes in Room-temperature Storage Treatments *389*
24.7.2 Functional Group Changes of UF Resins in 7–50-day Room Temperature Storage *392*
24.7.3 Test Results of Particleboards Bonded with Various UF Resins *393*
24.7.3.1 Internal Bond and Bending Strengths of Particleboards (PBs) *393*
24.7.3.2 F/U_1 Mole Ratio Effects on Formaldehyde Emission Values of Particleboards *394*
24.7.4 Applications to the UF Resin and PB Manufacturing Industry *396*
24.7.5 Summary *397*
24.8 Effects of Mild Heat/Stirring and Room-temperature Storage Treatments of UF Resins Synthesized with Various F/U_1 Mole Ratios by ^{13}C NMR and Formaldehyde Emission Tests *397*
24.8.1 Resin Synthesis and Treatment Procedures *398*
24.8.2 Collection of Data *398*
24.8.3 Heat Treatment Effects of UF Resins Made with Various F/U_1 Mole Ratios *402*
24.8.4 Heat Treatments at 60 °C for 2.5 hours Followed by Storage at Room Temperature for up to 27 days *405*
24.8.5 Reaction Mechanisms of Methylene and Methylene-ether Group Forming Ractions *406*
24.8.6 Viscosity and Phase Changes of UF Resins During Heating/Storage Treatments *408*
24.8.7 F/U_1 Mole Ratio and Posttreatment Effects on Resin Polymeric Structures and Compositions *409*
24.8.8 F/U_1 Mole Ratio and Heat Treatment Effects on Particleboard Strength Properties *411*
24.8.9 Particleboards Bonded with Resins of 2.5-hours Heat and Followed Room-temperature Storage Treatments *415*
24.8.10 Summary *417*
References *418*

25 Advanced Reading Materials on UMF Resins Modified with 6–12% Melamine *420*
25.1 Introduction and Synthesis and Analysis Results of UF and UMF Resins *420*
25.1.1 Background and Known Resin Synthesis Methods of UMF Resins *420*
25.1.2 Reaction Chemistry of MF and UMF Resins *421*
25.1.3 Resin Syntheses of UF and UMF Resins *422*

25.1.4 Characteristics of UMF Resins and Resin Intermediates Determined by ^{13}C NMR Spectra *422*
25.1.5 General Resin Characteristics of Synthesized UMF Resins *429*
25.1.6 Conclusion *432*
25.2 DMA Curing Properties of UF and UMF Resins *432*
25.2.1 Past Research Using Dynamic Mechanical Analysis (DMA) Method on Wood Adhesive Resins *432*
25.2.2 DMA Sample Preparation and Experiments *433*
25.2.3 Three Curing Segments of DMA Curing of Resin $UFA_{1.15}$ *433*
25.2.4 Curing Chemistry of Resin $UFA_{1.15}$ *435*
25.2.5 Effects of Catalyst Level and Mix Time and Temperature on DMA Curing of Resin $UFA_{1.15}$ *438*
25.2.6 Comparison of Resins $UFA_{1.15}$ and $UMF_{1.15}$ by DMA at Three Different Temperatures *439*
25.2.7 Effects of Three Different F/(U + M) Mole Ratios on Curing Rates at 150 °C *441*
25.2.8 Summary *441*
25.3 Bond Performance of UMF Resins as Particleboard Binders *443*
25.3.1 Preparation and Tests of Laboratory Particleboards *443*
25.3.2 Test results of Particleboards Bonded with Resins at an F/(U + M) Ratio of 1.15 *443*
25.3.3 Summary *445*
References *445*

26 Advanced Reading Materials on UMF Resins Modified with 2.5% and 5.0% Melamine *447*
26.1 UMF Resins Synthesized with 2.5% and 5.0% Melamine Levels *447*
26.1.1 Experimental Procedures *447*
26.1.2 Resin Chemical Structures Determined by ^{13}C-NMR *449*
26.1.3 Storage Stability of UF and UMF Resins *451*
26.1.4 Properties of Catalyzed Resins *451*
26.1.5 Particleboard Test Results *453*
26.1.6 Summary *456*
26.2 UMF Resins Synthesized by Adding Melamine at Different Points *456*
26.2.1 Syntheses of UMF Resins *457*
26.2.2 Physical Properties of Synthesized UMF Resins *458*
26.2.3 Resins' Chemical Structures From ^{13}C NMR Results *458*
26.2.4 Storage Stability of Un-catalyzed Resins *459*
26.2.5 Properties of Catalyzed UMF Resin *461*
26.2.6 Particleboard (PB) Test Results *462*
26.2.7 Summary *466*
References *467*

27 Advanced Reading Materials on Diethylene Tricarbamide-Formaldehyde Resins *468*
27.1 Introduction *468*
27.2 A Higher Functionality Urea Analogue – Diethylene Tricarbamide *469*

27.3 Chemical and Physical Properties of Diethylene Tricarbamide *469*
27.4 An Efficient Synthesis Method of Diethylene Tricarbamide *470*
27.5 Synthesis Chemistry and Procedure and Properties of D-formaldehyde (DF) and Copolymer Resins *470*
27.6 Synthesis Formulations of DF Resins vs. UF Resins *471*
27.7 Room Temperature Storage Properties of DF and Copolymer Resins *472*
27.8 Pot-lives and Curing Characteristics of DF and Copolymer Resins *472*
27.9 Preparation of Laboratory Particleboards and Testing *473*
27.10 Test Results of Prepared Laboratory Particleboards *473*
27.11 Summary *475*
References *475*

28 Advanced Reading Materials on PF Resole Wood Adhesive Resins *476*
28.1 Chemical Structures of PF Resole and Novolac Resins by ^{13}C NMR Spectroscopy *476*
28.1.1 PF Resin Model Compounds *476*
28.1.2 An Example Spectrum of a PF Resole Resin *479*
28.1.3 A Second Example Spectrum of a PF Resole Resin *480*
28.1.4 A Third Example ^{13}C NMR Spectrum of a Novolac PF Resin *480*
28.1.5 Summary *481*
28.2 Reaction Rates and Structures of a PF Resole Resin Synthesized at 70 °C *482*
28.2.1 Introduction *482*
28.2.2 Peak Intensity Changes of Aromatic Carbons Due to Polymerization During Resin Synthesis *483*
28.2.3 Peak Intensity Changes of Aliphatic Carbons due to Polymerization Reaction *484*
28.2.4 Polymerization of PF Resole Resins Observed by Gel Permeation Chromatography (GPC) *487*
28.2.5 Summary *488*
28.3 Polymer Structures of a PF Resole Resin Synthesized at 102 °C vs. a Commercial Resin *490*
28.3.1 Introduction *490*
28.3.2 Acetylation of PF Resin Samples and Fractionation *491*
28.3.3 ^{13}C NMR Chemical Shift Values of Model Compounds and PF Resin Fractions *492*
28.3.4 ^{13}C NMR Results of PF Resin Model Compounds and Resin Fractions *492*
28.3.5 Intrinsic Viscosity ($[\eta]$), Vapor Phase Osmometric (VPO) Analysis, and M–H Equation *496*
28.3.6 Conclusion *499*
28.4 Polymer Structures of High Molecular Weight Fractions of a PF Resole Resin *500*
28.4.1 Synthesis of an Oriented Strand Board Binder-type PF Resin and GPC Analysis *500*
28.4.2 Acetylation and Fractionation *501*
28.4.3 Instrumental Analyses of Acetylated PF Resin Fractions *501*

28.4.4 Results and Discussion *502*
28.4.5 Conclusion and Summary *508*
28.5 Polymer Structures of Cured PF Resole Resins by Solid-state ^{13}C NMR *508*
28.5.1 PF Resin Synthesis and Curing Procedures *508*
28.5.2 Curing Mechanisms of PF Resole Resins and Loss of Some Formaldehyde? – a Review *510*
28.5.3 Solid-state CP/MAS ^{13}C NMR Spectra *511*
28.5.4 Solid-state CP/MAS ^{13}C NMR Chemical Shift Assignments of Aliphatic Carbons *511*
28.5.5 Solid-state CP/MAS ^{13}C NMR Chemical Shift Assignments of Phenolic Ring Carbons *513*
28.5.6 Solid-state CP/MAS ^{13}C NMR Chemical Shift Assignment of Quinone Structures *513*
28.5.7 Results and Discussion *514*
28.5.8 Conclusion and Summary *517*
References *518*

Index *521*

Preface

This book originated from the lecture notes prepared for undergraduate and graduate students in the course titled "FP 4113/6113 Adhesives and Finishes for Wood," by the author during his tenure at the Department of Forest Products, Mississippi State University, from 1988 till 2014. The consensus was that the lecture notes were a great help for students, and a textbook was envisioned but not materialized then but recently it became a reality. This book is designed for college-level courses with an emphasis on gaining the basic knowledge of the subject at the introductory, advanced, and practicing levels. Therefore, technologists who start careers in the wood adhesives and forest products industries with science and engineering backgrounds would also benefit from this book. Familiarity with properties, manufacturing methods, and application practices of various adhesive and finish formulations would be invaluable. Suggestions for a revised edition as a textbook will be truly appreciated.

In this book, the basic polymer chemistry and properties of polymer materials and the thermosetting and thermoplastic principles of wood adhesive resins and finishes are introduced in Chapters 1–4. Then, the chemistry and technology of urea-formaldehyde (UF) resins, melamine-urea-formaldehyde (MUF) resins, phenol-formaldehyde (PF) resins, phenol-resorcinol-formaldehyde (PRF) resins, isocyanate (Iso) resins, polyvinyl acetate (PVAc) resins, and other specialty adhesive resins are discussed in detail, along with their uses as binders for particleboard, medium density fiberboard, hardwood plywood, softwood plywood, oriented strandboard, and various furniture assemblies in Chapters 5–17. Then, the theory of wood adhesion and the evaluation methods of wood adhesive bonds are discussed in Chapters 18–20. Then, uses of polymers as coating binders and methods of application on wood as finishes were presented based on a well-prepared literature from a coatings company. Finally, advanced reading materials for major wood adhesive resins were presented in Chapters 24–28, designed especially for upper-level graduate students and practitioners in order to give more hands-on knowledge on the underlying chemistry and curing processes of these resins, with emphasis on the methods of reducing the formaldehyde emission levels of UF-resin-bonded wood composite boards and of improving the performance properties of PF resole resins.

Over the many years of my tenure at Mississippi University, the author is greatly indebted to Georgia-Pacific Corporation's resin plants in Louisville and Taylorville, MS, for generously providing 50% formaldehyde solution; to Georgia-Pacific Corporation's OSB plant in Grenada, MS; and to Norbord Corporation's OSB plant in Guntown, MS, for supplying wood strands, to Roseburg Lumber Company's particleboard plants in Louisville and

Taylorsville, MS, for supplying wood particles; and to Weyerhaeuser Company's plywood plant in Millport, AL, for supplying softwood veneers. The author is also greatly indebted to the Department of Sustainable Bioproducts of Mississippi State University for providing a conducive environment for research and teaching and also to his former colleagues, Dr. Terry Seller, Jr., and Mr. George D. Miller, for their continuous help in laboratory classes and collaborative research. The author is also greatly indebted to his industry colleagues for delivering invited class lectures and participating in collaborative research, Dr. Larry Amos, Mr. Charles R. Davis, and Dr. Joe Marcinko. Thanks also go to our former undergraduate and graduate students and research associates, whose research results are cited in the advanced reading materials sections. My gratitude to my parents is immeasurable. My thanks also go to Wiley-VCH, Inc. and its staff for their help and encouragement in publishing this book.

June 5, 2023

Moon G. Kim, Ph. D.
Laguna Woods, California

Author Biography

Moon Geu Kim, PhD, is the fifth child of ten born to Mr. Bok-sung Kim and Mrs. Bun-sim Lee in Boosoo-ri, Hoeinn-myon, Boen-gun, Choongbook-do, Republic of Korea. He attended Hoeinn Primary School and Cheongju Middle and High Schools. He has completed his BS in Chemistry at Kyung Hee University, Seoul, 1963, and PhD in Organic Chemistry at Kansas State University, Manhattan, Kansas, 1973. He served as a postdoctoral research associate at Oregon State University, Department of Chemistry, Corvallis, OR, between 1973 and 1974; a senior scientist at Weyerhaeuser Company, R&D, Tacoma, WA, between 1974 and 1984; a senior development chemist at Georgia-Pacific Corp., R&D, Decatur, GA, between 1984 and 1988; a professor at Mississippi State University (MSU), MS between 1988 and 2014. He taught *Adhesives and Finishes for Wood* course for undergraduate and graduate students and led research courses for graduate students' theses and dissertations at MSU. Authored or coauthored numerous papers on UF/UMF resins, PF resins, and bonding of PB, OSB, and softwood plywood. He lives in Laguna Woods, CA, USA. He is a retired, Emeritus Professor.

1

Introduction to Polymers, Wood Adhesives, and Wood Finishes

1.1 Good Wood Adhesives Must Be Optimum Polymers with Optimum Secondary Forces

Many structural materials used in this world, including wood adhesives, are hard, strong solid materials, i.e. their softening points (temperatures) are appreciably higher than ambient temperatures. If they become soft or liquid at use temperatures, their structural uses would be limited. Stone, wood, plastics, metals, glass, brick, cotton, etc., are used as structural materials. Liquid adhesives that bond these materials should be transformed into structural solids when the bonding process is completed; otherwise, the bonded elements will separate under a load. Furthermore, among solid materials, certain materials such as sugar or salt have some physical strength under dry conditions, but the strength disappears quickly when they are contacted with water because they dissolve in it, i.e. the molecules become disengaged from each other. Molecular units in structural materials do not disengage from each other easily even when contacted with water or heated to high temperatures since they consist of, firstly, covalently bonded (primary bonding) atoms in the form of high molecular weight polymer molecules (macromolecules), and secondly, these macromolecules also have certain optimum molecular geometry and secondary bonding forces among the atoms and molecules, resulting in solid structures that give strength and stability against exposure to water and/or high temperatures.

For example, the basic chemical structural units in a strand of cotton or wood pulp fibers are cellulose molecules, each having about 5000 or more glucose molecules covalently bonded. It should be reminded that glucose, the basic unit of cellulose polymers, is the main constituent of corn syrup with a molecular weight of 180 and dissolves readily in water due to its low molecular weight and high affinity for water. The major reason why cellulose is cellulose is that the glucose units are bonded together covalently to a high degree of polymerization and also that the secondary forces, exerted by the many hydroxyl groups present on cellulose molecules, are optimally arranged to make the cellulose molecular strands set into crystallites (highly ordered state) that do not get easily disturbed by water. Continuation of these crystallite formation leads to the formation of cellulose fiber strands. Although these fiber strands are further reinforced by lignin and hemicellulose molecules in wood (polymeric glue of wood), much of the strength of wood is derived from the polymeric, structural nature of cellulose molecules and their secondary forces. Thus, it is readily recognized that structural adhesives have to be of (covalently

Adhesives and Finishes for Wood: For Practitioners and Students, First Edition. Moon G. Kim.

bonded) polymers that have additional molecular structures and secondary bond forces that impart the strength properties under the common use conditions of moisture and temperature.

1.2 Polymeric Materials

A polymer is defined as a high molecular weight material (macromolecule) composed of many repeating units of a *monomer* or monomers joined by covalent bonds. *Poly* means "many" and "*mer*" means "part," i.e. "polymer" literally means "many joined parts." Polymeric materials can be organic or inorganic and natural or manufactured, as illustrated in the following:

- *Inorganic*: clay, brick, cement, pottery, sand, glass, glass fiber, carbon fiber, metallic carbides.
- *Organic/natural*: polysaccharides (cellulose, starch), proteins (hide, silk), natural rubber from rubber trees (polyisoprene).
- *Organic/synthetic/semi-synthetic*: polyethylene, nylon, polyester, rayon, etc. Monomer materials of synthetic polymers are derived mostly from coal, petroleum, natural gas, and some from wood, like rayon. The diverse properties and many different uses of polymers, as we will see in the next several chapters, can be traced to their high molecular weights and inter- and intramolecular secondary forces (molecular structures) present in the polymer molecules as well as different types (atoms) of monomers. Low molecular weight materials are not useful in most structural applications unless they are transformed into high molecular weight materials during the final forming (curing or molding) steps.

1.3 Synthetic Polymer Preparation Methods

Synthetic polymers are prepared by *polymerization* of monomers [1]. Evidently, the monomers must have two or more hands, i.e. functional groups (that can make covalent bonds), to form a repeating polymer molecular chain. When the monomer (M) has two functional groups, i.e. two free (bonding) electrons, the polymer formed will have a linear chain structure:

$$n\,\mathrm{M} = n\,(\ast\,\mathrm{M}\,\ast) \longrightarrow -\mathrm{M}_1-\mathrm{M}_2-\mathrm{M}_3-\mathrm{M}_4-\mathrm{M}_5-\mathrm{M}n-$$

Carbon–carbon bond

There are currently many different carbon-based monomers having two functional groups. When monomers having three or more functional groups are used, the polymer chain will be branched and, at the end of polymerization, the polymer molecules will be cross-linked. Also, aside from the common C—C bonds, the C—O bonds (ethers and esters), and C—N bonds (amides) are common chemical bonds formed in polymers. The end groups in polymers are not normally defined clearly in equations, but they are only slightly different groups than those in the middle of the chain, formed by some side reactions at the end of the polymerization process. The end groups are of minor importance in polymer technology because the value "n," *degree of polymerization*, reaches one thousand or more, and, therefore, the proportion of end groups in the entire polymer molecule becomes exceedingly small.

1.4 Typical Synthetic Polymer Materials

- *Polyethylene*: One ethylene molecule (gas) is added to another ethylene molecule, another ethylene molecule, etc., often called an "addition polymerization," with the help of a catalyst and under pressure. Ethylene itself is a gas at ordinary temperatures, and polyethylene is a thermoplastic solid material; the difference being the high molecular weight of the latter, attained by polymerization. Uses and properties: bottles, containers, films for packaging (shopping bags), and high-strength fiber. Polyethylene softens when heated to 95–120 °C and finally melts at higher temperatures (a thermoplastic polymer); at room temperature, it is stretchable, flexible, hydrophobic, etc. It has been the largest volume polymer among all synthetic organic polymer materials.

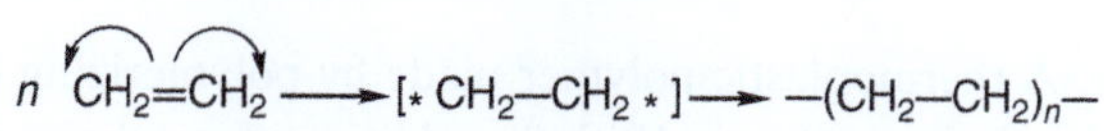

- *Polypropylene*: A thermoplastic addition polymer made by (head to tail) polymerization of propylene as monomer. Uses and properties: laboratory beakers, bottles, ice chests, rope, etc.; it softens when heated to about 150 °C (thermoplastic), similar to polyethylene, but is stiffer and the softening point is higher.

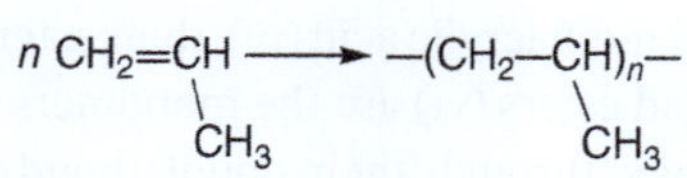

- *Polystyrene*: A thermoplastic addition polymer made by (head to tail) polymerization of styrene as monomer. Commercial polystyrene polymers normally have an average molecular weight in the range of 50–200 k Da and are extensively used as expanded materials, such as cups and dishes, and for insulation and package sheeting.

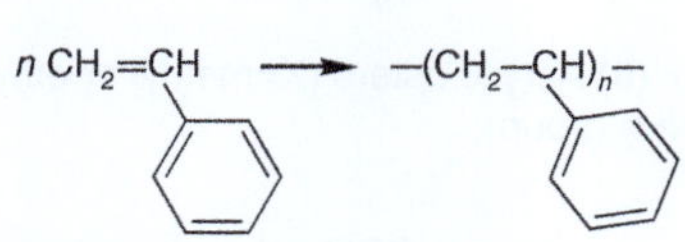

- *Polyvinyl acetate (PVAc)*: A thermoplastic addition polymer made by (head to tail) polymerization of vinyl acetate as monomer, commonly done as an emulsion polymerization process to result in latices. The glass transition (softening) temperature of commercial solid PVAc polymers is about 28 °C, and they are modified with 10–15% plasticizer (dibutyl phthalate) for use as wood binders and in coatings. Uses and properties: wood glue, interior use paint vehicle (binder). It melts when heated (thermoplastic) and has the polar characteristics needed for wood adhesive polymers due to the presence of polar acetate (—O—Ac) groups in comparison to polyethylene or polypropylene.

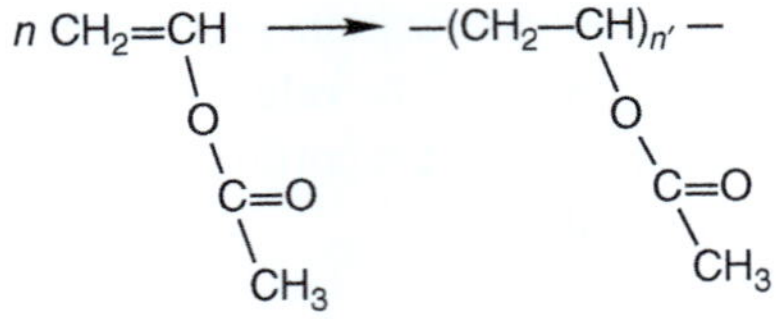

- *Ethylene-vinyl acetate (EVA) co-polymer*: A thermoplastic "co-polymer" made by co-polymerizing ethylene (5–20% by weight) and vinyl acetate monomers. The resultant co-polymer materials show high clarity, low-temperature flexibility, and crack resistance. The polymers are more flexible than PVAc. The co-polymers made with about 20% ethylene contents are used as hot-melt adhesives used in wood and paper bonding.

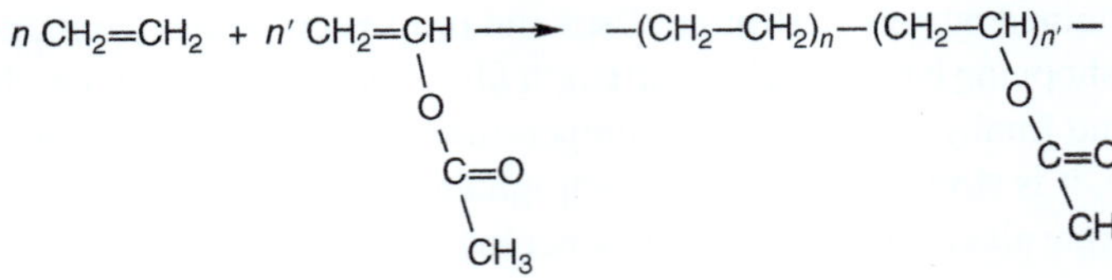

- *Polyvinyl chloride*: A thermoplastic polymer made by polymerizing (head to tail) vinyl chloride as a rigid, colorless polymer. With the addition of stabilizers and fillers, they are used extensively as water pipes, coatings, etc.

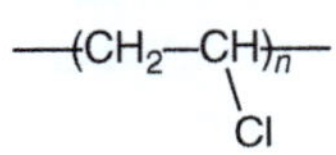

- *Acrylics*: Acrylic acid (i) and methacrylic acid (iv), their ester and amide derivatives (ii, iii, and v), and cyanoacrylic acid esters (vi) are the monomers for acrylic resins. Monomers are polymerized, respectively, through their double bonds to result in high molecular weight thermoplastic emulsion (latex) polymer materials, which are widely used as coatings, adhesives, and pane glasses.

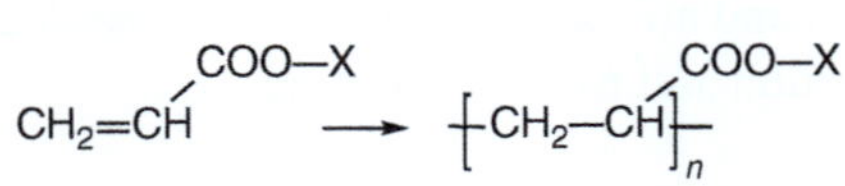

(i) Acrylic acid (X=H): **(ii)** Acrylic esters (X=methyl or ethyl groups):
(iii) Acrylamide (X=NH_2 group).

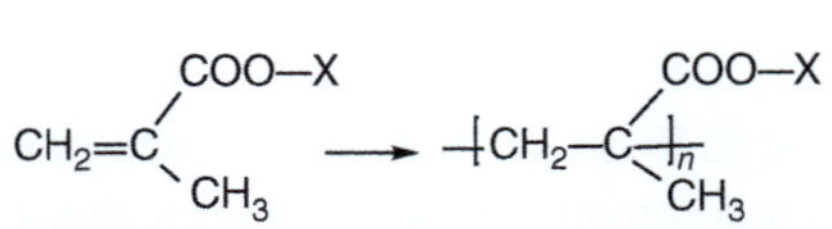

(iv) Methacrylic acid (X=H): **(v)** Methacrylic esters (X=methyl or ethyl groups)

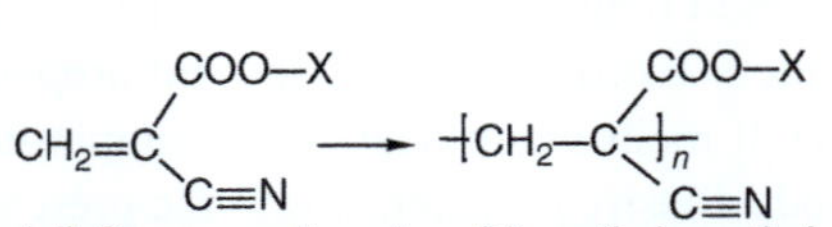

(vi) Cyanoacrylic esters (X=methyl or ethyl groups)

- *Nylon*: A polyamide, thermoplastic, made by condensation polymerization of a dicarboxylic acid and a diamine as monomers: the two monomers are condensed in exactly 1.00 : 1.00 mol ratio, where a water molecule is expelled to form amide bonds (condensation polymerization reaction). Uses and properties: filament yarn (nylon, clothes) and engineering composites and moldings. They melt when heated and are flexible, stretchable, and strong.

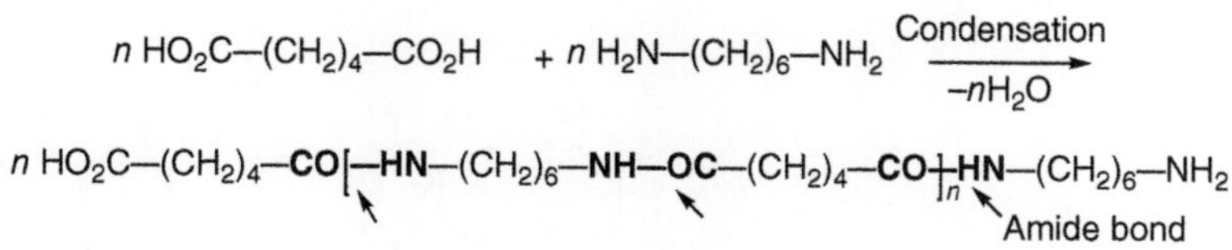

- *Polyesters*: A thermoplastic polymer made by condensation of a diol and a dicarboxylic acid as monomers. Uses and properties: filament yarn (polyesters, clothes, and polyester fibers, bottles, etc.) They melt when heated and are less hydrophobic than polyethylene or nylon. There are a variety of diols and diacids used to make polyesters and one class of them is alkyd resins used extensively as coating vehicles for wood.

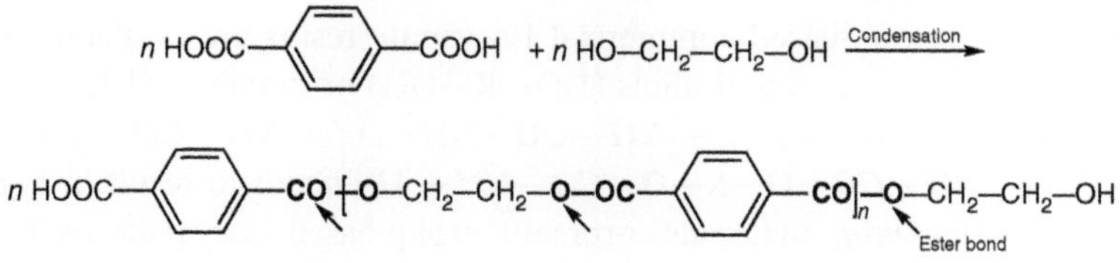

- *Phenol-formaldehyde (PF) resins*: Condensation polymers of formaldehyde and phenol as monomers; phenol has three functional groups and formaldehyde has two, use of less than 1.0 mol of formaldehyde leads to the novolac-type resins as shown, which need more formaldehyde in the curing step (two-step PF resins), and use of more than 1.0 mol of formaldehyde in resin synthesis leads to resole-type resins that cure directly (one-step PF resins). Uses and properties: wood adhesives for softwood plywood and oriented strand board, molding compounds for pot handles, etc.

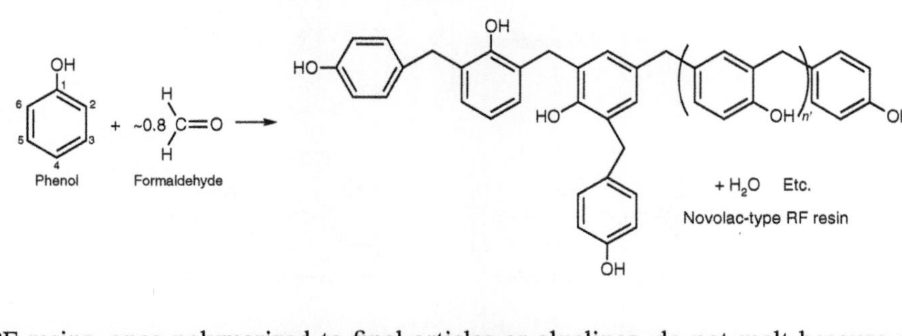

PF resins, once polymerized to final articles or gluelines, do not melt because of the branched (three dimensional) and cross-linked structures of polymer chains. In order to make an article using PF novolac resins, a source of formaldehyde is added, and it is then put in a mold, and heat and pressure are applied until the polymerization reaction is complete. PF resole-type resins do not need the second formaldehyde.

- *Resorcinol-formaldehyde resins*: Condensation polymers of resorcinol and formaldehyde are synthesized with about 1.0 : 0.7 mol ratios, novolac-type resins, as shown. Addition of some more formaldehyde and curing leads to cured RF resin polymers. They are thermosets similar to novolac PF resins in formalism, but their reactivity is greater, and they are used as room temperature-curing wood adhesives and other applications such as tire cord-rubber binders.

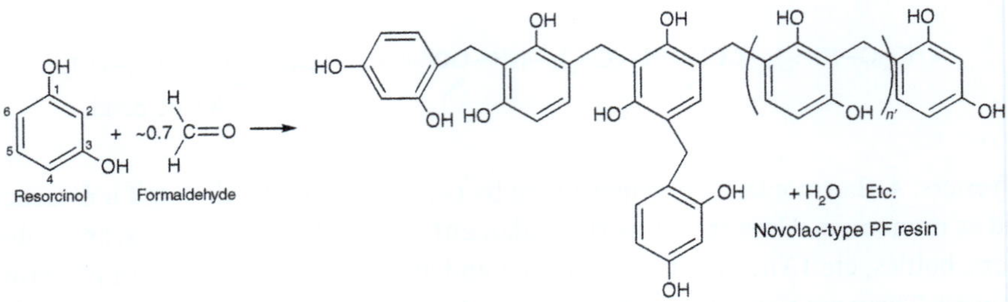

- *Isocyanate resins*: Isocyanate resin monomers are manufactured in industry to have two or more isocyanate groups (—N=C=O) per monomer molecule, and several aromatic (AR) and aliphatic (R) group-based commercial isocyanate resins are available. Isocyanate monomers are then reacted with diols (HO—R—OH) or diamines (NH_2—Ar′—NH_2) of various kinds to form *urea (AR—**NH—CO—NH**—Ar′′—**NH—CO—NH**—AR—) or urethane (AR—**NH—CO—O**—R—**O—CO—NH**—AR) bonds* to result in high molecular weight *thermosetting* materials. Aromatic group-based isocyanate resin materials such as diphenylmethane 4,4′ and 2,4′di-isocyantes (dimers) are used in large volumes in foam manufacture, and the higher oligomeric (trimer ~ hexamer) isocyanate materials (pMDI) are much used as core-layer binders of oriented strand boards and in preparation of various coating vehicles.

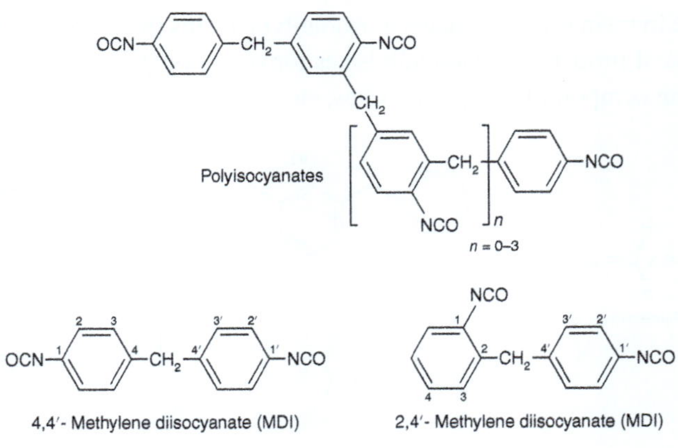

4,4′- Methylene diisocyanate (MDI)

2,4′- Methylene diisocyanate (MDI)

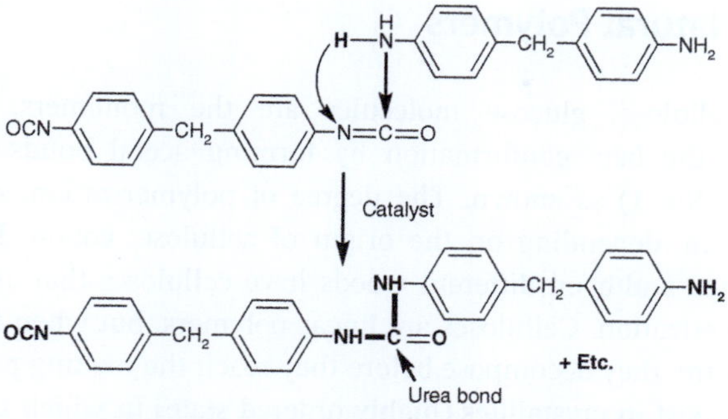

Reaction of isocyanate groups with amine groups in polymer formation

- *Urea-formaldehyde (UF) and melamine-formaldehyde (MF) resins*: Urea or melamine is reacted with formaldehyde in 1.0 : 1.1–1.8 mol ratios to make thermosetting resin materials and then cured by catalysts and heating to cross-linked polymers as adhesive binders for wood composite boards and wood moldings. They are used in the manufacturing of interior-grade particleboard, fiberboard, and laminated beams and are also used as cross-linkers in some thermosetting coatings.

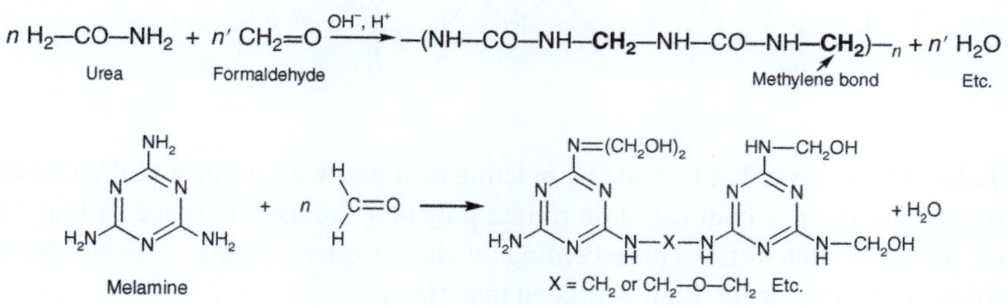

- *Epoxy resins*: Epoxy resins have two or more epoxy groups in a monomer molecule that react with primary and secondary amino groups of hardener parts as well as with hydroxyl groups generated in the process to result in thermosetting cross-linked polymers. They are used extensively as matrix resins in engineering composite materials and as specialty adhesives.

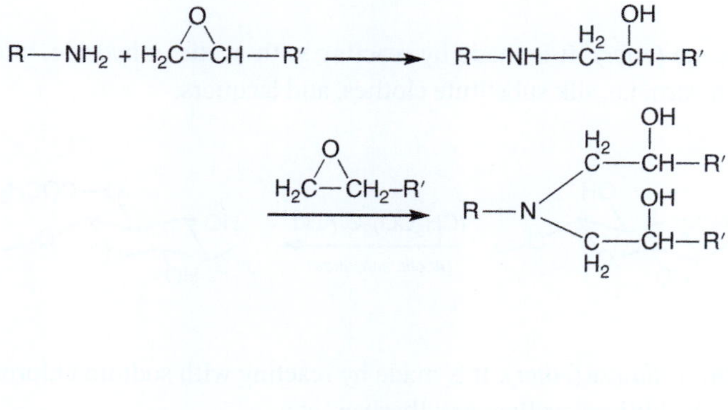

1.5 Typical Natural Polymers

- *Cellulose*: In cellulose, glucose molecules are the monomers, polymerized via 1,4′-linkages in the beta conformation by forming acetal bonds on the aldehyde groups (Carbon No. 1) as shown. The degree of polymerization, *n*, reaches several thousand or more, depending on the origin of cellulose: cotton, linen, wood pulp, rayon, ramie, etc., and even different woods have celluloses that are different in the degree of polymerization. Celluloses are linear polymers, but when they are heated to a high temperature, they decompose before they reach the melting point. The cellulose chains in wood exist in crystallites (highly ordered state) in which the many hydroxyl groups exist in tightly hydrogen-bonded structures (secondary bonding forces), and some portions are also in amorphous forms. Various derivatized cellulose polymers are currently manufactured by reacting various mono-functional chemicals on the cellulose hydroxyl groups to form ester or ether groups, as shown in the following examples:

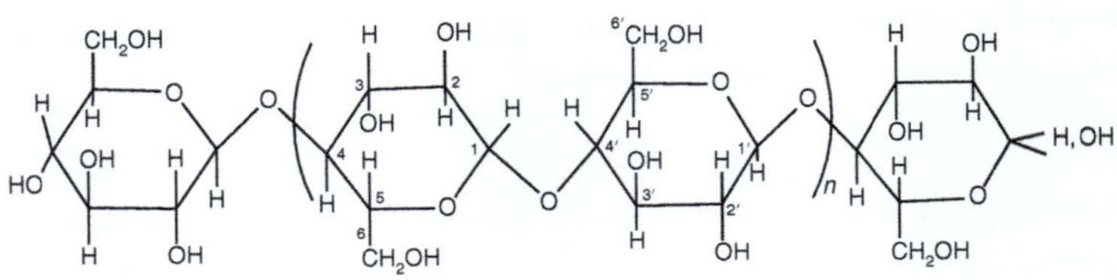

- *Cellulose nitrate (ester)*: It is made by reacting cellulose with a nitric acid/sulfuric acid mixture, and the resultant cellulose nitrate polymers are used to make various "celluloid" products, such as films for recording movies, lacquer (coating) polymers for wood finishes, and gun powder (highly nitrated materials).

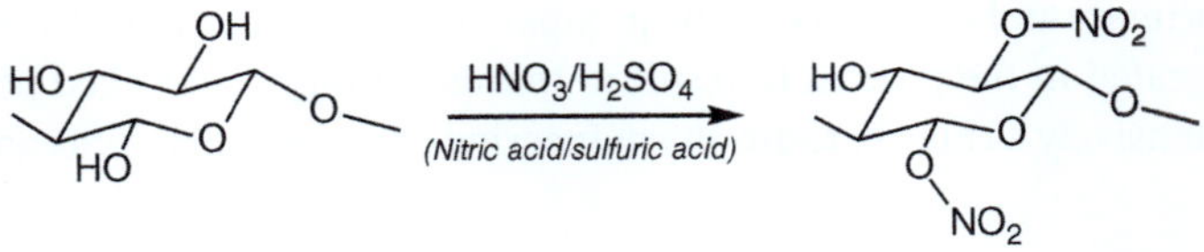

- *Cellulose acetate (ester)*: It is made by reacting with acetic anhydride, and it is used as acetate rayon yarn, i.e. silk substitute clothes, and lacquers.

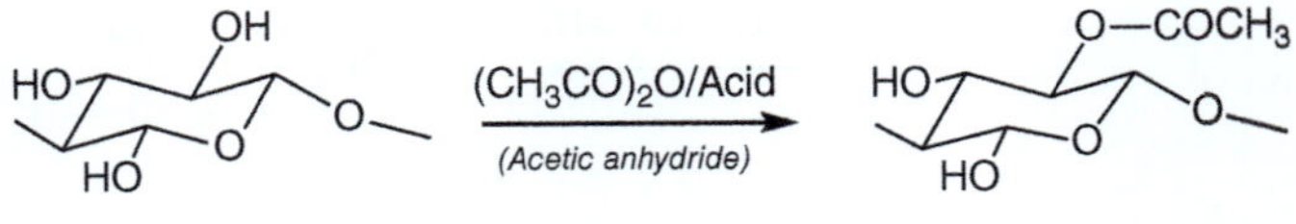

- *Carboxymethyl cellulose (ether)*: It is made by reacting with sodium chloroacetate, and it is used as food additives, wallpaper adhesives, etc.

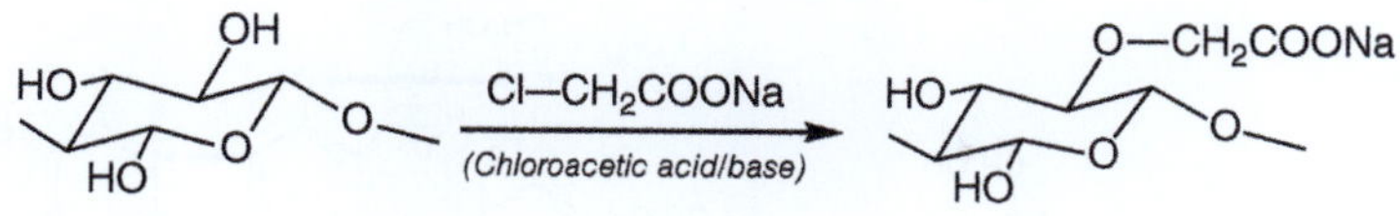

- *Hydroxyethyl cellulose (ether)*: It is made by reacting with ethylene oxide, and it is used as thickeners in paints and adhesives, etc.

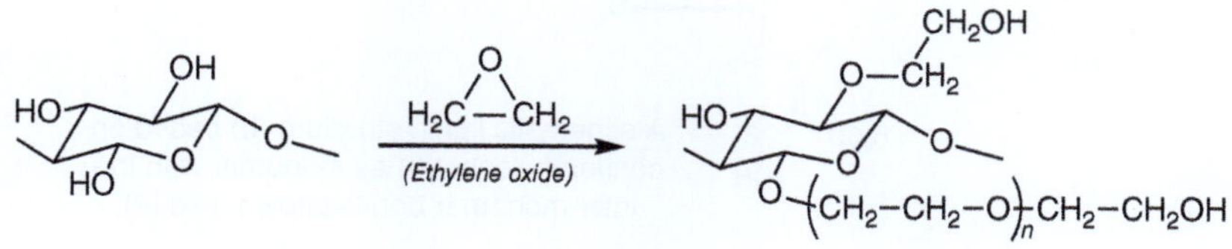

- *Starch*: It occurs in nature, in which glucose molecules are polymerized into 1,4′- and 1,6′-linkages with alpha configuration different from that of celluloses. They are obtained from plants as mixtures of (linear) amylose and (branched) amylopectin. The degree of polymerization is high and variable depending on the origin: potato, corn, wheat, tapioca, etc., and they swell in water when cooked. Used in food, interior use adhesives for corrugated paper products, and others.

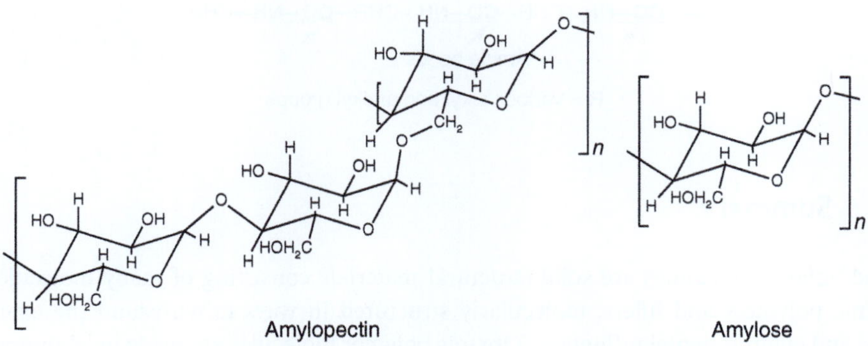

Amylopectin Amylose

- *Lignin*: "Lignin" components, together with hemicellulose components, are the adhesive polymers of cellulose fibers in wood. Wood pulps manufactured in the wood pulping industry are primarily composed of undigested cellulose molecules, and the lignin polymers and hemicellulose molecules are dissolved and partially decomposed to come out as black liquor in the Kraft pulping process. Black liquors are used as an energy source for the pulping plant. In lignin molecules, phenolic rings are present in polymerized forms, in a somewhat irregular, branched manner, connected through the 3-carbon aliphatic side chains, forming "ether" bonds with phenolic oxygens of another ring. Intact natural lignin polymers are difficult to isolate, and therefore the accurate structures of lignins have been difficult to determine, and they vary depending on the origin of wood. "Industrial lignin" products from the sulfite pulping processes are lignosulfonates, available for use as additives in wood adhesives and in other applications.

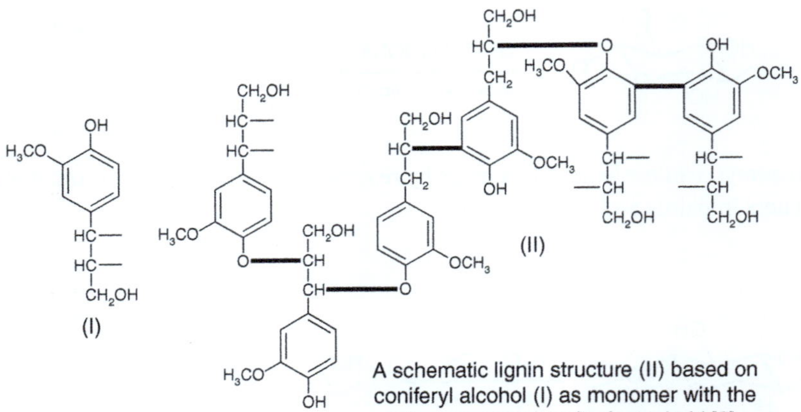

A schematic lignin structure (II) based on coniferyl alcohol (I) as monomer with the inter-monomer bonds shown bold [2].

- *Proteins*: Most proteins are natural *polyamide polymers* of various "alpha-amino acids," as shown below. The functional groups involved in polymerization are the amine and carboxylic acid groups, as in nylon-type polyamide polymers. Twenty or so different amino acids occur in nature (with various R groups), and many combinations are possible depending on the origins of protein materials, but industrial proteins useful as adhesives are casein from milk, animal tissue proteins (collagen), blood proteins (albumin), and soybean proteins.

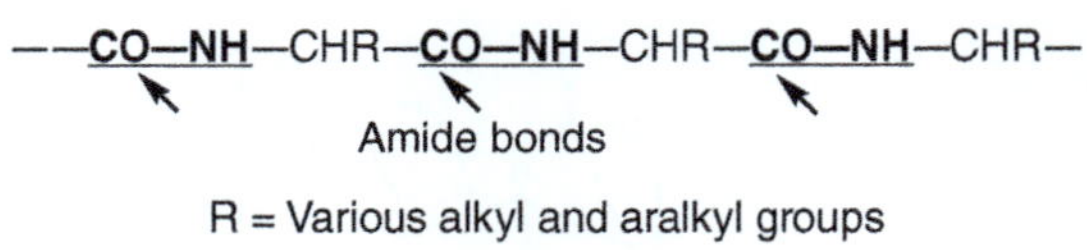

1.6 Summary

Wood adhesive gluelines are solid structural materials consisting of many molecules of organic polymers and fillers, molecularly structured in ways to withstand the external loads and environmental influences. Organic polymer molecules are made in laboratory or by nature, where monomers are first created and then polymerized to very high molecular weight polymers. Monomers are commonly gases, liquids, or solids with molecular weights lower than about 200 Da, and polymers are mostly solids having high molecular weights of up to 100 000 Da or higher, or unmeasurable. Various monomers and mechanisms of polymerization are presented. Many monomers are relatively simple molecules that have a carbon–carbon double bond (C=C), one of which is opened by catalysts to form carbon–carbon (C—C) covalent bonds with other monomers almost endlessly (addition polymerization) reaching to high molecular weight materials. Thus, ethylene gas is polymerized to polyethylene, propylene to polypropylene, styrene to polystyrene, etc. On the other hand, condensation polymers are formed between two different monomers such as —N=C=O groups of the first monomer reacting with a hydroxyl or an amino group of the second monomer to form polyurethane and polyurea polymers, respectively. Similarly, dicarboxylic acid groups of the first monomer condensed at exactly 1.00 : 1.00

group ratios with dihydroxyl or diamine groups of another monomer to form condensation polymers such as polyesters or polyamides with the expulsion of water molecules. Proteins are a special case of condensation polymers where many different amino acids, each having a carboxylic acid and an amino group, co-polymerize to form polyamides, i.e. protein polymers. Similarly, monomers having an aldehyde group and hydroxyl groups, such as glucose monomer, open the aldehyde C=O double bond and condense with two hydroxyl groups to form acetal bonds, resulting in poly-glucoses, such as cellulose and starch. Cellulose and starch can be derivatized on their hydroxyl groups by ether or ester bond formation to nitrocellulose and other derivative polymers. Monomers having two or more epoxy groups polymerize by opening the triangular epoxy bonds to form polyether polymers or polymerize with diamines to form thermosetting cross-linked epoxy polymers. Further, a monomer is formed by reacting two different components, such as formaldehyde with phenol, resorcinol, urea, or melamine, to form hydroxymethyl derivatives, which are then activated by an acid or base catalyst to polymerize into high molecular weight cross-linked polymers. These latter-group polymers are the major polymer materials used as thermosetting adhesives in bonding of wood and wood composite boards.

References

1 Saunders, K.J. (1988). *Organic Polymer Chemistry*, 2e. London and New York: Chapman and Hall.

2 Sarkanen, K.V. and Ludwig, C.H. (1988). *Lignins, Occurrence, Formation, and Structure and Reactions*. New York: Wiley.

2

Principles of Polymer Chemistry for Wood Adhesives and Finishes

2.1 Degree of Polymerization and Molecular Weight

A polymer is a covalently bonded large molecule, commonly obtained by polymerization of small monomer molecules. The properties of a polymer greatly depend, in part, on how many monomers are polymerized into the polymer molecule. This number is defined as the *degree of polymerization* (n), and the molecular weight (MW) of the polymer is obtained by multiplying $\boldsymbol{n}$ with the monomer molecular weight, for example, a polyethylene material produced has a degree of polymerization (n) of 1000, then,

$$\text{MW} = 1000 \times (28) = 28\ 000\ \text{Da}$$

$$(\text{Ethylene } CH_2 = CH_2; H = 1, C = 12; \text{monomer MW} = 28)$$

The high molecular weight of polymers, first of all, is reflected as their increased glass transition (~melting) points from the low-boiling gases or liquids of monomers to values anywhere up to about 200 °C or higher, although for thermoplastic polymers the melting point is not sharp like those of crystalline monomeric materials but shows a range of values from the glass transition (softening point) to the completely liquid stage (melting point). Of course, polymers are structurally used below the glass transition points but are often processed at above the glass transition or melting temperatures.

2.1.1 Determination of Molecular Weight or Degree of Polymerization

The molecular weight or degree of polymerization of a polymer material cannot be determined just by doing the polymerization reaction nor by visual inspection. They must be determined using analytical measurements. In addition, when a polymer is synthesized, the molecular weight or degree of polymerization is not just one high value but a range of values. The calculation shown above for polyethylene is a simplified representation. A piece of polymer sample contains many polymer molecules whose molecular weights spread over a range of values, and therefore the *average molecular weight value* is normally determined and used as the representative MW value. Further, due to certain restrictions in the analytical methods used in determining MW or degree of polymerization, there are different average MW values: *number-average* (Mn) and *weight-average* (M_w) molecular weights; these values are derived from using different analytical methods.

Adhesives and Finishes for Wood: For Practitioners and Students, First Edition. Moon G. Kim.

Vapor pressure osmometry (VPO) method gives the number average molecular weight because the method is based on counting *the number of polymer molecules* (particles) present in a given volume of polymer solution, naturally giving the number-average molecular weight.

Light scattering (LS) method gives the weight-average molecular weight value because the method is based on the amount of light scattered (reflected) by a given amount (weight) of polymer material dissolved in a given amount of solvent. As can be visualized, larger molecules (particles with larger mass) reflect more light than smaller ones, and measuring the amount of reflected light, therefore, gives the weight-average (size) molecular weight.

Gel permeation chromatography (GPC) method uses a column of tightly packed gel particles; each gel particle has many small channels (holes) of various different (measurable) diameters. The polymer molecules dissolved in solution are then forced to pass through this column of gel particles, such as by pumping with pressuring, and the polymer molecules in solution pass through the gel channels and finally are eluted at the end of the column according to the sizes of polymer molecules. A mass or UV detector measures the polymer concentrations in the eluting solution in real time. The result of polymer molecules passing the gel-packed column is that the smaller the size of polymer molecules, the longer time they take to travel through the given length of column. It is because longer paths are available for the smaller molecules and on the other hand, larger molecules *bypass* the small diameter channels to elute faster. The result is that larger polymer molecules are eluted first, and the smaller polymer molecules elute later, to give a spread of the time of elution (molecular weight) versus the polymer mass concentration (intensity). Thus, the chromatogram shows the molecular weight distribution of the polymer. By doing similar experiments with a series of known molecular weight standard materials, the molecular weight distribution curve of the unknown sample can be calibrated and the MW value calculated, either the number- or weight-average molecular weight values. The method can be done in a short time and is therefore used widely, in comparison to the VPO or light scattering procedures.

2.2 Properties of Polymer Solutions or Suspensions

Polymers are often manufactured, processed, or used as solutions or suspensions in organic solvents or in water before they are made into the final products of various shapes such as film, sheet, rod, fiber, or yarn [1]. Polymers dissolved in a solvent impart various properties to the solution in comparison to a solution made with similar weights of monomers or low molecular weight materials. One of the solution properties is the viscosity, an important processing parameter, which increases depending on the kind of polymer, molecular weight, kind of solvent, and concentration of polymer materials in solution, and also on the temperature of solution.

2.2.1 Viscosity of Polymer Solutions

Viscosity (or viscosity index) of a liquid is defined as the ratio (coefficient) of the force applied for the flow to occur (stirring, tilting, etc.) and the shear rate, i.e. the number of breaks of molecular contacts occurring in unit time effected by the force applied. Definition of the (dynamic) viscosity of liquid is in line with the mass (m) and acceleration of Newton's second law of motion and with the modulus (E) and strain of solid materials subjected to the stretching force. Thus, the viscosity of a solution is a measure of "how much the molecules are sticky to each other," or a measure of the resistance to the flow of liquid, such as by tilting, stirring, or shaking, arising from secondary molecular forces, which are needed to be broken (shear) during the molecules' movements to conform to the force applied (tilting, for example). The force applied (energy) to the liquid is thus expended in breaking the secondary molecular bond forces between solvent molecules as well as between solvent and polymer molecules, and the ratio of force to the number of all secondary bond breaks (shear rates) is the viscosity (index). The molecules then continuously re-make the secondary force bonds with next neighboring molecules and give up the energy to surrounding molecules as heat and take up a new shape or position. If the stirring (force) is continued, these molecular breaking and remaking of secondary forces bonds go on, and the overall force of stirring thus applied is consumed to generate heat, resulting in increasing the temperature of the solution. If the force applied is increased, for example, a faster stirring is applied, the breaking of bonds per second, i.e. shear rate, also increases, and the viscosity (index) remains unchanged, as the equation suggests. This type of solution is called a *Newtonian liquid or solution*. Many common solutions of monomers or low molecular weight organic materials and solvents behave as Newtonian solutions.

$$\text{Viscosity (index)} = \text{Force/shear rate.}$$

- Compare this equation with: **Mass** = Force/**a**cceleration (Newton's second law), or **E** = Force/strain (%); where **E** is modulus for solids.

Solvents like water and many low molecular weight organic solvents have low viscosity values since neighboring molecules in the liquid can be sheared (slip) easily in comparison to materials like thick honey or cooking oils. The dynamic viscosity (index) units are P (poise) or cP (centipoise, cps); other relative scales are also used in practice for convenience of measurement. Example: Gardner-Holdt bubble scale of A to Z (low to high) is used widely in wood adhesive technology. *Viscosity index* is normally referred to as "viscosity" for simplicity. Many polymer materials, such as polyvinyl acetate (PVAc) and uncured liquid PF or UF resins, are solids when they are in their pure state (100% solids content), but they are often processed and transported more easily in solutions or emulsions. When an amount of a polymer is dissolved or suspended as fine particles in a solvent, the resulting solution or emulsion becomes more viscous (thick) than the solution obtained by dissolving a comparable weight of monomer in the same solvent due to the larger molecular (particulate) sizes of polymers.

2.2.2 Dependence of Viscosity on the Shear Rate (Stirring Rate)

Depending on the nature of polymers and their interactions with the solvent molecules in which they are dissolved in, polymer solutions behave as a *Newtonian solution* or *non-Newtonian solution*. For Newtonian solutions, as mentioned above (Section 2.2.1), similar to Newton's second law (**F** = **ma**), as the stirring force (F) is increased, the shear rate of molecules in solution increases proportionately, resulting in a constant viscosity index value. For non-Newtonian solutions, however, the viscosity index value decreases as the shear rate increases for *shear-thinning materials*; or the viscosity index increases as the shear rate increases for *shear-thickening materials*. The nature of materials is more complex, and most organic adhesives, paints, and other coating materials show the *shear-thinning* effects on the stirring rate as well as on *the length of stirring time*: a longer time of stirring at the same rate makes the viscosity lower than a shorter time of stirring. This stirring time dependency of viscosity along with the dependency on the shearing rate is called "*thixotropy*." Many organic adhesives show this thixotropic behavior, and this property has important implications in manufacturing, transporting, and application (use) of polymer solutions. Therefore, in measuring viscosity values using various viscometers, the shear rate (the turning rate of the spindle) of the viscometer must be specified, because the result can change depending on the shear rate and time length of shearing.

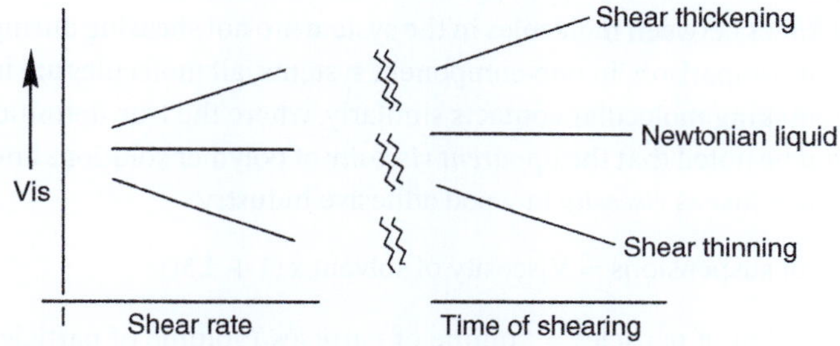

2.2.3 Temperature Dependence of Viscosity of Polymer Solutions

A liquid means its molecules are attracted to each other, by secondary forces, strong enough to hold them in the liquid state or within limited distances from each other. Increasing the temperature of a liquid increases the mobility of molecules in the solution and also the volume; in other words, the intermolecular (secondary) forces are weakened, and, therefore, less force is needed to break apart the intermolecular bonds, i.e. lowers the viscosity of solution. This applies also to liquid adhesives and coating systems. An adhesive (solution) made at the same viscosity in the plant (25 °C) becomes more viscous when it is sitting outside in winter and less viscous in summer: an increased viscosity can cause difficulties in pumping and spraying, *etc.* Heating is sometimes applied to lower the viscosity.

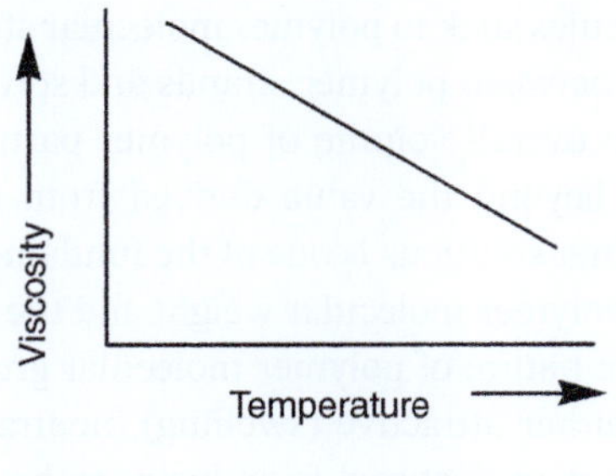

2.2.4 Dependence of Viscosity on the Concentration and Molecular Weight of Dissolved Polymers

A liquid that contains dissolved or finely dispersed insoluble polymer solid particles in solvent, such as a polymer solution or milk, shows an apparent viscosity value in comparison to one-component systems [1]. Albert Einstein worked out the calculation for the apparent viscosity of this type of mixture, observable with increasing levels of solid particles (concentration) in the system. When the particles are assumed to be spherical and not swelling, the equation shown below applies, stating that by adding solid particles to a liquid (or a resin solution) and stirring long enough to make the particles remain uniformly distributed, the viscosity of the resultant system will increase by $(1 + 2.5f)$ times the viscosity of the solvent, where **f** is the volume fraction of the particles in the solution. For example, addition of one (volume) percent (%) of clay or any polymer, not dissolving, to water will increase the viscosity by a factor of 1.025 $(1 + 2.5 \times 0.01)$. This increase in viscosity occurs because the solvent molecules are diverted to flow around the solid particles in stirring. Solvent molecules are far smaller than solid particles, and breaking of intermolecular secondary forces within the solid particles does not occur under the stirring force applied. This is also why we call the viscosity "the *apparent viscosity*," to indicate that some secondary chemical bond forces between molecules in the system are not shearing during the viscosity measurement. In comparison, in one-component systems, all molecules are independently breaking and remaking molecular contacts similarly, where the true definition of viscosity applies. It should be noted that the *apparent viscosity* of polymer solutions and suspensions is commonly called just as *viscosity* in wood adhesive industry.

$$\text{Viscosity of suspensions} = \text{Viscosity of solvent x } (1 + 2.5f).$$

f = volume fraction of particles = volume of particles/(volume of particles + volume of solvent).

- *Example*: (a) A 5% clay by volume added to a 300-cps adhesive as a filler at 25 °C: Vis. = $300\,\text{cP} \times (1 + 2.5 \times 0.05) = 337.5\,\text{cps}$.

On the other hand, the polymer particles suspended in liquids can wholly dissolve into the solvent or liquid to result in one-phase systems to naked eye. In these cases, for low molecular weight materials, such as sugar, the viscosity will not change much from that of the suspended particles, as shown above with clay particles. On the other hand, if the suspended material is a polymer that dissolves in the (good) solvent, the polymer molecules of compact solid structures first separate into individual molecular strands to exert higher resistance to the force of flow of solvent molecules. Also, the volume of dissolved polymer molecules can further increase because the polymer strands can be solvated with solvent molecules, i.e. solvent molecules stick to polymer molecular strands and not shearing (due to stronger attractive forces between polymer strands and solvent molecules), resulting in bigger molecules. Thus, the overall volume of polymer particles can increase, resulting in increasing the viscosity beyond the value derived from the *Einstein equation*. This increase in viscosity of polymer solutions is one of the fundamental properties of polymers and it depends a lot on the polymer molecular weight and the degree of solvation, and the latter depends greatly on the nature of polymer molecular groups and solvent molecules, namely, whether they are either attractive (swelling), neutral, or repelling (balling up). Also, polymer materials are manufactured in industry to have various (high) molecular weight (MW) values by extending or controlling the polymerization reaction condition.

When a polymer is dissolved in a solvent at a specific weight concentration, the viscosity of the polymer solution generally increases with the molecular weight (M) increases, and this relation is described by the Mark–Houwink equation, where the exponent "a" is the expansion factor of the polymer that can change depending on the kinds of solvent. The intrinsic viscosity ($[\eta]$) is the viscosity value of polymer solution at the infinite dilution and its meaning can be extended to mean the common viscosity values of the polymer solution. i.e. viscosity increases with molecular weight. When a polymer solution is to be applied on a substrate using a spreader or sprayer, the viscosity of the solution must be within a suitable range irrespective of the polymer molecular weight. For example, when a polymer of certain high molecular weight is to be dissolved to have a solution of low viscosity, the concentration of polymer must be lowered, i.e. diluted. Adding more (same) solvent reduces the viscosity because of the lowered concentration. On the other hand, when a certain high concentration of polymers in the solution is desired, a lower molecular weight polymer of the same kind must be used. Both these remedies can change operational parameters in the application and the performance properties of resultant materials.

Mark–Houwink Equation of polymer solutions [1]

$$[\eta] = K \times M^{a}$$

K, constant; M, molecular weight; a, expansion factor; and $[\eta]$, intrinsic viscosity.

2.2.5 Methods of Viscosity Measurement of Liquid Materials

Gardner-Holdt bubble tube method: the test solution is filled in a standardized test tube until a small space (air bubble) remains on the top. After closing the test tube with a stopper and equilibrating it to a standard temperature, normally 25 °C, the test tube is inverted to cause the air bubble to rise or the liquid to fall down through the space between air bubble and wall of test tube. Then, the rising speed of air bubble is matched against a series of same-size tubes containing known-viscosity standard (liquid) materials. The matched viscosity value (same rising rate of air bubble) is the (kinematic) viscosity (in units of stokes or centistokes or from A to Z values) of the sample, which can be converted to centipoise (cps) units by multiplying with the density of test solution, which is measured separately.

Brookfield viscometer: the viscometer rotates a vertically positioned (stainless) spindle of a disc in a beaker with a measured force and rate, and the test solution is poured into the beaker until the disc is immersed in the solution. The viscometer measures the resistance (force) exerted by the sample solution on the disc and shows a reading (0–100). The rotating speed and the surface area of the spindle disc determine the shearing rates of molecules in the test solution. There are available several sizes of spindle discs and various rates of rotation, and their shear rates are calculated and provided in the form of a sliding chart. The spindle (number) and rotation speed can be changed in a stepwise manner for a particular test solution until the reading comes into the middle range at 40–60 scale, and use the reading to obtain the viscosity of the solution in cP or P from the provided sliding chart made of the spindle number and rotation speed. A wide range of viscosity at varying shear rates can be measured. This viscometer also allows non-Newtonian behaviors of solutions to be determined due to the varying rotation speeds and spindle disc sizes available. The spindle number of measurement has to be specified for measured viscosity values. The viscosity values can vary slightly when the shear rate becomes different. This measurement is also carried out at a specified temperature, commonly at room temperature or in a water bath of known temperature.

Cup methods: There are methods of using cups of various sizes having a hole at the bottom. Picking up one full cup of a liquid and measuring the time of passing down of the liquid until empty measures the viscosity according to the times measured with known viscosity standard materials. The method is useful for high-viscosity materials measured at elevated (cooking) temperatures. Zahn Cup and Ford Cup viscometers.

2.3 Polymer Solids Level and Specific Gravity of Polymer Solutions

Many polymers, such as liquid phenol-formaldehyde (PF) and urea-formaldehyde (UF) wood adhesive resins, are traded in the market either as solution volumes or weights and the price is based on the actual polymer solids content (concentration), and it is normally determined by evaporating the solvent, which includes curing reactions, and weighing the residues (*non-volatiles*). Inorganic materials, such as sodium hydroxide, are also present and these are measured by burning the non-volatile residues and weighing the ash residues. The ash weight should be subtracted from the nonvolatile value to obtain the true organic polymer solids level value. Also, in transportation and other handling of polymer solutions, the specific gravity of polymer solutions is a parameter often used because it reflects the polymer solids content. Specific gravity (s.p.) values of liquid resins depend on their non-volatile solids content because most organic polymer materials have s.p. values near about 1.4–1.5 and water has a value of 1.0, and a polymer solution's specific gravity value changes in this range depending on the polymer concentration. S.p. gravity is measured using a refractive index meter by comparing the readings with the graphic chart that relates to specific gravity values of the polymer solution, with which volumes of adhesive samples can be converted to weights, *vice versa*.

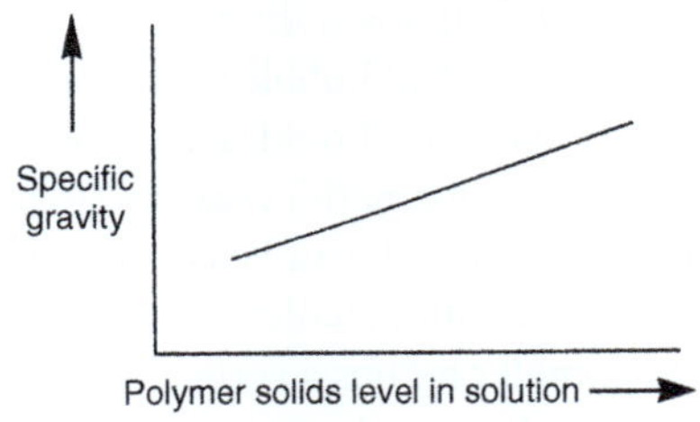

2.4 pH of Polymer Solutions and Buffers

2.4.1 pH

Acids and bases neutralize each other. Solutions of polymers that have acidic or basic groups or contain acidic or basic additives as a result of their manufacturing procedure show certain pH values: acidic, neutral, or basic pH values. Post-modification of these original pH values by neutralization, *etc.* sometimes results in drastic changes in the polymer solution, such as precipitation of polymers from the solution. Furthermore, acids and bases are often used as catalysts in polymer manufacturing and also as catalysts for curing reactions in the application stage. Most industrial polymer solutions especially aqueous polymer solutions and emulsions need to be made with a well-defined pH value and (acid or base) buffering capacity.

The pH value of a solution is defined as the negative logarithm value of the molar concentration of hydrogen ion (H^+) in the solution:

$$pH = -\log [H^+] \; : \; [H^+] = \text{Hydrogen ion concentration in moles per liter.}$$

In dilute solutions of an acid, the hydrogen ion concentration is the same as the molar amount of acid dissolved in a liter of water but, as the acid concentration increases, the actual proton concentration deviates slightly from the actual acid concentration. The hydrogen ion concentration of water solutions is at the same time related to the hydroxide ion (OH^-) concentration by the equation: $[H^+] \times [OH^-] = 1.0 \times 10^{-14}$ (in mole/l); (at 25 °C). Therefore, the pH value of a solution also defines the alkalinity of the solution. pH values are conveniently measured using a pH meter in the laboratory. Examples are shown in the table:

H^+ concentration (mol/l)	pH	Alkaline/Acidic	
1.0×10^{-14}	14.00		Strongly alkaline
5.6×10^{-10}	9.25		Mild alkaline
1.0×10^{-7}	7.00		Neutral
5.6×10^{-4}	3.25	Mild acidic	
1.0×10^{-1}	1.00		Strongly acidic
$1.0 \times 10^{+1}$	−1.00	Strongly acidic+	

2.4.2 Acidic Buffers and Buffer Capacity

A solution or suspension (emulsion) of materials in water, such as of UF resins, PF resins, or wood particles, is said to have an acidic buffer capacity if the material contains some amounts of any (weakly) basic materials such as amine (R-NH_2) groups or conjugates of weakly acidic materials as salts, namely the anionic conjugate components of mild acids, such as: CH_3COO^- Na^+; R—COO^- Na^+; and C_6H_5—O^- Na^+. The presence of common strong bases such as sodium hydroxide and sodium carbonate is also a part of the acidic buffer capacity but it is uncommon. When a strong acid is added to a solution or suspension of materials that contain these groups or materials, these (weakly) basic components would capture the hydrogen ion to form the corresponding weak acids, for example, RCOOH, thus consuming the strong acid added, resulting in making the solution only weakly acidic, not strongly acidic, to the extent of the amount of conjugate anionic materials present. Weak acids dissociate less, and the hydrogen ion concentration does not increase as much as when the same amount of strong acid is added to pure water. This capturing of the strong acid added is continued until the acid conjugate anions in the material are all depleted. These delayed pH decreases are the *characteristics of acidic buffer capacities of materials or adhesives*. In the absence of the buffer species, the solution pH value will decrease rapidly. Therefore, the concentration of these weakly basic (anionic) components in a solution is the (total) acidic buffer capacity of the solution.

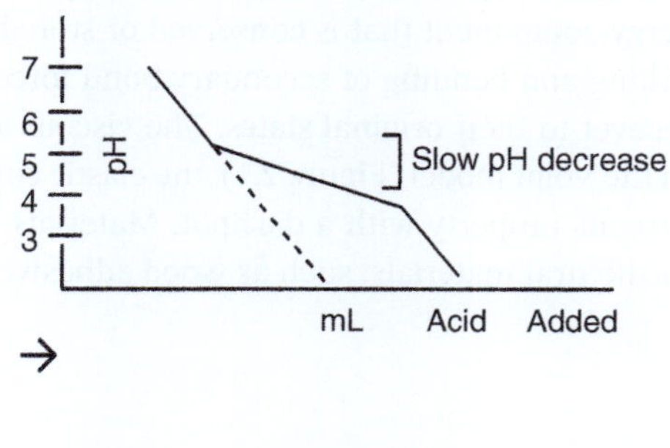

2.5 Solid Properties of Polymer Materials

2.5.1 Adhesive Property of Polymer Solids

A wood adhesive must be a liquid for application, and the glueline formed between wood pieces must be cured to a solid. The adhesive property (adhesion) of solidified adhesives is difficult to measure directly without breaking them apart. They are therefore mostly evaluated by testing (breaking) the glue bonds for strength and durability performances. The adhesive layer's polymer molecules' good compatibility with the surface molecules of wood substrates is an important criterion for a successful bonding, in terms of polar and nonpolar interactions. Details of adhesion phenomena of wood with various solidified adhesives will be discussed in later Chapters (Chapters 18 and 19).

2.5.2 Cohesive Property of Adhesive Polymer Solids

Cohesive property of a polymeric adhesive material means the strength of bulk polymer material, such as in stretching and bending. In successful bonded wood products, obviously, *the cohesive strength* of cured adhesive must be greater than both *the adhesion strength* and the *substrate strength*. Otherwise, the bonded product in testing will fail in the middle of the gluelines. The cohesive strength of a polymer material depends on the kind of polymer, molecular weight, and molecular structure, *etc.* For *thermoplastic polymers*, crystalline structures occur, i.e. regularly repeating chemical structures that result in the cohesive strength, while for *thermosetting polymers*, amorphous cross-linked polymer structures occur to give the cohesive strength, depending on the kind of polymer and the extent of cure or the molecular weight attained.

2.5.3 Cohesive Strength Measurement and Viscous and Elastic Responses

Cohesive strengths of polymer materials [2] can be measured, for example, by applying *tensile* or *bending* forces on the sample. When stressed, most organic structural materials experience a strain, i.e. dimensional increases or bending angle changes. When the stress is removed, the dimensional changes would recover if the stress was stopped within the elastic proportional limit of strain. In stressing of polymer samples, the *viscous and elastic* responses (properties) are involved. *Viscous response* occurs through the positional changes of atoms, i.e. involving actual moves and rearrangements of polymer molecules and atoms against neighboring components, i.e. actual breaking of secondary bond forces, thus dissipating (consuming) a part of stress energy into heat (loss response). It is often defined as the *loss modulus*, i.e. the lost energy component. This loss modulus component is often coming from the amorphous region of polymer structure. In comparison, the *elastic response* represents the stress energy component that is conserved or stored in the polymer molecular structures (in stretching and bending of secondary bond forces) and later causes the polymer structures to recover to their original states. The viscous and elastic components are often illustrated with the Voigt model (Figure 2.1): the elastic component is represented with a spring and the viscous property with a dashpot. Materials with high elastic properties constitute useful structural materials, such as wood adhesive polymers. The viscous

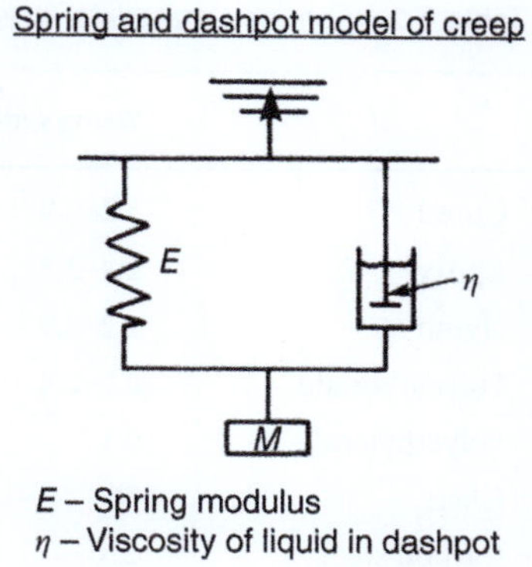

Figure 2.1 Spring and dashpot model of creep. Source: [3] Reproduced with permission from the American Chemical Society.

component is far more pronounced in thermoplastics than in cured thermosetting materials because of the lack of cross-linking bonds in the polymer structure. Also, the viscous property increases with temperature for most thermoplastic polymers, which soften and melt in the end to become 100% viscous. Most cured structural wood adhesives are based on thermosetting polymers that have only minor viscous components.

2.5.4 Measurement of the Elasticity of Solid Polymer Materials

Among various modes of strains used in testing for elasticity values, the tensile (stretching) mode is illustrated, (Figure 2.2) where the pulling force (F) applied at the two ends of a sample is related to the strain through the constant modulus (E):

$$\text{Stress (Force)} = \text{Modulus } (\mathbf{E}) \times \text{strain, or } \mathbf{E} = \text{Stress/Strain, where strain} = \Delta l / l_0.$$

The strain is the fraction of the stretched length over the original length of the sample. Modulus of elasticity (MOE) is a set value for a given solid material, unchanging within the proportional limit of stretching. Thus, the equation states that the strain is proportional to the stress within the proportional (elastic) limit of the bulk sample. Stretching the sample further than the proportional limit will start to cause a flow (slippage) of material components within the sample and eventually lead to a rupture. Rupture means breakage of secondary as well as primary chemical bonds between atoms and molecules. The elastic modulus (slope) measured in the stretching mode is commonly called *Young's modulus*. Materials having high Young's moduli are stiff materials. Similar relationships exist for compression, bending, and shearing forces, with respect to the strains. In Table 2.1, Young's moduli and the maximum strain values are shown for typical structural materials.

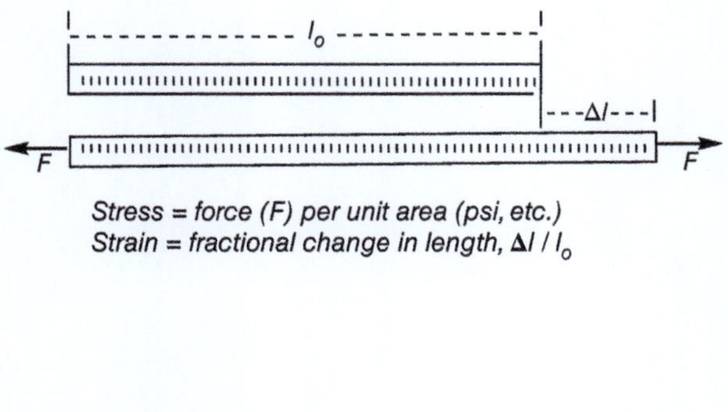

Figure 2.2 Measurement of elasticity of solid materials.

Table 2.1 Common materials' strength properties [3].

	Young's modulus ($\times 10^3$ MPa)	Max strain (%)
Cured PF	5.2–6.9	1–1.5
Epoxy	2.4–3.5	3–6
Nylon 66	1.2–2.9	60–300
Polycarbonate	2.1–2.4	100–130
Polyethylene	0.1	90–800
Glass	70	
Carbon steel	210	
Southern Pine	10	

References

1 Flory, P.J. (1953). *Principles of Polymer Chemistry*, 308. Ithaca, New York: Cornell University Press.

2 Ferry, J.D. (1961). *Viscoelastic Properties of Polymers*, 66. New York: Wiley.

3 Pocius, A. and Dahlquist, C. (1986). *Adhesion and Adhesives - Audio Course*. Washington, DC: ACS p. 88, 95.

3

Thermosetting and Thermoplastic Wood Adhesives and Practices

Wood adhesives have been used since the ancient times, but since 1940s the gluing of wood has become an increasingly more important component in the conversion of timber into marketable forest products, as well described by Alan Marra in 1956, [1] due to (i) development of new synthetic adhesives that offered durable bonds and became producible in large quantities with modern manufacturing methods, (ii) depletion of choice timbers that had been used as higher grade structural materials, (iii) advantages of glued products such as improved strength and better dimensional stability attained by selection, location, and orientation of different wood components, (iv) versatility of glued products in design with respect to sizes and shapes, and (v) many other advantages such as utilizing inferior wood species in non-critical parts in the construction of wood composites. Until recently, glue and gluing practices of wood had been an art than a science, partly because it encompasses three wide areas of science, i.e. chemistry, engineering, and wood technology. Many advances have been made since then, but gluing must be carried out with utmost care with prescribed procedures since there is no fool-proof adhesive or gluing process available, as will be found throughout this course. The adhesive bonding operation for wood includes: (i) adhesive selection based on the composition, (ii) adhesive application on wood substrates, (iii) bringing the mating surfaces into contact, (iv) applying pressure, (v) curing or hardening of adhesive, and (vi) evaluation of resultant wood bonds. In 1998, Terry Sellers, Jr. and C. R. Davis reported the wood adhesive consumption volumes and costs of North America as shown in Table 3.1 [2].

3.1 Selection Criteria of Wood Adhesives

There are about six classes of wood adhesive resins: urea-formaldehyde (UF), melamine-urea formaldehyde (MUF), ISO, PF, phenol-resorcinol-formaldehyde (PRF), and polyvinyl acetate (PVAc) resins. Adhesive selection for a wood composite manufacturing plant is made in the beginning based on the following considerations: (i) Wood element size and wood product type: plywood, lumber, OSB, PB, FB, furniture, etc. (ii) Type of stresses to put on the bonded products: structural or semi-structural. Most thermosetting wood adhesives belong to structural uses, and thermoplastic wood adhesives are for semi-structural uses. (iii) Durability requirements: cyclic dimensional changes in temperature and humidity, chemical and biological attack; classified commonly as "exterior" or "interior" uses. (iv) Bonding condition and bonding time: high temperature or low (room) temperature

Adhesives and Finishes for Wood: For Practitioners and Students, First Edition. Moon G. Kim.

Table 3.1 Estimated 1998 consumption of major wood adhesives in North America [2, 3].

Polymer type	North American consumption	Cost ($/lb) (delivered)
Amino	1060 kt (59.5%)	0.30 (UF)
Phenolic	658 kt (31.9%)	0.45 (liquid)
Isocyanate	90 kt (5.1%)	0.65 (Powder)
Polyvinyl acetate	50 kt (2.8%)	0.85
Resorcinol type	7 kt (0.4%)	
Soy-casein	5 kt (0.3%)	
Total	1780 kt (100%)	

Source: Society of Plastic Industries, industry sources, and other calculations. Reproduced with the permission of the Forest Products Society and Mr. C. R. Davis of Georgia Pacific Corp.

bonding process and bond warranty. (v) Cost of adhesive. (vi) Color; colorless adhesives (UF, MUF, ISO, PVAc) are preferred for interior and furniture applications, and PF and PRF resins give dark or dark brown glueline colors.

Wood composite manufacturing plants select adhesives based on these considerations made at the start of the plant's construction. Later, adhesive modifications and sometimes adhesive changes can occur for various reasons in some minor ways; also, the resin suppliers frequently make minor changes in adhesive formulations for lowering the cost of resin, for better performance, and for adapting to new incoming wood materials. In general, a drastic change in the kind of adhesive for a wood product plant is rather rare but can occur as the recent introduction of isocyanate (ISO) resins as core layer binders in OSB plants. Development of new or improvement of existing wood adhesive resins requires a lot of effort and can be mostly carried out by resin manufacturing companies. Most wood adhesives currently used in the industry have gone through such development processes, and even now the adhesive resins are being improved or adjusted continuously for better overall performance or for adjusting to new raw wood materials and mat moisture content-related parameters.

3.2 High Temperature-Curing Thermosetting Adhesives and Curing Conditions

UF resins are used for bonding interior-grade wood composite boards such as hardwood plywood, particleboard, and medium-density fiberboard. UF resins are usually cured at about 300–350 °F under pressure with press times ranging 2.5–3.0 minutes for ½″-thick boards. Resin loading levels run at about 8–10% based on the oven-dry wood weight for PB and MDF. Wood particles' moisture content before resin application runs at about 2–4%.

PF resins are used in bonding of softwood plywood, OSB, and hardboard. PF resins are more expensive than UF resins on weight basis, but PF resins' loading levels are lower than UF resins. The curing temperature is somewhat higher than for UF resins at 350–400 °F. Resin loading levels run 3–4% for OSB based on the oven-dried wood weight and cured in 3.0–5.0 minutes. Wood moisture contents for ½″-thick boards before (liquid PF) resin application run at about 4% for OSB and slightly higher for softwood plywood.

Isocyanate resins (ISO): ISO resins are often used as core layer binders in OSB manufacturing in place of PF resins because somewhat lower press temperatures can be used and greater moisture-forgiving characteristics are obtained. Various handling and toxicity problems are associated with ISO resins due to their high chemical reactivity and water insolubility.

3.3 Room Temperature-curing Thermosetting Wood Adhesives and Processes

PRF and MUF resin adhesives are applied on wood substrates and cured at room or slightly elevated room temperatures. Room temperature curing is convenient and low cost in comparison with the hot-pressing processes since no large hot press nor the energy of heating is needed. However, the method generally requires longer press (or clamp) times (six hours or more) and therefore the methods are useful only in manufacturing of large (long) solid wood assemblies such as laminated beams and trusses. The laminated beam curing presses consist of a zig and a series of clamps with pressure gauges operated with torque wrenches. In this process, the wood substrates are freshly planed, and the liquid adhesive is applied, normally, in parallel lines as beads or by brushing. Then the pieces are mated and clamped to a certain pressure (tightened), and the assembly is allowed to stand in a warm room while the clamp pressure is monitored and adjusted often to compensate for the compaction of wood occurring slowly in the early clamping period. In the case of manufacturing of arched beams, clamps are arranged in a *jig* of pre-determined curve form, and individual adhesive-applied wood pieces are forced (bent) into the zig, followed by tightening of the glue-applied assembly. Due to the high reactivity of the resorcinol component in PRF resins, the adhesive adequately cures at room or slightly elevated temperatures. After unclamping, a maturing period of a few days is normally followed, and sanding and dimension cutting can be performed before shipping.

Casein (protein from milk) and MUF resin adhesives are also used similarly at room temperature in the manufacturing of wood laminated beams, but their lower water resistance property restricts to the bonding of beams for interior structural applications. End-jointed and edge-bonded wood products, such as tabletops and wooden benches, are also bonded with these resin adhesives and used in semi-structural, interior uses. The bonding processes using PRF, MUF, and casein adhesives are also done continuously at elevated temperatures by passing the wood assembly through a radiofrequency (RF) heating unit.

3.4 Room Temperature-curing Thermoplastic Wood Adhesives

PVAc resin adhesives are thermoplastic in nature and used in general wood bonding in furniture manufacturing. They are more aesthetically appealing due to colorless gluelines in comparison with PRF resins. They are also more convenient to use since no heating is needed, and they are low-toxicity materials. The durability of PVAc resins is poor than PRF,

MUF, and casein resin adhesives and therefore they are recommended only for interior, low-load uses.

Hot-melt resin adhesives are thermoplastic in nature and used in limited wood bonding in furniture manufacturing and other applications, such as corrugated paper board bonding. The adhesives are applied on substrates in a molten form and let them harden to form bonds under pressure from cooling by loss of heat to the substrates.

3.5 Adhesive Application Methods and Loading Rates and Costs

Adhesive application methods: depend on the wood elements and adhesive kinds:

- *Plywood*: Formulation of a PF resin with fillers and extender, followed by curtain coating, roll coating, or spraying.
- *OSB, PB, FB*: Neat liquid resin is sprayed on wood elements using a spinning disk sprayer or pressure nozzle in the tumbling or mixing type blenders. Powder PF or UF resins are applied by blowing into tumbling blenders using compressed air sprayer.
- *Lam beam, I-beam*: Adhesive mixes are continuously extruded in bead form on the lumber.
- *Furniture*: Adhesive mixes are brushed or extruded in bead form.

Adhesive application (loading) rates in wood composites affect the performance of bonded products and the bonding cost. The application rate is monitored and controlled by measuring the total weights of wood and resin going into the blending step. When comparing various wood composite products, the general rule on the adhesive application rates per unit volume of products is that the larger the surface area per unit volume of wood elements, the higher the adhesive application rate needs to be. Thus, the order is: laminated lumber < plywood < OSB < PB ≤ MDF. Hardboards, although the wood elements are smaller, generally use lower (PF) adhesive loadings than OSB or PB due to the extra bond formation occurring by the lignin present on the fiber surfaces and by the higher board densities used. However, since the cost of adhesives, the selling price of the bonded wood products, and the cost of wood raw materials used for these different wood composite products are all different, such a simple comparison of adhesive costs is less meaningful in practice.

Adhesive cost is a substantial component in the manufacturing of most bonded wood composite products, reaching 20–25% of the total production cost for OSB and 10–15% for softwood plywood. The resin application (loading) rate is continuously monitored and adjusted to obtain the best possible, acceptable products at minimum adhesive loading levels or costs. Attainment of adequate bonding performance at the lowest adhesive application rate is a continuing goal for most wood composite manufacturing plants.

3.6 Adhesive Curing by Hot Presses and Other Heating Methods

Multi-opening hot presses are used in the manufacturing of plywood, OSB, PB, and FB. Multi-opening presses have various numbers of openings that simultaneously open and close; usually 14–16 openings. The platen size ranges up to 8 × 24 ft for the multi-opening presses.

Continuous presses or long hot-presses (up to 50 ft) are used in manufacturing of thin particleboard and more recently for OSB, MDF, and laminated veneer lumbers (LVL). The presses are heated by circulating steam or hot oil, heated by burning wood wastes or fuel oil.

Steam-injection presses are used for manufacturing thick oriented strand board, particleboard, and medium-density fiberboard, which subsequently cut into dimension lumbers, moldings, and other products. Steam injection presses are designed for injecting high-temperature steam into the mat through many small holes bored in the hot press platens. Various steam pressures (temperatures) can be used, and platens are maintained at a desired temperature using a separate heating system. In this process, the wood mat core is heated very quickly by injected steam, resulting in short hot-pressing times for thick board mats than standard hot presses. However, the injected steam introduces, by steam condensation, extra moisture into the mat and this becomes another variable that often interferes with the curing of binder resins, especially for PF resins, which are currently judged to be unsuitable in this application for OSB. UF and ISO resins show less of such problems if the starting mat moisture is kept at normal levels. Steam injection pressing is currently used for manufacturing of oriented strand lumber (OSL) and door-core particleboard by Weyerhaeuser Co., and medium-density fiberboard by a small manufacturer. This pressing method is not used for plywood or LVL because stream penetration into solid wood veneers does not occur efficiently.

Radiofrequency (RF) heating processes: Alternating electric fields in the RF range of the electromagnetic wave spectrum, applied through radio tubes, cause (electric) dipoles of polar molecules in the wood substrate and adhesive layer to vibrate at the same frequency. The vibrating dipoles in wood/adhesive molecules relax to release heat, thus converting electric energy into heat. This method is also called induction heating. Since adhesives normally contain more polar molecules (water) than wood substrates, the RF heating results in a preferential heating of adhesive layer. This method is used in spot-welding in the manufacturing of end-jointed lumber and also in the glueline-heating in face-bonding of laminated wood beams in a continuous manner using PRF and MUF resin binders. This method of heating allows the bonding of large or long objects without using large presses or ovens that would require a large capital investment. The RF heating method consumes expensive electric energy in operation, but it is compensated for by short hot-pressing times. In RF curing processes of I-beams, the adhesive-applied wood members enter the first section of the process, where they are consolidated with (plastic) pressure hands and then pass through a RF heating unit at a predetermined speed. The glueline and nearby area of wood substrate are preferentially heated during this passage due to the higher moisture content in the glue-applied sections. The line speed and the adhesive curing rate are adjusted in such a way that the mechanical pressure that holds the wood assembly together can be relieved soon after the bonding assembly passes the RF heating unit. The adhesives used are specifically formulated for fast curing speeds, and the curing of glueline in the RF process is normally made to advance just enough to hold the wood member together and the cure continues until the heated section cools down.

3.7 Evaluation of Adhesive Bonds

Evaluation of cured wood adhesive bonds are normally done by mechanical testing of bonded wood composite products. Cured adhesives in the gluelines are difficult to separate from wood for observation or testing for strength. Adhesion occurs between the wood surface and glue layer surface and the adhesion quality is more dependent on the surface-surface bonds rather than adhesive layer. Mechanical tests are also done after aging of samples for durability evaluation of gluelines. The adhesion theories and the evaluation methods will be discussed in later lectures (Chapters 18–21).

3.8 Summary

Advantages of gluing, gluing methods, selection criteria of thermosetting and thermoplastic wood adhesives, and current usage levels of various adhesive resins are presented. UF, PF, and ISO resins are high-temperature curing thermosetting resins, and PRF resins, casein resins, and MUF are room or elevated room temperature curing thermosetting resins. PVAc emulsion resins and hot-melt resins are thermoplastic resins, and the former are room temperature curing, and the latter are melt/cool curing adhesives. The heating methods include multi-opening hot press, continuous or long hot press, steam injection press, and RF heating continuous press. For strength and durability reasons, thermosetting resins are more extensively used for bonding lumber and wood composite products. Thermoplastic resins are mainly used for interior grade, low-strength wood bonds, such as in furniture manufacturing. Adhesive costs or spread rates are relatively low if calculated based on wood weight but significant when calculated based on the selling prices of bonded products, and therefore, lowering adhesives costs and improving adhesion procedures is an ongoing task of many wood composite product plants.

References

1 Marra, A.A. (1956). Developments in glues and gluing. *Forest Products Journal* 6 (2): 54.

2 Sellers, T. Jr. (2001). Wood adhesive innovations and applications in North America. *Forest Products Journal* 51 (6): 13.

3 Davis, C.R. (2005). Presentation in class lecture. Georgia-Pacific Resins, Inc.

4

Principles of the Curing of Thermosetting and Thermoplastic Wood Adhesives

As discussed in Section 3.3, phenol-formaldehyde (PF), urea-formaldehyde (UF), and isocyanate (ISO) wood adhesive resins are "low molecular weight, live polymers" and finally polymerized (reacted) and cured (cross-linked) at elevated temperatures to high molecular weight cross-linked polymers in the heating (hot-pressing) step, also called as "thermosetting" or curing step, and this type of resins are called *thermosetting resins* or sometimes "*thermosets*." The bond quality of wood composite boards made with these resin adhesives depends on how rigorously the adhesive curing and wood consolidation processes are carried out. This thermosetting and consolidation process entails physical/chemical transformation of adhesive layers and movements of wood elements, as well as expelling of air and moisture from the wood mat and adhesive layer. The resultant cured PF, phenol-resorcinol-formaldehyde (PRF), and ISO resin gluelines have high strength and high durability and are useful for exterior structural uses, while those of UF, melamine-urea formaldehyde (MUF), and casein resins have limited durability and are useful for interior uses.

On the other hand, polyvinyl acetate (PVAc) and hot-melt wood adhesives are *thermoplastic resins*, which are manufactured as high molecular weight solid materials in the resin manufacturing step. Since adhesives must be in a liquid form to function, *PVAc adhesive resins* are commonly available as an emulsion in water, which needs to be dried in the bonding process. *Hot-melt adhesives* are available as solid rods to be melted before application. The curing steps for thermoplastic resins involve mainly physical changes of adhesive polymer molecules to solids by evaporation of the water of emulsion or by cooling to room temperature of molten hot melt. No major polymerization or crosslinking reactions occur during the bond formation and no external heating of wood elements is needed, and their glueline performances are limited in strength and durability, making them useful only for interior, ordinary temperature, and nonstructural uses. In general, the solidification process of *both* thermosetting and thermoplastic adhesives is often termed as "curing."

4.1 Principles of Curing of Thermosetting Adhesive Resins at Elevated Temperatures

In manufacturing of wood composite boards, such as oriented strand board (OSB), plywood, particleboard (PB), and medium density fiberboard (MDF), a liquid (50–60% water content) or powder-type (no water content) thermosetting PF or UF adhesive resin is applied

Adhesives and Finishes for Wood: For Practitioners and Students, First Edition. Moon G. Kim.

to wood substrates, which are pre-dried to 2–8% moisture contents. For core layers of OSB, neat (100% solids, liquid) ISO resins are also used. Mat is then formed and transported to the hot press. The time taken from the resin application step to the hot-pressing step can vary from about 30 minutes to an hour or longer. Loss of moisture from adhesive layer to wood substrate often occurs as well as some thermosetting curing processes of resins in this time span depending on the transit time length and the room temperature.

In the hot-pressing step, the mat is consolidated by the pressure and heat of press platens, maintained at 300–430 °F, during the three to five minutes of pressing time for ½ thick boards. Longer press times are needed for thicker boards. In this hot-pressing process, the moisture in the mat (moisture of wood substrates and water-borne resins) moves from the surface layers of mat to the core layer of mat and, from there, is slowly expelled through the edges of mat during the latter part of hot-pressing period. *The adhesive molecules lose almost all water, further polymerize, cross-link to high molecular weight polymers, and at the same time form adhesive bonds to wood substrates.* Various physical and chemical changes occur in the adhesive layer as well as in wood substrates, simultaneously and/or sequentially, as discussed below.

4.1.1 Viscosity Changes of Adhesive Layer Before Hot-Pressing

The viscosity of liquid PF or UF adhesive layers or droplets on wood substrates increases due to loss of water by absorption into wood as well as by evaporation into the environment. The drying of resins results in stickiness (tack) that dies off as moisture loss goes further. This tackiness can hinder the resin blending and board matting process, but it also helps keep the formed mats from breaking. Then, the adhesive layer reaches a semi-solid consistency at the time of hot-pressing. Powder PF or UF resins or ISO resins would change little in viscosity in this stage. Overall, chemical (polymerization) reactions would proceed only minimally in this stage, and in most cases an adequate lay-up or matting time is available, or adhesives are so adjusted in the resin manufacturing.

4.1.2 Spreading and Penetration of Adhesive Layers in Hot-Pressing

In the hot-pressing, the semisolid-like adhesive layer initially becomes fluid again due to the increasing temperature. With the help of compaction pressure, the fluid adhesive layer or droplet spreads out and also gets transferred to surfaces of opposite wood pieces and can also penetrate into wood substrates due to the lowered viscosity. This movement of adhesive layers during the liquid stage is very necessary in achieving good bonds, especially because the adhesive resin loading level is limited for minimizing the resin usage. In the case of OSB manufacture, the wood surfaces are only partly covered with adhesive droplets, resulting in spot bonding of wood strands. Adhesive layers that optimally penetrate into subsurfaces of wood substrate, i.e. into cut lumens and broken cell-wall structures would result in a good "anchoring" and "repairing" effects of damaged wood structures incurred in the wood (strand) preparation step. This movement of adhesive materials occurs only until the adhesive layer remains fluid. However, an over-penetration of adhesive can also occur to result in a starved glueline, i.e. a lack of adhesive layer in the glueline proper if the adhesive layer becomes too fluid. On the other hand, a poor adhesive movement can also occur if the adhesive viscosity becomes too high at this stage of pressing, either by a wrong resin formulation or from resin pre-curing occurring in the lay-up period. The result will be a poor coverage of glueline with low bond strength.

4.1.3 Viscosity Decreases by Moisture Content Changes

In addition to the viscosity decrease by temperature rise, the viscosity can also decrease by layer-by-layer dilutions due to movements of the moisture in wood mat during the hot-pressing. The mat moisture (about 10% based on wood weight) moves from the face layers to the core layer as an advancing zone of (accumulated) high moisture content. The moisture moves to the core layer and is then expelled into the environment. The moisture front may help the adhesive spreading but also can result in diluting the resin layer to result in a washed-out glueline resulting in poor bond strength. The probable explanation for the poor bond strengths occurring when bonding wood with higher moisture contents than normal would be this dilution effect of adhesive layers. Ideally, an ample polymerization needs to be attained in the resin manufacturing step. In general, over-penetration, resin dilution, and resin squeeze-outs from the glueline are undesirable in wood bonding operations, which can be controlled to some extent by adjusting the degree of polymerization of adhesive polymers in the resin manufacturing step and by formulating resins with fillers and extenders, when applicable. PF resole resin components are water-soluble until cured and, therefore, they are more sensitive to the moisture movement discussed above than water-insoluble, faster-curing UF or ISO resins. In all cases, this moisture effect is significant and should be minimized as much as possible by adjusting such resin parameters as molecular weight and sodium hydroxide content (for PF resins) as well as by limiting the initial mat moisture content.

4.1.4 Viscosity Increases and Curing of Adhesive Layer by Polymerization Reaction

In this hot-pressing period, thermosetting resin polymers in adhesive layers polymerize further, resulting in viscosity increases to the point where the adhesive spreading is greatly reduced. This viscosity increase has to be synchronized with the press closing schedule, that is, *the major viscosity increases have to occur after all wood elements are pushed and placed in their proper consolidated positions.* Too early or too fast increases of viscosity can restrict the resin mobility (resin flow) to result in lower resin efficiency or lower bond strength, often called as pre-cure. On the other hand, if the viscosity of adhesive layer increases too slowly, the resin flow would be good, but since the total hot-pressing time is limited, the cure of adhesive may not be completed at targeted press opening.

4.1.5 Flow and Wetting of Adhesive Layers and Molecular Adsorption

Thermosetting adhesive resins consist of low molecular weight polymers (oligomers), which polymerize to high molecular weight polymers as hot-pressing progresses. Also, the adhesive layer becomes ~100% resin-solids, i.e. most water is dissipated. These two processes result in increasing the viscosity of adhesive layer, but it needs to be understood that *the adhesive layer needs to flow even when it becomes 100% resin-solids for a good bonding to occur.* Simultaneously, the (uncured) 100% resin-solids adhesive layer molecules have to "wet" the molecules of surface layers of wood substrate for a good bonding to occur. "Wetting" means that adhesive molecules and molecules of wood surfaces come into atomic contacts, within ~4.0 Å distances. Air, water, or any extraneous small molecules should be expelled for wetting to effectively occur. Also, a good flow of adhesive molecules, after most water is gone, is necessary for good wetting to occur and it is often called "good" or "poor" flow of adhesive resins, which can be discerned by observing the broken test pieces of boards. A poor flow of adhesive resins often results from "dry-outs" – the resin failed to flow by drying up (by pre-curing) before making the needed close atomic contacts. Adhesives (molecules) must be so selected to have *the chemical affinity* to the molecules of wood surface, such as good *polar–polar* interactions for a good wetting to occur ahead of gelation of resin, and it needs to be maintained throughout the life of glued products for the adhesive layer to perform adequately as a structural component. Most thermosetting wood adhesive resins, such as UF, PF, and PRF resins, contain many hydroxyl or amide groups and wood surfaces also have many hydroxyl groups, a good condition for a good wetting to occur.

4.1.6 Gelation of Adhesive Layer

Gelation of adhesive layer is defined as the stage in the curing "process where the adhesive layer stops flowing." A gel is a "weak" solid having an established internal molecular structure that would *break apart* if sufficient pressure is applied. A gel ("gello") can be cut and even be chewed. The gelation of thermosetting adhesives occurs when the viscosity of adhesive increases to an extremely high value and then some crosslinking of polymer molecules has occurred in such way that the whole adhesive layer structure is loosely interconnected by polymer chains. The typical liquid character, such as flow, disappears and some physical strength of the adhesive layer arises (yield strength). In terms of thermosetting polymer chemistry, the gelation means that at least one (long and branched) polymer molecular strand is formed in the resin mass that spans through the entire adhesive layer, an irreversible point in the transformation of thermosetting resin adhesives. Gelled gluelines, if broken, such as by a belated press closing, will not get repaired by subsequent heating or pressuring because the gelled and broken glue layer had already lost the capacity to flow and cannot be molecularly re-mixed to re-make a homogenous phase. Therefore, the gelation of glueline has to occur in the curing process only after all the bonding elements are put into proper places, i.e. after the board mat is pressurized to the final thickness. And, after the gelation point, no wood element should move against each other in position too much, or the weak-gelled glueline will break and cannot be repaired. Therefore, a premature gelation will ruin the adhesion process, and it is normally observed in wood bonding as pre-cure; an example being the loose flakes or particles most often found in the top-most surface layers of OSB or PB. The poor flow of adhesive layers can occur also by dryouts, i.e. the resin layer is dried up, but its melting point is higher than the prevailing temperature and not melting. An ideal adhesive resin needs to have a long flow at the curing temperature, but this is often difficult to obtain for most thermosetting adhesive resins because the resin also should cure fast (short press time) for optimum productivity.

4.1.7 Vitrification of Adhesive Layer

After gelation, the polymer molecules further undergo polymerization and crosslinking (or thermosetting or curing) reactions, leading to three-dimensional cross-linked macromolecular structures. Initially, the gelled glueline begins to gain in physical strength and starts to show the typical "rubbery" characteristics (Young's modulus up to 10^8 dynes/cm^2) [1]. Then, as the curing continues, the rubbery structure transforms into a typical solid structure with Young's modulus value of 10^{10} dynes/cm^2 or higher at the curing temperature. This transformation from rubbery state to the real solid state is termed as the "vitrification" of the resin. The original meaning of the word is "to become a solid glass" from molten glass in the glass manufacturing industry. The hot-pressing of boards can be stopped soon after the vitrification point. If the press is opened before this point, the adhesive bond, still having the rubbery characteristics, may not withstand the expanding pressure of the mat, resulting in a spring-back (increase in board thickness) or even blow of the mat by the internal steam pressure. In manufacturing of OSB in the laboratory, this inadequate vitrification of glueline at press opening is often seen as a delamination of board, often accompanied with a blowing noise depending on the amount of gas (mostly water vapor) entrapped in the board at the time of press opening. Often, OSBs would show internal bond strengths of about 25 psi or less if the vitrification process of the glueline has not proceeded adequately at the time of press opening. A longer hot-pressing time would make the vitrification process complete to result in a whole board.

4.1.8 Post-Curing of Adhesive Layer

Even after the vitrification point, the curing reaction in the glue layer normally continues due to the remaining, unreacted chemical functional groups available for further reaction. The curing after the vitrification point can continue after the hot press is opened, i.e. normally during the "post-cure" or "hot-stacking" period, a term used for keeping the hot boards stacked together overnight or longer in PF resin-bonded OSB and softwood plywood manufacturing plants. The process often results in improved internal bond strengths of boards. However, in the manufacturing of UF resin-bonded particleboard or MDF, the resin curing speed is relatively fast, and the hot-stacking is not recommended, rather some cooling of the hot boards is followed to prevent the over-curing, i.e. the breakdown of chemical bonds in cured UF resin polymers.

4.2 Relationship of Temperature and Moisture Content in Hot-Pressing of Wood Composite Mats

In hot-pressing of wood composite boards, the consolidation of wood elements, expulsion of air and moisture, compaction of mat to a higher overall density, and curing of gluelines are accomplished in a relatively short time, where the temperature, pressing pressure, internal gas pressure of mat due to water and other gaseous products, board thickness, and moisture content in the mat continuously change [2–4].

4.2.1 Platen Pressure, Mat Thickness, and Density Profiles

In the beginning of hot pressing, a relatively high platen pressure is applied, and the target mat thickness (0.5″ for example) is reached in a predetermined closing time (about 15–45 seconds for OSB). The initial high platen pressure is maintained for a short period of time, and then it gets lower as a result of the compaction of wood elements going on, and this lower pressure is maintained until near the press opening. As the wood elements are plasticized by the increased temperature and moisture, the mat's resistance to the pressing pressure decreases, resulting in lowered pressure in the later period of pressing. The curing of adhesive layers begins as the temperature rises. In most industrial processes, the thickness gauge on the hot press continuously measures the mat thickness, and the target value is maintained by adjusting the platen pressure. As the curing of adhesive layer progresses and the moisture in the mat is expelled as steam (water vapor) from the edges of the middle layer, the vapor pressure in the mat decreases but remains significant until the end of hot-pressing. As the mat's consolidation and adhesive bond strength increase the vapor pressure of mat approaches a value low enough to allow the press to open. In practice, there is often some pressure remaining due to the residual steam in the board. If the vapor pressure is too high, boards can have blown spots, usually in the center of boards. For this reason, it is important not to have excess moisture in the mat to begin with.

The mat consolidation time can be adjusted or shortened, to a limited extent, by varying or increasing the platen pressure, but a maximum pressure exists in practice due to the cost of necessary hydraulics. Also, the mat consolidation time cannot be too short in order not to cause physical damage to wood elements by a fast compaction occurring ahead of plasticization of the wood. Also, the trapping of steam inside the mat needs to be avoided from fast closing. The press closing rate is an important parameter for board characteristics such as the density profile across the board thickness, i.e. a higher surface-layer density for fast closing rates, etc. Also, a press closing that is too slow would result in pre-curing of resins in the mat surface layers because of the heat transfer occurring to adhesive layers before wood elements make complete contacts. Fast closing is generally desirable, and surface layers of commercial boards have higher densities than middle layers.

4.2.2 Temperature Changes and Moisture Movements in Board Mat

In practice, hot-pressing is done at ~300 °F for UF resins and 350–400 °F for PF resins. The heat flows from platens toward the core-layer of mat during hot-pressing. The mat surfaces quickly attain the temperature of platens and then the heat transfer occurs from the surface layers toward the core-layer of mat largely by moisture movements. Normally, the overall mat moisture content ranges up to about 8–10% for PF resin-bonded OSB or UF resin-bonded particleboard. The moisture in the surface layers vaporizes, moves into the next inner layers, and then condenses to give back the heat, and this heat transfer process continues, layer by layer, until the two high moisture fronts meet in the core layer. Wood is a poor heat conductor and direct heat transfers from face-layers to the core-layer occur minimally at least in the beginning. In the end, a steady-state temperature gradient is established from the surface layers to the core layer and this state continues as long as the moisture exists in the surface layers. The moisture in the core-layer moves horizontally to the edges of mat and gets expelled into the air, and toward the end of pressing time, the face-layers would become devoid of moisture and the core-layer would harbor the most of moisture remaining in boards. Also, the core-layer temperature can remain significantly lower than the face-layer temperature, but it has to reach a significantly higher temperature (>250 °F) than the boiling point of water. Also, the bond strength of core-layer must be high enough to allow the press to be opened at an average mat moisture content of about 2–3% for PF and

UF resin-bonded wood composite boards. The moisture at the press opening stage resides mostly in the narrow core-layer and therefore, the core-layer tends to become the weakest layer, breaking along with the internal bond strength testing of boards. Also, toward the end of press cycle, the moisture content in the core-layer would be lowest in the center region and highest in the edge regions of board, often resulting in lower internal bond strength values for board edges, or edge splitting might occur when the cycle time is too short.

4.2.3 Extents of Resin Cure in Hot-Pressing

Thermosetting resins, such as PF and UF wood adhesive resins, start to cure when heat is applied with the curing rate increasing with increasing temperatures but decreasing with increasing moisture contents. Because of these temperature and moisture dependency of curing reactions, the resin in face-layers cures first and the resin in the core-layer cures in the later stage. In industry, the hot-pressing step is the bottleneck of production so the hot-pressing time is minimized by using a faster-curing resin in the core-layer and a slower curing resin in the face-layers. However, it is still true in most cases that the face-layers tend to over-cure while the core-layer tends to under-cure in daily manufacturing with given press times. The hot press in wood composite plants is the most expensive equipment, and therefore their production capacities are built to be just enough to produce the targeted production volumes. Therefore, using a resin with fast curing time, especially in the core-layer is preferred to obtain the shortest hot-pressing time, and also, in the case of PF resin-bonded OSB and softwood plywood, the hot press time is adjusted to be just enough to hold the boards together and the curing is augmented by post-curing, i.e. hot-stacking of boards.

4.3 Curing of Thermosetting Adhesives at Room Temperature

The general curing mechanism described above for the hot-pressing of wood composite boards can be applied to the room-temperature curing of laminated beams, etc. using PRF or catalyzed MUF-type thermosetting wood adhesive resins. First of all, the adhesives are formulated with fillers to have relatively high viscosities, although the resin solids levels are kept at about 50%. The adhesive is applied on wood substrates by brushing or beading. Also, the room-temperature curing process is carried out over a longer period of time (>6 hours), and, therefore, the initial viscosity reduction occurs slowly because of the absence of external heating. A small extent of temperature increase occurs when the curing reaction starts and generates some reaction heat, which causes some viscosity reduction to result in increasing the adhesive spread or exudation. The moisture in the adhesive layer dissipates mostly through absorption into the wood substrates, and the overall changes in total substrate moisture content are very small. Adhesive spread, wetting, and bonding occur at room or slightly elevated temperatures. Often, PRF resin adhesives contain some organic solvent additives, such as ethanol or ethylene glycol, to help improve adhesive's wetting and penetrating properties. The curing room is maintained at warm temperatures through all four seasons not only to speed up the resin cure but also to help improve other resin curing characteristics. In some plants, these adhesives are also cured under the radiofrequency (RF) heating condition in a continuous press at a relatively good speed.

The adhesives are formulated to have relatively high viscosities and also with necessary spreading and wetting properties under the room temperature condition, so that the adhesive layer is contained within the glueline against the relatively low pressing pressure (100–200 psi) applied on wood substrates. Relatively speaking, the adhesive cure progresses more slowly at room temperature than in the hot-pressing or RF heating processes, and therefore the adhesive wetting, penetration, flow, and curing occur in a more orderly manner. Therefore, consistently good bonding results are usually obtained for rightly formulated adhesive resins. Commercial PRF resin-bonded laminated beams are known to be durable for exterior uses, while MUF resin-bonded laminated beams or wood blocks are semi-durable and therefore used in interior applications.

4.4 Curing of Thermoplastic Emulsion Adhesives

This process is typically represented by PVAc emulsion adhesives used in the furniture manufacturing industry [5]. Again, the adhesive spreading, penetration, and wetting have to occur at room temperature. These processes are initially promoted by the water in the adhesive formulation (~50% resin solids). Other factors also play important roles in the later stages of curing because, in time, most of the water dissipates from the glueline and only the thermoplastic resin mass and fillers remain in the glueline. The polymer's molecular weight and polymer chain structures and the monomer unit's chemical compositions all must have optimum values for both *the cohesion and adhesion* to take place efficiently. The uniqueness of PVAc emulsion resins as a good interior use wood adhesive, although they are for low-load nonstructural uses, arises from the polymer molecular composition that amply promotes the cohesion and adhesion processes at room temperature tailored by controlling the molecular weights to optimum values in the resin manufacturing step.

PVAc wood adhesives are supplied as an emulsion, composed of small swollen particles of PVAc polymer molecules suspended uniformly (not settling) in water with the help of small amounts of surfactants and colloids. (It is likely that one particle consists of one polymer molecule.) When the adhesive layer applied on wood surface is dried sufficiently, the solid PVAc polymer particles would first get stacked up in the glueline to form a solid-like structure that is similar to clay particles slurried in water, settled down, and drained of water. Then, PVAc polymer molecules have such chemical structures and group components that the particles collapse from the swollen structure and are consolidated. The inter-particle bonds (secondary bond forces) then begin to arise in a process termed as "coalescing," where a single solid phase is established out of many particles and results in acquiring the cohesive strength of the dried glue line. The polymer particles interact with each other first through their peripheral, loose chain groups, by secondary atomic bond forces such as hydrogen bond and van der Waals forces, and by the vacuum (pressure) arising from the evaporation of water molecules existing between polymer particles, as schematically shown in the figure below. Often, *a coalescing agent*, (organic solvents) that evaporate slower than water, is formulated as a component in PVAc-type emulsion glues and paints. The coalescing agent remains in the polymer layer after evaporation of water and starts to work, i.e. making the polymer chains move more freely for coalescing. The coalescing agents finally evaporate out of the solidified adhesive region. This coalescing process needs some thermal energy supplied by keeping the room at slightly higher temperatures, and there are also chances that all water and coalescing agents are evaporated out with the coalescing process remaining incomplete, as when gluing is carried out at cold temperatures.

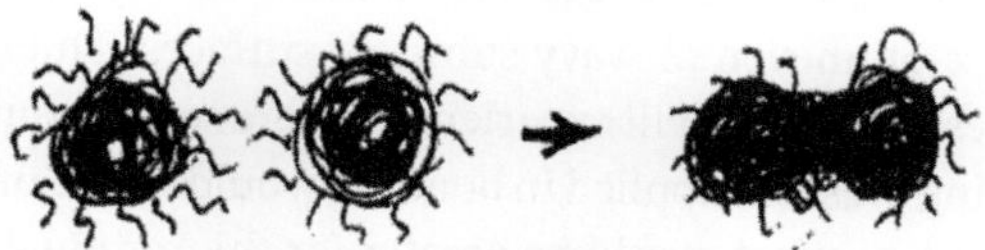

A schematic representation of the coalescing process.

This coalescing process is different from the normal melting/cooling process (as in hot melt) since the core portions of polymer particles would not be involved much. The coalescing process is also the main mechanism in the formation of a continuous film for latex (emulsion) paints. As we often see, however, these paint films or gluelines are not as durable or strong as those made from thermosetting resins or by melting process of thermoplastic resins because the coalescing process, although it is low cost and very convenient to carry out, does not truly mobilize all the polymer structures in the glueline to form a unitary structure throughout. Coalescing boundaries are to be formed, and they often constitute the weakest points in the solidified glueline structure. Therefore, the cured adhesives are not so resistant to water nor are they stable at high temperatures, and their uses are limited to ordinary temperatures and under relatively dry conditions and light loads. PVAc adhesives are widely used in the furniture manufacturing industry and will be discussed in detail in Chapters 14 and 15.

4.5 Volume Contractions of Adhesive Layers upon Curing

For most wood adhesives, there are physical and chemical reasons that cause the adhesive layers to contract in volume upon curing or cooling down, often manifested as a shrinkage of glueline thicknesses that can affect the bond strengths negatively [6]. The gelation of thermosetting resin adhesives and the initial solidification of thermoplastic adhesives means loss of water and resultant shrinkage of adhesive layer thicknesses, but these shrinkages are mostly compensated by the external pressure applied. However, after the gelation stage, the polymerization reaction further progresses, resulting in compaction of polymer molecules' volume and generation of some reaction water. Then, the water and solvents remaining in the adhesive layer would diffuse out as the cure progresses resulting in creating voids in the polymer adhesive layer that can be filled in by neighboring polymer chain strands, resulting in further contraction of adhesive layers. In case of high-temperature curing thermosetting resins, the contraction also occurs from the cooling of cured hot adhesive layers to room temperature.

The contraction is often observed when wood composite boards are hot-pressed and the adhesive is cured adequately, i.e. the board thicknesses are often thinner than the platen gaps used. Of course, this results in increasing the board density above the targeted value albeit the thickness may recover to some extent by re-equilibration of moisture. In practice, of course, some intra-particle voids could remain unfilled or filled by impurities to become defects, which can be the initiating points of breakage in strength testing of boards. Almost all adhesives, i.e. water-borne, solvent-borne, or 100% solids, have this volume contraction tendency, the extent being greater with adhesives having low solids content at the time of gelation and curing, and with the condensation-type adhesives, such as PF and UF resins. Further, the glueline volume contraction is a special concern for laminated beam and plywood manufacturing, where the glueline is more clearly defined and the glueline thicknesses vary because of uneven or wavy substrate surfaces. Thicker glueline spots are inevitably present in these cases and will experience a more severe contraction than thinner glueline spots. The mating pressure applied in bonding would not be uniformly transmitted, and the glueline contraction effect would be greater for thicker glueline area, resulting in harboring more voids or defects. This glueline contraction is also the reason why all-purpose gap-filling glues are difficult to formulate and most of the time they do not work well. Probably you experienced this phenomenon of an ineffective bonding when trying to glue wood pieces with Elmer's glue for a mating arrangement that is loose or has a wide gap. The failure is partly due to this contraction effect as well as to the fact that thick gluelines tend to show lower strength values than thinner gluelines. Gluelines mean *thin* gluelines in the first place, which will be discussed in Chapters 18 and 19.

4.6 Thermal and Moisture Expansion/Contraction of Adhesive Layers

Other factors that can weaken the gluelines are the different extents of moisture and thermal expansion and contraction between adhesive layers and wood substrates. The coefficients of thermal expansion of gluelines and wood substrates are often different. When a glued piece made at a higher temperature is cooled to room temperature or to any cold temperature, the wood substrate and glueline dimensions change differently to result in strains in the interphase layer, but this thermal problem is minimal for wood composites. On the other hand, wood expands and contracts greatly upon moisture content changes, and gluelines of most wood adhesives do not change as much as wood substrates, and herein lies the cause of many facets of durability problems of bonded wood products, to be discussed in Chapter 20.

4.7 Summary

Principles of the curing of thermosetting and thermoplastic wood adhesive resins are presented. Thermosetting wood adhesive resins are low molecular weight, *reactive* materials available as low-viscosity liquids. When a thermosetting adhesive is applied on wood, the low molecular weight reactive resin molecules are activated by catalyst and heating. The adhesive resin then goes through the stages of curing under pressure: loss of carrier solvent if any, spreading and penetration, viscosity decreases and wetting, and then viscosity increases by the polymerization of reactive functional groups, gelation, vitrification, and post-curing to reach to the cross-linked thermoset polymer gluelines. During these resin curing processes, the moisture in the mat is only slowly escaping, often counteracting the resin curing processes to the last minute. Thermoplastic wood adhesive resins, after being applied on wood substrates, the loss of water occurs through absorption into wood, and the high molecular weight polymer particles get packed up against each other and go through a process called *coalescing*, where the peripheral polymer chain strands interact to form a unity phase by forming secondary bond attracting forces. Hot-melt thermoplastic wood adhesive resins are applied as a melt at about 200 °C and the curing process lets the resin adhesive cool to room temperature, leading to solid thermoplastic polymer gluelines.

References

1 Ferry, J.D. (1961). *Viscoelastic Properties of Polymers*. New York: Wiley.

2 Wolcott, M.P., Kamke, F.A., and Dillard, D.A. (1990). *Wood and Fiber Science* 22 (4): 345.

3 Kamke, F.A. and Casey, L.J. (1988). *Forest Products Journal* 38 (6): 38.

4 Bolten, A.J., Humphrey, P.E., and Kavvouras, P.K. (1989). *Holzforschung* 43: 265.

5 Lambourne, R. (ed.) (1987). *Paint and Surface Coatings, Theory and Practice*. West Sussex, England: Ellis Horwood Ltd.

6 Pocius, A. and Dahlquist, C. (1986). *Adhesion and Adhesives – Audio Course*. Washington, DC: ACS.

5

UF and MUF Wood Adhesive Resins (Manufacturing and Resin Chemistry)

Urea-formaldehyde (UF) resins, including melamine-urea-formaldehyde (MUF) resins, are currently the largest volume wood adhesive resins: 3.3 billion pounds in North America and 15.6 billion pounds worldwide in 2006 [1], followed by phenol-formaldehyde (PF), isocyanates, phenol-resorcinol-formaldehyde (PRF), polyvinyl acetate (PVAc), and some other specialty resins. UF resins are used as binders for particleboard, medium-density fiberboard, hardwood plywood, and other wood composite products. UF resins have the advantage of low cost and ready availability of raw materials, good adhesion for many different wood species, and a wide range of curing temperatures (10–150 °C). Also, cured UF resins have acceptable durability for interior uses and are colorless, light-fast, and arc-resistant compared with PF resins.

Reaction of urea with formaldehyde was first reported by K. Hoelzer in 1884 [2] and 1885 [3], and K. Goldschmidt reported the results of reacting urea (U) and formaldehyde (F) at various F/U mole ratios in 1896 [4]. Then, the polymeric nature of reaction products and their potential use as adhesives, moldings, etc., were first patented by H. John in 1918 [5]. Wood bonding with UF resins was first proven and industrial production began in 1931–1934 in Germany. UF resins were one of the first synthetic adhesives, but research on UF resin chemistry went through many twists and turns in the beginning due to the complex reversible chemistry and associated analytical difficulties [6, 7], and finally in 1950s, several fundamental research results were reported that laid the groundwork for much of the current understanding of UF resins. Then, the particleboard industry started and began to use large amounts of UF resins. A comprehensive book on UF and melamine-formaldehyde (MF) resin technologies was published by Wirpsza and Brzezinski in 1972 [8]. In the late 1970s, ^{13}C NMR analysis results on UF resins began to appear, and thus much of the complex chemistries of UF resins were cleared up by 2000s.

5.1 Raw Materials of UF Resins

$$CH_3OH + {}^1\!/_2O_2 \rightarrow CH_2O + H_2O + 38\ \text{Kcal} \qquad (5.1)$$

$$CH_2O + {}^1\!/_2O_2 \rightarrow HCOOH \qquad (5.2)$$

$$CH_2O + H_2O \leftrightarrow HOCH_2OH + 15\ \text{Kcal} \qquad (5.3)$$

Adhesives and Finishes for Wood: For Practitioners and Students, First Edition. Moon G. Kim.

$$n\,HOCH_2OH \leftrightarrow HO(CH_2O)_n - H + (n-1)H_2O \quad n\text{ (degree of polymerization)} \tag{5.4}$$

$$CH_2O + CH_2O ----> CH_3OH + HCO_2H \tag{5.5}$$

$$HOCH_2OH + CH_3OH \leftrightarrow HOCH_2OCH_3 + H_2O \tag{5.6}$$

$$n\,HOCH_2OH \leftrightarrow HO(CH_2O)_n - H + (n-1)H_2O \tag{5.7}$$

Formaldehyde was discovered in 1859 by A. Butlerow [9]. It is a highly reactive chemical widely used in the manufacturing of various thermosetting polymers. Formaldehyde is a gas at room temperature with a pungent odor detectable in air by human nose at a concentration of 0.001 ml/l. Currently, methanol is the raw material in manufacturing of formaldehyde. Methanol is a volatile organic chemical/solvent (b. p. 64.7 °C), produced from natural or oil-field gas (methane, ethane, propane, etc.) in a partial oxidation reaction process. It is a large-volume chemical produced worldwide and used as solvent and fuel additive. A large volume of methanol is used to produce formaldehyde, with the plant commonly standing as a separate unit in PF or UF resin manufacturing plants. Formaldehyde is produced in a catalytic oxidation process (Eq. (5.1)), where a mixture of methanol vapor and air (oxygen) is passed through a bed of catalyst (silver or molybdenum oxide) at 350–450 °C. The methanol is (partially) oxidized to formaldehyde and water [10, 11]. The exiting hot gas of formaldehyde and water is scrubbed with cold water to obtain the water solution of formaldehyde. A minor side reaction in this process is a further oxidation of the generated formaldehyde to formic acid (Eq. (5.2)).

Formaldehyde in water exists as (hydrated) methanediol (Eq. (5.3)) and its **$CH_2{=}O$** (Formaldehyde Molecule) oligomeric forms, i.e. polyoxymethylene glycols (Eq. (5.4)). This equilibrium is almost unchanging between pH 2–10. The reaction rate of this equilibrium reaction is slowest in the pH range between 5 and 7. Formaldehyde, even in pure water, easily undergoes the Cannizzaro reaction (Eq. (5.5)), often catalyzed by metallic cations such as Na^+, Al^{+3}, and Fe^{+3} ions present as impurities. Formaldehyde solutions made initially in the neutral pH tend to gradually go to the acidic pH due to this reaction. Formaldehyde solutions in water are stable at room temperature only up to 30% concentrations, and higher concentrations when kept at room or lower temperatures result in forming precipitates of polyoxymethylene glycols with $n > 3$ (Eq. (5.4)). Addition of methanol helps stabilize those formaldehyde solutions by hemi-formal formation (Eq. (5.6)). In wood adhesive industry, 50% solutions of formaldehyde in water, without methanol addition, is stored and used by keeping the solution at 50–60 °C. Formic acid content in formaldehyde solutions needs to be lower than 0.05%, and metal salts such as aluminum and iron are undesirable because they catalyze the oxidation reactions. Iron salts also cause resultant UF resins to appear with some colors. Therefore, all equipment is made of stainless steel. Formaldehyde is also marketed without methanol as pre-condensates of urea, such as UF-85, which has 59% formaldehyde, 26% urea, and 15% water, kept at pH 6.5–7.5. Use of this formaldehyde source in UF resin manufacture can be more economical compared with other methods, such as using methanol-stabilized formaldehyde. UF-85 type pre-condensates are stable at −20–40 °C for up to six months.

Industrial formaldehyde solutions often show white precipitates of paraformaldehyde, which makes it necessary to correct for the actual concentration of formaldehyde present in the liquid part used in resin manufacturing. Also, the formic acid present in formaldehyde solutions needs to be neutralized in the resin manufacturing step. The formaldehyde solutions in UF resin manufacturing industry are normally kept at a 50% concentration, a value chosen for economy and convenience in UF resin formulations, in order to have no water additions nor evaporative removal of water necessary from the resin synthesis mixture for resins having about 60–64% resin solids contents and at F/U mole ratios of 1.10–1.30. Fifty percent formaldehyde solutions are also used in the manufacturing of PF and PRF wood adhesive resins. *Formalin* is formaldehyde solution made stable at room temperature by adding about 7.0% methanol as stabilizer, available from chemical suppliers at ~37.0% concentration [10]. The methanol needs to be removed in manufacturing of wood adhesive-type UF resins.

Formaldehyde is currently classified as a health-hazardous material and a potential carcinogen [12]. *Polymeric forms of formaldehyde* are manufactured from aqueous formaldehyde solutions by polymerization and dehydration as white solid flakes called paraformaldehyde or paraflakes (Eq. (5.7)), which have a strong formaldehyde odor. The degree of polymerization of these precipitates reaches up to 8 with melting points in the 80–120 °C range, and when heated or mixed with water they readily de-polymerize to give back formaldehyde. Paraformaldehyde dissolves in water slowly but quickly in the presence of base and is therefore used as a source of formaldehyde in the hardener component of PRF resin adhesive systems.

Urea is a large volume chemical used widely as nitrogen-source fertilizers [10]. Natural gas, crude oil, or coal are the raw materials, from which hydrogen and carbon dioxide are obtained. Nitrogen is extracted from the air. In manufacturing of urea, the hydrogen is reacted with nitrogen in the presence of catalyst under a high pressure to produce ammonia in the well-known *Haber process* (Eq. (5.8)). The ammonia is reacted with carbon dioxide at 160–200 °C under a high pressure, which pushes the equilibrium reaction to the right side to form ammonium carbamate, which is then dehydrated to urea (Eq. (5.9)). Commercial urea is supplied as white granules with a melting point of 133 °C and it is nontoxic to the skin. The urea used in UF resin manufacture is higher grade material compared to the fertilizer grade urea. In overall, formaldehyde and urea are both large volume, low-cost materials that are also used in various other industrial and agricultural applications thanks to their functionality and low raw material costs, which also leads to the low cost and good availability of UF resin adhesives used for bonding various low-cost wood composite boards throughout the world.

$$\tfrac{1}{2}\,N_2 + 3/2\,H_2 \leftrightarrow NH_3 \tag{5.8}$$

$$2\,NH_3 + CO_2 \leftrightarrow (NH_2COO^-\;NH_4^+) \rightarrow NH_2CONH_2 + H_2O \tag{5.9}$$

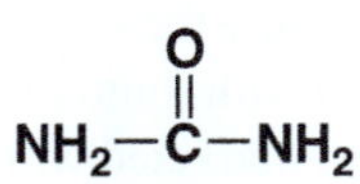

Urea Molecule

5.2 Urea-Formaldehyde (UF) Resins

UF resins are thermosetting pre-polymer resins manufactured often in a liquid form in water in the resin manufacturing plants. They are transported to wood composite board

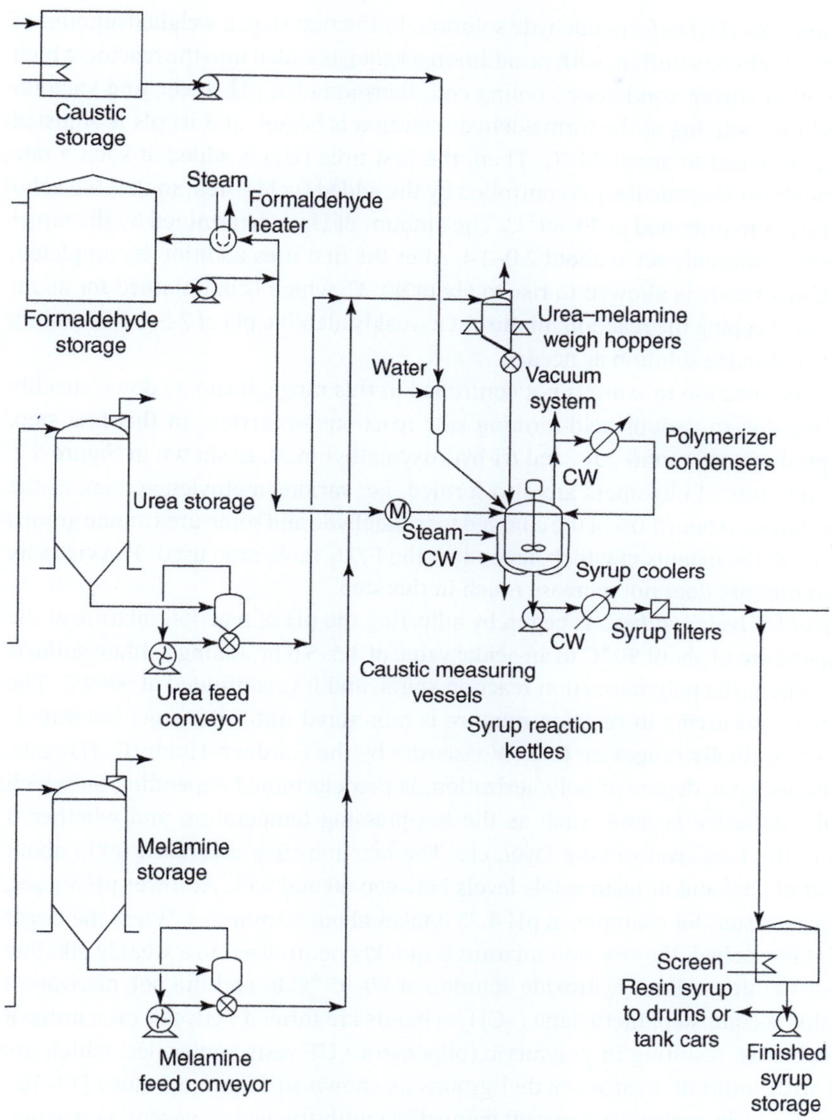

Figure 5.1 A schematic layout of UF and MUF manufacturing plant. CW = cold water. Source: Courtesy of Encyclopedia of Chemical Technology, 4th ed. Volume 2, Wiley-Interscience 1995. p. 614.

plants and, after being added with a catalyst, sprayed into the wood furnish at about 8–12% resin solids loading levels and cured (thermosetting) at an elevated temperature and under pressure in the hot-pressing process. A schematic layout of UF and MUF resin manufacturing plant is shown in Figure 5.1. The reaction kettle is where the reaction is going, and heating is done with steam and cooling is done with cold water. Urea or melamine and formaldehyde and caustic solutions are introduced in separate lines.

First step of resin synthesis procedure: Wood adhesive UF resins are typically synthesized in three consecutive steps [13], with urea being added in two parts, often described as first urea

(U_1) and second urea (U_2) to formaldehyde solution. In the first step, a weighed amount of warm 50% formaldehyde solution, with no additional water, is added into the reactor, which is equipped with a stirrer, condenser, cooling coil, thermometer, pH meter, and vacuum capability, and then stirring of the formaldehyde solution is begun, and its pH is adjusted to about 8.0 and heated to about 70 °C. Then, the first urea (U_1) is added at such a rate that the reaction's exothermic heat is controlled by the added (cold) urea, so that the solution temperature is maintained at 70–80 °C. The amount of U_1 is determined by the target F/U_1 mole ratio commonly set at about 2.0–2.4. After the first urea addition is completed, the reaction temperature is allowed to rise to about 90 °C, which is maintained for about 30 minutes while keeping the reaction mixture at a weakly alkaline pH of 7.5–8.5 by adding dilute sodium hydroxide solution as needed.

If the pH of the reaction mixture is not controlled in this range, it can go down steadily to about pH 6.2 due to various acid-forming side reactions occurring. In this first step, the reaction products are mono-, di-, and tri-hydroxymethylureas, as shown in Figure 5.2. Some minor amounts of oligomers are also formed, i.e. various methylene-ethers of the hydroxymethylureas. About 7.0% of the charged formaldehyde and some urea amide groups remain unreacted, the extents of which depend on the F/U_1 mole ratio used. The viscosity of the reaction mixture does not increase much in this step.

Second step of UF resin synthesis is begun by adjusting the pH of reaction mixture at the reaction temperature of about 90 °C to an acidic value of 4.5–5.0 by adding a dilute sulfuric acid solution. Then, the polymerization reaction starts, and it is continued at 90–95°. The viscosity increase occurring in reaction mixture is monitored until the target viscosity is reached, which normally ranges up to T–W viscosity by the Gardner–Holdt (G–H) scale. The target viscosity, i.e. degree of polymerization, is predetermined depending on which application the adhesive is used, such as the hot-pressing temperature and whether it will be used in the face-layer or core-layer, etc. The reaction time can reach up to about 110 minutes at pH 5.0 and at resin solids levels between 60 and 65%. At lower pH values, the reaction goes faster; for example, at pH 4.75 it takes about 80 minutes. When the target viscosity point is reached, the reaction mixture is quickly neutralized to a weakly alkaline pH by adding a dilute sodium hydroxide solution at 90–95 °C to end the polymerization reaction. In this second step, methylene ($-CH_2-$) bonds are formed between urea units of hydroxymethylureas, resulting in polymeric (oligomeric) UF resin molecules, which are also bonded with multiple hydroxymethyl groups as shown in Figure 5.3 (top) [13–16]. These oligomeric resin molecules are well compatible with the water present as reaction medium due to the hydrophilic hydroxymethyl groups, and therefore the reaction mixture is commonly not separated into the resin-rich and water-rich phases and looks clear or

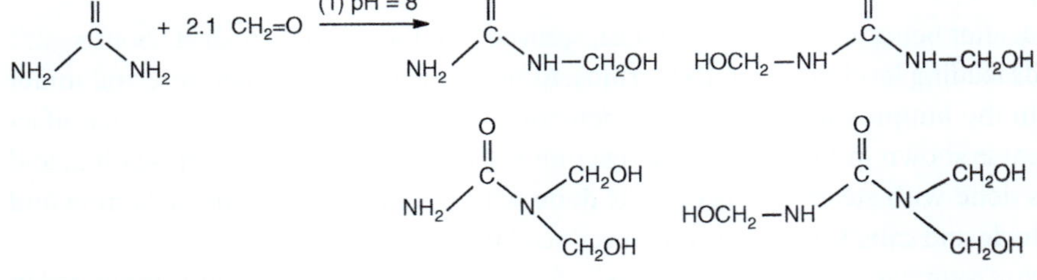

Figure 5.2 First step of UF resin synthesis [13].

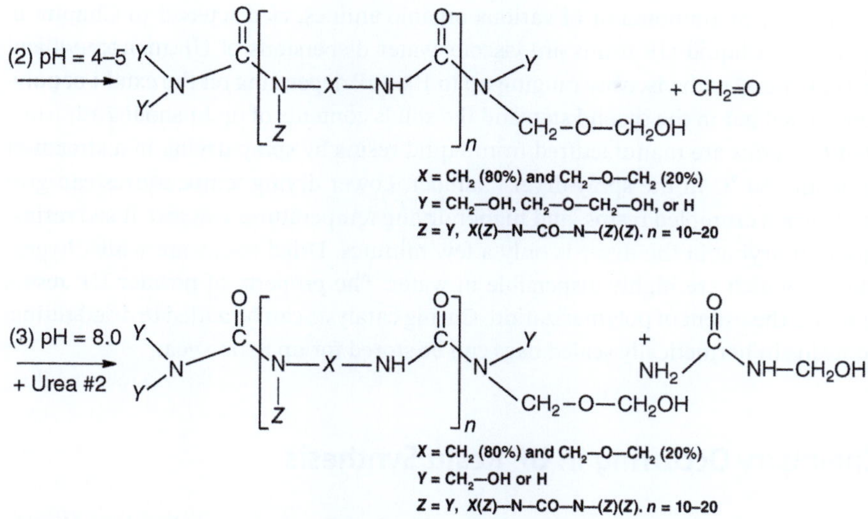

Figure 5.3 Second and third steps of UF resin synthesis [16].

may show a slightly milky appearance due to the formation of minute insoluble oligomer particles that are uniformly distributed throughout the liquid resin phase. In this reaction step, the viscosity of reaction mixture increases from about A_1 continuously and can be allowed to go up to about T–W by G-H scale.

Third step of UF resin synthesis begins by cooling down the reaction mixture, and when the temperature reaches to about 70 °C, the second urea (U_2) is added with vigorous stirring. The amount of U_2 is predetermined by the target (combined) formaldehyde/urea mole ratio values of resins $[F/(U_1 + U_2)]$, currently at about 1.10–1.25 or lower for particleboard or MDF adhesive resins. The cooling is continued until the resin mixture reaches to room temperature, which normally takes about two hours or more in the resin manufacturing industry. In this cooling period, the second urea reacts with the free formaldehyde present in the reaction mixture to form various monomeric hydroxymethylureas, as shown in Figure 5.3 (bottom). Some free urea groups remain unreacted, and a small amount of free formaldehyde (0.5% or less based on liquid resin weight) also remains unreacted in finished resins [13–16]. Freshly manufactured UF resins look clear or slightly milky in appearance and have 60–65% resin solids contents and pH value of about 8.0. The viscosity of finished resins will be lower than the viscosity attained in the second step reaction but could be above the target viscosity values for the resin, depending on the history of cooling procedure. Normally, the viscosity may come out at around O-P at the end of cooling step and then, it can slowly go lower to reach to the final (target) viscosity. This initial, gradual viscosity lowering is due to the mixing of monomeric urea into the polymeric resin components and migration of hydroxymethyl groups from the first urea (U_1) to the second urea (U_2) components by reversibility. Afterwards, during the transportation and storage period, the viscosity of resins can decrease and then increase, i.e. storage time is limited. The viscosity of resin at the point of use is a rigorously adhered parameter due to accurate pumping and spraying (atomization) operation required, and therefore UF resins are manufactured to reach to the target viscosity and transported in cooled containers, especially for long-distance shipments and in summer months. Finished UF resins can be added with curing catalysts as needed, which are often acid salts of ammonia or of various organic amines, as discussed in Chapter 6. Thus, the resultant liquid UF resins are viscous water dispersions of UF monomers and oligomers (polymers) with viscosity ranging up to 1000 cP depending on the extent of polymerization carried out in the second step and the solids contents of up to about 70%.

Powdered UF resins are manufactured from liquid resins by spray-drying in a stream of hot air at about 150 °C in the spray-dryer chamber. Lower drying temperatures can give incompletely dried crumpled resins, and higher drying temperatures can give fused resins. The duration of drying in the dryer is only a few minutes. Dried resins are white, hygroscopic powder, which are highly dispersible in water. The property of powder UF resins also depends on the extent of polymerization. Curing catalysts can be added to dried resins. Powdered resins in hermetically sealed bags can be stored for up to one year.

5.3 Chemistry Occurring in UF Resin Synthesis

Hydroxymethylation reactions of UF resins in the alkaline pH were reported by several early researchers, who also identified various monomeric (=methylolureas) and dimeric hydroxymethylureas of UF resins [17–22]. Later, De Jong and De Jonge reported their pioneering work on the kinetics of reactions between urea and formaldehyde at 35 °C and various pH values, summarized in Figure 5.4 [23–27], where the methylolurea-forming reactions (upper curve, UF) occur relatively fast in all pH ranges. In industry, in the first step of resin synthesis, the weak alkaline pH range of 7–8 is preferably used because of complications that can occur in the acidic pH, as will be discussed Section 5.12. The first-step alkaline reactions proceed rapidly and are completed in less than 30 minutes at 90 °C, and further reactions do not occur much more [13].

In current resin manufacturing practices, the F/U_1 mole ratio is kept at about 2.0 or slightly higher or lower and, therefore, the product composition at this point, when cooled to room temperature, would correspond to a di-methylolurea on average. This standstill of reaction, as reported by De Jong and De Jonge in Table 5.1 [23], shows that the hydroxymethylation reactions are **reversible reactions** of forming mono-, di- or tri-hyroxymethylureas from urea and formaldehyde. The reverse reaction rates (k_{-1}) for

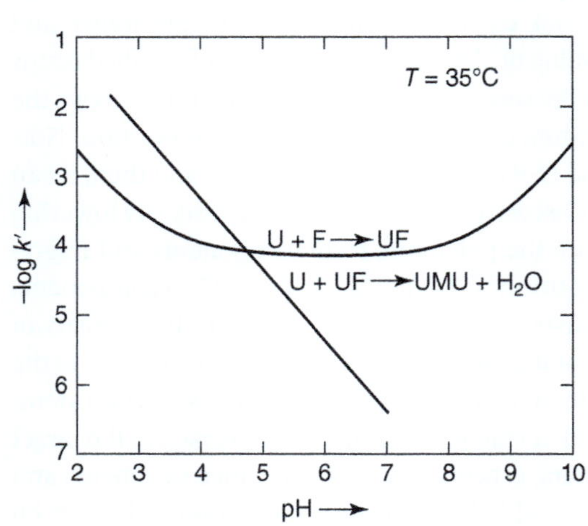

Figure 5.4 Influence of pH on the addition reaction rates of U and F by general acid and base catalysis and on the condensation of UF and U by specific acid catalysis, by J.I. de Jong and J. de Jonge [24].

Table 5.1 Forward (k_1), reverse (k_{-1}) reaction rates, and formation constant (K) of mono-methylolurea [23].

$$NH_2CO{-}NH_2 + CH_2{=}O \leftrightarrow NH_2{-}CO{-}NH{-}CH_2{-}OH$$

Temperature (°C)	pH	k_1 (l/moles.sec.)	k_{-1} (l/moles.sec.)	$K = k_1/k_{-1}$ (l/moles)
25	7	0.60×10^{-4}	1.20×10^{-6}	50.0
35	7	1.10×10^{-4}	3.20×10^{-6}	34.4
42	7	1.90×10^{-4}	6.80×10^{-6}	27.9
59	7	5.70×10^{-4}	31.0×10^{-6}	18.4
35	4	4.40×10^{-4}	14.0×10^{-6}	31.4

reactions with an F/U mole ratio of 1.0 are significantly slower than the forward reaction rates (k_1), resulting in significant (positive) formation constants (K) at various temperatures and pH values. First, it should be noted that this type of reversible reaction is ongoing at all ordinary temperatures. Thus, the forward reaction is about 50 times faster than the reverse reaction at 25 °C and pH 7.0, and both reaction rates increase with increasing temperature, but the formation constant decreases because the reverse reaction rate increases faster with the temperature. It is noted that this reversibility is also effective in acidic pH, for example at pH 4.0, and therefore also in the second synthesis step of polymerization and acidic curing steps, as well as in the neutral or alkaline pH of resin transportation and storage periods. Therefore, the reversibility becomes the cause of formaldehyde emission problems of UF resin-bonded wood composite boards, an inherent deficiency of UF resins, i.e. the only way to reduce the free formaldehyde content or emission potential is to increase the urea amide group content, which means lowering of the resin's F/U mole ratio.

$$\begin{aligned} &R'{-}NH{-}\mathbf{CH_2{-}O{-}H} + H^+ \rightarrow [R'{-}NH{-}\mathbf{CH_2{-}O = H_2}]^+ \\ &\quad \rightarrow \left[\mathbf{R'{-}NH{-}CH_2{}^+}\right] + \mathbf{H_2O} \end{aligned} \tag{5.10}$$

$$\begin{aligned} &\left[R'{-}NH{-}CH_2{}^+\right] + NH_2{-}R'' \rightarrow \left[R'{-}\boldsymbol{NH{-}CH_2{-}NH_2{}^+}{-}R''\right] \\ &\quad + A^- \rightarrow R'{-}\mathbf{NH{-}CH_2{-}NH} - R'' + H^+{-}A^- \end{aligned} \tag{5.11}$$

5.3.1 Second Step of Resin Synthesis

Polymerization reaction mechanism of UF resin monomers occurring in the acidic pH of 4.0–5.0 and at 90–95 °C in resin manufacturing arises from the methylol groups' tendency to form active cationic species (carbenium ion, $-CH_2{}^+$) by the action of acid catalyst (H^+) (Eq. (5.10)). The acid catalyst attacks on the oxygen atom and the resultant reaction complex [$-CH_2-OH-H^+$] breaks, giving off a water molecule (Eq. (5.10)). The resultant (electrophilic) cationic carbon species, then, attacks on a nucleophilic electron pair of an amide nitrogen of another methylolurea molecule in a relatively slow bimolecular reaction step, resulting in forming a bimolecular complex that has a **nitrogen–carbon–nitrogen (methylene)** bond (Eq. (5.11)). This molecular complex finally reacts with an anion (A^-) and loses the hydronium ion to form a dimer and regenerates the acid catalyst species (H^+-A^-) (Eq. (5.11)), which can keep working as catalyst for further reaction.

Polymerization reaction rates of monomeric species in the second step resin synthesis:

$$\begin{aligned} &NH_2{-}CO{-}NH{-}CH_2{-}\mathbf{OH} + \mathbf{NH_2{-}CO{-}NH_2} \\ &\rightarrow NH_2{-}CO{-}NH{-}\mathbf{CH_2{-}NH}{-}CO{-}NH_2 \\ &\boldsymbol{k = 3.30 \times 10^{-4}} \end{aligned} \tag{5.12}$$

$$\begin{aligned} &HO{-}CH_2{-}NH{-}CO{-}NH{-}\mathbf{CH_2}{-}OH + \mathbf{NH_2{-}CO{-}NH_2} \\ &\rightarrow HO{-}CH_2{-}NH{-}CO{-}NH{-}\mathbf{CH_2{-}NH}{-}CO{-}NH_2 \\ &\boldsymbol{k = 2.10 \times 10^{-4}} \end{aligned} \tag{5.13}$$

$$\begin{aligned} &NH_2{-}CO{-}NH{-}CH_2{-}\mathbf{OH} + \mathbf{NH_2}{-}CO{-}NH{-}CH_2{-}OH \\ &\rightarrow NH_2{-}CO{-}NH{-}\mathbf{CH_2{-}NH}{-}CO{-}NH{-}CH_2{-}OH \\ &\boldsymbol{k = 0.85 \times 10^{-4}} \end{aligned} \tag{5.14}$$

$$\begin{aligned} &NH_2{-}CO{-}NH{-}CH_2{-}OH + \mathbf{HOCH_2}{-}NH{-}CO{-}NH{-}CH_2{-}OH \\ &\rightarrow HO{-}CH_2{-}NH{-}CO{-}NH{-}\mathbf{CH_2{-}NH}{-}CO{-}NH{-}CH_2{-}OH \\ &\boldsymbol{k = 0.50 \times 10^{-4}} \end{aligned} \tag{5.15}$$

$$\begin{aligned} &HOCH_2{-}NH{-}CO{-}NH{-}\mathbf{CH_2{-}OH} + HOCH_2{-}\mathbf{NH}{-}CO{-}NH \\ &{-}CH_2{-}OH \rightarrow HOCH_2{-}NH{-}CO{-}NH{-}\mathbf{CH_2{-}N} \\ &= (CH_2{-}OH)CO{-}NH{-}CH_2{-}OH \\ &\boldsymbol{k = 0.03 \times 10^{-4}} \end{aligned} \tag{5.16}$$

(pH 5.0; 35°C; in mole/l – sec.)

Reaction rates for several dimer-forming reactions at 35 °C and pH 5.0 were reported by De Jong and de Jonge, as shown in Eqs. (5.12)–(5.16) [23–27]. It is noted that the reaction is fastest when free urea is involved (Eqs. (5.12) and (5.13)), but free urea is not much present in the second step of resin synthesis. The next fast reaction is the one involving one **amide** group (Eqs. (5.14) and (5.15)). This type of reaction would be the major reaction in the second step of resin synthesis, and the rate constant values are in accord with the reaction rates calculated from the viscosity increases observed in the UF resin synthesis reaction [13]. The reactions involving the **imide** nitrogen (Eq. (5.16)) go at significantly slower rates, and this type of reaction would be progressing in the later part of the second resin synthesis step. These reaction rates show that the higher the degree of substitution of urea with methylol or methylene groups, the lower the nucleophilic reactivity of urea's nitrogen atom, or an increased crowding is present in the molecular structure.

Reaction times in the second step resin manufacturing: The viscosity of resin mixture starts from "A_1" and increases gradually, at pH 4.0, to reach to T in about 105 minutes at 95 °C and in about 144 minutes at 70 °C, reflecting the growth of resin molecular size, or **methylene group** content, which increases to about 31%, as shown in Table 5.2 [14]. The extent of

Table 5.2 Progression of polymerization of UF resins with F/U_1 of 2.10, and after the alkaline first step reaction [14][a].

pH 4.0 and 95 °C	Reaction time (min)	45	60	85	105
	Viscosity	C	F	M	T
	Methylene group (%)	*19.95*	*28.40*	*27.47*	*29.70*
pH 4.0 and 70 °C	Reaction time (min)	64	94	134	144
	Viscosity	C	F	M	T
	Methylene group (%)	18.51	25.17	29.31	31.49

a) Methylene group content is based on the total amount of formaldehyde (F) used in resin synthesis.

viscosity or molecular weight increase attained in this second step of resin synthesis is a key parameter for industrial UF resins in achieving the necessary tack property and curing speed, commonly, the higher viscosity resulting in higher tack and higher curing speed. Also, the molecular weight increase goes through the point of no water dilutability, where some resin molecules start to get precipitated when the resin is diluted with water. Finally, these methylene bond-forming polymerization reactions stop nearly completely when the pH is adjusted back to 7 or higher, as shown in Figure 5.4. Further, in Figure 5.4, the methylene bond-forming reaction rate increases as pH gets below about 5, and it does so rapidly at lower pH values. At pH 4–5 the reaction rates are in a convenient range where an accurate control of cooking operation is possible, neither too slow to take a long cooking time nor too fast to have a risk of going into gelation in the kettle. In industrial manufacturing of UF resins, the batch is large and it is important to control this polymer-forming reaction rate right in order to arrive at the target viscosity (or molecular weights) accurately without under- or over-reaching, because the resin's binder performance, such as curing speed, etc., is changing with increasing molecular weights, and for operational safety reasons. This can be difficult if the reaction goes too fast and, therefore, the pH of reaction mixture needs to be controlled accurately: 5.0 is relatively safe and 4.0 can be somewhat fast. Fortunately, in many cases when using a good quality urea and formaldehyde, pH values in this range once set to a target value do not change much by themselves during the whole acidic cooking period, although attention needs to be paid to local gelation that can occur when adding the dilute acid to adjust the pH at the beginning of the second step. Such gelation often occurs and can be severe if the acid concentration is too high or the addition rate is too fast. Also, it is a good practice to arrive at the target pH value in the early part of the acidic step, since another shot of acid at a middle resin viscosity value, to lower the pH value further to speed up the synthesis reaction, which can result in local gelation. In industry, some local gelation often occurs and most manufactured UF resins go through a filtration step before shipping in order to remove these gel articles and any other extraneous materials that often come with bulk urea. When the target viscosity point is arrived, adding sodium hydroxide solution leads quickly to pH 7 or slightly higher and stops the polymerization reaction.

The free formaldehyde level increases in the acidic cooking step from about 7%, in the beginning, to reach at the end of the second step to about 14% of charged formaldehyde [13], which arises because the urea molecules in the reaction (1.0 mole) cannot accommodate all the formaldehyde present (~2.0 moles) in the molecular structure of the resultant oligomeric UF polymer structure, i.e. the two amide groups of a urea unit cannot take up all 2.0 moles of formaldehyde as methylene and hydroxymethyl groups in the polymer structure. It is reported that the reacted formaldehyde amount reaches to a final F/U_1 mole ratio value of about 1.81 vs. the reacted F/U mole ratio of about 1.95 in the alkaline first step with the charged mole ratio of 2.10 [13–16]. Thus, the reaction mixture at the end of second step contains a considerable amount of free formaldehyde, which is reduced to less than about 0.5% after the addition of second urea in finished resins. It is also useful to keep in mind that the reverse hydroxymethylation reactions are also ongoing during this acidic reaction step. The reaction products at this stage are (oligomeric) UF polymer molecules consisting of methylene and methylene-ether groups as the inter-urea bonds and hydroxymethyl groups bonded to the polymeric urea groups, plus free formaldehyde existing as methylene glycol and its oligomers as well as hemi-formal groups bonded to the hydroxymethyl groups.

In the third step of resin synthesis, the addition of second urea (U_2) during the cool-down period also expedites the cooling process because of the (cold) urea and its dissolution absorb heat (endothermic). The viscosity of reaction mixture also decreases significantly due to the monomeric urea added to polymeric UF resin components, i.e. a dilution effect. The free formaldehyde present reacts with second urea to form mostly mono- and di-methylolureas similarly as in the first alkaline step of resin synthesis. These reactions may not go to completion in the short cooling/discharging period. Also, some of the methylol groups bonded to the polymeric UF components start to dissociate by the reversibility, and the freed formaldehyde makes bonds to second urea, and these reactions, at ordinary temperatures, continue but slowly toward reaching to the equilibrium value even during the transportation and storage period. Thus, UF resins require ripening times to reach to the stage of more consistent composition and consistent bonding performance [13–16]. Also, when stored for longer times or at elevated outdoor temperatures, UF resins, at weak alkaline pH, are known to advance in viscosity as well as increase in turbidity, and in the end, the resin can become useless. This topic is discussed further in Chapter 24. (Advanced reading materials section.)

5.4 Polymer Chain Branching in UF Resins

Formation of branch polymer structure: It has been difficult to prove the existence of branched UF polymer chain structures in liquid or cured UF resins using the traditional wet chemical methods. In past research, some linear oligomeric UF reaction products formed during UF resin synthesis reactions, apparently as side reaction products, were readily separated out as compositionally pure crystalline materials and identified to have linear methylene-urea-type molecular structures (Eq. (5.17)) up to about 7-mers and most of them are only sparingly soluble in water even in the presence of excess formaldehyde [17–22], indicating that these linear molecular materials may not necessarily be the "real" constituents of UF adhesive resins proper. Therefore, the investigators have kept the view of the branched polymer structures for major components of UF resins [17–22]. More recently, much of the various reaction pathways and reaction products proposed by the early researchers turned out to be in accord with recent ^{13}C NMR results, and it appears that the linear oligomers were only minor side-reaction products in typical UF resins, and some 60% methylene and methylene-ether groups are involved in the branching of UF polymer structures [13–16].

Linear: $$\text{—CO—NH—CH}_2\text{—NH—CO—NH—CH}_2\text{—NH—CO—NH—CH}_2\text{—NH—CO—NH—CH}_2\text{—NH—} \quad (5.17)$$

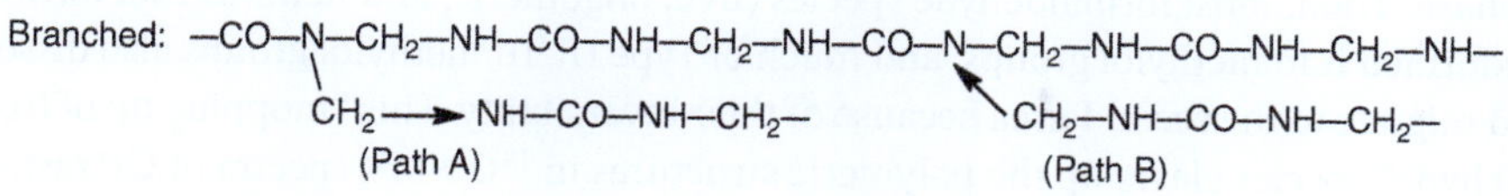

(5.18)

To form branched methylene bonds (Eq. (5.18)), either Type IIi methylol groups need to react with urea amide groups to transform themselves into type II (or III) methylene bonds (*Path A*) or Type I methylol groups should attack on imide nitrogens to form Type II or type III methylene bonds (*Path B*). Similar reaction paths would be applicable to the formation of methylene-ether-based branch chains.

Branched bond-forming reaction rates would be relatively slow, occurring mostly in the later part of the acidic polymerization reaction step. One such indication is that the reaction mixture's viscosity is increasing continuously without much phase separation from formation of precipitates. Linear UF polymers (Eq. (5.17)), if formed to some extent during the synthesis, are likely to form precipitates, possibly before gelation. Branched UF polymers would be more resistant to precipitation. Also, branched UF polymers naturally have more urea amide-end groups that are apt to be bonded with Type I or Type II methylol groups, which, because of their hydrophilic properties, would make the polymer chains more compatible with water medium and remain in the liquid phase longer than linear polymers.

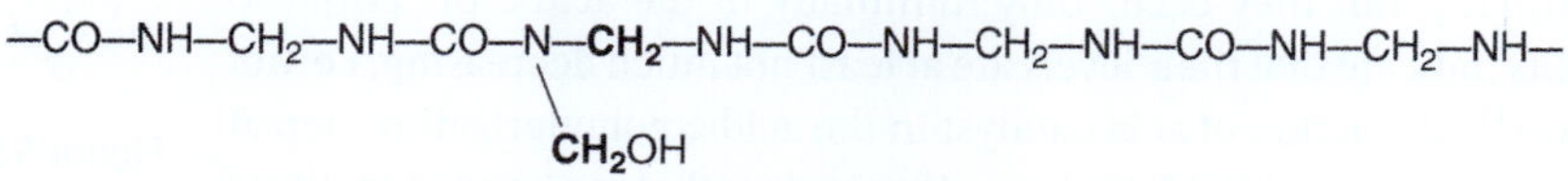

(5.19)

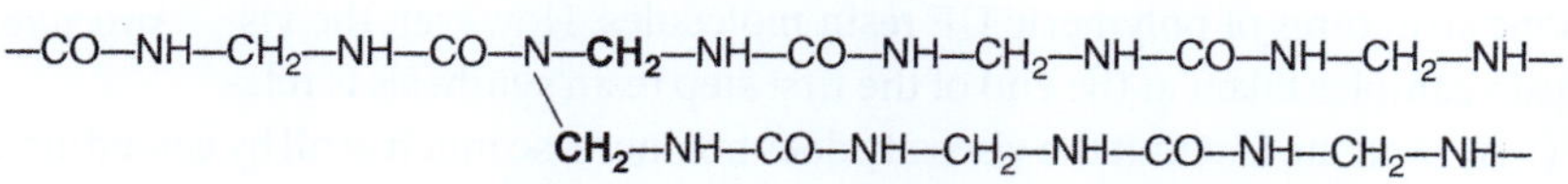

(5.20)

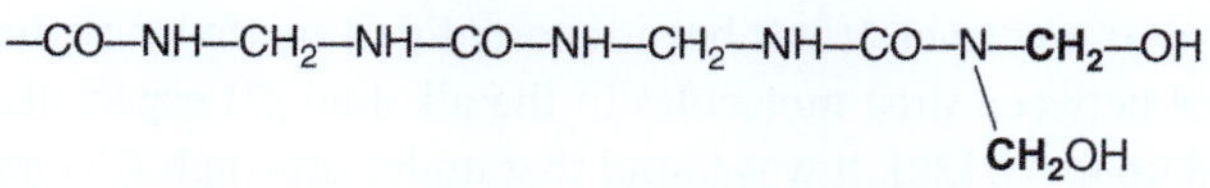

(5.21)

Branched chain question has been a somewhat complicated problem in ^{13}C NMR analysis studies because Types II and III methylene groups classified as such due to the presence of Type IIi methylol groups (Eq. (5.19)) also occur and overlap with those of Type II methylene groups (Eq. (5.20)). Furthermore, Type IIi methylol groups' position (chemical shift) is overlapped with those of Type IIt methylol groups (Eq. (5.21)) and cannot be measured separately. Type II and IIi methylol groups are themselves potential branching points, but they can also be lost as free formaldehyde because of their reversibility when second urea is added and also during the storage/transportation or during the hot-pressing steps of boards, i.e. they do not necessarily make branch points that can last to the cured structures of UF resins. This means that the branching extent measured can be an over-estimate to the extent of Type IIi methylol group content values, which cannot be measured directly. Approximate values of branching can be reached by adding the second urea and heating for an extended period of time or an excess urea is added, for example, to an F/U ratio of 0.8 or lower, which is possible in laboratory because urea itself is highly soluble in water without disruption of resins' phase. Then, most formaldehyde species (free, oligomeric, and hemi-formal forms) are transformed into methylol groups, and much of Type IIt/IIi methylol groups also dissociate and migrate to the second urea because of their reversibility. This "mopping up of free formaldehyde" species clears up the polymeric structures in ^{13}C NMR spectra of UF resins to give close to the "real" polymer molecular chain structures based on mostly methylene (M) and methylene-ether (ME) bonds. The UF resin polymer structures were thus proved to consist of about 60% Type II/III (branched) M/ME groups and about 40% Type I (linear) M/ME groups [13–16].

5.5 Other Reactions Occurring in UF Resin Synthesis

$$\begin{aligned}&\mathrm{HO{-}CH_2{-}NH{-}CO{-}NH{-}}\mathbf{CH_2{-}OH}\\&\quad+\mathbf{HO{-}CH_2}\mathrm{{-}NH{-}CO{-}NH{-}CH_2{-}OH}\\&\quad\rightarrow\mathrm{HO{-}CH_2{-}NH{-}CO{-}NH{-}}\mathbf{CH_2{-}\underline{O}{-}CH_2}\mathrm{{-}NH{-}CO}\\&\quad\mathrm{{-}NH{-}CH_2{-}OH}+\mathrm{H_2}\mathbf{O}\quad\text{Methylene–ether bond}\end{aligned}\tag{5.22}$$

Methylene-ether bonds are formed in significant amounts in the first alkaline reaction step, as mentioned above, from reactions of two methylol groups (Eq. (5.22)) [13]. But they occur only minimally in the acidic pH [16]. ^{13}C NMR results indicate that their levels are at least not much decreasing, i.e. not breaking up by the action of acid catalyst in the acidic polymerization step at temperatures of up to 95 °C. Methylene-ether and methylene bonds constitute the backbone structures of polymeric UF resin molecules. However, the viscosity of resin samples taken at the end of the first step resin synthesis is relatively low (~A_1) and the fact that the viscosity does not increase much at all by extending the cooking time at the alkaline pH indicates that the ether bond formation (oligomerization) does not proceed beyond that low level. It has been reported that methylene-ether bonds are the only bonds formed between urea molecules in the alkaline pH especially at high F/U_1 mole ratios from 2.5 up to 4.0 [28]. It was found that under this high F/U mole ratio condition and acidic pH, the methylene-ether bonds also occur in the form of cyclic uron structure (Figure 5.5) [18, 28]. ^{13}C NMR results indicate that methylene-ether bond contents in UF resins amount to about 17–20% of the total formaldehyde used in resin synthesis [13–16].

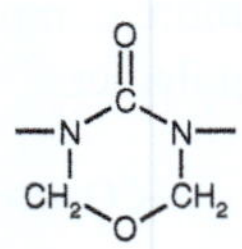

Figure 5.5 Uron structure.

Methylene-ether bonds have been considered to be breaking up in the curing reaction [8], likely giving rise to one methylene bond and one free formaldehyde molecule (Eq. (5.23)) or two formaldehyde molecules (Eq. (5.24)). If this breakage happens before the gelation of resin, formaldehyde will have a good chance of assimilating back into the resin phase to contribute to the bonding. However, if these breakages happen after the gelation, the formaldehyde molecules would have little chance of assimilating back into the resin phase and later get emitted from boards as free formaldehyde [8]. This transient nature of methylene-ether groups, on the other hand, may give rise to a long flow in resin curing or a stepwise progression of curing of UF resins, although it is currently not well understood what roles they play in the curing of resins. Methylene-ether bonds are also reported to remain in certain cured UF resins at significant levels [29, 30]. Thus, methylene-ether bonds appear to have some important implications, especially with respect to the curing process and formaldehyde emission potential of UF resin-bonded wood composite boards.

5.6 F/U_1 Mole Ratio Effects in the Polymerization Step

Upper limits of F/U_1 mole ratio: In UF resin manufacturing industry, the F/U_1 mole ratio ranges from 1.8 to 2.4. In theory, higher values would result in resins with higher extents of reaction, and lower values would result in lower extents of reaction, and each chosen value may offer some specific resin performance characteristics. Further, it is interesting to know why there exist the upper and lower limit values. In practice, with F/U_1 values higher than about 2.6, the polymerization reaction step in acidic pH 4–5 does not proceed much at all and no significant viscosity increase is obtained, i.e. UF polymeric components are not much formed, and if the reaction is pushed further by lowering the pH further, much uron-type intramolecular cyclic methylene-ether groups are formed. This phenomenon arises from the steric constraint present in the molecular structure of urea, i.e. not allowing more than three substituents on one urea molecule, although one urea molecule has four hydrogen atoms that could be replaced in theory. Tetra-substitutions on urea molecules are observed at extremely low levels in UF resins by ^{13}C NMR [13].

$$\begin{aligned}&\text{R–NHCO–}\underline{\text{NH–}\mathbf{CH_2\text{–}O\text{–}CH_2}\text{–NH}}\text{–CO–NH–R}\\ &\rightarrow \text{R–NH}_2\text{–CO–}\underline{\text{NH–}\mathbf{CH_2}\text{–NH}}\text{–CO–NH–R} + \underline{\text{CH}_2 = \text{O}}\end{aligned} \tag{5.23}$$

$$\begin{aligned}&\text{R–NHCO–NH–}\underline{\mathbf{CH_2\text{–}O\text{–}CH_2}\text{–NH}}\text{–CO–NH–R} + \text{H}_2\text{O}\\ &\rightarrow \text{R–NH–CO–}\underline{\text{NH}_2} + \underline{\text{NH}_2}\text{–CO–NH–R} + \underline{2\,\text{CH}_2 = \text{O}}\end{aligned} \tag{5.24}$$

This fact has led the industry to utilize urea as a formaldehyde carrier in the form of products known as *UF-85*. As mentioned in Section 5.1, 50% formaldehyde solutions used in industry need to be stored and transported at about 50 °C, which can be costly. UF-85 type products are made by reacting urea and formaldehyde in the weak alkaline pH, similarly in the first step of resin synthesis, with F/U_1 mole ratios of about 4.0–5.0, resulting in a mixture of various mono-, di-, and tri-methylolureas and some excess formaldehyde. Cooled to room temperature, this mixture is stable, i.e. no polymerization reaction occurs, and the excess formaldehyde does not form polymeric formaldehyde precipitates as 50% formaldehyde solutions would do. This formaldehyde product can be stored and shipped at ordinary temperatures and used as formaldehyde source, i.e. by adding more urea to obtain target F/U_1 mole ratio values in resin manufacture.

Lower limits of F/U_1 mole ratio: There also exists a lower limit for F/U_1 values in the acidic polymerization step at about 1.8, where the resin synthesis reaction starts to show turbidity that gets worse as the F/U_1 mole ratio is lowered further, resulting in unusable resins. The formation of white precipitates in UF resin synthesis has been long known to be arising from the formation of various linear methylene-urea molecules [3, 18]. This problem in using low F/U_1 mole ratio values in UF resin syntheses appears to be arising from the lack of formaldehyde in the resin system, i.e. some of dimeric or higher oligomeric UF resin components, as soon as they are formed, start to lose hydroxymethyl groups by reversibility to transform themselves into *methylene-di-urea* (Eq. (5.25)) or *tri-methylene-tetra-urea*-type products (Eq. (5.26)), which have a high tendency to crystallize out as sparingly soluble, high melting, **finely dispersed** precipitates [13–16]. Once formed during the acidic polymerization reaction step, these white precipitates do not go back into the resin phase even when more formaldehyde is added, a reflection of their high insolubility. In milder cases, the turbidity problem gets worse as the polymerization reaction progresses and the viscosity increase becomes irregular, resulting in missing of the endpoint of cooking compared with resins made with higher F/U_1 values. This type of turbid UF resins were found to perform less satisfactorily in bonding of wood composite boards, probably due to inefficient resin spraying or inadequate hot-pressing temperature because the temperatures of core-layers of particleboard in industry would not reach to the high melting points of these precipitates, resulting in their functioning only as fillers.

$$\begin{aligned} &\mathrm{HOCH_2}\mathbf{-NH-CO-NH-CH_2-NH-CO-NH-}\mathrm{CH_2-OH} \rightarrow \\ &\quad 2\,\mathrm{CH_2=O} + \mathbf{NH_2-CO-NH-CH_2-NH-CO-NH_2} \downarrow \\ &\quad (\mathrm{mp}\ 218\,^\circ\mathrm{C}) \end{aligned} \tag{5.25}$$

$$\begin{aligned} &\mathrm{HOCH_2}\mathbf{-NH-CO-NH-CH_2-(NH-CO-NH-CH_2)_2} \\ &\quad \mathbf{-NH-CO-NH-}\mathrm{CH_2-OH} \rightarrow \\ &2\,\mathrm{CH_2=O} + \mathbf{NH_2-CO-NH-(CH_2-NH-CO-NH)_2} \\ &\quad \mathbf{-CH_2-NH-CO-NH_2} \downarrow \\ &\quad (\mathrm{mp}\ 230\,^\circ\mathrm{C}) \end{aligned} \tag{5.26}$$

Mid-range F/U_1 mole ratios: The F/U_1 **mole** ratio values of about 1.8–2.5 currently used in industry determine the amounts of the second urea (U_2) to be added in resin synthesis. For example, with an F/U_1 value of 2.10 and target total F/U ratio of 1.15, the second urea value needs to be 0.82 mole vs. 1.00 mole of first urea. One result that can be derived from the U_2/U_1 ratio of resins is that the finished resin would consist of ~45% monomeric urea and methylolureas and ~55% branched oligomeric UF polymers. This ratio can change by using different F/U_1 ratios in resin synthesis. Since UF resins with more polymeric urea components and higher degrees of polymerization would have faster curing rates, the F/U_1 ratio would be the one key parameter that determines various resin-curing rates and bonding performance characteristics. Thus, *F/U_1 mole ratio, final F/U mole ratio, and extent of polymerization attained in the second step of resin synthesis are the three key variables* used for manufacturing various different UF resins, including the core- and face-layer resins of wood composite boards.

5.7 Final F/U Mole Ratios and Formaldehyde Emission Problem

Due to the formaldehyde emission problems of UF resins, the formaldehyde to urea ($F/(U_1 + U_2)$) mole ratio has been lowered in the UF resin industry, mostly by increasing the amount of second urea (U_2), to as low as *about 1.05 for face-layer resins and about 1.15–1.25 for core-layer resins* in particleboard binder applications. The mole ratio is slightly higher for MDF and hardwood plywood binder resins. In the early years of PB industry, UF resins having an F/U mole ratio of about 1.50 were used for fast curing rates and high strength values of boards [31]. In theory, the mole ratio of 1.50 means every urea molecule in cured resins is to be bonded to three methylene bonds, i.e. perfectly three-dimensionally linked polymer networks are formed. However, even at that time, formaldehyde emission levels of boards were noted to be very high during hot-pressing and from finished boards, indicating that the urea functionality would be less than three. The reason behind this dilemma of formaldehyde emission is that, as shown by De Jong and De Jonge [23] discussed above, the hydroxymethylation reaction is **reversible** and the forward/reverse reaction rates increase significantly with temperature, more so in the acidic than in alkaline pH. Thus, at the curing temperature of about 170–200 °C and pH of 4.0 or lower in boards, both forward and reverse reaction rates will increase significantly. Since about 50% of charged formaldehyde is present as hydroxymethyl groups in *pre-cure* UF resins, the amount of methylol groups that dissociate into formaldehyde would be significant for resins made even with F/U mole ratio values of 1.0–1.25. Some of the dissociated formaldehyde would remain in the uncured resin phase and participate in the curing process, but those that escape from the resin phase would have little chance of returning, becoming the source of formaldehyde emissions during hot-pressing and from boards. Of course, when low F/U mole ratios are used in resin manufacture, there are less methylol groups and less dissociation of formaldehyde, but lowering the F/U mole ratio results in longer hot-pressing times and lower board strength properties, i.e. a lack of the methylol functional groups. Until recently, boards bonded with UF resins made with an F/U mole ratio of about 1.15 gave emission values of about 0.18–0.20 ppm by the Large-Scale Chamber method (ASTM E1333-1990) [31], significantly higher than regulation limits, and a further lowering of F/U mole ratio would lead to lower board strengths and longer hot-pressing times.

5.8 Physical and Chemical Tests and Properties of Industrial UF Resins

Cold storage and transportation: After manufacture, the cooled, finished resin is filtered, transported, and pumped into a storage tank, which is often maintained at a cool temperature, especially in summer months. Some stirring can be applied to minimize side reactions that can occur during the storage [11]. Transportation of UF resins to board manufacturing plants is also done in cooled tanks, and it is also customary to use up resins within some prescribed storage times. Industrial UF resins are commonly clear or milky, viscous liquids having 58–70% resin solids contents and shelf-lives of about two to three weeks of time after arrival. Some slow but significant reactions are ongoing during the storage period and, therefore, the wood bonding/formaldehyde emission performances of a UF resin can vary slightly with storage times and conditions. UF resins stored too long could reach to a high viscosity and make the spraying operation inefficient to result in poor bonding performances. Commonly, a number of physical and chemical tests are done by resin users to assure resins' uniform quality, as in the following:

Viscosity: An indicator of the degree of polymerization, resin solids level, and a criterion of sprayability (atomization). For PB or MDF manufacturing, it can range at about 200 cps, and for hardwood plywood manufacturing, it can go up to 500 cps or higher. When the resin is to be stored for longer periods of time in board manufacturing plants, especially at hot temperatures, the viscosity needs to be monitored for any increases. It is commonly determined by using a Gardner–Holdt (bubble tube) viscometer or a Brookfield viscometer.

pH: Another value to be checked from time to time during storage to detect any changes occurring from the delivered values to assure storage stability. Determined by using a pH meter. The pH value of UF resins ranges 7–8 in the industry.

Resin-solids level: An indicator of the accuracy of raw material charges used in resin manufacturing. Resin manufacturing involves weighing large amounts of materials, and operator errors or calculation inaccuracies can occur and, if occurred, it is not easy to find them in finished resins. Resin's solids level is measured with catalyst, which must be specified, commonly done also at board manufacturing plants, by putting about 1.0 g of resin and 1.0 g of distilled water with a drop of catalyst on an aluminum pan, shaking to mix, and then heating at 115 °C for an hour, followed by weighing the residual solids. The value commonly is in 60–64% ranges, and higher values are preferred if the viscosity can be within the sprayable range because less water goes into wood mat. This resin-solids level determined with catalyst addition is the basis of pricing between the supplier and user.

Refractive index: An indicator of the resin-solids level of resins, tested on a refractor meter and the reading is compared with the standard chart prepared with several similar resins of different resin-solids levels, to read the resin-solids level. It is a simple alternative to the direct resin-solids determination method.

Free formaldehyde content: Another indicator of the accuracy of raw material charges and cook procedures as well as of the resin's final F/U mole ratio. Formaldehyde content of UF resins is usually determined by the **hydroxylamine hydrochloride** method [32]. Formaldehyde reacts readily with the reagent and liberates hydrochloric acid, the amount of which is determined by titration using a standardized sodium hydroxide solution. Current industrial UF resins, due to their low F/U mole ratios used in resin manufacture, have less than about 0.5% formaldehyde content based on the liquid resin weight. The free formaldehyde present in UF resins can evaporate during blending, matting, and hot-pressing of boards, so it is necessary to have a ventilation system in board manufacturing plants.

F/U mole ratio: A useful value to obtain for competitive UF resins or when the material charge values used in resin synthesis are suspected. The wet chemical analysis method involves separate measurements of total nitrogen content by the Kjeldahl method and total methyleneic group ($-CH_2-$) content by the hydrolysis method [29]. ^{13}C NMR methods can be used for comparative purposes for different resins [13–16]. Both procedures are somewhat lengthy, and therefore the test is not routine, mostly done by resin manufacturers. The F/U mole ratio of UF resins is a technological decision: higher F/U mole ratio values give faster hot-pressing times (curing speed) and higher board strength values, but the formaldehyde emission values can be higher, so that a compromise is commonly made, and face-layers use slightly lower values than core-layer resins. Also, UF resins used for bonding of MDF and hardwood plywood have slightly higher F/U mole ratio values than resins used for particleboard.

Gel times: The test is performed with and without catalyst. Without catalyst, it can show the resin storage stability and other manufacturing quality control parameters such as consistency in a series of cooks. With catalyst, the test value is an indicator of curing speed in hot-pressing of boards and is used as a quality control test with respect to the hot-pressing time and temperature. In gel time tests, about 10 g of resin is taken in a test tube and placed in a boiling water bath (100 °C) and the resin is kept stirred with a glass rod until it is set. The resin sample loses some small portion of the water of resin (in open air) and the setting of resin is mostly due to chemical curing reactions occurring in the presence of water of resin. Thus, gel time is a partial story of resin curing versus the curing in hot pressing of boards, where most of resin water is lost. Gel time test is simple to carry out in the resin user's laboratory.

Molecular weight: It can be measured by gel permeation chromatography [33, 34]. ^{13}C NMR analysis method can also be used as a comparative method for different resins [14, 15], with the number-average molecular weight values reported in range 5–20 urea units for the polymeric components of PB and MDF binder-type UF resins. This test is not routine and is mostly done by resin manufacturers.

Storage life is generally a parameter known to be limited to two to three weeks after delivery. For this reason, UF resin manufacturing plants tend to be located at a central point in a cluster of wood composite manufacturing plants. Resin storage life is commonly measured by viscosity increases during storage. Optimum viscosity values for PB and MDF plants are in the range of about 200 cps and for hardwood plywood plants it goes up as high as 500 cps. Another indicator for the storage life is the phase separation of the resin-rich phase (oil phase) from the water-rich phase, which would limit useful handling operations. This problem does not occur often unless another component, such as melamine, is added in the wrong way, or drastically different polymerization scheme is used in the resin manufacturing stage, or the latent catalyst is added wrongly, which is all uncommon. No special measure is commonly done such as adding methanol to resin to increase the storage life because any organic solvent added to resins has to be evaporated off in the hot-pressing stage, resulting in increasing the organic air pollutants of the plant, the amount of which is normally regulated by local laws. Buffers such as borax can be used to keep pH changes at a minimum and therefore increase the storage life, but this method is not generally used because more curing catalysts are needed.

Tack property: Pressure-sensitive adhesives are the typical tacky materials that form strong bonds when brought into contact with another surface for a short time (seconds) under light pressure (a few psi). Tack is also an important parameter for UF resin adhesives used in particleboard and hardwood plywood manufacture. It indicates the adhesive's holding capability of wood particles or plies in mats being formed, as a result of gradual loss of moisture and advancing of UF resins. An ideal tack development graph is shown in Figure 5.6 [35]. Tack property of UF resin adhesives starts to arise as soon as it is mixed with wood particles but ideally, it needs to be minimal in the resin blending and mat-forming stations and then increase to the maximum value shortly after the mat formation and pre-pressing. The initial low tack helps blending of resin with wood particles, and a good tack later prevent the formed mat from breaking especially on mat edges, which is necessary during transport to hot-press. Tack property of liquid UF resins sprayed on wood particles increases and reaches to the maximum as the moisture loss increases, and it starts to gradually decrease and disappear when the resin is completely dried, but once

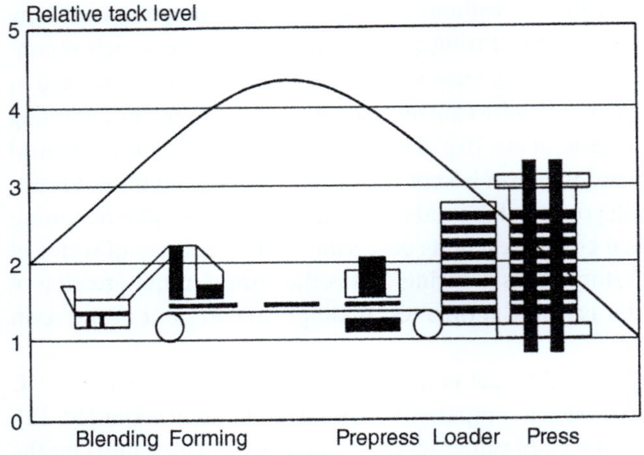

Figure 5.6 An ideal tack development curve of UF resins in particleboard manufacturing process [35]. Source: Reproduced with the permission of the Composite Panel Association.

formed, the tack bonds can remain till hot-pressing. Tack property of uncured, liquid UF resins arises from the polymeric adhesive molecules' attracting forces among themselves (viscosity) as well as their adhering forces to the molecules of wood surface, both of which are adequately present in current industrial UF resins due to the resin molecules' hydrophilic and oligomeric nature (appropriate degree of polymerization). Moisture loss of resin increases with the elapsing time during blending and matting by evaporation into the air and adsorption into the wood. Also, the polymerization reactions of UF resin molecules initiated either by external catalysts or by wood acid catalysts can play into accelerating or changing the tack curve. Also, the rate of water dissipation and catalyst effects depend on the room temperature and humidity as well as the wood moisture content with the tack curve changing accordingly. The degree of polymerization, F/U mole ratio (hydroxymethyl group content), and water content of incoming UF resins are commonly constants for a particleboard plant, but they could change according to seasons or other changes in board manufacturing conditions to result in different, sometimes ineffective tack curves. Also, high molecular weight and high F/U mole ratio resins often show a good tack while resins having less methylol groups, such as current low F/U mole ratio resins, can show poor tack properties. For low-tack resins, remedies such as minimizing the shaking of mats during transport to hot press are needed, which is not easy. Tack property is not directly related to the bond strengths of finished boards.

5.9 Resin Changes Occurring After Manufacture of UF Resins

UF resins' storage stability: Liquid UF resins are generally stable as a one-phase (resin-in-water) system with respect to the various handling operations, such as transporting, pumping, and spraying. One major concern is the viscosity of resin that is changing or more often increasing as the storage time is extended or resin is exposed to elevated temperatures in hot weather. UF resins are a mixture of UF monomers and polymers, water, and with or without added catalysts. UF resin molecules having methylol groups are relatively well compatible with the water medium, so that the one-phase system is maintained, but as the polymer molecular weight is increased in resin manufacturing for faster cure or better strength properties, their compatibility with water becomes poorer and the dilutability of some resin components in water decreases. Overall, the one-phase (clear) resin system requirements are all met in the current resin synthesis schemes with the viscosity value (200–500 cps), extent of polymerization (5–20 mer for U_1), and resin solids level (60–64%). Also, the chosen resin synthesis procedure and qualities of urea and formaldehyde are all important factors that determine the clarity and storage time of resultant resins. Turbid UF resins are two-phase resin systems and they tend to age faster. Also, increasing the extent of polymerization, followed by adding water to lower the viscosity can lead to a point where the resin separates into the resin-rich and water-rich phases.

Viscosity increases and turbidity changes also occur for liquid UF resins during storage. Even at room temperature and neutral pH the viscosity of UF resins can decrease initially and then increase, and turbidity can also arise as storage time is extended. Turbidity can also arise during resin manufacturing. These **viscosity and turbidity changes** will be discussed in detail in Chapter 24, the advanced reading section, but it can be summarized here as arising from the interplay between methylol groups bonded to the first urea and the amide groups of the second urea, which could be at mis-equilibrium in freshly manufactured UF resins in terms of the reversibility of methylol groups. Thus, methylol groups bonded to the first urea slowly dissociate to free formaldehyde, which then reacts with second urea over a period of one to two weeks at room temperature. The average molecular weight is decreased by this migration and therefore the viscosity decreases. After this period, the viscosity can start to increase due to certain polymerization reactions occurring, effected by the (slow) general acid catalysis of water molecules and hydroxyl groups, etc. This slow polymerization reaction, progressing more readily without agitation, also results in increasing the turbidity of resins [13] that indicates the resin is a two-phase system, but it is okay in use as long as the fine solid particles do not agglomerate and precipitate out, i.e. if the resin remains as an "apparent one-phase" system. Though, the turbidity can increase further to cause additional viscosity increases and changes in the spraying pattern. Turbidity of UF resins often arises, as discussed above in Section 5.6, when the F/U_1 mole ratio in the typical resin synthesis procedure is set to be below about 1.8, and it gets severe as the mole ratio is reduced further, even though the final $F/(U_1 + U_2)$ mole ratio is set equally to 1.15 or similar values. On the other hand, many industrial UF resins are clear in the beginning, but freshly made resins can also show some turbidity because of some ill-defined catalytic effect of metallic impurities present in the formaldehyde or urea or by other synthesis procedures. Some experimental UF resins show severe turbidity because of the synthesis procedures chosen to obtain certain specific resin performance properties.

Formation of methylene di-urea (insoluble).

$$\begin{aligned}&\mathrm{HO{-}CH_2{-}NH{-}CO{-}NH{-}CH_2{-}NH{-}CO{-}NH{-}}\\&\mathrm{CH_2OH}\\&\quad\text{Dimethylol methylene di-urea (soluble)}\\&\leftrightarrow \mathrm{NH_2{-}CO{-}NH{-}CH_2{-}NH{-}CO{-}NH_2} + 2\,\mathrm{CH_2=O}\nearrow\end{aligned} \tag{5.27}$$

The reason behind the turbidity phenomenon appears to be the formation of methylene-di-urea (Eq. (5.27)) and tri-methylene-tetra-urea-type linear oligomers that have no methylol groups and tendency to precipitate out as fine solid particles (white color) that have high melting points and poor solubilities in water [23, 36]. In 1950s, Staudinger and Wagner and De Jong and De Jonge studied these insoluble products extensively assuming that they might be the high molecular weight UF polymers to be formed as the ultimate products of UF polymerization/curing reactions [26, 36]. However, the insoluble particles were found to be short-chain linear methyleneureas formed as side-reaction products. Currently, these types of insoluble UF resin products are manufactured for use as UF-based white pigments or slow-releasing fertilizers by reacting at F/U mole ratios of around 0.6–1.0 [37, 38]. Also, the high melting points of these precipitates present in resins suggest that they might not be easily assimilated into the resin phase during hot-pressing of boards. i.e. they could well act only as a filler possibly not much contributing to bond strength. In common UF resins, the amount of the white precipitates present is exceedingly small, and its bond strength effect would be minimal, but if the turbidity progresses too far, such as in old resins, the viscosity increases due to the precipitates will be significant, and lower bonding performance can result not only from the poor sprayability, but also from the loss of functional resin molecules that are used only as fillers.

pH dependence of storage stability: one major factor that determines the storage stability of stored UF resins is the pH, and the resin stability is highest in the pH regions of 7.0–7.5, which is the common value practiced in the industry. At lower pH values, methylene bond formation proceeds to cause viscosity increases. At pH values significantly higher than 7.5, UF resins can show lower stability by polymerization through the formation of methylene-ether bonds by base catalysis. Since pH values of UF resins can change during transportation and storage due to side reactions such as Cannizzaro reactions, acidic buffers such as borax are often added, which needs to be minimized because an additional acidic catalyst is required to overcome it in the blending step of board manufacture.

5.10 Bond Performances and Durability Tests of UF Resins in Industry

Bond performances of experimental UF resins are obtained mostly by laboratory board making carried out with well-defined parameters of target density, wood moisture content, layer constructions, resin spraying time and pattern, wax and catalyst additives, mat forming and dwell time, hot-pressing parameters, total hot-pressing time, etc., followed by testing the boards according to the standard procedures. It is often difficult to do every step the same way as is done in the industry, but data obtained in the laboratory are the basis for one or a few mill trial experiments using 5–20 drums of resins. The most important criteria is whether the resin allows the plant to run smoothly at desired speeds and resultant boards also show adequate strengths and formaldehyde emission levels. In the case of low board strength values or high emission values, catalyst level adjustments, longer hot-pressing times, and higher pressing temperatures can be experimented. If not successful, the resin may need a re-formulation with respect to the deficiencies observed. The board manufacturing plant may also need to adjust some other processing parameters, such as wood moisture contents. If the resin runs well, its performance must be validated by quality control tests. Any seasonal temperature/moisture changes or wood species mix changes must be taken care of by resin/catalyst formulation changes. I marvel when looking at the sanded, perfect-looking loads of 8′ × 4′ boards coming out of the finishing line, a result of resolving so many conflicting processing parameters and, above all, with possible profits to everyone involved. Durability tests of UF resin-bonded boards are commonly limited to thickness swelling and water absorption percentages in 2 and 24 hours water-soaking tests at 20 °C. The methylene bonds formed in cured resins are the backbone of adhesive layer in manufactured boards, but they are also reversible, i.e. can be broken, going back to methylol and urea amide groups when soaked in water for long times, but this reverse reaction rate is exceedingly slow and expedited only at high temperatures and high moisture contents. UF resin-bonded boards are adequate for interior uses.

5.11 UF Resins vs. Various Operating Parameters in PB, MDF, and Hardwood Plywood Industry

The wood composite board industry is supplied with UF resins tailored to meet each plant's own set of operational parameters regarding the wood species, press temperature, and press time, as well as the required minimum resin loading level, formaldehyde emission level, and all strength-related board properties.

Particleboard plants use wood particles dried to about 4% moisture content, and the face-layer and core-layer particles are slightly different in size and quality and bonded with each type of resin, the former relatively slower curing, with lower advancement, lower F/U mole ratios, and lower catalyst levels than the latter to compensate for the temperature differences between the two layers in hot-pressing. Of course, the hot-pressing speed of boards is mainly determined by the core-layer resins' extent of polymerization arrived in the second step of resin synthesis, final F/U mole ratios, and catalyst levels, as well as by the hot-pressing temperatures (300–350 °F) used in the board manufacturing plant.

For MDF plants, there are two different processes with respect to the mode of adhesive resin application. The first process uses UF resins that are similar to particleboard plants and they are added to dried fibers in the paddle-type blenders with separate face- and core-layer lines. In the second process, the binder resin is injected into the hot blow-line that goes from the digester to the dryer tube, wherein the resin and undried (wet) fiber are mixed rapidly and pass through the high-temperature drying tube in a few seconds together and, therefore, the UF resin has to be relatively greener (less polymerized and water-dilutable) than the resins used in the dry process. The resulting resinated fibers are then matted and hot-pressed to boards using relatively short hot-press times.

Hot-pressing parameters for UF resins: In PB and MDF plants, the hot presses are large, up to 8′ × 24′ and up to 8–12% resin solids are applied based on wood weight to result in about 10–12% mat moisture contents, which need to be expelled to about 4% or less during the hot-pressing. The usable press temperatures are often limited due to UF resins' pre-curing or over-curing tendencies and because of limited boiler capacity; therefore, the press temperature in a plant tends to be a constant, not easily increasing or decreasing for hot-press time control. On the other hand, the extent of polymerization of resin in synthesis can be maxed out but is also limited by the viscosity requirement for spraying operation at the required resin-solids level and, also by the required maximum moisture level in board mats. The F/U mole ratio of core as well as face layer resins is also limited because of the formaldehyde emission level to be achieved, resulting in limited strength values of boards. Currently, the F/U mole ratio value for face-layer may go down to~1.0 and for core-layer the value goes up to 1.15, and the board may still emit some level of formaldehyde.

Hardwood plywood industry uses UF resins made with higher viscosities (500 cP or more), and the resins are mixed with extenders such as rye flour and catalysts to result in formulated resin adhesives having viscosities in the range of 2000 cps or higher, which are then applied on veneers with roller-type applicators. The plants have their own operational parameter requirements, such as short press times and moderate pressing temperatures, and on-site bonding requirement of no separation of veneers out of hot-press or no strike-through of adhesives, especially through the top veneers. The formaldehyde emission requirement is similar as for particleboard or MDF plants, and methods of lowering the F/U mole ratio of resin have been known to be similarly limited. On the other hand, it is reported that use of high-viscosity adhesives with roller-type applications allows "trade-secret" soy-based or polyamide-based resins that are manufacturable only as high-viscosity materials to be applicable with very little or no formaldehyde emissions, although the detailed bond quality and hot-pressing parameters have not been published.

5.12 Other UF Resin Synthesis Procedures

In literature there are different kinds of bases and acids reported to be used in the methylolation and polymerization steps of UF resin manufacturing: the bases are sodium hydroxide, triethanolamine, trimethylamine, ammonia, etc., and acids are hydrochloric, sulfuric, phosphoric, formic, oxalic acids, etc. Also, the pH values can be anywhere between 7.0–9.5 or basic pH and 3.5–7.0 for acidic pH. Also, there are various cooking temperatures and time lengths employed for each step. The various UF resins resulting from using different combinations of basic pH and acidic pH values and cooking temperature and times cannot be the same theoretically, i.e. can result in different adhesive performances. There are also the F/U_1 and $F/(U_1 + U_2)$ mole ratios, the first in the range between 1.8 and 2.5 and the latter anywhere between 1.25 and 1.00. Overall, however, the typical resin synthesis procedures discussed in this chapter would represent these many "slightly" different UF resins relatively well.

There are also drastically different resin synthesis procedures reported in literature, aiming to obtain better bonding or other better performance properties. *One reported method* includes the first acidic reaction step with an F/U_1 mole ratio of about 3.0 and a strong acidic pH value, such as pH 1.0, followed by reaction in a neutral or alkaline pH, and then advancing the resin in the mild acidic pH, followed by adding second urea in the alkaline pH [39–41]. This type of procedure can result in different methylene/methylene-ether bond ratios as well as different chain branches in resin structure, such as uron-type ether bonds, to result in different performance properties, possibly hydrolytically more stable or lower formaldehyde emission values as claimed in comparison with typically synthesized resins [30]. In these procedures, resultant UF resins tend to be turbid probably because linear UF polymers having no methylol groups, such as methylene-di-urea and tri-methylene-tetra-urea are formed and come out as crystalline insoluble precipitates [15].

In extreme cases, highly cloudy resins are obtained with unpredictable viscosity behaviors and short storage lives. Then, is it okay to use somewhat cloudy resins, as is often the case with industrial UF resins, performance-wise? As discussed above, the answer would depend on whether the solid materials can be adequately assimilated into the liquid resin phase to participate as effective binder components in the curing step. In general, low levels of cloudiness will be of little concern, but highly cloudy resins can show lower bonding effectiveness, especially in the core-layers of boards where the maximum temperatures can be significantly lower than the melting points of these solid materials.

5.13 Polymeric Chemical Structures of UF Resins

Conclusively, typical UF resins, such as those made with an F/U_1 ratio of 2.10 and finished to have an overall F/U ratio of 1.15, are represented in Figure 5.7. Poly(hydroxymethyl-methylene-methylene-ether ureas) amount to about 55%, and monomeric hydroxymethyl ureas and free urea amount to about 45% based on the urea weights used in resin synthesis. The viscosity may range often in 200 cP, resin solids content at 60–64% with a clear or slightly milky appearance, the pH value in the 7–8 range, and usable lives of about two or three weeks when stored at cool temperatures.

5.14 Melamine-Urea-Formaldehyde (MUF) Resins

Melamine was discovered by J. Liebig in 1834 [42] and is currently manufactured by heating urea at 250–350 °C in the presence of ammonia and under high pressure, a method discovered in 1943 (Eq. (5.28)). This melamine manufacturing process from urea is a complex chemical process with carbon dioxide and ammonia by-products being required to be recycled [10]. Therefore, melamine's cost is high, higher than phenol by about 50–70% but less expensive than resorcinol. Melamine is a colorless crystalline solid material with a high melting point of 345 °C, one of the highest melting and resistant to burning organic compounds. It has a low solubility in water (0.3 g/100 ml water at 20 °C). These molecular

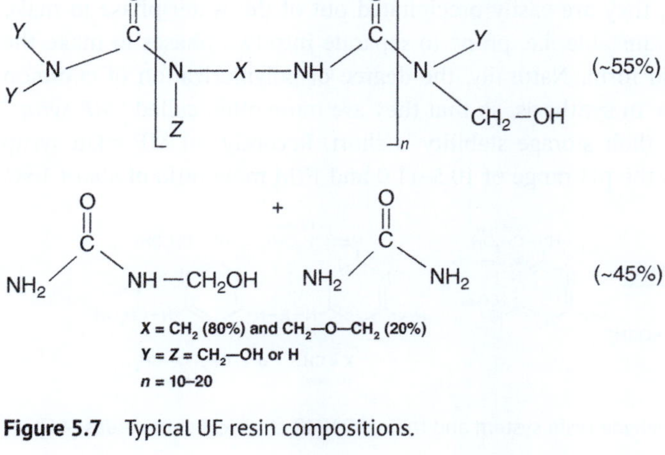

Figure 5.7 Typical UF resin compositions.

properties are thought to be the source of the good strength and good heat stability properties of MF resin products compared to UF resin products. Melamine molecule has three reactive amino groups; each amino group can react with two formaldehyde molecules to form mono-, di-, tri-, tetra-, penta-, and hexa-hydroxymethyl-melamine in the presence of excess formaldehyde. These hydroxymethylation reactions are reversible similarly to the UF resin system but to a lesser extent. The higher functionality and lower reversibility of hydroxymethyl groups means that MF resins would have lower formaldehyde emission potentials than UF resins.

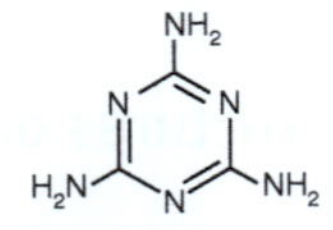

Melamine molecule

$$6CO(NH_2)_2 \rightarrow C_3N_6H_6 + 3CO_2 + 6NH_3 + 100\ \text{Kcal/mole} \tag{5.28}$$

MF resins are manufactured by reacting melamine with formaldehyde in a mildly alkaline solution of pH 7.5–8.5 and at temperatures above about 70 °C, similarly to the first step reaction in the UF resin manufacturing [13]. In this process, various **methylol-melamines** are formed in a reversible manner similarly as UF for resins, [43] and they can undergo condensation reactions to form dimers, trimers, etc., by forming methylene or methylene-ether bonds between melamine rings, as illustrated in Figure 5.8 with ^{13}C NMR chemical shift values assigned to the various carbon groups [43–46]. The liquid MF resin sample used for the ^{13}C NMR was made in the laboratory with an F/M ratio of 2.10, but it was unstable and in order to obtain the spectrum, the resin was modified by adding urea to F/(M + U) = 1.15 and reacting further at pH 7.0–8.0 to make a MUF resin, in which the urea mostly reacts to form methylol groups with little methylene and methylene-ether groups formed. For MF resins, methylene bonds tend to form predominately when pH value is below 7.0,and methylene-ether bonds are formed predominately at higher pH values. Methylene-ether bonds are known to break down when the resin is heat-cured in the acidic pH. The MF resin synthesis procedure commonly ends at an average degree of polymerization of only 1.5 in the slightly alkaline pH. No acidic polymerization step is needed, since melamine dimers, trimers, etc., are not much soluble in water and, when their proportion increases to a certain level, they are easily precipitated out of the water-phase to make the resin cloudy or storage-unstable, i.e. prone to separate into two phases to make the resin unusable in the liquid form. Naturally, the degree of polymerization of common liquid MF resins is kept low in synthesis, so that they are more often called "*MF syrup*" rather than MF resins and their storage stability is short. Recently, an MF resin syrup was reported to be made in the pH range of 10.5–11.0 and F/M mole ratio of about 1.90,

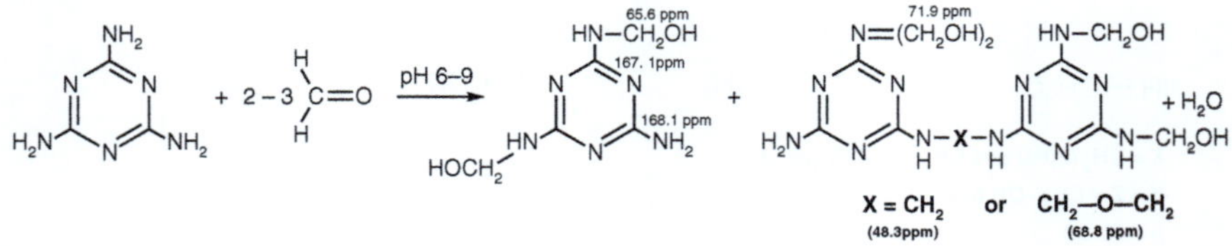

Figure 5.8 Melamine-formaldehyde resin system and their ^{13}C NMR chemical shift values [44].

stable up to about 60 days [47]. MF resin syrups are more often spray-dried for long-term uses. Pure MF resins are used in Formica-type laminate products and rarely used as wood adhesives because of the high cost, and therefore they are copolymerized with urea as MUF resins.

Synthesis of MUF resins: The synthesis of copolymer MUF resins is commonly carried out by adding melamine as a part of the second urea in the typical synthesis procedure of UF resins, i.e. a UF resin is synthesized through the alkaline and acidic steps to the desired viscosity and neutralized to pH 7–8 and then add melamine and some necessary urea and formaldehyde and cook further in the slightly alkaline pH until the target viscosity is attained. The ratio of melamine to urea (M/U) by weight goes close to 1.0 or slightly lower, and the F/(M + U) mole ratio for wood adhesive applications ranges in 1.7–2.0. Several other resin synthesis variations are possible, such as varying the extent of polymerization of UF resin components. The resultant liquid MUF resins are also unstable and therefore spray-dried to powder MUF resins. The pale yellow powder MUF resins store relatively well and after mixing with fillers they are shipped to wood manufacturers to be used after mixing with water and acid catalysts. In the resin-forming reactions some melamine-urea copolymer bonds (Figure 5.9) are likely to form, i.e. either methylene or methylene-ether bonds between melamine and urea molecules, [48] a desirable reaction path in theory.

Curing of copolymer MUF resins and formaldehyde emission: The curing of MUF resins is accomplished either by catalyst only at room temperature or by catalyst and heating, although the latter method is known to give more durable bonds. For heat curing, the radiofrequency (RF) heating method is a common method. Technologically, MUF resins can be used for bonding hardwood plywood, but the cost is high and there is little advantage because the formaldehyde emission level would be high at the current F/(U + M) mole ratios used in resin syntheses, although the durability of gluelines would be improved significantly. Durability of adhesive bonds of MUF resins measured by cyclic boiling tests or exposing to elevated temperatures (120–158 °F) was reported to be poor but increasing proportionately with melamine/urea ratio of MUF resins [49, 50]. The curing mechanism of MUF resins would be similar to that of UF resins, discussed in Chapter 6, by forming methylene bonds between amino groups of melamine and urea molecules. The overall F/(U + M) mole ratio of MUF resins would be the key parameter that determines [51] the curing speed and quality of gluelines, such as the bond strength or durability, which has been known to be optimum at about 1.7–2.0. The formaldehyde emission problem can be severe with these mole ratios [57], but their current application is not in bonding of particleboard or MDF, but in specialty applications where the formaldehyde emission problem does not attract much attention as of now.

Uses of MUF resin adhesives in wood composite products: MUF copolymer resins are quite more expensive than UF resins on weight basis because of the melamine contents. Furthermore, the powder form MUF resins need to be formulated with a catalyst, filler, and water in a slurry form, which cannot be used directly in bonding of PB or MDF boards because

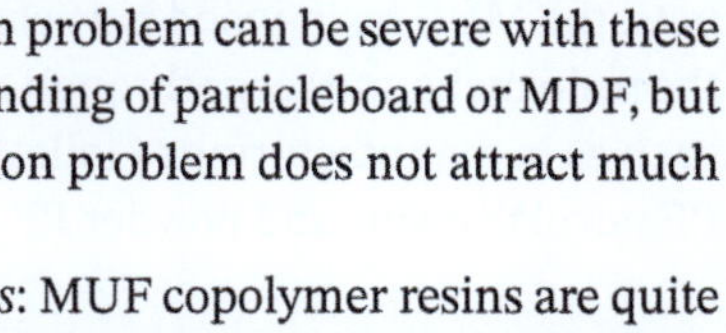

Figure 5.9 Copolymer formation of urea and melamine in MUF/UMF resins [44].

of the difficulty of obtaining a uniform distribution of adhesive mix among wood particles or fibers in the current resin blending/matting systems. Mixed MUF adhesives are applied with brush or beading applicators and, are currently, used where PRF resin adhesives are competitively used. It is reported that MUF resins are more widely used in Europe since the introduction of liquid forms of MUF resins and catalysts in 1987 with cure times of two to three hours at 20 °C [31]. Gluelines of MUF resins are nearly colorless and show relatively good resistance to water exposure, significantly better than UF or PVAc resin adhesives. In the United States, MUF resins have been used in lumber lamination, end-jointing, and some face-gluing operations such as in bonding of tabletops, truck and rail car flooring, chopping boards, school furniture, curved hardwood plywood, and molded wood products such as toilet seats.

MUF resins are not as durable as PRF resins and are not recommended in long-term exterior or high-humidity exposure uses. In contrast to PRF resins, however, MUF resins must be heat-cured at 240–260 °F for better durability, and this is more conveniently done by using the RF heating method, to which MUF resins have a better response than PRF resins, i.e. a lower arcing tendency. End-jointing is also an operation where the fast-curing speed is desired for productivity, i.e. the long, just-bonded lumbers need to be moved, lifted, or turned around, and this requires relatively high bending or shear strengths right after the bonding. In lumber lamination plants of long wooden beams, individual lumbers are made by initially end-jointing (mostly finger jointed) using MUF resin binders under the RF curing condition, and then the end-jointed pieces are face-glued with PRF resins for lamination, followed by RF heating or by curing overnight in a warm room. In this type of exterior-grade beam construction, the proportion of the less durable MUF resin gluelines in the whole beam assembly is small compared to the PRF resin gluelines. Also, these PRF resin-bonded beams, although classified to have exterior use qualifications, are mostly used under some protection from direct rain, and, therefore, the lower durability of MUF resins in the end-joints is a minor concern.

MUF resins are also used as moisture-barrier binders for wood-fiber ceiling tiles, applied by spraying the slurry MUF resins on the tile surface during manufacturing. Subsequent heat-curing of tiles leaves thin cured MUF resin film on the surface that acts as moisture barrier and also prevents the sagging of ceiling tiles that often occurs in use due to absorption of moisture and weight gains and expansion of wood fibers. In manufacturing of molded products, the MUF resin-wood particle mixes are compressed for densification under heat, and the resin melts and seeps through the wood particles and polymerizes to bind them together, such as in manufacturing of toilet seats and school furniture. In these applications, powder PF novolac resins and powder PF resole resins are also competitively used.

5.15 Urea-Melamine-Formaldehyde (UMF) Resins

Current UF resins have formaldehyde emission problems, as will be discussed in Chapter 6. Industry efforts to address the formaldehyde emission problem have led to use of resins with low F/U mole ratios of 1.00–1.15, which resulted in low bond strength and high water-absorption/thickness-swell values of boards as well as longer hot-press times. Therefore, modifying UF resins by adding low levels of melamine (2–10%) in the form of UMF resins has been widely investigated and our laboratory also reported some positive

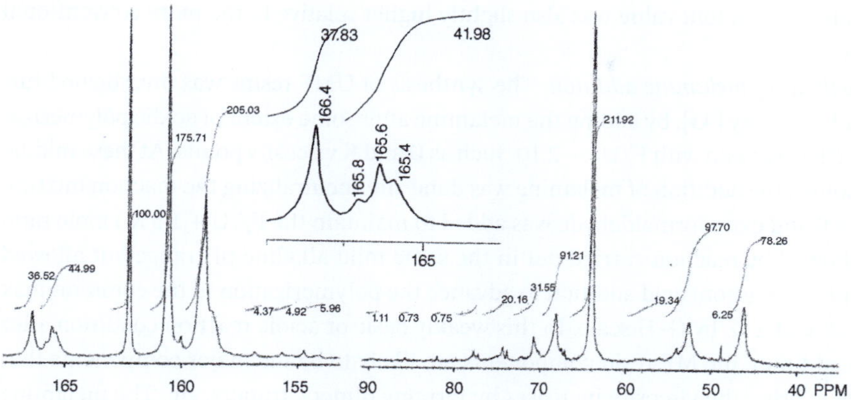

Figure 5.10 ^{13}C NMR Spectra of Resin 5.0% UMF1.05 in water (Method D) [52].

results [44, 52]. Melamine molecule has three amino groups compared to two similar groups of urea for reaction with formaldehyde, and, therefore, UMF resins are expected to give boards with lower free formaldehyde contents.

Four different methods of melamine addition: The modification of UF resins with low levels of melamine was commonly done in the third step of UF resin synthesis, where the melamine is added first, followed by adding the second urea (Method #4). This type of melamine addition was found to give adequately stable UMF resins. An example ^{13}C NMR spectrum of this type of resin is shown in Figure 5.10 where the base UF resins were made with an F/U_1 mole ratio of 2.10 similarly with the typical UF resin synthesis and advanced to P-Q viscosity in the acidic polymerization step, followed by neutralization, addition, and reaction of melamine (5.0% based on liquid resin weight), and then the second urea was added to an overall $F/(U + M) = 1.05$ [52]. The resultant UMF resin entails reacted (~165.5 ppm) and unreacted (~166.4 ppm) melamine molecules due to the limited availability of formaldehyde at the point of addition.

Other methods of melamine addition were also explored: melamine is added in the very beginning of resin synthesis with the first urea (Method #1), added in the middle of the first urea by dividing the first urea into two parts (Method #2), and added at the end of the first alkaline resin synthesis step just before the acidification step (Method #3) [53–55]. Resins cooked by Methods #2 and #3 were cloudy and found to show poor storage stability, probably because the melamine components could not survive the second synthesis step of acidic polymerization by forming precipitates of dimers, trimers, etc. On the other hand, it is probable that the Method #1 allowed to form ample methylol groups on melamine molecules, resulting in helping MF resin components survive through the acidic polymerization step, i.e. somewhat more stable resins. In this acidic polymerization step of Method #1, copolymers between urea and melamine molecules are likely to form (Section 5.14), a desirable reaction in theory. In conclusion, UMF resins made with 5.0% melamine made by Methods #1 and #4 resulted in laboratory particleboards having better IB strength and better water absorption values than the control UF resin, and the formaldehyde content of boards measured by the European standard test method [56] ranged from 6 to 8 mg/100 g board in comparison to 10–12 mg/100 g boards for control UF resin, at the same mole ratios of 1.05 [57]. Further, the UMF resin of Method #1 gave slightly higher strength values, but the formaldehyde content value was also slightly higher relative to the more conventional Method #4.

Other methods of melamine addition: The synthesis of UMF resins was investigated further in our laboratory [43], by adding the melamine after some extent of acidic polymerization of the UF base resin with $F/U_1 = 2.10$, such as D and K viscosity points. At these middle viscosity points, the addition of melamine was done after neutralizing the reaction mixture to pH 7.0–8.0 and extra formaldehyde was added to maintain the F/(U + 1.5 M) mole ratio at 2.10, followed by reaction carried out in the same mild alkaline pH range but allowed to drift to pH 6.2 without acid addition to advance the polymerization to the common max viscosity value of Y–Z by G–H scale. In this weakly basic or acidic reaction condition after melamine addition, the MF resin components are likely to be the major components that polymerize to bring the viscosity increases by forming dimers, trimers, etc. The melamine level studied was from 6.0% to 12.0% based on the liquid resin weights, which is from about 12.0–25.0% based on the combined weights of urea and melamine. The synthesized resins were also tested by making laboratory particleboards by hot-pressing at 325 and 365 °F for four and five minutes press times. In conclusion, UMF resins made with 5.0% melamine resulted in laboratory particleboards with better IB strength and better water absorption values than the control UF resin, and the formaldehyde content of boards by the European standard test method [56] ranged in 10 mg/100 g board ranges in comparison to 12 mg/100 g boards ranges for control UF resin, at the same mole ratio of F/(U + 1.5M) = 1.05. Further, the UMF resins made with melamine addition at D and K viscosity points showed differences in resins' storage stability and physical strength values of boards, arising from the different extents of polymerization of UF components in the pH range of 4–5 and of MF resin components in the pH range of 8.0–6.7. Details and resin structures studied by ^{13}C NMR are discussed in the advanced reading materials section in Chapter 25. Various other methods of melamine addition in UMF resins are also reported in various patents [58–60].

Handling properties of UMF resins and prospects: Thus, some synthesized UMF resins can be handled similarly to liquid UF resins, and they improved strength values and lowered the formaldehyde content values of boards. The cost increase due to melamine addition is one major reason the industry has been reluctant to adopt these UMF resins. On the other hand, the resultant formaldehyde content values of these UMF resins are just on the borderline of the required emission levels, leaving the industry to work on additional devices and methods in order to comply with the current emission regulations. Another potential problem to overcome is the slower curing speed of UMF resins in comparison with UF resins due to the added melamine as well as the lower F/(U + M) mole ratio values. Melamine is more basic than urea and a stronger acid catalyst is reported to be needed [53–55]. How much catalyst is to be used and how much hot-pressing time is to be lengthened is the question that cannot be answered adequately in the laboratory. Then, the industry needs to evaluate the stronger acid catalyst's effect on the long-term durability of boards because strong acids tend to degrade wood. The hot-press is the bottleneck in the PB board manufacturing plants, and for ½ thick boards the press time is normally 180 seconds, and even a slowdown of 10 seconds can reduce the production volume by 5%, and similarly, adding melamine at about 5% level will increase the resin cost by almost 10% or more. For these various reasons, adopting the UMF resin technology can be a challenging task for the industry.

5.16 Summary

Synthesis chemistry and industry manufacturing practices of typical UF resins are presented. In the first step of resin synthesis, the first urea (U_1) is reacted with 50% formaldehyde at pH 7–8 and at 70–90 °C with an F/U_1 mole ratio between 2.0 and 2.4 for about 30 minutes, resulting in a mixture of mono-, di-, and tri-hydroxymethylureas and their methylene-ether oligomers. Then, in the second step of reaction, the mixture is acidified to pH 4–5 and reaction is continued, where the hydrogen ions (H^+) catalyze the hydroxymethyl groups to methylenic cations that readily react with urea amide groups of other hydroxymethylurea molecules to form methylene bonds between urea molecules, resulting in dimers, trimers, and higher molecular weight oligomers (or polymers) that are still bonded with some hydroxymethyl groups. Also, some formaldehyde is formed back from the dissociation of hydroxymethyl groups. The viscosity of reaction mixture increases, commonly reaching up to P–Z viscosity, and the synthesis reaction is ended at the target viscosity (or degree of polymerization) by neutralizing the acid catalyst with a base to a neutral pH of 7–8. Then, in the third step of reaction, while cooling down the reaction mixture, the second urea (U_2) is added and mixed until the mixture reaches to room temperature. The second urea reacts with the free formaldehyde present to form hydroxymethyl groups, resulting in UF resins having about 60–63% resin-solids contents and viscosity values anywhere between about G and V. Thus, UF resins consist of polymeric (hydroxymethyl-methylene-methylene-ether-ureas) with a number-average degree of polymerization about 12–20 derived from first urea (~55%) and monomeric hydroxymethylureas and free urea (~45%) derived from second urea. Further, some of the hydroxymethyl groups bonded to the polymeric resin components slowly dissociate to formaldehyde and re-form hydroxymethyl groups on the second urea molecules, resulting in various lower viscosity values for resultant resins. When stored at room temperature for about two weeks or longer at alkaline pH, hydroxymethyl groups go on to further react to result in viscosity and turbidity increases that can be marked as a measure of the storage life of the resin. Most hydroxymethyl groups of UF resins remain as potential functional groups in the alkaline pH waiting to be activated by acidic catalysts. Reaction mechanisms and kinetics of hydroxymethylation and methylene bond forming reactions in resin synthesis and resultant branching of polymer chains and various resin properties used for quality controls in using UF resins as wood composite binders are discussed. Also, different resin synthesis procedures reported in literature are discussed on their possible advantages and disadvantages. Also, the resin synthesis and polymer structures of MUF and UMF resins are discussed with their potential for reducing the formaldehyde emission potentials and increasing the strength/durability properties of boards. The curing chemistry of UF, MUF, and UMF resins is discussed in Chapter 6 and more advanced reading materials are presented in Chapters 24–27.

References

1 The Innovation Group (2006). *Report on UF Resins*, 07960. Morristown, NJ.
2 Hoelzer, K. (1884). *Berichte* 17: 659.
3 Hoelzer, K. (1885). *Berichte* 18: 3302.

4 Goldschmidt, K. (1896). *Berichte* 29: 2438.

5 John, H. (1918). Manufacture of aldehyde condensation products capable of technical utilization. German Patent 392283; 1920. U.S. Patent 1,355,834.

6 Vogel, R.E. (1941). *Kunststoffe* 31 (9): 309–313.7.

7 Vogel, R.E. (1954). *Kunststoffe* 41 (8): 335.

8 Wirpsza, Z. and Brzezinski, J. (1973). *Aminoplasty*. Moscow: Khimia.

9 Butlerow, A. (1859). *Annalen* 145: 242.

10 Saunders, K.J. (1988). *Organic Polymer Chemistry*, 2e. New York: Chapman and Hall.

11 Gollob, L. (1990). *Proceedings of the NPA Resin and Blending Seminar*, 9–14. National Particleboard Association.

12 U.S. Senate Bill: S-1660 (2011). American Jobs Act of 2011.

13 Kim, M.G. and Amos, L.W. (1990). *Industrial and Engineering Chemistry Research* 29: 208.

14 Kim, M.G. (1999). *Journal of Polymer Science Part A: Polymer Chemistry* 37: 995.

15 Kim, M.G. (2000). *Journal of Applied Polymer Science* 75: 1243.

16 Kim, M.G. (2001). *Journal of Applied Polymer Science* 80: 2800.

17 Einhorn, A. and Hamburg, A. (1908). *Berichte* 41: 24.

18 Kadowaki, H. (1953). *Bulletin of the Chemical Society of Japan* 11: 1027.

19 Kadowaki, H. (1956). *Bulletin of the Chemical Society of Japan* 11: 248.

20 Dixon, A. (1918). *Journal of the Chemical Society, Transactions* 113: 246.

21 Pollak F. (1921). Improvements in or relating to the manufacture of condensation products from formaldehyde and urea, thiourea derivatives. Great Britain Patent 171,094.

22 Van Lauer, M. (1919). *Bulletin des Societes Chimiques Belges* 28: 381.

23 De Jong, J.I. and De Jonge, J. (1952). *Recueil des Travaux Chimiques des Pays-Bas* 71: 643.

24 De Jong, J.I. and De Jonge, J. (1952). *Recueil des Travaux Chimiques des Pays-Bas* 71: 661.

25 De Jong, J.I. and De Jonge, J. (1952). *Recueil des Travaux Chimiques des Pays-Bas* 71: 891.

26 De Jong, J.I. and De Jonge, J. (1953). *Recueil des Travaux Chimiques des Pays-Bas* 72: 139.

27 De Jong, J.I. and De Jonge, J. (1953). *Recueil des Travaux Chimiques des Pays-Bas* 72: 1027.

28 Zigeuner, G. (1955). *Fette,-Seifen-Anstrichmittel* 57 (14): 100.

29 Williams, J.H. (1982). Presentation, Forest Products Society. 36th Annual Meeting, June 20. Forest Products Society.

30 Williams, J.H. (1983). Hydrolytically stable urea-formaldehyde resins and process for manufacturing them. U.S. Patent 4,410,685.

31 Grostad, K. and Ormstad, E.B. (2005). Melamine and melamine-urea adhesives for European market. In: *Wood Adhesive 2005* (ed. C.R. Frihart), 191. Forest Service, FPL.

32 Roffael, E. (1982). *Adhaesion* 3: 18.

33 Ludlam, P.R. and King, J.G. (1984). *Journal of Applied Polymer Science* 29: 3863.

34 Hope, P. and Stark, B.P. (1973). *British Polymer Journal* 5: 363.

35 Rammon, R. (1990). *Proceedings of the NPA Resin and Blending Seminar*, 15. National Particleboard Association.

36 Staudinger, H. and Wagner, H. (1955). *Makromolekulare Chemie* 12: 168.

37 Guy, J. and Solange, S. (1975). Application of precipitates of methylene-ureas and products obtained. U.S Patent 3,869,311A.

38 Renner, A. (1976). Process for the manufacture of porous solids consisting of crosslinked urea-formaldehyde polycondensation products. U.S. Patent 3,931,063A.

39 Sun, Q.N., Hse, C.Y., and Shuppe, T.F. (2011). *Journal of Applied Polymer Science* 119: 3538.

40 Camaloni, D. (1981). Process for the preparation of urea-formaldehyde condensates and a binder obtained by the process. European Patent 0.038,180.

41 Paiva, N.T., Hendriques, A., Cruz, P. et al. (2012). *Journal of Applied Polymer Science* 124 (3): 2311.

42 Liebig, J. (1834). *Annalen* 10: 48.

43 Koeda, K. (1959). *Kobunshi Kagaku* 16: 62.

44 No, B.Y. and Kim, M.G. (2004). *Journal of Applied Polymer Science* 93: 2559.

45 No, B.Y. and Kim, M.G. (2005). *Journal of Applied Polymer Science* 97: 377.

46 No, B.Y. and Kim, M.G. (2007). *Journal of Applied Polymer Science* 106: 4148.

47 Van Langenberg, K. and Reilly, M. (2005). Influence of storage of on the stability of melamine-formaldehyde resins. In: *Wood Adhesive 2005* (ed. C.R. Frihart), 513. Forest Service, FPL.

48 Yanagawa, R., Mizuno, T., and Sasaki, A. (1962). *Nippon Mokuzai Gahkaishi* 8 (6): 231.

49 Rider, S.H. and Kozdemba, S.E. (1954). *Forest Products Journal* 4 (5): 283.

50 Blomquist, R.F. and Olson, W.Z. (1957). *Forest Products Journal* 7 (8): 266.

51 Blank, W.J. (1979). *Journal of Coatings Technology* 51: 61.

52 Mao, A., Hassan, E.B., and Kim, M.G. (2013). *BioResources* 8: 2453.

53 Mao, A., Hassan, E.B., and Kim, M.G. (2013). *BioResources* 8: 2470.

54 Mao, A., Hassan, E.B., and Kim, M.G. (2013). *BioResources* 8: 4659.

55 Mao, A., Hassan, E.B., and Kim, M.G. (2013). *BioResources* 8: 5733.

56 IN EN 120-92, Wood-Based Panels - Determination of Formaldehyde Content-Perforator Method.

57 Margosian, R. (1990). *Forest Products Journal* 40 (6): 19.

58 Shiau, D.W. and Smith, E. (1985). Low formaldehyde emission urea-formaldehyde resins containing a melamine additive. U.S. Patent 4,536,245. Boden Inc. Columbus, OH.

59 Druet, B. and Rochet, G. (1991). New process for the manufacture of urea-formaldehyde resins containing melamine. U.S Patent 5,008,365. Orkem Group, Paris LaDefense, France.

60 Breyer, R.A., Hollis, S.G., and Jural, J.J. (1997). Low mol ratio melamine-urea-formaldehyde resins. U.S. Patent 5,681,917. Georgia-Pacific Resins, Inc., Atlanta, GA.

6

Urea-Formaldehyde and MUF/UMF Wood Adhesive Resins (Curing)

Urea-formaldehyde (UF) resins can undergo a rapid polymerization reaction of curing when acidified to pH 4–5 or below and heated. For example, the gel time of particleboard UF resins measured at 100 °C without catalyst is about 25 minutes or longer, and it shortens to less than 2 minutes with a catalyst. Catalysts are acids or acid-generating materials. Catalyzed UF resins will increase in viscosity and eventually solidify. Since most southern U. S. wood species are acidic and have low acidic buffer capacities, the wood acidity can be enough to shift the resin's pH to the acidic side, and, therefore, catalysts are normally not needed in board manufacturing. In other regions, acid catalysts are often needed, especially for high pH and/or high buffer capacity wood materials. The acidic catalysts, often called latent catalysts, are designed to act slowly at room temperature and are added to resin by the resin manufacturer or by board manufacturer by spraying the catalyst solution and resin simultaneously onto wood particles in the resin blender. Latent acid catalysts used are normally ammonium or organic amine salts of strong acids.

6.1 Typical Latent/External Catalysts Based on Ammonium Salts for UF Resins

$$\begin{aligned} &NH_4^+ - X^- \rightarrow NH_4^+ + X^- \leftrightarrow NH_3 + H^+ + X^- \\ &\quad X = Cl\ (\text{Ammonium chloride}); \\ &\quad X = (SO_4)_{\frac{1}{2}}\ (\text{Ammonium sulfate}) \end{aligned} \tag{6.1}$$

$$\begin{aligned} &6\,CH_2 = O + 4NH_3 \rightarrow (CH_2)_6N_4 + 6H_2O \\ &\quad (\text{Hexamethylenetetramine}) \end{aligned} \tag{6.2}$$

$$\begin{aligned} &\text{UF Resin} \rightarrow \text{UF Resin} + CH_2 = O \\ &\quad \text{Heat} \end{aligned} \tag{6.3}$$

Catalysts are added to UF resins at various levels (0.25–2.0% based on liquid resin weight) to obtain a faster, controlled cure rate in the board manufacturing process

Adhesives and Finishes for Wood: For Practitioners and Students, First Edition. Moon G. Kim.

[1–3]. When an ammonium salt is dissolved in (aqueous) UF resins, reactions shown in Eqs. (6.1)–(6.3) occur. In the first step, the ammonium salt dissociates into ammonium ion (NH_4^+) and conjugate anion (X^-) and then the ammonium ion further dissociate into *ammonia* and *hydrogen ion*, the latter being the (strong) acidic species (Eq. (6.1)). However, the extent of dissociation is small, and the solution shows only a weak acidic pH because the reaction proceeds to the right-hand side only slightly at room temperature in water. This dissociation increases as the temperature of the solution goes up, but only to a limited extent, and as such, the amount of acid generated by this mechanism is not enough.

Then, the UF resin comes in to play a part in catalyzation equation. UF resins contain some free formaldehyde, which reacts readily with the freed ammonia to form hexamethylenetetramine (hexamine) (Eq. (6.2)), which does not dissociate back to formaldehyde and ammonia. This depletion of formaldehyde by ammonia results in pushing Eq. (6.1) to the right side as long as there remains free formaldehyde to react in the resin, forming more acid (H^+). UF resins normally contain a small amount of free formaldehyde (<0.5%) and the acid generation at room temperature is still limited, but in hot-pressing of boards the resin is heated, and some additional formaldehyde is produced due to the breakage of some hydroxymethyl groups of resin molecules (Eq. (6.3)), resulting in generation of more hydrogen ions. Thus, the available hydrogen ion concentration in curing of resins increases with the amount of formaldehyde available in the first place, and if there is an enough amount of formaldehyde, then also on the catalyst level, which is added usually at about 0.5–2.0% by weight of resins. The catalyst effect on resin curing is proportional to the amount of the acid (H^+) produced, i.e. the amount of catalyst added within the capacity of formaldehyde generation of resins. This type of catalyst is called "latent catalysts," since the catalyst activation occurs mainly in the hot-pressing stage.

This type of ammonium salt has been the main catalyst used in bonding of particleboard (PB), medium density fiberboard (MDF), and hardwood plywood, mostly in the Eastern and Northern United States, especially in the winter seasons. Fast hot-pressing times are desired in most plants, but limitations arise for this type of catalyst due to the low F/U mole ratios of UF resins currently used in board manufacturing, adopted to minimize the formaldehyde emission levels of boards. For low F/U mole ratio resins, the free formaldehyde produced in the hot-pressing is proportionately less than those of higher mole ratio resins, and therefore, the catalyzation effect is less, and adding more ammonium salt catalysts does not help as the reaction mechanism would indicate. Therefore, current UF resins often need additional catalysts, and this is met in limited extent by other acidic materials, as discussed below.

6.2 Organic Tertiary Amine Salts of Strong Acids and Other Catalysts

$$(CH_3CH_2)_3N - HCl \xrightarrow{\text{(Heat)}} (CH_3CH_2)_3N + H^+ + Cl^- \tag{6.4}$$

$$(HOCH_2CH_2)_3N - HCl \tag{6.5}$$

Tertiary amine and other salts: Salts of strong acids of organic tertiary amines, such as triethylammonium hydrochloride, are used as additional catalysts [2], which dissociate similarly as shown in Eq. (6.4), but the triethylamine generated from the dissociation does not have the reactivity for free formaldehyde because it has no hydrogen atom to be replaced on its nitrogen atom. The dissociation proceeds at a low degree and is not helped by presence of free formaldehyde. At hot-pressing temperatures, however, the dissociation proceeds at higher levels than with ammonium salts due to the alkylamine's chemical structure, generating the free acid. Overall, tertiary amine salts are also slow-acting catalysts and are therefore used mostly as an augmenting component to the ammonium salt catalysts to enhance the curing speed of resin. Tri-ethanolamine hydrochloride (Eq. (6.5)) and other tertiary amine salts are also used [4]. Another particular material used in the U. S. industry as a catalyst and filler, up to about 5% of resin solids, is the common salt (sodium chloride) [5]. Also, magnesium chloride is claimed to be acting similarly, making the resin cure faster and at the same time adding to the resin-solids level [4]. The reaction mechanisms of these additives have not been reported, but the latter (magnesium) was suggested to be making complex-type bonds with the amide groups of urea. These types of catalysts are used only by some board manufacturers mostly added to resin at the time of resin manufacturing.

Aluminum sulfate or chloride and free acids: For low F/U mole ratio UF resins, however, stronger catalysts might be needed when the curing speed (hot-pressing time) is to be increased close to those of higher mole ratio resins used in the previous times. These external catalysts used are the stronger acid-generating materials such as aluminum sulfate or aluminum chloride, or other free acid materials such as formic acid, citric acid, sulfuric acid, and hydrochloric acid. The addition level is important for these strong catalysts. With the aluminum sulfate catalyst, as shown in Figure 6.1 [6], the 100 °C gel times are greatly reduced in comparison with ammonium sulfate catalyst, and they are also nearly halved by increasing the level from 0.5% to 2.5% for all mole ratios of resins, while with ammonium sulfate catalyst, the gel time changes far less with additional levels. It is noted that resin mole ratio effects for ammonium salt catalysts are far larger due to the availability of higher levels of free formaldehyde in higher mole ratio resins. It is apparent that aluminum sulfate-type strong acid catalysts can drastically shorten the gel times of resins, especially for low F/U mole ratio resins, which do not respond well to the increasing levels of ammonium sulfate catalyst. In this approach of using strong acidic materials as catalysts, buffering materials are often used at the same time to control the pot lives of mixed resins; ammonia, urea, or hexamine are used. One question that remains is whether the stronger free acids will degrade wood in boards more than ammonium sulfate-type catalysts.

6.3 Catalyzation of UF Resins by Wood Acids

All catalysts discussed above have to be used judiciously with regard to the acidity (pH) and acidic buffer capacity of incoming wood materials in board manufacturing. Wood acids (organic carboxylic acids and phenolic hydroxyl groups of lignin components) are mild acids. In general, the statement that wood is acidic means the carboxylic acid or phenolic hydroxyl groups are present as acid forms in raw woods and are ready to dissociate into hydrogen ions to some small extents.

$$R - COOH \rightarrow R - COO^- + H^+ \tag{6.6}$$

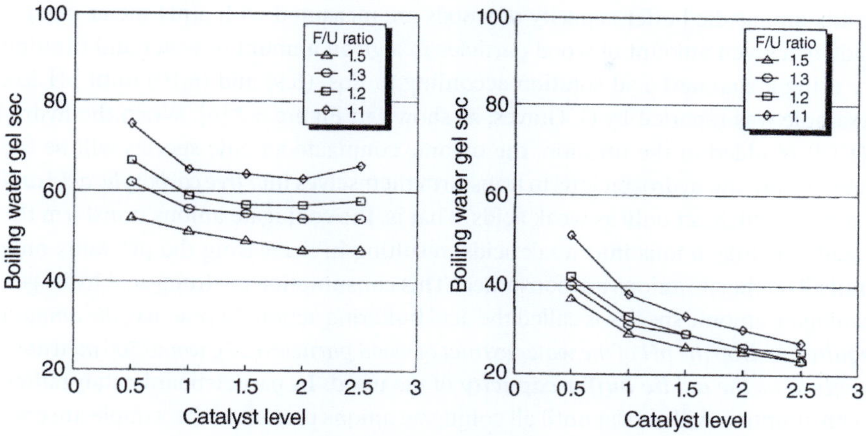

Figure 6.1 Gel times of various F/U mole ratio UF resins at 100 °C with ammonium sulfate (left) and with **aluminum sulfate** (right) addition levels (%. [6]. Source: Reproduced with permission of the Composite Panel Association.

$$\text{Ph} - \text{O} - \text{H} \rightarrow \text{Ph} - \text{O}^- + \text{H}^+ \tag{6.7}$$

$$\text{Hemicellulose} - \text{O} - \text{COCH}_3 + \text{HO} \rightarrow \text{Hemicellolose} - \text{OH} + \text{CH}_3\text{COOH} \tag{6.8}$$

Origin of wood acidity: We can eat foods that contain these acidic groups: acetic acid (vinegar), citric acid, aspirin, etc. These weak acids are present in all wood materials in significant amounts: lignin components have carboxylic acid (Eq. (6.6)) and phenolic hydroxyl (Eq. (6.7)) groups, and hemicelluloses contain acetate groups (Eq. (6.8)) that can hydrolyze relatively easily to acetic acid. Wood acids are present in higher amounts in hardwoods than in softwoods, i.e. different amounts in different species, and also the wood aging factor comes into play in board manufacturing because the wood acidity and acidic buffer capacity would change as wood particles age during the storage and particle preparation periods. Although wood acids are weak acids, they constitute the acid catalysts in the curing of UF resins in most PB manufacturing in the Southern United States. In other areas where woods of higher pH or higher buffer capacity are used, the kind and amount of catalysts need to be chosen and increased according to the acidic buffer capacities of woods and seasons.

$$\text{R} - \text{COO}^-(\text{Na}^+) + \text{H}^+(\text{X}^-) \rightarrow \text{R} - \text{COOH} + \text{NA}^+ + \text{X}^- \tag{6.9}$$

$$\text{Ph} - \text{O} - (\text{Na}^+) + \text{H}^+(\text{X}^-) \rightarrow \text{Ph} - \text{O} - \text{H} + \text{Na}^+ + \text{X}^- \tag{6.10}$$

The acidic properties of woods are defined in terms of their acidity and acidic buffer capacity: In woods, both the free carboxylic acid form and the conjugate anionic form (*salt*) are present, and their ratio in a wood sample varies depending on the wood species and the aging history of wood particles. Thus, Eqs. (6.6) and (6.9) play out in a sample of wood in the presence of water, where the former reflects acidity (pH) and the latter the acidic buffer capacity. In the conjugate anionic salt form, the counter ions are some metallic cations such as sodium or potassium ions, which are naturally present in wood or get into the wood during wood processing and aging processes.

The acidity and acidic buffer capacity of woods are measured with a pH meter using a slurry made of a given amount of wood particles in a given amount of water and titrating the slurry using a standard acid solution according to Eqs. (6.9) and (6.10) until pH 3.0. A good example was reported by G. Graves, as shown in Figure 6.2 [6]. When the hydrogen ion (HCl) is added in the titration, the organic conjugate anionic species will be the first ones to capture the hydrogen ions to transform themselves into *free carboxylic acid and phenolic species*, which act only as weak acids. That is, the conjugate anions transform the strongly acidic hydrogen ions into weak acids, resulting in decreasing the pH value only slowly until all conjugate anions are consumed. This consumption of strong acid hydrogen ions by conjugate anionic species is called the acid buffering action. In practice, *the amount of acid required to drop the pH of the water extract of wood particles (25 g wood/500 ml water) to 3.0 is defined as the* ***acidic buffer capacity*** *of the wood*. In particleboard catalyzation, this acid consumption will persist until all conjugate anions present in the sample are consumed, and only then the hydrogen ions present and generated from catalysts will begin to catalyze the curing reactions of resins.

Acids and acidic buffer capacity of various woods: In Figure 6.2, various wood materials measured for acidic buffer capacities are presented: initial pH values between 4 and 4.9

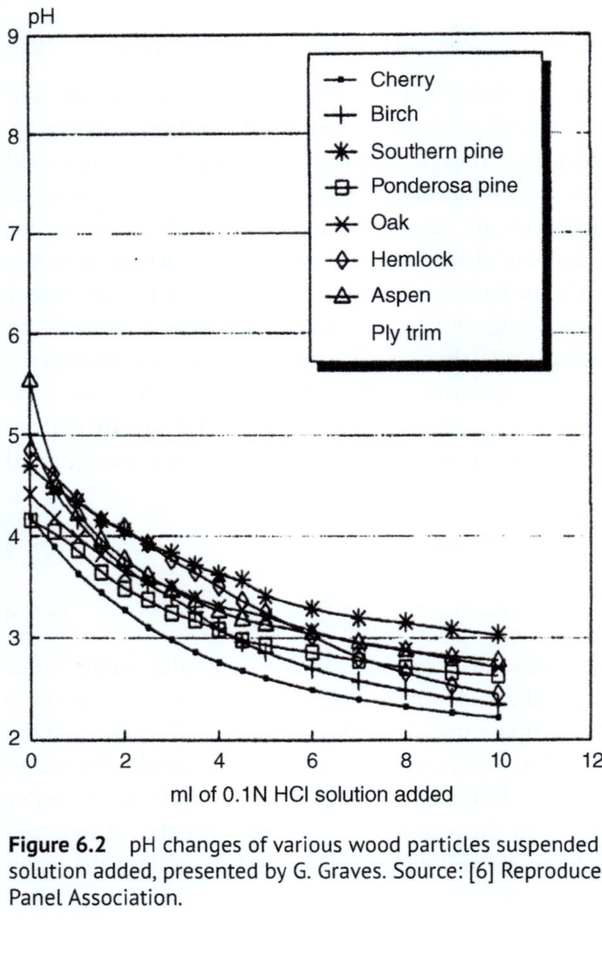

Figure 6.2 pH changes of various wood particles suspended in water vs. ml of standard acid solution added, presented by G. Graves. Source: [6] Reproduced with permission of the Composite Panel Association.

except aspen at pH 5.5, the least acidic wood [6]. Cherry, ponderosa pine, and oak are more acidic with the least buffer capacities, while southern pine shows the highest buffer capacity value. (Fortuitously, the curve for plywood trims missed printing in the publication, but it would be starting from pH 8 to 9.) In particleboard industry, various wood species are used, and different aging histories of wood particles are also common so the acidity and buffer capacity can vary significantly. Therefore, the pH and buffer capacity of incoming wood particles need to be measured, and wood mixes made accordingly to maintain the acidity and acidic buffer capacity values within a constant and narrow range over time to avoid the changing the catalyst level and/or hot-pressing time.

Softwood plywood trims that are often incorporated into particleboard manufacturing in the Southeastern U. S. region have high pH values and high acidic buffer capacity due to the caustic in the PF resin binders and can result in problems, i.e. requiring more catalyzation or longer hot-pressing times.

White fir and sycamore are known to have high pH and buffer capacity values, while *red cedar* has a high acidity. All need to be considered when these woods are used.

Burner ash: Another source of high buffer capacity for all particleboard plants is the *burner ash* that is often flown into the wood particle drying chamber. Wood contains small amounts of metallic components such as sodium, potassium, or calcium, and when wood wastes and bark are burned to generate hot air for drying of wood particles, the metallic components transform into carbonates or oxides (ash), all strongly basic materials, and some ash particulates are entrained in the hot air and deposited on the drying wood particles. Thus, the incoming wood and burner operation need to be closely monitored, and any changes need to be noted for possible catalyst level adjustments. Overall, the natural wood acidity allows many particleboard/MDF plants in the Southern U. S. region to run the plants without using any catalyst or at most small amounts of catalysts most of the year, while in other areas the catalyst kinds and levels need to be adjusted seasonally, stronger catalysts in winter and less strong catalysts in summer, etc. It should be noted also that catalysts by themselves do not compensate for low F/U mole ratio values used in resin manufacture, i.e. do not increase board strength values by themselves.

6.4 Buffering of Acidic Catalysts

Board manufacturing plants often use another catalyst-type material such as *ammonia*. When it is expected to take a longer time than normal from the blending to the hot-pressing station, or the catalyzation needed for the target short hot-pressing time would be too strong in comparison with the transit time during the hot weather, or when pre-cure problems are noticed, free triethylamine or ammonia solution (ammonium hydroxide) is often added as a buffer to capture the acidity as salts [6, 7] thus delaying the action of acid generated from acid catalysts. One convenient method showing this effect is the differential scanning calorimeter (DSC) analysis as shown in Figure 6.3 [8], run at an isothermal temperature of 60 and 70 °C. The thermal analysis of catalyzed resins, measuring the amount of heat generated from the curing reaction, shows the maximum heat generation at about 10 minutes at 60 °C, but it is delayed to about 24 minutes when buffered by adding ammonia. This method might be mostly needed in hot summer days in particleboard or MDF manufacturing plants, as well as in room temperature curing of *melamine-urea-formaldehyde*

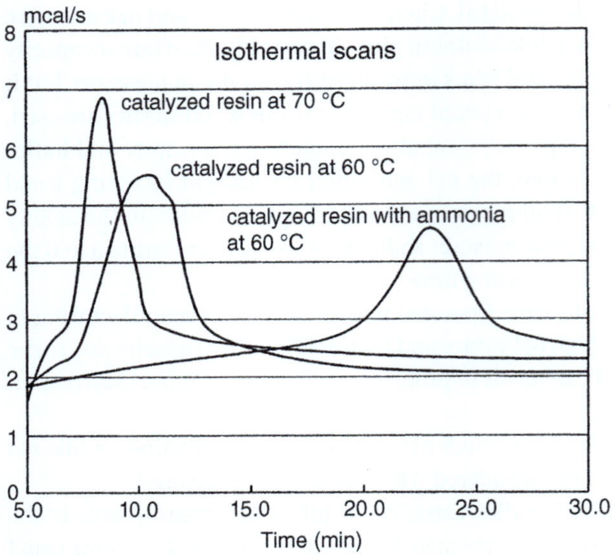

Figure 6.3 Generation of the reaction heat in DSC of a catalyzed and buffered UF resin reported by R. Rammon. Source: [8] Reproduced with permission of the Composite Panel Association.

(MUF) resins catalyzed with aluminum sulfate-type stronger catalysts and for longer assembly times.

6.5 Thermosetting Curing Reactions, Cured Resin Structures, and Formaldehyde Emission Problem

$$\begin{aligned} &R-NH-CH_2-O-H+H^+ \rightarrow [R-NH-CH_2-O^+=H^2] \\ &\rightarrow [\mathbf{R-NH-CH_2^+}]+H_2O \end{aligned} \tag{6.11}$$

$$\begin{aligned} &[\mathbf{R-NH-CH_2^+}]+NH_2-R \text{ or } NH=R_2 \rightarrow R-NH-\mathbf{CH_2}-NH-R \text{ or } R-NH \\ &-\mathbf{CH_2}-N=R_2 \quad (R=-CONH-, H-, -CH_2OH, \text{etc.}) \end{aligned} \tag{6.12}$$

Curing reaction mechanisms: When UF resins are catalyzed and cured at elevated temperatures, the methylol groups present on the polymeric and monomeric resin species are activated by the attack of hydrogen ion to become methylene carbenium ions ($-CH_2^+$) by losing a water molecule (Eq. (6.11)). These electron-deficient reactive cations, present only as a transient form, attack initially on the (electron-rich) nitrogen atoms of free amide groups ($-NH_2$) of monomeric and polymeric urea species, resulting in forming methylene bonds between urea molecules (Eq. (6.12)), resulting in extending the polymer chain, similarly as in the acidic resin synthesis reaction. These acidic polymerization/curing reactions can go on as long as there are reactive methylol functional groups remaining in the resin system. The F/U mole ratio of UF resins is currently in the range of 1.10–1.25, and urea has two amide groups (two or three functional hands) and formaldehyde has two functional hands ($-CH_2-$). Therefore, the F/U mole ratio of one (1.0) means in theory that the formaldehyde in the resin system can bond with all free amide groups (two hands) of urea to form methylene bonds, forming very long linear UF polymer chains.

$$
\begin{array}{l}
-\mathrm{NH}-\mathrm{CO}-\mathrm{NH}-\mathrm{CH_2}-\underline{\mathbf{NH}}-\mathrm{CO}-\mathrm{NH}-+-\mathrm{NHCO}-\mathrm{NH}\underline{-\mathbf{CH_2}-\mathbf{O}-\mathrm{H}}\\
\rightarrow\\
-\mathrm{NH}-\mathrm{CO}-\mathrm{NH}-\mathrm{CH_2}-\mathbf{N}-\mathrm{CO}-\mathrm{NH}-\\
\qquad\qquad\qquad\qquad\quad |\\
\qquad -\mathrm{NHCO}-\mathrm{NH}-\mathbf{CH_2}+\mathrm{H_2O}
\end{array}
\tag{6.13}
$$

Cured resin structures: However, as discussed in Chapter 5, the polymeric UF resin components are partly branched structures *via* methylene and methylene-ether bonds, i.e. the branch point urea nitrogen atoms have three functional hands (Eq. (6.13)). These branched structures might be considered not be formed much in curing reactions since the imide group participation in methylene bond formation would be relatively sparse under the low F/U mole ratio conditions, but those branched structures of methylene groups formed in the resin manufacturing stage will mostly remain unbroken during the curing stage, to result in a mixture of (1) branched or crosslinked structures (a), (2) linear structures (b and d), and (3) amide group-terminated structures (c) in cured UF resins as shown in Figure 6.4, where there are 15 urea and 18 methylene units resulting in an F/U ratio of 1.20. Also, the methylene groups consist of 12 branch methylenes (*a* type) and 6 linear methylenes (*b* and *d* types), resulting in a ratio of 2.0 for the branch/linear methylenes groups. (A crosslinking structure involves three methylene groups according to the ^{13}C NMR structure classification.) The author considers the structure in Figure 6.4 is a reasonable one for a cured UF resin with an F/U mole ratio of 1.20. The same ratio value calculated for a typical uncured UF resin turns out to be 0.45 (Section 24.4.2). Thus, it appears that the curing process of common UF resins involves forming more crosslinking bonds and it, therefore, indicates the importance of curing conditions. The ratio of these three different structures can vary and, in current low F/U mole ratio UF resins, high (*a*) structures mean high (*c*) structures, and since (*c*) structures need to be minimized for better bond performance, (*a*) structures need to be minimized in uncured resins, for which there appears to be some room to juggle the F/U_1 mole ratio in the resin manufacturing step. In theory, crosslinking is possible when the overall F/U mole used in the resin system is greater than one (1.0) and its extent would increase with increasing F/U mole ratio up to 1.50. In general, higher degrees of crosslinking means better strength properties, i.e. higher internal bond (IB) strength values of boards. Thus, although the detailed molecular structures of cured UF resins are currently difficult to determine, the polymer chains in cured resins are considered to be branched/cross-linked by the 0.10–0.30 mol of formaldehyde per mole of urea present in excess of 1.0 mol. The result of branching/cross-linking would be that the whole resin mass is cured in the end to become a rigid solid material that does not melt at the curing temperature or on further heating, a thermosetting resin system.

$$
\mathrm{R}-\mathrm{CO}-\mathbf{NH}-\mathbf{CH_2}-\mathbf{OH}\leftrightarrow\mathrm{R}-\mathrm{CO}-\mathbf{NH_2}+\mathbf{CH_2}=\mathbf{O} \tag{6.14}
$$

Formaldehyde emission in curing of UF resins: However, this thermosetting process of UF resins occurring in hot-pressing of boards involves various unfavorable reaction parameters

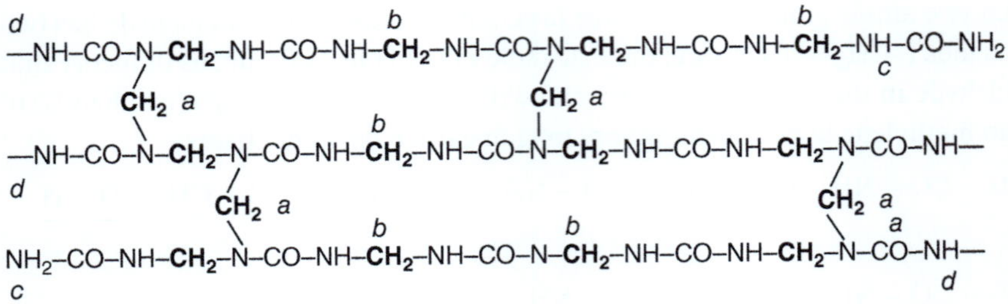

Figure 6.4 A schematic cross-linked UF resin polymer structure.

for the methylol groups in comparison with the similar reactions occurring in the low viscosity resin synthesis processes: no stirring, high temperature, short reaction time, high viscosity, gelation, and vitrification, i.e. nonequilibrium, kinetic processes. Therefore, even for a "fully" cured resin system, some methylol groups can remain unreacted, kind of orphaned, bonded to the polymer chains, which can affect little on the adhesive performance properties *per se* but later decompose, by reversibility, and defuse out as free formaldehyde, a major source of formaldehyde emission from boards. Further, as discussed in Chapter 5, the reverse reaction of methylol groups becomes faster at higher (curing) temperatures than the forward reactions (Eq. (6.13)). It is because the curing reactions occur in the nonequilibrium condition where the freed formaldehyde molecules cannot easily find their way back or find other amide groups to react with, especially after the resin system crosses the gelation stage and then the vitrification stage. Then, little freed formaldehyde would go back to form methylol groups. During or at the end of curing and cooling down period, the freed formaldehyde molecules also have to find their places: remain trapped in the cured resin/wood matrix or get expelled out of the board with steam into the air. The trapped and adhered to wood formaldehyde species, i.e. those stabilized by hydrogen bonds and/or other secondary forces on polar surfaces or as hemi-formal groups on hydroxyl groups of wood, will eventually break off and diffuse out slowly as formaldehyde emission.

Because of the formaldehyde emission problems, the final F/U mole ratios of resins currently run at low values, and the cured resins are cross-linked at significantly lower levels than the totally cross-linked system with F/U = 1.50. For current core-layer UF resins, one urea molecule makes bonds on average with about 2.3 (=2 × 1.15) bonds. Assuming the amount of formaldehyde lost in the curing process is small, this value means that 15 urea units among 100 urea molecules are forming branch points in cured resins. On the other hand, if the mole ratio is reduced further, say to 1.05 of current face-layer resins, it means only 5 urea units out of 100 are forming branch points, indicating a drastic reduction in the cross-link density of adhesive layer and resultantly decreasing board strength properties. In the end, the reversibility of methylol groups in UF resins, the source of formaldehyde emission, appears to be an inherent molecular property of the carbamide groups of urea molecules, and one limited but apparent way to suppress it is to lower the F/U mole ratio further and use higher resin-solids loading levels in boards in order to overcome the decreasing crosslink densities of cured resins.

Methylene-ether–cellulose bonds: Another route of formaldehyde emission from UF resin-bonded wood products that has not been much considered until now is the possible formation of methylene-ether bonds with cellulose hydroxy groups in the hot-pressing of boards: $-U-CH_2-O-$ Cellulose bonds, shown to be formed in a ^{13}C NMR model compound study [9], and their slow breakdown to emit formaldehyde. This route cannot be excluded, but the extent of such bond formation would be low, and it also has to be questioned whether such bonds would really break down under the ordinary use conditions.

6.6 Control of the Curing Speeds of UF Resins

Board manufacturing plants aim to attain the highest productivity (short hot-pressing time), adequate board strength values, and low formaldehyde emission values. Resin kind, mole ratio and loading level, catalyst kind and level, wood acidity, tack of mat, hot-pressing temperature and time, and formaldehyde emission values are the processing variables to consider. First of all, the higher the hot-pressing temperature, the faster the resin curing speed/shorter hot-pressing times, but the hot-pressing temperature is commonly set at about 300–350 °F and some manufacturers use lower temperatures and others use higher temperatures (400 °F), especially in the Northern United States and Canada. Also, the temperature differential between the face-layer and core-layers is inevitable: the core-layer attains temperatures of about 25–30% lower than the face-layer temperatures. Therefore, the hot-pressing temperature and time chosen invariably result in some over-curing of face-layers, i.e. weak flaky surfaces that need to be sanded off in the finishing station. This sanding means loss of wood and resin as well as the costs incurred, and this leads to the upper limit in choosing the hot-pressing temperature.

With a hot-pressing temperature set in a manufacturing plant, using shorter hot-pressing times increases the productivity and can decrease the surface over-curing, but this method is limited because of possible under-curing, i.e. sub-par strength values and increased formaldehyde emission levels. The face-layer/core-layer board construction method allows the face-layer to incur less over-curing and lower formaldehyde emission levels by using lower F/U mole ratio resins and less catalyst in the face-layers, while higher F/U mole ratio resins and more catalysts are used in the core-layers for faster curing and better board strength values. Commonly, the face-layer is less of a problem in controlling the curing speed, although the resin mole ratio cannot go much lower than 1.05 for board strength values. On the other hand, the cure speed of core-layer is the key to the productivity and board strength qualities; the catalyst kind and level need to be controlled to the maximum, and the resin mole ratio needs to be controlled to a value high enough for fast curing and good strength values but not too high for formaldehyde emission values of boards. The core resins' mole ratio in industry ranges from 1.15–1.25 depending on the various conditions of board manufacturing plants. Another UF resins' F/P mole ratio-related parameter in board manufacturing plants is the *tack property* of resins as discussed in Chapter 5.

6.7 Side-Reactions Occurring in the Curing of UF Resins

Methylene groups' limited heat stability: The methylene bonds formed in the curing of UF resins are somewhat unstable in the common curing temperature ranges of 300–400 °F, i.e. the bonds partially break down (decomposition) back to methylol and amide groups (Eq. (6.15)).

$$-\mathrm{NH}-\mathrm{CO}-\mathbf{NH}-\mathbf{CH_2}-\mathrm{NH}-\mathrm{CO}-\mathrm{NH}- \rightarrow -\mathrm{CO}-\mathbf{NH^-}+{}^{+}\mathbf{CH_2}-\mathrm{NH} \\ -\mathrm{CO}-\mathrm{NH}- \rightarrow +\mathbf{H_2O} \rightarrow -\mathrm{CO}-\mathbf{NH_2}+\mathbf{HO}-\mathbf{CH_2}-\mathrm{NH}-\mathrm{CO}-\mathrm{NH}- \quad (6.15)$$

This reverse reaction of curing and cross-linking becomes significant when the hot-pressing temperature is high or the hot-pressing time becomes long, i.e. *over-curing* of resin. This breakdown of the polymer backbone structures leads to weakening of board strength as will be discussed in Chapter 7. Because of this limitation of UF resins, the hot-pressed boards are normally cooled rapidly to minimize the strength degradation, such as by standing boards sideways separately and sometimes by fanning them. This is also the reason why shorter press times are difficult to obtain just by increasing the hot-pressing temperature to increase the productivity of board manufacturing.

Breakdown of methylol groups to formaldehyde: As discussed above in Section 6.5, methylol groups are the functional groups of UF resins needed to form methylene groups in curing, but some of them fail to react in that way and just break down back to formaldehyde (reversibility of methylol groups). However, it is important to note that the reaction is a minor side reaction. Commonly, boards made with a UF resin with F/U mole ratio of 1.15 have free formaldehyde content of about 10–15 mg/100 g of dry board [10]. This formaldehyde value amounts to approximately 0.2–0.3% of the formaldehyde used in resin manufacture. Even if the formaldehyde emitted during the hot pressing is included, the formaldehyde involved in this errant pathway would be less than 0.5% of the amount used in the resin manufacture, although this causes many emission problems due to the toxicity of formaldehyde.

6.8 Phenomenological Changes of UF Resins in Curing

Drying and tack development: When a UF resin is sprayed on wood particles, the resin droplets begin to lose their moisture to air and wood substrates, resulting in increasing of viscosity. At some point, the resinated particles begin to show some useful tackiness. As discussed in Chapter 5, *tack* is an important processing parameter [8]. In the particleboard industry, especially in the caul-less hot process, the beginning of tack is ideally set to coincide with the end of mat-forming station, i.e. after the particles are laid down on the belt and subsequently pre-pressed, so that the mats, especially the edges, retain their form. Tack is a resin property modulated by catalyst, wood acidity, mat moisture content, and room temperature and, therefore, can be controlled only to limited extents. In theory, tack is not related to the final bond strength of boards. *Dry out* is more often referred to as the lowered strengths of boards as a result of not only the evaporation of water from the adhesive resin but also of pre-curing reactions occurring during the mat-forming stages.

Gelation, rubbery stage, vitrification, and full cure: As discussed in Chapter 4, after most water in the resin is gone, the curing of thermosetting resins starts from the liquid (molten) stage and goes to gelation, rubbery stage, vitrification, and high elasticity stage. The *dynamic mechanical analysis (DMA)* method shows these changes in terms of *rigidity (G″)*, *loss modulus (G″)*, and *tan delta* curves against the elapsed time as shown for a typical

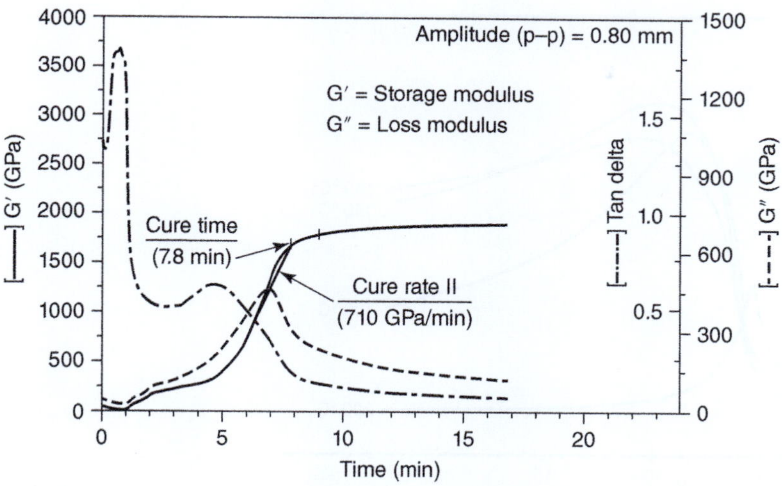

Figure 6.5 DMA results of a catalyzed typical UF resins obtained at final temperature of 120 °C [11].

catalyzed UF resin having an F/U mole ratio of 1.15 and curing temperatures of 120 °C in Figure 6.5 [11]. In this DMA diagram, a piece of fiberglass cloth impregnated with resin is cyclically flexed and heated at 120 °C, and the rigidity curve (solid line) starts to rise at the gelation point at about 1.0 minute of elapsed time, i.e. showing some strength or rigidity, it then reaches to the rubbery stage at about 2.5 minutes that lasts until about 4.5 minutes, i.e. a low strength value of the rubbery stage and, then the rigidity curve increases rapidly and passes the vitrification point at about 7.0 minutes where the loss modulus curve (dotted line) reaches to the maximum. This two-stage curing process was also observed in the torsional braid curing analysis of UF resins [12]. The rigidity curve goes up further to reach the plateau level where the resin is fully cured. The maximum rigidity attained remains unchanged until the end of the DMA run (17 minutes), indicating that the cured UF resin structure is stable at the curing temperature, i.e. not degrading [13]. The loss modulus (dotted line) and tan delta (loss modulus/rigidity ratio, dot-and-solid line) curves reach low values, indicating the test sample became a true elastic solid material. The loss modulus represents the liquid-like materials remaining in the sample that absorb the flexing energy and dissipate it into heat energy, and the curve reaches to the very low value at the end, indicating that there is not much of such materials left in the sample. Thus, the rise of rigidity curve reflects the growth of polymer chains and formation of cross-linked structures due to the formation of methylene bonds, a typical behavior of "thermosetting process."

Degradation of cured UF resins. In Figure 6.6 [13], the DMA results of the same UF resin show the rigidity curves obtained at higher curing temperatures of 140, 160, and 180 °C, respectively, where the maximum rigidity is reached at about 10 minutes but soon starts to degrade quickly at 180 °C while it remains somewhat stable at lower temperatures. At 200 °C, the rigidity curve did not even rise much and degraded quickly. The results show the instability of cured UF resins at higher temperatures, or the optimum curing temperature ranges to be used in the curing of UF resin-bonded wood composite boards, and also the need for cooling of boards soon after the hot-pressing.

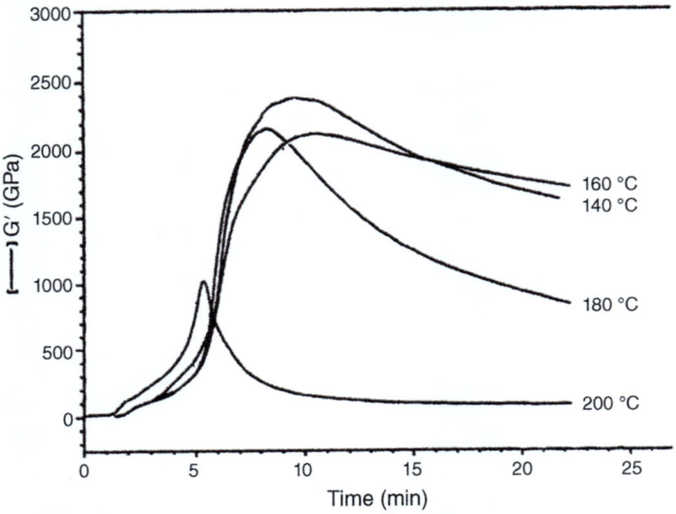

Figure 6.6 DMA results of the Rigidity of a laboratory-catalyzed UF resin (F/U = 1.15) at various temperatures. Source: [13] Reproduced with permission of the Forest Product Society.

6.9 Chemical Curing Mechanisms, F/U Mole Ratio, and Cured Resin Structure

$$\mathbf{HO-CH_2-NH-CO-NH-poly-NH-CO-NH-}$$
$$\mathbf{CH_2-OH} \quad \underline{(\sim 55\text{ mol\% of charged Urea})};$$
$$\mathbf{NH_2\text{-}CO\text{-}NH\text{-}CH_2\text{-}OH//NH_2\text{-}CO\text{-}NH_2}$$
$$\underline{(\sim 45\text{ mol\% of charged Urea})}$$
$$\text{Polymeric/monomeric composition of common UF resins} \tag{6.16}$$

Various chemical curing stages: The curing chemistry of UF resins is quite more complex than discussed above. Typical UF resins consist of polymeric and monomeric components, as discussed in Chapter 5. Further, the methylol groups can in part move around among the various molecular species in a time and temperature-dependent manner. Thus, the resin sprayed on wood substrates will form a layer consisting of water-swollen balls of about 55% branched UF polymeric molecules and the balance consisting of monomeric UF resin species, as indicated in Eq. (6.16). As the curing reaction progresses, the polymeric species will grow in size by reacting with the monomeric species to form polymer balls, and the less reactive low molecular weight resin species and the water generated from reactions and freed formaldehyde will get pushed into the boundaries of these balls of polymer molecules. Then, polymer balls grow further and start to loosely touch each other and then start to form (methylene) bonds between them to form strings of balls, and then the glueline would soon reach a point where a few such strings of polymer balls span throughout the glueline layer, the *gelation point*. Then, the polymer networks grow further by making more methylene bonds *within* polymer balls and also at their boundaries to pass *the rubbery stage* of glueline. Finally, the polymer balls would pass the *vitrification stage* and grow denser to reach to the final elastic stage.

During these various stages, the water, freed formaldehyde, and unreacted UF molecules will get pushed into the boundary areas, and then most of the water and freed formaldehyde will get pushed out of the adhesive layer while unreacted UF resin fragments and molecules can remain in the boundaries. Therefore, it is quite possible that the boundaries become the weakest links in the network of balls of polymeric UF molecules, especially in under-cured gluelines. To obtain strong adhesive layers, it is considered important that a good amount of boundary-to-boundary methylene bonds be formed. The level of methylol functional groups remaining on the surface of the balls of polymeric species in the last curing stage would be the key to forming a good bond to wood surface molecules. With current resins with low F/U mole ratios of about 1.05 for face-layers a thorough curing is mostly possible, but higher mole ratio resins in core-layers need to be cured equally thoroughly, albeit under a lower temperature and higher moisture content conditions. Further, as the F/U mole ratio of resins is lowered further due to high formaldehyde emission levels of boards, the boundaries of balls of polymeric molecular species would be lacking in the methylol functional groups, resulting in weak boundary links. In the end, a weak adhesive layer would result if the mole ratio went below 1.00.

Role of methylene-ether bonds: The relatively high level of methylene-ether bonds in uncured liquid UF resins (about 20% of urea-urea bonds) presents an intriguing question since ether bonds have been considered to be breaking up in acid-catalyzed curing processes. The acidity and temperature of breaking up and its rate would determine the role it plays in the cured resin's strength or quality. There are some indications that the uron bond, one kind of methylene-ether bond discussed in Chapter 5, if its level is increased, results in better IB strength values of boards [9]. One key to a good performance of thermosetting resins is to have a good "flow," i.e. flow of resin in the absence of water just before the gelation. Since the breaking up of methylene-ether bonds means not only an emission of formaldehyde but also a pushback of polymerization extent, and if it occurs before gelation, the flow will be increased. If it occurs after gelation, a delayed curing, longer rubbery stage, and an increased formaldehyde emission may be the results. This subject has never been investigated. The thermal breakdown of strength values of UF resins shown in DMA curing and UF resin-bonded particleboards out of hot-pressing in Section 6.8 have not yet been explained regarding whether it is due to methylene or methylene-ether bonds that are degrading.

6.10 Composition of Cured UF Resins

Direct measurements of cured UF polymer structures such as by using X-ray have not been possible because of the amorphous nature of cured structures. Neither the solid-state ^{13}C NMR has given much quantitative information thus far because of the lack of resolution. On the other hand, the results of wet chemical analyses for formaldehyde-derived methylenic groups of four different F/U mole ratio, un-cured and cured resins as well as two-hour desiccator formaldehyde emission values of boards were reported in 1982 by J. H. Williams, shown in the following Table 6.1 [14]. The methylenic group values of uncured UF resins are generally in line with more recent values determined by ^{13}C NMR analyses, although they fluctuate much among the four resins, indicating the wet analysis methods available at that time probably need some improvements. For cured UF

Table 6.1 Analysis results of methyleneic groups of un-cured and cured UF resins.

	Mole ratio of un-cured UF resin				Mole ratios of cured UF resin			
Functional group	**1.0**	**1.2**	**1.3**	**1.6**	**1.0**	**1.2**	**1.3**	**1.6**
Free formaldehyde	—	0.79%	0.52%	1.47%	—	—	—	—
Methylol	51.08%	45.23%	43.83%	62.03%	—	1.12%	2.83%	5.49%
Methylene-ether	21.09%	9.44%	28.01%	12.95%	28.07	45.37%	44.34%	32.43%
Methylene	27.83%	44.54%	26.64%	23.55%	71.93%	53.51%	52.83%	62.08%
Total formaldehyde	*100.00%*	*100.00%*	*100.00%*	*100.00%*	*100.00%*	*100.00%*	*100.00%*	*100.00%*
			Average emission UG/ML		0.5	1.0	1.4	6.0

Source: Williams [14] Reproduced with permission of the Forest Products Society.

resin samples, the methylol group content of 1.0 mole ratio resin is none, but for other resins, the value is low but increases as the mole ratio increases, which was interpreted to be the main cause for the increasing levels of formaldehyde emission values of boards. The two-hour desiccator formaldehyde emission (ASTM D 5582) level for 1.0 mole ratio resin is low but still sizable. The methylene–ether group content of cured resins increases with the resin mole ratio, and the high values reported are somewhat surprising in view of the suggestions in the literature that they are formed in the resin synthesis stage but breaking up in the acid-catalyzed curing stage [15]. Maybe the resins were cured without or with inadequate amount of catalysts, and it therefore appears useful to reexamine the compositions of cured, low F/U mole ratio UF resins.

6.11 Uses of UF Resins

UF resins are used as binders for particleboard, medium-density fiberboard, and hardwood plywood in the United States. Most hardwood plywood imported from Asia is also bonded with UF resins.

6.12 Properties of UF Resin-Bonded Wood Composite Boards

The colorless gluelines of cured UF resins are ideal for interior use applications. The low cost of UF resins also makes them particularly useful for production of low-cost wood composite products, and the low cost also allows to use higher resin solids loading levels to obtain better physical performance properties of boards from exceedingly small wood particles or fibers. (Imagine that about 10% weight of PB or MDF panels consists of cured UF resins.) The dry strength and stiffness properties of UF resin-bonded wood composite boards have been adequate for interior, semi-structural applications. Cured UF resins do not give water-resistant properties to boards in long-term exterior uses, as often seen with a water-soaked piece of PB furniture that expands and eventually disintegrates. A short contact with water does not cause problems due to the UF resin bonds' limited water resistance properties, aided by the boards' water repellency due to petroleum wax added in board manufacture. Above all, the UF resin-bonded boards can have long useful lives when used under dry conditions, especially with coatings or overlay applications. On the other hand, the natural degradability of UF resin-bonded boards can be a positive in the present consumption-oriented economy, i.e. disposed board pieces would degrade soon when buried underground.

6.13 Formaldehyde Emission Mechanisms, Mole Ratios, and Board Strength Properties

One major drawback of UF resin-bonded boards is the emission of formaldehyde gas during the hot-pressing and from manufactured boards during use. Formaldehyde is classified as toxic material and a potential carcinogen, and the emission values in the recent past ranged at about 0.15–0.20 ppm by the standard Large- or Small-Chamber test methods [11, 16], but the emission regulation limits were recently lowered to about 0.08 ppm by OSHA, HUD, EPA, state governments, and U. S. Senate Bill S-1660 (2011), which require all UF resin-bonded wood products to comply with the regulation by March 22, 2019. The new law also introduced new definitions of "*no-formaldehyde-added resins*" and "*ultra-low formaldehyde-emitting resins*" classes that aim at boards that emit formaldehyde less than 0.04 ppm in the future. European and Japanese governments are also aiming at the ultra-low formaldehyde-emitting wood composite boards [17]. Formaldehyde emission levels depend approximately on the F/U mole ratio of resin used, which is currently near 1.15 on average, and further lowering of mole ratio of UF resins would lead to marginal physical properties of boards and the hot-pressing times could be lengthened.

Chemical mechanisms of formaldehyde emission: Raw woods by themselves give off formaldehyde at extremely low levels (~1 mg/100 g wood) [18, 19]. The majority of formaldehyde emitted from UF resin-bonded boards has been considered to be from (i) *breakdown of methylol groups* during hot-pressing, which is trapped in board and come out slowly during use; (ii) *methylol groups that remain unreacted in the cured gluelines of boards* would break off slowly to give off formaldehyde during use; (iii) *methylene-ether groups* that break up during hot-pressing or during use of boards [15, 20]; and (iv) *methylene groups*, which are the major backbones of cured UF resin polymers but can also breakdown to give off formaldehyde. However, breakdown of methylene groups would be *very slow* in comparison to the other mechanisms of formaldehyde emission, and it is expected that only high humidity and high-temperature conditions would result in a significant emission by this mechanism.

Mole ratio vs. board strength and formaldehyde emission levels:

It has been reported that the mole ratio of UF resins is intimately proportional to the formaldehyde emission levels and strength values of boards, as reported by A. T. Go in Figure 6.7 [21]. The IB values of boards reach to the maximum at mole ratio of 1.20–1.25 and decrease continuously with mole ratio decreases as expected. Also, the perforator (DIN EN 120) and 2-H desiccator values (ASTM D 5582-94) of formaldehyde emission decreased with resin mole ratios. IB values need to be about 90 psi minimum for the density of boards and perforator (formaldehyde content) values of less than about 8.0 and desiccator values of less than about 0.5 are required by federal emission regulations. Thus, data in Figure 6.7 indicate that even 1.10 mole ratio resins will be difficult to meet the required formaldehyde emission

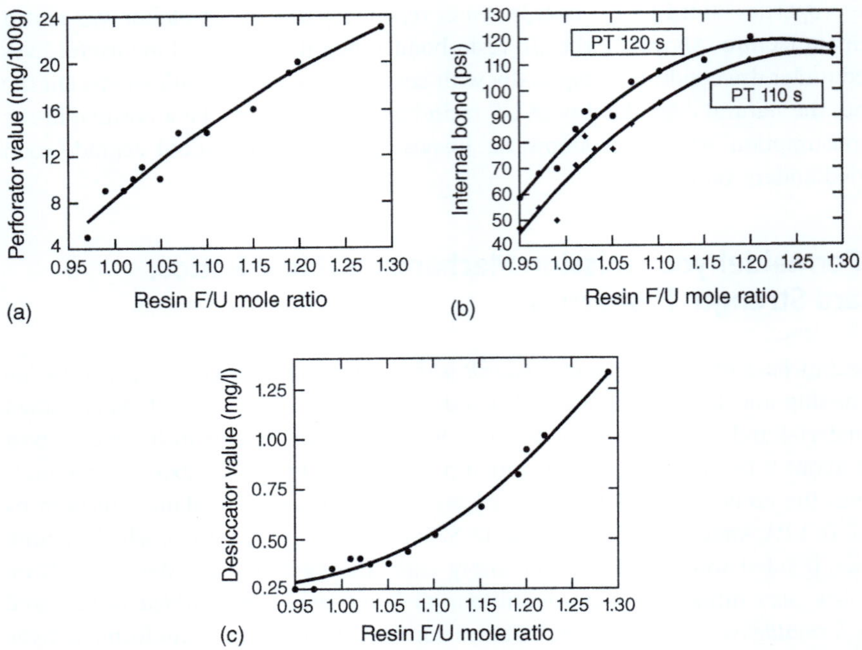

Figure 6.7 Relationship of mole ratio of typical UF resins and (a) Perforator values (PV), (b) internal bond (IB) strengths, (c) 2-Hour desiccator values of laboratory particleboards, reported by A. Go in 1991. [21] Boards: 5/8″ × 2′ × 2′; same resin in both face and core; resin loading of 10.5% in face with no catalyst and 8.5% in core with 1–3% ammonium sulfate catalyst; press temperature 200 °C; Press time 120 seconds; mat moisture 8–10%; target board density 42 lb/ft^3. Averages of 9 samples for IB, 6 samples for DV, and 4 samples for PV. Source: Reproduced with permission of Washington State University Wood Composite Center.

values and the IB strength values would be far lower than needed. Until recently, PB manufacturers used UF resins made with F/U ratios in the range of 1.05–1.25 the higher one in the core-layer and the lower one in the face-layer of boards to maintain the short hot-pressing time capability [21]. Lowering the mole ratios further appears to be a big challenge if the hot-pressing condition and board performance properties are to be maintained.

Other variables: Other variables that increase the formaldehyde emission values are [22, 23]: (i) curing catalysts, too fast or too slow; (ii) wood particles' high moisture content; (iii) under-cure or over-cure; (iv) board density and face-layer density values also modulate formaldehyde gas diffusion rates; and (v) higher temperatures and higher moisture contents of board give higher emission values; (vi) formaldehyde emission levels of boards decrease only slowly with aging of boards with half-lives reported to range about 216 days at room temperature [24].

6.14 Various Methods Proposed/Practiced for Reducing the Formaldehyde Emissions of Boards

Various formaldehyde emission reduction methods have been reported [25, 26]: (i) treatment of boards with ammonia gas or ammonium carbonate or bicarbonate solutions;

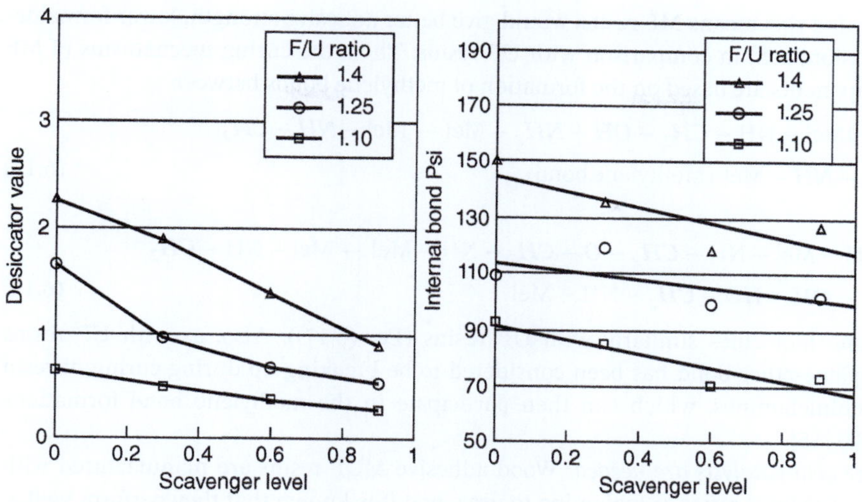

Figure 6.8 Urea scavenger levels vs. (a) 2-h desiccator values and (b) internal bond values in lab boards for various F/U mole ratio resins. Source: [6] Reproduced with permission of the Wood Composite Panel Association.

(ii) treatment of boards with surface coatings or barrier membranes; (iii) injection of scavengers in board manufacturing using a separate line from the resin application line; (iv) optimization of board hot-pressing conditions, resin formulation, catalyst, and F/U mole ratio; and (v) airing or heating of boards. Each of these methods has certain limitations of its own because either the technology is less than adequate or the cost is too high. Though the scavenger method of using urea solutions has been reported [27] and practiced in some particleboard plants as reported by G. Graves and shown in Figure 6.8 [6]. It is noted that the formaldehyde emission level is reduced by increasing the urea scavenger level, but the effect gets smaller for low F/U mole ratio resins, and the IB strength values of boards also decrease as the scavenger level increases. However, the author indicated that higher internal bonds can be obtained by using higher mole ratio resins along with higher levels of scavengers than by using lower mole ratio resins without using scavengers at the same combined F/U mole ratio levels. The advantage of this method may be found also in obtaining shorter hot-press times by using the higher mole ratio resins and high scavenger levels. In the end, however, the IB strength of boards has to be in certain ranges, about 100 psi for 50 pcf density boards, while the emission level has to be below about 0.5 ppm by the two-hour desiccator test, and a fully satisfactory solution seems to be somewhat difficult to achieve by this method. More manufacturers are looking to use polyisocyanate (pMDI) resins [28].

6.15 Curing of Melamine-Urea-Formaldehyde (MUF) Resins

Methylene bond formation in curing:

In general, the curing chemistry of melamine-formaldehyde resins is similar to that of UF resins. However, melamine molecule has three amino groups, and each amino group can make bonds with one or two methylene, methylene-ether, or methylol groups. Thus, one melamine molecule can be cured to form three to six methyleneic bonds around its molecule, and this means MF resins would give better adhesive strength, lower formaldehyde emission, etc., in comparison with UF resins. Thus, the curing mechanisms of MF resin components are based on the formation of methylene bonds between

$$\text{MelMel} - \text{NH} - \boldsymbol{CH_2} - \boldsymbol{OH} + \boldsymbol{NH_2} - \text{Mel} \rightarrow \text{Mel} - \boldsymbol{NH} - \boldsymbol{CH_2} - \boldsymbol{NH} - \text{Mel (Methylene bond)}. \tag{6.17}$$

$$\text{Mel} - \text{Mel} - \text{NH} - \boldsymbol{CH_2} - \boldsymbol{O} - \boldsymbol{CH_2} - \text{NH} - \text{Mel} \rightarrow \text{Mel} - \text{NH} - \boldsymbol{CH_2} - \boldsymbol{OH} + \boldsymbol{HO} - \boldsymbol{CH_2} - \text{NH} - \text{Mel} \tag{6.18}$$

melamine molecules similarly with UF resins (Eq. (6.17)). Also, as with UF resins the methylene-ether bond has been considered to be breaking up during curing of resin to methylolmelamines, which can then participate in the methylene bond formations (Eq. (6.18)) [15].

Stronger acid catalysts are needed: Wood adhesive MUF resins are manufactured with about 4 : 6 (weight) ratios of melamine to urea, and it is known that they perform well at F/(M+U) mole ratios of about 1.8, utilizing less than half of its functionality [4]. Further, MUF resins, because of the more basic nature of melamine, require stronger acidic catalysts than those used for UF resins, such as aluminum chloride or sulfate. Apparently, these strong acidic catalysts have not yet revealed wood degradation problems in the industry, probably because these resins are not used in bonding of particleboard-type wood composite boards. For MUF resins, the curing/polymerization reactions involve the curing reactions of UF resins and MF resins separately as well as some co-polymerization reactions between the two monomer types [20].

Industry bonding practices: In the wood products industry, MUF resins are used in the form of slurry made by mixing the powder MUF resin/catalyst/filler in water. The mixes have pot-lives of a few hours, and their cure rates can be adjusted with acid catalyst levels. They are cured either at room temperature or by heating, but the heating method has been reported to give better bonding results [29]. MUF adhesives also cure well and fast with the radiofrequency (RF) heating method, and therefore, they are used in some limited extent for bonding laminated wood beams designed for interior uses and more extensively in the finger-jointing of lumber in these plants due to their adequately high curing speed, while the competing phenol-resorcinol-formaldehyde (PRF) resins, used mostly in the face laminations in the same plants, do not provide such a curing speed under the RF curing condition. MUF resins are also the preferred adhesive binders for bonding of hardwood lumber blocks such as tabletops and truck beds [30], because of their colorlessness and good RF curing capability with less arching problems and reasonably good water resistance properties of the gluelines albeit not as good as PRF resins. The general gluing operation is similar to PRF face gluing operation discussed in Chapter 10. MUF resins do not bond well to the CCA preservative-treated laminated beams, but the adhesion was reported to be improved significantly by priming the lumber using the hydroxymethylated resorcinol (HMR) coupling agent [31].

6.16 Curing of Urea-Melamine-Formaldehyde (UMF) Resins

Recent research results on formaldehyde emission reduction: Some PB plants are reported to be considering using *urea-melamine-formaldehyde* (UMF) resins made by adding 2–10%

melamine to UF resins for lower formaldehyde emission values. It should be noted that with these low levels of melamine in UF resins, the melamine effects, such as higher water resistance properties, would be minimal. However, UMF resins made by using low F/(U + M) mole ratio values in resin syntheses were reported to achieve some lower formaldehyde emission values of boards. These resins were also found to need stronger catalysts to fully utilize the melamine's strength improvement potentials of boards [32]. In this study, UF resins were first synthesized using the typical three-step procedure, and melamine was added in the beginning of third step, followed by reaction for 60 minutes at 90 °C and pH 8.0. Further, Catalyst *A* was the commonly used 25% ammonium sulfate (AS) solution, and Catalyst *B* was a 25% AS solution added with 5% sulfuric acid. Tables 26.5 and 26.6 show the PB preparation parameters and test results of the boards. First of all, for UF and 2.5% and 5.0% UMF resins, the formaldehyde content levels (Perforator value) of boards greatly depended on the F/U mole ratio values of core-layer resin and using 1.05 mole ratio resins in both core and face-layers brings the PE value to 6.0–8.0 mg/100 g of board, just approaching the Large-Chamber emission levels allowed by the federal law [32].

Board strength properties were improved significantly with UMF resins having 5.0% melamine levels in the core-layer and better results were obtained when using these resins in both core and face-layers, with higher levels of the stronger catalyst B. These boards also showed lower PE formaldehyde content values of 5.2–7.0 mg/100 g of board. It is also noted that the longer hot-pressing times (3.5 minutes) show improvements in the formaldehyde content values. Melamine has 3–6 functionalities per molecule while urea has on average 2.3 functionalities and, therefore, the melamine addition provides more functionality for capturing the formaldehyde and increasing the strength properties. The increased bond strengths and reduction of formaldehyde emission potentials of particleboard or medium-density fiberboard using various UMF resins were also reported by various other authors [33–35]. Furthermore, the methylene bonds formed between melamine and melamine or melamine and urea molecules in curing of resins are considered to be stronger bonds than those bonds formed between urea molecules only. This may be why melamine resins give higher strength and higher water resistance values than UF resins, as well as lower formaldehyde emission levels. On the other hand, *UMF resins were found to need higher levels of sulfuric acid-containing catalyst B* and *also the melamine level of resin needs to be at about 5.0%* to have a definite advantage in the board strength values over UF resins. Therefore, as the emission regulation gets stricter in the future, UMF resins may become useful more widely in industry, although stronger acidic catalysts would be required.

6.17 Cost Increases Estimated for Boards with Very Low Formaldehyde Emission Values

Various adhesive approaches have been investigated aiming at manufacturing PB or MDF panels with extremely low formaldehyde emission values, i.e. F**** grade (<0.3 mg/l according to JIS A5905 method), which includes the formaldehyde emission level of wood itself that equals to about 10% of the limit value [36, 37]. Four possible methods and the estimated cost increases and production capacity losses from current levels were reported by M. Dunky [38]: (i) Improve UF resins with melamine, 30–70% cost increase including resin load increase and capacity loss of 10–20% due to the slower curing rate; (ii) Use of modified phenolic resins, about 100% cost increase including resin load increase and capacity loss of 20–33% from the slower curing rates; (iii) Use of tannin resins, 100–200% cost increase including resin load increase and capacity loss of about 30%; and (iv) Use of isocyanate resins or in combination with amino resins [28], five to six times higher cost and about 50% capacity reduction. Among these approaches, tannins are in limited supply, isocyanate resins undergo steep cyclic price changes, and the reduction in board production capacity means current board manufacturing plants must be expanded or additional plants need to be built to meet the demand. These problems cast a big uncertainty on the profit picture for the large-scale board production plants. A further discussion on UF and UMF resin syntheses and curing process is presented in Chapters 24–27

6.18 Summary

Curing mechanisms of UF, MUF, and UMF resins and industry curing practices using ammonium sulfate and other catalysts are presented. In curing UF resins, an ammonium salt added at room temperature dissociates into hydrogen ion and ammonia by itself, and then the free formaldehyde present in the resin reacts with the ammonia to form hexamethylene tetramine. The reduction of ammonia in the resin results in pushing the equilibrium to generate more hydrogen ions from ammonium salt. The amount of formaldehyde present in resins at room temperature is small, and therefore, the resin catalyzation is minimal at room temperature. In hot-pressing, the resin generates more formaldehyde and leads to further catalyzing the polymerization reaction of curing. The total amount of formaldehyde available in the hot-pressing is lower for lower F/U mole ratio resins to give slower hot-press curing speeds and increasing the catalyst level results in no effect. In practice, UF resins are made with F/U mole ratios between 1.05 and 1.25 and 0.5–1.5% ammonium salt is added, and the formaldehyde emission is still significant. Wood acids and buffer capacities of various woods as well as softwood plywood trims and burned ashes are discussed in relation to catalyst effects on curing speeds. Further, curing mechanisms, cured resin structures, and causes of formaldehyde emission problems are presented. Curing of UF resins occurs from further growths of methylene bonds between urea units, reaching to the cross-linked state of polymer chains using all available hydroxymethyl groups in the resin. Cured resin structures are discussed in terms of branched, cross-linked molecular chains formed by methylene groups, and side-reactions, including the breakage of hydroxymethyl groups in curing, are discussed in relation to the formaldehyde emission problems. The formaldehyde emission decreases with decreasing F/U mole ratios. Current F/U mole ratio values are low, but the emission levels are still above the regulation limits. Using urea-based scavengers and other procedures for formaldehyde emission reduction purposes are discussed. The curing and use of UMF are specifically discussed as an approach to solving the formaldehyde emission problem. A significant cost increase due to the necessary resin and process modifications is discussed as reported in literature.

References

1 Thackston, W.J. (1940). Resin Emulsion. US Patent 2,196,367.
2 Cordier, D.E. (1948). Thermosetting Urea-Formaldehyde Composition. US Patent 2,414,025. Monsanto Chem. Co.

3 Scott, M.J. (1949). Curing Catalysts for Aminoplasts. US Patent 2,466,744. Borden Co. British Patent 938,786 (A).
4 Shiau, D.W. and Smith, E. (1985). Low Formaldehyde Emission Urea-Formaldehyde Resins Containing Melamine Additive. US Patent 4, 536, 245 A.
5 Black, W.R. (1972). Urea-Aldehyde Adhesive Process. US Patent 3,697,355. Borden Company.
6 Graves, G. (1990). *Proceedings of NPA Resin and Blending Seminar*, 23. Gaithersburg, MD: NPA.
7 Nevers, A.D. and Lehmann, W.F. (1969). *Forest Products Journal* 19 (2): 53.
8 Rammon, R. (1990). *Proceedings of NPA Resin and Blending Seminar*, 15. Gaithersburg, MD: NPA.
9 Philbrook, A., Dunlop, N., Easton, C.J. et al. (2005). *Wood Adhesive* (ed. C.R. Frihart), 143. FPL: FS.
10 Mao, A., Hassan, E.B., and Kim, M.G. (2013). *BioResources* 8: 2470.
11 No, Y. and Kim, M.G. (2004). *Journal of Applied Polymer Science* 93: 2559.
12 Steiner, P.R. and Warren, S.R. (1987). *Forest Products Journal* 37 (1): 20.
13 Kim, M.G. (2005). *Wood Adhesive* (ed. C.R. Frihart), 149. FS, FPL.
14 Williams, J. H. (1982). Presentation, Forest Products Society 36th Annual Meeting, 20 June 1982.
15 Wirpza, Z. and Brezezinsky, J. (1973). *Aminoplasty*. Moscow: Kimya (Chemistry).
16 Margosian, R. (1990). *Forest Products Journal* 40 (6): 19.
17 Sigvarten, T. and Dunky, M. (2006). *Wood Adhesives* (ed. C.R. Frihart), 23. Madison, WI: Forest Products Society.
18 Myers, G.E. (1983). *Forest Products Journal* 33 (5): 27.
19 Roffael, E. (1982). *Adhaesion* (3): 18.
20 Higuchi, M., Tajima, S., Irita, H. et al. (1991). *Mokuzai Gakkaishi* 37: 1041.
21 Go, A.T. (1991). *Proceedings of the 25th International PB and Composite Materials Symposium* (ed. T.M. Maloney), 285. Pullman, WA: WSU.
22 Marutzky, R., Roffael, E., and Ranta, L. (1979). *Holz als Roh-und Werkstoff* 37: 303.
23 Christiansen, R.L. (1972). *Forest Products Journal* 22 (4): 17.
24 Zinn, T.W., Cline, D., and Lehmann, W.F. (1990). *Forest Products Journal* 40 (6): 15.
25 Myers, G.E. (1986). *Forest Products Journal* 36 (6): 41.
26 Barry, A. and Corneau, D. (2006). *Forest Products Journal* 56 (9): 90.
27 Seller, T. Jr., Miller, G.D., and Nieh, W.L.-S. (1991). *Forest Products Journal* 41 (1): 53.
28 Ruffing, T.C., Smith, P.M., and Brown, N.R. (2010). *Forest Products Journal* 60 (2): 119.
29 Selbo, M.L. (1965). *Forest Products Journal* 15 (2): 475.
30 Blomquist, R.F. (1958). *Forest Products Journal* 8 (2): 45.
31 Vick, C.B. (1997). *Forest Products Journal* 47 (7/8): 83.
32 Mao, A., Hassan, E.B., and Kim, M.G. (2013). *BioResources* 8: 2453.
33 Hse, C.-Y., Fu, F., and Pan, H. (2008). *Forest Products Journal* 58 (4): 56.
34 Hse, C.Y. (2009). *Forest Products Journal* 59 (5): 19.
35 Lee, S.-M., Park, J.-Y., Park, S.-B., and Hwang, W.-J. (2012). *Forest Products Journal* 62 (3): 207.
36 Meyer, B. and Boehme, C. (1997). *Forest Products Journal* 48 (5): 45.
37 Weigl, M., Wimmer, R., Sykacek, E., and Steinvender, M. (2009). *Forest Products Journal* 59 (1/2): 88.
38 Dunky, M. (2005). *Wood Adhesives* (ed. C.R. Frihart), 343. FS, FPL.

7

Particleboard, MDF, and Hardwood Plywood Bonding with UF Resin Binders

7.1 Particleboard

(Much of this chapter is adapted with permission of the publisher and minor revisions from notes taken during class visits to PB manufacturing plants of Roseburg Lumber Co. in Louisville and Taylorsville, MS and from proceeding papers presented by referenced authors at the Composite Panel Association's Resin and Blending Seminar, in May 1990).

Particleboard (PB) production in the United States increased rapidly between 1963 and 1973 and replaced softwood plywood in many applications because of its lower cost and adequate performance. Since then, the production volume generally followed the housing starts and case goods demands. PB is manufactured by bonding wood particles with binder under heat and pressure, and urea-formaldehyde (UF) resins have been the dominant binder because they are less costly and cure faster than phenol-formaldehyde (PF) resins. General principles in using UF resins for PB and medium density fiberboard (MDF) manufacture are well summarized in recent conference proceeding papers [1–3]. UF resin-bonded PB cannot withstand heat and moisture well, but the major use of PB in the United States is in interior applications, where the lower moisture resistance is accepted. Low-level melamine-modified urea-formaldehyde (UMF) resins might be introduced to improve the moisture resistance and lower the formaldehyde emission values in some manufacturing plants. Seventy-six PB and thirty-eight MDF manufacturing plants were operating in 1998 in North America [4].

7.1.1 Wood Furnishes in Particleboard Manufacture

Various wood mill residues, such as shavings, hog-trim, ply-trim, PB-trim, and also raw wood chips are used for the production of wood particles collected from nearby wood products plants. Planar shavings and raw wood are preferred materials, but their quantity is limited. Ply-trim, because of contamination with alkaline PF resin adhesives, can cause problems of increased buffer capacity. PB-trims also are less desirable because they mostly result in generating superfine particles, but some intra-plant PB-trims have to be reprocessed rather than thrown away as (costly) wood wastes. Also, the density of wood species coming in needs to be considered since the compaction ratio of board is one key parameter to PB manufacture. It is important for PB manufacturing plants to prepare

Adhesives and Finishes for Wood: For Practitioners and Students, First Edition. Moon G. Kim.

consistent face and core furnishes from these mixtures of varying wood sources and qualities [5]. Since UF resins cure adequately only at the acidic pH, testing the pH and acidic buffer capacities of incoming wood materials and recording their aging histories are necessary, and appropriate mixing and separating methods are also necessary to produce consistent core and face-layer furnishes.

The incoming sorted raw wood particles are refined by passing them through hammer-mills or attrition mills to smaller particles, followed by screening to remove extraneous materials as well as the superfine wood particles. Over-refining needs to be prevented since the superfine particles, if they get into the face-layers, pick up high levels of resin-solids that can be wasted in the sanding of finished boards. The remainder is then separated through sieves into fine and coarse particles for use, respectively, in the face- and core-layers. The wood particles are then dried by passing through a rotating-type or suspension-type dryer, into which the hot combustion gas (800 °F) generated in the burner is passed through counter currently. The combustion gas needs to be clean, i.e. the fly ash flown into wood particles has to be minimized, or it will increase the buffer capacity to retard the curing of UF binder resins. The dwell time, hot gas temperature, and the amount of wood particles put through the dryer are controlled so that the exiting wood particles retain an average moisture content of 3–6%. The moisture content of particles has to be in this low range since excess moisture will cause under-curing of resin or generate high steam pressure in boards to result in weak internal bonds or blown boards. Particles that are too dry can also cause bonding problems since some moisture is needed to help the resin flow in curing. The fine wood particles are often used in the core-layer and coarse particles are used in the face-layers for optimum bending strength properties of boards, but in recent years some manufacturers have also used fine, preferably flat, particles in the top-most surface-layers to achieve smooth board surfaces that reduce the sanding in the finishing operation. The consistency of particle preparation is commonly checked by screen analysis, from which the distribution of particle sizes and moisture content are obtained [2]. The prepared particles are transferred to a bin and then to the blending station.

7.1.2 Blending of Resin, Catalyst, and Wax Emulsion in Particleboard Manufacture

Short-retention time blenders are widely used separately for core and face-layer particles (Figure 7.1) [6]. Littleford Brothers blenders are more typical blenders, and Grenco and Keystone blenders are also used in industry [5]. Short-retention blenders consist of a relatively small-volume clam-shell drum, available in various sizes (30–75 cm diameter) with throughput capacities ranging from 1 to 14 tons per hour for low bulk density wood

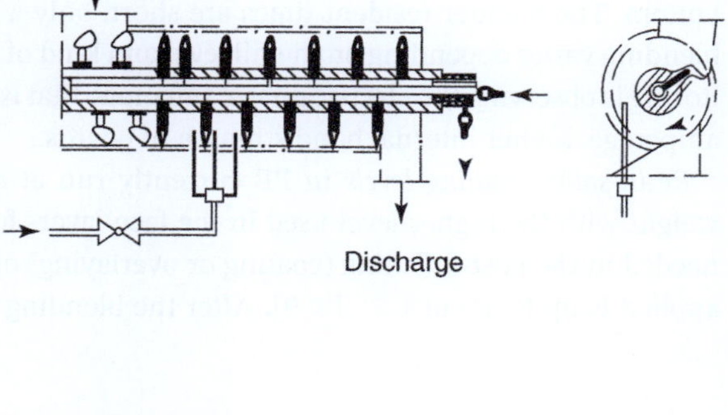

Figure 7.1 Horizontal short-retention time blender. Source: [6] Reproduced with permission of the Composite Panel Association.

particles and from 2 to 35 tons per hour for higher bulk density wood particles. The mixing zone is roughly divided into the beginning **distribution zone and** the **finishing dispersion zone**. Different machine manufacturers have put into certain designs that put different emphasis on these two zones. The impellers set to have a narrow clearance of 10–18 mm from the inside of wall rotate at about 800 RPM for core-layer particles and 1050 RPM for face-layer particles. The distribution zone is designed to *uniformly distribute* the resin/wax on wood particles, and the dispersion zone is designed to force the wood particles *to rub each other* and also to break apart the agglomerates of wood particles into smaller particles so that the resin/wax can spread further on the entire surfaces of the particles. Littleford blenders are known to have put a greater emphasis on the dispersion mixing, and Grenco and Keystone blenders are more on the distribution mixing. The objective of these in-line blenders is to attain the best distribution efficiency of resin and wax components as well as the highest throughput. The impellers run at high speeds with only small clearances, and the whole machine, including the impeller shafts, needs to be circulated with cold water for dissipation of heat generated in the mixing in order to minimize the advancement of resins and also to reduce the buildup formation of hardened resins inside the blender that needs to be cleaned out regularly.

In operation of blenders, as shown in Figure 7.1, the wood particles are continuously weighed and pushed into the inlet chute with the feed plows that hurls them at a high peripheral speed of 4–7 kft/min [6, 7]. The feed rate is controlled by the angle of the feed plows, which results in forming a concentric, continuously flowing ring of wood particles inside the blender. The paraffin wax emulsion, resin, and catalyst/buffer solution, as needed, are injected into the concentric flowing ring of wood particles peripherally through the multiple entering tubes from the bottom. The high rotating speed of wood particles makes the distributive mixing of additives to be completed quickly. The wood particles then get pushed into the high-shear, high-speed dispersion mixing zone, which consists of multiple rotating "horn-shaped" impellers that literally "mix" the wood particles. Finally, the wood particles advance toward the pneumatically controlled discharge gate that is also synchronized with the inlet feed gate through a separate mechanism. The task of the blender is to get the maximum throughput as well as the best resin/wax efficiency in terms of internal bond (IB), modulus of rupture (MOR), and modulus of elasticity (MOE) values of the resultant boards. Therefore, the weights of wood particles going in and those getting discharged are equalized, and then the operation is controlled to maintain the highest fill level in the blender and also let the whole operation go continuously and smoothly. Too high fill levels would lead to an inefficient distribution/dispersion mixing due to lack of particle movements between the inner and outer rotating rings of wood particles, resulting in lower internal bond strength values [7]. Too low fill levels lead to low dispersion mixing due to lack of interparticle collisions, as well as low throughputs. The resin/catalyst solution and wax emulsion are separately injected into the drum or sprayed through nozzles from the bottom. The blender resident times are short, only a few seconds. The efficiency of resin blending varies depending on the fill level and kind of wood particles, and one way of monitoring is observing the electric power amperage that is going into the blender, [7] i.e. higher amperage, higher internal bond strength of boards.

Resin-solids loading levels in PB currently run at 8–10% based on the oven-dry wood weight with the higher level used in the face-layers for the tight surface quality of boards needed in the post-finishing (coating or overlaying) operations [1–3]. The wax solids level applied is up to about 1.5% [8, 9]. After the blending step, the resin-blended particles are sent to the holding bin and then to the former. The overall distribution of resin and wax has been known to be adequate with these short retention time blenders due to the good blender design as well as the relatively large amounts of resin being used. On the other hand, the resin distribution among different size particles is often not uniform: small particles tend to attract more resin than large particles (on wood weight basis) because of the greater surface area of smaller particles per unit weight. For example, particles in the range of 10–50 mesh sizes were found to have 1–15% of resin loads at an average application rate of 6–7%. One natural thing to do is to blend the coarse core and fine face particles separately [6, 7]. Also, different UF resins and different catalyst levels are used for face- and core-layers to optimize the resin cure speeds, with faster curing speed for the core-layer and slower one for the face-layer.

7.1.3 Mat Formation and Transport

The PB mat is formed by felting, i.e. continuously dosing the resin-blended particles through a forming machine onto the moving belt. There are at least two types of forming machines. The first is *the wind sifting-type former*, where the wood particles are thrown by carding rolls into the air stream that is moving counter to the moving belt so that they settle down on the moving belt according to their particle sizes, finer ones move further to make up the bottom-most surface-layer. The second pass is for the core-layer of coarse particles. The third pass is for the top layer in the opposite direction of the first layer to finish the mat formation with the fine particles on the top-most layer. In this process, the fine top layer particles that tend to collect more resin-solids in the blending process are partly eliminated in the sanding step. The second method is *the mechanical forming machines* consisting of carding rolls, a bank of rollers, and other devices to result in uniform mats where the fine particles are distributed throughout the entire face-layers. Accurate and uniform matting (by weight) is important since lower-density areas of board will invariably show up as deficient spots, lowering the overall board quality that can often be viewed as a result of poor adhesive bonding. The mat moisture content is again an important parameter. UF resins contain water at about 40% of their weights, and this water is added to wood particles, resulting in about 8–10% moisture content in the mat. Higher mat moisture contents would result if the resin had to be diluted for some reason before spraying, and this will retard the resin cure rate in the hot-pressing, possibly lowering the board quality. The mat finally passes through a nuclear gauge that monitors the weights of mat over the whole mat area continuously to assure the uniformity of mat weights.

The mat is then cut, pre-pressed (at a high pressure in the case of caulless pressing), and transferred on a caul plate or directly (caulless process) to the loader, which loads the mats into the hot press all at once. The tack of UF resins on wood particles arises in the blending and matting period, and its strength arises, maxes, and then disappears, as discussed Ref. [2] in Chapter 6. It is important to have the necessary tack for formed mats since some juggling occurs during transit to the hot press. Breaking of mats, especially on the edge, due to poor tack can result in defective boards. The pre-curing problem that reduces the board quality can also arise during the blending and matting period. As discussed in Chapter 6, UF resin is often catalyzed to activate the resin curing ideally after the mat is placed and consolidated in the hot press, but pre-curing problems can arise from over-catalyzation, warm surrounding temperatures in the blender or forming box, waiting too long in the loader or hot press, etc. [1–3]. The resin activation in these stages will advance the resin to higher degrees of polymerization or even to the gelation stage. This pre-advancement of resin can make the resin binder flow less in hot-pressing, resulting in lack of coverage and wetting, etc., which are essential to form strong resin bonds between wood particles.

7.1.4 Hot-pressing

Multi-opening and other types of hot presses are used in industry with a temperature range of 300–350 °F for UF resin-bonded PB manufacturing operations, as shown in Table 7.2. In the typical multi-opening hot press, the loader loads mats into all openings at once and the hot-pressing starts soon. Press closing speed, total press time, and pressing pressure profile are the important hot-pressing parameters in successful board manufacturing. The board thickness is attained and maintained by press pressure and spacing gauges. The hot press is often the most expensive equipment in the wood composite plant and therefore needs to be built with a minimal size to meet the target production volume. Often, therefore, the hot press in the plant constitutes the major bottleneck of production line and needs to be operated most efficiently: a short press time as practically as possible (currently about 17 seconds per 1/16″). Radio-frequency heating presses have also been used, especially for thick board mats, to shorten the hot-pressing time as an auxiliary [10]. The hot-pressing has to be carefully handled, in a way that the resin curing progresses in concert with press-closing and completes within the total press time. Use of a slower curing surface-layer resin and faster curing core-layer resin is widely employed to help reduce the surface pre-cure and over-cure problems.

Mat compaction and density profile of boards: In practice, however, the heat transfer to the core-layer, occurring by steam movement, is slower than the mat compaction speed, and the surface-layers attain higher temperatures for longer duration than the core-layers, as shown in Figure 7.2a [2], resulting in a higher plasticization effect on the surface-layers. Thus, surface-layers result in a higher degree of compaction (higher density) than the core-layer, as shown in Figure 7.2b [10], at three different average board densities. A lower density

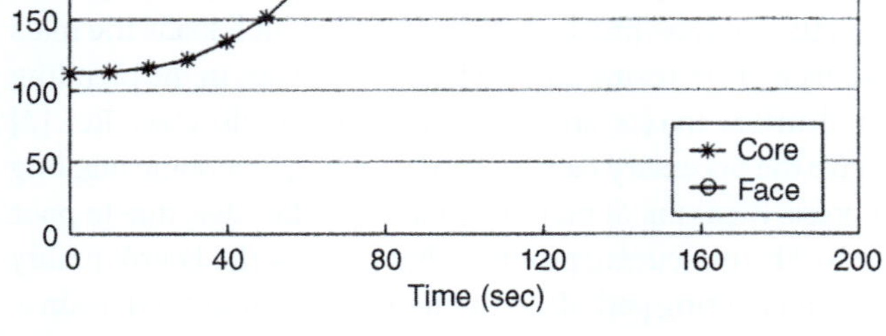

Figure 7.2a Typical temperature changes of the face- and core-layers in the hot-pressing of particleboard at 300 °F over 200 seconds. Source: [2] Reproduced with permission of the Wood Composite Panel Association.

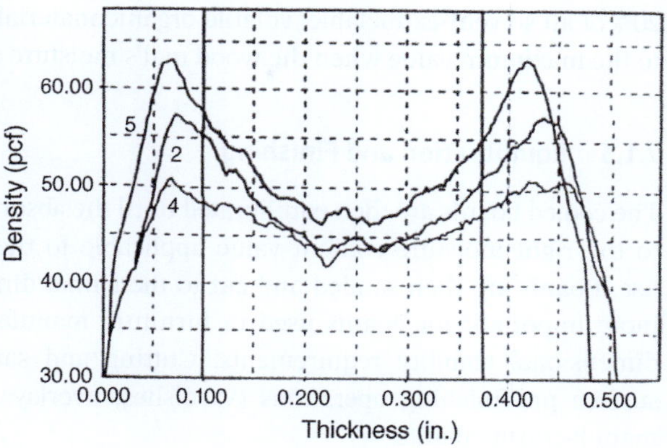

Figure 7.2b Typical laboratory particleboard density profiles with the average densities: No. 2: 44.9; No. 4: 49.6; No. 5: 54.5 pcf. Source: [11] Reproduced with permission of the Forest Products Society.

in the core-layer results in lower internal bond strength and lower edge screw-holding strength, but the higher density of the face-layers improves the board's stiffness (MOE) and surface quality in comparison to an imaginary case of boards having a uniform density profile across the board's thickness. The density gradient increases with higher press temperatures or faster closing times. The importance of the resin cure speed control, i.e. slower curing resin for face-layer, must be recognized to obtain a good quality board. If pre-curing of resin occurs in the surface layers, such as by a slower press-closing speed, a higher density of surface layers could still result, but the board bending strength will suffer because of the precure of the face-layer resin. In industry, the uppermost surface layers turn out to have very low densities, and they are removed by sanding.

The total press time should be controlled carefully, so that it is not extended any longer than necessary for UF resin binders to adequately cure but not to over-cure because of the high over-curing tendency of UF resins, especially in the face-layers. Over-cure will degrade the strengths of boards [1–3]. Apparently, the over-cure occurs because methylene bonds formed in the curing of UF resins are also, in part, broken down at board-pressing temperatures, as discussed in Chapter 6. The bond breakage arises from the fundamental chemistry of methylene groups in UF resin polymers, and no other recourse is available except for lowering the temperature. Although the bond-breaking rate is slower than the bond-forming rate during the hot-pressing, the rate of bond-breaking can increase toward the end of the curing cycle.

Cooling of boards: Thus, boards just out of the hot-press are customarily cooled by placing them separately and vertically on a turning wheel and often air is blown on them, a practice that is quite the opposite to the practice of hot-stacking of boards in the PF resin-bonded softwood plywood or OSB manufacture. If the cooling of boards is inefficient, lower internal bond and bending strengths of boards would be obtained. Thus, hot-pressing parameters influence the density profile, MOE, MOR, IB, thickness, board surface quality, board durability, etc. Hot-pressing of PBs also emits various volatile organic compounds (VOC) [12], including formaldehyde and methanol, the major components originating from UF resin binders with emission levels increasing with hot-pressing temperature, hot-pressing duration, resin loading level, and resins' F/U mole ratio values [12]. There are many other VOC components coming out in hot-pressing, such as hexanal and α- and β-pinenes from southern pine PBs, and the total VOC amount was measured to be about

20% of all solvent-extractable, volatile organic materials present in wood, and it increases to the maximum value when the wood mat's moisture content reaches 10–14% [13].

7.1.5 Equilibration and Finishing

The cooled boards are then equilibrated until the absorption of moisture from air proceeds to the right moisture content value applicable to the intended geographical regions of use. Boards are then sanded and cut to their final dimensions. This equilibration step is more important for boards used in furniture manufacturing due to the more stringent dimensional stability requirements. Cutting and sanding as well as some edge and surface pre-finishing operations (varnishes, overlays, etc.) are often performed in PB manufacturing plants.

7.1.6 Bison–Mende process

Thin PBs (1/8–1/4 in. thick) are produced by the continuous Bison–Mende process. Wood particles as well as wood fibers are used with UF resin binders. The boards are used for interior wall paneling and furniture manufacturing, as well as for door jambs, etc., after surface finishes are applied.

7.1.7 Properties of Commercial Particleboards

Various grades of PB with specific property requirements have been established in the industry, and manufacturers also supply minimum specifications for their boards to buyers, but board properties on the market vary substantially because of different wood materials, different adhesive loading levels, and other manufacturing techniques [14]. A research report shows the properties of seven different manufacturing plants' PBs (four different companies), as shown in Table 7.1 [14]. Plant No. 4 is located in the West using ponderosa pine and all other plants are located in the Southeastern regions using southern yellow pine as major sources of furnishes. Recommended specification values of ANSI grades 1-M-1 and 1-M-2 are shown [15], and all boards passed, and all boards were accepted by users except No. 6, not because of IB but because of rough surface properties (not tested). Substantial variations from one supplier to another existed except for IB, which is the one property that suppliers and users tend to go by closely. Coefficient of variance (COV) values for each PB shown in parentheses would be a good indicator of the variability useful for users, such as cabinet manufacturers. Even though there are some variabilities from company to company, the variability within a company is relatively low except for the linear expansion, 28% and 35%, measured from changes in relative humidity (RH) from 39% to 68% and 98%, respectively. The authors indicated that the difficulty in accurately measuring the linear expansion properties was a factor in addition to the physical property variations of boards [15].

7.1.8 US Particleboard (PB) Industry

The PB industry in the United States keeps the high quality of their products through *the American National Standard for Mat-Formed Wood Particleboard (ANSI A208.1-1987)*, established by the American National Standard Institute (ANSI) and National Particleboard

Table 7.1 Minimum property values of two different ANSI standard grades (1) of particleboard and average properties of particleboards tested [14].

Grade or surface	Density (pcf)	MOE (10^3 psi)	MOR (psi)	IB (psi)	Surface bond (psi)	Screw withdrawal (b)		Linear expansion (%)		Moisture content		
						Face	Edge	39% to 68% RH	39% to 98% RH	RH	RH	RH
1-M-1	—	250	1600	60	—	—	—	0.35[a)]				
1-M-2	—	325	2100	60	—	—	—	0.35[a)]				
1	48.7	317	1860	157	249	362	126	0.19	1.1	9.7	12.7	24.1
	(4.1)[b)]	(16)	(15)	(11)	(14)	(8.5)	(19)					
2	49.6	409	2380	156	311	422	139	0.22	1.1	8.8	11.7	23.4
	(2.6)	(5.3)	(8.1)	(13)	(11)	(8.2)	(12)					
3	52.1	420	2410	152	327	418	135	0.22	1.4	8.2	11.4	23.6
	(3.3)	(8.3)	(7.6)	(17)	(9.7)	(15)	(19)					
4	48.3	434	2590	144	272	364	123	0.17	1.0	8.6	11.6	22.7
	(1.7)	(10)	(11)	(11)	(11)	(12)	(14)					
5	52.2	421	2440	142	244	413	160	0.34	1.8	8.6	11.9	23.9
	(2.5)	(13)	(15)	(14)	(15)	(9.5)	(19)					
6	46.5	349	2110	141	255	316	129	0.14	0.7	8.5	11.3	23.3
	(2.5)	(11)	(12)	(10)	(16)	(14)	(9.8)					
7	47.9	400	1970	130	233	293	95	0.15	1.2	9.2	11.9	22.9
	(3.8)	(13)	(18)	(10)	(9.1)	(6.5)	(13)					
Average	49.3	392	2250	147	270	370	129	0.21	1.2	8.8	12	23
	(5.1)	(15)	(16)	(14)	(17)	(17)	(21)	(35)	(28)	(5.3)	(3.8)	(3.3)

a) Minimum linear expansion between 50 and 90 percent RH(1).
b) Numbers in parentheses are coefficients of variation.
Source: Reproduced with permission of the Forest Products Society.

Association (NPA). *Type 1* PB is the common UF resin-bonded product and is classified in eight different grades with respect to various uses, differentiated by IB, MOR/MOE strength values, hardness, linear expansion, and screw-holding strength values. The Type 1 product grades are from high to low strength values of the board: 1-H1–3, 1-M1–3, and 1-LD1–2, which are mostly controlled by the board density values, H for high, M for middle, and LD for low-density values. Type 1 PB is typically used in the manufacturing of kitchen cabinets, office and residential furniture case goods, and other demanding applications, such as floor underlayment and stair treads. PB is also used with overlays. Formaldehyde emission levels are also included in the standard so as to meet the emission regulation limits. *Type 2 PB* are in the ANSI Product Standard as exterior-use grade materials but not much manufactured in the US industry.

7.1.9 Summary

In the manufacturing of PBs, various wood mill residues are collected from nearby wood processing plants and refined into particles, dried to 3–6% moisture contents, and separated into face- and core-layer particles of consistent qualities. These particles pass through the short-retention resin blender, where a consistent flow of circulating wood particles is maintained at full level by the fast-rotating impellers, and the injected (catalyzed) UF resin and wax are distributed on wood particles. Accurate and uniform distribution of additive materials and high throughput of the blender are essential for production volume and bond quality. The wood particles then go through the matting and pre-pressing stations. As soon as it is sprayed on wood particles, the UF resin begins to lose water and increase in viscosity and tackiness, the timing of which is often controlled by resin manufacturers to have it minimal in matting and maximum in pre-pressing steps, so that mats stay firm till the hot-pressing. Loss of water and curing reactions of resin are completed in the hot-pressing commonly carried out at 300–350 °F with the core-layer reaching about 225–250 °F, but the curing reactions of resin can continue to the over-cure stage, so the boards need to be cooled immediately, often by airing. Also, the hot-pressing time needs to be as short as possible, currently 17 seconds per 1/16″ board thickness, for adequate production volumes and to avoid over-curing. Currently, industrial PBs show adequate strength and moisture stability values, but the formaldehyde emission values have been high, and the problem needs to be resolved without compromising the strength values.

7.2 Medium Density Fiberboard (MDF) Bonded with UF Resins

(This section is adapted with permission and minor revisions from the proceeding paper; Production variables, blowline, and alternative blending systems for MDF given by Mr. Ron Frashour at the National Particleboard Association Seminar) [16].

7.2.1 Fiber Manufacture and Blowline Resin Blending

Fiber manufacture and wax addition in the blowline resin blending systems shown in Figure 7.3a, the wood chips are first steam-conditioned at an elevated temperature and under pressure in the digester and then fed directly into the connected pressurized refiner [16]. The emulsion form of wax is commonly added into the digester or refiner, although molten wax would be okay at the prevailing temperatures. Some loss of wax efficiency is to be observed when added in these harsh digester/refiner processes, but this method is preferred in the blow-line resin-blending processes. The refiner consists of a single rotating disk having special refining plates and a matching plate on the fixed disk. Under the same elevated temperature and pressure condition, the treated wood chips are fed into the refiner and fiberized, and the refined fiber is then blown out of the refiner through a small blow valve with a discharging speed of 600–1700 ft/sec into the blowline.

Addition of UF resins and drying of the fiber and resin mix: The UF resin is injected through a tube into the blow valve or blowline and get mixed with the fiber stream, as shown in Figure 7.3a (top). The blowline commonly consists of a pipe about 18-ft long and 4–6 in. in diameter and is connected to the blow valve on the refiner, and the other end is connected to the first bend of the dryer tube, as shown in Figure 7.3b. The fiber/resin mix travels the bowline further and then gets thrown into the drying tube to meet the drying air at about 0.3 vol% of fiber in the air steam at a temperature of about 300 °F or higher. The wood fiber/resin mixture then travels straight for about 100 ft to meet the bend of the

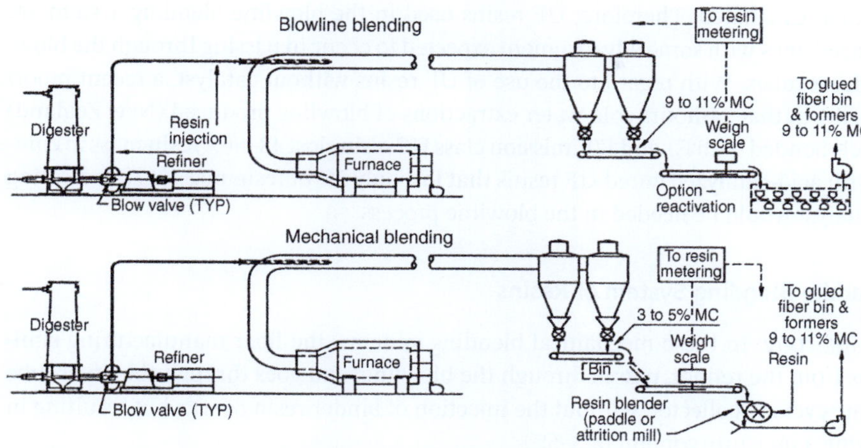

Figure 7.3a Blowline vs. mechanical blending of resin binder in MDF manufacturing process. Source: [16] Reproduced with permission of the Composite Panel Association.

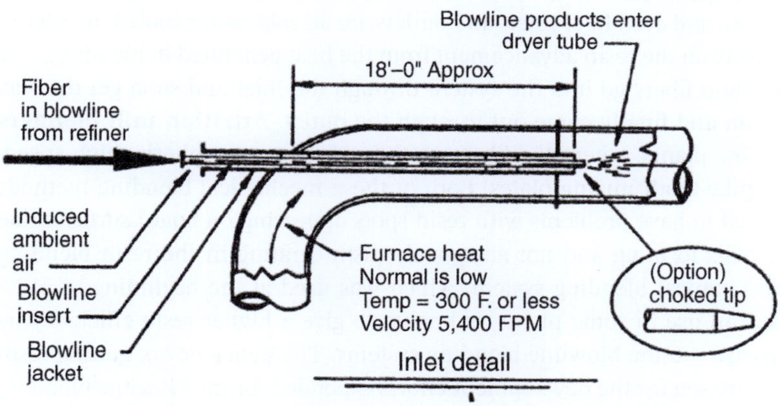

Figure 7.3b Details of the blowline where the resin binder can be added. Source: [16] Reproduced with permission of the Composite Panel Association.

dryer tube and travels further for about 200 ft to meet the cyclone separators, where the dried, resinated fiber is deposited onto the bottom, and the dryer air is exhausted through the vent together with most of the moisture from the fiber. The moisture content of the resinated fiber commonly runs 9–11% based on dry wood weight, while the resin loadings run at 8–10% based on the wood weight. In the beginning of the MDF industry in the United States, PF resins were also blended this way, but most production lines of PF resin-bonded MDF were discontinued.

UF resin requirements: UF resins used in MDF manufacture are basically similar to those used in PB manufacture. In the Southeastern United States, most plants use UF resins without the acidic latent catalyst, relying on the wood acidities that can vary significantly depending on the wood species and qualities. The flow characteristics of resins, i.e. the viscosity and dilutability, are important for fast-delivering systems, such as the blowline blending systems, similarly to the short retention time blenders in PB manufacture, since a slightly higher viscosity, for example, due to a temperature change, can lead to lower resin loadings than calculated. Therefore, UF resins used in the blowline blending system are low-advanced ones with some advancement expected to occur in passing through the blowline and dryer tubes. With regard to the use of UF resins without catalyst, a recent report is noteworthy in that 24-hours cold water extractions of blowline produced (New Zealand) MDF panels bonded with E1 and E0 emission class UF resins lost 48–66% resin mass in comparison with acid-catalyzed cured UF resins that lost only 2% of resin mass [17], indicating an acid catalyst would be needed in the blowline process.

7.2.2 Paddle Blending System of Resins

Fiber manufacture: In these mechanical blending systems, the fiber manufacturing similarly starts from the refiner, passes through the blowline, and goes through the dryer tube to reach the cyclone collector, without the injection of binder resin or wax but resulting in fibers having a moisture content of 3–5%.

Blending of resin and wax emulsion: The cool dry fibers are then sent to the high-speed paddle or attrition mill blenders for treatment with the binder resin and wax emulsion [16]. A typical **high-speed paddle blender is** shown in Figure 7.4, of which the outer shells, shafts, paddles, and even fiber inlets and outlets are all cold-water cooled in order to assure minimal effects on the resin advancement from the heat generated in blending. The paddles rotate fast, and fibers go into the system through the inlet and soon get the wax and resin sprayed on and finally come out through the outlet. **Attrition mill blenders** are also used in some plants, where the fibers go through two atmospheric, high-speed, counter-rotating, spike-tooth mixing plates. Both of these mechanical blending methods have been recognized to have problems with resin spots appearing on board surfaces due to the resin/fiber balls forming and not adequately disintegrating in the resin blending processes. This mechanical blending system, which was used at the beginning of MDF industry and is still in use in some plants, is known to give a higher resin efficiency by 0.5–2.0% in comparison to the blowline blending systems. The deficiency of getting resin spots was the main reason for the development efforts expended for the blowline blending method. The mechanical blending systems require more frequent, time-consuming cleaning in comparison with blowline blending systems, and therefore, a tandem blender setup is often necessary, so that the board production can be continued by switching off on one and starting on the other. Another variation of mechanical blending is the inclusion of the

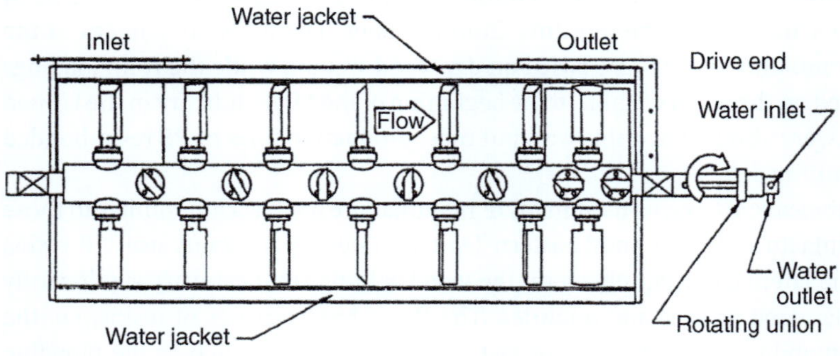

Figure 7.4 Details of water-cooled paddle blender. Source: [16] Reproduced with permission of the Composite Panel Association.

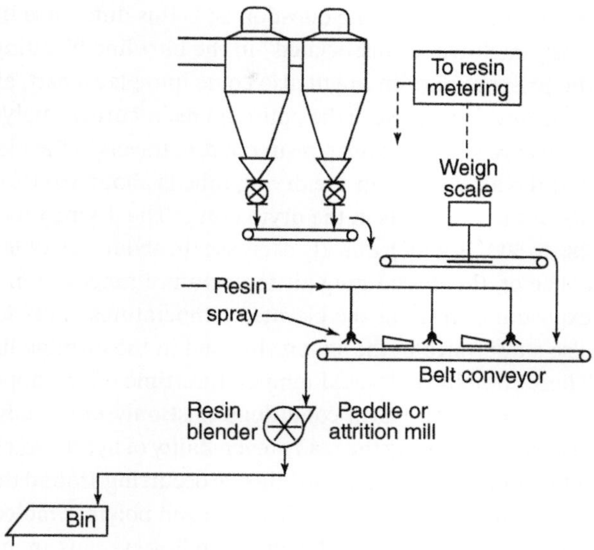

Figure 7.5 Details of resin binder application on fiber on moving conveyer belt. Source: [16] Reproduced with permission of the Composite Panel Association.

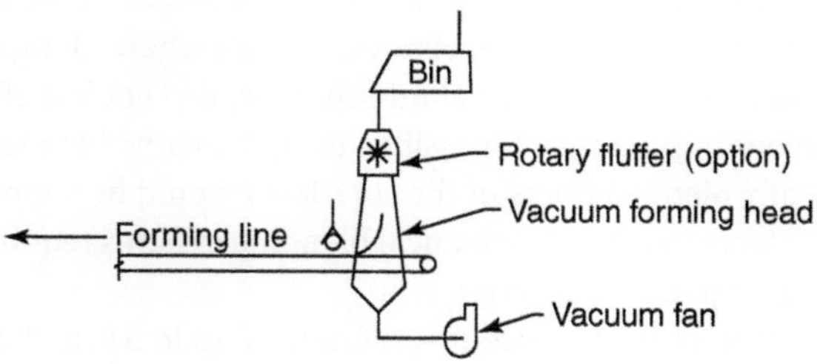

Figure 7.6 Details of the vacuum former in MDF manufacture. Source: [16] Reproduced with permission of the Composite Panel Association.

belt blending step [16], where the fibers are spread on a wide moving conveyer belt and sent through the resin sprayer booth to get resin application directly on fibers and then sent further to a paddle or attrition mill blender, as shown in Figure 7.5.

Mat formation: The resinated fibers are then stored in the storage bin and sent through a rotary fluffer, where mat is formed with the help of vacuum suction on the moving belt, as shown in Figure 7.6.

7.2.3 Comparison of MDF Resin Blenders

Dry-out and re-activation: the blowline blending methods have been considered to result in some dry-out of resins, and some plants spray water on fibers for re-activation, but the fiber moisture content of 9–11% attained in this process is advantageous in comparison with 3–5% MC of the fiber used in the mechanical blending process [16]. Though, it has been known that the blowline resin blending process results in significantly increased formaldehyde emission levels from the dryer stack. In some regions, the stack emission amount of volatile organic compounds (VOC) is strictly limited.

Fire hazard: Mechanical blending processes use more energy in drying of fibers and have an increased fire hazard condition.

Resin usage levels: It has been recognized that blowline blending systems require 0.5–2% more resin solids than mechanical blending systems for attaining the similar board strength properties. The question is: Is this difference in resin usage related to resin being "lost, modified, or dissociated" in the blowline blending process? All seems possible, and the pre-cure of resin might also come into play a part, all due to the high temperature and high air velocity. Use of the optimum resin-curing catalyst discussed earlier may be needed.

Fiber velocity: The fiber/resin mixture travels in the blowline tube of about 80 ft in length, and the air velocity in the drying tube is about 5500 ft per minutes, and the dwell time is about three seconds in the drying tube. The drying tube inlet air temperature ranges from 280 to 500 °F, which quickly decreases to about 220° or less in the drying tube. In the cyclone collector, the discharging air temperature ranges from 120 to 160 °F. Overall, the time of exposure of resin at the elevated temperatures is brief. However, the fiber/resin mixture also stays warm in the storage bin and in the forming line until delivered to the hot-press. These combined elevated temperature/time effects appear to underlie the resin efficiency loss, while such warm conditions exist only minimally in the mechanical blending processes. Considering the ready reversibility of hydroxymethyl groups in UF resins at elevated temperatures and the air turbulence occurring around the resinated fibers, the physical separation of freed formaldehyde from resin polymer molecules appears quite possible.

Resin spots in boards: Blowline blending systems are not without some maintenance, and resin spots also can appear in boards due to the resin buildups often occurring in the first few feet of dryer tube, elbows, and anywhere along the dryer tube and breaking off to give resin spots. Cleaning of buildups is needed but less often in comparison with the mechanical blending systems. If possible, using blowline blended fibers in the face-layer and mechanically blended fibers in the core-layer would be appropriate.

Maintenance: Mechanical blending systems require frequent cleaning to avoid resin spotting on the board surface.

Refiner sizes: Currently, refiners of up to 54-in. diameter are available that deliver 40 000 pounds of oven-dry fiber an hour. Refiners supplying 17 000 to 20 000 pounds of oven-dry fiber an hour would allow plants to use the separate face- and core-layer board manufacturing systems for a higher resin efficiency in the core-layer and lower resin spots in the face-layer with a more economical outcome.

Blowline configuration: Blowline consists of a 4 to 6-in. diameter pipe with the resin injection tube of 1/4-in. diameter welded at 90° or at a slanted angle so that the injected resin merges with the flow of fiber with less turbulence. Blowline itself also can have a neck section of about 3-in. diameter and of 3-ft long or longer, to where the resin injection tube also can be welded. In this restricted section, the fiber flows faster and increases in turbulence, possibly resulting in a better mixing of resin, steam, and fiber.

Economics: For an MDF plant where 40 000 pounds/hour of oven-dried fiber operating 22 hours a day and using a UF resin cost of 13.85 cents per solid pound at 9% loading level, if the resin level can be reduced by 0.5–8.5% from the average of 9.0%, the savings per year would amount to US$201,000.

Reducing the resin loading level by 1.5–7.5% would increase the savings to US$603,000 [16]. These are significant cost savings and worthy efforts to achieve by researching both the blowline blending system and the resin technology. Therefore, despite the resultant resin spot problems, the mechanical blending systems are worthy of further development in view of their better resin efficiencies [16]. MDF is higher-value material than PB because of the higher uniformity attainable by the fibrous nature of wood that results in better strength, better water-soak properties, better workability (cutting, sawing), etc. Both PB and MDF

Table 7.2 Hot-pressing parameters of PB and MDF in industry and market trends [18].

	Particleboard	MDF
Resin loading (%)[a]	5–8	8–10
Press temperature (°F)		
Standard press	320–350	320–350
Continuous press	340–370	350–370
RF cure press[b)]	280–300	280–300
	7.5–10.5	7.5–10.5
Mat MC (%)		
Press time (min) ¾-thick mat		
Standard press	4.0–5.0	4.0–5.0
Continuous press	3.5–4.5	3.5–4.5
RF cure press[c)]	2.5–3.5	
Standard press[d)]	2.0–3.0	

a) Resin % N.V. in panel-based o.d. dry wood wt.
b) Radio-frequency heating.
c) Button to full open.
d) Catalyzed UF resins.

Market trend:

- *Liquid PB UF resin growth rate: positive.*
- *PB/MDF industry driving force:*

Building industry + Formaldehyde emission – Furniture industry + Canadian imports – Exports+

boards are extensively used as core stock in furniture and case goods, but it is probable that furniture made of MDF is slightly higher priced than those made of PB.

7.2.4 Mat-Forming and Hot-Pressing Parameters of PB and MDF in Industry

Because of the bulky nature of fibers, the mat-forming step for MDF is different from that of PB manufacture. The resinated fiber goes through the air fluffer and is laid down on the forming belt in a continuous manner, as shown in Figure 7.6. Some plants have the face/core separate mat forming systems and some others have one-layer systems. Also, the orientation of fibers has been investigated by passing fibers in the forming box at about 50–100 ft/min speed through a 60-cycle electric field of 1500 to 3750 V/in., and fibers are collected at the end through a screen. The bending strength increases, and the linear expansion in water soak tests decreases significantly in the direction of alignment. Also, there are multi-opening hot-pressing and continuous hot-pressing systems [18]. The current industry hot-pressing parameters for PB and MDF are reported in Table 7.2, presented in our "Adhesives and Finishes for Wood" classes for several semesters around 2005–2010 by Mr. Charles Davis, R & D Manager, Georgia-Pacific Resins, Inc., Crossett, AR. It is noted that the resin loading levels are generally higher for MDF than PB, reflecting

Table 7.3 Average test value results for all test procedures for each MDF manufacturer.

Mill No.	Density (g/cm^3)	MOR (psi)	MOE 1000 (psi)	IB (psi)	Screw withdrawal, edge (lb)	Screw withdrawal, face (lb)	Thickness swelling 47–93% RH (%)	Residual thickness swelling (%)	Linear expansion 47–93% RH (%)
1	0.73	4837	466	125	257	326	9.59	4.36	0.360
2	0.90	4932	576	136	325	407	6.32	1.61	0.391
3	0.79	3366	432	282	330	445	6.26	2.48	0.611
4	0.82	5703	635	121	252	326	8.88	2.83	0.346
5	0.95	3565	517	133	405	509	11.44	5.45	0.577
6	0.80	5278	578	103	315	404	10.49	4.52	0.376
7	0.77	5421	572	179	360	464	8.74	3.18	0.440
8	0.71	5107	858	158	324	416	8.17	3.03	0.413

Source: [19] Reproduced with permission of the Forest Products Society.

that the fiber surface area is greater than the particle surface area for a given volume of boards, and this results in overall better board quality for MDF. It is also noted that the MDF industry often uses the radiofrequency (RF) heating method as a pre-heating device just before the mat goes into the hot press.

7.2.5 Physical Properties of Commercial MDF

Various grades of MDF panels with specific property requirements have been established by the industry, and manufacturers also supply the minimum specifications to board buyers. Compared with PB, MDF shows exceptional surface and form stability. However, board properties vary substantially because raw materials, adhesives, and manufacturing techniques can vary. A paper published in 1978 by O. Suchsland, D.E. Lyon, and P.E. Short shows board properties of eight different MDF manufacturing plants' as shown in Table 7.3 [19]. In overall, the authors indicated that higher-density boards have better physical properties but with higher manufacturing costs. For example, board No. 5 required 1/3 more furnish than board No. 8. Further, MOE values strongly correlate with board *surface* density values but not much with the *average* board density values. Also, there is no clear relationship between average density or core-layer density values and IB values. Wood species, resin content, fiber quality, etc. are of minor importance with regard to MOE values. But the core-layer density values did affect the screw-holding power. Neither thickness swelling nor linear expansion is related to board density or density gradient. *Surface stability* is a measure of resistance to deterioration of surface smoothness caused in the water-soaking and redrying test, measured by a profilometer, and board No. 2 showed the highest stability and No. 1 the lowest. *Form stability* is defined as the ratio of the volume of moisture absorbed vs. the volume expansion of boards based on MC equilibration values changing from RH 47% to 93%, and this value is found to correlate strongly with the unrecoverable component of swelling [19]. Overall, the authors indicated in 1978 that the MDF industry would expand into the PB market due to its better properties.

7.2.6 Medium Density Fiberboard Industry

The MDF industry in the United States keeps the high quality of their products through *the American National Standard for Medium Density Fiberboard for Interior Application (ANSI A208.2-2002)*, established by the American National Standard Institute (ANSI) and National Particleboard Association (NPA). MDF is graded with numbers as: *110*; *120*; *130*; *140*; *150*; and *160*, identified by their increasing MOR, MOE, IB, and screw-holding strength values. In addition, the standard includes the moisture resistance designation: MR10 for reduced thickness swell values; MR30 for increased MOR values; and MR50 for both reduced thickness and increased MOR values; thus, a board can be graded as *110-MR10*, etc. The standard also classifies thinner MDF panels (3/8 in. or less) into four grades with numbers 210, 220, 230, and 240, with all grades having a thickness swell requirement of 0.080 in. or less. MDF is used similarly with PB in manufacturing kitchen cabinets, office and residential furniture case goods, and other demanding applications, such as floor underlayment and stair treads. MDF is also used with overlays. Formaldehyde emission levels are also included in the standard so as to meet the emission regulation limits.

7.2.7 Summary

In the manufacturing of MDF, wood chips are first produced from low-grade whole trees and go through the digesting and refining processes under pressure and high-temperature steam with wax often added. Then, the refined fiber with wax is allowed to exit into the blowline under pressure, into which the UF binder resin is injected in those plants where the *blowline blending method* of resin is used. The fiber loaded with wax and resin is then shot from the blowline into the long drying tube at about 0.3 volume percentage of hot air of about 300 °F that travels at a high velocity, gets dried, and is carried to the collecting cyclones and get precipitated with a moisture content of 9–11%. The glued fiber then goes to the air-fluffer-type former and then to the hot-pressing station. Although this blowline blending method is more of recent development, in other older plants, a similar process is carried out without resin injection and with some extra drying to result in fibers having a lower moisture content of 3–5%, followed by blending with wax and UF resins in a paddle or attrition type blender to yield resinated fibers having a similar moisture content of 9~11%, followed by the similar forming and hot-pressing stations. Advantages and disadvantages of the two methods are presented. The paddle blending system has a better resin efficiency due to the fact that the resin does not go through the harsh drying tube, but the process tends to produce boards with (dark) resin spots on board surfaces because of inefficient fiber-resin mixing in the blender, which is the main reason why the blowline resin blending system was developed. The hot-pressing condition of MDF is similar to that of PBs but MDF panels have generally better board properties due to the use of refined whole fibers and higher resin loadings, and they are used in higher-grade furniture in comparison with PB.

7.3 Hardwood Plywood Bonding with UF Resin Adhesives

Hardwood plywood industry entails various decorative plywood panels and laminated veneer lumber (LVL) products in North America, manufactured using UF or MUF resin

Table 7.4 A typical adhesive mix for hardwood plywood.

UF resin (60.0% RS)	56.8%
Wheat flour (95.0%)	12.0%
Water	17.8%
Wheat flour (95.0%)	12.0%
Catalyst	1.4%
Marker (color)	*D*
Resin solids	34.08%
Extender solids	22.80%
Catalyst	1.4%
Total solids	58.28%
Target viscosity	~5 k cps

binders. One hundred fifteen hardwood plywood manufacturing plants were operating in 1998 in North America [4]. Various hardwoods are used, and the face-layer veneers are of higher quality, and the cross-band and bottom-layers can be of lower quality hardwood veneers or even of some thin wood composite panels. Hardwood plywood panels are mostly used for interior, and decorative applications, such as in wall paneling, cabinets, furniture, and flooring. Birch, walnut, oak, gum, and maple are the most often used hardwood species. Also, some hardwood plywood made of lauan is imported from Asia and used mostly in wall paneling.

Manufacturing procedure: The hardwood plywood panel construction procedure is similar to that of softwood plywood, using cross bands in the core-layer to obtain a better overall stability and strength. Another class of hardwood plywood is the hardwood LVL, where veneers are all laid up in one direction and the bonded panels are cut into lumber pieces for use as frames for interior upholstered furniture, *etc.* The adhesive binders for hardwood plywood are mostly UF resins formulated with extenders, such as wheat flour, and catalysts, as shown in Table 7.4, so that adhesive mixes have about 5k cps viscosities for gap-filling capability and good control of spreading and flow, similarly to the softwood plywood manufacturing with PF resin adhesives. The adhesive color is often minimal, only deriving from the extender materials used, and this low color of UF resin adhesives is a positive attribute to hardwood plywood because they are used mostly as decorative panels, which are often finished with coatings of desired color and texture. Also, a marker color is often added, which is identified only by instrument, for tracing to the panel manufacturers in case of a consumer complaint.

In the hot-pressing of hardwood plywood, a good tack is also needed to hold veneers together after pre-pressing, since this allows the hot-pressing to be done in a more orderly and efficient manner. The tack is mostly attained from the interaction of wheat flour, water, and UF resin. In recent years, because of the formaldehyde emission problem of UF resins, the F/U mole ratio of resins is decreased and also the tack and bond strength properties decreased. To ameliorate these bonding and formaldehyde emission problems, fortification of UF resins with low levels of melamine resins in the form of UMF resins is often done in some segments of the industry, where the increased binder cost needs to be accommodated. The UMF resin addition would also improve the moisture resistance of boards.

Various troubleshooting methods: Various troubleshooting methods in hardwood plywood manufacturing plants have been reported: [20] (i) *Bleed throughs* from too thin viscosity of the adhesive mix, short assembly time, high moisture content of veneer, and an improper portion of filler in the glue. (ii) *Blows or blisters* from the similar reasons and also from too high press temperature or too short press time or defects such as knots, worm holes, or checks in the stock. (iii) *Dimpling* from defects in the stock, too-thin crossbands, defects in the lumber core, nonuniform moisture content in the PB cores, or board surfaces that are pitted or have voids. (iv) *Face checking* from improper conditioning in the assembly station, too long press times, or too high press temperatures. (v) *Indentations* from cauls or platens that are pitted or damaged. (vi) *Low water resistance* from too high extensions of glue mix or incomplete cure. (vii) *Pre-cure* from too long assembly times, using the stock directly from the kiln or dryer before it has cooled, too high press temperatures, too light glue spreads, or variability in thickness of the stock. (viii) *Spotty bonding* from too long assembly times, worn or damaged spreader rolls, careless veneer handling, face veneers, which are sized or glazed, insufficient or nonuniform pressure, improper mixing, rough stock, or nonuniform glue spread. (ix) *Starved joints* from too short assembly times, too thick adhesive mixes, faulty extension, or a foamy mix. (x) *Warpage* from defects in core stock, improper conditioning of panels, nonuniform spread, over-curing in hot press, too high press temperatures, or unbalanced construction. (xi) *Weak bonds* from adhesive mix not reactive enough, too heavy spreads, too low moisture contents in the stock, too short press cycles, or too low press temperatures. Stability of hardwood plywood products (flooring, etc.) against warping is important, the lack of which is often the major cause of consumer complaints, and it can become more acute for low mole ratio UF resins and also when using different species or different quality veneers. Some manufacturers also use polyvinyl acetate (thermoplastic) adhesives.

Hardwood plywood grades: Hardwood plywood used for furniture and flooring is graded as Technical, Type I, Type II, and Type III with the water resistance properties increasing in that order, and each grade is specified to be bonded with particular grades of adhesives, possibly with higher F/U mole ratios or higher levels of UF resin or PF resin solids. Another class of hardwood plywood is thin (1/4″) panels manufactured mostly of lauan-type hardwood veneers imported from Asia and finished locally with grooving and coating applications for interior wall paneling. Hardwood panels are normally higher-priced than similar-use panels made from thin PB or fiberboard.

Summary: In the manufacturing of hardwood plywood, veneers are produced by peeling or slicing, and various hardwoods are used. Better grade veneers are used for top layers, and lower grades are used in the crossband and bottom layers. Also, thin PB or lumber is used in middle layers. The adhesives are all UF resins formulated with an extender and acidic curing catalysts, commonly comprising ~34% UF resin solids, ~23% wheat flour, and 1.5% catalyst. The adhesive is applied with a roll-coater and a good tack property is needed. The hot-pressing is done at ~300 °F and various adhesive-related or substrate-related possible problems are discussed. Hardwood plywood is used mostly as decorative wall or flooring panels and traded with grades: Technical, Type I, Type II, and Type III with the water resistance properties increasing in that order, made mostly through adjusting the resin/extender ratio as well as the F/U mole ratio of resin. The formaldehyde emission is similarly an important issue for hardwood plywood.

7.4 Paraffin Wax and Uses in Wood Composites

7.4.1 Use of Paraffin Wax

In the manufacturing of most wood composite boards (PB, OSB, and MDF), paraffin wax is applied up to 1.5% based on oven-dry (OD) wood weight to improve the water repellency, so-called, the temporary moisture resistance of boards, effective for some duration in contact with water or humid air, [8] not designed for continuous water contact, but the temporary moisture repellency is a technically important parameter for board manufacturers. Paraffin wax is a relatively low-cost material, and its addition interferes little with adhesive bonding when the usage level is kept in the low range. Paraffin wax molecules are low molecular weight hydrophobic materials, and they are not permanently bonded to wood surfaces; therefore, boards lose water repellency when high humidity conditions are encountered for an extended period of time. Waxes added more than the level mentioned above can cause lower bond strengths and increase the flammability.

7.4.2 Properties of Paraffin Wax

Paraffin wax is obtained in the petroleum refining process as a solid fraction between the heavy oil and tar residues. A commercial wax consists of linear and slightly branched low molecular weight, saturated (no double bond) hydrocarbon molecules, with the number of carbons in molecules ranging anywhere from about C_{18} (Mw = 256) to about C_{40} (Mw = 560) or slightly higher. These aliphatic hydrocarbons, containing no oxygen or nitrogen, are highly hydrophobic (nonpolar) and have low molecular affinity to polar wood surfaces, and therefore the bonding on wood is relatively weak. **Melting point** of commercial waxes is a property that depends on the average carbon number and the heavy oil content. Waxes from different crude oils or different manufacturing lots can show different melting points and different heavy oil contents. Melting points of industrial waxes range from 120 to 150 °C, which is generally within the range of the mat core-layer temperatures found in the hot-pressing of most wood composite boards. But it is noted that the particular wax chosen for a process must have the right melting range in order to be effectively melted under the hot-pressing condition, and this has to be obtained with a minimum heavy oil content. Recalling that PB and MDF manufacturing plants use 300–350 °F and OSB plants 400–430 °F for hot-pressing and their core-layer temperatures are quite lower, the wax choice needs to be made accordingly. Further, commercial paraffin waxes invariably contain some **heavy oil**, i.e. the lower molecular weight hydrocarbons of up to about 17 carbons (heptadecane, mp 22 °C or 71.6 °F) at less than about 10% of total weight.

The heavy oil components, sort of impurities of wax, coming from an incomplete separation in the distillation process, lower the melting point of the wax. The **volatility index** of waxes, obtained as the weight loss observed by heating a wax sample at 325 °F (162 °C) for three hours, is a measure of the heavy oil contents, reported to be normally less than about 3.0% [8, 9]. Most volatile fractions are hydrocarbons with carbon numbers around decane (C_{10}), which has a boiling point (bp) of 344 °F. Most of the heavy oil components deposited in the board with wax will remain, but their water repellency effect would be minimal since they would disappear as the board ages because they would migrate to board surface and evaporate off. Also, some volatile heavy oil components would vaporize during hot-pressing to be deposited as buildups on vent stacks, and this can lead to fires [9].

7.4.3 Analyses of Wax

Waxes have been analyzed using the gas chromatographic (GC) method. In this method, the wax sample vapors in the carrier gas are sent through a thin column packed with fine inert particles that have a low affinity for wax molecules. Under an optimum condition of carrier gas speed and column temperature, the wax molecules come out of the column in the order of their volatility, i.e. the more volatile molecules (lower carbon numbers) exit ahead of the less volatile ones (higher carbon numbers). Thus, the early peaks are of low molecular weight and the later ones are of higher molecular weight hydrocarbons. Two wax examples with reported melting points thus analyzed are shown in Figure 7.7a and 7.7b with the elapsed times recorded from the sample injection time [8]. From the resultant chromatograms, the mean or middle carbon numbers can be identified, and the overall population of peaks can be correlated to the melting points. In these gas chromatograms, each sharp peak represents a linear hydrocarbon molecule of one particular carbon number, say a hydrocarbon having 25 carbons, i.e. $CH_3—(CH_2)_{23}—CH_3$. There are also peaks present in between the sharp peaks with lower intensities that represent the branched hydrocarbons, or isomers of similar carbon numbers. It has been known that the linear (normal) hydrocarbons afford better water repellency than branch isomers, and the higher molecular weight hydrocarbons afford better water repellency [9]. The value of linear hydrocarbon content is referred to as the "**normality**" of a wax, which can be calculated from the chromatograms. Normality index value is often a fixed value depending on the source of the crude oil.

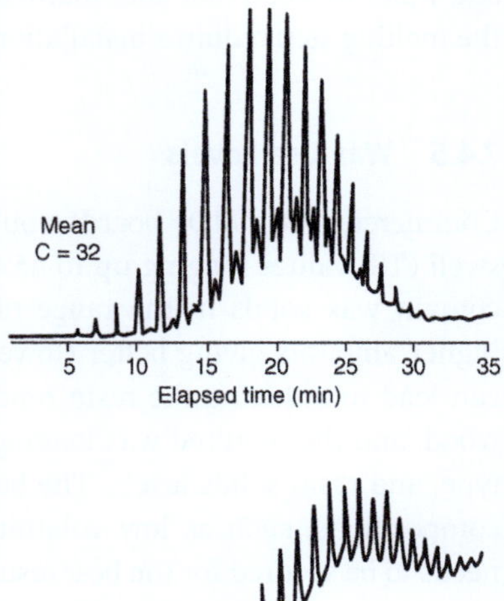

Figure 7.7a Chromatograms of a low melting (120 °C) wax [8].

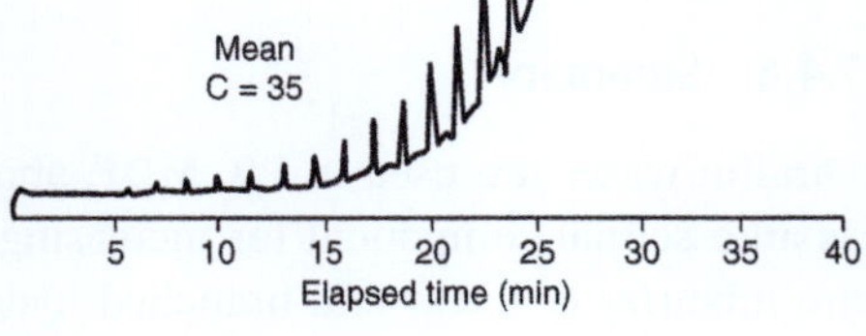

Figure 7.7b Chromatograms of a high melting (140 °C) wax, indicating the increasing melting point with increasing amounts of later eluting components [8].

7.4.4 Wax Application Methods

Wax is applied on wood particles or fibers by spraying the **emulsion form of wax** in the manufacturing of PB and MDF. A separate air-atomizer-type sprayer is set up in the blender. In the blowline MDF manufacturing processes, the wax emulsion is added to the digester or to the refiner. Some manufacturers premix the wax emulsions with binder resin and spray them, but this method tends to cause incompatibility problems. Wax emulsions, normally produced by wax emulsion companies, contain about 50–65% wax solids in water. The emulsification process involves melting the slack wax in hot water and dispersing it into finely suspended liquid wax particles, and the suspension is stabilized with emulsifiers, surfactants, or colloids, followed by stirring while cooling to room temperature. The surfactants are nonionic, anionic, and cationic surfactants, among which the emulsions based on nonionic surfactants are most common because of their greater compatibility with resins and stability against the pumping-type shearing forces.

One typical commercial wax emulsion has the following specifications: wax solids content of 46–48%; specific gravity of 0.94–0.96; viscosity of 20–80 cps; and pH of 6.5–7.0 [5]. They do not normally reveal the kind of surfactant or chemical/physical quality of the wax itself. Wax emulsions have a low viscosity and are easy to spray, but one disadvantage of using wax emulsion in comparison to slack wax is the extra water that goes into the mat and the additional cost of making the emulsion. Also, the surfactants tend to show affinity to water and can result in reducing the efficiency of wax emulsion as a water repellency additive. PB and MDF manufacturing plants mainly use wax emulsions because the blenders are more amenable to wax emulsions. If *molten slack wax* is to be applied in these processes, the wax particles would cool down quickly and would not disperse adequately in the blender. In OSB plants, molten wax is used by first melting the slack wax, and the whole line and sprayer are kept heated above the melting point of the wax. Molten wax application does not add water to wood mat and, therefore, is widely used in OSB manufacturing, even though the melting step requires installation of the heating/delivery equipment.

7.4.5 Wax Use Levels

Commercial PB or MDF boards would show 24 hours water-absorption (WA) and thickness swell (TS) values ranging up to 45% and 15%, respectively, and the boards are known to contain wax solids in the range of 0.75–1.50% based on OD wood weight, commonly higher amounts giving better (lower) water absorption properties. But excess wax levels can lead to reducing the resin binder's performance. Wax costs significantly more than wood, and the required wax loading level can vary depending on the wood species, resin type, and resin solids levels. The best possible distribution of wax and the best chemical compositions, such as low volatility, optimum melting point, and optimum normality, needs to be secured for the best results.

7.4.6 Summary

Paraffin waxes are used in PB, MDF, and OSB at 1.0–1.5% levels based on wood weight as an essential component for increasing the water repellency of boards. Paraffin waxes are mixtures of linear and branched hydrocarbons of C18 carbons and up to about C40 carbons, derived from crude oils, and contain some heavy oil components up to about 10% by weight. Typical slack waxes purchased from suppliers have a melting point between 120 and 150 °C. Wax components with lower carbon numbers have lower melting points and *vice versa,* and the melting point of a wax is affected by heavy oil contents. OSB manufacturers use solid waxes and have the option of using a wax based on its melting point. Since the water repellency of wax hinges on its melting and spreading in hot-pressing, the melting point of wax needs to be lower than the hot-pressing temperatures prevailing, especially in the core-layers. Lower melting waxes would appear preferable, but the level of heavy oil components needs to be taken into consideration since they can lower the melting point but do evaporate off during hot-pressing and also during the use of boards. PB and MDF plants use ~50% wax emulsions, and their wax components' melting points or heavy oil contents would be worthwhile to be made known. Analyses of waxes can be carried out by analytical service companies using the gas chromatography method to determine the distribution of hydrocarbons of various carbon numbers, including the heavy oil components, and also the ratio of the linear/nonlinear hydrocarbons. Linear hydrocarbons are suggested to be more effective water repellents, although the ratio would be decided mostly by the source of crude oil.

Advanced Reading Materials

7.5 Effects of UF Resins' Curing Catalysts on Tool Wear in Machining of Boards – An Example of Laboratory PB Manufacturing and Testing

(Adapted with permission of the publisher and minor revisions from Tool- wear effects of anionic components of UF resins used as particleboard binders. Kim. M. G., H. A. Stewart, and H. Wan. 1999. Forest Products J. 49(4):60–65.)

Tool wear is an important economic parameter in the secondary wood manufacturing industry, where wood composite boards, such as PB and MDF, are machined extensively. Increased tool wear has been observed in various wood machining industries. Dirt incorporated during the board manufacturing process has been known to increase tool wear due to its increased abrasiveness. Boards with higher surface-layer densities also have been known to cause excessive tool wear due to more materials present in the given cutting volume. Furthermore, certain chlorine and sulfur anionic components present in wood composite products have been known to cause increased tool wear, ascribed to the facilitation by these chemical species on the oxidation/corrosion-wearing processes of tool materials occurring during machining [21–24]. The cutting zone in wood machining reaches temperatures of more than 800 °C and with the oxygen and wood materials present, a highly oxidative/reductive environment is established. It appears that chlorine and sulfur anionic components form metal chlorides that easily volatilize or metal sulfidation products that are easily removed by abrasion. This active involvement of chlorine and sulfur in accelerating tool wear was specifically noted in the machining of MDF with tungsten carbide–cobalt tool materials (C2) (WC-6% Co) [24]. PB, MDF, and

Table 7.5 Experimental parameters and methods used in the laboratory PB manufacture.

Wood particles: southern pine with about 25% mixed hardwoods
Board size: 63.25 mm by 55.66 mm by 12.65 mm
Replication: 6 boards, 3 boards from one blending
Resin blending: 15 min
Matting: 15 min after resin blending for 10 min
Mat moisture content: 9–10%
Target board density: 801 kg/m^3 as standard, and 720 and 870 kg/m^3 as density controls
Resin and wax loads: 8.0% and 1.0% solids, respectively, on oven-dry wood weight
Hot-pressing temperature: 1490 °C for UF-resin-and 1710 °C for PF-resin- PF-resin-bonded boards
Hot-pressing: 1.0 min compression to −2760 kPa (−400 psi), holding for 3.0 and 0.5 min decompression for UF resin-bonded boards; the holding was lengthened to 6.0 min for PF resin-bonded boards
Catalyst and additives: the levels (Table 2) are based on liquid resin weight
UF resins had 64.0% resin solids; PF resins had 44.0% resin solids

Table 2 = Table 7.6.
Source: Reproduced with permission of the Forest Products Society.

other wood composite boards can contain chloride, sulfate, and other anionic chemical components derived from the latent acid catalysts and also from the acidic components used in the syntheses of UF resins [25]. Most of the acids will remain in the boards after board manufacture, probably as salts of weak bases, but they will easily dissociate back into free acids when contacted with tool surfaces in the hot wood cutting zone. Since strong acids lower the oxidation potentials of various metals [26], the freed acids would result in increased tool wear by facilitating the oxidation process. Thus, the tool wear potential of these anionic components in UF resins was investigated.

Resin syntheses and PB preparation: In this work, UF resins (F/U = 1.15) were synthesized according to a known 3-step procedure starting with an F/U_1 mole ratio of 2.10 using a small amount of sulfuric, nitric, or oxalic acid as the polymerization catalyst, [25] and formulated to have various levels of chloride, sulfate, formate, and nitrate ions, and laboratory PBs were manufactured using the parameters shown in Table 7.5 under the typical laboratory PB manufacturing conditions. Bark-free wood particles were obtained from the Georgia-Pacific Corporation PB plant in Louisville, MS. An alkaline PF resin, OSB core-layer binder type, synthesized in the laboratory [27], was also included in the study for comparison.

Physical Tests of PBs: IB, MOR/MOE values were obtained in accordance with ASTM Procedure D1037 for each set of PBs. Eight IB and two MOR/MOE samples were cut from each board using a pattern, and the average test values were obtained. Only IB values were reported since all MOR/MOE values were within the normal range. The board density values of all tool wear test circles were directly measured from the weight and dimensions, and the averages and standard deviations were obtained for each set of PBs. Samples for density profiles were cut from the center pieces of boards left from testing of tool wear test circles, and density profiles were measured using a QMS Density Profiler (Quintek Measurement Systems, Inc., Oak Ridge, TN), generously provided by Georgia Pacific Corporation,

Louisville, MS. The maximum surface-layer densities were read from the curves, and the averages and standard deviations were calculated and reported as surface-layer densities.

The results reported in Table 7.6 indicated that the overall bonding process for PBs was relatively well controlled, and the levels of added catalysts did not adversely affect the board's properties much. The average density values of test circles also showed good consistency within target values, as indicated by the small standard deviation values. The average surface-layer density values derived from the density profile curves indicated that they are also relatively consistent and have acceptable standard deviation values ($\leq$4.5%). In Figure 7.2b, the density profiles of PBs made at average target densities of 721, 801, and 881 kg/m^3 are shown as examples. The IB strength values of PBs ranged from 110–150 psi for standard 800-kg/m^3 density boards, 100 psi for 719-kg/m^3 density boards, and 180 psi for 873-kg/m^3 density boards, which are all within the ranges of typical commercial boards. Overall, the prepared boards showed a consistency that appears to be sufficient to give an acceptable comparison of tool wear effects in relation to additive levels.

Elemental analyses of mineral elements, sulfur, and chloride contents: To assure that sands and other extraneous materials were not inadvertently incorporated into boards, the raw wood particles and selected density profile test pieces were analyzed for mineral elements and sulfur and chloride contents using ion-coupled plasma spectroscopy and appropriate wet chemical test methods (Iowa Testing Laboratory, Eagle Grove, Iowa). The results reported in Table 7.6 indicated that the amount of dirt present was too small to distort the tool wear test results, except for Board #1, which was therefore not used in this discussion. The sulfur and chloride contents were found to be generally consistent with the samples' history, but they were very low, although, for some unknown reason, the chloride contents were several times higher compared to the sulfur contents for all samples, including the wood particles used in board preparation. Metallic elements were present in levels comparable to reported values for various woods [29] (data not shown). The metallic elements have not been known to have any direct effects on tool wear.

Test methods of tool wear effects: The tests were done using a lathe-type cutting machine known to give good, consistent tool wear results when cutting MDF [28, 30, 31]. Briefly, in this test method, the board circle (0.5″ thick, ~10″ diameter) is held on a lathe and rotated at 550 rpm against the knife to continuously cut a thin slice of board along its circumference. The parallel and normal forces exerted on the knife are continuously recorded, and the end values are read. It was earlier shown that the tool forces measured in this test method increase linearly as the actual wearing of the tool edge increases [31]. The knife was 12.6-mm wide tungsten carbide (C2), with a 15-degree rake angle and 10-degree clearance angle. The C2 tungsten carbide with standard cobalt binder used in this test is a common material of current wood working knives in the industry. One test consisted of consecutively cutting *four* 233-mm-diameter board circles with one knife, which resulted in a total cutting length of approximately 304 m. The test was replicated three or five times for each additive variable. The normal forces increased more rapidly than parallel forces; therefore, only normal force values were reported with averages and standard deviations calculated.

Test results of tool wear effects: The results reported in Table 7.6 indicated that the tool force curves obtained from PBs showed a somewhat wider oscillation than those obtained with MDF due to the more granular nature of the wood elements used [30]. However, the oscillation was minimal, and the median values were found to be consistent to adequately represent the tool force increases incurred. Further, in general, the tool force increased as

Table 7.6 UF resins, catalysts, cutting forces, and other test results of particleboards.

Acid[a]	Catalyst[b]	Cat. (%)[c]	Board no.	Average density (kg/m^3)[d]	Surface density (kg/m^3)[e]	Normal cutting force (N)[h]	IB (kPa)[f]	Elem. (%)[g]	
								Cl	S
UF-H_2SO_4	$(NH_4)_2SO_4$	0.1	1[i]	807(20;19.7)[j]	1059(6;39.4)	171(5;14.3)	1089	0.10	0.03
		0.4	2	795(20;22.4)	985(6;21.9)	87(5;9.2)	1151	0.12	0.02
		0.8	3	799(20;11.4)	990(6;24.6)	86(5;5.1)	779	0.12	0.04
		0.1	4	719(12;10.7)	795(6;27.7)	82(3;3.9)	772		
		0.1	5	873(12;13.0)	1006(6;30.6)	159(3;11.0)	1131		
	Na_2SO_4	2.0	6	814(20;13.0)	1000(6;20.0)	86(5;6.2)	1585		
		5.0	7	803(12;8.3)	977(6;33.3)	108(3;2.6)	772	0.10	0.14
	NH_4Cl	0.1	8	801(20;14.6)	974(6;28.8)	82(5;4.5)	1096		
		0.4	9	817(20;14.6)	1004(5;14.1)	136(5;9.2)	1324	0.17	0.02
		0.8	10	814(20;8.2)	1024(6;26.3)	121(5;22.4)	896		
	Nacl	2.0	11	804(20;14.4)	956(5;34.3)	84(5;5.8)	972	0.20	0.02
		5.0	12	806(12;10.6)	971(4;15.5)	165(3;10.3)	979	0.20	0.01
UF-HNO_3	NH_4HCO_2	0.0	13	869(20;23.1)	990(6;27.2)	128(5;6.0)	1661		
		0.1	14	843(20;21.5)	996(6;12.0)	121(5;12.1)	1310		
	$NaNO_3$	5.0	15	798(12;6.2)	982(6;23.1)	152(3;10.0)	876		
UF-$(COOH)_2$	NH_4HCO_2	0.0	16	795(12;8.8)	891(6;17.0)	111(3;4.5)	800		
		0.1	17	795(12;9.5)	873(6;63.1)	108(3;2.6)	848		
		0.4	18	796(12;10.7)	940(6;39.9)	117(3;3.4)	917		
		0.8	19	798(12;6.1)	922(6;34.8)	119(3;6.4)	1193		
Alkaline PF	NaOH	5.2	20	805(12;21.5)	1045(5;36.0)	128(3;2.6)	1138	0.10	0.01

a) Acids used in resin syntheses.
b) Acid salts and bases added to resins before board preparation.
c) Catalyst level added in resins based on liquid resin weight.
d) Average overall board density for tool-wear test circles.
e) Average surface-layer density from PBs from board density profiles.
f) Internal bond strength of PBs adjusted to the standard board density of 801 kg/m^3.
g) Elemental analysis results of PBs for chlorine and sulfur.
h) N = Newton.
i) Board #1 was found to be contaminated, by elemental analysis, and not considered in the discussion.
j) The first value in parentheses represents the number of test samples; the second value indicates the standard deviation.
Source: Reproduced with permission of the Forest Products Society.

the board density increased, as shown clearly by tool forces of boards made at average board densities of 719 kg/m^3 (Board #4) and 873 kg/m^3 (Board #5) both by using 0.1% ammonium sulfate catalyst: 82 and 159 Newtons (N), respectively. The different surface-layer densities, 795 and 1006 kg/m^3, respectively, could have also contributed to the tool force difference. The higher surface-layer density of boards would show increased tool force since knife dulling occurring in the fractional lengths of knife in contact with board surface layers could cause an inordinately greater increase in tool force. However, in wood composite boards made at given overall densities, the increased surface-layer densities will result in comparatively lesser densities in the core-layers, partly canceling out the higher density surface-layer effect. In practice, the surface-layer densities of boards are not easy to vary systemically. The surface-layer densities obtained in this work were relatively uniform, their variations were small, and they could be separated out in most cases as factors that are insignificant when comparing the additive effects of tool forces. Overall, it appears that the **average board density** is more pertinently related to tool force increases.

Effects of additives on tool forces: (i) PBs bonded with 0.4% and 0.8% levels of ammonium sulfate as catalyst and 2.0% sodium sulfate as a diluent additive showed relatively low tool forces of 86–87 N. PBs with 5.0% sodium sulfate showed an increase in tool force of 108 N, indicating that sulfate ions are a tool wear accelerating component when present at high levels. (ii) PBs made with 0.1% ammonium chloride and 2.0% sodium chloride showed low tool forces of 82–84 N, but boards made with 0.4% and 0.8% ammonium chloride and 5.0% sodium chloride showed very high tool forces of 121–165 N, indicating that chloride ions are very potent tool wear materials, and more potent than sulfate ions on a weight basis. Sodium chloride is known to be intentionally added to some UF resins [32, 33]. (iii) The results also indicate that ammonium salts cause more tool wear than sodium salts. (Ammonium salts are likely to remain in boards as hexamine salts, and the acid components are far easier to dissociate than from sodium salts.) (iv) PBs bonded with UF resins made with nitric acid as a polymerization catalyst gave high tool forces with and without 0.1% ammonium formate catalyst, and the tool force increased further, with an addition of 5.0% sodium nitrate, to the levels of PBs made with high ammonium chloride or sodium chloride levels. Nitrates appear to be a very potent agent of tool wear. (v) PBs bonded with UF resins made with oxalic acid as a polymerization catalyst and formulated with zero to various levels of ammonium formate catalyst, containing no chloride or sulfate components added, all showed relatively high tool forces between 108 and 119 N. All these boards had an average board density of nearly 801 kg/m^3, but their surface-layer densities were relatively low and mildly fluctuated in the 873–940 kg/m^3 range. The relatively high tool forces observed for these low surface-layer density boards clearly contrasted with, for example, the PBs bonded with 0.4% and 0.8% ammonium sulfate catalysts. The latter boards have the same average density, but far higher surface-layer densities in the 956–1000 kg/m^3 range with low tool forces of 87 N and below. Thus, these results indicate that formate ions are not, but oxalate ions are a very potent agent of tool wear and that the surface-layer density itself plays an insignificant role for tool wear. (vi) Sodium chloride, sulfate, and nitrate added at 5.0% levels showed significantly increased tool forces, even though these salts are neutral salts, i.e. not easily hydrolyze to free acids. The reasons behind the high tool wear may be traced to the high temperatures in the cutting zone, accelerating the oxidation/corrosion reactions involving these ions. Various (pure) metals are known to degrade fast when mixed with chlorides and nitrates and subjected to high temperatures [23]. Overall, the acidity and high-temperature oxidation/corrosion tool wear mechanisms would be operative. (vii) Alkaline PF resin-bonded PBs also showed considerably increased tool forces, indicating that strong bases may also participate in the accelerated wearing of tool materials.

7.6 Summary

A publication on the tool wear effect of the curing catalysts of UF resin binders used in PB manufacture is presented to familiarize readers with the laboratory board manufacturing and testing procedures. The experimental parameters used show a detailed composition of materials and processing parameters for a PB having the targeted dimensions and densities. Curing catalysts for UF resins selected are ammonium salts of strong acids, such as sulfuric, hydrochloric, nitric, and oxalic acid. Sodium chloride is also used since some PB manufacturing plants use it as an additive to UF resins with claims of cure acceleration. An alkaline PF resole resin was included as a comparison. Also, the manufacturing of UF resins needs an acid catalyst in the polymerization step, and sulfuric acid is common, and nitric and oxalic acids are also included. The manufactured boards were tested for surface and average densities, dry IB strength, and chlorine and sulfur contents by elemental analysis. The boards were tested for tool wear by measuring the normal resistance force exerted on a fixed tungsten carbide knife blade set at a 15° rake angle to slice a thin layer of board continuously from a 23.3 cm-diameter circular specimen of board that is rotating on a lathe at a constant speed. *Four* circular boards were consecutively cut, resulting in a total cutting length of 304 m to read the final normal resistance force. Overall, sulfuric acid and ammonium sulfate were found to be the least tool-wearing acidic materials for use in synthesizing and catalyzing UF resins for PB. Boards bonded with UF resins having hydrochloric, nitric, oxalic, and formic acid components and PF resins containing sodium hydroxide showed significantly higher tool-dulling effects.

References

1 Gollob, L. (1990). *Proceedings of the NPA Resin and Blending Seminar*, 33. Gaitherburg, MD: NPA.

2 Rammon, R. (1990). *Proceedings of the NPAP Resin and Blending Seminar*, 15. Gaitherburg, MD: NPA.

3 Kennedy, H. (1990). *Proceedings of NPA Resin and Blending Seminar*, 20. Gaithersburg, MD: NPA.

4 Sellers, T. (2001). *Forest Products Journal* 51: 12.

5 Setera, S. (1990). *Proceedings of the NPA Resin and Blending Seminar*, 9. Gaitherburg, MD: NPA.

6 Resing, T. (1990). PB blending. In: *Proceedings of the NPAP Resin and Bending Seminar*, 49. Gaitherburg, MD: NPA.

7 Halloway, T. Blender efficiency. In: *Proceedings of the NPA Resin and Blending Seminar*, 52. Gaitherburg, MD: NPA.

8 Mecham, R.M. and Davis, C.R. (1991). Private Communication. Georgia-Pacific Corp. Crossett, AR.

9 Press, W.A. (1990). *Proceedings of NPA Resin and Blending Seminar*, 29. Gaithersburg, MD: NPA.

10 Carll, C. (1980). *Proceedings of 1980 Symposium of Wood Adhesives*, 213. FPL and WSU.

11 Kim, M.G., Stewart, H.A., and Wan, H. (1999). *Forest Products Journal* 49 (4): 60.

12 Wolcott, J.J., Motter, W.K., Daisy, N.K. et al. (1996). *Forest Products Journal* 46 (9): 62.

13 Wang, W. and Gardner, D.J. (1999). *Forest Products Journal* 49 (3): 65.

14 Cassens, D.L., Bradmueller, J.P., and Picado, F. (1994). *Forest Products Journal* 44 (10): 50.

15 American National Standards Institute (1987). ANSI/A208. 1-1987. Particleboard. NPA, Gaithersburg, MD.

16 Frashour, R. (1990). *Proceedings of the NPA Resin and Bending Seminar*, 62. Gaitherburg, MD: NPA.

17 Grigsby, W.J., Carpenter, J.E.P., Thumm, A. et al. (2015). *Forest Products Journal* 65 (1/2): 15.

18 Davis, C.R. (1995–2005). Invited Class Lectures. Georgia-Pacific Corp.

19 Suchsland, O., Lyon, D.E., and Short, P.E. (1978). *Forest Products Journal* 28 (9): 45.

20 Booth, C.C. (1959). *Forest Products Journal* 9 (6): 13A.

21 Black, W.R. (1975). Urea aldehyde resin adhesives and process. US Patent 3, 905,847. Borden, Inc., Columbus, OH.

22 Ellis, E.L. (1965). *Cellular Ultrastructure of Woody Plants* (ed. W.A. Cote), 181. Syracuse University Press.

23 Glasstone, S. (1946). *The Elements of Physical Chemistry*, 434, 482. New York: Van Nostrand Co.

24 Jacobson, N.S., McNallan, M.J., and Lee, Y.Y. (1986). *Metallurgical Transactions A* 17A: 1223.

25 Kim, M.G. and Amos, L.W. (1990). *Journal of Industrial and Engineering Chemistry Research* 29: 208.

26 Moeller, T. (1952). *Inorganic Chemistry*, 212–232. New York: Wiley.

27 Kim, M.G., Amos, L.W., and Barnes, E.E. (1990). *Journal of Industrial and Engineering Chemistry Research* 29: 2032.

28 Padilla, M.H., Rapp, R.A., and Stewart, H.A. (1991). *Forest Products Journal* 41 (10): 31.

29 Reid, A.S., Stewart, H.A., and Rapp, R.A. (1991). *Forest Products Journal* 41 (11/12): 12.

30 Stewart, H.A. (1985). *Forest Products Journal* 35 (11): 4.

31 Stewart, H.A. (1988). *Proceedings of the 19th International Wood Machining Seminar* (ed. R.L. Lemaster), 320. Berkley: UC.

32 Stewart, H.A., Shantynski, S.K., Harbinson, B., and Rabin, B. (1986). *The Carbide and Tool Journal* 18: 2.

33 Black, W. R. (1972) Urea-Aldehyde Resin Adhesive Process. US Patent 3, 697, 355. Borden Company.

8

PF Novolac Wood Adhesive Resins (Manufacturing and Chemistry)

The facile reactivity of phenol toward formaldehyde in the presence of acid or base was first discovered in 1872 by Adolf Bayer, and the reaction was named as Lederer-Manasse reaction [1, 2]. Based on this reaction, Leo Baekeland first patented the *Novolac*-type phenol-formaldehyde (PF) resins in 1909, the first synthetic plastic material [3, 4]. In the early period, the main aim of research was on producing artificial shellac materials and met the difficulty of brittleness, which was overcome by adding fillers, curing under pressure, and heating, resulting in the first man-made plastic material. The underlying chemistry of PF resins was elucidated thereafter in 1940s through the concerted research carried out by Zincke and Hultzsch group in Germany, Euler group in Sweden, and many other contributors. The results of these pioneering works were summarized in several books and journals published since 1950s [5–13]. The *Resole*-type PF resins were also conceived in this period and practical applications were reported in several patents [14, 15]. Thus, there are two types of PF resins: *Novolacs and Resoles*. The current major uses of novolac PF resins are as heat-resistant molding compounds used in manufacturing of pot handles, etc., and as bonding and filling materials in the manufacturing of light bulbs and electric switches. Some PF novolac resins have been used as binders of wood composite boards, but their use has been declining due to resole PF resins, which began to be used widely in bonding of softwood plywood [14, 15]. In this chapter on PF novolac resins, we will discuss the general principles of the novolac PF resin chemistry and technology.

8.1 Raw Materials of PF Resins

Phenol is manufactured by oxidizing benzene or toluene [16], which are obtained as aromatic components from crude oil in the petroleum refining process, where gasoline, diesel oil, paraffin wax, and many other products are also obtained. The aromatic components, like benzene and toluene, are important components of gasoline, especially in premium grades. The price of phenol, therefore, fluctuates according to the price of crude oil. The manufacturing of phenol from benzene or toluene is currently done on large scales because phenol is widely used in the manufacturing of not only PF resins but also epoxy resins, polycarbonate resins, etc. Polycarbonates are widely used as engineering plastic materials in molded forms (Lexan and soft drink bottles), and epoxy resins are used as matrix binders for high-performance composite materials. Currently, the world production volume of phenol is large due to these various large volume uses, and the

Adhesives and Finishes for Wood: For Practitioners and Students, First Edition. Moon G. Kim.

supply and demand are normally well balanced worldwide as with many other commodity chemicals. Occasionally, a wide price fluctuation of phenol occurs when an unexpected plant shutdown occurs such as by a safety accident. Phenol is relatively cheaper in the United States than in Europe or Japan due to our petroleum import policy and therefore more extensively used in wood adhesives in the United States than in these other countries. Africa, Australia, and South America import most of phenol they use, and therefore phenol is rather expensive as a wood adhesive raw material in these regions.

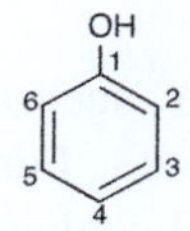

Phenol is a severe poison. It is a solid at room temperature with a pungent odor, and it is handled more conveniently in industry as a molten liquid form by keeping it above the melting point of 42 °C, at about 50–60 °C, or as 90% solutions in water (liquefied phenol). Molten phenol or any phenol solutions in water are similarly dangerous materials for bodily contact, and their handling requires rubber gloves, protective clothing, and breathing and eye protection masks. It should be noted here that most PF resole and novolac resins, although made from poisonous phenol and formaldehyde, are not a direct poison in skin contact and this somewhat benign characteristic is one of the reasons PF resins are widely used as wood composite binders.

Formaldehyde, the second component in PF resin manufacture, is discussed in Chapter 5 (Section 5.1.1). One additional comment to make is that the current price of natural gas or methanol, the precursors of formaldehyde, became linked to the price of crude oil because major uses of natural gas and crude oil are the same: transportation and heating fuel.

$$CH_2{=}O$$

8.2 Reaction Chemistry of Phenol

Phenol molecule has one hydroxyl (–OH) group on the 6-carbon benzene ring, which is sometimes called an "aromatic" ring because this ring is the basic structure of many organic chemicals that are aromatic, indicating strong odors, sometimes pleasant, associated with them. This benzene ring structure occupies a special position in organic chemistry due to its structure's sturdy nature; the ring does not break or open up easily and acts as one unit in many organic transformation reactions. The six carbons of phenol molecule are numbered, as shown in Figure 8.1, carbon 1 being the one that is bonded to hydroxyl group. Carbons 2 and 6 are also called *ortho*-carbons, carbon 4 as *para*-carbon, and carbons 3 and 5 as *meta*-carbons. The hydroxyl group (–OH), or phenolic hydroxyl group, shows a weak acidic character, different from those hydroxyl groups bonded to aliphatic carbons, due to the benzene ring's *electron-withdrawing* nature in combination with the phenolic hydroxyl group oxygen atom's *electron-releasing* tendency toward the benzene ring. The electronic charge on phenolic oxygen atom thus delocalizes into the benzene ring and accumulates mainly on the *ortho- and para-carbons* but not on *meta*-carbons. The accumulated electronic charge makes both *ortho*- and *para*-carbons

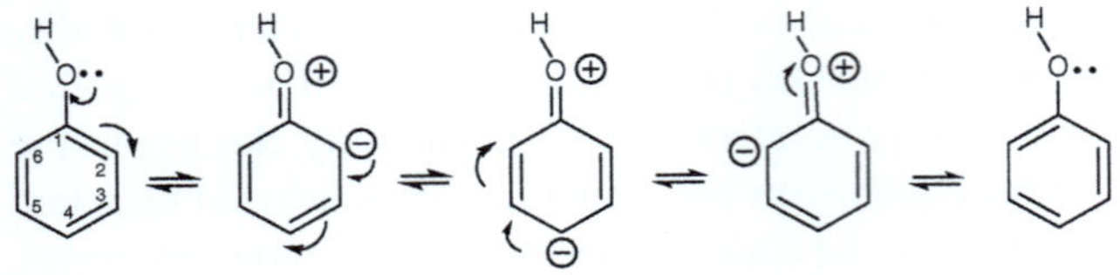

Figure 8.1 Phenol molecular activation structures: carbons 2, 4, and 6 become electron-rich both in the acidic and basic media, and the carbons react selectively with electrophilic formaldehyde molecule. Carbons 3 and 5 are not activated.

accept the incoming electrophilic agents, such as formaldehyde, in both neutral/acidic and alkaline media, as shown in Figure 8.1. Furthermore, phenolic hydroxyl group can ionize to produce hydrogen ion, H^+, and conjugate phenolate anion, prominently in the basic medium, such as by addition of sodium hydroxide (NaOH), where the hydrogen ion is neutralized to form water. In this case, the anionic charge on the phenolate anion is more extensively delocalized into the benzene ring, i.e. *ortho*- and *para*-carbons become more electron-rich carbon sites, as shown in Figure 8.1 (bottom), i.e. become more reactive nucleophiles toward incoming electrophilic agent, such as formaldehyde. The reaction with formaldehyde proceeds more efficiently in the presence of a base and somewhat more slowly in the neutral or acidic medium. Another electrophilic agent that arises in PF resin chemistry, except formaldehyde, is the *o*- or *p*-benzylic carbenium ion formed in the second stage of reaction, and they similarly go into reactions only with *o*- and *p*-carbons of another phenolic molecule (discussed below). These base-catalyzed reactions are the basis of manufacturing of PF resole resins, while the acid-catalyzed reaction is the basis of manufacturing of novolac PF resins. Furthermore, the *para*-carbon is appreciably more reactive than *ortho*-carbons, and this reactivity difference and the fact that *ortho*- and *para*-carbons are "structurally" different in positions of benzene ring leads to PF polymers having molecular structures and molecular weights that are too complex to depict in a simple formula.

8.3 Synthesis and Chemistry of Novolac Phenol-Formaldehyde (PF) Resins

Synthesis procedure of novolac PF resins consists of reacting formaldehyde and phenol in the presence of an acid catalyst at an F/P mole ratio less than 1.0 in theory, but the ratio most often used in practice is less than about 0.85 [17]. The reaction is normally done in a reactor shown in Figure 8.2 and starts with charging phenol and some water in the reactor, and stirring and heating are initiated to go to about 90–100 °C in the presence of a strong acid catalyst such as sulfuric acid at about 0.5% level based on the batch weight. Then the formaldehyde solution is added in small amounts over a period of time in such a way that the exothermic heat from reaction can be kept under control at the reaction temperature with the help of external cooling coil and a condenser. After the formaldehyde addition is completed, the reaction temperature is *slowly* increased up to about 180 °C while slowly distilling out the water and water of condensation reaction. The unreacted phenol begins to distill at around 180 °C and then, a vacuum is applied to remove much of the unreacted phenol remaining in the reaction mixture.

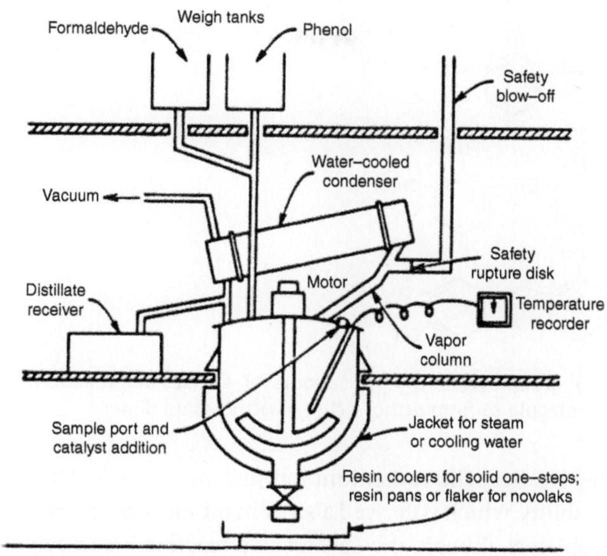

Figure 8.2 Phenol-formaldehyde resin manufacturing unit. Source: Kirk-Othmer Encyclopedia of Chemical Technology, 1985. Courtesy of Wiley-Interscience, Inc.

In this initial reaction period, the phenol's electron-rich *ortho-* and *para*-carbons react with the electrophilic formaldehyde to form *o-* and *p*-hydroxymethyl-phenols (methylolphenols), as shown in Figure 8.3. Reaction heat is evolved (exothermic) and it must be timely dissipated by cooling, or an unsafe situation can arise. Adding formaldehyde in a continuous manner over a period of time at a temperature where the reaction occurs quickly, usually between 90 and 100 °C, helps control the rate of heat evolution. The reaction heat, thus generated in a controlled manner, is removed both by the cooling coil that goes around inside the reactor and cooling condenser connected on the top of the reactor, through which cold water is passed whenever cooling is needed. Also, the reactor is normally kept under a mild vacuum to facilitate the evaporation of water in the kettle. The water vaporizes goes to the condenser, captured and cooled, and returned to the reactor or siphoned off for removal.

Polymerization of hydroxymethylphenols: Further, in this acid-catalyzed regime of cooking, as shown in Figure 8.3, the hydroxymethyl groups of *o-* and *p*-hydroxymethylphenols become reactive by forming cationic intermediate ion species, benzylic carbenium ion. The acid (H^+) acts as a catalyst that initially attacks the oxygen atom of benzylic hydroxyl group and forms a complex of a water molecule, which is broken off to leave the reactive benzylic

Figure 8.3 Activation of ortho and para positions of phenol and reaction with formaldehyde to form *o-* and *p*-methylolphenols.

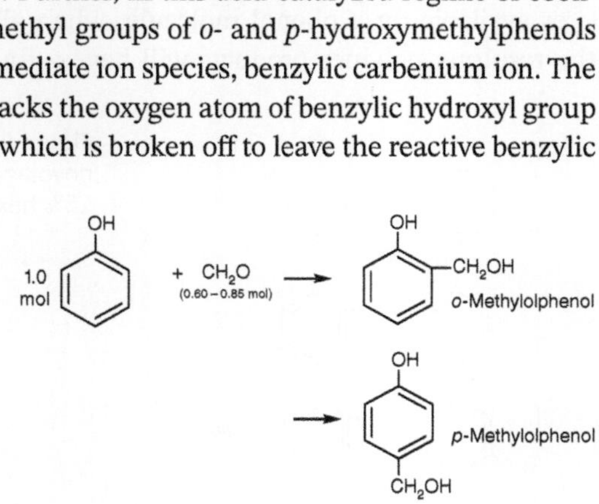

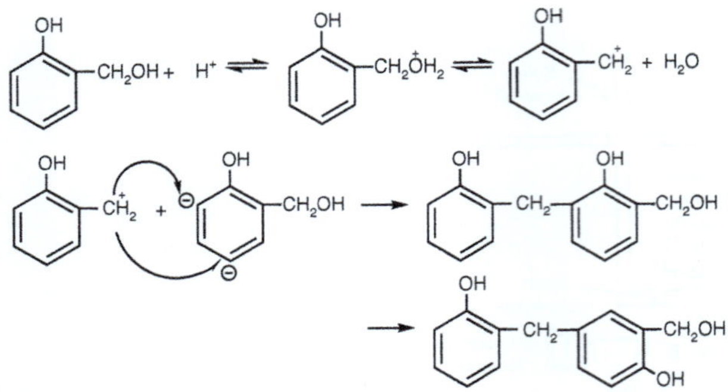

Figure 8.4 Activation of *ortho*-methylol group by acid to form cationic ion species and reaction with *ortho*- or *para*-carbon of another molecule to form ortho–ortho or ortho–para dimer.

cation. The facile formation of reactive benzylic carbenium ion intermediates is derived from the resultant ion's enhanced stability, which is derived also from the electron-releasing effect of phenolic oxygen atom. In practice, though, this second-step reaction is somewhat sluggish at 100 °C in presence of water but accelerates at higher temperatures and in the absence of water. By continuously increasing the reaction temperature up to about 180 °C in resin manufacturing, the intermediate cationic ions thus formed keep attacking *ortho*- or *para*-carbons of other phenolic molecules, again due to the activation of the *ortho*- and *para*-carbons by the phenolic oxygen atom, to form methylene bonds between two phenolic rings by replacing the hydrogens, as shown in Figure 8.4.

Further polymerization: The resulting molecules initially formed are *dimers*, i.e. two phenolic rings are bonded via a methylene group, and there are three different dimers: ortho–ortho (2,2′-), ortho–para (2,4′-), and para–para (4,4′-) methylene bonds between two phenolic rings as shown in Figure 8.5 [19, 20] with their gelation times measured at 160 °C after mixing with 15% weight of hexamethylene-tetramine [18]. In resin synthesis, the dimeric molecules can have another hydroxymethyl group on them because the F/P mole ratio used is higher than 0.5, i.e. 0.60–0.85, which allows to repeat the formation of the cationic intermediates and attacking another phenolic ring to form trimers, tetramers, etc. This is *the polymerization reaction*, which is reflected in increases in viscosity of reaction mixture. Due to the F/P ratio used in the synthesis and due to the three reactive sites available on a phenol molecule, when all methylol groups have reacted this way, the resulting polymer products will have a finite (average) degree of polymerization or

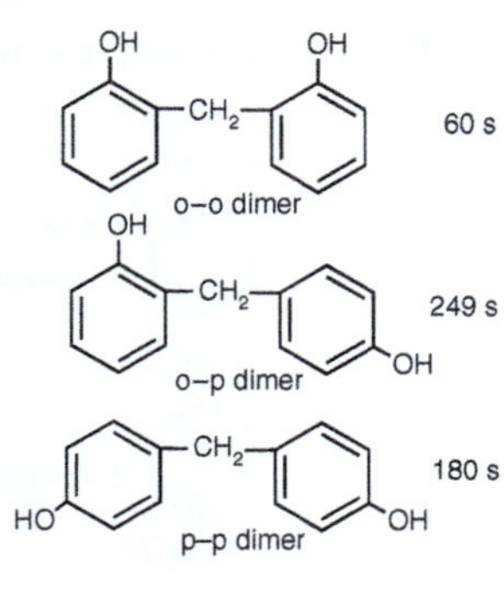

Figure 8.5 Three different dimer structures of novolac PF resins and their gel times reported with 15% hexamine at 160 °C [18].

molecular weight. The resulting resins would be stable (nonreactive) and melt without chemical reactions occurring, although they still entail some unreacted, potentially reactive *ortho-* and *para*-carbon sites on the polymeric phenolic rings.

Finishing of resin manufacturing: Stopping the reaction before the complete consumption of all hydroxymethyl groups can happen in practice, but this must be minimized to avoid resin instability problems in subsequent processing steps. It is also useful to recognize that use of a higher F/P mole ratio leads to more hydroxymethyl groups and higher extents of polymerization, resulting in resins of *higher melting points*. Melting points become higher as the degree of polymerization increases. Higher melting point resins can give shorter cure times, but they can also show limited flow capability in molding process and vice versa. Also, since some unreacted phenol is distilled out in the resin synthesis process, actual reacted F/P mole ratio values in finished resins would be higher than the charge values, and the extent of phenol removal can depend on the reaction procedure and therefore the molecular weights or melting points need to be measured experimentally. Another characteristic of polymerization is that one final resin product is composed of polymer molecules having a range of molecular weights, i.e. a normal distribution of molecular weights with its median value increasing with the charged F/P mole ratio values. This molecular weight distribution can be known only by measurements, such as by *gel permeation chromatography (GPC)*. Thus, the melting point is also an averaging result of molecules of varying molecular weights. The resin manufacturing is completed by pouring out the hot molten resin in the reaction kettle rapidly onto a pan and cooling it, resulting in solid novolac resin chunks, which are then crushed and pulverized.

Properties and structures of novolac PF resins: Although the resin synthesis is to be carried out until all hydroxymethyl groups are consumed and all free phenol is distilled out, small amounts of unreacted phenol can remain in the resins. Reported molecular weights of *novolac PF resins* range from a few hundred to a few thousand Da [17]. ^{13}C NMR spectroscopy gives information on the chemical structures of novolac PF resins, such as molecular weight and reacted mole ratio. A schematic example is shown in Figure 8.6 [19] taken in dimethylformamide (DMF) solution: C_1 (phenolic) carbons at about 157 ppm are separated into four groups of peaks: I_4, that of phenol itself; I_3's, those of mono-substituted phenols; I_2's, those of di-substituted phenols; and I_1, those of tri-substituted phenols. It is noted that there are some tri-substituted phenolic rings in this relatively low-advanced novolac resin, indicating the presence of branched polymers. Peak k_2 at about 133 ppm is assigned to C_4's of 4,4′-substituted phenol rings, and peak k_1 is assigned to C_4's of 2,4′-substituted phenol rings. In the regions represented by peak i's at about 130 ppm are asigned to C_3's and substituted C_2's. Peak i_4 was identified as C_3's of phenol or of 4,4′-dimer. All substituted C_2's appear in a very narrow range, 127.70–128.22 ppm. Unreacted C_2's and C_4's are readily assigned to peaks at about 116.0 ppm (peak f) and 120.9 ppm (peak g) ppm, respectively. The methylene peaks are at about ~30 ppm for ortho–ortho, ~35 ppm for ortho–para, and ~40 ppm for para–para methylene bonds (as shown expanded and integrated), well separated and quantifiable. There are no hydroxymethyl functional groups found in the 60–72 ppm region as a novolac should be. Thus, the final structures of novolac resins formulated from the ^{13}C NMR data, includes mono-, di-, and tri-substituted phenolic rings connected with o—o, o—p, and p—p methylene bonds, and can be represented as "poly-methylene-phenols" as shown in Figure 8.7 [19].

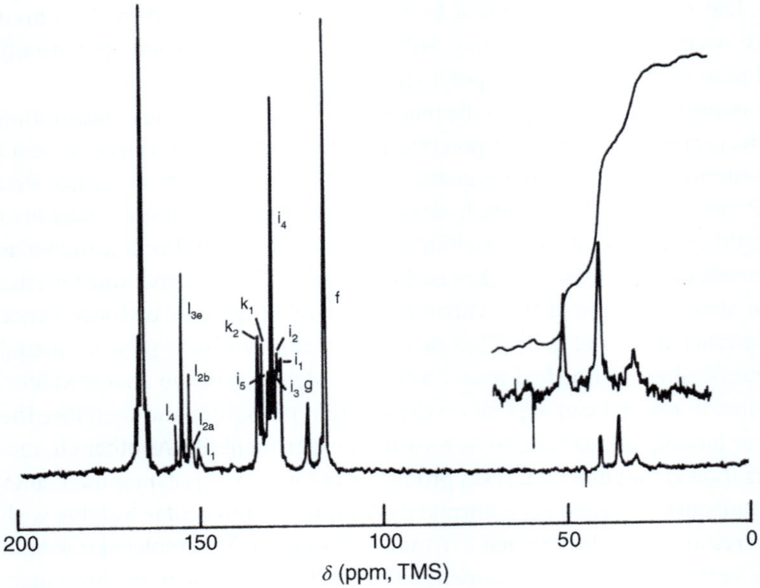

Figure 8.6 ^{13}C NMR spectrum of a 0.76 mol ratio novolac PF resin prepared with phosphoric acid catalyst, in dimethylformamide (DMF) solution at 166.2 ppm [19].

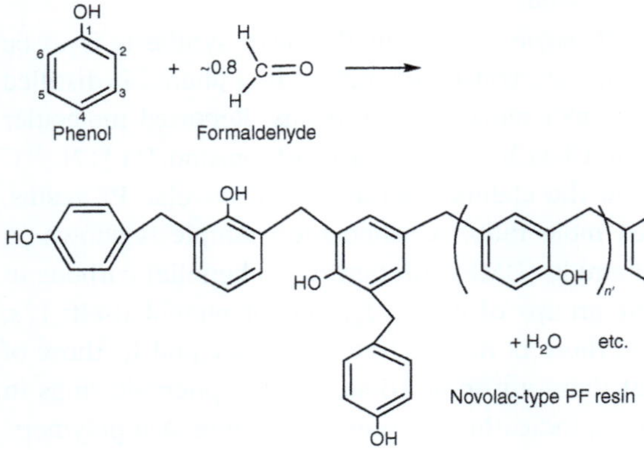

Figure 8.7 Formation of PF novolac resin molecules.

Calculation of degree of polymerization for PF resins

$$\text{Degree of polymerization} = \frac{1}{(1 - \text{Reacted F/P ratio})} \tag{8.1}$$

Number-average molecular weight of novolac PF resins: Detailed quantification of mono-substituted phenolic rings vs. di- and tri-substituted phenolic rings can give the number-average molecular weight or degree of polymerization values. Similarly, quantification of total methylene and total phenolic groups gives the *reacted CH_2/Phenolic (F/P) mole ratio values*, i.e. degree of polymerization, as shown in Eq. (8.1). Further, quantification of three different methylene groups would result in a general description of the shapes of PF resin polymer chains. In the past, there was a good deal of effort aimed at finding synthesis procedures for novolac PF resins having more o–o methylene groups, since the unreacted C_4 carbons then could result in fast-curing resins [20]. Also, as an example, use of Eq. (8.1) is illustrated: for the molecule in Figure 8.6, the reacted F/P mole ratio value of 5/6 measured by ^{13}C NMR, then the degree of polymerization becomes 6. This type of number-average degree of polymerization (DP) value is affected steeply by the reacted F/P mole ratio values, becoming 100 at F/P ratio of 0.99. Here, we find the difficulty of controlling the DP values accurately in resin syntheses only by the charge F/P ratio values used, although the reacted F/P ratio value is generally proportional to the charged F/P ratio; thus, the melting point of novolac resins increases with charged F/P ratios and a series of trial-and-error type experiments can lead to desired reacted F/P ratio values and desired melting points. Melting point is an important variable in the compounding operation of novolac resins with hexamine and in subsequent molding and curing operations. Conclusively, one key point in the resin manufacturing procedure is that the finished resins consist of only low molecular weight polymers (oligomers) similarly as for most other thermosetting resins. The high molecular weight, cross-linked resin products, and adhesives are obtained only by further polymerization and cross-linking with additional amounts of formaldehyde in the subsequent molding or hot-processing steps of boards or other products.

8.4 Compounding and Curing of Novolac PF Resins

Compounding of novolac PF resins: In novolac PF resin molecules, there are numerous *ortho-* and *para*-carbons remaining unreacted on the polymeric phenolic rings and they are the potential reaction sites for final curing and cross-linking reactions. Since there is no methylol group present in the resin, the resin is stable and melts upon heating without any further polymerization reaction to occur. Of course, the melt viscosity, which is an important parameter in the subsequent compounding and molding processes, is affected greatly by the molecular weight, i.e. F/P ratio used in resin manufacturing. Though, it is apparent that an additional amount of formaldehyde is needed to cure the novolac PF resins. Making novolac resins curable by adding a formaldehyde-releasing material and fillers constitutes the *compounding process*. Practical considerations such as water insolubility of novolac resins and the fact that the resin is a solid material do not allow using aqueous formaldehyde solutions. Hexamethylene-tetramine {hexamine or hexa, $(CH_2)_6N_4$}, a solid material, produced by reacting formaldehyde and ammonia (see Chapter 6), is suited for this purpose, because of its relatively good stability at temperature below about 160 °C, and then above that temperature it decomposes to formaldehyde and ammonia. Thus, a mixture of powdered novolac resin, powdered hexamine, in an overall F/P mole ratio of up to about 1.40, and filler, dye, and lubricant, etc., as shown in Table 8.1 as an example, are mixed and passed through a heated roll mill or an extruder repeatedly at about up to 140 °C until the mixture becomes a semi-fluid, well-dispersed uniform mixture. The reaction of novolac resin with hexamine is allowed to occur to some extent, so controlled that the compounded product can be later molded with a short, target molding time. The fluid uniform mass is then cooled to solid chunks, which are then crushed and ground using a milling machine (usually ball mill) into fine powders. Use of varying novolac resins, control of resin level with respect to hexamine and filler levels, and adjustment of compounding temperature and time leads

Table 8.1 Mixing components in parts of a typical novolac molding compound [18].

Novolac resin	100
Hexamine	12.5–15.0
Wood flour	110–125
Lime or MgO	2.0–2.5
Lubricants	2.0–2.5
Colorants/pigments	2.0–2.5

to molding compounds having desired molding speed and strength performances. Some unreacted phenolic components and hexamine still remain in the molding compounds, which are then shipped to molders, who put them in molds and apply heat to about 200 °C or higher under pressure to fully cure the molding compounds into desired shapes of articles.

Reactions occurring in the final molding step consist of the hexamine reacting with phenolic molecules to form intermediates that eventually decompose to give off ammonia, and the residual formaldehyde makes bonds with unreacted *ortho*- and *para*-carbons of the novolac resin molecules to form methylene bonds, resulting in advancing/curing of the resin/filler system. The ammonia escapes out, but the molding pressure keeps the article from blistering. The number of reactive sites available on phenol molecule (3) makes the polymer chain grow to form branches and crosslinks, resulting in products that are insoluble in water or in any organic solvents and they are infusible and highly resistant to heat. Although detailed structure of cured PF novolac resins is not completely determinable presently, the schematic crosslinked structure in Figure 8.8 entailing intramolecular methylene bonds that form rings of five or six phenolic molecules was proposed by K. Hultzsch and earlier researchers [22], an approximation toward the perfectly well-cured novolac PF resin structure, which is convenient to visualize but considered difficult to be attained in practice [5]. There could be numerous missing methylene bonds and kinks in polymer chains that can act as defects, potential initiating points of breakage when the sample is stressed.

8.5 Curing Speed and Testing Methods of Molding Compounds of PF Novolac Resins

The curing speeds of novolac molding compounds depend much on the degree of polymerization and the polymer chain structures, i.e. the ratio of o—o (2,2′) methylene bonds relative to the o—p (2,4′) and p—p′ (4,4′) methylene bond contents and the molecular weights, resulting in various short, medium, and long hardening times as shown in Figure 8.9 [18]. It is noted that resins having high 2,2-methylene bond contents would have more of the faster-reacting *para*-carbons available to result in faster hardening speeds. Fillers normally constitute more than 50% of the mix, and cellulose and wood, and nutshell flours are the general-purpose filler materials, which serve to reduce the shrinkage of curing and improve the impact strength. Inorganic fibrous materials, such as glass fiber, clay, talc, and mica, are also used for imparting heat, impact, and electrical resistances. Asbestos fibers were

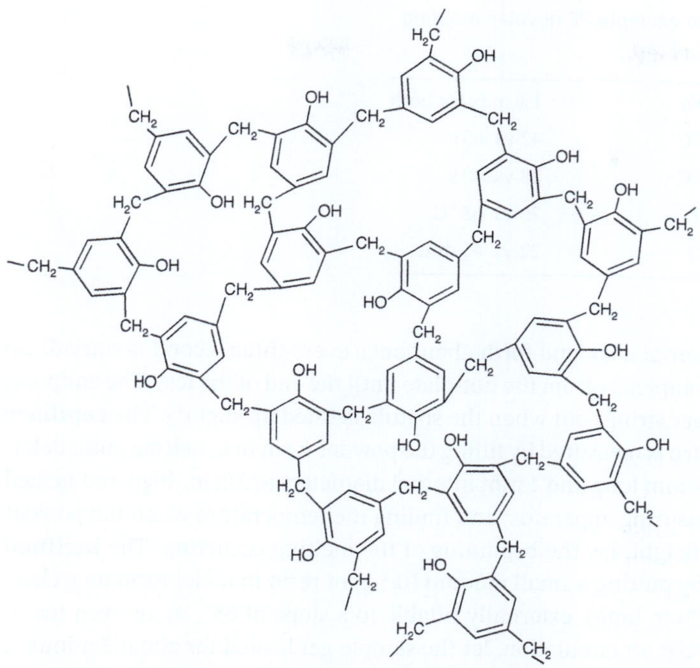

Figure 8.8 Schematic representation of cured PF novolac resins fully reacted and by forming five- and six-membered rings of phenol molecules. Source: Adapted from Hultzsch [22].

Figure 8.9 Hexa-hardening speeds of solid novolac PF resins, fusible enough to flow. Time in seconds required for resins to harden at 160 °C relative to the 2,2′ methylene bond content vs. combined 2,4′ and 4,4′ methylene bond content [18].

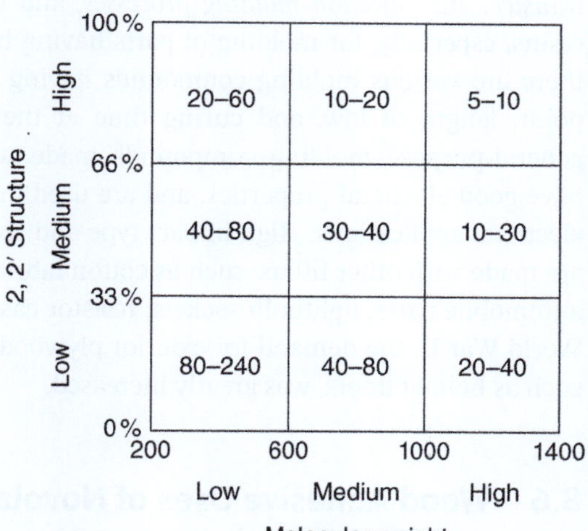

also used in the past but were discontinued for health reasons. These cured novolac resin materials have particularly good heat and water resistance properties.

Testing methods of molding speed: The curing and other processing characteristics of two example commercial PF novolac compounds are shown in Table 8.2 [18]. The **hot-plate cure time**, often called as stroke-cure time, is measured by putting about 0.5 g of solid resins or about 1.0 g of liquid resin on a chromium surfaced hot-plate, maintained at given temperature, and spreading in an area of 1.5 in. × 3 in. with a stainless spatula. The clock is started

Table 8.2 Properties of two example PF novolac molding compounds [18].

Screen analysis on 200 mesh	1% max for both
Hot plate cure time at 150 °C	42 vs. 60 s
Hot plate cure time at 165 °C	28 vs. 40 s
Capillary contraction point	70 vs. 65 °C
Inclined plate flow (125 °C)	22 vs. 45 mm

and stroking of resin material back and forth about once every three second is started. Do not remove the spatula completely from the hot-plate until the end of the test. The endpoint is when the resin no longer strings out when the spatula is lifted up slightly. The **capillary contraction temperature** is measured by filling the powder resin in a melting point determining capillary tube (40 mm long and 1 mm internal diameter) to 3/8 in.-high and heated on the melting point measuring apparatus, and finding the temperature when the powder resin starts to shrink in height, i.e. the beginning of the melting occurring. The **inclined plate flow test** is done by putting a small amount (0.5 g) of resin in tablet form on a clean horizontal glass plate (25 cm long), externally tiltable to a slope of 65°, in an oven maintained at 125 °C with forced air circulation, let the sample get heated for about 3 minutes, and then tilt the plate to 65° and continue heating for exactly 20 minutes, remove the glass plate and cool, and measure the length of flow of molten resin.

Various molding processes: Three molding processes are commonly used: *compression*, *transfer*, and *injection molding processes*, and the latter two methods require long flow resins, especially, for molding of parts having both thick and thin-walled sections. Thus, there are various molding compounds having varying curing characteristics of melting point, length of low, and curing time at the (high) molding temperatures [12]. The general-purpose molding compounds made with wood flour fillers are low cost and have good electrical properties, and are used in molding utensil handles and low-voltage electrical applications. High impact type and special electrical grade molding compounds are made with other fillers, such as cotton fabric, glass, rubber, and mica, and are used for automobile parts, lightbulb sockets, resistor cases, and special electrical switches. During World War II, the demand for exterior plywood and other phenolic molding compounds, such as helmet liners, was greatly increased.

8.6 Wood Adhesive Uses of Novolac PF Resins

Use as oriented strand board (OSB) Binders: Finely powdered novolac PF resin molding compounds made with minimal amounts of fillers are used as binders in the waferboard and toilet seat manufacturing industries. The novolac resin binders need to be selected based on the temperature of hot-pressing and curing times, etc. Currently, the novolac adhesive powder is dosed into the rotating blender wherein the dried and wax-sprayed wood wafers or particles are tumbling around. The novolac resin particles are fine particles and they are attracted to the surface of dry wood substrates by static (electric) forces and onto the sticky spots of sprayed wax deposits. In most cases, a uniform adhesive distribution on wood surface is obtained, but the applicable amount of adhesive is limited, and in case of waferboard manufacture, up to about 2% based on wood weight, because the adhesive particles do not adhere to each other. Excess resin particles, if applied, would fall to the bottom of the blender. In the blender, the tumbling of wafers needs to be random as much as possible for a good distribution of powder adhesive, and the randomness is more feasible when the wood wafers are square rather than rectangular strands. It was soon found that square wafers give less stiff boards than oriented rectangular strands and therefore most waferboard plants were transformed into strand board plants also with their name changed to oriented strand board (OSB). Some OSB plants use powdered novolac or *powdered resole PF resins*. The average curing temperature of novolac resins is 160–180 °C [21]. The resin loading level, up to about 2% based on dry wood weight, is somewhat low, and strengths of these boards are somewhat inferior to boards bonded with *liquid* resole-type PF resins, which are loaded at levels of 3% or higher. Also, the novolac powder adhesives are being replaced by spray-dried resole-type PF resins because the latter can be manufactured at lower costs. Remember that the compounding process of novolac resins includes dry-mixing, melting, cooling, and grinding steps, which are generally messy and therefore expensive. Powder-form PF novolac and resole resins have an advantage of carrying no water in them, allowing a better control of mat moisture content in OSB manufacturing and a lower cost in transporting to long distances, and the binders have longer shelf-lives than liquid PF resins.

Other uses of novolac PF resins: A development reported in a patent literature indicated that using un-dried, liquid novolac PF resins as an added component in manufacturing of spray-dried PF resole resins imparts better flow/curing properties [23]. In this approach, a novolac PF resin is made but stopped when the distillation of water would begin and then cooled, and it is added, at an appropriate ratio, to a PF resole resin being cooked in alkaline pH (see Chapter 9). The resultant mix will become one phase (uniform) due to the sodium hydroxide content in the resole resin part and, this uniform resin mix is spray-dried to obtain powdered PF resins. In this case, the curing of the novolac resin components is possible because of the hydroxymethyl groups supplied from the resole resin components. By adjusting the mix ratio of novolac and resole resins, resin molecular weights, and spray-drying temperature and time, the patent claims, the resultant PF powder resins perform better in bonding OSB under the higher mat moisture content conditions than normal.

8.7 Summary

The acid-catalyzed manufacturing and basic chemistry of novolac PF resins and molding compounds made with hexamethylene tetramine (hexamine) for uses in molding of various thermoset articles are presented. Phenol is activated on its *ortho-* and one *para*-carbons, and the reaction is commonly carried out with dropwise addition of less than 0.85 mol of aqueous formaldehyde to one mole of phenol in the presence of 0.5–2% sulfuric acid and at temperatures between 90 and 100 °C, where *ortho-* and *para*-hydroxymethyl phenols are first formed. The reaction is then continued by increasing the reaction temperature up to about 180 °C while removing the water and unreacted free phenol as distillates, where ortho–ortho, ortho–para, and para–para methylene bonds are formed between phenol molecules (polymerization) to form dimers, trimers, etc. until all hydroxymethyl groups are consumed, followed by cooling to room temperature, resulting in solid PF novolac resins consisting of lightly branched oligomeric "poly-methylene-phenols" having a range of molecular weights but distinctive melting points that depend much on the F/P mole ratio and the history of cook temperature. Then, the novolac resin chunks, fillers, and hexamine are melted together at an elevated temperature, thoroughly mixed and advanced to the desired extent of polymerization, and cooled to room temperature, resulting in solid chunks that are again milled to fine powders, yielding the "novolac molding compounds," available to molders in various grades of curing speed and applications. During the compounding, the combined F/P mole ratio is often made to go up to about 1.40 with hexamine, and the mixing temperature and duration are controlled in such way that a limited extent of reaction occurs between resin's unsubstituted *ortho-* and *para-*carbons and formaldehyde, so that the compounded material increases in molecular weight to have the target molding time at the molding temperature. The remaining reactivity of the molding compounds is cured completely in the final molding step, used in manufacturing various thermoset articles including binders for wafer boards at a loading level of up to about 2% based on dry wood weight.

References

1 Bayer, A. (1872). *Bericht* 5: 25–56, 280–282, 1094–1100.

2 Schnell, H. and Krimm, H. (1963). *Angewandte Chemie International Edition in English* 2: 373.

3 Baekeland, L. H. (1909). Method of making insoluble products of phenol and ormaldehyde. US Patent 942,699.

4 Baekeland, L.H. *Journal of Industrial and Engineering Chemistry* 1: 149.

5 Carswell, T.S. (1946). *Phenoplasts: Their Structure, Properties, and Chemical Technology*. New York: Interscience.

6 Hultzsch, K. (1950). *Chemie der Phenolharze*. Berlin: Springer-Verlag.

7 Robitschek, P. and Lewin, A. (1950). *Phenolic Resins: Their Chemistry and Technology*. London: Iliffe.

8 Reese, J. (1955). *Kunststoffe* 45 (4): 137–145.

9 Martin, R.W. (1956). *The Chemistry of Phenolic Resins*. New York: Wiley.

10 Megson, N.J.L. (1958). *Phenolic Resin Chemistry*. New York: Academic Press.

11 Whitehouse, A.A.K., Pritchett, G.K., and Barnet, G. (1967). *Phenolic Resins*. New York: Elsevier Press.

12 Knop, A. and Pilato, L.A. (1985). *Phenolic Resins*. Berlin: Springer-Verlag.

13 Pizzi, A. (1983). *Wood Adhesives, Chemistry and Technology*. New York: Marcel Dekker.

14 Van Epps, C.F. (1944). Water-soluble phenolic adhesive resins and method. US Patent 2,360,376.

15 Redfern D.V. (1953). Production of thermosetting phenol-formaldehyde resin condensation products. US Patent 2,631,098.

16 Saunders, K.J. (1988). *Organic Polymer Chemistry*, 2e, 317. New York: Chapman and Hall.

17 Whitehouse, A.A.K., Pritchett, G.K., and Barnet, G. (1968). *Phenolic Resins*, 149. New York: Elsevier Press.

18 Bender, H. 1954. Modern Plastics. March 115.
19 Kim, M.G., Tiedeman, G.T., and Amos, L.W. (1981). *Weyerhaeuser Co. Science Symposium*, 263. Tacoma, WA: Weyerhaeuser Co.
20 Kopf, P.W. and Wagner, E.R. (1973). *Journal of Polymer Science Part A: Polymer Chemistry* 11: 939.
21 Chow, S.Z. and Steiner, P.R. (1979). *Forest Products Journal* 29 (11): 49.
22 Hultzsch, K. (1950). *Chemie der Phenolharze*, 7. Springer-Verlag.
23 Udvardy, O.D., Titus, P.E., and Navratil, M. (1984). Powdered phenol-formaldehyde resins. US Patent 4,424,300. Borden, Inc., Columbus, OH.

9

PF Resole Wood Adhesive Resins (Manufacturing and Chemistry)

9.1 Alkaline PF Resole Wood Adhesive Resins

Wood adhesive PF resole resins are made by reacting phenol and formaldehyde in water in the presence of a base catalyst with an F/P mole ratio in the range of 1.4–2.5, which is different from novolac PF resins discussed in Chapter 8. Acid catalysts are not used because the polymerization reaction with these high F/P mole ratios can lead to gel formation even if the reaction is to be conducted at low temperatures. In the alkaline pH, once the initial hydroxymethylation reactions are completed, the reaction goes relatively slowly and can be stopped at an appropriate extent of polymerization simply by cooling the reaction mixture to room temperature. The resulting resins are called *phenol-formaldehyde (PF) resole resins*, and the polymerization reaction is resumed in the curing step by heating back to a high temperature without adding another shot of formaldehyde. For this reason, PF resole resins are called *one-step PF resins* in comparison to the two-step novolac PF resins. They began to be used widely as binders for softwood plywood after several patents were published in the 1940s [1, 2]. PF resole resins have limited but useful shelf-lives at room temperature and are normally shipped and stored at relatively low temperatures of 10–25 °C. The raw materials of PF resole resins represent a significant portion of the total production cost of resins, as much as 90%, and a one-cent increase in the price of phenol results in 0.5–0.6 cent increase in the cost of liquid PF resole resin-solids. Currently, most PF resole resins are produced by high-volume resin manufacturers, and the production volume in the United States generally follows the national GNP and housing starts since the phenolic resins' largest outlet is wood composite products and glass wool insulation materials [3, 4]. PF resole and novolac resin chemistry and technology have been written up in several excellent books and journal articles [5–14] from which much of the description in this chapter is derived. The author's personal experiences gained in these industries and research are also included without referencing.

9.2 Synthesis Chemistry and Manufacturing Practices of Alkaline PF Resole Resins

Wood adhesive PF resole resins are synthesized using one mole of phenol and 1.4–2.5 moles of formaldehyde, which was found to be a necessary condition in arriving at fast-curing resin materials [1, 2]. Also, a strong base catalyst, usually sodium hydroxide, is added

Adhesives and Finishes for Wood: For Practitioners and Students, First Edition. Moon G. Kim.

initially at up to a NaOH/Phenol mole ratio of about 0.20, and the level is often increased in the course of resin synthesis up to 0.60 NaOH/P mole ratio depending on what products the adhesive resin binder is destined to be used for [15]. The industrial reactor is similar to the reactor used in manufacturing of UF or novolac PF resins (Section 8.3). It is a big stainless-steel drum, connected to phenol and formaldehyde charging pipes and jacketed with insulation and steam-heating/cooling coils around inside or outside the wall of the reactor. It is also equipped with a stirrer, thermocouple, and sampling and discharge ports. A cooling condenser is connected to the top of the reactor, which is further connected to a vacuum system. The cooling coil and cooling condenser are circulated with (cold) tap water when a cooling of the reaction mixture is needed. Cooling coil directly cools the reaction mixture, and cooling condenser cools the hot water vapor coming off from the reaction mixture by forming condensates that flow back into the reaction mixture. Vacuum is often applied through the condenser when the evaporation and condensation process is expedited to achieve a faster cooling.

Initial Condensation Step: In the beginning of resin manufacturing, the agitation is started, all phenol and a part of total sodium hydroxide are added, and the mixture is heated to about 60–70 °C. Then, after applying a slight vacuum in the reactor and with some cooling on, 50% formaldehyde solution is added in small portions over a measured period of time, normally 30 minutes to an hour or more depending on the batch size, to prevent the exothermic heat of reaction from arising too much too fast. Cooling is applied through the cooling coil and cooling condenser during most of the formaldehyde addition period. After the formaldehyde addition is completed, the reaction mixture is held in the same temperature range for a while to assure the completion of the exothermic condensation reaction step. In this initial condensation step, as shown in Figure 9.1, phenol is partially neutralized by the sodium hydroxide catalyst and transformed into phenolate anion, and the negative charge on the oxygen atom of phenol delocalizes (moves) by resonance to the ortho- and para-carbons and activates them to have an increased nucleophilic character for the attacking formaldehyde molecules. Formaldehyde has the (positive) electrophilic character and reacts to form carbon–carbon bonds with ortho- and para-carbons of phenol molecules. Initial reaction products include two mono-methylolphenols, two di-methylolphenols, and a tri-methylolphenol, and some unreacted phenol remains in the reaction mixture [16, 17]. The reason behind this uneven reaction is that an introduced hydroxymethyl group on the phenolic ring results

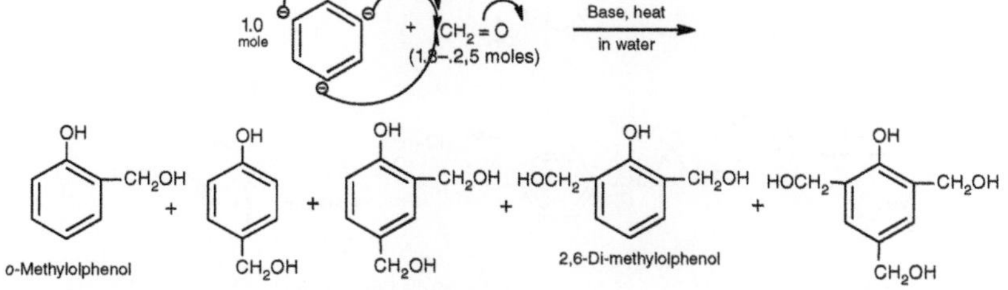

Figure 9.1 Base-catalyzed reaction of phenol and formaldehyde to form various hydroxymethylphenols at 60–100 °C. Source: Adapted from De Jong and de Jong [16].

in increasing the reactivity of the phenolic ring further, i.e. formaldehyde reacts faster with the formed mono-methylolphenols than free phenol, resulting in the formation of di-methylolphenols, and similarly, di-methylolphenols react faster with formaldehyde than mono-methylolphenols to form tri-methylolphenol [17]. Naturally, some phenol can remain as free phenol until late, where there is no formaldehyde left to react with. At this stage of resin manufacturing, the reaction products consist mostly of these monomeric species, which are insoluble in water but solubilized by the sodium hydroxide present [18], and their polymerization to dimers and trimers, etc., occurs only minimally.

A large amount exothermic heat, about 25.0 Kcal per mole of formaldehyde (30 g), is evolved in this stage [19]. In industry, one batch of PF resole resin cook can range up to about 100 000 pounds of resin, and the amount of exothermic heat is large, which is controlled both by adding the formaldehyde solution over an extended period of time and also by removing the heat by circulating (cold) water through the cooling coil inside the reactor as well as through the cooling condenser with vacuum application. Failure to control the exothermic heat can result in an explosion-type reaction that can destroy the reactor and its facilities. For example, this type of accident can happen when the reaction temperature is accidentally kept too low (below about 60 °C) initially and therefore the formaldehyde added is not reacted fast enough to result in an accumulation of free formaldehyde, and later when the reactor temperature catches up, the reaction goes all at once and evolves a large amount of heat.

Polymerization Step: In the second step of resin synthesis, after the exothermic reaction period is passed, the resin synthesis reaction is continued by heating the reaction mixture to a higher temperature between 80–100 °C, which is known as the resinification step of hydroxymethylphenols [18]. In this higher temperature and basic regime, the base catalyst promotes the formation of cationic intermediate form from the phenolic hydroxymethyl groups, as shown for *p*-hydroxymethylphenol in Figure 9.2 [20, 21]. This methylene cationic intermediate (**II**) of p-hydroxymethylphenol (**I**) can transform into its

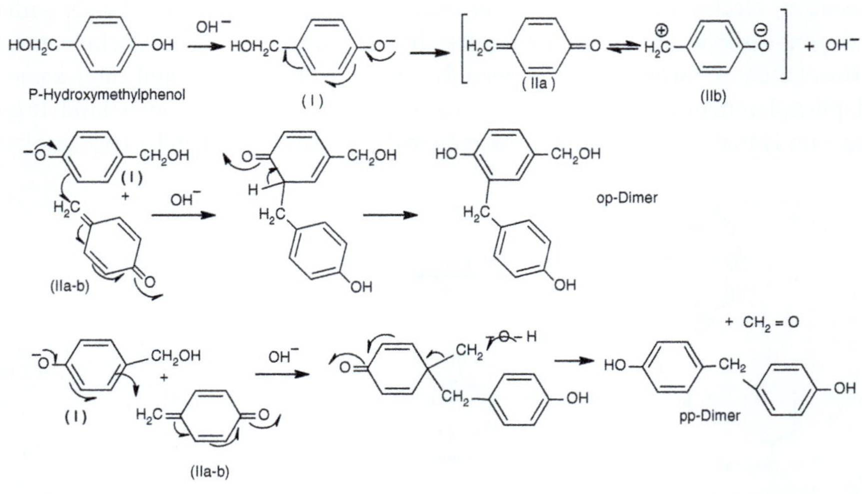

Figure 9.2 Activation of *o*- and *p*-carbons as carbanionic species which attack on quinonoid-form of *p*-hydroxymethyl group to form *o*–*p* and *p*–*p* methylene bonds, resulting in dimers. Source: Yeddanapall and Francis [20].

quinone methide form (**III**) for molecular stabilization, therefore gaining a longer lifetime for reaction as an electrophile (positive ion character), and is attacked by an ortho- or para-carbon (nucleophile) of another hydroxymethylphenol (**I**) molecule, as shown in Figure 9.2. The attacking ortho- or para-carbon of the second phenolic molecule has either a hydrogen or hydroxymethyl group on it, which is replaced by a methylene group of the first molecule, and the hydrogen ion or hydroxymethyl group is kicked out as hydrogen ion or as formaldehyde molecule.

These reactions lead to formation of methylene bonds (carbon–carbon bonds) between two phenolic rings, resulting in formation of dimers. *Dimers are formed with ortho–ortho (o–o), ortho–para (o–p), and para–para (p–p) methylene bonds between two phenolic rings.* Since a dimer molecule has another or more hydroxymethyl groups on it, the methylene bond-forming reactions can go on as long as the second phenolic ring has an available hydrogen or hydroxymethyl group to be replaced, but not a methylene group on its ortho- and para-carbons, resulting in various trimers, tetramers, etc. This is the *polymerization* reaction of PF resole resins, which results in increasing the viscosity of reaction mixture, measurable by taking a sample as the resin manufacturing progresses. The polymerization step is commonly carried out by keeping the reaction mixture at a temperature between 80–100 °C and more sodium hydroxide and water can be added and reacted until the resin reaches to the target resin-solids level, sodium hydroxide content, and viscosity or degree of polymerization (advancement), normally in 5–6 hours. The resin polymerization reactions go faster among various methylolphenols than between methylolphenols and free phenol [20]. Also, because the reactions occur between molecules of different ionization extents, the reaction rate is reported to go fastest at a NaOH/phenol molar ratio of 0.3 (pH ~ 10) and slows down with either higher or lower NaOH/Phenol mole ratios [22]. Also, the polydispersity of resultant PF resins is highest at a NaOH/phenol ratio of 0.4, decreases at 0.6 and 0.8 mol ratios, and increases as the molecular weight (M_n) increases [22–24]. The polymerization progression and chemical structures of resultant PF resole resins are discussed further in a later lecture (Chapter 28). The resin batch is then cooled to room temperature, which takes about two hours or longer for large batches. The polymerization reaction of resin molecules nearly stops when cooled to room temperature, and therefore, manufactured liquid PF resole resins are "live" resins but can be shipped and stored until use, normally up to two to three weeks with little property changes.

Different Synthesis Procedures of PF Resole Resins: There are various other procedures used in synthesizing PF resole resins. A more common procedure often used in the past is initially mixing the phenol, formaldehyde solution, and 1–2% sodium hydroxide while keeping the mixing temperature under control at about 40 °C, and then starting to raise the temperature according to, for example, one degree per min to 80 or 90 °C under a tightly controlled heating/cooling schedule and then hold till the exotherm is expired. Then, the addition step of more water and sodium hydroxide can be carried out similarly to the procedure described above. Any (small) differences in the synthesis procedure, such as temperature schedule, sodium hydroxide split ratio and addition time, and resin-solids level, can lead to a "different resin," i.e. it could behave differently in the curing process or in wood bonding performances. Producing "the exactly same" resin batch after batch is an important target in the large-scale PF resin manufacturing industry.

9.3 Typical Synthesis Procedures of Alkaline PF Resole Resins for Various Uses

PF resole resins are used in various binder applications, such as for glass wool, paper laminates, oriented strand board (OSB) face-layer, OSB core-layer, softwood plywood, and many others. These resins have different F/P mole ratios, sodium hydroxide contents, resin-solids contents, and curing speeds, pre-tested by trial-and-error experiments for optimum performance and better economy in their respective uses. The performance requirements of resins, especially the cure time, are met by advancing the resin to the "right" extent of polymerization in the resin manufacturing step. Polymerization reaction in resin synthesis results in increasing the viscosity of reaction mixture, and when it arrives at a point where the stirring becomes difficult or inefficient, some water and sodium hydroxide are added, resulting in lowering the viscosity and easing of stirring, and the polymerization reaction is continued further. This step also results in lowering of resin-solids content and increasing of resins' sodium hydroxide content and can be repeated one or more times. The reaction temperature also can be changed, often decreased toward the end of resin synthesis for more accurate control of viscosity toward the target value. The viscosity of reaction mixture is monitored, and the reaction temperature/viscosity profile is constructed as the reaction progresses and, finally, a fast cool-down operation is carried out to about 60 °C or below as the target viscosity (advancement) is approached. In industry, the cooling speed becomes slower as the temperature of reaction mixture goes down due to the limited cooling capacity of tap water. The cooling step normally takes about 2 hours or longer for large batches, during which the viscosity also advances albeit slowly. In the end, the cooled, finished PF resin needs to have various target resin properties, i.e. target degree of polymerization, viscosity, resin-solids level, and sodium hydroxide content. The F/P mole ratio of resins is normally preset at the beginning with one-shot addition of formaldehyde. The finished, cooled resin is filtered to eliminate foreign materials and gel particles.

Typical Synthesis Schemes of PF Resole Resins: The degree of polymerization (advancement), i.e. how much polymerization is effected during the resin synthesis procedure, determines the curing speed of PF resole resins, becoming faster as the resin's degree of polymerization becomes higher, while the *resin's flow capability* gets poorer in the curing step, possibly to give poorer bond qualities. Resins' flow capability means how long the resin-solids, after evaporation of most water, remains as a flowing (molten) liquid at the curing temperature until it gels. During the hot-pressing of wood composite boards, the wood particles are consolidated, and the adhesive resin has to flow for spreading or adjusting to the shapes according to movements of wood particles, all before the adhesive resin starts to gel. An ample flow time of resin is desired, but it also means the total curing (or hot-pressing) time is lengthened. Lack of the flow capability of resin during the mat consolidation (or pre-curing) can ruin the adhesion process. Therefore, the resin advancement in resin manufacturing needs to be ended before this over-polymerization occurs for intended use. In industry, PF resole resin technologies evolved to optimize resins for the various different wood composite manufacturing parameters, while the wood composite manufacturers developed their machinery and operating procedures to be able to effectively use the optimized resins. The more important parameters are the hot-pressing temperature and hot-pressing time and the assurance of high bond strength values. Softwood plywood, OSB, hardboard (FB), paper laminates, and glass wool are the typical products bonded with PF resole resins.

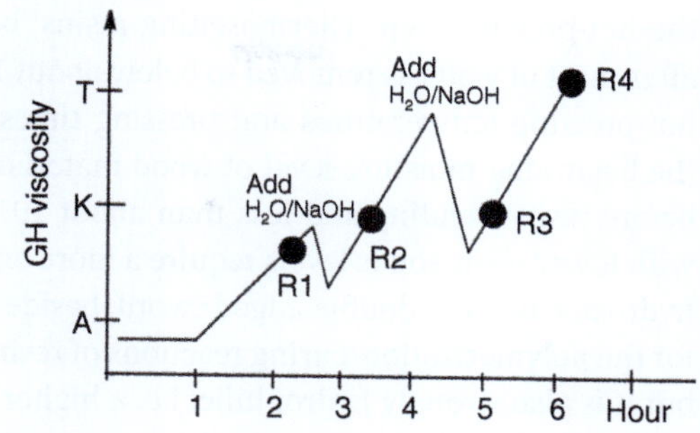

Figure 9.3 Schematic representation of PF resole resin cooking time (hour) and viscosity changes at 65–85–95 °C. Resins R_1–R_4 can be taken as representative resins used as various wood binders.

Resin-solids Level, Sodium Hydroxide Content, and Viscosity: The extent of polymerization entails three major parameters: *resin-solids level* (how much water is added), sodium hydroxide content, and viscosity values that fit for spraying or other application methods of resins. The schematic resin synthesis diagram in Figure 9.3a shows the increases/decreases in the viscosity of reaction mixture with cook times at an elevated temperature, normally 85–100 °C, with the addition of sodium hydroxide and water at indicated max viscosity points as commonly practiced in manufacturing of resins. In the beginning, 90% phenol and 50% formaldehyde solution are reacted at a mole ratio of about 2.0 in the presence of about 1–2% of first sodium hydroxide based on the total resin charge weight. When the initial condensation reaction is completed at 60–70 °C, the reaction mixture shows a viscosity value of A or lower by Gardner-Holdt (GH) scale. As the polymerization progresses at higher temperatures, the reaction mixture increases in viscosity and when it reaches to about J-K, the mixing of reaction mixture becomes inefficient, and the risk of local gelation arises if a further advancement is to be carried. A further advancement of resin is made possible by adding some water and sodium hydroxide, which lowers the viscosity of the reaction mixture. The polymerization reaction is then continued to another viscosity point of J–K and then, similarly, some water and sodium hydroxide can be added. The reaction is then carried to the final viscosity point, often T or higher for softwood plywood-type adhesive resins.

Viscosity of polymer solutions decreases with decreases of polymer-solids level in the solution. Water is a solvent for PF resole resin polymers, and the addition of water lowers the resin-solids level. Sodium hydroxide also has the viscosity-reducing effect due to increasing ionization of phenolic hydroxyl groups of PF resin polymer molecules [22], as shown in Figure 9.4. The ionized phenolic molecules, i.e. anions, repel each other or reduce the attraction forces between molecules, lowering the shearing forces of molecules in stirring. Viscosity is a measure of shearing forces between moving molecules and, thus, both water and sodium hydroxide contribute to lowering the viscosity of reaction mixture. On the other hand, the addition of water results in decreasing the resin-solids level or increasing the water content, otherwise an unwanted component of resin, since it increases the moisture content of wood composite mats that needs to be removed in

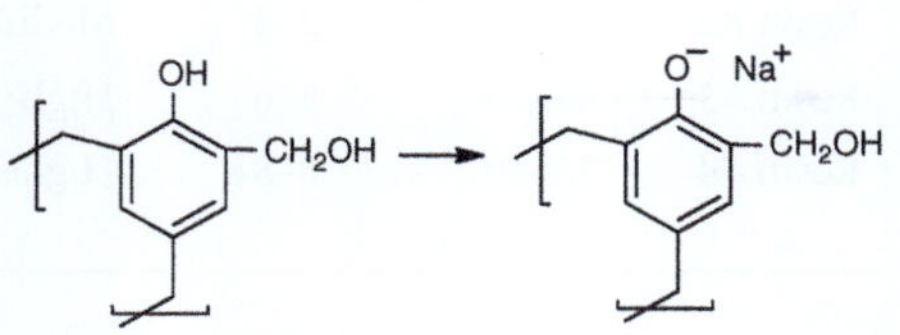

Figure 9.4 Ionization of phenolic resole resin molecules with addition of sodium hydroxide.

the hot-pressing step. Thermosetting resins' bond formation is normally completed only after most of water is removed to below about 1–2% of mat weights. In general, for current hot-pressing temperatures and pressing times used in the OSB manufacturing industry, the beginning moisture level of wood materials needs to be controlled to below about 4% before resin blending and less than about 10% after resin blending, and therefore resins with lower resin-solids levels require a more exhaustive drying of wood substrates. Sodium hydroxide is also a double-edged sword, beside its viscosity-reducing effect, it is the catalyst for the polymerization/curing reactions of resin, i.e. needed for faster curing rates of resins, but it is also strongly hydrophilic, i.e. a higher tendency of absorbing water, and the more sodium hydroxide in the resin the stronger the hydrophilic character of cured PF resins or boards [24]. Therefore, the wood composite board's water absorption capacity and thickness swelling values can increase as the sodium hydroxide level of the adhesive is increased.

PF Resole Resins vs. Degree of Advancement, Resin-solids Level, and Sodium Hydroxide Content: Any sample taken during the cook in the schematic resin synthesis diagram of Figure 9.3 could be *a PF resole resin* that will cure to a decently cross-linked PF resin product, but its curing time and flow character or bonding quality will be significantly different from a resin taken at another point, i.e. the PF resin industry chooses a resin synthetic procedure that best matches the required press time and bond quality at the available hot-press temperature. The resin samples taken at R#1, #2, #3, and #4 points are schematically classified with respect to their resin-solids levels, sodium hydroxide contents, relative molecular weights and curing speeds, and application areas as shown in Table 9.1. It is noted that the resin-solids level decreases, and the sodium hydroxide content increases in the given order as the resin's curing rate and molecular weight increase [25]. This resin property-performance relationship also shows that there could be many different PF resole resins due to different extents of polymerization, resin-solids levels, and sodium hydroxide content in addition to differences in cooking schedules. Another note to make is that although the viscosity of PF resole resins needs to be in the usable (sprayable, etc.) range of about below 500 cP (cps), its meaning with respect to the curing speed is limited unless the comparative resins have the same resin-solids and sodium hydroxide content values. Thus, *Resin #4 types* (softwood plywood, etc.) are among the most advanced resins [26], in which resin-solids levels are the lowest and sodium hydroxide catalyst contents are the highest. *Resins #3 types* (OSB core-layer) have medium-high MW values, medium-low resin-solids levels, and medium-high sodium hydroxide contents. *Resin #2 types* (OSB face-layer layer)

Table 9.1 Various kinds of wood adhesive PF resole resins according to Figure 9.3.

	Resin solids (%)	NaOH (%)	Rel MW and cure speed	Uses
Resin #1	60–65	1–2	Low	Glass wool; paper overlay; acid-cure resin
Resin #2	55–60	2–3	Medium	OSB face-layer; dry process hardboard
Resin #3	45–50	5–6	High	OSB core-layer; particleboard
Resin #4	35–40	6–8+	Highest	Softwood plywood; wet-process hardboard, laminated veneer lumber

Table 9.2 Synthesis procedure of an OSB core-layer binder PF resole resin.

- 1044.4 g 90% Phenol; 120.0 g 50% NaOH. 1500.0 g 50.0% Form.; F/P ratio 2.50.
  1. Mixed phenol and sodium hydroxide and heated to 70 °C,
  2. Added 50% formaldehyde dropwise over 1.0 h while keeping the temperature at 70 °C,
  3. The reaction is continued at 80 °C until the viscosity reaches to 275 cps measured by GH viscometer,
  4. Add 80 g 50% NaOH and 245 g water and continued the reaction until the viscosity reaches to 375 cps,
  5. Add 80 g of 50% NaOH and 460 g of water and continued the reaction until the viscosity reaches to 600 cps,
  6. Cooled.
- Total batch 3529.4 g.
- NaOH/P mole ratio 0.35.
- PF resin-solids level 36.6%.
- 5 ~ 10% urea is commonly added, resulting in ~45% resin solids level and 250 cps viscosity.

Source: Kim et al. [27].

have medium-low MW values, medium-high resin-solids levels, and medium-low sodium hydroxide contents. Resin #1 types have the lowest molecular weights among PF resole resins, highest resin-solids, and lowest sodium hydroxide contents.

An Example Synthesis of PF Resole Resin: An OSB core-layer binder-type PF resole resin is synthesized as shown Table 9.2, where three sodium hydroxide and water additions are used. The NaOH/phenol ratio of the resin is not as high as that of Resin #4 type shown in Table 9.3, and the final viscosity reached to a little high value in consideration of adding 5–10% urea after synthesis to capture the free formaldehyde and lower the viscosity for OSB binder-type PF resins [28]. Commercial OSB PF resole resins are red/dark, viscous liquids having 43–58% resin-solids content, dilutable with water 25 : 1, specific gravity of 1.10–1.30, and storage stability of about 10–20 days.

9.4 Properties and Polymer Structures of Alkaline PF Resole Resins and Measurements

Molecular Weights of PF Resole Resins Measured by GPC: The average molecular weights of PF resole resins, like other common thermosetting resins, are relatively low (up to a few thousand Da) in comparison to thermoplastic polymers that can go up to several hundred thousand Da. Polymerization reactions in PF resin syntheses occur among monomers and oligomers of similar or different molecular weights to form bigger oligomers and, therefore, PF resole resins have broad molecular weight distributions, as shown by the gel permeation chromatogram (GPC) in Figure 9.5 [29]. This analytical method uses a column of compacted gel particles and an eluent solvent that passes through the column under pressure. A small amount of polymer sample is injected into the head of the column and detects the polymer molecules coming out at the end of the column. The eluent is a good solvent for the polymer so that all molecules in the sample dissolve and move freely, i.e. no strong association to form bigger particles. The gel particles need to be in a very well-compacted state so that most polymer particles and eluent molecules have to pass through various size holes in *the gel particles* with the help of pressure applied on the entering eluent.

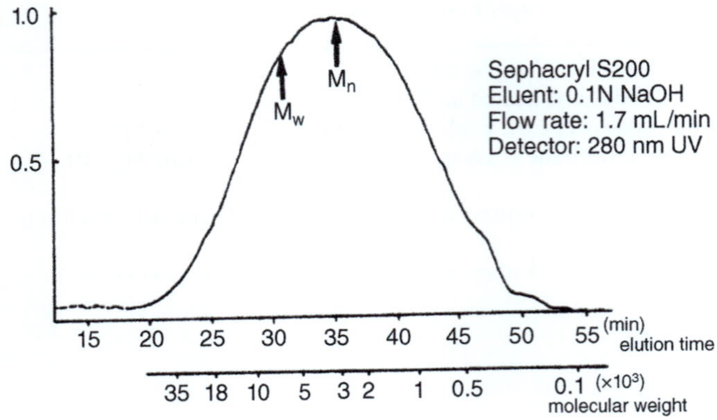

Figure 9.5 Aqueous GPC of the PF resin synthesized in Table 9.1 calibrated with polystyrene sulfonates and number- and weight-average molecular weight positions indicated. Source: [27, 29].

Calculation of average Mn and Mw values from a GPC result [28]

$$\text{Number-average, } M_n = \sum N_i \times M_i / \sum N_i$$
$$\text{Weight-average, } M_w = \sum N_i \times M_i^2 / \sum N_i \times M_i \qquad (9.1)$$

N_i: number of molecules.
M_i: mol. Weight of each molecule.

When a polymer sample consisting of varying molecular weights is injected at the head of the column, since the amount of sample is very small compared to the amount of gel particles, all polymer molecules meet the gel particles. Then, larger polymer particles can go through only larger holes with shorter path lengths and come out of the column first. Smaller polymer molecules can also move through larger holes fast but most of them soon meet more smaller holes, and since there are many more smaller holes, they will meet the holes of their molecular sizes to pass through the many layers of gel particles consecutively. Thus, larger polymer molecules win the race over smaller polymer molecules, resulting in a fractionation of the polymer sample according to the molecular sizes, which is continuously detected at the end of column using a refractive index or UV concentration detector to give chromatograms consisting of elution time (molecular weights) vs. intensity (amount of materials) values, displayed from the high molecular weights to low molecular weights. A series of similar experiments with narrow molecular weight standard materials with known molecular weight values then allow the elution times to be assigned to the molecular weights of the chromatogram. The resultant chromatogram of molecular weight (M_i) vs. population (N_i) is analyzed to give the number- and weight-average molecular weights (M_n and M_W) values, calculated using the Eq. (9.1) [29]. Of course, the ΣN_i in the equation is set to 1.00 to obtain consistent graphs of MW values from sample to sample. The M_n value of about 3000 Da and M_w value of about 6300 Da are obtained from the GPC of Figure 9.5 with an M_w/M_n ratio of about 2.10 and the highest MW fraction reached to about 36 000 Da. ^{13}C NMR, light-scattering, and vapor phase osmometry methods have also been used to measure MW values of PF resin fraction samples [27].

Viscosity Values of Alkaline PF Resole Resins: As a measure of molecular weights of PF resole resins, the viscosity values have a relatively narrow meaning as mentioned above because viscosity values have a greater dependency on resin-solids levels and sodium hydroxide contents than molecular weights. On the other hand, if two resins have the same solids levels and same sodium hydroxide contents, one resin that has a higher viscosity would have a higher degree of advancement and faster curing speed. Furthermore, a PF resole resin at 50% resin-solids content can be made to have the same viscosity as that of another PF resole resin at 45% resin-solids content. Then, the molecular weight will be higher and curing speed will be faster with the latter resin. Overall, the viscosity of PF resole resins is an important parameter for spraying operations or in monitoring of stored resins for stability. Lower the viscosity the higher the spraying efficiency, in general.

Approximate calculation of pH values for PF resole resins

$$\text{P—O—H} \leftrightarrow \text{P—O}^- + \text{H}^+$$

$$\frac{[\text{P—O}^-][\text{H}^+]}{[\text{P—O—H}]} = \text{Ka} = 1.28 \times 10^{-10}$$

$$\text{pH} = -\log[\text{H}^+] = -\log 1.28 \times 10^{-10} \times \frac{[\text{P—O—H}]}{[\text{P—O}^-]} \tag{9.2}$$

pH values of alkaline PF resole resins are an indicator of the sodium hydroxide (base) content and its determination is frequently done in quality control operations. Although the base contents of PF resole resins can be high, ranging up to about 8–9% based on liquid resin weight, the phenol component in the resin, a weak acid, is in excess and neutralizes the strong base, resulting in a buffered solution of PF resole resin molecules. Therefore, the pH value of PF resins increases only slowly and in a narrow range as their base content is increased. The pH calculation is shown in Eq. (9.2). Let us represent the phenol as P—O—H and the dissociation constant of phenol in aqueous solution as Ka, which is a known constant. The bracket $[\text{P—O}^-]$, etc., means the concentration of phenolate anion, etc., in moles per liter. The molar concentration of phenolate anion, formed as the result of the neutralization by sodium hydroxide, is initially equal to the concentration of sodium hydroxide added in resin synthesis. The ratio of sodium hydroxide to phenol present in the resin can be calculated using the respective amounts of raw materials charged in resin manufacture. As an example, if the sodium hydroxide to phenol mole ratio in the resin is 0.1/1.0, then: $[\text{P—O—H}]/[\text{P—O}^-] = 0.9/0.1$, and $\text{pH} = —\log 1.28 \times 10^{10} \times \log 0.9/0.1 = 8.93$. If the sodium hydroxide to phenol mole ratio is 0.6/1.0, then, $\text{pH} = -\log 1.28 \times 10^{10} \times \log 0.4/0.6 = 10.1$. The first example approximately corresponds to about 2% of sodium hydroxide content, and the second example to about 8% sodium hydroxide content in PF resins with 50% resin-solids levels. Thus, the pH value ranges between ~9.0 and ~10. The actual resin pH value will be slightly higher due to other effects, but this approximate calculation shows the slow increases in pH value in a narrow range. However, it is noted that pH is in the logarithm scale, i.e. the pH change from 9 to 10 means a 10-fold increase in the hydroxide ion content. The pH values of PF resins are low in comparison to cases where a similar amount of sodium hydroxide is dissolved in pure water, with pH value of 13.7 and 14.3, respectively, for the same amount of sodium hydroxide used in the above examples. On the other hand, PF resole resins show alkaline pH values despite the excess acidic phenol present because phenol is a weak acid and the phenolate anion equilibrates back to partly produce hydroxide ion and free phenol.

Polymer Structures of PF Resole Resins: In syntheses of base-catalyzed PF resole resins, the formaldehyde molecules freed in the polymerization step as a result of breaking of methylol

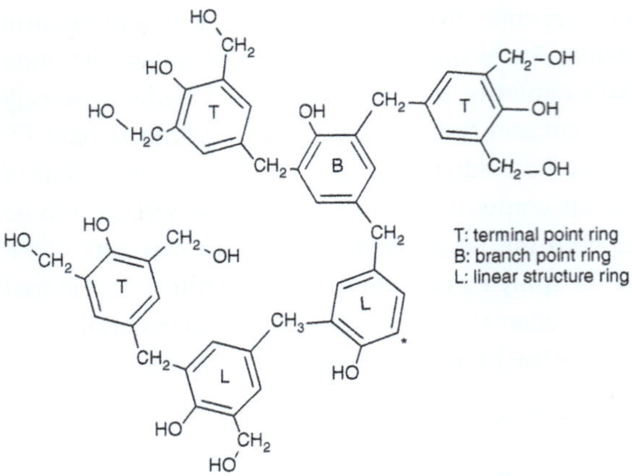

Figure 9.6 Polymeric structure of PF resole resin molecule. Asterisks (*) carbons indicate no hydroxymethyl group nor methylene group is bonded.

groups, shown in Figure 9.2, can react again with unreacted ortho- or para-carbons to form methylol groups and, ideally, for resins made with an F/P ratio of 2.0, this process can go on until all ortho- and para-carbons are bonded to a hydroxymethyl or methylene group. In practice, however, some formaldehyde remains in the finished resin, especially when the F/P ratio is higher than 2.0, and it is also common to have some formaldehyde remaining even if the mole ratio is lower than 2.0. Also, some ortho- and para-carbons of phenolic rings can remain unreacted depending on the temperature schedule of synthesis. On the other hand, side reactions such as Cannizzaro and other side reactions do occur to some extent to consume some formaldehyde during the reaction, but the extent appears minimal since the head space in the reactor is under the water vapor and the stirring rate is kept minimal so as not to allow much air to be whipped into the resin phase. A schematic PF resole resin molecular structure synthesized with an F/P mole ratio of 2.0–2.5 and advanced to where most ortho- and para-carbons of phenolic rings are bonded to a methylene or hydroxymethyl groups, is shown in Figure 9.6.

Relationship of reacted FP mole ratio and number-average degree of polymerization for PF resins with a full substitution. F_f: Free formaldehyde

$$F/P = 2.0 + 1.0/n + F_f/P; \text{or},$$
$$n = 1.0/\{(F/p) - 2.0\}, \text{where } F/p \geq 2.0 \tag{9.3}$$

One ortho-carbon (asterisk) is left unreacted to show such a possibility. On the other hand, in the case of a full substitution, *the reacted F/P mole ratio* is related to the degree of polymerization (n) by Eq. (9.3), i.e. for F/P ratio of 2.5, $n = 2$; F/P of 2.25, $n = 4$; F/P of 2.10, $n = 10$; F/P of 2.0, $n = \infty$. Thus, high molecular weight PF resins can be made by using F/P mole ratio values of about 2.0, where the free formaldehyde content can be very low. Use of formaldehyde more than corresponding amounts means some free formaldehyde is remained and can evaporate in the spraying operation in board manufacturing step to cause health and environmental problems. On the other hand, these PF resole resins will keep polymerizing and cross-linking when they are heated, such as in curing in boards, by making methylene bonds between phenolic rings and expelling formaldehyde molecules derived from de-hydroxymethylation, as discussed in *Polymerization Step*. Since phenol molecule has three reaction sites, any F/P mole ratio used in resin synthesis above 1.50 will have to be expelled or dissipated as oxidation products.

F/P Mole Ratios of PF Resole Resins: In practice, the optimum F/P mole ratio for a particular resin is found by trial-and-error experiments, i.e. a series of laboratory resins are made by varying the F/P ratio and testing them for curing speed and free formaldehyde content. The curing speed of PF resole resins generally becomes faster with increasing the F/P mole ratio, but the free formaldehyde level in resins is normally kept to below about 0.3% in practice. It is known that F/P mole ratios lower than 2.0 are also utilized, likely to result in PF polymer structures where some unreacted ortho- or para-carbons remain in the resin structure, as shown in the schematic structure in Figure 9.6 with asterisks (*). Then, how is the F/P mole ratio value of resins decided in relation to the different classes of PF resole resins? It is apparent that in going from Resin #1 to Resin #4, the F/P mole ratio needs to be lowered to about 2.0 or lower to keep the free formaldehyde content at a minimum as well as to avoid the crowding of resin molecular structures. In bonding of softwood plywood, the Resin #4 type PF resins with F/P ratios of 1.6–1.9 have been known to give better delamination values than resins made with 2.2–2.5 F/P mole ratios [26]. For *Resin #1 ~ Resin #3 types*, F/P mole ratios up to 2.5 are used, and minimization of the free formaldehyde content with respect to the degree of advancement is often attempted. In the case of OSB core-and face-layer binder PF resins, there is room to use more formaldehyde since some urea, up to 5–10%, can be added to lower the cost of resin-solids as well as the free formaldehyde content [30, 31]. The urea reacts to form UF resin products, and the UF resin-solids apparently is added to PF resin solids to give some room for price negotiation for resin suppliers.

9.5 Resin Characteristics and Uses of Various PF Resole Resins

Resin #4 type PF Resole Resins: The highest molecular weight faction of this class of PF resole resins is reported to reach to molecular weights of about 40 000 Da and some of these molecules are considered to be pushed into the formation of intramolecular cross-linking (methylene) bonds and turned into solid-like components that may still remain in the liquid resin phase, undifferentiated to naked eyes or in normal handling procedures [28]. These materials would be of limited thermosetting capability due to their already cross-linked nature. Since this tendency of intramolecular reaction is expected to be higher for higher F/P mole ratio resins, for highly advanced resins such as *Resin #4 type* the F/P mole ratio in syntheses is often lowered to near 1.8, and also thickeners such as hydroxyethylcellulose [32] are added to prevent the formation of such highly advanced MW fractions. On the other hand, with these low F/P mole ratio resins, there are some unreacted ortho- and para-carbons present on the phenolic rings, and the free formaldehyde content is low. These types of PF resins cure fast and are used in bonding softwood plywood panels, allowing the hot-pressing of panels to be done at relatively low temperatures with short pressing times, near 300 °F and 3–5 minutes for 0.5′-thick panels. The low resin-solids level of these types of PF resins is ameliorated by adding fillers and extenders. Also, more sodium hydroxide or sodium carbonate is often added in the formulation step to help shorten the hot-press time further.

Resin #2 and #3 type PF resole resins are used as face-layer and core-layer binders of OSB, respectively, and the F/P mole ratio ranges from 2.0–2.3 with higher values selected for core-layer resins for a higher curing speed. The top-most surfaces of boards tend to experience pre-curing of binder resins, and therefore a lower degree of polymerization and lower F/P mole ratio are used in comparison with core-layer resins. One additional constraint for core resins is the moisture level of core-layer after resin application, which needs to be less than about 10%, and therefore the resin-solids level of core-layer resins cannot be lowered further. In manufacturing of wood composite boards, the hot-pressing time needs to be as short as possible and the bond strength be as high as possible; further, the least extent of pre-curing in face-layers and the least amount of free formaldehyde emissions are desired. However, it is often difficult to closely match the degree of polymerization with the F/P mole ratio of resin according to Eq. (9.3), inevitably leaving some free formaldehyde in the resin. As discussed above, addition of 5–10% urea based on the liquid resin weight is usually done to reduce the free formaldehyde content and reduce the odor problem. The urea addition results in only a minimal reduction of internal bond (IB) strength values [30, 31].

Resin #1 types have the lowest degree of polymerization and are used in at least three different applications: glass-wool binders, acid-curing wood binders, and paper-saturation matrices, which are discussed in Section 9.11.

9.6 Molecular Weights (Sizes) of Alkaline PF Resole Resins and Wood Cell Wall Penetration

The curing speed of PF resins is related to the molecular weight, i.e. shorter curing times for higher molecular weights and longer curing times for lower molecular weights. This relationship or concept cannot be carried out too far to make faster-curing resins in practice due to decreasing of the flow property. Good resin flow during the final bonding stage would allow glue molecules to make good (molecular) contacts with wood surface molecules and also to spread to wider wood surface area, including filling of cut lumens and crevices generated in the cutting steps of wood. The flow capability of resins decreases as the molecular weight of resin increases, and molecular weights rapidly increase in *Resin #4 types*. Poor bond qualities can result in a resin that is advanced beyond the optimum range. For example, using Resin #4 types in place of Resin #3 types in OSB core-layers is likely to cause a poor bonding.

Another adhesive property in wood bonding proposed in the past and being relied on until today by many practitioners of PF resole resins is that the cut wood cell walls must be penetrated by at least some significant fractions of resin molecules and polymerized *in-situ* in order to attain a good exterior-grade bond quality [33]. This thesis appears plausible since wood would expand a lot more than cured PF resins when exposed to moisture and, if there are no imbedded chemical bonds in the interface, the adhesive layer and wood surface are likely to separate from each other, i.e. no or low durability. For a good penetration to occur, the resin molecules should first of all make an intimate molecular contact with cell wall structures and also need to be small enough, or low molecular weight species. One result of the molecular penetration is the more gradual continuum of adhesive structure in the interphase of gluelines, i.e. the "**diffusion theory of adhesion**," discussed in Lecture 20. Under the normal gluing condition, the cell wall has a very tight structure of cellulose and lignin molecules but might expand for resin molecules that are small enough, probably monomer types, to penetrate. High molecular weight PF resins tend to have less of these low molecular weight polymer species and can result in poor bonding due to an ineffective penetration. Thus, advancing the degree of polymerization in PF resin synthesis must be done within an appropriate range for each application type, even though a short hot-pressing time is desired in practice and is obtainable within a certain range by advancing the resin. One area of application where the cell wall penetration of resins is very essential is in the manufacturing of paper laminates for various uses, including air, oil, and gasoline filters and acid battery separators. Resin #1 types with mostly low molecular weight polymer species and low F/P mole ratios are used for the flexibility of cured products. In these highly durable products, the cell wall penetration capability of PF resin molecules and formation of non-hydrolyzing methylene bonds appear to be the key.

9.7 Powder PF Resole Resin Manufacturing by Spray-drying

In industry, some PF resole resins are spray-dried to fine powders. The spray-drying plant consists of a large circular insulated chamber, in which the hot drying air is slowly circulating around and partly allowed to exit through the vent. The liquid resin is sprayed as a fine mist from the top center of the chamber and the resin droplets soon dry to solid particles in the air and settle to the bottom, which are collected and bagged. The spray-drying process advances the PF resin's degree of polymerization to some extent. Normally, *Resin #2 and #3 types* are spray-dried for use as binders for OSB face or core-layers and they are also used in bonding of wood moldings, such as toilet seats. Properties of typical commercial powder PF resins are pinkish powders with bulk densities of 34–38 pcf, screen analysis results of about 30% on 200 mesh screen, stroke cure times of 12–16 seconds, and storage life of about four months. The cost per resin-solids weight is higher than liquid resins due to the spray-drying process. However, the powder PF resole resins, as with powdered novolac PF resins discussed in the previous chapter, do not add water to the wood mat and thus make it possible to dry the wood substrates less, allowing a better control of the mat moisture contents. This moisture advantage has been reported to be improved further by adjusting the molecular weight distribution of resin, i.e. by combining a high molecular weight resin (5600 Da, acetylated, 80%) with a low molecular weight resin (1630, acetylated, 20%), followed by spray drying [34]. Also, liquid PF resole resins were shown to result in improved bonding values for higher moisture content veneers or OSB when they are mixed with a high molecular weight powdered PF resins as a dispersion [35]. Powder PF resole resins are also more stable than liquid resins and shipping cost is lower on the basis of resin-solids weight, making them more economical to use for board manufacturing plants located remotely. Resins #1 and #4 types cannot be spray-dried since the former has a too much liquid character, and the latter has too much (cured) solid character to make it difficult to disperse in water in formulating of softwood plywood binder-type adhesives.

9.8 Curing of Alkaline PF Resole Resins

9.8.1 Pre-dry and Pre-cure, Resin Spreading, and Flow

In OSB manufacturing, when an (aqueous) PF resole resin is applied on wood strands by spraying, the water in the resin droplets begins to evaporate into air and also penetrate into wood substrates. This water dissipation continues during the mat lay-up, resulting in increasing the uniformity of moisture content throughout the mat. This movement of moisture results in partially drying of resin droplets, the extent of which depends on the incoming substrates' moisture content, temperature and humidity in the air, and length of lay-up time. Resin droplets of 50–60% resin-solids contents dry to semi-solids to the touch in the matting station. In the subsequent hot-pressing step, the resin droplets melt and spread initially without significant polymerization reactions taking place. The residual moisture in the resin droplets expedites this melting/spreading process, so it is seen that the mat moisture content, including the moisture in resin droplets, needs to be not too low. If the mat moisture content is too low and resin molecular weight is too high, the resin droplets may not be melting adequately and fail to cure properly. This can often happen in the top-most face-layers to cause pre-cure, i.e. wood strands tend to flake off in use of boards, and it is common practice to spray some water mist on the mat in transit to the hot-press in order to minimize the pre-drying.

In softwood plywood manufacturing, the lay-up time includes open and close assembly times, which, if they are too long, often give rise to "dry out" problems. Plywood PF resins are advanced to the highest extent for fast curing, so the resins' polymer molecules tend to lose mobility when they are dried. In this case, the flow of resin molecules in hot-pressing can be hampered if the curing temperature happens to be lower than the melting point of the dried resin. The molecular mobility would be better if some moisture is present in the adhesive, which means the length of assembly time needs to be controlled to be not too long. Another way of improving the resin flow is to synthesize the resin at a lower molecular weight so that the melting point of the dried resin becomes lower. This should be done by the resin supplier if it is warranted, i.e. such a gluing condition remains the same for an extended period of time. On the other hand, when the molecular weight of a PF resin is lowered, it will result in a longer press time, resulting in lowering the plant productivity. When all other factors are in good order, this is what is normally observed. However, if the hot-pressing time is too short or the press temperature is too low, the PF resin glueline will cure incompletely (under-cure). Or, if the mat or veneer moisture content is too high, the moisture gets attracted to resin molecules more tenaciously when the molecular weight is low in comparison to when it is high. This excess moisture can cause dilution of adhesive droplets in hot-pressing, resulting in starved gluelines. The molecular weights of PF resins used in OSB are appreciably lower than those of softwood plywood PF resins, and in bonding of softwood plywood, the moisture in adhesive layers is controlled to some extent with the help of fillers and extenders to obtain adequate flow properties for the higher molecular weight PF resin molecules.

In manufacturing of paper laminates, Resin #1 types with low F/P mole ratios are used to expedite the penetration of low molecular resin molecules into cellulose structures. *Low F/P mole ratio resins* are also preferred to have a better flexibility of cured products, and therefore the resins contain even some free phenol as a useful component. In this process, first, a web of kraft paper, etc., is impregnated with PF resin at about 30–40% weight pickup of resin-solids, and then sent through a hot chamber to dry off the solvent water and allow the resin to advance further in the so-called *B-staging* step. B-staged sheets can be stored, shipped to users, or used directly in manufacturing of laminates after cutting and layering, and by hot-pressing. B-staging means partly curing (polymerizing) of resin to such an extent that the resin still retains an ample polymerization reactivity, flow capability, and also a good storage stability, so that the laminating process can be done with a short hot-pressing time and minimal flashing (exudation) of resin in the final hot-pressing. In the past, softwood plywood was bonded using this type of B-staged PF resole resin films (Teco film).

9.8.2 Final Curing of Alkaline PF Resole Resins in Wood Bonding

Growth of Molecules in Liquid Stage: The curing of PF resins involves resumption of the polymerization reactions of resin molecules, which were stopped by cooling of resin at the end of resin manufacturing, to reach to the cross-linked, high molecular weight resin stage, commonly at a high temperature of about 150 °C or higher and under pressure. In the curing, due to the tri-functionality of phenol molecules and the high F/P mole ratio used in resin synthesis, the polymer chain can grow three-dimensionally, and at some point, some intramolecular methylene bond linkages are likely to form. The current model of the polymer growths is the iso-colloidal model: Figure 9.7(a) → (d), rather than the isocell model by Houwink (c) or spherocolloidal model (b). Thus, fluid spheres of resin molecules keep growing in number and to various sizes till they finally reach to a point where polymer spheres pack the volume of glueline space [6]. Much of the resin molecules would be incorporated into these polymer spheres (grain) of various sizes, but the water molecules remaining in resin spheres and those formed from resin polymerization reactions and some low molecular weight resin molecules that did not have a chance to get incorporated into the resin spheres are gradually expelled to the boundary regions of spheres to be evaporated out as steam and low molecular weight volatile materials [6]. The resin spheres (grains) are still in the liquid stage, i.e. the adhesive can flow to a wider area and wet the wood surfaces. One note to make here is that the size of macromolecular spheres can vary depending on the resin compositions and also on the prevailing temperature of curing, likely to get smaller at higher temperatures.

Gelation of Resin: As the water in the boundary regions gets evaporated out, the resin spheres are touching each other and start making chemical (methylene) bonds between them across the grain boundaries, as shown in Figure 9.7d [6], as proposed by R. A. Barkhuff and T. Carswell [7]. This is *the gelation point* and the beginning of developing a unity structure of the glueline, the rubbery stage. The glueline stops flowing and if a gelled glueline would break up, for example, by a delayed pressuring of board mat, the broken glueline cannot be properly recouped in the subsequent pressuring and heating, and the chance of attaining a cured, unity glueline structure is lost, resulting in a pre-cured glueline.

Vitrification of Resin: The rubbery stage develops further, gaining more elasticity (stiffness) from methylene bonds that are cross-linking the resin spheres, i.e. (inter-sphere bonds.) Also, more intra-sphere methylene bonds continue to be formed to reach to the transition region from the rubbery stage to the true solid stage of high elasticity, i.e. *the*

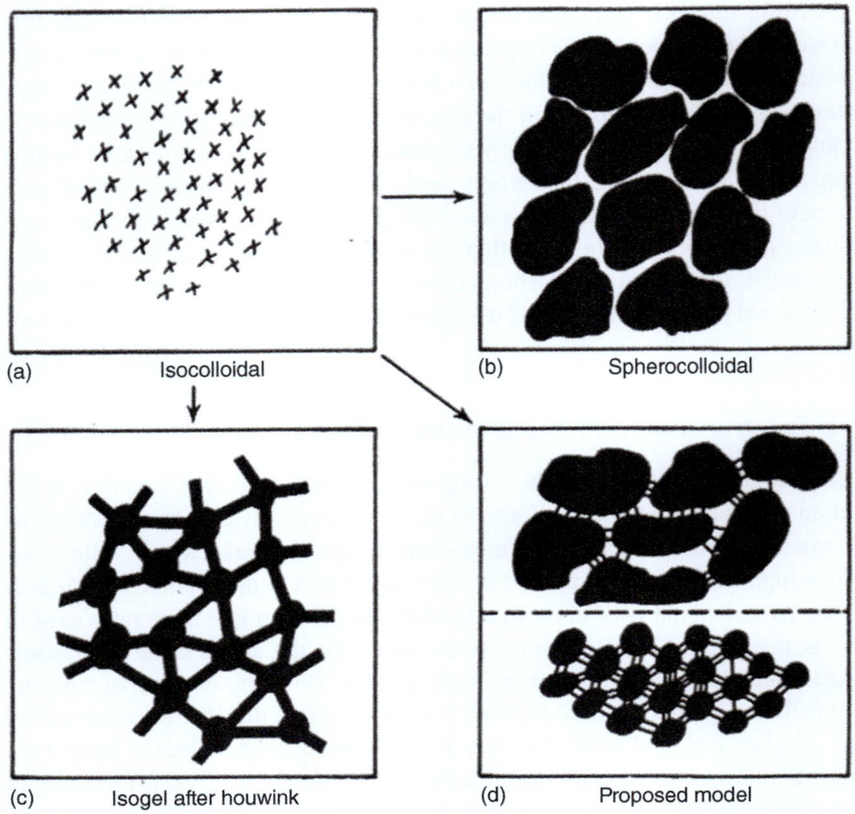

Figure 9.7 Diagrammatic states of resin cure according to various points of view. Source: Carswell [6].

vitrification point of the resin. From the point of gelation to the vitrification stage, the volume of the glueline, or mostly the thickness of glueline, shrinks significantly due to the loss of water and low molecular weight volatile resin materials. Thus, *gaps are to be formed* within the boundary regions of resin spheres and also between the gluelines and substrates unless the substrates are externally pressured, which is one reason for the consolidation pressure needed in board manufacturing. Passing the vitrification point means that the glueline became a true solid material that can at least hold the shape or the thickness of glueline in pressed boards, and the board can be taken out of the hot-press.

Post Curing of Resin: In practice, the curing of resin is allowed to go further for PF resins, *such as by hot-stacking of boards*, often resulting in further strengthening of boards. Thus, curing of PF resole resins means the hydroxymethyl groups are transformed into methylene bonds, and this has been confirmed by solid-state ^{13}C NMR analysis.

Chemical Structures of Cured PF Resole Resin: In this curing process of PF resins, the water molecules and unreacted or impurity materials keep evaporating out, but some of them can remain in the boundary regions of the polymer spheres of 'cured' gluelines in boards, a consequence of forming a unity structure starting from very small polymer molecular spheres in the course of curing process, and therefore, *the boundary regions* can be the weakest points in the breakage of cured gluelines or products in testing or in use [6, 7]. It is believed that this is one of the reasons why high mat moisture contents can be deleterious to bonding of boards when the hot-pressing temperature or time is inadequate for expelling the mat moisture to a very low level, commonly 1–2%. But since the presence of some water would be also essential until the gelation stage, the mat moisture content has to be neither too high nor too low, and the appropriate level has to be found by experiment [15]. A further discussion on cured resin structures is presented in Chapter 28.

9.8.3 Gel Times and Stroke Cure Times of Liquid PF Resins

In OSB and softwood plywood manufacturing, some water remains in the gluelines and reaches to the hot-press. This water would help adhesive spreading in the initial phase of curing, but too much water can cause the adhesive polymers to penetrate into wood too much and reduce the bond strengths due to lack of resin materials in the glueline, i.e. starved gluelines. Water in gluelines also tends to retard the resin polymers' curing speed since the polymerization reaction itself is a process that generates some water, especially in the later stages of curing. This dynamic interaction of moisture level of mat and curing speed of resin is often investigated by hot-pressing of boards with board mats having various moisture contents and at various hot-pressing temperatures and pressing times. Extreme conditions are easier to visualize. For example, in the *gelation* time measurements, a resin sample is put in a test tube and heated in a boiling-water bath (100 °C) and the resin is stirred with a glass rod continuously until the resin gels, i.e. until when the stirring becomes impossible. In this test, the resin does not lose much water, and the gelled resin becomes semi-solid when cooled and shows some (low) physical strength even though it contains a good amount of carrying water. PF resole resins for bonding plywood and OSB gel in about 20 to 30 minutes and it is apparent that this gelation time reflects one kind of cure time but would not directly predict the actual hot-pressing time of board because the resin gelation occurs in the presence of excess water. On the other hand, *the stroke cure time* of PF resins is measured with a resin sample of about 1.0 g, put on a hotplate at about 125 °C, and stirred with a flat laboratory spatula while stroking the boiling resin within an area of about one square inch. A small sample of the resin is then often picked up with the spatula to a height of about a one-half inch and returned to the hot resin while observing how stringy the resin becomes as the polymerization reaction progresses. The endpoint is when the resin string is no longer stringy but breaks on picking up, commonly after twenty to forty seconds. In this test, the interference effect of water is minimal, and the results also do not directly indicate the required hot-pressing times of boards. Though, these tests are generally useful in comparing the curing rates of different PF resins for quality control purposes in resin manufacturing and use.

9.8.4 Dynamic Mechanical Analysis (DMA) Method of Curing Tests for Thermosetting Resins

DMA mechanism and procedure: One good way of observing the curing phenomenon of PF resole resins from the liquid to the high-strength solid state under the dry condition is the dynamic mechanical analysis (DMA) method using, for example, DuPont 982 model instrument, the mechanism of which is shown in Figure 9.8 [36, 37]. In the test of liquid PF resole

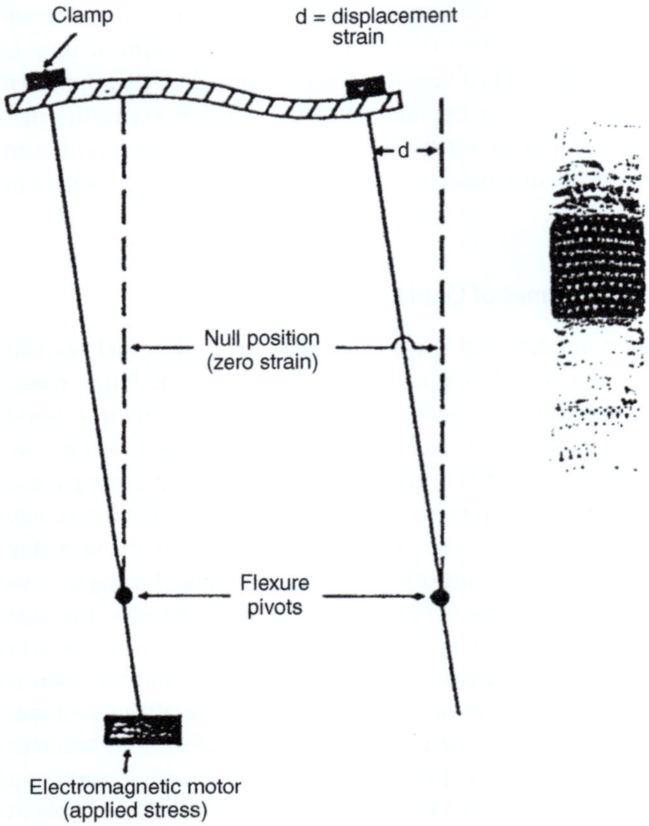

Figure 9.8 Mechanism of Du Pont 982 DMA analyzer in the flexure mode and a cured PF resin specimen of 15.0 × 0.30 × 12.50 mm run in the shear mode. Source: [36, 37].

resins using this instrument, a piece of glass-fiber cloth is held tight between the ends of two stout parallel arms with an exposed area of, for example, 15.0 mm × 12.7 mm dimensions, and is impregnated with a known amount of resin, for example, to obtain 0.10 mm thickness of resin-solids after curing. The assembly is put in the heating chamber, followed by rocking (stressing) the sample at a fixed frequency of about 1.0 Hz and with an amplitude of about 0.8 mm. In Figure 9.8, the sample is set to move in flexure mode, but for liquid PF resole resins, the authors chose to run in shear mode by orienting the holder horizontally, as shown on the right side with a cured sample, with the result obtained in rigidity (shear modulus) values [37]. Then, the chamber temperature is jumped to the curing temperature within about one min. The two arms sit on pivots of minimal friction outside the chamber, and one arm is the main arm operated by a transducer sitting at the end of the arm, which shakes and detects the mechanical forces experienced by the sample electromechanically. The second arm is freestanding. The two arms are connected only through the glass cloth and when the run is initiated, the two arms start rocking at the set frequency and with a small amplitude. The force of rocking is monitored as the stiffness of resin sample increases progressively with the curing of resin. The small amplitude of rocking limits the stress/strain of the sample to occur only within the elastic limit. As the liquid resin

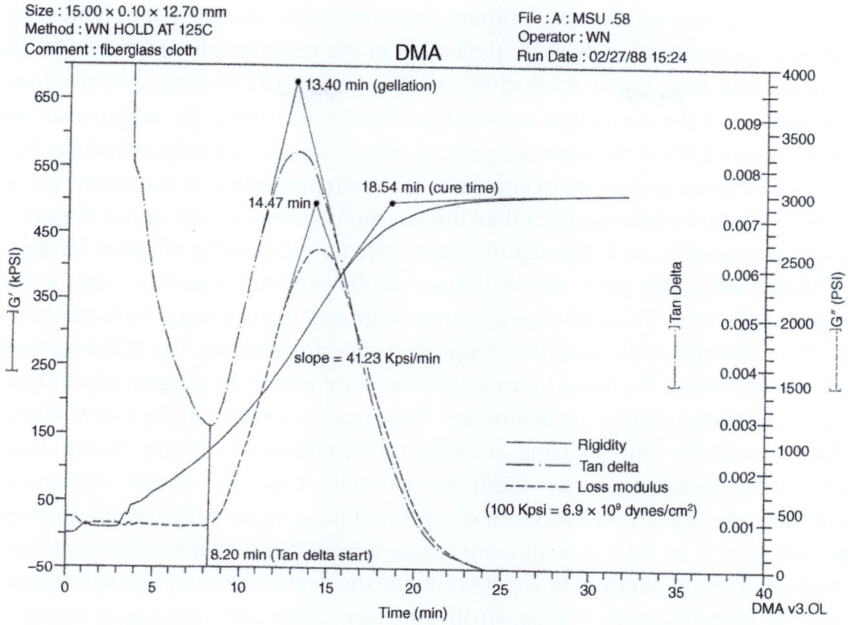

Figure 9.9 Typical DMA run with PF resole resin at 125 °C showing the rigidity, loss modulus, and tan delta curves. Source: Kim et al. [37]/American Chemical Society.

goes through the gelation and vitrification stages, the instrument continuously displays the increasing (complex) rigidity of sample and, after the run is over, the actual thickness of cured sample is measured and put into calculation. The result is three temporal curves: rigidity (G′), loss modulus (G″), and tan σ (loss modulus/rigidity ratio) curves.

DMA results of an OSB core-layer binder PF resole resin run at 125 °C [37] are shown in Figure 9.9. *Rigidity curve* (solid line) represents the strength of emerging polymer networks, that is, the conserved dynamic force (*G*′); *loss modulus curve* (dotted line) represents the uncured resinous phase that is physically coupled with the emerging polymer networks and thus able to dissipate the rocking energy into heat (*G*″), and *tan delta curve* (solid and dotted line) is their ratio (*G*″/*G*′). (*With this early version of DMA 982, the G″/G′ ratio values in the graph were erroneously calibrated to be about 100 times smaller, which was corrected in the DMA 983 version to range from 0.00–1.00.)* In the test done at 125 °C [37], both moduli are adjusted to be at zero in the beginning, and therefore the starting tan delta value is indeterminate in the beginning, and then the rigidity curve started to increase at about 0.8 minutes elapsed time, indicating that the resin already passed *the gelation point* to begin the rubbery stage and it remained flat until about 3.0 minutes. Gelation point is typically defined where tan σ value reaches to 1.0, but in Figure 9.9 it lies out of graph scale. Then, the rigidity curve starts to increase sharply to reach to a small value, the loss modulus stays small and flat, and tan σ curve keeps decreasing fast to reach to a low point at about 8.2 minutes. In this early period, the rigidity increased significantly (or the gel gets stronger) ascribable to the formation of polymer networks, while the flat loss modulus curve indicates that the emerging polymer networks have little resinous phase to participate in the dissipation of rocking energy, perhaps indicating the rigidity growth occurring mainly among the higher molecular weight components. The loss modulus curve then started to sharply increase at about 8.2 minutes, indicating that the formed polymer networks began to engage increasingly more molecules of the resinous phase for dissipation of rocking energy into heat, and it reached to a maximum at 14.47 minutes, perhaps indicating the formation of the methylene bonds that interconnect the colloidal particles, as shown in Figure 9.7 a–d. Then, the resinous phase molecules began to diminish and reached to an extremely low value at the end of cure. The tan σ curve reached to the maximum at 13.40 minutes, about 1.07 minutes earlier than the loss modulus maximum, and at this point there is a slight change of slope in the rigidity curve with a rigidity value of about 250 kpsi, or 6.95×10^9 dynes/cm^2. This point can be defined as *the vitrification point of resin at the curing temperature* [37, 38]. Then, the rigidity curve increased further and soon turned into a plateau at about 500 kpsi, indicating the completion of curing process. The DMA method thus gives *the curing time* of the resin, for example, the point where the tangent of the rigidity curve meets the plateau line at 18.54 minutes. The tan σ curve follows the loss modulus curve to a low value at the end of curing, showing the formation of a highly elastic cured resin mass that contains low amounts of resinous materials. When the sample is cooled to room temperature, the rigidity will increase sharply and tan σ value will decrease further. Typical tan σ values of 0.10–0.15 at room temperature were indicated for a fully cured thermosetting polymer [38]. In addition to vividly showing the development of strength values, the DMA method gives the gelation time, vitrification time, cure time, and cured strength. Also, the area value under the tan delta curve has been proposed to be a useful measure of the amount of uncured resinous materials left in B-staged PF resin materials [39]. Also, in comparison with DMA cure times of 10 minutes or longer observed for OSB binder PF resins, the stroke cure times of less than 1.0 minutes at the same temperature of 125 °C discussed above indicate that the curing progressed only to a small extent in the stroke cure tests, telling a part of the curing story.

DMA Results at Different Curing Temperatures: Results of DMA experiments of a typical OSB binder PF resin cured at three different fixed temperatures 90, 110, 125 °C, are shown in Figure 9.10 as an example [37]. The stiffness increases faster at higher temperatures and slower at lower temperatures, and the results also show that final rigidity values are higher for higher cure temperatures, indicating that there is a minimum or optimum temperature to cure a particular PF resin for best strengths. Also, the maximum tan σ values are higher for lower temperature curing, indicating that the area value under the tan delta curve and time of its maximum depends on the curing temperature, and therefore its use as a cure potential of B-staged resin samples would need a careful control of temperature or the rate of temperature increasing schedule. Overall, the DMA results appear to also indicate that higher number of polymer spheres are initially involved in the strength development when cured at higher curing temperatures (Figure 9.7 a–d, bottom) than at lower curing temperatures (Figure 9.7d-top), resulting in leaving less materials to act as loss modulus components. In the hot-pressing of OSB, for example, the middle layer can reach to about 150 °C toward the end of hot-pressing cycle. The rigidity values obtained in DMA measurements can be related to the IB strength values of boards. Overall, however, it needs to be kept in mind that DMA runs are made under the dry condition and without external pressure or containment. Also, the DMA rigidity values are related to the cohesive strength values of resin, while IB values of boards are much related to the adhesion between wood and cured resin surfaces. The adhesion cannot be studied directly on DMA, but the high rigidity is one necessary condition for any good adhesives for wood composite boards.

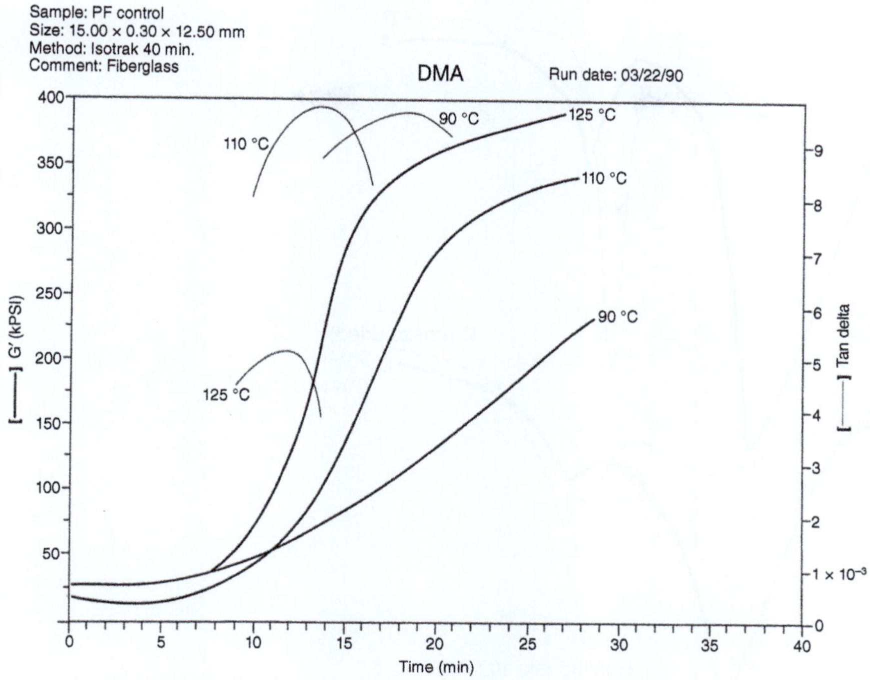

Figure 9.10 Dynamic mechanical analysis (DMA) test results of an OSB core-layer binder-type PF resole resin: showing the development of rigidity and Tan delta values vs. elapsed time (min) at 90 °C (bottom), 110 °C (middle), and 125 °C (top). Permission was obtained from the publisher and authors. Source: Adapted from Kim et al. [37].

Thermoplastic Resins on DMA: The stiffness would decrease to a low or zero value if a *thermoplastic polymer*, such as polyethylene, is to be likewise tested, since the DMA test temperatures can be higher than the melting point of the polymer and there is no cross-linking reaction to occur during curing. Naturally, most thermoplastic polymers cannot be used as binders for hot-pressed wood composite boards.

9.8.5 Differential Scanning Calorimetric (DSC) Analysis of PF Resole Resins

Differential scanning calorimetric (DSC) analysis method is also often used to study the curing of PF and other thermosetting resins, often using a Perkin–Elmer DSC instrument. A small sample of liquid or dry resins (~4 mg) is taken in a cell and put under the ambient pressure or increased pressure, and then the chamber is heated to 200 °C or higher at a controlled rate and continuously detects the heat coming out (exotherm) or going into the sample (endotherm) relative to the chamber temperature, resulting in a graph of endothermic and exothermic changes as shown in an example in Figure 9.11 [40, 41]. With a 10 °C/min heating rate, it is seen that the endothermic peak between 60–100 °C is assigned to the evaporation of water, and the exothermic peaks at 130–180 °C are assigned to the polymerization reactions of PF resin molecules. The positions and peak heights are characteristic of the PF resins' F/P mole ratio, degree of polymerization, water content, and sodium hydroxide content, and the methods have been used to evaluate various PF resins with respect to hot-pressing temperatures used for board manufacturing.

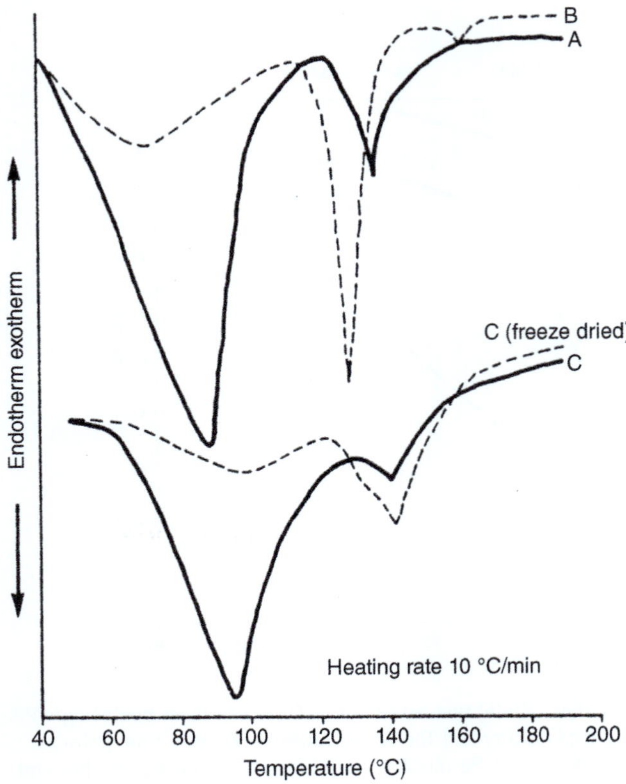

Figure 9.11 DSC thermograms of liquid PF resole resins A, B, and C, at ambient pressure. Resin A : F/P ratio, 2.2; pH, 9+; resin-solids, 43%. Resin B : F/P ratio, 1.7; pH, 9+; resin-solids, 65%. Resin C : F/P ratio, 2.1; pH, 5.0; resin-solids, 77%. Source: Chow et al. [40]/John Wiley & Sons.

9.9 Good Wood Adhesion and Exterior Durability of PF Resin Adhesives

Thus far, we have considered PF resins' curing phenomenon in the form of hardening, but the adhesion aspect cannot be recognized in this hardening process. The adhesion quality attainable in the curing of boards with PF resins can be determined only by gluing and testing the bonds' strengths and durability. Cured PF resin gluelines' adequate bond strengths and good exterior durability for various wood composite products have been found by many experiments and field tests and in uses. This fact indicates that well-cured PF resole resins are durable, and the interphase structure of gluelines formed with wood surfaces is likewise good and durable. The durable adhesion indicates that the cured PF resin's chemical structures and polarity are well compatible and highly attracted (molecular adhesion) to surface molecules of wood. The high polarity is expected from the (polar) phenolic hydroxyl groups present in the whole of polymer structure. The durability of wood adhesives has been theorized to require a good *diffusion of resin molecules* into cut cell wall structures, which can be possible for PF resole resins since there are ample amounts of low molecular weight resin fractions that can molecularly *diffuse* into cut wood cell wall structures and polymerize in-situ to high molecular weight polymers. An interphase of gradual structure changes from wood to cured adhesive polymer layer would thus result, in minimizing the expansion stresses occurring in the glueline during water exposure. On the other hand, the *penetration of adhesive* means the movements of adhesive resin/filler materials into cut lumens and other damaged cell structures, the extent of which is dependent on the viscosity and curing speed of resin, pressing temperature, wood moisture content, and above all on the structures of wood surface, such as whether they are damaged cells or not, radial or tangential cuts, heartwood or sapwood, and latewood or earlywood [42, 43]. A good repair of damaged cells and filling any open voids on wood surface will also be required to obtain a durable glueline. Of course, the extent of adhesive penetration is related to the efficiency of adhesive: too much penetration leads to a starved glueline and poor bond strength, or rougher surfaces need more glue spread, etc. The PF resins that best meet these various wood bonding parameters were found by making various resins and adhesives and testing their wood bonds rather than by designing the molecular system from scratch or by calculations.

9.10 Formaldehyde Emission Problem of PF Resin-bonded Products

Most PF resin-bonded wood composite products emit exceedingly small amounts of free formaldehyde and are not required to abide by the protocols of formaldehyde emission levels required for UF resin-bonded products.

9.11 Various Binder Uses of PF Resins in the US Wood Products Industry

Softwood Plywood and Laminated Veneer Lumber (LVL) Binders: discussed in Chapter 11 [5].

Oriented Strand Board (OSB): Liquid and powder PF resole resins are used and some isocyanate resins are also used in the core-layers; discussed in Chapter 13.

Hardboard panels are made from lignocellulosic fibers bonded with PF resole resins under heat and pressure and used as *door skins (panels)*, sidings, and many other products [5]. Hardboard manufacture was pioneered by the Masonite Corporation in Laurel, MS, and twenty-one hardboard manufacturing plants were operating in 1998 in North America [44]. In this process, wood fibers were made by the explosion process for many years, but they are now made by grinding and refining wood chips under high-pressure steam similarly as in the MDF manufacture. There are two different hardboard manufacturing processes in industry.

Hardboard Wet Process of Manufacturing: Wood fibers and a Resin #4 type highly advanced PF resin are mixed with water where the resin is precipitated on wood fiber, and the slurry is fed onto a screen to form wet mats, which are then pressed in a multi-opening hot-press into bone-dry panels at high temperatures as shown in Table 9.3. Boards have one or both surfaces smooth, referred to as S1S or S2S boards. The rough side of boards results from using a screen in hot-pressing to expedite the steam escape. The smooth side is also made to have a broad range of specialty surfaces such as perforated, striated (ridges),

Table 9.3 US industry's PF resin-bonded hardboard manufacturing (hot-pressing) parameters and market trends.

	Dry process	Wet process
Resin usage (%)	1.5–5.0	1.5–3.0
Press temp. (°F)	400–475	430–480
Mat MC (%)	1–8	65–75

PF resin usage (non-volatiles) is based on oven-dry wood weight.

- Market growth trend:
  Declining on conventional panels and siding products.
  Increasing on molded door skin-type products.
- Driving forces:
  Negatives: Imported boards, inefficient plants, environmental.
  Positives: uses in building industry, furniture, and automotive parts.

Source: Davis [51].

grooved, tiled, and embossed with wood grain patterns such as for door skins, and also sold as prefinished panels after a primer coat.

Hardboard Air-suspension Process of Manufacturing: used in the Masonite plant in Laurel, MS, using a low molecular weight *Resin #l type* PF resin, injected at about 2–3% loading level into the wet fiber in blowline and sent through a long drying tube of high-velocity hot air, where the water in the fiber/resin mixture is evaporated. The dry, resinated fiber is collected at the end of the drying tube in a cyclone collector. The dwell time of the drying step is short, less than 2 minutes. The resinated, dried fiber is then matted and hot-pressed using a patterned platen at about 400–475 °F with a short press time, as shown in Table 9.3. The board density of hardboards tends to be high, since the panel stability requirement is high in applications, such as door skins. The panel is often applied with linseed oil on both surfaces and cured in an oven, a process called "tempering." The panel is then cooled and humidified, painted with thermosetting polyester-type enamel paints, cured, trimmed, and humidified again before shipment. Hardboards are often overlaid, printed, and painted (often enamel) and used as various interior or exterior decorative panels for furniture, doors, and lap sidings in house construction. Hardwood panels have many outstanding characteristics: uniformity, workability, low cost, non-corrosiveness, compatibility with other materials, excellent dimensional stability under common temperature and humidity conditions, exceptional resistance to weathering, abrasion, denting, and cracking, ability to take and hold coatings, high impact resistance, and excellent dielectric properties. Thickness of hardboards ranges from 1/12 to 1/2 in. and panels of 4 ft. × 16 ft. length are common. However, hardboard markets have been significantly reduced due to their replacement by thermoplastics and aluminum (sidings).

Molded wood products are bonded with novolac PF or spray-dried PF resole resins. The mix of wood particles and resin is melted by passing through heated rolls or extruders and cured in molds under heat and pressure. Lubricants and colors are often added. One high-volume product is wood toilet seats, of which market share has been invaded by thermoplastic products. However, the all-around performance of phenolic binders ensures a substantial market share of molding products to remain competitive. Also, particleboard can be bonded with PF resole resins to produce exterior grade products for use as mobile home decking where the improved weathering characteristics and low free formaldehyde emission potential are needed. Use of phenolic resins would cost more, and production speed would be reduced significantly compared to UF resin-bonded particleboard.

Glass-wool Insulation: Glass wool is produced by melting specially formulated glass, and the molten glass is pushed through small openings of platinum dies. The emerging molten threads are blasted with steam and compressed air in the forming chamber. A resole resin of Resin #1 type, usually made at about 40% solids and F/P ratio of about 2.5 and added with urea, is further diluted, and sprayed on molten glass wool, followed by pressuring the glass wool into a low-density blanket and curing. Ammonia is often added for pH adjustment to the resin prior to use, which also ties up the excess free formaldehyde in the resin. The PF resin binds the glass wool and the ammonia and urea additions result in giving pink or yellow colors. The loading of PF resin binder solids on glass wool insulation products runs at about 7%, and the products are used for insulating homes and commercial and industrial buildings.

Wafer boards are the forerunner of OSB. The industry was pioneered in Idaho in the late 1950 and started production in Canada in the late 1960. The U. S. production began in 1974 by Blandin Wood Products, in Grand Rapids, MI. The adhesive binder was initially novolac PF resins, but OSB became recognized as an improvement and, also, liquid and spray-dried PF resole resins were introduced as binders.

Laminates are manufactured by impregnating fibrous substrates such as paper, cotton cloth, glass cloth, or nylon with PF resole resins of Resin #1 types with relatively low F/P ratios and low sodium hydroxide contents and often made into alcohol solutions. The resin-solids pickup runs 30–70%, and for decorative laminates, this value usually runs at 32–34%. Advancing the resin to the B-stage is done after the impregnation and during the drying, which is a very important step in manufacturing of quality finished laminates. Pressing together a number of B-staged layers necessary for desired thickness is followed at up to 375 °F and under pressure up to 2 kpsi to effect the final cure into the desired shape. There are decorative and industrial laminates [5, 12].

Decorative (high pressure) laminates [5, 12, 13] are exemplified by Formica® panels produced by Formica Corporation, used as table and counter tops, paneling, and cabinet faces. A decorative laminate is made with a backer that is composed of several layers of phenolic resin-impregnated kraft paper sheets and a surface layer consisting of a cellulose pulp sheet impregnated with melamine-formaldehyde resins, and another sheet often printed with a design (print sheet) appropriate for end use is added. Another layer of pulp sheet (top sheet) with MF resins is overlaid on top of the print sheet, and the assembly is cured at elevated temperature of about 130 °C under a high pressure of about 1000–1400 psi. Both print and overlay sheets are impregnated with melamine-formaldehyde resins at a rate of 35–40 pounds per 1000 square feet, chosen for clarity, no color, and light stability, as well as good physical properties such as abrasion resistance and chemical, water, and heat stability.

Low-pressure laminates: are often called "decorative or melamine component paneling (MCP)," used in vertical surfacing of wood, such as drawer fronts and paneling for furniture, cabinets, and partitions. The boards are produced by directly bonding (curing) melamine-impregnated sheets to both faces of a (manufactured) particleboard core, at about 150–175 °C and under pressure of about 200–500 psi. The melamine resins are often modified (etherified) for better flow, and the resin loading level is slightly higher at about 50–55 pounds per 1000 square feet [5, 12, 13].

Industrial (high pressure) Laminates: PF resole resins are used for high-pressure laminates for the same reasons they are used in high-pressure decorative laminates, such as low cost, good strength-to-weight ratio, hardness, and resistance to heat and moisture. The substrates used depend on the properties required of the final products: paper sheets based on wood pulp for electrical properties, cotton cloths for mechanical properties, and glass cloths for both mechanical and electrical properties and also for good heat resistance. The manufacturing method is similar to that of decorative high-pressure laminates. Resin-impregnated layers are dried and advanced to the B-stage in the manufacturing process, and then stored/transported in rolls or sheets, and later cured under heat and pressure to produce sheets, tubes, rods, or molded into various desired shapes. After curing, the products can be machined by metalworking techniques, such as turning, drilling, and sawing. Paper laminate products include oil, air, and gasoline filters and acid battery separators in which low molecular weight PF resole resins (nearly monomeric) are used for better penetration, or saturation, into cellulose cell wall structures [5, 12, 13]. Resin loading is high, and cured PF resin molecules completely coat the cellulose fibers to protect them from contacting with and eroding by the acid.

9.12 Acid-curing PF and PMF Resole Resin and Dispersion Wood Adhesives

Acid-curing PF resins were used in the 1940s and 1950s in bonding of wood, but they were mostly abandoned because the strong acid catalysts needed for curing turned out to be degrading wood. The resins are commonly made with F/P mole ratios in the range of 2.0–2.5 with sodium carbonate or hydroxide catalysts at resin-solids levels of 50% or higher. The resin synthesis reaction is carried out at 80–100 °C until the formaldehyde content diminishes to a few percent and is commonly stopped before the reaction mixture separates into water-rich and resin-rich phases when a resin sample taken is cooled and neutralized with a mild acid, such as citric acid, to pH 4.5 or lower. The extent of polymerization is relatively low like with Resin #1 type. The resin batch is then cooled, partly neutralized with an organic acid, and tested for curing rates with a strong acid catalyst, such as p-toluene-sulfonic acid, to determine the amount of catalyst needed for desired curing speed. In order to obtain decent curing rates, the amount of acid catalyst normally needs to be increased to about 1.0% of sulfuric acid or higher based on the liquid resin's weight, which brings the pH of mix down to about 1.0–1.5. This resin mix cures relatively fast, almost like catalyzed UF resins, and can be used as bonding or molding resins for acid-resistant woods, such as cedar veneer or other hardwoods, under the radiofrequency heating methods at a significantly lower cost than PRF resins [45], but the possibility of wood degradation has dissuaded its use in the industry, especially for products designed for exterior exposure [46–48]. Also, this type of PF resole resins and those modified with melamine-formaldehyde resins are reported, where the resins are made into dispersions (in water) to avoid the resin separation problem or spray dried and tested for higher-moisture curing OSB binder resins [35, 49, 50], but the acidity would still effect wood deterioration. Also, these types of phenol-formaldehyde resins modified with melamine (PMF resins)) are reported to be adequately cured by using a weak acidic catalyst, such as ammonium chloride, to give panels with improved durability vs. melamine-formaldehyde resins only [14].

Acid-curing PF resole resins are also used in manufacturing of rigid and friable PF resin foam products used as base plates in flower arrangements. Novolacs and acid-curing resole resins are used as binders for friction materials, such as brake linings, clutch facings, grinding wheels, and emery paper, where the acid catalysts cause little problems [5, 12, 13].

9.13 Summary

Reaction chemistry, industrial and laboratory manufacturing, various grades and applications, and curing of alkaline PF resole wood adhesive type resins are presented. Phenol molecule is activated on its two ortho- and one para-carbons and it is commonly reacted with 1.5–2.5 moles of aqueous formaldehyde in the presence of sodium hydroxide at about 65–70 °C to form various hydroxymethyl phenols and then further reacted at up to 100 °C to form methylene bonds between phenol molecules (polymerization) until the targeted degree of polymerization is attained, resulting in branched poly-(hydroxymethyl-methylene-phenols) having a range of molecular weights. Sodium hydroxide is added stepwise to reach up to NaOH/phenol mole ratio of about 0.8 along with some additional water, which keeps the viscosity of resin mixture within an adequately stirrable range and lowers the resin-solids level. The polymerization reaction of PF resin components increases the viscosity, and the sodium hydroxide and water additions decrease it, and the reaction is ended by cooling to room temperature, resulting in PF resole resins having a target viscosity, resin-solids level, and sodium hydroxide content. The curing speed of PF resole resins is decided by the NaOH/P mole ratio and degree of polymerization, the higher the faster. The wood bonding industry uses PF resole resins in four curing-speed grades: paper saturation type, OSB face-layer type, OSB core-layer type, and softwood plywood-type binders. PF resole resins can be usefully cured only at elevated temperatures above about 120 °C with the curing speed becoming faster with increasing temperatures up to about 200 °C. Bonding performance of thermosetting adhesives depends much on whether the adhesive polymer solids go through the melt stage to wet the substrate. As the degree of polymerization of PF resins increases, the polymer's melting point increases. In curing of liquid PF resin adhesives, the water evaporates first, and the remaining resin-solids need to melt to cure adequately, but if the resin polymer melts but needs to advance further to cure, the cure time would lengthen. Various types of PF resole resins with their relative cure speeds, application areas and testing of resin properties are presented as practiced in the wood composite board manufacturing industry.

References

1 Van Epps, C.F. (1944). Water-soluble phenolic resin adhesive resin and method. US Patent 2, 360,376.

2 Redfern, D.V. (1948). Art of making phenol-aldehyde reaction products and product thereof. US Patent 2, 457, 493.

3 Redfern, D.V. (1953). Production of thermosetting phenol-formaldehyde resin condensation products. US Patent 2, 631,098.

4 Van Epps, C.F. (1944). Water-soluble phenolic adhesive resin and method. US Patent 2, 360,376.

5 Stanford Research Institute (1982). *Chemical Economics Handbook*. Menlo Park, CA: SRI International.
6 Carswell, T.S. (1946). *Phenoplasts: Their Structure, Properties, and Chemical Technology*, 69. New York: Interscience.
7 Barkhuff, R.A. and Carswell, T.S. (1944). *Industrial & Engineering Chemistry Research* 36: 461.
8 Hultzsch, K. (1950). *Chemie der Phenolharze*. Berlin: Springer-Verlag.
9 Robitschek, P. and Lewin, A. (1950). *Phenolic Resins: Their Chemistry and Technology*. London: Iliffe.
10 Megson, N.J.L. (1958). *Phenolic Resin Chemistry*. New York: Academic Press.
11 Whitehouse, A.A.K., Pritchett, G.K., and Barnet, G. (1967). *Phenolic Resins*. New York: Elsevier Press.
12 Knop, A. and Pilato, L.A. (1985). *Phenolic Resins*. Berlin: Springer-Verlag.
13 A. L Wooten. 1988. Personal communication. Mississippi State University 1988.
14 Pizzi, A. (1983). *Wood Adhesives, Chemistry and Technology*. New York: Marcel Dekker.
15 Northcott, P.L., Hancock, W.V., and Colbeck, H.G.M. (1962). *Forest Products Journal* 12 (10): 478.
16 De Jong, J.I. and de Jong, J. (1953). *Recueil des Travaux Chimiques des Pays-Bas* 72: 497.
17 Freeman, J.H. and Lewis, C.W. (1954). *Journal of the American Chemical Society* 76: 2080–2087.
18 Granger, F.S. (1937). *Industrial & Engineering Chemistry Research* 29: 860.
19 Debing, L.M., Murray, G.E., and Schatz, R.J. (1952). *Industrial & Engineering Chemistry Research* 44: 354.
20 Yeddanapalli, L.M. and Francis, D.J. (1962). Makromoleculare. *Chemie* 35: 74.
21 Jones, R.T. (1983). *Journal of Polymer Science: Polymer Chemistry Edition* 21: 1801.
22 Haupt, R.A. and Waago, S. (1995). *Wood Adhesives 1995* (ed. A. Christiansen and A. Corner), 220. FS, FPL.
23 Sprengling, G.G. and Lewis, C.W. (1953). *Journal of the American Chemical Society* 75: 5700.
24 Roffael, E. and Schneider, A. (1980). *Zum Holz als Roh-und Werkstoff* 38: 151.
25 Gollob, L., Krahmer, R.L., Wellons, J.D., and Christiansen, A.W. (1985). *Forest Products Journal* 35 (3): 42.
26 Hse, C.Y. (1972). *Forest Products Journal* 22 (9): 104.
27 Kim, M.G., Nieh, W.L., Sellers, T. et al. (1992). *Industrial & Engineering Chemistry Research* 32: 973.
28 Flory, P.J. (1953). *Principles of Polymer Chemistry*. Ithaca, NY: Cornel University Press.
29 Wooten, A.L., Prewitt, M.L., Sellers, T. Jr., and Teller, D.C. (1988). *J. Chromatography* 345: 371.
30 Lee, S.M. and Kim, M.G. (2007). *Journal of Applied Polymer Science* 105: 1144.
31 Watt, C., Kim, M.G., and Davis, C.R. (1996). *Journal of Wood Chemistry and Technology* 16: 21.
32 Freeman, H.G., Baxter, G.F., and Howell, J.T. (1972). Phenol-formaldehyde adhesive resin containing hydroxyethylcellulose. US Patent 3, 663, 481, Weyerhaeuser Co., Tacoma, WA.
33 Tarkow, H., Feist, W.C., and Southerland, C.F. (1966). *Forest Products Journal* 16 (1): 61.
34 Ellis, S. (1993). *Forest Products Journal* 43 (2): 66.

35 Andersen, A. and Troughton, G.E. (1996). *Forest Products Journal* 46 (10): 72.

36 Lofthouse, M.G. and Burroughs, P.J. (1978). *Journal of Thermal Analysis and Calorimetry* 13: 439.

37 Kim, M.G., Nieh, W.L.-S., and Meacham, R.M. (1991). *Industrial & Engineering Chemistry Research* 29: 798.

38 Ferry, J.D. (1961). *Viscoelastic Properties of Polymers*. New York: Wiley.

39 Geimer, R.L., Follensebee, R.A., Christiansen et al. (1993). *Proceedings of 27th International PB/Composite Materials Symposium* (ed. T.M. Maloney), 65. Pullman, WA.

40 Chow, S. and Steiner, P.R. (1979). *Journal of Applied Polymer Science* 23: 1973.

41 Chow, S., Steiner, P.R., and Toughton, G.E. (1975). *Wood Science* 8: 343.

42 Gillespie, R.H. and River, B.H. (1976). *Forest Products Journal* 26 (10): 21.

43 Brady, D.E. and Kamke, F.A. (1988). *Forest Products Journal* 38 (11/12): 63.

44 Sellers, T. (2001). *Forest Products Journal* 51 (6): 12.

45 Vick, C.B. (1988). *Forest Products Journal* 38 (11/12): 8.

46 Vick, C.B. (1984). *Forest Products Journal* 34 (10): 37.

47 Vick, C.B. (1987). *Forest Products Journal* 37 (10): 43.

48 Little, G.E. and Pepper, K.W. (1947). British Plastics, October: 430.

49 Steiner, P., Troughton, G.E., and Andersen, A.W. (1993). *Forest Products Journal* 43 (10): 29.

50 Nakamura, N. (1977). *Journal of The Adhesion Society of Japan* 13 (1): 33.

51 Davis, C.R. (2005). Invited presentation in class lecture at Mississippi State University.

10

PRF Novolac Wood Adhesive Resins and Lumber Lamination

Laminated lumber and wood I-beams are manufactured in North America mostly using phenol–resorcinol–formaldehyde (PRF) (copolymer) resin binders with an annual usage of about three million pounds, including some resorcinol–formaldehyde (RF) (homo-polymer) resins. Because of the higher cost of resorcinol, PRF and RF resin adhesives are more expensive than PF resins on a weight basis. Resorcinol (R) is about \$3.00 vs. phenol (P) is about \$0.45 per pound. RF resins were first developed in 1943 [1], and then PRF resins were developed by incorporating phenol to lower the cost [2] while maintaining similar curing characteristics of RF resins. PRF and RF resin adhesives have been the sole adhesives certified for bonding exterior structural laminated beams (glulam). Forty-three glulam manufacturing plants, 43 I-joist manufacturing plants, and 50 lumber end-jointing plants have been operating in 1998 in North America [3]. In this chapter, we will principally discuss PRF resins, and one advanced reading material on RF resins is included. Tannin–formaldehyde resins are discussed briefly, which are similar to PRF resins in the preparation, application methods, and areas of use but not much used in North America.

The P/R mole ratio of common PRF adhesive resins ranges 0.25–1.25 and therefore PRF resins can come with varying grades and prices, curing speeds, and performance characteristics. The price of common PRF resins is about three times higher than PF resins on weight basis. Fillers are used in PRF resins, and their overall adhesive cost is currently at about \$1.5 per pound, which is reasonable in comparison with many other specialty adhesives available on the market, such as epoxy resin adhesives. PRF resins are unique in that they can be cured at room temperature and their durability is unmatched by other adhesives. The capability of curing at room temperature or by radiofrequency (RF) heating method makes it possible to manufacture large dimension wood composite products using only jigs and clamps without requiring large expensive ovens or hot presses. The lumber laminating industry has been operating in the Northwest using Douglas fir and western hemlock and, in the South and Southeast using various southern yellow pines [4]. Hardwoods can also be bonded satisfactorily using PRF resin adhesives but are not much used in laminated lumber manufacturing [5, 6].

Adhesives and Finishes for Wood: For Practitioners and Students, First Edition. Moon G. Kim.

10.1 Manufacturing and Chemistry of PRF Resins

High reactivity of resorcinol toward formaldehyde in comparison with phenol: PRF resins' room temperature curing reactivity is derived from the resorcinol, which has two hydroxyl groups, on C_1 and on C_3 positions on the benzene ring, in comparison to one hydroxyl group in phenol, as shown in Figure 10.1. In resorcinol, the 2, 4, and 6 ring carbons (ortho and para carbons with respect to the first hydroxyl group) are activated doubly by the second hydroxyl group in comparison with phenol. The activation is further increased by sodium hydroxide catalyst as discussed in Chapter 9 on PF resins. Because of this double activation, the reactivity of resorcinol toward formaldehyde and that of the resultant hydroxymethyl resorcinol molecules is far greater (about 12 times) than those of corresponding carbons of phenolic ring discussed in Chapter 9 [7]. This high reactivity of resorcinol allows RF resins to bond wood at room temperature as reported early in a U.S. patent [1]. Later, PRF resins were developed by incorporating the lower cost phenol into the backbone structure of PRF resin polymers where resorcinol molecules are bonded as peripheral groups, for maximizing the fast-curing capability [2, 8]. (On the other hand, PF resins can be cured usefully only at elevated temperatures as discussed in Chapter 9.)

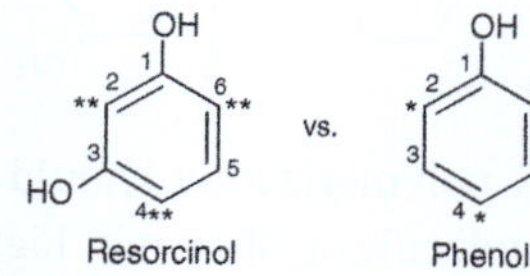

Figure 10.1 Molecular structures of *resorcinol* and *phenol* with carbon numbers shown.

Copolymer novolac PRF resin chemistry and manufacturing procedure: PRF resins are novolac-type *copolymers of phenol and resorcinol* made by reacting a limited amount of formaldehyde with respect to the combined amount of phenol and resorcinol: $F/(P+R)=$ 0.6–0.8 [8, 9]. In order to achieve an optimal high extent of copolymerization between the slow-reacting phenol and fast-reacting resorcinol components, the synthesis of PRF resins is conducted in two consecutive stages as shown in Table 10.1 as an example.

In the first stage of resin synthesis, phenol and formaldehyde are reacted in the presence of water and 1–2% sodium hydroxide catalyst based on the charge weight at a relatively low F/P mole ratio of up to about 1.10–1.25 to form a low molecular weight PF resole prepolymers as shown in Figure 10.2. Due to the relatively low F/P ratios used in this step, the resultant PF resole polymer structure is mostly linear and the extent of polymerization is low, but in theory, each PF prepolymer molecule needs to have one or two terminal hydroxymethyl groups for reaction with resorcinol to obtain "resorcinol-capped" PRF resin molecules. For example, in the resin of Table 10.1 with an F/P mole ratio of 1.08, the degree

Table 10.1 Typical synthesis procedure of a PRF resin (1000 g batch).

1. 250.0 g of 90% phenol (2.39 moles), 44.0 g of water, and 15.0 g of 50% NaOH are charged in a reactor and heated to 70 °C.
2. Begin to add 155.0 g of 50% formaldehyde solution (2.58 moles) dropwise and finish in 15 minutes and hold for 20 minutes at 70 °C ($F/P = 1.08$).
3. Add 15.0 g of 50% NaOH and heat to 90 °C and hold until G viscosity by G–H viscosity scale.
4. Add 99.0 g of water and 180.0 g of 100% resorcinol (1.63 moles) and heat to reflux until W viscosity. {$F/(P+R) = 0.64$}.
5. Add 200 g of water and reflux until X viscosity.
6. Cool to room temperature and add 42.0 g of ethanol.
   - Resin solids content = 46.5% (calculated).
   - P/R ratio = 1.21 by weight and 1.47 by moles.

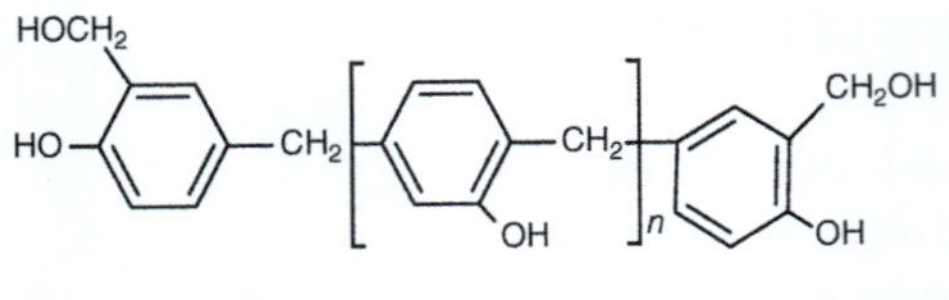

Figure 10.2 A schematic representation of PF base resins' oligomeric structure (n = about 12).

of polymerization should reach about 12 in order to have one hydroxymethyl group per molecule, as shown in Figure 10.2. Thus, the F/P mole ratio and the extent of reaction have to be at such a level that: (i) the free phenol content must be low since it is likely to remain unreacted in the final PRF resin and also in the final cured PRF resin. There is not much chance for free phenol to react with formaldehyde from hardener in subsequent curing steps at room temperature; (ii) the molar amount of methylol groups in the prepolymer should be significantly less than the molar amount of resorcinol to be added in the second step to result in efficient PRF resins; and (iii) the free formaldehyde content of the prepolymer should be low enough to minimize the formation of resorcinol homopolymers as well as to assure no violent reaction occurring when resorcinol is added. In practice, the resin synthesis formula and reaction procedure and condition need to be found by trial and error-type experiments.

In the second stage of synthesis, resorcinol, and some water is added, and the mixture is reacted under the same alkaline condition at 80–105 °C. In the initial period, it is common to have some free formaldehyde remaining in the prepolymer resin, and it reacts fast with resorcinol to form various mono-, di-, and tri-hydroxymethyl resorcinols [10] and then to form resorcinol dimers and trimers to result in a rapid increase in viscosity of reaction mixture.

This is commonly minimized in practice by minimizing the free formaldehyde content of prepolymer resins. Afterward, the methylol groups of PF prepolymers, catalyzed by the sodium hydroxide present, form the benzylic cations (I and II), which are stabilized as quinone methide intermediates (Ia and IIa) as shown in Figure 10.3, react mostly on the C_4, and C_6 carbons of resorcinol molecules, which are also further activated by sodium hydroxide, resulting in forming of *methylene bonds between phenol and resorcinol molecules*, i.e. formation of copolymers of phenol and resorcinol. This reaction between phenol and resorcinol molecules is sluggish because, under this basic reaction condition, the reactive hydroxymethyl groups belong to "phenol" rings that determine the reaction rate, not to the doubly activated resorcinol rings. When the initial reaction of resorcinol with residual formaldehyde is completed, the reaction temperature is usually raised up in a stepwise manner to the refluxing temperature near 100 °C or slightly higher. The reaction becomes somewhat sluggish because the hydroxymethyl group content of prepolymer, the only functional group in the system, becomes scarcer as the synthesis reaction advances. The progression of reaction is monitored by measuring the viscosity of samples taken at appropriate intervals. As the target viscosity endpoint of resin synthesis is approached, the viscosity increase needs to be minimal or none at all since the hydroxymethyl group content is designed to go nearly to zero in order for a pure novolac resin to be attained. The target viscosity value is commonly S–Z by G–H scale at 50–60% resin-solids content levels.

Resultant novolac PRF resins: The amount of resorcinol used in PRF resin syntheses should be more than enough to consume all hydroxymethyl groups so that a good amount of unreacted reactive sites of resorcinol molecules remain in the resin system as shown in Figure 10.4, including those of free resorcinol molecules, to ensure the final curing reactions to take place adequately (with formaldehyde in hardener) in the gluing step. When PRF resins with high levels of resorcinol are needed for special curing/performance

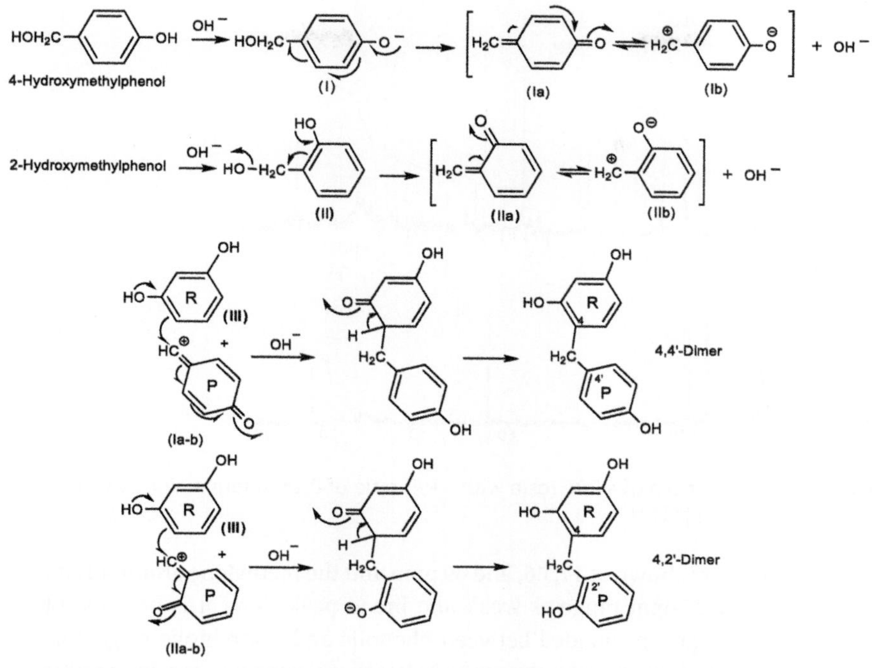

Figure 10.3 Dimer formation by attacking of the 4-carbon site of resorcinol (III) on the *p*- and *o*-hydroxymethyl phenols in the basic media.

Figure 10.4 A schematic representation of PRF co-polymer resin molecules with potential reactive sites indicated with * for carbons of phenol and with ** for carbons of resorcinol.

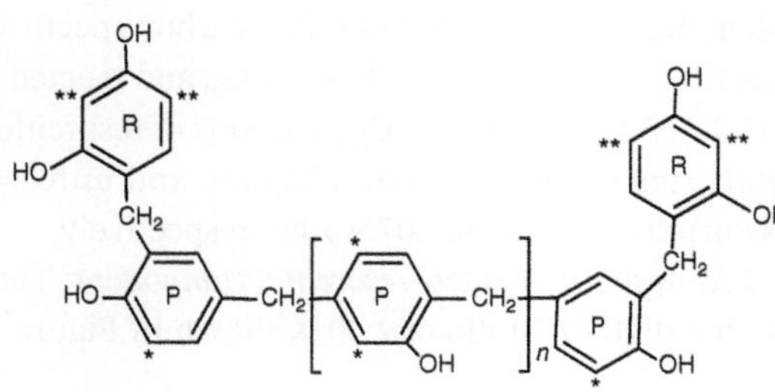

characteristics such as of the fast curing finger-jointing adhesives, the F/P molar ratio and extent of reaction in the prepolymer synthesis stage could be adjusted in such a way that some RF homopolymers are formed. Thus, the overall F/(P + R) molar ratio is usually in the range of 0.6–0.8 as mentioned above, i.e. novolac PRF resins that have no reactivity by themselves. Well-made PRF resins, therefore, are stable and have very long storage lives, often six months or longer. They are dark brown liquids and contain 2–3% sodium hydroxide that acts as catalyst and often contain some ethyl alcohol or ethylene glycol type organic solvents added at the end of manufacturing to improve the wetting property on wood substrates.

10.1.1 Chemical Structures of PRF Resins

PRF resins with a high resorcinol component: The ^{13}C NMR spectrum of a PRF resin having a P/R ratio of 0.25 and F/(P + R) ratio of 0.70 is shown in Figure 10.5 with expanded spectra and integration results (unreported results). In this spectrum, small amounts of

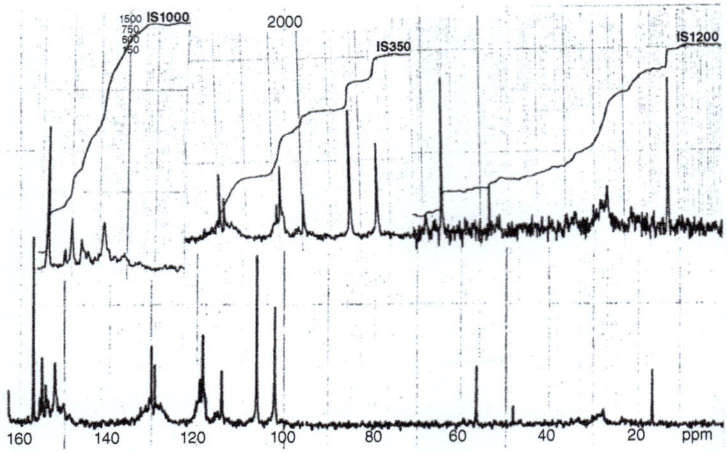

Figure 10.5 ^{13}C NMR spectrum of a PRF resin with a P/R ratio of 0.25 in formamide solvent at about 166 ppm (not shown) [11].

ethanol and methanol are shown at 17, 66, and 49 ppm and the methylene groups of copolymers appear in the 18–35 ppm range as weak and broad peaks, and it is not possible to pinpoint the methylene groups bonded between phenolic and resorcinolic rings. But the major methylene peaks at 28–29 ppm appear to indicate the presence of such methylene groups considering the methylene group chemical shift values of monomers of PF resins discussed in Chapter 9 and RF resin monomers discussed in Section 10.10 [12]. Further, phenolic C_1 and resorcinolic $C_{1,3}$ carbons occurred at 148–158 ppm, C_5 carbons of resorcinol ring, $C_{3,5}$ carbons of phenol ring, and reacted $C_{2,4}$ carbons of phenolic ring all occurred at 127–134 ppm, reacted $C_{2,6}$ carbons of resorcinol ring and unreacted $C_{2,4}$ carbons of phenolic ring occurred at 116–122 ppm, and unreacted C_2 and C_4 carbons of resorcinol ring occurred at 103.5 and 107.9 ppm, respectively.

PRF resins with a low resorcinol component: The ^{13}C NMR spectrum of a PRF resin made with a high P/R ratio of 2.40 is shown in Figure 10.6, which clearly shows the lower level

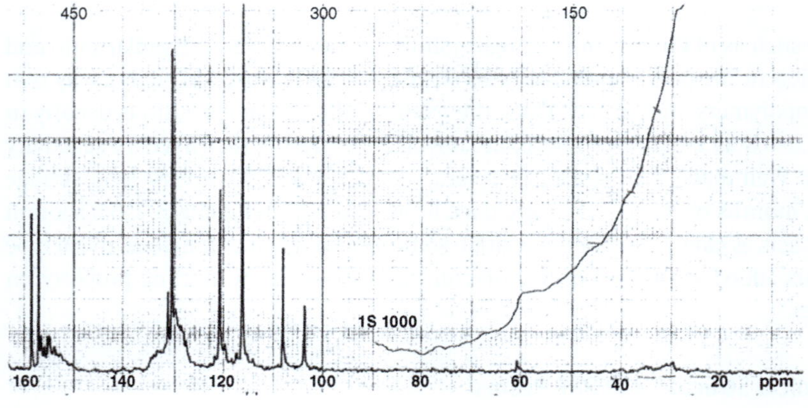

Figure 10.6 ^{13}C NMR spectrum of a PRF resin with a P/R ratio of about 2.40 in formamide solvent at about 166 ppm (not shown) [11].

of resorcinol (by unreacted C_2 and C_4 carbons at 103.5 and 107.9 ppm) and the higher level of phenol (by unreacted C_2 and C_4 carbons at 116.0 and 119.6 ppm), all as sharp peaks (unreported results). One unexpected peak is the small peak at 60.61 ppm that indicates the presence of a small amount of phenolic hydroxymethyl groups, which would affect the resin storage stability. These chemical shift assignments of PRF resins are in good agreement with chemical structures of (pure) RF resins [12] discussed in Section 10.10. Overall, the ^{13}C NMR analysis results reflect reasonably well the general novolac molecular structures of PRF resins schematically depicted in Figure 10.4.

10.1.2 Relationship Between Target Viscosity, Resin-Solids, and Storage Stability of PRF Resins

Finishing of resin manufacture: As discussed in Section 10.1, cooking a PRF resin to the point of no viscosity increase is an important criterion in PRF and RF resin manufacturing due to the fact that one shipment of resin would be used over a long time period of time, several weeks and longer, i.e. a good viscosity storage stability is needed. The viscosity attained at the point of no viscosity increase in resin synthesis can come short or long in comparison with the target viscosity, possibly to result in variation of adhesive mix viscosity. When it is short, small amounts of formaldehyde can be successively added and cooked to increase the viscosity to reach the desired viscosity endpoint, which would little affect the resin-solids levels. In case of long, addition of some water or solvent could be a way, which lowers the resin-solids level and may require a modification of adhesive mix components later on, but this erring needs to be minimized in the next cook by adjusting the F/P ratio. Also, when the resin viscosity appears to keep increasing beyond the target point, the batch can be cooled and called the synthesis finished, but this resin would contain some phenolic hydroxymethyl groups and show a shorter storage stability. When its extent is excessive, the adhesive bonding performance has been shown to be negatively affected, presumably because the unreactive hydroxymethyl groups hung on the polymer chains interfere in the setting/curing of adhesive.

Molecular weights of PRF resins are distributed in a wide range up to 10K Da and higher as shown by gel permeation chromatography in Figure 10.7, where dimers (~200) and tetramers (~400) show separate peaks in the low molecular weight region, and the high molecular weight region overreach the column capacity of 10K Da [13].

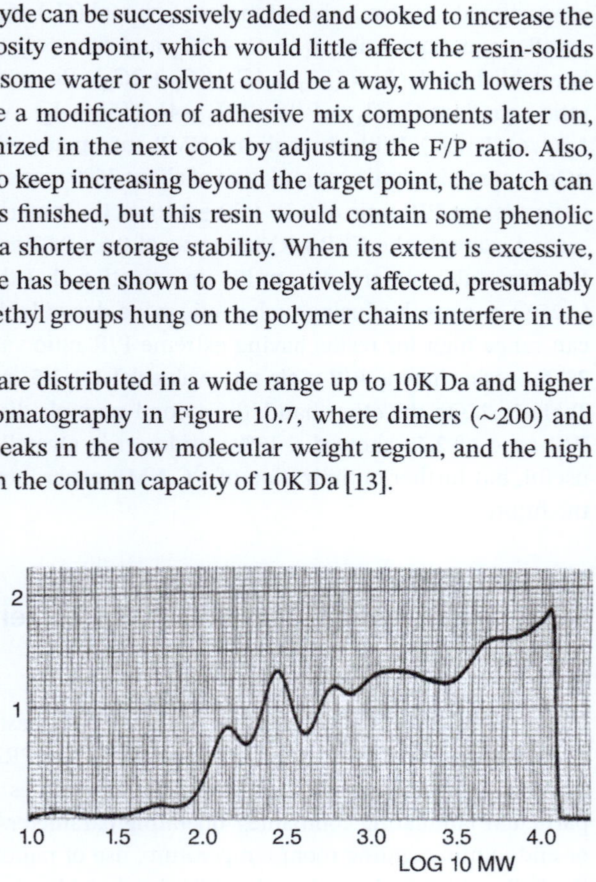

Figure 10.7 Molecular weight distribution of commercial PRF resins by GFC on a Sephacryl 200 column in 0.10 N NaOH [13].

10.1.3 P/R Ratio Measurements from ^{13}C NMR Spectra

PRF resins were developed to lower the higher cost of RF resins, and this has been achieved without significant loss of wood bond quality in most cases [8, 9], and therefore, the commercial volume of PRF resins used in the wood lamination industry far exceeds that of RF resins. Then, there are different grades of PRF resins depending on the resorcinol contents, normally higher resorcinol content resins for finger-jointing type radio-frequency heat-curing resins and lower resorcinol contents for beam lamination-type room temperature curing resins and, in each kind, the resorcinol level can also vary to make the resin cure faster in cool weather and cure slower in warm or hot weather, etc. The phenol/resorcinol ratio of commercial PRF resins is never let known to adhesive users albeit it can be usually reflected in the pricing.

One way of measuring the P/R mole ratio is using the quantitative ^{13}C NMR spectroscopy [11]. Larry W. Amos of Weyerhaeuser Co. of Tacoma, Washington, worked out an indirect equation in a unpublished report: for example, for Figure 10.5, let $\boldsymbol{I_1}$ be the integrals of combined C_1 carbons of phenolic ring and $C_{1,3}$ carbons of resorcinolic ring, $\boldsymbol{I_2}$ be the integrals of all other combined aromatic carbons, and I_3 be the integrals of combined methylene carbons. Luckily, $\boldsymbol{I_1}$ and $\boldsymbol{I_2}$ peaks are well separated in chemical shift values and their integrals can be separately measured. Since $\boldsymbol{I_1}$ and $\boldsymbol{I_2}$ intensities are each proportional to the number of moles of phenol (C_P) and resorcinol (C_R) components, two parallel equations arise: $I_1 = C_P + 2C_R$ and $I_2 = 5C_P + 4C_R$, and a ratio (I_2/I_1) equation is established: $I_2/I_1 = (5C_P + 4C_R)/(C_P + 2C_R)$. Rearranging this equation results in this ratio equation: $C_P/C_R = \mathrm{P/R} = (2I_2 - 4I_1)/(5I_1 - I_2)$. Also, the F/(P + R) ratio is given by $I_3/(1/6)(I_1 + I_2)$. With six replicate NMR runs made with an ample pulse-delay time and nOe suppression and accumulation of about 3600 scans for the first example above resulted in an average P/R ratio of 0.20 + 20 ± 0.10 vs. the materials charge ratio of 0.25, and for the second example the F/(P + R) ratio was of 0.6 ± 0.07 vs. the material charge ratio of 0.70. Dr. Amos also remarked from his investigation that the downside error range was high (−25%), as a result of using such a ratio value-dependent equation, indicating that the error can range high for resins having extreme P/R ratio values. The error ranged only about 3% for resins having P/R mole ratio values of 1.0–1.5, such as the example resin cooked in Table 10.2. On the other hand, the second example discussed above (Figure 10.6) having P/R ratio of 2.40 showed a 13% upside error. Overall, this method of analysis appears useful, but further optimization of ^{13}C NMR experimental parameters could be useful in the future.

10.2 PRF and RF Resin Adhesives Currently Available in Industry

Properties of industrial PRF resins: There are at least three large resin manufacturing companies in the U.S., supplying various grades of PRF and RF resins. Properties of one typical PRF resin are shown in the table. The grades of PRF resins are optimized to fit particular laminating companies' operating parameters such as wood species, laminating or end-jointing, gluing room temperature, use of radiofrequency heating, etc. In general, the R/P and F/(P + R) mole ratios, filler kind and level, sodium hydroxide content, organic solvent kind and content as well as the price will vary for different grades of PRF resins. In practice, the resin company, based on the laboratory evaluation data, will present two or three different grade resins with offer prices to the laminator, who will run them in a trial run and the bond quality including the exterior durability will be evaluated. Replacement of existing adhesive with a new adhesive is then decided based on whether the adhesive is easier to use with minimum control, the bond quality including the durability is satisfactory, and the cost is competitive.

Physical properties of a typical commercial PRF resin

Appearance – red/purple liquid.
Nonvolatiles (%), 53.5–54.5.
Density (lbs./gal. @ 25 °C), 9.58–9.66.
Viscosity (cps @ 25 °C), 1200–2000.
pH, 8.8–9.8.
Flash point (°F), >200.
Free Formaldehyde (%), <0.1.
Free phenol (%), 9.0–9.5.
Storage life @ 25 °C, 3 mo.

PRF resins for use with a powder or liquid hardener: For room temperature-curing PRF or RF resins, besides the long-term storage stability, the target viscosity and resin-solids level are the important variables since the resins need to be formulated with hardeners into adhesive mixes that have certain target viscosity and resin-solids levels. Commercial PRF resins are manufactured roughly in two grades: *the first types* have a higher viscosity and higher resin solids levels (~65%) to be used with a liquid hardener using the continuous meter-mixing setups, and *the second types* have a lower viscosity and lower resin-solids levels (~50%) to be used with a powder hardener in the batch mixing process [14, 15]. The batch mixing process using solid hardeners had been the common method in the past. On the other hand, in the meter-mixing process, the liquid form hardener is first made in the laminating plant by slurrying a powder hardener in water and pumped into a line that meets the resin line and they go through the mixing zone and applied on the lumber with an option to stop and go, as necessary. This meter-mixing method thus avoids the wasting of mixed adhesives when the production line has to be stopped unexpectedly. In the batch mixing process, the mixed adhesive is made by mixing the resin and solid harder shortly before a shift begins, and when the plant is stopped unexpectedly the unused mixed adhesive has to be discarded because of limited pot-lives of about two hours. The batch mixing method can be more convenient for small-scale operations, including research work. The mixed adhesives' viscosity and resin/filler solids levels of the two methods are in the same range, 7k–10k cps at adhesive resin/filler solids level in 50–60% ranges. In the mixing process, some free formaldehyde present in the hardener reacts soon with resin molecules and generates some heat and it results in advances in mix viscosity, necessitating some cooling of mixing tank.

10.2.1 Some Nonideal Aspects of PRF Resins

Shelf life: Some commercial PRF resins will have limited shelf lives due to the minor aberration in charge mole ratios arising from small weighing errors and resultant incomplete cooking extent, i.e. some hydroxymethyl groups on the phenolic rings remain in the finished resin while the target viscosity is already reached, resulting in some advancement of resin viscosity during the storage period. The storage period of resins often extends to several months in industry. Commonly, the resin manufacturing industry takes good care of this aspect of resin technology and the extent of viscosity increase is often small and does not affect the gluing operations too much. For example, one commercial PRF resin indicates a nine-month storage life. However, excessive viscosity increases can affect the gluing process including changes in the adhesive application rate and open/closed assembly times. Also, the residual hydroxymethyl groups can interfere with the curing process to result in poorly cured polymer structures.

Free phenol content: PRF resins often contain some free phenol due to the low F/P molar ratio used in the first step of resin synthesis. Use of higher F/P mole ratios would lower the free phenol content, but this would require higher amounts of resorcinol in order to arrive at a stable resin viscosity. On the other hand, although the free phenol content commonly reaches 5% or higher, this currently does not seem to adversely affect the gluing operations or performances, although the phenol causes odor problems in the gluing room and is potentially hazardous for bodily contact. The free phenol will eventually dissipate into the air, polluting the air and wasting valuable material.

Free resorcinol content: the resorcinol charge levels in PRF resins are normally kept at relatively high enough to assure the curing reactivity of adhesive mixes and, therefore, there is some free resorcinol left un-reacted in PRF or RF resins. This does not adversely affect the gluing procedure or performance since free resorcinol readily reacts with the hardener and can participate in the curing process as with PRF copolymer resin molecules. However, free resorcinol is also hazardous for bodily contact and can cause odor problems as with the free phenol present in resins.

10.3 Hardeners for PRF Resin Adhesives

$$\text{HO–(–CH}_2\text{–O–)}_n - \text{CH}_2\text{–O–H} \xrightarrow[\text{NaOH/H}_2\text{O}]{\text{Depolymerize}} (n+1)\,\text{CH}_2\text{O} \tag{10.1}$$

Paraformaldehyde as hardeners: PRF resins are novolacs that need to have extra formaldehyde, i.e. hardeners, to cure. For various reasons, the industry adopted hardeners made of paraformaldehyde (a solid) as the formaldehyde donor, mixed with wood flour filler for control of mix viscosity and penetration. Paraformaldehyde is a relatively low molecular weight polymer of formaldehyde obtained by polymerization of monomeric formaldehyde available in aqueous solutions. Paraformaldehyde depolymerizes readily in the basic aqueous solution of mixed PRF adhesives to give off free formaldehyde (Eq. (10.1)). Technical grade, ground paraformaldehyde, often called para powder, as supplied by manufacturers, is a fine white powder or flakes with a mild formaldehyde odor due to the presence of small amounts of residual monomeric free formaldehyde.

Preprocessing of paraformaldehyde: This para powder or flakes need to be ground as necessary and often de-fumed prior to formulate into hardeners to minimize the workers' exposure to free formaldehyde in the gluing area and also to obtain consistent adhesive mix viscosity values in the mixing and initial application periods.

Amount of paraformaldehyde: For a proper cure of PRF resin adhesives the amount of paraformaldehyde currently stands at a combined F/(P + R) ratio of 1.2–1.4, including the formaldehyde used in the resin synthesis step. For example, if the PRF resins were made with a F/(P + R) mole ratio of 0.75, the amount of formaldehyde in the hardener needs to be 0.35–0.65 mole per mole of phenol and resorcinol combined.

Fillers: Hardeners also have fillers such as walnut shell or wood flour and clay-type materials for viscosity increase and control. The clay-type fillers also give the adhesive mixes some thixotropic character that helps the applied adhesive flow minimally on vertical surfaces and at the same time gives an optimal flow in pumping and an adequate adhesive transfer between substrates when they are mated and pressured [15].

Powder hardeners are made by mixing ground and de-fumed paraformaldehyde powder and fillers together commonly using the ribbon-type blender and packed tight in plastic bags for storage and shipment. Normally, the mixing weight ratio of the resin to the powder hardener is 4 : 1 or 5 : 1 and adhesive mixing is conducted in batches.

Liquid hardeners: Powder hardener preparation procedure is the same except an acid component is added for stabilizing the hardener mix in water while waiting in the supply tank perhaps to the end of a gluing shift. The weight of the hardener to be mixed with water at the laminating plant needs to be specified, and the mix is made at the laminating plant. Liquid hardeners have certain advantages over the powder hardeners, such as a fast meter-mixing, longer working lives, and less adhesive wastes [14, 15].

10.4 Mixing of a PRF Resin and Hardener, Pot-Lives, Working Life, and Assembly Time

Batch mixing: A powder hardener is used in the batch mixing mode, which involves about 5–10 minutes of vigorous mixing. The polymeric paraformaldehyde of hardener partially dissolves into the resin (which has a basic pH) and soon gives off the free formaldehyde present in the para powder. The free formaldehyde reacts with resin (resorcinol components) generating some heat. Cooling may be needed at times. Soon after, the depolymerization of para powder ensues slowly, the rate of which depends on the character of para powder, i.e. its degree of polymerization, and the adhesive mix starts to increase in viscosity toward the gelation without input of external heat. The rate of depolymerization also depends on the sodium hydroxide content of mix and temperature of mixing tank. The heat generation is often controlled by external cooling during the mixing operation, especially in summer months to obtain the specified *pot-lives.* The adhesive mix's viscosity needs to be in 7k–10k cps range, which is mostly attained with the solid hardener components added to the base resin of 0.5k–2k cP viscosity. Because of the continuous generation and reaction of free formaldehyde from hardener the adhesive mix increases in viscosity and has a limited pot-live, commonly, about 4.0 hours at 60 °F, 2.5 hours at 70 °F, and 1.0 hour at 80 °F. Resin suppliers also give the adhesive mix's "*working-life,*" which is often 10–15 minutes shorter than pot-life. During this pot-life period, the viscosity of the mix advances and it can be checked by measuring the viscosity using the Brookfield viscometer, and it reaches to the point where the spreading and wetting become inefficient, i.e. the end of pot-life. Also, resin suppliers have different grades of powder hardeners, i.e. with increasing levels of para powder for increasing the curing speed of adhesive, which can also shorten the pot-life, working-life, and assembly time. The batch mixing process tends to generate wasted resin mixes from operation interruptions, and this has been mostly minimized by introducing the meter-mixing, liquid hardener technology.

Continuous meter-mixing technology: A liquid hardener is used in the meter-mixing mode, where the limited pot-life problem of adhesive mix is mostly eliminated [14, 15]. The PRF resin is made in a more concentrated form (higher resin solids) and higher viscosity to compensate for the water going in from the liquid hardener. The (liquid-type) hardener is supplied by resin manufacturers and made into a slurry with a given amount of water. Then, the slurry and PRF resin are meter-pumped separately into one small-diameter mixing channel in a given ratio, and the mixed adhesive coming out of the channel is delivered directly on the wood substrates. The meter-mixing operation can be stopped and resumed, as necessary, and the mixing channel can be flushed out with water if the gluing operation needs to be stopped for a period of time longer than the working life of adhesive mix. This continuous meter-mixing system eliminates much of the pot-life problem and minimizes generation of adhesive wastes.

10.5 Applying and Curing of PRF Adhesives in Wood Lamination

Room temperature curing procedure: The mixed PRF resin adhesive is commonly formulated to have a viscosity range of 7k–10k cP and applied on one surface of moving wood substrates by extruding using a positive-displacement pump through a bank of nozzles as a ribbon form [16]. The positive-displacement pump is known to deliver a more accurate amount of adhesive than the air-pressure pump for high-viscosity adhesives [16].

Moisture content (MC) of wood substrates: it needs to be in 12–17% ranges for most softwood species. Southern pines are used in the South and Southeastern regions and D. fir and western hemlock in the Northwestern region for laminated beam productions, but some hardwoods can be successfully laminated using PRF resin adhesives within these moisture ranges [5]. In finger-jointing of Norway spruce with RF resins, good joint strength values were obtained with lumber at MC levels of 14–26% [17]. At lower MC values, the adhesive mix can lose water quickly to result in partially losing molecular mobility needed for curing reactions, and at higher MC values the adhesive layer would cure slower and some water can remain in the cured glueline even after unclamping to result in a porous glueline structures after evaporation of the excess water.

Knife-planed lumber is preferred in most cases and sanded lumber does not bond well [18]. Sanded wood surfaces often retain partially broken and dislodged cell walls on the very top surface layer that cannot be fully reinforced by cured adhesive, becoming weak points of glueline. The dislodged cell walls are also reported to be blocking the adhesive penetration into cut lumens resulting in weakening of the overall adhesive strength.

End-grain to end-grain joints are a necessary component in wood lamination, and scarf joints and various finger joints are used. Butt joints are not often used [19]. More often, the end joints must be placed only in the middle layers in laminated beams with their number and spacing in a beam regulated by certification rules of the industry. In recent years, finger joints are used most often with more fingers and deeper cuts preferred and, since these joints go into middle layers, the PRF adhesive application and curing step can be done in the same lamination process with pressure applied also from the ends of the beam. The strength of finger joints depends in part on the amount of end pressure applied [20]. A clean cut of joints by using sharp knives is similarly necessary for good end-joint bond strength. Even with sharp knives, the damaged crosscut cell walls have been reported to be the main reason for poorly performing butt end joints, where there can be more dislodged cell walls that not only block the penetration of adhesives into cut lumens but also become the weak link of the cured bond line [20]. Scarf (1 : 6) and various finger joints involve less crosscut cell walls. Commonly, end joints are bonded more efficiently using the faster curing RF or MUF resins under the radio-frequency heating method. In the case of using MUF resin-bonded end joints in PRF resin-bonded laminate beams there remains the question of wood degradation due to the acidic MUF resins in bonded finger joints.

Laying up of laminated beam: A set number of adhesive-applied lumber is then assembled in a jig of target dimensions and shapes (straight or curved) within the specified *assembly time* or working life of adhesive mix, which also depends on the temperature of the room and the adhesive mix's spread rate. The spread rate goes from 50 to 90 pounds of adhesive mix-solids per 1000 square feet of bond area single glueline. At the common temperature of 70 °F, the allowed assembly time goes from 30 to 80 minutes, longer for higher spread rates, and at higher temperatures, it becomes shorter and shorter. Also, this assembly time is divided into the open assembly time and closed assembly time, and the open assembly time is to limit the drying of adhesive mix in open air and losing of wetting capability on the mating surfaces. The closed assembly time is also to limit the adhesive from losing the liquid character before the glue lines get the pressure and attain the target thicknesses. Then, the pressure is applied through a series of belt-clamps along the length of the assemblage, at about 100–200 psi in case of straight beams. Over-pressuring must be avoided to prevent too much flashing of adhesive mix that can result in starved gluelines, and under-pressuring is also avoided to ensure optimally thin gluelines formed. Flashing of adhesive mix also depends on the adhesive mix viscosity and thixotrophy, less for higher viscosity and more for lower viscosity, and since flashing, first of all, means wasting of adhesive, the mix viscosity needs to be well controlled to the right value, i.e. until only a tad bit of flashing is observed.

Curing of PRF adhesive in a laminated beam: Over the curing time, or often called the *clamping time*, the clamping pressure is monitored and adjusted (tightened) as necessary to maintain the same pressure because the clamping pressure decreases (loosens) as the wood members are compacted from the curing of adhesive and dissipation of water from the gluelines. The clamping time is normally 6–12 hours at room or at slightly elevated temperatures, commonly at about 80 °F [21]. After the clamping time, clamps are removed, and the bonded beams are moved to a warm maturing room and kept for a day or more to complete the curing of gluelines. MC equilibration and dimension cutting are followed before shipment. The cured gluelines of PRF or RF resin adhesives are dark brown and highly resistant to water and other environmental elements [22]. Thus, it is noted that the whole gluing operation goes clockwise according to the properties of adhesive mix, i.e. PRF resin and hardener components. *The quality control department* of a lamination beam plant has to carry out surveillance of bond quality in all phases of fabrication and also the completed products (trims) are sampled and tested for acceptance or rejection. The glue preparation and glue application including the end- and edge-bonding and the clamping steps are the more critical quality control activity areas [23].

Power generation in dielectric heating

$$\text{Power in watts} = 1.415 \times 10^{-6} f v^2 \varepsilon'' \,(A/d) \tag{10.2}$$

f = frequency in MHz, v = volts between electrodes, ε'' = loss factor – energy conversion to heat, A = square inch of adherend exposed, d = gap in inches between electrodes.

Radiofrequency (RF) heat-curing of PRF and RF resin wood adhesive gluelines: The radiofrequency (RF) heating/curing method, also called dielectric heating method, has been used to cure various resin adhesives in large, laminated beam manufacturing plants for faster curing operations [24]. This heat-curing method is also preferred in the finger-jointing and webbing of I-beam manufacturing operations. In these processes, the adhesive/adherend beam assembly *under the zig pressure* passes slowly between the two electrodes of an RF generator, where the gluelines are preferentially heated instantly often to a temperature higher than 100 °C. The power absorbed by the adherend is given by Eq. (10.2) and the RF frequency lies between 10 and 27 MHz [23], that rapidly acts on the polarity (dielectric moment) of adhesive/water molecules to result in generating heat by friction of molecules in the glueline. The cured beams are then sent to the maturing room.

In the continuous wood *I-beam* manufacturing plants, the PRF adhesive is applied into the grooves of two parallel flanges (lumber) and the web piece is fed to fit into the grooves with pressure hands. The assembly is continuously made and pushed through the RF heat curing unit, where an instant, short-duration heating is obtained to result in curing of adhesive to a degree that the operator soon can lift and stack the large, bonded products. PRF resin adhesives cure faster at elevated temperatures as shown in Figure 10.8 with rigidity-increasing curves in DMA tests [25]. The requirement for PRF resin adhesives to cure by RF heating is more demanding than that for the room temperature curing procedure: the adhesive should cure faster to avoid arching (or burning) through the beam due to the electric current passing through directly, which is often observed when the adhesive cure rate is slower by such reasons as lack of catalyst or lower resorcinol contents. It is probable that the PRF resin adhesives contain higher amounts of resorcinol or are (pure) *RF resin* adhesives. Acid-curing UF and MUF resins have been known to cure well under this RF heat curing procedure [26] and are more often used for bonding interior use laminated beams. RF and PRF resins are known to cure slower but produce wood bonds

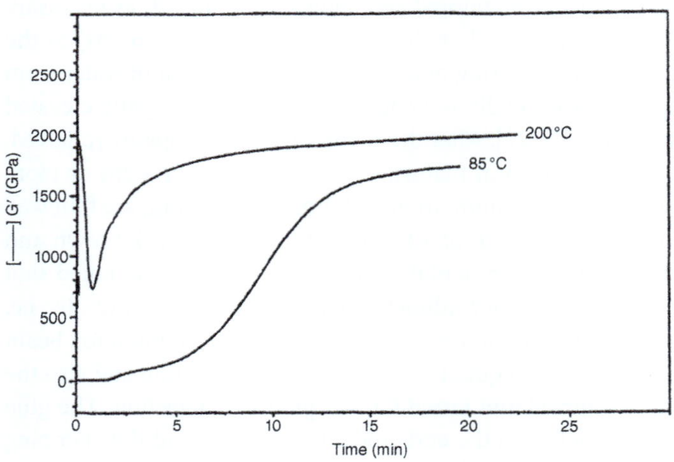

Figure 10.8 DMA curing of a typical PRF resin wood adhesive at two elevated temperatures [25].

that are durable in exterior applications [27]. In general, an excessive glue spread may also cause arching and under-curing [26].

10.6 Curing Chemistry of PRF Resin Adhesives and Industry Practices

The curing chemistry of PRF resin adhesives has not been well elucidated using the more modern instruments like the solid-state NMR method, but it has been considered the curing occurs between resorcinol and paraformaldehyde to form methylene bonds between resorcinol molecules, as shown in Eq. (10.3) from the observation that the exothermic heat of curing in DSC analysis occurs at about 75 °C [28], not at 120–150°that corresponds to the curing of phenolic components. Curing of phenolic components would not occur much, considering that PF novolac or resole resins do not cure to any useful extent at room temperature or by heating for a short period time in the presence of extra formaldehyde. It should be remembered that the PRF resin synthesis procedure was targeted to incorporate the phenol components into the middle sections of PRF resin polymers. But this synthetic scheme is not fully realized, and some phenol components would have only singly bonded to resorcinol or even free phenol remains especially for resins made with high P/R ratios for lower resin costs. Therefore, the question of *free phenol* in PRF resins needs to be raised as to the role and safety of the workers, known to be at 5% or higher on liquid resin weight basis. Another concern is the *formaldehyde gas* emitted from gluelines during the bonding and maturing stages. Decreasing the amount of para powder in the hardener to resolve this issue entails the possibility of under-curing due to lack of formaldehyde in the adhesive layer. Laminated beams are used as main structural components in relatively large buildings such as strip shopping malls and churches, and the liability from a failed beam can be high, therefore the formaldehyde gas problems must be handled without the possibility of compromising bonding performances. One recent report also indicated that the ethanol often added to PRF resins for controlling the flow properties during curing comes out mostly as acetaldehyde (an oxidized form), a potentially toxic material [29].

Curing of PRF resin molecules with formaldehyde, forming hydroxymethyl groups on resorcinol rings and then forming methylene bridges between resorcinol rings.

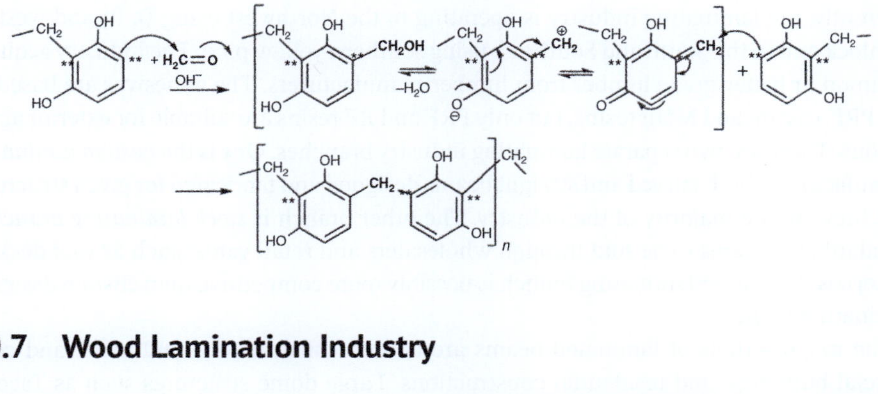

10.7 Wood Lamination Industry

Wood laminating industry sectors: Compared to solid and mass timber, laminated lumber has various advantageous properties, such as longer length, curved structure, balanced and stronger structure, and various shapes. The industry started early in 1800s in the U.S., the earliest example being the laminated arch beam structure of the Trenton Bridge built by

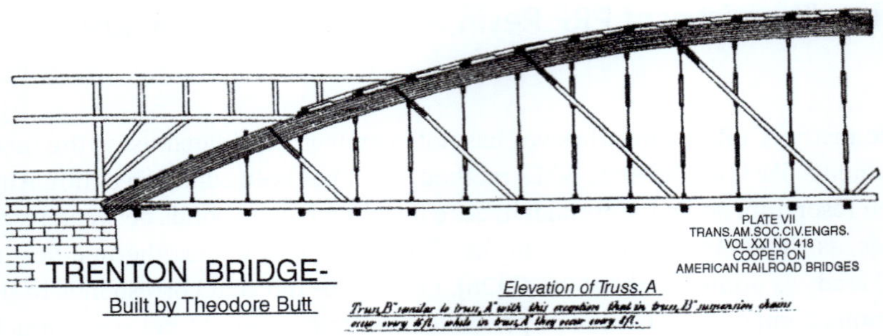

Reproduced with permission of the Forest Products Society from paper by M. Steinhaus [30].

530-ft diameter Tacoma Dome (Tacoma, Washington) in construction (c. 1982) using plywood and laminated beams manufactured from downed trees in the 1980 eruption of Mount St. Helens (Photo courtesy of Wester Wood Structures, Inc., Tualatin, OR) [31].

Theodore Burr, which served more than 70 years for vehicular traffic, shown on the left [30]. Currently, the laminating industry is operating in the Northwest using D. fir and western hemlock and in the South and Southeast using southern yellow pine. The industry acquires common or better-grade lumber from lumber manufacturers. The adhesives are based on RF, PRF, casein, and MUF resins, but only PRF and RF resins are suitable for exterior applications. There are two separate laminating industry branches. One is the *custom laminators* manufacturing both curved and straight beams designed and fabricated for given structures and they are the majority of the industry. The other branch is *stock laminating branch* of standard-size beams to be sold through wholesalers and retail yards, such as roof decking materials. This latter laminating branch is possibly more competitive than custom designed laminating branch.

The major outlets of laminated beams are church constructions, industrial and commercial buildings, and residential constructions. Large dome structures such as Tacoma dome (above) were constructed using PRF resin-bonded D. fir laminated beams and PF resin-bonded plywood [31]. Another industry sector using RF and PRF resin adhesives is the *house log industry* that produces various house logs by laminating kiln-dried 4 × 4 or 8 × 4 wood materials and machining on two or all four sides of the lumber into various log profiles. The adhesive bonding process includes the usual application of PRF resins and clamping, but also the continuous process is used where the lumber is first heated and then fed into the adhesive application chamber and pressuring zones. The produced laminated house logs would be more dimensionally stable and stronger on average compared with whole logs, and the industry appeared to have a significant potential for growth [32].

Wood lamination industry practices: The laminated wood manufacturing industry in the U.S. keep its high quality of products through *the American National Standard for Wood Products – Structural Glued Laminated Timber*, initiated by the American Institute of Timber Construction (AITC) with associated Test Methods for Structural Glued Laminated Timber (ANSI/AITC A190-1992). The standard and test methods define for manufacturers: wood species, lumber qualities and grades, adhesive qualification test (ASTM D 2559-89), adhesive-type designations for dry-use (casein type) and wet-use (PRF, melamine-urea-formaldehyde resins with melamine contents 60% or higher), thickness of adhesive bond line, all manufacturing procedures for face gluing and end-jointing, number and spacing of knots and proof-loaded end-joints in a beam for tension and other layers, and quality control measures with test procedures, and certification by a testing agency. Furthermore, all lumber and end-jointed lumber need to be graded and qualified and so marked in a beam. Also, the required daily quality control tests include dry bond strength tests, and for wet use face and end-joint bonds they include the Cyclic Delamination Test (T110-AITC) with the passing grade attained with the delamination extent less than 5% in the first cycle for softwood and less than 8% for hardwood, and if the values are slightly higher, the values must be less than 10% in the second delamination test cycle. The high level of manufacturing standardization by the industry appears to underlie the wide use of the products and high acceptance.

T110-AITC cyclic delamination test procedure [33]: This test is done for wet use face and end jointed bonds with a sample block cut to be thicker than 6″ and 3″ long and the bondlines are evaluated for delamination on the sawn (end) faces of the samples. The weight of each test specimen shall be measured to the nearest gram and recorded. (i) The specimens are placed in an autoclave and submerged in water at a temperature of 65–85 °F, (ii) A vacuum of 20–25 in. (510–640 mm) Hg shall be drawn and held for 30 minutes then released, (iii) A pressure of 75 ± 5 psi shall be applied for a period of two hours, (iv) The test specimens shall be dried using air at a temperature of 160 ± 5 °F to within 12–15% of their original weight in 10–15 hours, (v) During drying, the specimens shall be placed at least 2 in. (50 mm) apart, with the exposed bond lines on the end grain surfaces parallel to the direction of the airflow. (vi) When the weight has returned to within 12–15% of the original test specimen weight prior to the beginning of the test, the specimens shall be removed from the oven, and the delamination is immediately measured and recorded. (vii) The delamination is measured along the bond lines and is reported as a percentage of the sum total of bond line lengths on both sawn faces of the test sample block or both sections of a single end joint that has been sawn in two parts prior to testing.

10.8 Recent Developments in PRF Resin Adhesives

A good gap-filling capability for PRF resin adhesives is highly desired because the knife-planed lumber surfaces can still harbor waviness and roughness so that a uniform gap thickness is often difficult to achieve. As used presently, PRF resin adhesives have an adequate gap-filling capability for knife-planed wood substrates, attained by controlled viscosity and consistency of mixed adhesives, i.e. through optimizing the filler and thixotropy parameters [9, 16]. Generally, the type of surface roughness generated from sawing is too coarse for the current PRF gluing process. Thick cured bond lines can be inherently weak, as will be discussed in Chapter 20. Similarly, the *abrasive-planed (sanded)* surfaces of D. fir and southern pine were known to give weaker bonds with PRF resin adhesives due to the presence of damaged wood fibers resulted from sanding and not fully repaired in the bonding process [34].

Bonding of wood preservative and fire-retardant treated lumber is another area of current research activity. RF and PRF resin adhesives have been known to bond well for CCA-type chemically treated woods when knife-planing is done before bonding [35–38]. Chemically treated lumber surfaces can be variable since morphological changes occur on the surface from chemical treatments, and these chemically treated lumber need to be dried and knife-planed before gluing. However, knife-planing can be expensive because of not only fast dulling of knives, but also the resultant planing debris' toxic nature, costly to dispose of. Adhesives that do not require the knife-planing step appear to be desirable [39]. On the other hand, it is possible to bond laminated beams first with PRF adhesives and followed by preservative or fire-retardant chemical treatments, although it means treating larger pieces and possibly for longer times. In the case of preservative-treated hardwood species, PRF resin-bonded laminated lumber showed variable bond durability depending on the wood species, also, poor performances were observed for oil-borne pentachlorophenol or creosote-treated hardwoods [40]. On the other hand, CCA-treated, knife-planed laminated lumber or plywood was reported to bond effectively on *building foundations* when the PRF resin adhesive is formulated to have an enhanced gap-filling capacity and the wood is primed with dilute sodium hydroxide solution [41]. On the other hand, ACQ (alkaline copper quaternary) and copper azole-type treated woods were reported to give poor bonding with PRF adhesives even with using hydroxymethyl resorcinol (HMR) coupling agents discussed below [42]. Also, MUF and MF resin-bonded CCA-treated laminated beams do not normally pass the commercial standards of durability but the adhesive bond is improved significantly when wood is primed with HMR coupling agent [43].

Bonding of fiber-reinforced plastics: Another area of recent development in the laminated beam manufacturing industry, arising from the scarcity of good quality lumber, especially for the tension (bottom) members of laminated beams. Use of aramid or epoxy resin and carbon or glass-fiber-based reinforced plastic (RFP) bands has been used as tension layers. This method allows manufacturing of lighter, longer wood beams, up to 500 ft long [44–46], and this method is further improved by using prestressed (stretched) glass-fiber RFP [47]. Of course, the PRF resin adhesives have met the requirements and also further improved by using some primers [48], and recently by using the HMR coupling agent discussed below. Glue-laminated timbers have been recognized as one of the major engineering construction materials, and areas of continued research include improvements in the bending strength, i.e. longer beams, and exterior durability.

Coupling agents based on resorcinol-formaldehyde resole resin chemistry: HMR was reported to be useful [49] in improving the adhesive bonding of wood with fiber RFPs. HMR is made in dilute water solution by mixing 3.34% resorcinol, 3.79% (37%) formaldehyde solution, 2.44% sodium hydroxide, and 90.43% water (F/R = 1.54) and reacting them at room temperature for about four hours. Thus, a resole-type RF resin shown in Eq. (10.4) is formed with a limited pot-live due to their further polymerization to form crosslinked polymers. HMR have been reported to be a good primer (coupling agent) in bonding CCA-treated pine lumber with epoxy, MF, MUF, and moisture-curing pMDI resin adhesives [43]. Later, the authors improved the HMR preparation method to a two-step formaldehyde addition system where the resorcinol is reacted first with a part of formaldehyde to make the novolac-type resin material and the second part formaldehyde is added just before application on wood, where the pot-life is reported to be extended to up to about seven hours [50, 51]. The HMR coupling agents were reported to promote better bonding by stabilizing the wood surface, and not by covalent bond formation with epoxy resin molecules [52].

Resole-type resorcinol–formaldehyde resole resins having limited shelf lives prepared at room temperature [43]

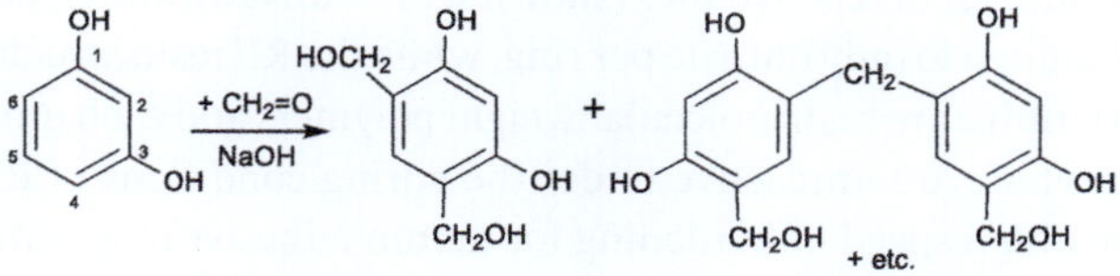

High-temperature resistance of PRF resin-bonded-laminated beams: One concern for face-lamination and finger-jointed wood products is whether the PRF resin bond lines would fail before the wood substrates in fire situations. Early on, fire tests at over 1000 °F indicated that wood joists resist fire significantly better than steel joists [53] and the concern has been also whether the PRF adhesives would soften and weaken before wood fails in fire, since such bond failures would endanger firefighters. Thermoplastic adhesives such as polyvinyl acetate resin adhesives often used for interior-grade finger-jointing of wood are likely to fail in fire. Glulam and LVL products are tested according to a new method ASTM D7247 for this purpose and PRF and RF resins adhesives passed the test handily but pMDI resin-based moisture-cure adhesives were found to be necessary to test case-by-case since there are pMDI adhesives in the market that have a high thermoplastic character, discussed in Chapter 12.

Cross-laminated timber (CLT): has been a recent development and PRF resin adhesives are the main adhesives, and pMDI adhesives are also being introduced. PF resin-bonded thick softwood plywood (LVL) can be similarly used after precut to sizes for floor, roof, and wall constructions. PRF and RF resins are somewhat expensive and can be at times in short supply, and pMDI can fill in but various durability and high-temperature tests need to be done to pass case-by-case.

10.9 Laminating Wood Adhesives Based on Other Resorcinolic Materials

Tannin–formaldehyde wood adhesive resins: Tannins are a class of natural, renewable materials composed of polyphenolic components (oligomers) present in limited but useful amounts in the core wood or bark of certain trees such as quebracho colorado and black wattle. In Africa, South America, and Australia, these tannin–formaldehyde resin adhesives are used as wood composite board binders, and also used to replace some of the resorcinol in PRF-type resin wood adhesives for room temperature curing lamination of lumber. The adhesive technology, including resin formulation and application methods, has been extensively researched and reported [54–59], but these tannin adhesives are currently not much used in North America. As shown in Figure 10.9, the chemical structure of tannins (A-ring) from wattle and quebracho is resorcinol-like materials while tannins from most southern pine barks are phloroglucinol-like materials. Phloroglucinol has one more hydroxyl group on the C_5 carbon of the benzene ring in comparison with resorcinol, and this makes the southern pine tannins even more reactive toward formaldehyde than resorcinol or wattle tannins. D. fir and a few pines have barks containing both resorcinol-like and phloroglucinol-like tannins.

Several important differences exist for wattle and quebracho tannins in comparison to resorcinol: (i) the number of reactive sites (shown with *, unsubstituted) on the resorcinolic ring in tannins is limited to only one site per ring, while the RF resin structures have 1.2–1.8 reactive sites; (ii) tannins are high molecular weight polymers and each unit has a bulky side groups (B-rings), which are unreactive under the curing conditions of PRF resins. These differences mean a high speed of hardening for tannin adhesives (by formation of methylene bonds between tannin polymer strands) in the presence of formaldehyde, or short pot-lives to require a stricter control of gluing operation; (iii) The bulky B-ring groups in tannin polymers would result in curing of the adhesive to a lower overall crosslink density in comparison to PRF resins and this can lead to require higher resin loading levels in adhesive uses; (iv) A recent report indicated that using the slowly formaldehyde-releasing hexamine instead of paraformaldehyde as the hardener component for mimosa tannin resins lengthen the pot-lives, making the resins useful for bonding of thick particleboards under the steam injection pressing condition [60]; (v) Since tannins are obtained by water extraction of tree barks or core wood, commercial tannins normally contain a significant amount of polymeric gums (water-soluble polymeric carbohydrates; hemicelluloses) extracted together, and commercial tannins therefore often show viscosity values that are high and variable at useful resin-solids levels in solution. Therefore, the viscosity of a particular tannin adhesive

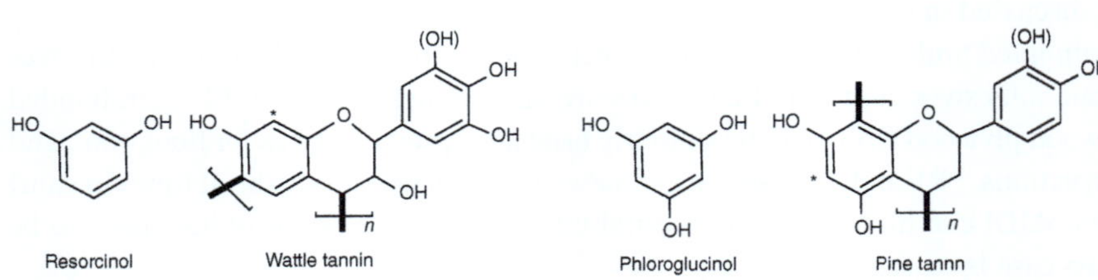

Figure 10.9 Chemical structures of tannins and related materials.

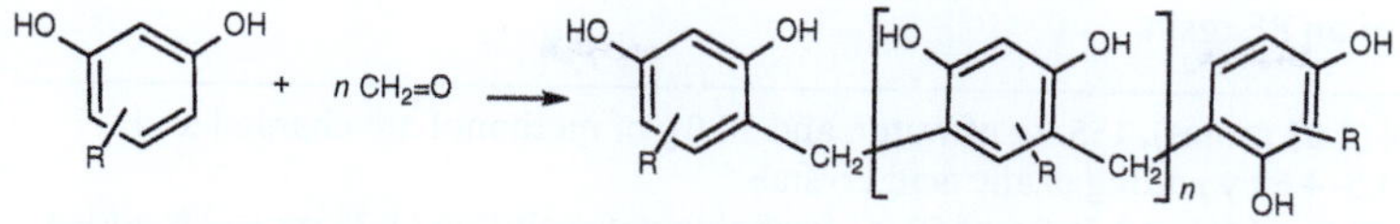

Figure 10.10 A schematic representation of alkyl–resorcinol–formaldehyde resin synthesis. R = methyl, ethyl, and propyl groups, etc. on 2-, 4-, 5-, 6-carbon positions.

formulation can fluctuate from shipment to shipment to make it necessary to adjust the mixing parameters; (vi) The price of wattle, quebracho, and radiata pine bark tannins is relatively high when shipped to North America [58], because the growing of tree, harvesting, and extraction of tannins are all labor-intensive processes, a limitation in scale-up production. In the U. S., therefore, some research has been going on for southern pine bark tannins [58, 59, 61, 62] but the fast-curing rates and other economic reasons made it difficult to bring pine or D. fir tannin-based resin adhesives to commercial uses.

Alkyl resorcinol-formaldehyde resins: Another PRF resin-type adhesive has been reported, made from a distillation fraction of Estonian shale oil [63, 64]. The fraction is composed of more than ten (10) different alkyl–resorcinols, including the major 5-methyl resorcinol as shown in Figure 10.10. Most of the components are capable of reacting with formaldehyde similarly to resorcinol. However, they are somewhat more reactive than resorcinol and each component has somewhat different reactivities and some have poor solubilities in water. Therefore, the resin synthesis procedure requires a complexation additive acetone and an additive ethanol. The resultant resins are reported to be useful in lumber laminations and other wood bonding applications, commercially available in Europe with a trade name "DFK Resins," possibly at lower costs than resorcinol-based resin adhesives. However, some components of the alkyl–resorcinol material are more volatile than resorcinol, producing a foul odor in use, a disadvantage. DFK resin adhesives are not known to be used in the U. S.

10.10 Advanced Reading Materials on Resorcinol–Formaldehyde (RF) Resins

(Adapted with permission of the publisher and minor revisions from: Kim, M. G., L. W. Amos, and E. E. Barnes. 1993. Investigation of a resorcinol-formaldehyde resin by ^{13}C NMR spectroscopy and intrinsic viscosity measurement. J. Polym. Sci.: Part A. 31: 1871-1877. [11])

10.10.1 Synthesis and Fractionation of an RF Resin and GPC Analysis

A RF novolac resin was synthesized according to the procedure shown in Table 10.2. The reactivity between resorcinol and formaldehyde is lowest at pH of about 3.0 and increases with increasing pH, and the reaction was conducted by increasing the pH in a stepwise manner from 4.5→4.6→5.6 and then increased to pH 7.5 to complete the reaction.

Fractionation of RF resin by the solvent/non-solvent procedure: A 100 g sample of the liquid resin was further stabilized by boiling further, followed by distilling out the water of

Table 10.2 Synthesis of an RF resin.

1. 210.0 g of resorcinol (1.91 moles), 155.0 g of water, and 10.0 g of methanol are charged and adjusted the pH to 4.5–4.6 by adding oxalic acid crystals.
2. The mixture is heated to 65 °C and 75.0 g of 50.0% formaldehyde solution (1.25 moles) is added over a period of 30 minutes (F/R = 0.65).
3. The mixture is then heated at a rate of 1 °C/min to reflux and held for two hours. $\eta = 75$ cps.
4. Adjusted pH to 5.6 using 50.0% NaOH solution and Reacted at reflux for one hour. $\eta = 90$ cps.
5. Added 0.3 g of 50.0% formaldehyde solution at a time while measuring the viscosity until the viscosity reaches 1000 cps, totaling about 5.0 g.
6. Adjusted the pH to 7.5 using 50.0% NaOH and reacted at reflux for 10 minutes and cooled. Resin-solids = 50.5%; F/R = 0.70.

resin. The residue was taken up in 660 g of ethanol and added 2635 g of water, followed by heating to 60 °C and cooling overnight while stirring slowly. The precipitates were collected as fraction RF_1 (6.70 g). This procedure of water addition to the solution and precipitation was repeated, resulting in fraction RF_2 (4.20 g with 575 ml water), fraction RF_3 (4.30 g with 1000 ml water), and fraction RF_4 (3.80 g with 1200 ml water). The residual solution was evaporated, the residue was dissolved in 500 ml of ethanol, and the undissolved material is collected as fraction RF_5 (3.90 g). To the residual ethanol solution, ethanol was added to 3l-volume and the undissolved precipitates were collected as fraction RF_6 (3.40 g). To the residue, added 200 ml ethyl acetate, 200 ml ethanol, and 910 ml ether to yield fraction RF_7 (1.70 g) as undissolved residues. To the residual solution, 200 ml ethyl acetate and 500 ml ether were added to yield fraction RF_8 (1.35 g). To the residual solution, adding 500 ml ether yielded fraction RF_9 (5.40 g). To the residual solution, addition of 100 ml ether and 50 ml heptane yielded fraction RF_{10} (3.70 g).

Isolation of model resorcinol resin compounds: The residual solution evaporated all solvents and the residues were sublimated under vacuum at 80 °C to yield free resorcinol (4.58 g) as sublimates. The sublimation residues were dissolved in ether and filtered to isolate inorganic salts (2.15 g), followed by evaporation of ether to yield fraction RF_{11} (4.32 g). Total fractionation yield was 47.14 g (94.3%). Also, a portion of fraction RF_{11} was further separated by silica gel column chromatography to obtain the 2, 4′ dimer (I), 4, 4′-dimer (II), and 4,4′; 6,4′-trimer (III), and a mixture of two other trimers (IV) and (V), as shown in Figure 10.11.

GPC analysis of fractions: All obtained resin fractions were analyzed first by gel filtration chromatography (GPC) using Sephacryl 200 and 0.1 N NaOH solution with a UV detector and the results are shown in Figure 10.12. The graphs generally are sharper for low molecular weight fractions than for high molecular weight fractions as expected, and the obtained (calculated) molecular weight values indicate the fractionation procedure was a reasonable one, separating resin polymers according to the molecular weight values: RF_1 the highest molecular weight and RF_{11} the lowest molecular weight. For the unfractionated whole RF resin, the majority of polymer molecules have molecular weights in the 1000–5000 Da range, which corresponds to 8-mers to 40-mers, quite low values but agreed with the commonly accepted concept of novolac resins. Though, these molecular weight values appear high in comparison with the average molecular weight value of about 500 Da or 3.6-mer that is expected for the resin synthesized with a charge F/R mole ratio of 0.72 assuming a uniform reaction. Thus, the results indicate, similarly as with PF resins, that the polymerization

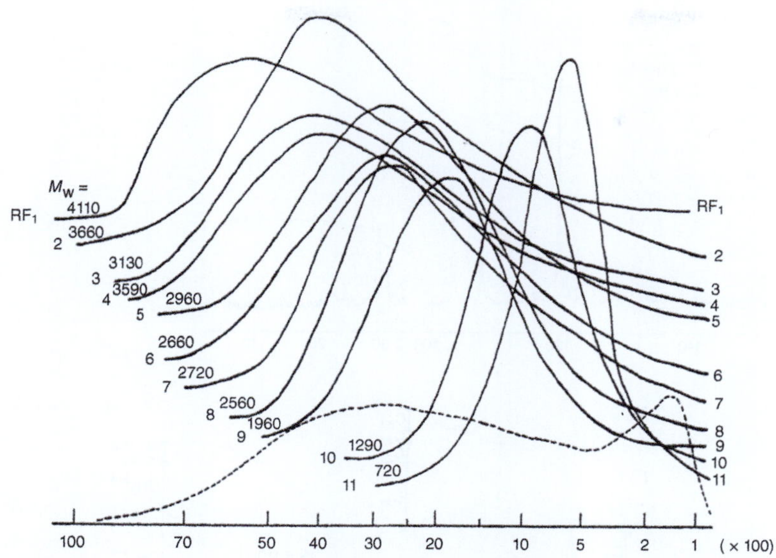

Figure 10.11 Gel filtration chromatograms of RF resin fractions 1-11 and unfractionated resin (---) with calculated weight-average molecular weights (M_w).

reactivity becomes greater as the resorcinol ring gets substituted with methylene groups, resulting in wide molecular weight distributions.

10.10.2 Assignments of ^{13}C NMR Chemical Shift Values of RF Resin Fractions

^{13}C NMR chemical shift values of various carbons of the RF resin model compounds obtained from the separation procedure, in methanol, are shown in Figure 10.12: (i) C_1 and C_3 carbons at 159.6 ppm for resorcinol; 157.4–156.1 ppm for monosubstituted resorcinol rings; 155.0–153.8 ppm for disubstituted resorcinol rings. Also, peaks at 152.0–151.1 ppm not observed in the model compounds but observed in resin spectra (Figure 10.13) were

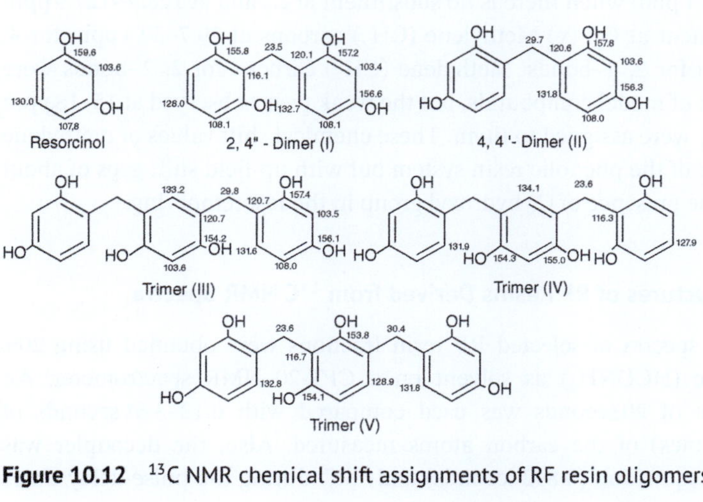

Figure 10.12 ^{13}C NMR chemical shift assignments of RF resin oligomers.

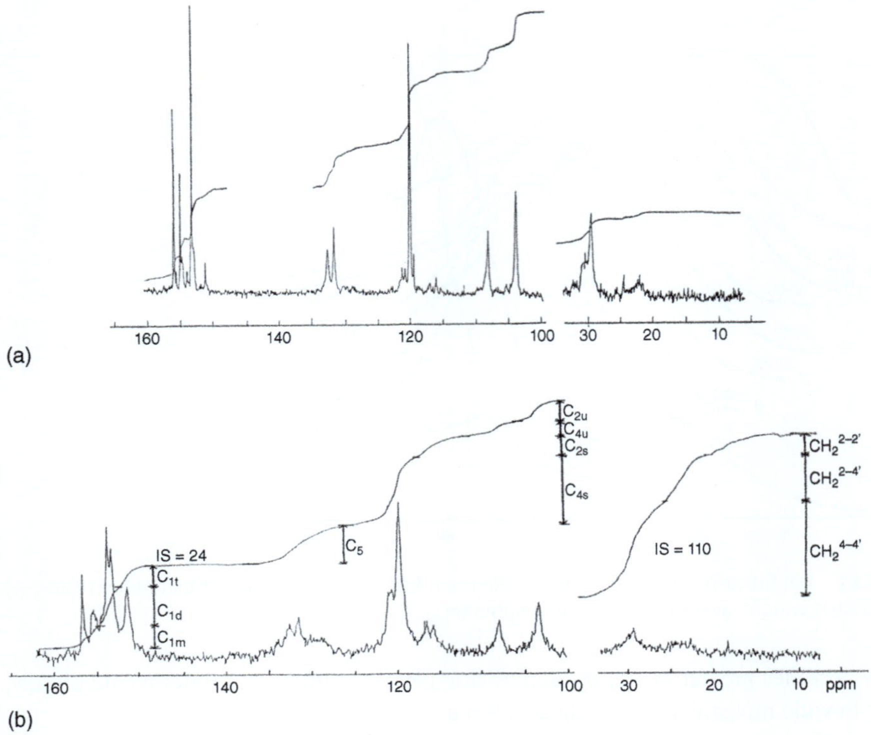

Figure 10.13 (a) ^{13}C-NMR spectrum of resin fraction RF_{11} in formamide (Integration is shown). (b) ^{13}C-NMR spectrum of resin fraction RF_1 in formamide (IS: integration scale). Carbons numbers are as noted in the text.

assigned, tentatively based on the substituent effects, to trisubstituted C_1 and C_3 carbons, which indicates the presence of branched polymer molecules. (ii) Unsubstituted C_2 carbons at 103.6 ppm; substituted C_2 carbons at 116.1–116.7 ppm; (iii) Unsubstituted C_4 carbons at 107.8–108.2 ppm and substituted C_4 and C_6 carbons at 120.1–120.8 ppm; (iv) C_5 carbons at 130.9–134.1 ppm when there is no substituent at C_2 and at 128.9~127.9 ppm when there is a substituent at C_2; (v) Methylene (CH_2) carbons at 29.7–30.4 ppm for 4, 4′-bonds; 23.5–23.7 ppm for 2, 4′-bonds. Methylene (CH_2) carbons for 2, 2′-bonds were not observed due to lack of model compounds, but the weak peaks observed at 15–18 ppm observed in fraction RF_1 were assigned to them. These chemical shift values of methylene carbons paralleled those of the phenolic resin system but with up-field shift gaps of about 8–10 ppm, ascribed to the presence of C_3 hydroxyl group in the resorcinol ring.

10.10.3 Polymer Structures of RF Resins Derived from ^{13}C NMR Spectra

Quantitative ^{13}C NMR spectra of selected RF resin fractions were obtained using 20% solutions in formamide ($HCONH_2$) as solvent on a CFT-20 NMR spectrometer. An ample pulse-delay time of 20 seconds was used compared with 0.12–3.61 seconds of T_1 (lattice-relaxation times) of the carbon atoms measured. Also, the decoupler was gated "on" during the acquisition (0.892 seconds) and "off" during the pulse-delay times

Table 10.3 ^{13}C NMR integrals and calculated values and intrinsic viscosity values of RF resin fractions.

RF#[a]	C_{1m}	C_{1d}	C_{1t}	C_5	C^s_2	$C^s_{4,6}$	C^u_2	$C^u_{4,6}$	CH_2	n_2	n_3	α	$[\eta]$[b] $(ml\,g^{-1})$
1	11.0	19.0	10.0	18.0	9.0	31.0	11.0	8.2	18.0	—[c]	80	0.36	10.6
2													8.0
3	10.2	20.0	8.0	20.2	8.2	31.5	8.1	0.0	14.3	—[c]	35	0.32	7.8
4	14.0	20.0	11.0	20.0	8.2	33.0	9.6	9.2	17.0	—[c]	30	0.37	6.7
5													6.9
6	13.5	19.5	9.0	20.0	7.0	32.0	13.0	8.0	18.0	40	19	0.29	7.3
9	10.0	20.5	5.5	17.0	7.0	27.0	12.5	7.0	15.2	21	16	0.21	6.2
10	13.5	18.5	3.0	17.5	7.0	27.0	14.0	10.0	15.4	6	7	0.10	5.2
11	17.0	19.0	3.0	17.5	3.0	23.5	15.5	9.0	16.8	2	6	0.08	4.3
Res[d]													2.7

a) RF#: RF resin fraction number. C_{1m} ,C_{1m}, and C_{1t}: Mono-, di- and tri-substituted c_1 carbons, respectively. s and u: Substituted and unsubstituted $c_{2,4,6}$ carbons, respectively. n_2 and n_3: Degree of polymerization calculated by Eqs. (10.5b) and (10.5c), respectively. α: Branching index calculated by Eq. (10.5d).
b) Intrinsic viscosity in ml g^{-1} in methanol at 30.0 °C.
c) Calculated n_2 values were indeterminate with values of $(1-1.04)^{-1}$.
d) Resorcinol.

(20 seconds) to suppress nuclear Overhauser (nOe) effect. The number of scans required for decent spectra increased from 4000 to 39 000 as the degree of polymerization increased. The peaks were integrated, and the analysis results are reported in Table 10.3 with the intrinsic viscosity values separately determined for the fractions. Spectra of fractions RF_{11} and RF_1 are shown in Figure 10.13a,b and the integral and calculated data are summarized in Table 10.3. In the spectrum of fraction RF_1 in Figure 10.13b, the chemical shift value assignments are also shown (from left to right): for resorcinol ring carbons, C_{1m} (monosubstituted), C_{1d} (disubstituted), and C_{1t} (trisubstituted), C_5, $C_4{}^s$, $C_2{}^s$, $C_4{}^u$, and $C_2{}^u$, and for methylene carbons, 4-4′ bonds, 2-4′ bonds, and 2-2′ bonds. As expected of novolac resins, no hydroxymethyl group carbon was observed in the 60–66 ppm regions. Thus, the spectrum shows the expected RF resin novolac structures consisting of resorcinol rings and methylene carbons only. The methylene group peaks broadened more for higher molecular weight fractions due to the ever so slightly different many chemical shift values for the same carbons in the polymer molecular structure. The degree of polymerization was calculated using the integration data of spectra, using Eq. (10.5a) [65].

$$n_1 = 1/(1-r), \text{ where } r = CH_2/(1/6) \times \text{total aromatic carbon intensity} \quad (10.5a)$$

However, the calculated values, n_1, were found to be quite variable, indicating that ^{13}C NMR quantitation parameters used were not good enough to compare the aromatic carbons directly with aliphatic carbons and the results are not reported. But the NMR spectrum indicated that $C_{2,4,6}$ carbons can be measured separately for the reacted (s) and unreacted (u) ones, from which the degree of polymerization can be calculated by using the intensities of only these aromatic carbon fraction values, i.e. the reacted C_2, C_4, and C_6 carbons, in

Eq. (10.5b), resulting in (n_2) values in Table 10.3:

$$n_2 = 1/(1 - 1.5xf_s), \text{ where } f_s = \left(C_2{}^s + C_4{}^s + C_6{}^s\right)/(C_2 + C_4 + C_6) \quad (10.5b)$$

Calculated n_2 values also showed out-of-range values for higher molecular weight fraction. Also, C_1 and C_3 carbons can be separately measured for mono, di, and trisubstituted ones to allow the calculation of the degree of polymerization, (n_3):

$$n_3 = 1/[1 - (0.5f_m + f_d + 1.5f_t)] = 2/(2f_m - f_t) \quad (10.5c)$$

where f_m is the fraction value of monosubstituted, f_d is the fraction value of disubstituted, and f_t is the fraction value of trisubstituted C_1 and C_3 carbons. Calculated n_3 values, 6–80, are shown in Table 10.3 and they appeared generally in accord with GPC molecular weight values discussed above.

Also, since the f_s value represents the extent of reaction and the f_t value the branched ring values, the branching extent of RF polymers, (α), was also calculated [63]:

$$\alpha = f_s \times f_t/[1 - f_s(1 - f_t)] \quad (10.5d)$$

The calculated α values are shown in Table 10.3 and appear to be in a reasonable range, increasing with fractions' degree of polymerization, or higher branching extents for higher molecular weight polymers. Another interesting result is the average unreacted/reacted carbon ratios: 1.70 for C_2 carbons and 0.25 for C_4 and C_6 carbons, indicating that the major reaction occurred in the resin synthesis is through C_4 and C_6 carbons, indicating that remaining unreacted C_2 carbons could play the major role in the curing step.

10.10.4 Relationship Between Intrinsic Viscosity and Molecular Weight Values from Equation n_3

Also, the intrinsic viscosity values of RF resin fractions were measured using a Cannon-Ubbelohde 4-bulb shear-dilution viscometer according to the procedure reported and the results are reported in Table 10.3, which indicated decreasing values from Fraction #1 to Fraction #11 as expected. The intrinsic viscosity values are in the similar range with those of typical wood adhesive (acetylated) PF resole resin fractions discussed in later lectures, of which molecular weights (acetylated) ranged up to 110 000 Da, measured by a light-scattering method. The intrinsic viscosity data and the molecular weight values derived from n_3 values were correlated to give the following Mark–Howink relationship [63]:

$$[\eta] = 0.406 \times M_n{}^{0.354} \text{ (in methanol)} \quad (10.5e)$$

The low exponent value of 0.354 indicates that the RF resin molecules have a compact solution structure expected of branched polymers.

In conclusion, this study on a typical RF resin revealed its relatively low molecular weight values and branched polymer structures formed by methylene bonds, of which their abundances are in the order of 4-4′>2-4′≫2-2′ bonds, between resorcinol rings. The average unreacted/reacted carbon ratios of 1.70 for C_2 carbons and 0.25 for C_4 and C_6 carbons indicate the resin synthesis reaction occurred mainly through C_4 and C_6 carbons and therefore remaining unreacted C_2 carbons could play the major role in the curing step. Also, the unreacted C_2 and C_4 and C_6 carbons are present on average in the range of 1.2–1.8 sites per resorcinol ring, to be reacted with the formaldehyde in the curing step. The present results also appear useful in understanding PRF resin structures.

10.11 Summary

Chemistry, manufacture, and practices of PRF resins are discussed where the bond curing is done at room or slightly elevated temperatures due to the significantly higher reactivity of resorcinol. RF resins were first used in the industry, and PRF copolymer resins were developed to lower the cost by replacing some of the resorcinol with phenol. In manufacturing of a typical PRF resin, phenol-formaldehyde resole resin oligomers are first synthesized at an appropriate low F/P mole ratio and resorcinol is added and reacted in the same basic pH until all hydroxymethyl groups on phenolic ring are transformed into methylene bonds, resulting in an aqueous PRF copolymer (novolac) resin of resorcinol-capped PF resin molecules. Presence of residual hydroxymethyl groups in the resin is avoided to have adequate storage stability and good bond performance. Resorcinol/phenol (R/P) mole ratio, formaldehyde/phenol (F/P) and F/(P + R) mole ratio, and targeted resin viscosity, resin solids level, and sodium hydroxide content all need to be accurately adjusted to obtain a storage-stable PRF resin that cures at the desired speed and also gives adequate bond strength values. Various grades of PRF resins are available mostly depending on the R/P mole ratio that determines the resin cost and curing speed, higher the ratio higher the curing speed. Liquid and solid hardeners based on paraformaldehyde and wood fillers are then discussed as practiced in the industry with respect to the resin solids levels, mixing procedures, and curing methods. The general bonding procedures used in the laminated beam manufacturing industry are also presented as well as some new developments in the industry such as bonding of fire-retardant and preservative-treated lumber and bonding of RFP tension layers in laminated beams. A resole-type RF resin product reported in literature as an effective coupling agent with a limited pot-life is also presented. Resins based on tannins and alkyl-resorcinols are briefly discussed and a research publication on a novolac RF resin is also presented.

References

1 Novotny, P.E. (1930). Synthetic resin and making the same. U.S. Patent 1,767, 696.

2 Rhodes, P.H. (1945). Production of resin. U.S. Patent 2,385,372.

3 Sellers, T. Jr., (2001). *Forest Products Journal* 51 (6): 12.

4 Sampson, G.R. and Fasick, C.A. (1967). *Forest Products Journal* 17 (11): 11.

5 Laposky, P. Jr., Blankenhorn, P.R., Janowiak, J.J., and Manbeck, H.B. (1993). *Forest Products Journal* 43 (2): 19.

6 Janowiak, J.J., Laposky, P. Jr., Blankenhorn, P.R., and Manbeck, H.B. (1993). *Forest Products Journal* 43 (9): 23.

7 Little, G.E. and Pepper, K.W. (1947). *British Plastics* October: 430.

8 Kreibich, R.E., Freeman, H.G., and Baxter, G.E. (1970). Method of producing an ortho-directed phenolic resin by condensing phenol and hcho in the presence of a bivalent metal ion and then adding resorcinol, and the resultant product. U.S. Patent RE. 26,681.

9 Gillern, M.F. (1977). Phenol-formaldehyde resorcinol resins and adhesives. U.S. Patent 4,061,620. Weyerhaeuser Co., Tacoma, WA.

10 Sebenik, A., Osredkar, U., and Vizovisek, I. (1981). *Polymer* 22 (June): 804–806.

11 Amos, L.A. and Kim, M. (1982). Unpublished report. Weyerhaeuser Company, Tacoma, WA.

12 Kim, M.G., Amos, L.W., and Barnes, E.E. (1993). *Journal of Polymer Science Part A: Polymer Chemistry* 31: 1871.

13 Barnes, E.E. and Kim, M. (1980). Unreported results. Weyerhaeuser Co., Tacoma, WA.

14 Grendon, W.C. (1984). Resin hardener and method of its use. U.S. Patent 4,425,178.

15 Kreibich, R.E., Freeman, H.G., and Baxter, G.E. (1972). Liquid hardener for phenol aldheyde resins. U.S. Patent 3,636,272.

16 Webb, D.A. (1970). *Forest Products Journal* 20 (4): 19–23.

17 Raknes, E. (1967). *Norske Skogindustrier* 10: 378–390.

18 Caster, D., Kutscha, N., and Leick, G. (1985). *Forest Products Journal* 35 (40): 45.

19 Schaeffer, R.E. and Gillespie, R.H. (1970). *Forest Products Journal* 20 (6): 39.

20 Madsen, B. and Littleford, T.W. (1962). *Forest Products Journal* 12 (2): 68.

21 Selbo, M.L. (1958). *Forest Products Journal* 8 (5): 145.

22 Freeman, H.G. and Kreibich, R.E. (1968). *Forest Products Journal* 18 (7): 39.

23 Rhude, M.J. (1957). *Forest Products Journal* 7 (3): 100.

24 Carruthers, J.F.S. (1963). *Forest Products Journal* 13 (5): 190.

25 Kim, M.G. (2005). *Wood Adhesives 2005*, 149. FS, Forest Products Laboratory.

26 Mann, J.W. (1954). *Forest Products Journal* 4 (6): 16A.

27 Estep, M.H. (1954). *Forest Products Journal* 4 (3): 138.

28 Ebewele, R.O., River, B.H., and Koutsky, J.A. (1986). *Journal of Applied Polymer Science* 31: 2275.

29 Tohmura, S., Miyamoto, K., and Inoue, A. (2005). *Wood Adhesives 2005.*, 411. USDA Forest Service, FPL.

30 Steinhaus, M. (1964). *Forest Products Journal* 14 (1): 40.

31 Kreibich, R.E. (1980). Personal Communication, Weyerhaeuser Co, Tacoma, WA.

32 Gorman, T.M., Hamanishi, C.M., and Callison, J.R. (1996). *Forest Products Journal* 46 (3): 80.

33 (1992). *American National Standard for wood products---Structural Glued Laminated Timber*. AITC.

34 Bryan, H.R., Murmanis, L., and Stewart, H.A. (1980). In: *Proceedings of 1980 Symposium of Wood Adhesives*, USDA Forest Service, Forest Products Laboratory, Madison, WI. p. 219.

35 Raknes, E. (1963). *Journal of Wood Science* 11: 24.

36 Schaeffer, R.E., Gillespie, R.H., and Blomquist, R.F. (1966). *Forest Products Journal* 16 (5): 23.

37 Sellers, T. Jr., and Miller, G. *Forest Products Journal* 47 (10): 73.

38 Lisperguer and Becker, P.H. (2005). *Forest Products Journal* 55 (12): 113.

39 Ropella, L.A. (1960). *Forest Products Journal* 10 (10): 513.

40 Janowiak, J.J., Manbeck, H.B., Blankenhorn, P.R., and Kessler, K.R. (1992). *Forest Products Journal* 42 (10): 68.

41 Vick, C.B. (1980). In: *Proceedings of 1980 Symposium of Wood Adhesives*, 142–153.

42 Lorenz, L.F. and Frihart, C. (2006). *Forest Products Journal* 56 (9): 90.

43 Vick, C.B. (1997). *Forest Products Journal* 47 (7/8): 83.

44 American Laminators and Prescott, R. (1994). *Chemical and Engineering News 1994* 11: 23.

45 Davids, W.G., Richie, M., and Gamache, C. (2005). *Forest Products Journal* 55 (1): 93.

46 Lopez-Anido, R., Gardner, D., and Hensley, J.L. (2000). *Forest Products Journal* 50 (11/12): 43.

47 Silva-Hendriquez, R., Gray, H., Dagher, H.J. et al. (2010). *Forest Products Journal* 60 (1): 33.

48 Loos, W.E. and Kent, J.A. (1968). *Forest Products Journal* 18 (3): 23.

49 Vick, C. B., Richter, K., and River, B.H. (1996). Hydroxymethylated resorcinol coupling agent and methof bonding wood. U. S. Patent 5,543,487.

50 Christiansen, A.C.B. and Vick, and E. A. Okkonen. (2003). *Forest Products Journal* 53 (2): 32.

51 Christiansen, A.W. and Okkonen, E.A. *Forest Products Journal* 53 (4): 81.

52 Christiansen, A. (2005). *Forest Products Journal* 55 (11): 73.

53 Anonymous. (1960). Wood outperforms steel. *Forest Products Journal* 10 (11): 617.

54 Plomley, K.F. (1966). Div. of Forest Products Tech. CSIRO Paper No. 39.

55 Saayman, H.M. and Oatley, J.A. (1976). *Forest Products Journal* 26 (12): 27.

56 Pizzi, A. and Roux, D.G. (1978). *Journal of Applied Polymer Science* 22: 2717.

57 Hills, W.E. (1981). *Weyerhaeuser Science Symposium* II: 171.

58 Pizzi, A. (1989). Tannin-based cold-setting adhesives for face lamination. In: *Adhesives from Renewable Resources. ACS Symposium Series 385* (ed. R.W. Hemingway and A.H. Connor), 203–216. Washington, DC: ACS.

59 Gornik, D., Hemmingway, R.W., McGraw, G.W., and Tisler, V. (2000). *Holz und Roh-Werkstoff* 58: 23.

60 Pichelin, F., Nakatani, M., Pizzi, A. et al. (2006). *Forest Products Journal* 56 (5): 31–36.

61 Kreibich, R.E. and Hemingway, R.W. (1985). *Forest Products Journal* 35 (3): 23.

62 Kreibich, R.E. and Hemingway, R.W. (1989). Syntheses and application of alkylresorcinol-formaldehyde DFK adhesives. In: *Adhesives from Renewable Resources. ACS Symposium Series 385* (ed. R.W. Hemingway and A.H. Connor), 203–216. Washington, D.C.: ACS.

63 Kiisler, K., Christiansen, P., Tanner, J. et al. (1981). *Weyerhaeuser Science Symposium* II: 223.

64 Christiansen, P. (2000). *Wood Adhesives 2000 Abstract*, 77. Madison WI: FS, FPL.

65 Flory, P.J. (1953). *Principles of Polymer Chemistry*, 350. Ithaca and London: Cornell University Press.

11

Softwood Plywood Adhesives and Manufacturing Technology

11.1 Softwood Plywood Manufacturing Technology

An excellent description on the softwood manufacturing is found in the booklet: "*Basic Plywood Processing*" written by Ted Dumas, published by the American Plywood Association. Another good reference is the book, *Plywood and adhesives technology*, written by T. Sellers, Jr. published in 1985. These books are recommended for reading in this course. The following subjects are included in these books:

Adhesive application	Assembly method	Assembly time	Pre-pressing
Pressing temp and time	Pressing pressure	Veneer thickness	Veneer drying
Veneer porosity	Extractive content	Veneer freshness	Veneer surface
Veneer smoothness	Veneer cleanness		

Softwood plywood panels are bonded with durable PF resole resin-based adhesives and used in exterior structural applications [1]. This industry started in 1920, using casein glues and later using soybean meal glues. In 1930, the demand for exterior durability led to the use of PF resole resin-based Tego films and soon-to-use liquid PF resole-based adhesives. The industry expanded greatly during World War II and mostly resided in the Northwest using D. fir, and it then expanded to the South and Southeastern regions (southern pines) with their exterior durability judged to be equal [2]. From the adhesion and gluing points of view, it is useful to make one distinction between softwood plywood and other wood composite boards such as oriented strand board (OSB) and various particle and fiber boards. Softwood plywood is manufactured in such a way that the original surface structures and qualities of wood are conserved while others come to extensively modifying those features, since in plywood manufacture the hot-pressing process uses less compaction pressure at about 150–200 psi and lower temperatures of about 300 °F than those of other wood composite boards. The retention of original wood surface structures is highly valued in some uses. Therefore, the adhesion process requires an entirely different principle, i.e. use of highly advanced PF resole resins, further formulated with additional caustics, fillers, and extenders to give the adhesive layer the gap-filling capability and fast curing rates at the relatively low hot-pressing temperatures. There are various species of southern pines, but the gluability with PF resin adhesives has been considered about the same except for differences derived from the specific gravity, amount of latewood present on gluing surface,

Adhesives and Finishes for Wood: For Practitioners and Students, First Edition. Moon G. Kim.

and amount of extractives [3]. Recently, OSB began to make inroads into traditional softwood markets, but softwood plywood market is still strong, and 103 manufacturing plants are operating in North America in 1998 [4].

11.2 PF Resole Resins for Bonding of Softwood Plywood

PF resole resins for bonding softwood plywood are *Resin#4 Types*, synthesized to the fullest extent in molecular weight, as evidenced by the lowest resin-solids level, highest sodium hydroxide content, and a high viscosity among various wood adhesive PF resole resins. Properties of a typical southern pine plywood adhesive PF resin are shown in Table 11.1. It is noted that fast curing rates at low hot-press temperatures are aimed for this class of PF resins, but fast cure rates are good only if the bond strength turns out good. The dry-out problem is an extreme case of fast-curing resins that gives a poor bond strength. As discussed in Chapter 9, PF resin polymer molecules are formed with a wide range of molecular weights and the cure rate becomes faster and the flow property becomes poorer as the resin is advanced further on. On the other hand, wood bonding with PF resole resins can be performed at about 125 °C (257 °F) or higher, up to about 232 °C (450 °F). For softwood plywood bonding, the current cure temperature in industry is at about 300 °F (149 °C), a relatively low temperature which is necessary to keep the extent of veneer surface degradation to a minimum. Therefore, the current extent of polymerization of softwood plywood PF resins attained in resin manufacturing can be considered to be at the balance point between the cure rate and flow property.

Then, what will happen if some further polymerization of resin is carried out, for example, by adding some more water and sodium hydroxide to current resins and cooking to the same viscosity as before the additions? It is likely that the cure rate or hot-pressing time may be shortened, but the bond strength may deteriorate due to the reduced resin flow capability. In industry, it was recognized that there appeared to be only a small leeway to further advancements of overall molecular weight to get a faster cure rate without reducing the flow capability because of the formation of high-end molecular weight fractions of resin [3]. The nature of the highest molecular weight fractions in highly advanced PF resole resins is not well understood, but they could be in 40 kDa ranges [5], possibly hard solid materials with high melting points and consisting of PF polymers having intramolecular methylene cross-linking bonds and therefore the molecules are less

Table 11.1 Properties of a typical southern pine plywood binder PF resin.

Non-volatiles	40.00%
NaOH content	5.11%
Free formaldehyde	0.07%
pH	11.2
Specific gravity, 25 °C, 1.1770	1.1770
Viscosity, Brookfield, 25 °C	790 cps
Gel time, 121 °C, Sunshine	~7.0 min

able to re-structure themselves in the formation of overall cross-linked cured PF resin structures. This tendency of intramolecular methylene bond formation in resin synthesis can be, in theory, minimized by lowering the synthesis F/P mole ratio, which also appears to have been optimized in current industry PF resins to *about* 1.8, significantly lower than about 2.1 for *Resin#3 Type* resins as discussed in Chapter 9. Another question is the cure temperature. The dynamic mechanical analysis (DMA) study of PF resole resins discussed in Chapter 9 indicates that PF resins cure faster at higher temperatures, i.e. shorter cure (vitrification) times, and a more advanced resin cures faster than less advanced resins at the same temperature [6]. It also indicated that curing of certain PF resins at lower temperatures may never reach to the full strengths obtainable at higher temperatures due to fact that the high molecular weight fractions in resins, once dried out at lower temperatures, would not reach to their melting ranges.

11.3 Veneer, Veneer Drying, and Adhesion Problems

Softwood plywood manufacturing plants have to use rough veneers, the extents of which depend on the wood anatomy and the mechanical processing factors. Wood cell size and orientation (in relation to the lathe knife), which are related to the tree growth rate, annual ring symmetry, and size and frequency of knots, are the wood anatomical factors for resultant degrees of roughness [7]. The knife severing alternately high- and low-density sections of wood causes strong vibrational forces to result in rough veneer surfaces. This veneer roughness, in terms of the number and depth of checks [8], can be minimized by heating peeler blocks to about 150–180 °F [9], knife sharpening, and tight veneering with correct lathe setup. The peeling process induces lathe checks to occur on the rough side of veneer and their depth has been known to reach to the smooth surface side to influence glue bond performances [8]. Rough veneers are difficult to glue because of difficulty of making intimate contacts between veneer surfaces, and they also promote conditions of dry-out and over-penetration by allowing the liquid adhesive to collect in cracks and valleys to result in starved joints with reduced bond quality. Rough veneers also mean an increased surface area of bonding and therefore the amount of glue application [7, 9].

The subsequent veneer processing operation in plywood manufacturing plants includes clipping and separating of heartwood and sapwood portions for different drying schedules as well as separating excessively rough and irregular veneers. Then, in the veneer drying step, surface inactivation of veneers often occurs to cause bonding problems such as delamination of panels, which is the source of most common customer complaints. Surface inactivation is a sort of "burning" of wood surface structures and often cannot be identified visually but by an instrumental method such as reflectance colorimeter [10]. Inactivated surface is an area of changed wood surface chemistry in varying degrees to have hydrophobic characters in comparison with the average wood surface. Certain abnormal temperature/relative humidity conditions occur in the veneer drying chamber to give rise to moving of *wood extractive chemicals* from inner wood and depositing on veneer surface, resulting in more hydrophobic properties. The extractive chemicals in softwood, the kind and amount of which are variable between wood species and also within species, are hydrolyzable and condensed tannins, lignins, fatty acids, resin acids, terpenoids, waxes, sugars, starch, etc. They are generally hydrophobic molecules and their concentration in wood is relatively small, but when moved and deposited on the veneer surface during drying, they lower the surface's wetting and permeability properties and result in poor bonding for PF resin adhesives [11]. This surface inactivation also occurs in long-term aging of lumber or veneers and therefore freshly prepared veneers or lumber should be used in adhesive bonding [12].

For good bonding, the veneer surfaces should be smooth and not too rough, or peeler checks should not be too deep, and the veneer thickness needs to be uniform, and moisture content (MC) should be within the prescribed range of 3–7% in southern pine plywood mills. Also, the patching and edge-gluing of veneers can leave spots that are too dry or too wet or too thick, or too thin to lead to weaker bonds or blows. Also, veneers should be cooled to room temperature before reaching the gluing station. Overall, the quality of veneer surfaces is an important parameter and, it was found that wood failure (WF) values increase with low roughness and breaking strengths increase with low depths of lathe checks [13].

11.4 Bond Performance Characteristics of Softwood Plywood Adhesives

Controlled penetration	Dry-out prevention
Pre-press bonding – tack	Fast curing at low temperature
Gap-filling, thick glueline	Low compaction pressure
Exterior durability	

The generally rough surfaces and lathe checks in veneers and the relatively low hot-pressing temperature requirement have made it necessary for PF resin adhesive formulations to possess the adhesive/adhesion characteristics as shown on the left. In the beginning of the softwood plywood industry in the United States, the unformulated liquid PF resins were found to have the required exterior durability, but they lacked in the other various adhesive/adhesion performance characteristics, and adhesive formulations using various fillers and extenders and mixing/application methods were developed over the years. The formulation research is continuing to respond to the changing raw materials, such as using different fillers/extenders as well as using veneers of different wood species, seasonal changes, and also to lower adhesive costs. Also, reducing the volatile organic compound (VOC) emissions from the drying process has been one of the major activities of the U. S. softwood plywood industry in recent years.

11.5 Softwood Plywood Adhesive Mixing Procedure

A typical adhesive formulation and mixing procedure for bonding of southern pine plywood is shown in Table 11.2. The adhesive mixing operation is carried out in a large tank equipped with a stirrer that can provide a high shearing action. More often, the mixing is done at the plywood manufacturing plant, but some plywood plants are supplied with mixed adhesives from resin manufacturers, to be used within a few days. Each component is sequentially added and mixed under a high shear for a length of time depending on the batch size until a smooth well-dispersed mix (to naked eye) is obtained. Keeping the

Table 11.2 A typical southern pine plywood adhesive mixing procedure.

Order of addition	Materials	Solids (%)	Function	Weight (%)	Mix solids (%)
1	Water	0	Diluent	16	0
2	Wheat flour	97	Extender	6	5.82
3	Furafil	97	Filler	6	5.82
4	PF resin	43	Binder	23	9.89
5	NaOH	50	Catalyst	3	1.50
6	PF resin	43	Binder	46	19.78
	Total			100	42.81

temperature of mixing low near the room temperature during and after mixing is necessary to obtain the target adhesive viscosity and maintain its viscosity during use. Often a mild cooling is applied during the mixing, especially in summer. A different mixing order, different splits of PF resin, or a higher mixing temperature can result in an adhesive mix with different viscosities with shorter/longer pot-lives, and other different gluing characteristics. Often, a surfactant is added to speed up the dispersion of the filler/extender and also to improve the bonding performance of hard-to-bond wood species. A de-foaming agent is often added to prevent the foaming from occurring during mixing. The target viscosity of the adhesive mix for a particular manufacturing plant depends on the adhesive application method: roll coating (high viscosity), curtain coating (medium viscosity), or coarse spraying (low viscosity) method. The target viscosity has to be attained in the adhesive mixing operation, or the adhesive cannot be applied efficiently.

11.6 Softwood Plywood Adhesive Formulation and Characteristics

11.6.1 Characteristics of PF Resin Adhesive Mixes

The plywood adhesive formulation includes PF resin, filler, extender, and sodium hydroxide. Also, mixing them in a given order is necessary to obtain a consistent, uniform adhesive mix with target viscosity values batch after batch. Properties of a typical southern pine plywood PF resin adhesive mix are shown in Table 11.3. Firstly, as discussed in Chapter 9, the PF resole resins of Resin #4 types are used in softwood plywood adhesives, the most advanced ones among PF resole resins used in the wood products industry. The synthesis procedure of a PF resin for a particular plywood plant is attained through a series of trial-and-error experiments by incrementally varying the F/P mole ratio, sodium hydroxide/phenol ratio, resin solids level, target viscosity, as well as by changing the temperature/time parameters in the cooking schedule. Commonly, caustic contents of 5.0% or more, resin solids level of about 40%, and resin viscosity ranges up to 500 cps are used, indicating a highly advanced nature of the PF resole resins. All these synthesis parameters affect the resultant resins' molecular weight and its distribution as well as branching structures to result in varying curing speeds and performances. Since the

Table 11.3 Properties of a typical southern pine PF resin adhesive mix.

pH	11.0
Non-volatiles	42.4
Viscosity, Brookfield, 25 °C	2900 cps
Specific gravity	1.1940

adhesive formulation, assembly times, curing methods, and wood properties also come into the bonding equation, delineating all these resin/adhesive and wood property parameters in plywood manufacturing or in other similar wood bonding processes becomes a highly complex task and exacting replications are the key to successful outcomes. The full story has never been published, and some such work done in industry tends to be kept confidential. Some bits and pieces were reported: in bonding of southern pine plywood, the shear strength and WF values were not much affected by F/P ratios in the range of 1.6–2.5 of PF resins, but the lower F/P values resulted in the least delamination values [14]. Also, the effects of too high or too low molecular weight and its distribution were also related to poor or better bond strength values of plywood or OSB [15, 16]. Overall, current PF resins meet the need for fast curing rates at a relatively low hot-pressing temperature of about 300 °F, which is lower in comparison to other PF resole resin-bonded wood composite products. It is necessary to minimize the heat degradation of veneer's natural appearance as well as surface wood characteristics needed in finishing with coatings in the secondary processing.

11.6.2 Gap-Filling Capability of Softwood Plywood Adhesives

PF resole resin adhesives are water-borne and both the carrier water and the water arising from resin polymerization reactions have to dissipate during the hot-pressing, resulting in volume (thickness) contraction of gluelines. Also, some water can remain in the gluelines at the end of hot-pressing and later dissipate from the adhesive layer, furthering the volume contraction. The volume contraction of adhesive layer occurring before gelation, i.e. during the period when gluing members as well as adhesive components can move freely against each other, have minimal deleterious effect because it is compensated by pressing. However, the volume contraction also occurs after gelation or after the gluing members are set in place to the final thickness position, so that the volume contraction results in a buildup of stretching stresses of adhesive molecules within the gluelines. These stresses, occurring due to shortage of materials within the given fixed volume of gluelines, can result in weakening of gluelines' solid structure, i.e. weakening of the cohesive as well as the adhesive strength. This volume contraction effect of adhesives, in general, is more deleterious for thicker gluelines than thinner gluelines. Thicker and thinner gluelines are present in the structure of softwood plywood because of the rough veneer surfaces and also due to the low hot-pressing pressures normally used in the industry (100–150 psi). These two contradictory parameters had required an extensive research/development work of formulating PF resin-based softwood plywood adhesives using fillers and extenders, and various lay-up and pre-pressing procedures were developed to meet this gap-filling adhesion requirement.

11.6.3 Viscosity Targets of Softwood Plywood Adhesive Mixes

Softwood plywood PF resin-based adhesives are normally formulated to have a relatively high viscosity in the range of 5k–10k cps. In this viscosity range, the actual value used in plants depends on the method of application: roll coating, curtain coating, or coarse spraying method. These methods of application are quite different than the fine mist spraying method used in other wood composite manufacturing processes. In the Southern United States, the coarse spraying method is widely used (GP, Louisville, MS). Caustic and water in the formulation act to reduce the viscosity of the adhesive system while the resin, fillers, and extender act to increase it, and therefore, the adhesive mix components are in such proportions that the adhesive viscosity reaches the target value at the end of the mixing period with the target resin/filler/extender solids levels. (Refer to the viscosity definition discussion in Chapter 2 regarding why fillers and extenders increase the apparent viscosity.) Then, the adhesive viscosity increases after the adhesive is applied on the veneer because of the loss of water from gluelines by absorption into wood and evaporation into the air, and the adhesive layer can reach to a high viscosity, near 20k cps at the end of the open and closed assembly times. If the veneers have abnormal moisture contents, either too high or too low, or the temperature or humidity of the surrounding environment deviates from the norm, the optimum high viscosity of applied adhesive may not be obtained.

11.6.4 Fillers in Softwood Plywood Adhesives

It is important to recognize that filler particles are doing the job of space-filling the glueline as well as blocking the crack propagation in cured adhesive layer and therefore, they should become an integral part (unity) of the gluelines' solid structure in the cured gluelines. Therefore, filler particles should be hard (strong) particles like cellulose particles found in nutshell flours. Also, the surface of filler particles has to make good atomic contacts (a good adhesion) with the PF resin and extender polymer matrix molecules, or if this atomic contact occurs poorly, the polymer matrix will behave like a porous, weakened structures. Fillers used in plywood adhesives are dried, often hardened, fine cellulosic particles, and each particle, as supplied, can be an assemblage of many smaller hard particles due to the electrostatic attractions arising between particles under the dry-grinding condition. "Furafil 100" is a typical filler supplied as fine dark-colored particles widely used in softwood plywood adhesive mixes. It is obtained as a byproduct in the commercial furfural production process, and the major component is undigested cellulosic residues in the thermal, acidic digestion (reaction) process of corn-cob at an elevated temperature. Also, fillers made of particleboard sander dust (70% through 270 mesh) are used in the southern regions [14] and various other organic materials such as walnut shell, oat hull, and pecan shell flours are used with their differences in performance known to be minor except their knife-wear tendency, which was found to be related to their (acidic) pH, and levels of sulfur, chloride, iron, aluminum, and silicon-type components [17, 18]. During the high-shear mixing of fillers in the adhesive preparation stage, the particle assemblages must be dispersed into each particle by wetting on each particle's surface with water, sodium hydroxide, and PF binder resin molecules. Ideally, each filler particle needs to be strong and not break further mechanically throughout the whole gluing/curing processes.

On the other hand, some swelling of the particle surfaces would be desirable because this would make the *adsorption interactions* occur more readily between PF resin matrix and filler particles. Thus, filler particles remain as solids, minimize the volume contraction, and contribute the most to the gap-filling characteristics. On the other hand, although filler particles are lower cost materials than PF resins or wheat flour (extender), they cannot be added too much in the adhesive mix, since the total adhesives solids level in adhesive mix is a fixed value and an increase of filler means a lower amount of PF resin or extender and lower overall adhesive cost but board performances may degrade because the filler level can reach to a value, beyond which a poorly structured polymer matrix would result in the cured gluelines. Also, the contribution of fillers to the adhesive mix viscosity is relatively minor because they do not dissolve in the adhesive mix, and therefore the high adhesive viscosity values required cannot be obtained by increasing only the filler level. Also, the viscosity contribution of PF resole resin component, although important, is not high, and a further increase of the molecular weight of PF resole resin for this purpose is not feasible due to the adverse effect of the high-end molecular weight components formed in increasing amounts. This dilemma finds an ally in the extender.

11.6.5 Extenders in Softwood Plywood Adhesives

The wheat flour used in adhesive mix formulations is often called an extender. First of all, it increases the adhesive viscosity far more than the cellulosic filler and PF resin because it swells and dissolves in the water of adhesive mix with the help of the sodium hydroxide and the shearing action affected on in the mixing procedure. Wheat flour consists of starch and some gluten, both high molecular weight materials and, as with other high molecular weight polymers, have a tremendous capability of increasing the viscosity when dissolved in a good solvent even at low usage levels of 1–6%. It is probable that wheat flour does not completely dissolve in the uncured adhesive stage, but the extent of dissolution can further increase at elevated temperatures during hot-pressing, resulting in further increasing the adhesive viscosity in the early stage of hot-pressing, contributing to minimization of glue squeeze-outs.

Another important contribution of wheat flour is strengthening of the cured gluelines. The swollen polymer structures of starch and gluten would collapse upon drying and curing of PF resin in hot-pressing, becoming solid materials dispersed throughout the cured glueline matrix. Thus, the resultant glueline is composed of cross-linked PF resin polymers as binder of matrix, filler particles as aggregates and gap-filling components, and starch/gluten polymers as a secondary binder. It is apparent that the matrix thus formed is a constructive one having a good durability against moisture, a result of an optimum combination of materials, their levels, and preparation procedures. That the wheat flour has been called as an extender means that the dispersed wheat flour components augment the cured PF polymer structures in a positive way, i.e. increasing the bond strength and durability. It appears that extender molecules themselves form uniform structures within the interstices of PF resin/filler components and also make intimate molecular contacts with them to result in strengthening the overall adhesive-layer structure. Another important function of the wheat flour molecules would be their optimum water-holding power as shown by their ready dissolution in the adhesive mix and also their capability of letting water molecules go off during the assembly and curing processes. It is also possible that a small amount of water is held by extender molecules until quite late in the curing process, promoting the adhesive layer to retain the flow property.

In addition, gluten, a protein and a minor component present in all varieties of wheat (9–12%), possesses ample amide groups, similarly as in urea molecules. These amide groups can react with formaldehyde to form hydroxymethyl groups and further react to form cross-linking methylene bonds within gluten molecules. Softwood plywood PF resins entail a small amount of formaldehyde and also can produce some more in the curing step. The reaction of gluten with formaldehyde can disrupt the adhesive mix viscosity if it occurs more than usual, possibly resulting in a higher mix viscosity, and also a disruption in the curing process may occur in the adhesive layer. Since the consistency of adhesive mixes is a key to the smooth operation of gluing process, the potential inconsistency problem is handled by extender manufacturers by blending various wheat varieties to produce a wheat flour of a constant (low) gluten content. Also, the wheat flour contributes to the tackiness of adhesive, which is necessary in pre-press bonding of veneer assembly. Since starch and gluten molecules are high molecular weight, hydrophilic polymers, they would easily make atomic contacts with hydrophilic wood surfaces (attractive forces) to give rise to high tackiness. Overall, the wheat flour in the adhesive mix is an essential component for pre-pressing and improving the bonding quality at costs lower than PF resins. In overall, fillers and extenders are essential components for controlling the gap-filling and other rheological properties, but they are not directly involved in crosslinking reactions with PF resins, and therefore they would contribute not much to the durability of gluelines. Therefore, their levels in adhesive mix have been known to be at low values [19], i.e. use of an excessive amount will degrade water resistance properties.

11.6.6 PF Resin Adhesives, Adhesive Spread Rates, and Application Methods

Standard adhesive spread rates for exterior grade plywood: The solids level of adhesive mix is the basis of setting the spread rate in use and cost calculation. Also, a certain level of PF resin adhesive solids is needed. Normally, the solids level of adhesive mix is a fixed value in a plant and also is used to calculate the amount of (liquid) adhesive mix to apply on a given area of veneer surface known as the spread rate, such as 40 lbs/1000 ft^2, single glueline basis, or 25 lbs/1000 ft^2, double glueline basis. These values form approximately a 0.005″-thick solid glueline and a 0.001″-thick penetration layer on either side [20]. Also, there is a constant desire to improve the bonding performance as well as to lower the cost of adhesive mixes, i.e. to use lower adhesive solids levels and/or lower PF resin solids levels since the glue cost makes up about 9% or more of total cost of manufacture [21]. Fillers and extenders cost less than PF resins and, theoretically, increasing the fillers and extenders and decreasing PF resins in the adhesive mix, or lowering the overall solids level of adhesive mix mean a lower bonding cost. This is to be pursued but with limitations: *PF resin is the durable matrix-forming material*, i.e. the cured PF resin polymer in the entire glueline consists of one PF resin molecule just like one automobile tire is composed of one rubber polymer molecule. For exterior use plywood this fact means that PF resin level in the adhesive mix cannot be lowered to the extent that allows filler particles to make direct contacts (no covalent bond) to other filler particles, a weak link in the gluelines. It has been reported that increasing the spread rate up to 75 lbs/1000 ft^2 and PF resin level above 28% in the adhesive mix significantly improves the strength and durability of southern pine softwood plywood [22, 23]. Most softwood plywood is produced for exterior applications with exterior durability classifications by using exterior grade adhesives.

Lower durability grade adhesive mixes: Some plywood is also produced for interior uses where the adhesive durability cis classified as interior glue, intermediate glue (Exposure 2) in the product standard, PS1-95 Construction and Industrial Plywood [1]. These adhesives can be based on PF resole resins but the level of PF resin-solids in the adhesive mix can be lower, determining the differing durability grades with savings on the bonding cost. Interior glues are adopted where there is little water or moisture exposure, and intermediate glues are used where there is some potential for water leakage or high humidity exposure. The interior, intermediate, and exterior durability classes of softwood plywood are defined in terms of the results of durability tests of different severity as discussed in Chapter 20.

Variable spread rates: Laboratory research on the possibility of using variable spread rates in plywood manufacturing based on the roughness of veneer surface and other processing variables, such as veneer moisture contents, assembly times, and veneer temperature, has been reported [24, 25]. The spread rates were set at 50 lbs/1000 ft^2 for control and varied between 25 and 68 lbs/1000 ft^2 or slightly narrower ranges depending on the roughness of veneers and other variables, intentionally controlled. Variable amounts of PF resin adhesive were manually applied, and other variables were also changed, and panels were hot-pressed at 300 °C for four minutes. Hot-pressed panels were tested, and analysis of test results led to a conclusion that an average 8.3% increase in WF value and 13.1% reduction in spread rate can be realized in comparison with the control, mostly through reduction of dry-out problems [24, 25], which often arise on rough surfaces of veneers. In order to utilize this procedure, sensors for each variable need to be installed and data be collected and analyzed to determine the prevailing spread rates and the results relayed to the adhesive dispensing unit.

Foaming application method: In addition to the roll spreading, spraying, and curtain coating methods commonly used in industry, the foam extrusion method is also used [26]. In this method, un-foamed, formulated PF adhesive is pumped at a controlled rate through an inline foaming device, and the foamed adhesive is extruded through 260 exiting tubes within a ~50″ span as parallel, flexible, coherent, ~1/10-in. diameter rods and deposited in a uniform pattern onto veneers as they pass through the layup process. The foaming increases the volume of adhesive mix by about five times. A foaming agent needs to be included in the adhesive mix formulation. It is reported that the coverage of adhesive is better, resulting in better bond strength or savings in adhesive usage. Overall, in the adhesive application and layup steps, careful attention should be paid to avoid core gaps, knots, splinters, etc. in veneers and also lapping of core veneers, and poor adhesive spreads on core veneers [27].

Wash water of PF resin plywood adhesives: In plywood manufacturing plants, adhesive storage tanks, adhesive mixers, transfer lines, and application equipment need to be cleaned regularly. In most cases, water is adequate for washing, but the produced wash water needs to be reused, or the leftover wash water needs to be treated internally at a cost. Sending phenolic resin components to municipal sewage treatment facilities is limited [28].

11.7 Open and Closed Assembly Times

The rate of water loss of adhesive mix applied on veneers occurring from the gluing station to the assembly station would also depend on the mix ratios of its components: PF resin, sodium hydroxide, filler, and extender components are different in their hygroscopicity. The optimum component ratio or choice of materials needs to be worked out in the adhesive formulation stage for optimum rate of water loss under the target veneer moisture content and weather condition, so that a significant increase in adhesive viscosity needs to be attained at the end of open assembly time to result in a good tack needed in the pre-pressing step. Adequate pre-press bonding is necessary in order for the loader to transfer veneer assemblies without fail into the narrow openings of hot-press. For a good tack of adhesive, its viscosity needs to be high enough at the end of pre-pressing but no drying out. Overall, the adhesive mix has to have the increasing viscosity characteristics arising from a gradual loss of water, i.e. the ideal tackiness profile is the adhesive mix starts with a relatively low viscosity for an easier application and then increases to a relatively high viscosity at the end of open and closed assembly times, occurring in an orderly manner.

11.8 Pre-pressing of Softwood Veneer Assembly

Pre-pressing is an important step in softwood plywood manufacturing in that a temporary bonding of veneers is attained with the help of increased tackiness of the adhesive layers as discussed above (Section 11.7).

The temporary bonding smooths out the bulky veneer assembly and makes their transport into the narrow openings of hot-press much more efficient and also shortens the press closing time, resulting in expediting the overall hot-pressing operation. Once the pre-pressing is completed, the loss of water by evaporation into the air is greatly reduced and this lowers the chance of adhesive dry-out. The temporary bond strength, or tackiness, is unrelated to the strength of cured gluelines or panel strengths, although adhesives having good tackiness profiles can lead to a more consistent hot-pressing operation to result in higher bond strength values over time.

11.9 Hot-Pressing Parameters of Softwood Plywood

Temperature and vapor pressure in mat: Hot-pressing is normally done in 20–30 multi-opening presses at about 300 °F for PF resin-bonded D. fir and southern pine plywood and this temperature is significantly lower than the hot-pressing temperatures employed in the manufacture of PF resin-bonded OSB or hardboard [29]. Also, the hot-pressing mat pressure is similarly lower (150–200 psi) because the compaction factor of plywood is not as high as for OSB or hardboard. When the hot-pressing begins, the steam pressure and temperature begin to increase within the mat, and the overall mat pressure and temperature increase.

Temperature and vapor pressure changes in mat: Changes in vapor pressure and temperature in hot-pressing of a mat consisting of 2.5 mm-thick veneers, five-layer D. fir plywood were reported, as shown in Figure 11.1. The steam pressure is highest in the middle layers

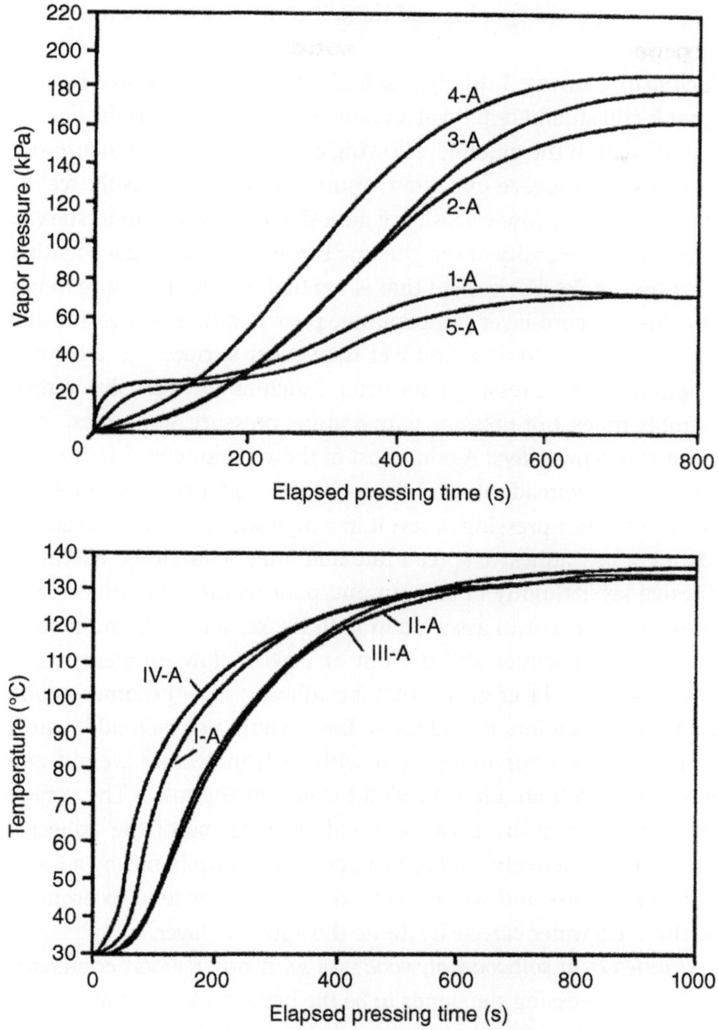

Figure 11.1 Variations of *vapor pressure* and *temperature* in hot-pressing of 2.5 mm, five-layer D. fir plywood (300 mm × 600 mm), hot-pressed at 150 °C (~300 °F). 1-A–5-A measured inside each veneer from the bottom to top layer; I-A–IV-A measured in each glueline from the bottom to top layer [29]. Source: Reproduced with permission from the Forest Products Society.

and somewhat lower in the outer layers only in later part of pressing since steam does not pass freely through veneers and platen heat moves mostly by conduction from outer layers to middle layers rather than steam evaporation and condensation as in the hot-pressing in OSB/PB type boards. The temperature is higher in outer layers than middle layer until about 10 minutes press time. Considering the common hot-pressing time is about five minutes for these thickness panels, it is noted that the core-layer temperature at five minutes press time reaches only to 100–120 °C (212–250 °F), which is relatively low for PF adhesive resins to cure, and the vapor pressure reaches to about 120 psi, which is within the platen pressure range commonly used. This low core-layer temperature is why the most advanced PF resins are utilized in softwood plywood bonding. The adhesive should cure adequately to just hold the veneers together at press opening and then, it continues to cure to the full extent in the hot-stacking period.

Viscosity changes of glueline and curing: Initially, the high viscosity of adhesive layer is lowered in hot-pressing, but it still should remain at a relatively high value in order for the adhesive layer to be contained within the glueline, allowing only a little or no squeeze-out from the mat edges. An excessive squeeze-out often results from the PF resole resins' molecular weight that is too low or the low viscosity of adhesive mix, resulting in starved gluelines. On the other hand, if the viscosity of the glueline before hot-pressing is too high or dried up due to the PF resin's molecular weight that is too high or pre-dried due to hot or dry environment, with the low core-layer temperature, poor bonding will result due to lack of fluidity needed for adhesive to flow and wet the veneer surface. An adequate bonding can arise from optimized hot-pressing parameters, such as the adhesive spread rate, open and closed assembly times, hot-pressing temperature, pressure, and times.

Interaction among hot-pressing parameters: Again, most of these parameters are interrelated. For example, a high adhesive spread rate and short open/closed assembly times can give a high adhesive fluidity in the hot-pressing to result in a high squeeze-out and starved gluelines. On the other hand, a low adhesive spread rate and long open/closed assembly times can give a low adhesive layer fluidity or dry-out and poor wetting due to too little water. These problems also tend to occur in association with the seasons of manufacture: under-cure problems more often in winter and dry-out or lack of flow problems more often in summer. A dried-out adhesive layer means that the adhesive layer becomes a solid layer before the hot-pressing and therefore the adhesive layer would not melt adequately in the hot-pressing, which tends to occur more often with high molecular weight and high caustic content PF resins, which are often needed for fast curing rates. The veneer moisture content also similarly affects the plywood bonding in terms of the adhesive dry-out or over-penetration: too dry veneers tend to promote dry-out problems by a faster water absorption rate and wet veneers and wet pockets in dried veneer tend to promote over-penetration because the extra water can easily dilute the adhesive layer.

Industry hot-pressing parameters of softwood plywood: As with other wood composite manufacturing processes, the hot-pressing step tends to be the bottleneck of the plant, and this greatly limits its hot-pressing time. The hot-pressing time should be long enough to assure an adequate cure of adhesive layers beyond the vitrification point. However, the hot-pressing time in practice tends to be set at the shortest possible time and then the adhesive cure is extended during the subsequent hot-staking period, similarly as in OSB manufacture. Thus, obtaining the shortest possible hot-pressing time in a plywood plant is related to the veneer moisture content and various adhesive formulation/application parameters. Currently, the softwood plywood industry in North America uses hot-pressing parameters shown in Table 11.4 presented by C. R. Davis, R&D Manager, Georgia-Pacific Corp., Crossett, AR, in invited lectures to the class [30]. It is noted that the industry continues research to find ways to use higher veneer moisture contents and shorter hot-press times by improving the adhesive formulations for the obvious economic and toxic air emission reduction reasons. The cure speed for PF core-layer resins cannot be increased too much due to the limitation of the resin chemistry involved, as discussed in Chapter 9, but certain additive catalysts appear to be used as discussed in the catalyzed adhesive systems.

Table 11.4 U. S. Industry's softwood plywood manufacturing (hot-pressing) parameters and market trends [30].

	PF conventional	PF high moisture	PF catalyzed high moisture
Resin usage[a)]	1.4–2.0	1.4–2.0	1.15–1.75
Press temp (°F)	300–330	300–330	270–320
Veneer MC (%)	3–7	9–12	9–12
Press time (min) 3/4″	6–7	6–7	5–6

a) Actual PF resin % (non-volatiles) in panels based on oven-dry wood weight.

- Market growth trend: Declining on conventional panels. Increasing on LVL type products.
- Driving forces: *Negatives*: environmental, high wood cost, alternate wood use, inefficient plants, and substitution by OSB. *Positives*: uses in building, furniture, and exports.

11.10 Laminated Veneer Lumber (LVL)

Laminated veneer lumber (LVL) is manufactured similarly to softwood plywood using softwood plywood-type PF resin adhesives. Veneers up to seven 1/4″-thick plies are laid up all aligned in the longitudinal direction, and the mat is made continuously by end-to-end lap-jointing of veneers at varying distances between layers and along the length of mat, cured by hot-pressing in a continuous hot-press. The cured board is cut into lumber in the finishing step. More uniform, higher strength, consistent quality, dimensionally stable, as well as longer length lumber are manufactured. The extra benefit is lower levels of sawdust produced out of the good wood, etc. in comparison with lumber made by cutting from the same kind of wood [31–33], i.e. LVL uses wood more efficiently than lumber. The thicker LVL mat requires hot-pressing at higher temperatures and longer times often in a continuous hot-press, and recently it was reported that the steam injection hot-pressing method is similarly applicable by incisings made parallel to the grain of all veneers with thin 0.625″ length knives at 0.5″ longitudinal and 0.375″ lateral spacings, so that the steam injection occurs efficiently [33–35]. The incising was reported not to degrade LVL strength values [36]. It is apparent that the formulated plywood adhesives cure well under the steam injection condition while neat PF resins do not as in the manufacture of oriented strand lumber (OSL) under the similar steam injection condition. LVL are used as beams, roof trusses, I-beam components, etc. for flat-wise and edge-wise loading situations. Twenty-one LVL manufacturing plants are operating in 1998 in North America [4].

11.11 Parallel (Veneer) Strand Lumber (PSL)

Parallel veneer strand lumber (PSL), shown on the left, is manufactured similarly to LVL using softwood plywood-type PF resin adhesives using veneer splits, by bundling up with pressure in desired dimensions and continuously compressing and curing by passing through a radiofrequency (RF) heating unit. The efficiency of wood use is reported to be better than in LVL because lower-quality veneers can be incorporated. Three PSL manufacturing plants are reported to be operating in 1998 in North America [4].

11.12 New Developments in the Softwood Plywood Adhesive Technology

Dry-out problems: In the early days of the southern pine plywood industry, the PF resin adhesives earlier developed and used for D. fir plywood in the Northwest were introduced, and the adhesive dry-out problems were found to occur excessively due to the greater absorption tendency of southern pine than D. Fir, especially when using drier veneers and longer assembly times. The problems have been ameliorated by (i) adding a small amount of thickener, such as corn starch or hydroxyethylcellulose (HEC), to the PF resin [37], which were shown by gel filtration chromatography (GPC) to decrease the high molecular weight fractions in the resin, and it was also noted that the additive needs to be added first into the formaldehyde solution and pre-react for best performances. It appears that the hygroscopic additives' higher water-holding power also plays a part in holding the water in the glueline longer to help the resin flow better in curing. (ii) Another resin synthesis method reported for dry-out resistant PF resins is based on the incorporation of more linear resin structures in PF resole resins by synthesizing one resin with a low F/P mole ratio (~1.0) and another with a high F/P mole ratio (~2.5) and mix them in an appropriate ratio and cook further [38, 39]. The more linear structure component appears to retain its structure in the final resin and help lengthen the liquid character and prevent dry-out problems. (iii) Also, cresols are reported to be useful as an additive component to phenol in resin synthesis for ameliorating dry-out problems [37], since cresols having only two reactive sites on the phenol ring in comparison to three reaction sites for phenol react and polymerize through only two reactive sites to form linear polymer chain segments partly in the three-dimensional polymer networks formed with the main component phenol. The linear polymers, although they will finally get incorporated into the PF resin structure, would similarly lengthen the liquid stage in the curing process and therefore result in better adhesive flow.

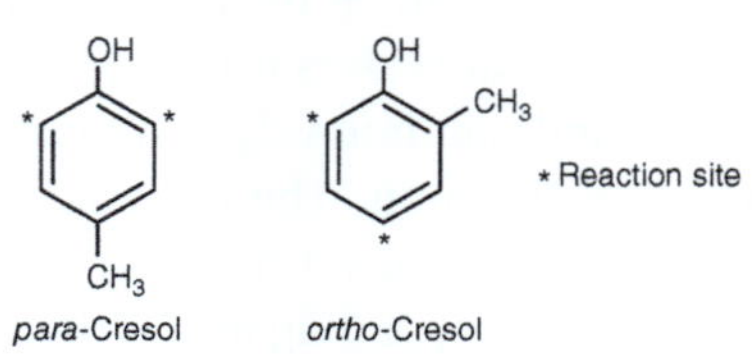

Bonding with higher moisture content veneers: In recent years, lowering the cost of plywood manufacturing has been a major concern of the industry by bonding higher moisture content veneers rather than the regular ~7% moisture levels without extending the hot-pressing time [30]. The approach has been reported to be somewhat successful with a catalyzed adhesive system, claimed to bond with southern pine veneers of up to 10% moisture content. The nature of the second component was not released to the public but the economic benefits of the new system are known to be significant, especially when the veneer drying capacity is the bottleneck of the plant [30]. Other reported methods such as adding potassium hydroxide to replace a part of the sodium hydroxide in the adhesive formulation or adding propylene carbonate or a small amount of resorcinol or guanidine carbonate have also been reported to be useful for bonding higher moisture content veneers [40–45]. Also, the bonding of fire-retardant and preservative-treated veneers has also been investigated. Thus, aspen veneers pressure-treated with (acidic) amino-resin fire retardants were successfully bonded by priming the surface with sodium carbonate or sodium hydroxide solution [46]. However, the PF resin adhesives did not develop strong bonds with the primed fire-retardant and DDAC preservative-treated veneers apparently because of the nonpolar alkyl chains of DDAC at the surface of veneers interfereing with penetration of the polar PF resin adhesives.

11.13 Phenolic Components Present in PF Resin-Bonded Softwood Plywood

Commercial southern pine plywood showed total methylene chloride extractable phenolic components in the range of 6.8–2.5 mg/kg (ppm) of panels and that of free phenol was 0.09–0.73 mg/kg (ppm) of panels, very low levels, indicating that softwood plywood is environmentally safe for various uses [47].

11.14 Softwood Plywood Manufacturing Industry

The softwood plywood manufacturing industry in the United States keeps the high quality of their products through the Voluntary Product Standard PS1-95, Construction and Industrial Plywood, initiated by APA – The Engineered Wood Association and developed under the Procedures for the Development of Voluntary Product Standards of the U. S. Department of Commerce [1]. In the newer PS1-19 version, plywood made of almost all soft and hardwoods is included and the durability grades are defined as (i) *Exposure* 1 and (ii) *Exterior use, bonded with exterior glue*. Most discussion in this chapter on softwood is for Exterior use grade. In contrast, OSB is classified as Exposure 1 and most hardwood plywood is Interior use grades. Exterior use grade is defined as that will retain glue bond when repeatedly wetted and dried or otherwise subjected to the weather or to exposure of similar severity and they must be proved accordingly as described in Section 20.2.3. The exterior grade softwood plywood is widely used as sheathings on roofs and walls as well as sub-flooring substrates with defined span ratings. The physical strength (wood species) and quality of veneers and the number of veneer layers in panels as well as the quality of adhesive bond are the determinant factors for the span ratings of panels, ranging from 16 to 48 in. Span rating numbers mean the distances between studs and 16 is the more common value in wall constructions. Since span rating numbers are measured by allowed minimum deflection extents on given static and impact loads, long span ratings mean stronger (stiffer) and often thicker plywood panels. Good adhesive bonds and good wood species and quality of veneers are the keys to get the desired span ratings with least board thicknesses or least materials costs.

11.15 Plywood Adhesive Fillers Made from Hydrolysis Residues of Municipal Newsprint Wastes

(Adapted with permission of the publisher Forest Products Society and minor revisions from: Oh, Y., T. Seller, Jr., M. G. Kim, and R. C. Strickland. 1997. Fillers for plywood adhesives made from acid hydrolysis residues of municipal newsprint wastes. Forest Products J. 47(10): 107-110.)

Biomass conversion to ethanol by acid hydrolysis process has been demonstrated as a feasible technology [48], and the potential to improve the process economy exists for the solid residues obtained as a byproduct by upgrading to higher-value materials. The Tennessee Valley Authority (TVA) explored the acid hydrolysis process of newsprint wastes and continued to study the residues as fillers for softwood plywood adhesives. In 1990, in the United States, various lignocellulosic particulate materials were used as fillers in wood adhesives and plastic moldings at about $266 per metric ton [49]. U. S. Environmental Protection Agency (EPA) in 1988 reported that US paper and paperboard accounted for an estimated 40% (64.6 million tons) of the municipal solid waste (MSW) stream [50, 51] and wood accounted for 3.6% (5.8 million tons) [51, 52]. In the United States the volume of lignocellulosic fillers used in adhesives to bond structural plywood is about 72 kt [53].

11.15.1 TVA Fillers and a Control Filler

Control filler: A furfural process residue (Cocob®) was selected. This filler was used extensively by the structural plywood industry in the southern region of the United States [49] and typically has a moisture content (MC) of 8–10% and a grind size of 75% passing through 75-μm (200-mesh) screen.

TVA acid hydrolysis residue fillers: At TVA, newsprint residues were prepared in a 20-L Monel-lined, electrically heated rotating digester. Coarsely shredded newspaper was hydrolyzed for 15 minutes at 160 °C using 2% (w/w) sulfuric acid at a liquid-to-solid (w/w) ratio of 5 : 1, determined to be optimal conditions for glucose production from newsprint. The digested contents were first cooled to 100 °C, filtered under vacuum, and washed three times with distilled water (four parts water to one part of newsprint, initial dry weight) to extract the majority of the solubilized sugars and acid used for hydrolysis. The filtered residues (70% moisture content) were dried in an oven at 50 °C to 6–8% MC and further processed in two ways: [54] (i) By grinding the dry material in a laboratory Wiley mill with a 200-mesh screen; (ii) By extracting the dried residues twice with methanol at a solid-solvent ratio of 1 kg: 5–6 L (33% solids) at 55 °C, followed by drying in an oven at 50 °C (4.6% of residues removed) and similarly grinding. The elemental analyses were done and found the fiber contents of about 95% in comparison with about 82% of the control filler (Cocob).

Table 11.5 Plywood adhesive mix characteristics using three fillers.

	Control filler	Unextracted filler	Extracted filler
Mix characteristics			
Resin solids (%)	29.7	30.3	30.3
Extender solids (%)	5.3	5.3	5.3
Filler solids (%)	5.5	5.5	5.5
Sodium hydroxide solids (%)	1.5	0.75	0.75
Total mix solids (%)	42.0	41.85	41.85
Total mix water (%)	58.0	58.15	58.15
Typical viscosity, Brookfield,			
Initial, 25 °C	5005	5100	5600
After 24 h in 4 °C storage	8800	5360	—
After 24 h ambient	10 740	6350	5740

11.15.2 Plywood Adhesive Mix

The plywood adhesive mixes were made following the recommendation of Georgia-Pacific Resins, Inc., using a phenol-formaldehyde resin with code number GP 5763. The mixing procedures used for TVA residue fillers were similar to the control, but only one-half of the recommended sodium hydroxide was used in order to obtain a desirable mix viscosity, as shown in Table 11.5. Once a usable mix viscosity was obtained (3.0k–10.0k cps), the mix viscosity stability was observed over 24 hours [53]. *The adhesive mixes* made with the residue fillers required less sodium hydroxide to yield desired mix viscosities. Once the proper viscosity mix was obtained, the formulation made with residue fillers was very stable in comparison to the control filler, measured by viscosity after standing for 24 hours at room temperature (25 °C). Excessively high mix viscosity cannot be applied with plywood adhesive coaters, and it is a common practice to adjust the amount of sodium hydroxide to reach a usable mix viscosity. The level of added sodium hydroxide does have an effect on resin penetration into the substrate surfaces and this penetration affects adhesive bonding, but it is noted that the adhesives made with TVA fillers at reduced levels of sodium hydroxide bonded the plywood very well.

11.15.3 Plywood Manufacture

Southern pine veneer, approximately 4.0 mm (0.159 in.) thick, was secured from a southern region structural plywood mill (Weyerhaeuser Co. Millport, AL). The MC of the veneer at the time of use was approximately 5.0%.

(1) *The veneer* was spread with adhesive by a Black Brothers double-roll coater (762 mm, 30 in. wide) at an objective spread rate of 214 g/m^2 (44 lbs/1000 ft^2), single glueline basis. *The construction* was 3-ply, 12.7-mm (1/2 in.) thick, 305 mm by 305 mm (12 in. by 12 in.) in size.
(2) *Panels were laid up* with 20-, 40-, and 60-minutes total assembly times, *consisting of 2 minutes layup, 10 minutes stand time before prepressing, 4 minutes prepress time (batch*

cold press), and *the balance stand time* before hot-pressing. The pre-press pressure was 1034 kPa (150 psi).

(3) *Panels were hot-pressed* for either four or five minutes at 150 °C under 1379 kPa (200 psi). *Four replications* per condition were made for total of 24 panels per residue filler type (2 × 24 = 48). *A control filler* was run with each residue filler for a total of 48 control panels. The total number of panels was 96.

(4) *Plywood tests*: Ten plywood tension shear specimens were cut from each panel per Product Standard PS 1-83 (now PS 1-95) [55]. The specimens were cut so that five specimens were pulled with closed lathe checks and five specimens were pulled with open lathe checks to average out any effects of this variable.

The specimens were subjected to standard vacuum-pressure (VP) accelerated aging procedure prior to testing with a Globe plywood shear test machine [55]. The shear strength of each specimen was recorded, and the wood failure (WF) percentages over the shear area was visually estimated after drying the specimen.

11.15.4 Plywood Test Results

A summary of the plywood shear test results is provided in Table 11.6 and the data were also compared in Figure 11.2a,b. The observations are as follows:

(1) The plywood made at 40- and 60-minutes assembly times using the unextracted residue filler was comparable in WF percentages to the plywood made with control filler at similar assembly times (95% vs. 97%, respectively).

(2) At the short assembly time (20 minutes), the plywood made with the unextracted filler had a considerably lower *WF* readings compared to control (75% vs. 94%, respectively), exhibiting adhesive *over-penetration* because of the slower dissipation of water from glueline. This problem could be solved probably by adjustment of mixed ingredients, resin changes, and/or by lower spreads.

Table 11.6 Plywood shear strength and wood failure percent estimates after a vacuum-pressure aging method for plywood bonded with three fillers.

		Shear strength			Wood failure estimates		
Assembly time	Press time	Control Cocob®	Unextracted TVA	Extracted TVA	Control Cocob®	Unextracted TVA	Extracted TVA
(min.)		(kPa)[a)]			(%)		
20	4	1120	1496	848	93	71	91
	5	1103	1427	1214	96	78	93
40	4	910	1103	958	98	99	98
	5	841	1269	1248	96	96	93
60	4	993	1330	1055	96	93	91
	5	1172	1434	1317	98	93	94
Average		1040	1343	1107	96	88	93

a) Multiply kPa times with a factor of 0.145 to obtain psi. The shear strength and WF values given represent averages of eight panels for Cocob® and four panels each for the residue fillers (10 specimens per panel).

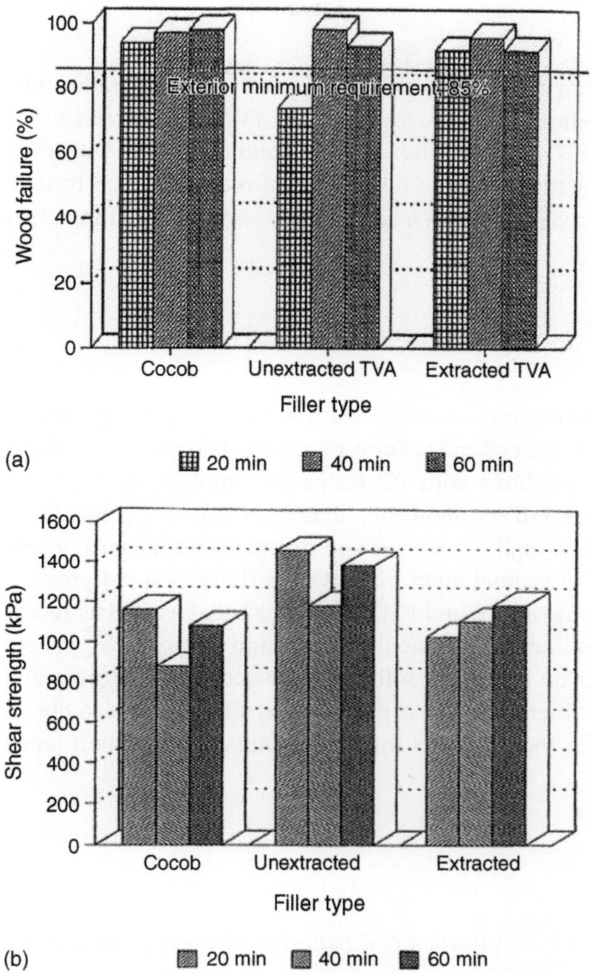

Figure 11.2 (a) Average wood failure results of plywood made with adhesives containing three fillers and bonded at three assembly times. The values in the figure include 2 press times and represent averages of 16 panels for Cocob® and 8 each for the residue filler at each assembly time: 10 specimens per panel. (b) Average shear strength test results of plywood made with adhesives containing three fillers and bonded at three assembly times. The values in the figure include 2 press times and represent averages of 16 panels for Cocob and 8 each for the residue filler at each assembly time: 10 specimens per panel.

(3) Plywood made with the extracted residue filler had *WF* results comparable to control (93% vs. 96%) for all assembly times.
(4) The average *shear strength* for the control filler was 1040 vs. 1343 kPa for the unextracted filler and 1107 kPa for extracted filler. These shear strength variations are typically due to veneer quality differences, although they may indicate different filler contributions to bonding.
(5) Hot-press times (four vs. five minutes) were not a significant factor in reaching 90% WF values (Figure 11.2a) except the 20-minute assembly time for the unextracted filler where higher WF values (78%) were observed with the longer press time compared to the shorter press time (71%).

11.16 Conclusion

The Product Standard PS 1-95 has set a minimum average of 85% WF value for exterior-type plywood construction. This study indicated that the residues from TVA acid hydrolysis process are good candidate fillers for PF resin adhesives used to bond *construction plywood* and *LVL* where high grades of structural glueline durability are required. Such residue uses would also add to the economic viability of TVA acid hydrolysis process for municipal newsprint wastes.

11.17 Summary

Industry and laboratory manufacturing procedures of softwood plywood are presented based on open literature and experiences obtained from numerous personal plant visits. Industrial adhesive formulation procedures with PF resins are emphasized regarding the filler and extender compositions and resultant mix adhesive's viscosity, solids level, gap-filling capacity, etc. The adhesive application rates and methods and the curing procedures, including open and closed assembly times, pre-pressing times, and hot-pressing times, were discussed. Manufacturing of LVL and PSL is also presented as well as recent research activities on using higher moisture content veneers and on reducing veneer dry-out problems. A research paper on laboratory softwood plywood manufacturing and testing carried out to evaluate new filler materials is presented to allow readers to obtain more realistic relationships between the manufacturing parameters and resultant panel properties.

References

1. *Voluntary Product Standard PS 1-95. Construction and Industrial Plywood*. Gaithersburg, MD: NIST/APA.
2. Blomquist, R.F. (1969). *Forest Products Journal* 19 (4): 36.
3. Fassnacht, D.L. (1964). *Forest Products Journal* 14 (1): 21.
4. Sellers, T. Jr., (2001). *Forest Products Journal* 51 (6): 12.
5. Kim, M.G., Nieh, W.L.-S., Sellers, T. Jr., et al. (1992). *Industrial and Engineering Chemistry Research* 32: 973.
6. Kim, M.G., Nieh, W.L.-S., and Meacham, R.M. (1991). *Industrial and Engineering Chemistry Research* 29: 798.
7. Neese, J.L., Reeb, J.E., and Funck, J.W. (2004). *Forest Products Journal* 54 (1): 67.
8. Fleischer, H.O. and Lutz, J.F. (1963). *Forest Products Journal* 13 (1): 39.
9. Koch, P. (1965). *Forest Products Journal* 15 (12): 495.
10. Chow, S.-Z. (1971). *Forest Products Journal* 21 (2): 19.
11. Hse, C.-Y. and Kuo, M. (1988). *Forest Products Journal* 38 (1): 52.
12. Stumbo, D.A. (1964). *Forest Products Journal* 14 (12): 582.
13. DeVallance, D.B., Funck, J.W., and Reel, J.E. (2007). *Forest Products Journal* 57 (1/2): 21.
14. Hse, C.Y. (1972). *Forest Products Journal* 22 (9): 104.
15. Gollob, L., Krahmer, R.L., Wellons, J.D., and Christiansen, A.W. (1985). *Forest Products Journal* 35 (3): 42.

16 Nieh, W.L.-S. and Sellers, T. Jr., (1991). *Forest Products Journal* 41 (6): 49.

17 Seller, T. Jr.,, Miller, G., and Nieh, W.L.-S. (1990). *Forest Products Journal* 40 (10): 23.

18 Seller, T. Jr.,, Miller, G., and Smith, W. (2005). *Forest Products Journal* 55 (3): 27.

19 Raymond, R.C. (1960). *Forest Products Journal* 205 (4).

20 Marra, A.L. (1980). *Proceedings of 1980 Symposium on Wood Adhesives*, WSU/FPL, Adhesives for wood composites. U. S. Forest Service/Forest Products Laboratory, p. 202.

21 Bauer, E. (1980). *Proceedings of 1980 Symposium on Wood Adhesives*, WSU/FPL, Recent advances in plywood technology. U. S. Forest Service/Forest Products Laboratory. p. 207.

22 Koch, P. (1965). *Forest Products Journal* 15 (9): 355.

23 Koch, P. and Jenkinson, P. (1965). *Forest Products Journal* 15 (1): 488.

24 Faust, T.D. and Rice, J.T. (1986). *Forest Products Journal* 36 (4): 57.

25 Faust, T. and Borders, B.E. (1992). *Forest Products Journal* 42 (11/12): 49.

26 Cone, C.N. (1969). *Forest Products Journal* 19 (11): 14.

27 Hine, J.M. (1960). *Forest Products Journal* 206 (4).

28 Haskell, H.H. (1971). *Forest Products Journal* 21 (9): 64.

29 Zavala, D. and Humphrey, P.E. (1996). *Forest Products Journal* 46 (1): 69.

30 Davis, C.R.. (1995–2005). R & D Manager, Georgia-Pacific Corp., invited lectures.

31 Bohlen, J.C. (1972). *Forest Products Journal* 22 (1): 18.

32 Koch, P. (1973). *Forest Products Journal* 23 (7): 17.

33 Laufenberg, T.L. (1983). *Forest Products Journal* 33 (9): 21.

34 Wang, B., Di, C., and Troughton, G. (2003). *Forest Products Journal* 53 (11/12): 99.

35 Troughton, G.E. and Lum, C. (2000). *Forest Products Journal* 50 (1): 25.

36 Koch, P. (1973). *Forest Products Journal* 23: 17.

37 Freeman, H.G., Baxter, G.F., and Howell, J.T. (1972). Phenol-formaldehyde adhesive resin containing hydroxyethylcellulose. US Patent 3,663,481. Weyerhaeuser Co. Tacoma, WA.

38 Chen, C.M. and Rice, J.T. (1976). *Forest Products Journal* 26 (6): 17.

39 Clarke, N.R., Steiner, P.R., and Anderson, A.W. (1988). Phenol-formaldehyde adhesive for bonding wood pieces of high moisture content and composite board and veneers bonded with such adhesive. US Patent 4,824,896.

40 Higuchi, M., Tomura, S.I., and Sakata, I. (1994). *Mokuzai Gakkaishi* 40: 604.

41 Tohmura, S.I., Higuchi, M., Hattoru, Y., and Sakata, I. (1994). *Mokuzai Gakkaishi* 40: 390.

42 Reuther, O.W., Graser, M., Mayer, J. et al. (1972). Use of formamide-based mixture as hardening accelerators for phenol-formaldehyde resin glues. German Patent 2,110,264.

43 Dormyer, D., Kochersperger, W., Wolf, F. et al. (1991). A curing agent mixture for alkaline phenol-formaldehyde resins containing fibrous fillers and carbonic esters. French Patent 1,550,847.

44 McVay, T.M. and Baxter, G.F. (1990). Phenolic resin containing a latent hardening agent comprising the carbamate reaction product. US Patent 4, 977,231.

45 Black, E.P. (1998). Phenol-formaldehyde resins modified with guanidine salts. US Patent. 5,763,559. Georgia-Pacific Corp., Atlanta, GA.

46 Vick, C.B. (1994). *Forest Products Journal* 44 (1): 33.

47 Tiedeman, G.T., Issason, R.L., and Sellers, T. Jr., (1994). *Forest Products Journal* 44 (3): 73.

48 Barrier, J.W. and Bulls, M.M. (1992). Emerging technologies for materials and chemicals from biomass. In: *ACS Symposium Series 476: Emerging Technologies for Materials and Chemicals from Biomass* (ed. R.M. Rowell, T.P. Schultz, and R. Narayan), 410. Washington, D. C.: ACS.

49 Sellers, T. Jr. (1990). Survey reveals use of lignins as partial substitute for phenol. *Panel World* 31 (26): 44.

50 Rowell, R.M.J., Youngquist, A., and McNatt, D. (1991). Composites from recycled materials. In: *Proceedings of the 25th International PB/Composite Materials Symposium* (ed. T.M. Maloney), 301. WSU, Pullman, WA.

51 U. S. Environmental Protection Agency (1988). *Characterization of Municipal Solid Waste in the United State: 1960-2000*. Washington, D.C.: EPA.

52 U. S. Environmental Protection Agency. (1990). Characterization of municipal solid waste in the United State: 1990 update. Report. EPA/530-sw-90-042. EPA, Washington, D.C.

53 Sellers, T. Jr., (1985). *Plywood and Adhesive Technology*, 661. New York: Marcel Dekker.

54 Sellers, T. Jr., and Gardener, D.J. (1989). *Forest Products Journal* 39 (3): 34.

55 National Standards Bureau (1995). *Product Standard PS 1-95*. Washington, D.C.: US. Department of Commerce.

12

Isocyanate Wood Adhesive Resins

12.1 Chemical Compositions of Isocyanate Wood Adhesive Resins

Chemical Compositions of Isocyanate Resins: Isocyanate (ISO) resins used as wood adhesives and coatings are manufactured mostly from benzene, which is obtained from crude oil, through a series of chemical transformation reactions by several large chemical manufacturers [1]. Isocyanate resins consist of 100% active resin solids and the molecules have two or more isocyanate groups, which are highly reactive to form covalent bonds with hydroxyl or amine groups of water, wood, cotton, and human body, etc. By limiting the amount of water available or reacting with organic materials containing poly-hydroxyl or poly-amine groups, the reaction of isocyanate resins results in forming high molecular weight, cross-linked (thermosetting) polymers. Because of this high reactivity, isocyanate resins are handled and transported in an anhydrous condition, usually in dry drums with a tight seal. Isocyanate resins as supplied are composed of dimeric 2, 4′- and 4,4′-diphenyl-methane diisocyanates (MDI) (~55%), trimers (~25%), and higher oligomers (~20%) as shown in Figure 12.1. In industry, the crude isocyanate product is commonly separated by distillation into fractions to obtain more consistent molecular properties for each fraction. The fraction used in OSB manufacture is what is often called polymeric-MDI (pMDI) composed of trimers and higher homologs up to hexamers (Figure 12.1, $n = 0$–3). They are dark brown liquids with a viscosity of around 250 cps, a flash point of about 220 °C , and vapor pressure of less than 1×10^{-4} mmHg, and available in various grades according to their isocyanate group content (%). There are also other types of isocyanate resins manufactured from toluene (TDI) and other aliphatic chain materials, which are used mostly in coatings.

General Uses of Isocyanate Resins: Isocyanate (mostly dimeric) resins are used with various polyols in the manufacture of flexible and rigid polyurethane foams used in furniture manufacturing and also as insulation materials in buildings and appliances. The good adhesive properties of isocyanate resins as thermosetting adhesives have been known for a long time, especially in the bonding of metals. The use of pMDI resins in industrial wood composite boards started only about 20 years ago, and they have progressively replaced PF resole *core resins* in OSB manufacture. The price of pMDI resins is higher than PF resin by about two times on weight basis, because of the more extensive chemical transformation reactions involved in their manufacture, but there are certain advantages in using pMDI resins as OSB core-layer adhesives.

Adhesives and Finishes for Wood: For Practitioners and Students, First Edition. Moon G. Kim.

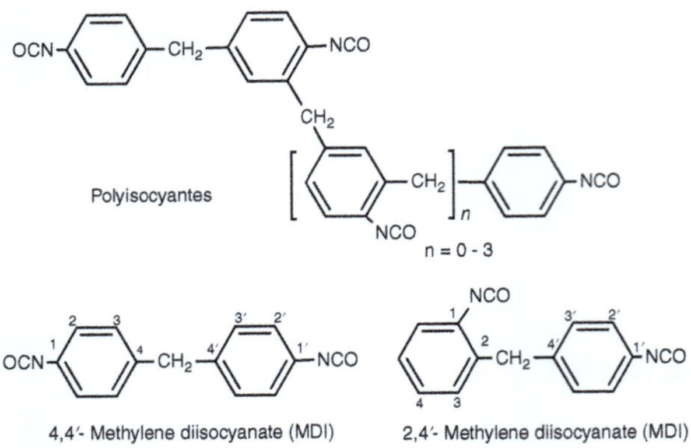

Figure 12.1 Structures of isocyanate resin molecules: MDI and other dimer and oligomer components.

Handling Procedures of Isocyanate Resins: The molecular weights of pMDI resins are relatively low compared to other thermosetting resins such as PF resins and their molecular association tendency in the liquid state is also low. Therefore, the viscosity of resins even at 100% resin-solid levels ranges in the relatively low ranges of about up to 250 cps, in the sprayable range. pMDI resins are supplied without any solvents, and since they are *insoluble* in but reactive with water, the handling of resins in wood composite manufacturing plants has to be done under anhydrous conditions. The resin curing reactivity will degrade by infiltration of water or moisture from the air. Cleaning of resin sprayers, etc., has to be done with organic solvents, which are mostly flammable and toxic, and the wash wastes need to be safely disposed of separately from the sewer system. Furthermore, isocyanate resins are highly toxic and allergenic to most people. Spraying of isocyanate resins in wood composite plants tends to form aerosol droplets, which can escape into other areas in the plant. Personal protective equipment has to be used in handling, especially where the resin is sprayed into the blender as fine mist in the OSB manufacture.

12.2 Reactivity, Polymerization, and Curing Reactions of Isocyanate Resins

Reaction of Isocyanate Resins with Water: The polymerization and cross-linking reactions of isocyanate resins should be carried out in the curing step, which is similar to other thermosetting resins, such as PF or UF resins [1]. *The reactive group is the isocyanate group (-N=C=O)*, which condenses with hydroxyl (-O-H) groups to form covalent bonds as shown in Eqs. (12.1) and (12.3). Each oligomeric molecule of isocyanate resin has **two** or more isocyanate groups (**O=C=N**-Ar-**N=C=O**), a basic requirement in forming of polymer molecules. Similarly, it is evident that the molecules that provide hydroxyl groups should have two or more such groups per molecule to be a member of polymer molecular chains, i.e. *they are polymeric diols or triols.* As shown in Eq. (12.1), however, the reaction with water (a diol) results in carbamic acid group (Ar-**NH-CO-OH**), which is unstable and breaks down to give off carbon dioxide (gas), leaving amine groups behind. Thus, when there is an ample amount of water available to react, all isocyanate groups will be transformed into amine groups, and the material becomes nonreactive or useless. Therefore, we use isocyanate resins in various reactions with restricted amounts of water.

$$O{=}C{=}N{-}Ar{-}N{=}CO + 2H{-}O{-}H \rightarrow [H{-}O{-}CO{-}NH{-}Ar{-}NH{-}CO{-}OH] \rightarrow\rightarrow NH_2{-}Ar{-}NH_2 + 2\,CO_2 \quad (12.1)$$

$$+2\,O{=}C{=}N{-}Ar{-}N{=}CO \rightarrow\rightarrow O{=}C{=}N{-}Ar{-}NH{-}CO{-}NH{-}Ar{-}NH{-}CO{-}NH{-}Ar{-}N{=}CO \quad (12.2)$$

$$O{=}C{=}N{-}Ar{-}N{=}CO + 2HO{-}R{-}OH \rightarrow\rightarrow HO{-}R{-}\mathbf{O{-}CO{-}NH}{-}Ar{-}\mathbf{NH{-}CO{-}O}{-}R{-}OH \quad (12.3)$$

$$+\,n\,O{=}C{=}N{-}Ar{-}N{=}CO \rightarrow\rightarrow HO{-}R{-}\mathbf{O}\{\mathbf{-CO{-}NH}{-}Ar{-}\mathbf{NH{-}CO{-}O}{-}R{-}\mathbf{O}\}_n{-}\mathbf{CO}{-}NH{-}Ar{-}N{=}CO \quad (12.4)$$

Reaction with Hydroxyl Groups of Polyols, etc: On the other hand, with diols or triols of aliphatic chain molecules, the condensation reaction with an isocyanate group leads to a stable carbamate ester (*urethane*) bonds (R-**O-CO-NH**-Ar), as shown in Eqs. (12.3) and (12.4), and in cases of mixing and reacting a *di-isocyanate resin* and a *diol* at a 1 : 1 mol ratio, one isocyanate resin molecule reacts initially with two diol (H-O-R-O-H) molecules to form two *urethane bonds*, resulting in a bigger molecule that has two terminal hydroxyl groups, i.e. a longer diol is resulted (Eq. 12.3). This longer diol keeps reacting further with another di-isocyanate resin molecule to form bigger molecules and in the end, the resulting high molecular weight molecules would have an isocyanate group at one end and a hydroxyl group at other end, as shown in Eq. (12.4). By presetting the 1 : 1 mol ratio accurately between isocyanate and hydroxyl groups in the beginning mix, the urethane bond formation reaction can be made to go to a very high extent to result in extremely high molecular weight urethane polymers. The end groups could be a hydroxyl or an isocyanate group, but their concentrations in the resultant polymer would be extremely low and do not affect the polymers' use properties too much. Such terminal isocyanate end groups could easily be changed to nontoxic amine groups when exposed to moisture, according to Eq. (12.1). This urethane-forming reaction often needs a catalyst to make it go at room temperature, and by selecting polyols having appropriate chain structures and molecular weights, thermosetting urethane engineering polymers (plastics) are manufactured, and by adding some water or a volatile freon-type gas to the reaction mixture, *flexible and rigid urethane foams* are manufactured, which are widely used in upholstered furniture manufacturing and as insulation materials. This type of urethane foam product is normally manufactured using the higher-priced di-isocyanate resin fractions composed of dimers or trimers, shown in Figure 12.1.

Wood Adhesives and Coatings Made by Premixing pMDI Resins and Polyols [1, 2]: The condensation reaction between hydroxyl and isocyanate groups occurs quickly only with a catalyst and without it the reaction goes only slowly and at above 40 ° [3]. This lower reactivity allows time for mixing the two components and applying them as wood adhesives or coatings, and the mixes can be cured at an elevated temperature [3]. Some of these mixtures are industrially manufactured: pMDI resins and polyols of appropriate molecular weight and chain structures are mixed at various mole ratios, for example, 1 hydroxyl to

12 isocyanate (1 : 12) groups, etc. Since polyols vary in molecular weight from low to high numbers, the mix ratio by weight could be different from the mole ratios. There are at least three different ways to make adhesives or coatings using pMDI and polyols.

Mix the Two Components with an Excess Isocyanate Group Ratio Without Any Water – The mix is applied soon and cured as binder of wood composites. In a reported example [4], pMDI was mixed with various molecular weight polyols and applied as particleboard binder. The mix viscosity would depend much on the polyol component, and the mixes commonly show several hours of pot-lives. The board mat (0.5″ × 18″ × 18″, 9% MC) was cured at 350 °F in a hot press for 6 minutes. The resin loading level was 5% based on dry wood. The best particleboard strength values were obtained with a hydroxyl:isocyanate group ratio between 1 : 6 and 1 : 12 using polyethylene glycol 1000 (Peg 1000) as polyol. Higher or lower hydroxyl:isocyanate group ratios performed less well. Also, ethylene glycol Peg 62 or Peg 1450 gave somewhat poorer values. Addition of a triol to the polyol part was found to give higher strength or higher thermal stability and durability values [4, 5]. Triols would enhance the crosslink density of cured polymers. Thus, the group ratio, molecular weight, and structures of polyols all affect the wood bonding performance and need to be optimized case by case. Poor performances were observed for higher hydroxyl : isocyanate group ratio values like 1 : 4 due to the lack of isocyanate groups left to form adhesive bonds. Poor performances were also observed when using higher molecular weight polyols due to too much thermoplastic character in the resultant cured polymers. It is interesting to note that these resin systems, even with a 1 : 12 mol ratio have a pot-life of several hours. This type of polymer system is often called *two-component adhesives or coatings*, meaning that it is necessary to mix two materials on-site before use. Coatings often contain some inert solvents as carriers, and this type of two-component urethane coating is often used as floor varnishes, known for their high toughness and durability.

Mix pMDI and an emulsion form of a polyol of known hydroxyl group content to result in an emulsion polymer/isocyanates (**EPI**) resin system – In a reported example [6], 15 parts of p-MDI were mixed thoroughly with 100 parts of a polyol in water emulsion, resulting in an EPI adhesive resin system that showed a lower viscosity but a pot-life of about one hour at room temperature due to the reactivity with water, which starts at above 40 °F, and when the mix is applied on wood at room temperature, the bond strength soon arose by dissipation of water, and the thermosetting properties developed more slowly from the condensation reaction of isocyanate groups with hydroxyl groups of emulsion polymer and perhaps also with wood hydroxyl groups if isocyanate groups are present in excess. This type of adhesive was reported to be useful for bonding COM-PLY lumber [5], i.e. as adhesives for bonding veneers for laminated veneer lumber (LVL) and also bonding LVL to PF resin-bonded particleboard. A closed-assembly time of up to 12 minutes, clamping time of up to two hours, clamping pressure of up to 175 psi, and an adhesive spread rate of 80–100 lbs/1000 ft^2 were experimented. The adhesive bonds needed to cure further for 7 days at 73 °F and 55%RH. The boards were tested for outdoor exposure for up to a year, and the results indicated that the higher spread rate and two-hour clamping time gave more than adequate dry and wet shear strengths values, but the wood failure values did not meet the 80% requirement for both unweathered and 6- and 12-month weathered samples. The control PRF resin bonds gave satisfactory 12 months outdoor exposure wood failure values. Overall, the EPI-bonded wood products appear to withstand direct weathering for up to one year without serious deterioration, but longer exposures appear not as good as PRF resin bonds. The short pot-life of this resin system is an inevitable outcome of mixing isocyanates with water emulsion of a polyol, making it necessary to mix ingredients on-site and use them quickly, although a sprayer application system would be possible to use due to the lower mix viscosity in comparison to using polyols directly. This approach has been explored in several reports [7, 8] and, in Japan, this type of water-borne pMDI resin adhesives have been reported to be in use, in which the polyol component consists of water and a polyvinyl alcohol (PVA), styrene-butadiene rubber or ethylene-vinyl acetate copolymer, and calcium carbonate as a filler [9]. The isocyanate component consists of pMDI that is designed to cross-link with PVA's hydroxyl groups on heating at 100–200 °C [10].

Mix the Two Components at a Hydroxyl:Isocyanate Group Ratio of About 1 : 16 or Higher – The mix is reacted at an elevated temperature without solvent until products of some mid-range high molecular weights are formed with a viscosity ranging up to about 25 k cps. The reaction stops at some point where there are no more hydroxyl groups left in the resin system, resulting in a thermoplastic polymer that is still a liquid or soft solid and contains a surplus of (reactive) isocyanate groups [11]. This type of intermediate reaction products has been explored as *one-component, moisture-curing polyurethane* binders for particleboard [3] and laminated beam [12], and they are commercially produced and used as the main components of reactive hot-melts (RHM) [11] and moisture-curing coatings (with solvents) [2, 13]. (RHM adhesives are discussed in detail in Chapter 15.) These adhesives are *moisture-curing systems* similar to the *pure* pMDI resins discussed in the next section, and therefore the wood substrates must have an appropriate level of moisture. At lower moisture contents, they may not fully cure to the high molecular weight structures. Also, since these adhesive molecules' extent of cross-linking is considered to be less than those of PF or PRF resin systems, their long-term creep resistance, durability, and fire properties need to be tested rigorously for certification for adhesives for structural laminated beams, *etc.* Various isocyanate formulations are currently trying to make inroads into the structural wood lamination markets, which are dominated by PRF and MUF resins. Therefore, the cured resins' creep resistance at elevated temperatures of fire (204 °C or higher) becomes an important issue, and some of these types of adhesives were reported to be showing higher strength degradation rates than solid wood [14, 15]. One difficulty for adhesive users would be that there are various isocyanate materials and there are more various polyol materials available, and glue suppliers would be hesitant to reveal the materials, or there could be too many candidates for the long, expensive certification tests. One report on using this adhesive system in wood lamination indicated good dry and wet strength values but low wet wood failure values, and this deficiency was found to be improved drastically by priming the wood with the hydroxymethylated resorcinol (HMR) priming agent discussed in Chapter 10 [16]. Also, it is noted that the presence of isocyanate groups in the adhesives or coatings makes it necessary to protect them from air moisture, commonly in thick paper cartridges for hot-melts or closed containers for coatings, but a long-term stability might be still poor. These types of one-component moisture-curing polyurethane adhesives are known to be used quite widely in Europe for wood laminations in non-load-bearing structures. Their various long-term performances have been investigated aiming at achieving performances that are competitive with those of PRF or RF resin adhesives [13].

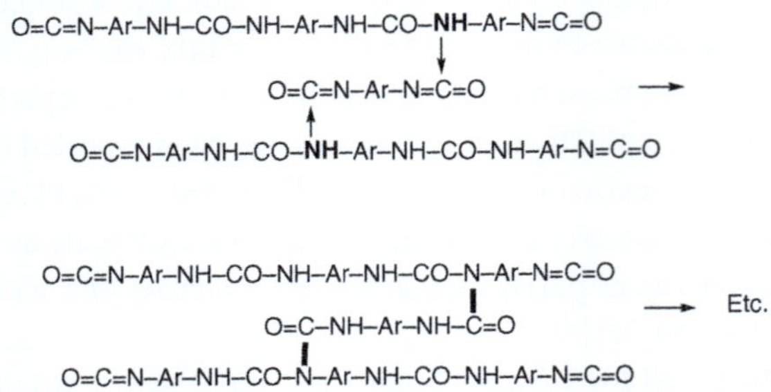

(12.5)

Cross-linking urea bonds
Ar = Diphenyl methane groups

Use of pMDI Resins in OSB and the Cross-linked Resin Structures: Neat pMDI resins are used as binders for OSB and other wood composite boards as moisture-curing systems, as reported in numerous successful reports [17–19]. When (pure) pMDI resin is applied on wood, as discussed above in Section 12.1, the isocyanate groups would react with the water available from wood to form amine groups and carbon dioxide gas since no external polyol is added except the hydroxyl groups of wood surface of which availability is indeterminate. And when the available water is exhausted, which is likely in OSB manufacture, the residual isocyanate groups would react with the generated amine groups to form urea-type amide bond group structures as shown in Eq. (12.2), extending the polymer chain growth.

Furthermore, all hydrogens of amine and amide structures (-NH-) are also reactive with isocyanate groups as shown in Eq. (12.5), resulting in cross-linked covalent *biuret-type bonds* among polymer strands; eventually leading to highly cross-linked thermosetting resin structures. The evidence of forming urea and biuret bonds has been observed by the solid-state NMR analyses [20]. Thus, when pMDI resins or moisture-curing intermediate MW (hot-melt) adhesives discussed above are applied on wood, the isocyanate groups are forced to undergo the reaction with water, followed by these chain-extension and cross-linking reactions until all isocyanate groups are exhausted. These reactions would make the resin system thermoset, similarly to PF or UF resins, not melting upon heating at elevated temperatures. The good adhesiveness of pMDI resins in bonding of OSB would also arise from the crosslinked isocyanate resin polymers' highly polar nature (nitrogen and oxygen atoms). Thus, the moisture available from wood substrates is a key, too much moisture can lead to stoppage of reactions at the amine forming stage, and too little moisture or presence of excess isocyanate groups leads to an incomplete reaction of isocyanate groups, remaining in the finished glueline, which may cause some unwanted effects later in the use of boards. Also, not all water in the wood substrates reacts, only some do, depending on the gluing or hot-pressing parameters including resin loading levels. So, the optimum moisture content ranges of wood substrates vs. resin loading levels have to be found by trial-and-error experiments under the bonding condition.

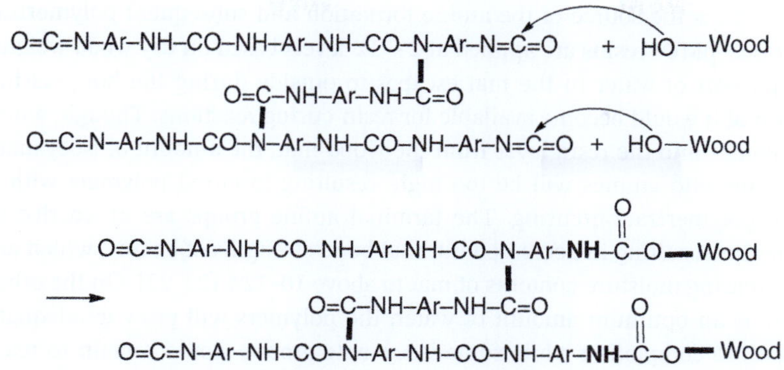

(12.6)

Ar = Diphenyl methane groups

Possible Reactions with Wood Hydroxyl Groups: As shown in Eq. (12.6), the wood surface hydroxyl groups appear to be able to react as diols or polyols with isocyanate groups to form urethane bonds, as discussed in Eq. (12.3) and (12.4). The reaction can occur *at elevated temperatures* without any added catalysts as in OSB manufacture, and it may compete with reactions with amine or amide groups discussed in Eq. (12.5). The polymerization and cross-linking reactions would occur firstly through the amine/biuret forming routes, and hydroxyl groups of the cellulose, hemicellulose, and lignin molecules of wood surface could come in later to react. There should be enough amount of isocyanate groups remaining after reactions with water in order to go on to form urethane bonds with wood components. The resultant urethane bonds with wood would greatly enhance the adhesion strength over the good adhesion strength already afforded by the highly polar nature of cured isocyanate polymers. However, the presence of such wood-isocyanate bonds has not yet been observed analytically, although pMDI resins have been known to give better IB strengths than PF resins even at lower loading levels in OSB manufacture. Therefore, better bond performances have been often ascribed to the possibility of forming such urethane bonds with wood hydroxyl groups. An intimate molecular association among the hydrogen atoms of adhesive molecules and of wood components in the interphase of isocyanate resin-bonded woods was shown to occur significantly by solid-state two-dimensional NMR studies [20, 21], indicating a good penetration of isocyanate resin molecules occurring into wood cell wall structures. A more uniform interphase between the wood and cured resin molecules appears to be formed in line with the diffusion theory of adhesion [22]. Although this result cannot be a direct proof of the covalent urethane bond formation with wood, it appears to indicate such bonds might be actually occurring to some extent.

12.3 Mat Moisture Levels in Using pMDI Resins for OSB Bonding

Analytical results indicate that Eqs. (12.1), (12.2), and (12.5) discussed above are the main reactions going on in the isocyanate resin bonding of OSB. This means that the water from wood mat, ~7%, is the source of the amine formation and subsequent polymerization/curing reactions. pMDI resins are applied at 2 – 3% levels based on dry wood weight. Of course, a major part of water in the mat escapes to outside during the hot-pressing and only a fraction of it would become available for resin-curing reactions. Though, when too much water comes into the resin layer from the wood mat, the amount of isocyanate resin that transforms into amines will be too high, resulting in cured polymers with a limited degree of polymerization/curing. The terminal amine groups are unreactive to wood hydroxyl groups resulting in decreased internal bond strengths of boards, which are observed with increasing moisture contents of mat to above 10–12% [22, 23]. On the other hand, when there is an optimum amount of water, the polymers will grow to adequate sizes and cure goes to completion while some isocyanate groups would remain to react to form urethane bonds with wood hydroxyl groups. On the other hand, in cases where there is a lack of water, some unreacted isocyanate groups may remain in finished boards, possibly expressing toxicity to board-handling workers. Thus, the optimum mat moisture content vs. the resin loading level has to be found by trial-and-error experiments, as is often done in the wood composite board manufacturing industries.

12.4 Bond Properties and Uses of pMDI Resins as OSB Binders

pMDI Resins Used as Core-layer Binder: Without any polyol component, pMDI resins are currently used as core-layer binders in OSB manufacturing at similar hot-pressing temperatures as those used for PF resole resins. They were also used in the past as entire binders for wheat straw boards in North America. It is also reported that they are widely used in Europe as particleboard binders. It has also been known that the resin-solids application rate in OSB (about 2.0–3.0%) is lower than those for liquid PF resins (3.5–4.0%) and comparable or slightly better physical strength properties are obtained. Overall, it is assumed that using isocyanate resins as binders at these levels still increases the overall resin materials cost in OSB manufacture, but it provides certain process and property advantages that can be translated into lowering the board manufacturing cost. An example of using pMDI resins as an OSB binder reported (Tables 12.1 and 12.2) shows better process parameters and enhanced board properties in comparison to PF resole resins [24, 25].

Advantages of using isocyanate resins in OSB manufacture are their more forgiving characteristics to mat moisture content (MC) fluctuations in comparison with liquid PF resins, as shown in Table 12.2 [24]. A wood MC of 7.0% or lower is needed for powder PF resins and 4% or lower MC is needed for liquid PF resins, and the latter gives better properties and is more widely used than the former in North America. On the other hand, higher wood MC values (7–10%) have been reported okay for pMDI resins, giving better board properties than liquid PF resins. Drying of wood strands is costly, and a uniform MC of wood strands can often be missed. Thus, isocyanate resins appear to give the advantage of saving energy during wood drying and also improve the overall board properties. In practice, however, it is not known whether the superior bonding quality is true or to what extent it is true when the resin application rate is lowered to the extent that the resin cost becomes equal to those of PF resins. Also, pMDI resins are known to bond oriented strand lumber (OSL) well under the steam injection pressing condition, while PF resins do not.

Table 12.1 Comparison of OSB manufacture with ISO and PF resole resin. Ex- press: out of hot-press. % Retention: 2-h boil test.

Comparison of MDI and PF boned 7/16 (11.1 mm) inch oriented waferboard		
Processing	**MDI[a)]**	**PF**
Dosage (%)	2.0–2.2	2.6
Press time (min:sec)	4:40	5:30
Press closing time (secs.)	38	70
Press temp., °F (°C)	356 (180)	410 (210)
Chip M.C. (%)	7.0–10.0	5.0
Ex-press M.C.(%)	3.5	1.0
Properties		
IB, psi (kPa)	100 (689)	93 (641)
MOR, psi (Mpa)	5680 (39.2)	4215 (29.1)
MOE, kpsi (Mpa)	704 (4854)	655 (4516)
Wet MOR, psi (Mpa)	2920 (20.1)	2360 (16.3)
% Retention	51.4	56.0
Thick. swell (%)	12.0	19.2
Moisture abs. (%)	21.1	20.3
Density, lb/ft³ (kg/m³)	41.6 (667)	41.3 (662)

No wax added to MDI boned board
a) Rubinate® MP-184 MDI isocyanate binder.
Source: [24] Reproduced with permission from the WSU Wood Composite Center.

Table 12.2 Thermal analysis results using differential scanning calorimetry predicting processing parameters for MDI and PF on wet and for MDI on very wet fibers. Source: [25] Reproduced with permission from the WSU Wood Composite Center.

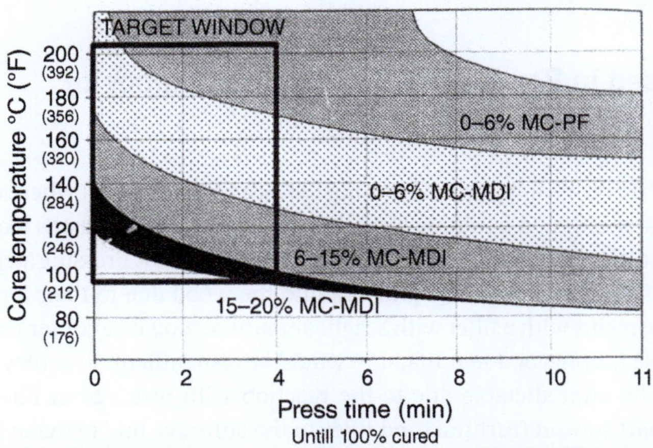

The higher moisture tolerance of isocyanate resins is apparently behind the OSL bonding capability. One Weyerhaeuser OSL plant at Elkin, NC, has been operating using pMDI resins under the steam injection pressing condition.

The disadvantages of using isocyanate resins are several, although not insurmountable in OSB manufacturing. The higher cost in comparison with PF resins (about two times per unit weight and fluctuating) is one disadvantage, but this is partially compensated for by the higher moisture tolerance, etc. Another disadvantage is the caul sticking tendency when pMDI resins are used in the face-layers of OSB, and this is the major reason they are currently used only in the core-layers. Various releasing agents have been tried, sprayed on hot-press platens, to solve this problem, but OSB plants are reluctant to allow even one caul sticking to happen in a day. Plant stoppage can eat away the productivity gain attainable by using isocyanate resins. Another disadvantage is its too-small volume that needs to be sprayed on wood strands (no solvent is involved) because the small volume makes it somewhat difficult to achieve a uniform resin distribution needed for maximum efficiency. Recently, emulsified pMDI resins, which are chemically modified to make them dispersible in water with only a minimal reaction with water occurring [26], but it appeared that the emulsified resins still showed limited storage lives and were not widely adopted in the industry. Another lingering question in using isocyanate resins is the health risk to the plant personnel since the isocyanate resins in general are toxic and allergenic. The resin spraying operation in OSB plants invariably generates isocyanate resin aerosols and resin-laden dust particles, which can circulate around inside the plant. Since maintaining a wood composite board plant free of dust and aerosols is a task that is almost impossible to achieve, the personnel health risk is difficult to eliminate completely. Some isocyanate resins can also escape from the mat during the hot-pressing along with escaping steam, or some unreacted isocyanate resin can remain in the pressed boards, especially when the mat moisture content is accidentally kept too low, resulting in expressing its toxicity to plant personnel in the cutting and loading areas, although the residual isocyanate resins would disappear in time by reacting with moisture gradually absorbed from the air. In OSB manufacturing plants where isocyanate resins are used, the air is collected from the blender, matting area, and hot-press vents and sent to the boiler burner to be used as combustion air or to RTO/RCO units.

12.5 pMDI Resins Used in Bonding of Other Wood Composite Products

Softwood Plywood: Use of pMDI resins in plywood bonding has been unsuccessful where a gap-filling capability is essential for good bonding. Neat, low molecular weight resins like pMDI resins would give difficulties in keeping the gluelines in place in the hot-pressing step since the resins can easily migrate away from the glueline into the wood due to their low viscosity. Mixing of isocyanate resins with a filler with a right amount of moisture content to make an adhesive mix suitable for plywood manufacture would be also difficult to achieve because the viscosity would be unpredictable due to the reaction with water from filler and air. Adding an organic inert solvent (furfural) and filler to the adhesive mix to make it suitable for softwood plywood bonding process resulted in a plywood product but it did not meet exterior requirements of Product Standard PS 1–83(12) for structural plywood [27]. The emission of the organic solvent would also be a roadblock to this approach.

Particleboard and MDF: Use of isocyanate resins in particleboard or medium density fiberboard manufacture would be possible and is known to be used in Europe to some extent with improved strength and durability of boards, but it has not been practiced in the United States due to the toxicity concern and higher cost. Also, a report of using an emulsified pMDI resin mixed with liquid UF resin (1% and 6%, respectively, based on dry wood weight) for bonding particleboard by hot-pressing at 175 °C was reported [18], which implicated an uncertain reaction pathway due to the water from UF resin during hot-pressing. The conclusion was that mixing the two components without ammonium sulfate catalyst and spraying gave slightly better IB and lower formaldehyde emission results than spraying separately with the UF resin catalyzed with ammonium sulfate. In the former case, it appears that the pMDI resin was cured in part with UF resin components in hot-pressing to give lower formaldehyde emission values, and in the latter case, the UF resin cured more completely to give higher formaldehyde emission values. Further research appears necessary since, recently, some board manufacturers began to consider this approach because of the federal formaldehyde emission reduction requirements.

12.6 Summary

Chemistry, reactivity, and adhesive uses of commercially available poly-isocyanate resins are presented. Isocyanate groups (R—N=C=O) are reactive with R′—O—H or R′—N—H_2 groups of polyols, polyamines, polyamides, or urethanes to form urea (R—NH—C[O]—NH—R) and urethane (R—O—C[O]—NH—R) bonds. When allowed to react with a limited amount of water, some of the isocyanate groups transform, by expelling carbon dioxide molecules, into amine groups (—NH_2) that react with other isocyanate groups to form polymeric urea-type groups that can further react with other isocyanate groups to form biuret groups {R′—N=$(CO—NH—R)_2$}, transforming the poly-isocyanate resins into cross-linked high molecular weight thermosetting polymeric materials or adhesives. Some oriented strand boards are bonded with pMDI resins in the core-layer by this reaction mechanism and a reported example is presented. Using polymeric diols or triols, the urethane chemistry (R—O—CO—NH—R′) is used in manufacturing of foam products and various adhesives and coating vehicles for wood in the form of hot-melts, one-part, two-part, or moisture-curing systems.

References

1 Saunders, K.J. (1988). *Organic Polymer Chemistry*, 2e, 358–387. New York: Chapman and Hall.

2 Lambourne (ed.) (1987). *Paint and Surface Coatings: Theory and Practice*. West Sussex, England: Ellis Horwood Lid.

3 Gallagher, J.A. (1982). Urethane bonded particleboard. *Forest Products Journal* 32 (4): 26.

4 Milota, M. and Wilson, J.B. (1985). *Forest Products Journal* 35 (7/8): 44.

5 Uemura, K., Takanashi, A., and Kawai, S. (1999). *Journal of Applied Polymer Science* 74: 1807.

6 Vick, C.B. (1984). *Forest Products Journal* 34 (9): 27.

7 Vrazel, M. and Sellers, T. Jr. (2004). *Forest Products Journal* 54 (3): 66.

8 Troughton, G.E. (1986). *Forest Products Journal* 36 (1): 59.

9 Hori, N., Takemura, A., and Ono, H. (2005). *Wood Adhesives 2005*, 459. FS: FPL.

10 Ling, N., Hori, N., Takemura, A., and Ono, H. (2005). *Wood Adhesives 2005*, 461. FS: FPL.

11 Higgins, E.D. (1995). *Forest Products Journal* 45 (10): 72.

12 Wan, H., Dahlen, J., Mao, A. et al. (2017). *Forest Products Journal* 67 (1/2): 112.

13 Gabriel, J. (2005). *Wood Adhesives 2005*, 179. FS: FPL.

14 Yeh, B., Herzog, B., Ipsen, P., and Brooks, R. (2005). *Wood Adhesives 2005*, 195. FS: FPL.

15 Richter, K., Barbezat, M., Pizzi, A., and Despres, A. (2005). *Wood Adhesives 2005*, 203. FS: FPL.

16 Vick, C.B. and Okkonen, E.A. (2000). *Forest Products Journal* 50 (10): 69.

17 Johns, W.E. (1982). *Forest Products Journal* 32 (11/12): 47.

18 Jackowski, J.A. and Smulski, S.J. *Forest Products Journal* 38: 49.

19 Sun, B.C.H., Hawke, R.N., and Gale, M.R. (1994). *Forest Products Journal* 44 (3): 34.

20 Bao, S., Daunch, W.A., Sun, Y. et al. *Forest Products Journal* 53 (6): 63.

21 Marcinko, J.J., Devathala, S., Rinaldi, P., and Bao, S. (1998). *Forest Products Journal* 48 (6): 81.

22 Gruver, T.M. and Brown, N.R. (2005). *Wood Adhesives 2005*, 123. FS: FPL.

23 McLaughlin, A., Alberino, L.M., Farrissey, W.J., and Waszeciak, D.P. *Proceed. of 1980 Symp. of Wood Adhesives*, 112. FS: FPL and Washington State Univ.

24 Galbraith, C.J. (1986). Presentation at the 20th Intern. PB/Composite Materials Symp. April 8-10.

25 Chelak, W. and Newman, W.H. (1990). *Proceed. 24th Intern. PB/Composite Materials Symp* (ed. T.M. Maloney), 205. Pullman, WA: WSU.

26 Wang, W., Zang, X., and Lu, R. (2004). *Forest Products Journal* 54 (9): 36.

27 Seller, T. Jr. (1989). *Forest Products Journal* 39 (11/12): 53.

13

OSB Manufacturing with PF and Isocyanate Wood Adhesive Resins

13.1 Oriented Strand Board (OSB) Manufacturing Processes

Oriented strand board (OSB) production research in the United States started in 1960s with the objective of making panels having the strength and dimensional stability of softwood plywood using the low-quality wood species. OSB has currently solidified its position as the preferred materials for wall sheathing and roof decking uses in the residential housing constructions, replacing much of the softwood plywood panels traditionally used. The success of OSB industry is based on the use of low-grade wood materials and higher levels of materials handling technologies (machinery) used in the manufacturing in comparison to the softwood plywood plants. The structural and durability performances of OSB are relatively good and comparable to those of softwood plywood in these applications, derived from the strand geometry and the crossband orientations as well as the durable phenol-formaldehyde (PF) resin binders used from the beginning of the industry. The increasing usage trend of OSB is expected to continue, and the industry is expected to expand further into the specialty, higher-grade panel markets by improving various performance properties, especially by minimizing the thickness swell values in the water-soak test. This evolution of the OSB industry includes the improvement of PF and pMDI resin technologies, i.e. optimizing the resin formulation, curing methods, and lowering the overall binder cost. Fifty-six OSB manufacturing plants were operating in 1998 in North America [1].

13.1.1 Wood Species Used in OSB Manufacture

Aspen is the major wood species used in the Great Lakes area and the eastern part of Canada, while southern pines are the major wood species used in the South and Southeastern U. S. regions with some soft hardwood species growing in the region are also used in admixture with the major species as they are available, such as sweet gum and yellow poplar up to almost 40% by weight. Pine logs used in OSB manufacture are of lower grades in comparison to lumber grades, used at lower costs. Raw logs are brought into the OSB manufacturing plants from within about 150–200-mi ranges. The advantage of using southern pine over soft hardwoods was reported: commercial OSB made with 100% southern pine and 50 : 50 southern pine/sweet gum mixture were compared, where the former was found to be better by 31% in MOE and 35% in MOR tested as is and by 69% in MOE and 62% in MOR tested after 48-hour water soak and drying [2]. Hard hardwood such as maple or beech

Adhesives and Finishes for Wood: For Practitioners and Students, First Edition. Moon G. Kim.

was reported to give stronger OSB panels if the compaction ratio is kept constant but oak panels performed poorly even though they were denser [3]. Aspen OSB showed a poorer creep behavior than southern pine OSB in test as a Sturd-I-Floor component under a high humidity condition [4].

13.1.2 Stranding (Flaking)

The manufacturing steps of OSB are shown in Figure 13.1 elegantly illustrated by John Lowood of Structural Board Association [5]. Logs are brought in and processed first through a de-barker, thawed ahead of time if necessary, and cut into short-length boles, and then sent through a flaking machine that slices them into strands. This disc flaking machine consists of rows of fixed-length knives with notching ends, tightly fixed by soldering on the inner side of a round disc, which is rotating fast while the wood boles are pushed sideways against the disc. In this stranding process, wood strands of 2–4″-long (longitudinal), about 20–30 mil (one mil = 1/1000 in.) thick, and varying width, and more of radial cut surfaces are first produced. The length and thickness values are set values and not often changed, and the cut strands come out with varying widths, and therefore they are lightly broken into nominally 0.5–1.0-in.-wide strands. In actuality, some small strands and fines are also produced, the amounts of which depend on the sharpness of knives, machine speed, wood quality, etc. At this stage, the very fine materials are separated and sent for fuel use. The strand thickness has been extensively investigated and came down to the current relatively low values of about 25–30 mil: thinner strands would give more uniform, less thickness-swelling boards, but it generates more fines (wastes) and larger strand surface area that necessitates higher resin loadings (higher cost). Also, the thicknesses of face-layer and core-layer strands are kept the same for process simplicity

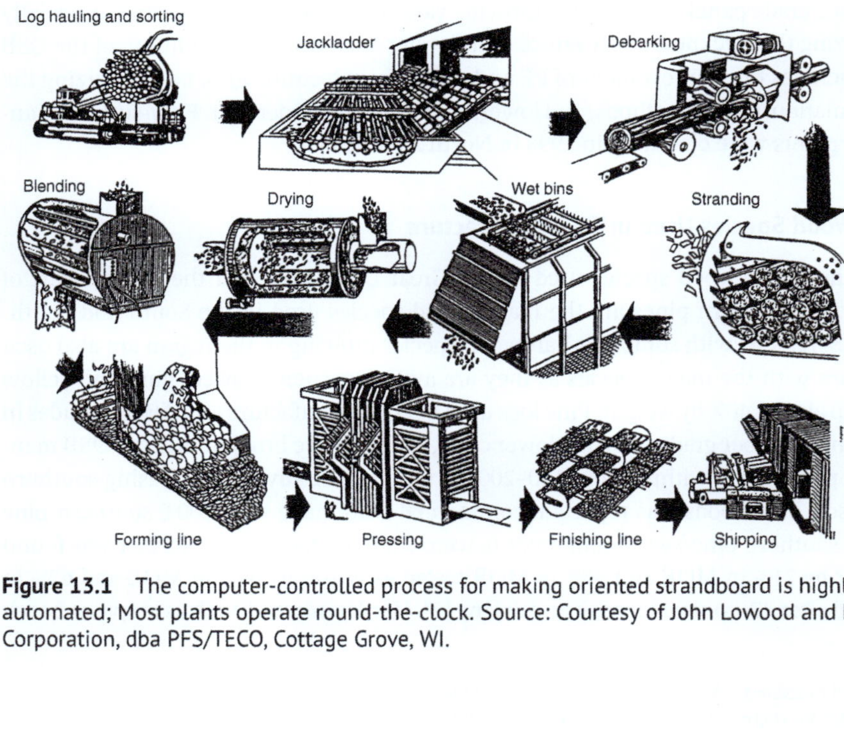

Figure 13.1 The computer-controlled process for making oriented strandboard is highly automated; Most plants operate round-the-clock. Source: Courtesy of John Lowood and PFS Corporation, dba PFS/TECO, Cottage Grove, WI.

(lower cost), although using a thinner face-layer and thicker core-layer combination might result in a more stable board. There are other types of flakers, such as ring or drum flakers, which produce flakes with more tangential cut surfaces, so that they are similar in performance overall but result in some different board properties, and therefore, a selection needs to be made depending on wood species to use [6]. It was also reported that disc flakers used on D. fir wood result in strands that have negligible internal damages and very smooth surfaces, irrespective of the moisture content at the time of flaking or the growth rates of wood, i.e. early wood and late wood [7].

13.1.3 Drying and Screening of Strands

Wood strands are then dried by passing through either a triple pass rotating-drum type or moving-belt type dryer, in which a stream of hot air (up to 870 °F for drum-type and 450 °F for belt-type), produced by burning mostly wood fines/bark wastes, is circulated. The strand drying condition, i.e. air temperature and dwell time (normally about 30 minutes or less) is controlled in such a manner that strands exit the dryer with a moisture content at between 4% and 8% and at a relatively cool temperature. The strands are then screened to remove fines and then separated into two streams, one stream having relatively good target strand dimensions, about 60% of the total, and the other stream having more broken-down dimensions, about 40% of the total. The first stream strands are used in the face-layers and the latter stream strands are used in the core layers, for overall better-bending strength and better face-layer qualities of boards. Most OSB plants have their board mats prepared to have moisture content of less than about 7% in order to have short hot-press times. Since powder PF resins or pMDI resin adhesives do not contain any carrying water, the users of these adhesives can dry wood strands to a less extent, to about 7%, while liquid PF resin adhesive users must dry to about 4% or less because of the water carried in the resin. In the OSB industry, significant resources are devoted to minimizing the emission levels of air pollutants and the lesser drying of wood strands leads to a lower emission of organic volatile components (VOC) due to the lower air temperatures used, and this can lessen the burning load on the electrostatic precipitator (EP), or regenerative thermal oxidizer (RTO), or regenerative catalytic oxidizer (RCO) [8], installed in most plants to capture VOC's and dust particles from dryers and boiler furnace stacks. The lesser emission of VOC is an environmental advantage for powder PF and p-MDI resin binders.

13.1.4 Binder Resin Types

Separate binder resins are used in the core and face layers in the manufacture of OSB as discussed in earlier chapters to achieve the optimum curing speed of boards within the given press time or to obtain the shortest press time. There are liquid PF resins and powder PF resins for both core and surface layers, and pMDI resins are used only in the core layers. There are about three PF resin suppliers and two or more pMDI resin suppliers in North America, and resins supplied by these companies would have their own curing characteristics, i.e. slightly different curing characteristics, derived from slightly different resin compositions. Since each OSB plant has its own characteristics with respect to wood species mix, stranding machine type, strand drying and hot-pressing conditions, etc., the binder resins are selected through a series of laboratory tests and trial runs. The unit costs of binders are also an important factor in the binder selection process due to the price fluctuations and competitive nature of the OSB market. The approximate concentration (%) of active solids in binder resins, current cost per resin-solids pound, and necessary loading level (%) based on the dry wood weight for commodity OSB panels are shown in the table. Furthermore, OSB plants in North America use one of the five or six binder combinations as shown in the table. Major differences between face and core PF resins are the sodium hydroxide content and extent of polymerization, the core-layer resin being the more advanced one as discussed in the previous chapter, and these values can be slightly different for different OSB plants depending on the wood species and other curing conditions. pMDI resins are used in core-layers, but not in face-layers because of the caul-sticking problems. Faster hot-pressing time, better mechanical board strengths, lower board thickness swell, and lowering of resin loading (costs) are the continuous objectives of all OSB plants, pursued mostly through adhesive choices.

Resin types	*Resin-solids level*	*Price/solids pound*	*Loading level*
Liquid PF core/face	45–55%	35 cents	3.5%
Powder PF core/face	100%	65 cents	2.0%
pMDI core	100%	85 cents	2.0%

Binder resin performance differences: On the other hand, the creep tests of commercial OSB as a component of Sturd-I-Floor component under a *high humidity* (95% RH) condition showed that southern pine OSB bonded with liquid PF resins in both face and core layers showed a significantly higher creep resistance than aspen OSB bonded with pMDI in both face and core layers [4]. Also, aspen boards bonded with all powder PF resins showed poor creep resistance, and the best creep resistance was obtained with southern pine flakes bonded with powder PF resin in face and pMDI in core layer. Overall, this research results showed that the creep tendency depends on wood species and adhesive kinds, and this is one reason OSB industry opts to use a particular set of core- and face-layer resin systems. Also, it was reported that extension of liquid PF resins with 25% clay or pecan shell flour with some water addition to lower the spraying viscosity resulted in adequate bonding of OSB at about 5% adhesive-solids loading levels [9].

Core-layer	*Face-layer*
Liquid PF	Liquid PF
Powder PF	Powder PF or liquid PF
Liquid pMDI	Liquid PF or powder PF

13.1.5 Blending of Binder Resin and Slack Wax

Resin blender types and operation: The blending of binder resin is carried out either in a liquid- or powder-type blender, and the kind of blender is determined early on when the plant is being built. Thus, OSB manufacturing plants normally do not have the flexibility of changing the blender types, for example, from powder to liquid, or liquid to powder.

The resin blender is a large drum rotating at a slanted angle at a low speed (~20 rpm) and weighed wood strands continuously go in from the top and come out from the bottom, while weighed amounts of resin binder and wax are also added continuously into the drum through sprayers.

Powder-type resin binder is sent into the blender through a pipe by using an auger-type device operating at a measured rate, and the delivered powder resin is then blown into the tumbling/swirling wood strands using compressed air. It is known that the particle sizes of current powder resins are so fine that strand surfaces are well covered by electrostatic forces uniformly at about 2% resin loading levels, and any extra loaded resin particles fall down to the bottom.

Liquid PF or isocyanate resin binder is sent into the blender using a metering pump to the spinning-disc type sprayer located inside the blender. The spinning disc rotates at an extremely high speed, and the liquid resin delivered onto the top center of disc moves uniformly to the circumference of the rotating disc and gets thrown into the air as fine mist particles. The swirling wood particles capture these (liquid) resin particles on their surface. At the current resin loading levels of 3.5% PF resin solids based on dry wood weight, the resin coverage on strands is not complete but adequate bond strengths are normally obtained. The spinning disc-type sprayer allows a more uniform application of liquid resins than the older-type air-pressured sprayers for resins having a viscosity of up to 1000 cps. It is reported that use of about 4% resin-solids of liquid PF resins based on wood weight establishes a continuous film of resin on all wood strand surfaces [10].

Efficiency of liquid resin binders vs. droplet sizes and resin levels: The efficiency of binder resins in terms of board strength and thickness swell values has been known to be the better the finer droplet size of liquid resins in the blending step [10]. For example, it was reported that in a study of the spraying efficiency of liquid PF resin at a 4% resin-solids loading level for bonding of D. fir OSB, the MOR and internal bond (IB) strength values increased with increasing resin distribution, finer the better, by 19% and 42%, respectively [11]. It has been also reported that use of higher resin loading levels in commercial southern pine OSB resulted in better MOR, MOE, and IB strength values and less dimensional changes in water soak test [12].

Resinated strands: Since the resin loading level is low relative to the amount of wood and wood is very dry, the liquid PF resin droplets lose the component water quickly, and the resinated wood strands appear almost dry to the touch at the matting station, visually indistinguishable from un-resinated strands. pMDI resins have no carrying water, but they also get readily adsorbed on wood surfaces to result in similarly dry-looking wood strands.

Wax is sprayed on the strands in the same blender at the same time with resin, up to about 1.5% based on dry wood weight. Most OSB plants currently use slack wax, wherein the wax is first melted by heating to about 120 °C or higher in a reservoir tank and meter-pumped into the wax sprayer located inside the blender. The wax sprayer is normally a compressed air type, and the lines and sprayer are kept heated with heating coils and insulation to prevent solidification of the molten wax.

Comparison of different binder resins: (i) *Powder PF resin* users have little ways of increasing the resin loading levels beyond about 2% because only so many resin particles can stick on the given strand surface area by static attraction forces. This restriction tends to limit the maximum strength values of boards obtainable. Customer complaints can arise when the adhesive bonding becomes inefficient, such as by using different wood species, higher hardwood contents, or poor-quality strands. Powder PF resin users also tend to have more dust particles, some are of binder resins, floating around inside the board manufacturing plant, originated from the blending and matting areas, and this can make the plants' air harder to keep clean. (ii) *Liquid PF resin binders* have the opposite characteristics, i.e. a greater extent of wood drying is necessary with more VOC emissions from the wood drying step, but the bonding quality can be improved significantly, if necessary, by increasing the resin loading level. Also, a somewhat cleaner environment in the board manufacturing plant is realized. One drawback of liquid resins is the buildup of solidified resin mixed with wood dust on the inside wall of blender that needs to be cleaned out regularly. The number of OSB plants using liquid PF resins is greater than those using powder PF resins in North America.

(iii) *pMDI resins* are liquids with 100% resin-solids contents and are used only as core-layer binders by some OSB manufacturers, and the boards might be stronger [13, 14] to be sold at a better price. The same spinning disc sprayers are used. The caul-sticking problem of pMDI resins, although it is a tell-tale sign of good bonding capability of the resins, currently prevents them from being used in the face layers of OSB. pMDI resins allow wood strands to be dried to a lesser extent, to about 7%, and therefore generally recognized as a binder that is more forgiving to mat moisture content in comparison with liquid PF resins. Being a liquid-type resin, the loading level can be also increased to higher levels to improve the bond quality, when necessary, but it is limited because of the increased cost and also chance of unreacted (toxic) resin remaining in manufactured boards to cause health problems. The pMDI resin users need a good ventilation system in the resin blending and matting area because of the toxicity and allergenic nature, specifically to the upper respiratory system and lungs.

13.1.6 Mat Forming with Strand Orienters

Core and face layer strands, separately blended with respective binder resins and wax, are sent to the matting station. Three-layer mats are normally made through a series of mechanical orienters; top face, middle (core), and bottom face layers in about 30 : 40 : 30 weight ratios. Wood strands in the face layers are oriented along the machine direction and those of core layers are aligned in the direction perpendicular to the face-layer direction. First, the wood strands in the hopper are dosed down onto the orienter, which consists of vertically spaced rows of discs in a rectangular box of mat width, and the entire orienter vibrates and also shakes up and down to expedite the falling of strands through the spacings between the discs [15]. The spacings between discs are just wide enough to catch and guide the strands to fall down while aligning them to the disc direction. Finally, the oriented strands fall down by gravity in the aligned direction continuously onto the moving belt to form a layer. Often four separate orienters are sequentially employed: first one for the bottom layer, second and third one for the core layer, and fourth one for the top surface layer. The orientation of face-layer strands was reported to increase the bending strength (MOR) and stiffness (MOE) in the aligned direction [16]. Overall, the current orientation methods result in about 43% of top surface layer strands aligned within about 20° of the longitudinal axis [17], enough to result in an appreciably higher bending strength values of boards [15] in comparison with un-oriented boards and also to waferboards, which are made with wood wafers of nearly square shapes. The three-dimensional mat uniformity and the accurate overall target mat density (weight per unit area) are the key parameters to be accurately obtained in the mat-forming station. In a study of relationship between the thickness swell (24-hour water-soak and drying) and IB strength values of commercial southern pine OSB, dry IB strength values were found to decrease linearly with increases in non-recoverable TS values, for which the in-plane density variation and nonuniform resin application due to curled strands were found to play a significant role [18]. The curls are derived from wood structure and drying process and need to be minimized. Furthermore, the strand dosing rate in the orienter and the belt moving speed need to be accurately controlled to achieve a three-dimensionally uniform mat formation at the target density value, which is monitored continuously using a nuclear gauze in transit to the hot-press.

13.1.7 Hot-Pressing of Mat

Loading of mats into hot-press: The formed mat, normally 8 ft wide and continuous, is cut on the way to the hot-press into the press lengths, normally 24 ft long, and loaded on a caul plate. In transit, the mat is often sprayed with a small amount of water mist on the top layer to compensate for the moisture loss of resin droplets occurring during the transit, to prevent the dry-out occurring on the top surface of boards. Mats with caul are then loaded into the multi-loader, one by one, and then the loader loads the mats into the multi-opening hot-press all at once. The multi-loader allows all mats to be loaded into hot-press at the same time; if mats are loaded into the hot-press one by one, the waiting period in the hot-press for some boards will cause pre-curing problems.

Press closing and effects of press closing rate: After loading, the multi-opening hot-press is closed to the target thickness in about 45–60 seconds, initially by using the main ram, and then with the help of jack rams for a finer control of mat thickness. The compaction of face layers occurs more quickly than that of core layer due to the preferential heating of the surface layers during the compaction and early part of the holding period. This hot-press curing process is similar to the hot-pressing of UF resin-bonded particleboards discussed in an earlier chapter. The rate of press closing, or press closing time, determines the shape and range of the density profile through the board thickness: [19] fast closing rates result in U-shaped density profiles and slower closing rates result in various M-shape density profiles. The shapes of density profile influences board's strength properties and also edge screw-holing strength and permeability [19]. In practice, the press-closing rate of OSB boards is kept as fast as possible with the mat pressure initially rising to as high as 850 psi or higher.

Temperature and moisture content variation in hot-pressing of boards: The hot-press platens are normally maintained at 400–430 °F by circulating heated oil through them. When the wood mat compaction begins, the platen heat starts to propagate to the surface layers of the mat and then gradually to the core layer. As discussed in Chapter 4, the heat transfer occurs mainly through the *steam transfer and condensation* mechanism. As shown in Figure 13.2 presented by Wolcottet al. [20] the steady-state temperature and moisture content differences from the surface to the middle layer are attained in the later part of pressing (for example, 430 °F → 350 °F). Similar graphs were also reported by Kamke and Casey [21] and by Bolten et al. [22]. The moisture in the surface layers moves to the core layer and, when the core temperature reaches the boiling point of water, the water vapor begins to escape from the mat through the edges, and therefore the mat's moisture contents in the face and core layers decrease. Also, the internal vapor pressure of mat first rises due

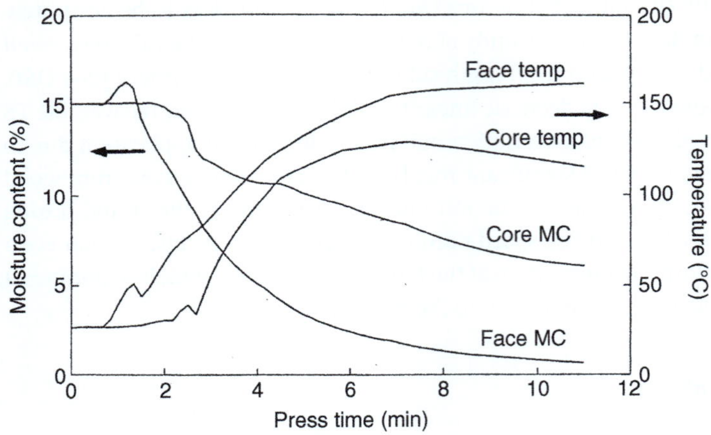

Figure 13.2 Average flake temperature and moisture content in hot-pressing of OSB mat of 15% MC and at 190 °C. Source: Walcottet al. [20]. Reproduced with permission of the publisher Wood and Fiber Science.

to the generation of steam and the mat compaction that restricts the steam from escaping. The internal vapor pressure finally decreases to a low value as the hot-pressing progresses [21] and steam escapes from the board edges. It is also noted that use of Fourdrinier-type screens either on the top or under the mat was reported to aid the venting of steam from the board surfaces [23].

Pre-cure or over-cure problems of boards can occur in the face layers due to the resin exposure too long at high temperatures before the full compaction of board mat, resulting in weakening of adhesive bond strength due to precure of resin. More often, such pre-cure problems occur in the face layers, especially on the top-most surface layers, and the result is seen as loose strands on the top surface layers that can be easily scratched off with a knife. This tendency can also show up as a surface-layer break in IB testing and as poor MOR and MOE values since the bending strengths are affected more acutely by the strengths of surface layers. A resin cure rate that is too fast, high platen temperature, or slow press closing speed can be traced to these pre-cure problems. These problems occurred more often in the beginning of the OSB industry when only one fast-curing PF resin was used in both face and core layers. In recent years, most OSB plants use separate resins and this problem diminished greatly, although more compact/strong surface layers than those of current boards would be desirable. Some OSB plants apply water mist on the top of the mat before it goes into the hot-press to lessen the pre-cure.

Under-cure problems often occur in the core layers. Either the press time that is too short, or the beginning mat moisture content that is too high, can result in weaker core layers. The under-cure effect on the IB strength is aggravated by the lower density of the core layer created by its slower compaction speed than in the face layers. In general, these layer compaction and bonding efficiency differences in different layers of board can be tested, by measuring the IB strength of each separate layer. In normal IB testing, it is customary just to note the breaking layers, as shown in Figure 13.3. Layer #3 is normal and other breaking layers would indicate hot-pressing modifications are needed. In the core layer, the temperature attained during the hot-pressing is lower than in face layers and the moisture

Figure 13.3 Various breaking layers in IB testing.

content and the vapor pressure remain high most of the hot-pressing time: with the platen temperature of 196 °C, which is closer to what industrial OSB plants are using currently, the core temperature reaches to about 125 °C in about four minutes of pressing time and the (vapor) pressure increases to about 50–60 kPa (1 psi = 6.9 kPa) in about five minutes as shown for a 1.9 cm thick board [21]. The moisture content of the core layer at this stage is too high since all moisture has gotten collected in the narrow core-layer and escapes only slowly. This is an unfavorable condition for PF resin curing when compared to the face layers, and therefore, as mentioned in Chapter 9, the core-layer PF resin is made to have a significantly higher degree of polymerization than the face-layer resin. In actuality, the hot-pressing time is adjusted to be just long enough to cure the core-layer resin to pass the vitrification point so that the board mat is held together within the target thickness. In case of very high mat moisture contents, i.e. 15% or higher, it is known that the core layer undergoes a saturated steam environment [21], which means that some liquid water is present in the core layer during hot-pressing time. The core-layer PF resins are very hydrophilic (higher sodium hydroxide content for faster curing) and can be readily diluted by this liquid water to result in penetration of resin into wood if it is not cured fast enough. This moisture content parameter will cause a poor adhesion in the core layer and is the reason for OSB plants to be run under a low and tightly controlled mat moisture content condition. pMDI resins used in the core layers are reported to be more forgiving to mat moisture contents and this is the main reason some OSB manufacturers have changed to this type of resins in spite of negative health effects and higher costs.

Industry hot-pressing parameters: Currently, the OSB industry in North America uses hot-pressing parameters shown in Table 13.1 presented in invited lectures by C. R. Davis, R&D Manager, of Georgia-Pacific Resins, Inc., Crossette, AR. [24] The industry also spend efforts to find ways to use higher mat moisture contents and lower hot-press temperatures for obvious economic and toxic air emission reduction reasons [25]. The cure speed for PF core-layer resins cannot be increased too much due to the limitation of the resin chemistry

Table 13.1 U. S. Industry's OSB manufacturing (hot-pressing) parameters and trends [24].

	PF liquid	PF powder	Isocyanates
Resin Usage[1]	2.6–4.0	1.2–2.5	1.8–4.0
Press temp (°F)	390–430	390–430	380–410
Mat MC (%)	5–10	3–8	5–10
Press time (min)[2]	4.5–7.0	4.5–7.0	4.0–5.0

(1) Resin usage – non-volatile solids (%) in panels based on oven-dry wood weight. (2) Press time – Button to full open, ¾″-thickness basis.
Positives for OSB industry: Penetration into siding and furniture, export, lower cost of wood, less labor intensive, substitution of softwood plywood, lower cost.
Negatives: Environmental rules, possible overcapacity, lower panel property, Canadian imports.

involved as discussed in Chapter 9. And, if the incoming mat moisture level is too high, the higher internal vapor pressure needs to be overcome with a longer press time, or a spring-back or blown boards will result due to the less than adequate adhesive strength reached in the given press time. The hot-pressing step also emits various volatile organic compounds (VOC) derived from degradation of adhesive resin, wax, and wood strands, and the amount of formaldehyde in the emission was found to depend on the PF resins' loading level, temperature, and other hot-pressing condition [25]. It is common practice to collect the air from hot-press, dryer, and matting area and send it for treatments.

Post-curing and finishing of OSB (hot-stacking): One favorable situation in practice is that the maximum temperature attained in the core-layer of boards is maintained for a while since, after the hot-press is opened, the board cools rather slowly over several hours due to the poor heat conduction of wood. Though, it is clear that the core-layer adhesive must be cured enough before press opening. IB strengths of boards are also determined mostly by the cure extent of core-layer resins. Thus, it is seen that the bottleneck of an OSB plant can easily find itself in the hot-pressing operation due to the limited cure speed of the core-layer resin. In OSB manufacturing plants, boards just unloaded from the hot-press are hot and directly sent to the cutting station where they are trimmed to 4′ × 8′ or other size panels and are stacked together up to 75 panels. Little cooling occurs during this short cutting/stacking period for panels. Stacked panels are then edge-painted with water-barrier paints (high wax content paints) having a certain distinctive color and logo, then strapped and stored until shipment. The OSB panel stacks stand in the plant and cool only slowly over time, and this practice is called the "hot-stacking" because the binder resins, especially of core layers, cure further during this period. This post-curing effect can be often shown by testing the IB strengths of samples taken before and after the hot stacking. On the other hand, recent reports indicated that long thermal treatments of manufactured OSB or softwood plywood at 76.6 °C and 60% RH, although some improvements were observed in thickness swell and water absorption properties, resulted in some deterioration of strength properties, and the deterioration effects were greater at higher temperatures above 140 °C [26].

13.1.8 Oriented Strand Lumber (OSL)

Oriented strand lumber (OSL) is a relatively recent development with one plant operating by Weyerhaeuser Company in Elkin, NC. The wood strands for OSL are slightly wider and thicker, and relatively longer up to 6″ or so for better stiffness of resultant (long) lumber. All strands are aligned in one direction and the hot-pressed boards reach up to about 3″ thick and quite long and wide, and they are cut into long lumber. The thick mat is hot-pressed by steam-injection pressing for a relatively short hot-pressing time and pMDI resins are known to work well. PF and MUF resins do not perform as well. An extensive discussion on why PF resins do not work and possible ways to improve the bonding in steam injection pressing was presented in recent reports [27, 28]. Overall, the steam injection pressing of OSB with pMDI resins resulted in improvement of dimensional stability of boards from reduced internal stresses due to stabilization of flakes [28]. On the other hand, the long-term creep tendency of the isocyanate resin-bonded OSB or OSL has not been adequately addressed yet [29], while PF resin-and PRF resin-bonded lumber products were extensively studied and proved to be adequate. The indication is that the creep tendency of pMDI resin-bonded OSB is slightly higher (poor) than PF resin-bonded OSB and problems might arise in future for OSL products, especially in cases of using at elevated temperatures and or higher humidity environments.

13.1.9 Adhesive Improvements Needed in the OSB Industry

OSB manufacturers are currently pursuing for several process improvements: (i) Lowering the panels' thickness swell, especially on the edges on exposure to water occurring in spite of the edges painted with a water barrier paint. The swelling of edges tends to give the so-called ripple effects in the roof. (ii) Lowering the overall thickness swelling of panels further, especially for uses in specialty applications such as for subfloor panels [30]. In this regard, the report that the combination of isocyanate resin in the core and PF resin in the face layers and use of thinner face layer strands (0.432 vs. 0.635 mm) results in lower thickness swell values than other combinations is noteworthy [30, 31]. (iii) Being able to use more hardwoods to lower the wood cost in the Southern U. S. region. (iv) Being able to make lower density boards, Currently, the density value is about 43 lbs/cft when using pines; it is desirable to use lower board densities and maintain the performance, i.e. savings of wood is desired. (v) Lowering the cost of PF resins, which has been realized to some extent by using urea-added PF resins [32, 33]. Some research works have been done on using various lignin additives to PF resins, but commercial uses have not been realized [34, 35].

13.1.10 History of Wafer Board and Flakeboard

Wafer board was invented by James D'Arcy Clarke, who pioneered the manufacturing technology of panels using the low-cost and abundant aspen wood in Idaho in late 1950s, and the industrial production was first started by Wisewood Ltd. in Hudson Bay, Canada in late 1960s using a 14-opening 4 ft by 16 ft hot-press [5]. The U. S. production began in 1974 by Blandin Wood Products in Grand Rapids, MI [5]. Wood wafers were roughly 1-in. wide and up to 1.5 in. length with similar thickness (0.03 in.) to those of strandboards, and the adhesive binder was initially powder novolac PF resins. The panels were initially called "Aspenites," which was transformed into the generic name of waferboard. Elmendorf Corp. further developed the waferboard technology by using longer strands and patented the OSB technology in 1965, and by 1980s the OSB became widely recognized as an improved product. It was also reported that orienting the strands in the face and core layers improves the IB, MOR, and MOE values of boards [15, 16], and therefore the boards were called oriented strandboard (OSB). Also, liquid PF resole resins were introduced alongside the spray-dried PF resole resins as binders. The term "*flake board*" was also used for OSB, but they are referred now for similar products made with wood flakes of about ½″ wide and thickness slightly thicker than those of strand boards.

13.1.11 Durability and Span Ratings of Oriented Strandboard in Use

The OSB manufacturing industry in the United States keeps the high quality of their products through the Voluntary Product Standard for Wood-based Structural-Use Panels (PS2-92), initiated by APA – The Engineered Wood Association. In this product standard, OSB is classified as *Exposure* 1, meaning boards can be used for partial exposure to weather or moisture, such as against the effects of moisture due to construction delays, high humidity, and water leakage, in contrast to softwood plywood, which is often classified as *Exterior* grade that can be used for permanent exposure to weather or moisture. The bond durability for the Exposure 1 grade for OSB is tested by the *Six-Cycle Vacuum Soak Test*, to be passed by a 50% strength retention (Section 20.2.3). Common PF and isocyanate resins at appropriate levels of loading have been known to result in boards that adequately meet this requirement. OSB panels are widely used as sheathings on roofs and walls as well as subflooring panels, replacing softwood plywood panels traditionally used in these applications. Therefore, MOR/MOE strength data of boards were used in deriving the span ratings of OSB panels, ranging from 16 to 60 in. for roof and 16 to 48 in. for subfloor or single floor. Span rating numbers mean the distances between studs and 16 is the more common value in wall constructions. Since span rating numbers are constrained by minimal deflection extents on given static and impact loads, long-span ratings mean stronger, often thicker boards. Good MOR/MOE and IB strength values of manufactured boards are the keys to get the desired span ratings with least board thicknesses or least materials costs.

13.2 OSB Bonded with PF Resins with Various Levels of Urea Added at the End of Resin Synthesis

(Adapted with permission from Taylor and Francis group and with minor revisions from Watt, C., M. G. Kim., and C. R. Davis. 1996. J. of Wood Chemistry and Technology 16(1) 21-34. Effects of urea addition to phenol-formaldehyde (PF) resin binders for oriented strand board. [32]*) (Advanced reading materials)*

Commercial OSB binder PF resins often contain urea at about 5–20% levels based on the resins-solids weight, added for lowering the cost. The urea addition is carried out at the end of polymerization step and is known to reduce the viscosity of resin apparently due to the hydrogen-bond breaking effect of urea [36]. Furthermore, it necessitates an additional amount of water to be added during the resin synthesis because of the adjustment of urea solids level to the targeted resin-solids level, currently being in the range of about 45–55%. Because of these reasons, it was claimed that the phenolic resin components can be advanced to higher molecular weights than when no urea is added [37]. The reaction of urea with formaldehyde is well-known as the first step reaction in the manufacture of UF resins [38]. OSB binder-type PF resins are manufactured only in the alkaline pH and a small amount of formaldehyde usually remains unreacted after the polymerization is completed. This residual formaldehyde similarly reacts with the added urea to form hydroxymethylureas [39], which do not polymerize among themselves in the alkaline pH [38] but may react at elevated temperatures with PF resin components to form PF resin-urea copolymer products [40, 41] similarly to such co-polymer products obtainable in the acidic pH [42, 43]. In the current resin manufacturing practice, the post-addition of urea carried out at low temperatures (about 60 °C and below) does not favor any significant copolymerization in the uncured resin stage [39], but the reaction is likely to proceed in the curing stage to positively affect the adhesive performance in comparison to the addition of non-reacting diluent additives. A preliminary study indicated that the post-added urea modifies the resins' curing characteristics [44], but its effects on adhesive performance or other resin-curing properties have not yet been reported. Therefore, OSB binder-type PF resins were made with four urea addition levels and at two F/P mole ratios, while keeping the resin-solids level constant at about 50%, and the resultant resins were evaluated by bonding laboratory OSB.

13.2.1 Syntheses of PF Resole Resins with Urea Added at the End of Synthesis

Eight PF resins with target resin-solids level of 50%, including the urea, were synthesized in duplicates, as shown in Table 13.2. The sodium hydroxide content was 4.6%; formaldehyde/phenol (F/P) mole ratios were 2.16 and 2.47, and urea addition levels were 0%, 10%, 18%, and 26% based on the resin-solids weight. The following typical PF resin synthesis procedure was used: (i) phenol, water, and one-half of the sodium hydroxide (50% solution) were charged in a stirred reactor having a heating mantle, cooling coil, and condenser, and the reaction mixture was heated to 65–70 °C. Formaldehyde (50.0%) solution was then added dropwise over a period of 30 minutes, while keeping the temperature in the same range by intermittent cooling. (ii) Then, the temperature was held for 10 minutes at that temperature and then gradually increased to 90 °C over a period of 30 minutes and maintained until D (Gardner-Holdt scale) viscosity is reached. (iii) Then, the temperature was lowered to 80 °C, and the remaining sodium hydroxide and water needed to attain the target resin-solids levels (after the urea addition) were added. The reaction continued at 80 °C until the resin reached H viscosity (approximately 200 cps) for the control PF resin that has no urea addition. (iv) For urea-added PF resins the reaction was extended until the viscosity reached 380, 550, and 850 cps, and added 10%, 18%, and 26% urea, respectively. The synthesized resins were cooled and stored frozen until analysis or use.

Table 13.2 Resin properties of eight synthesized PF and PUF resins.[a)]

Resin #	1	2	3	4	5	6	7	8
Urea	0%	0%	10%	10%	18%	18%	26%	26%
F/P ratio	2.16	2.47	2.16	2.47	2.16	2.47	2.16	2.47
pH	10.5	10.4	10.7	10.6	10.8	10.8	10.9	11.1
Viscosity cP	172	160	162	244	186	265	123	230
Solids level (%)	53.7	51.4	49.7	50.5	51.2	50.1	49.7	49.3
Gel time (min)	29.8	22.7	25.7	21.6	25.5	19.9	23.9	18.3
Free form (%)	0.31	1.95	0.0	0.0	0.0	0.0	0.0	0.0
Specific gravity	1.23	1.22	1.21	1.22	1.23	1.22	1.23	1.22
Mol. Weight	530	670	1230	1250	1740	2290	2910	8430
Mol. Weight-U	530	670	1110	1130	1440	1890	2170	6250

a) All values are averages of duplicate resins. Molecular Weight was determined by gel permeation chromatography on Sephacryl 400 gel and 0.10 N NaOH solution and UV detector at 280 nm. Urea in the resins was not detected and therefore recalculated by weighting the urea % values. Gel times were measured at 100 °C.

13.2.2 Testing of Synthesized Resins

Synthesized resins were tested according to the standard testing methods, and the results are reported in Table 13.2. All synthesized PUF resins showed low free formaldehyde contents as expected. Also, as the urea addition level is increased, the average molecular weight increased as expected from the synthesis schemes used. Furthermore, shorter gelation times were observed for resins made with urea additions, which appears to be due to the increased resin reactivity because of the reaction between PF resole components and urea [40]. PF resins made with the higher F/P ratio showed higher molecular weights and shorter gel times in comparison with the PF resins made with lower F/P mole ratio, reflecting the higher functional group contents. Possible benefits such as a shorter hot-pressing time is indicated, but the higher amounts of hydroxymethylureas may result in emitting increased amounts of formaldehyde.

13.2.3 Resin Curing Rates Determined by DMA

A DuPont DMA 982 dynamic mechanical analyzer was used to test the curing speeds of synthesized resins, in the fixed displacement mode and in the isothermal mode at 110, 125, and 150 °C. The oscillation of test sample was set to run at its natural frequency that increased from about 2 Hz to 25 Hz at 125 °C, as reported in Figure 13.4 with graphic averaging done for duplicate runs. The rigidity curves increased rapidly from about 10 minutes and reached the high final values by about 23 minutes, and the slopes (B) are defined as the cure rate, and the time to the final point (A) as the cure time [44, 45]. The cure rates and cure times of synthesized resins are collected in Table 13.3. First of all, (i) the higher temperatures made the cure times shorter and cure slopes steeper for all eight resins. (ii) Higher urea levels made

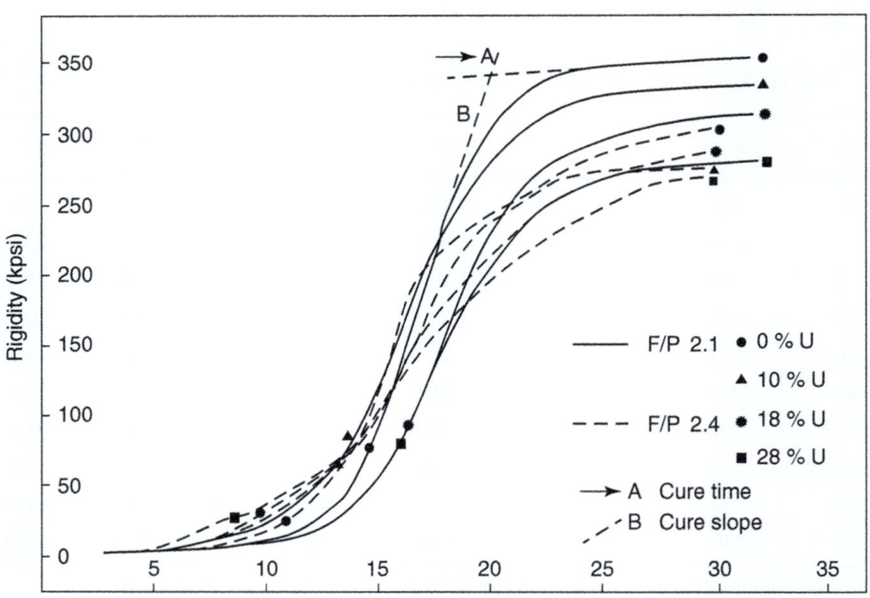

Figure 13.4 DMA Rigidity development vs. urea addition levels at 125 °C for OSB-type PF resins synthesized with F/P ratios of 2.16 and 2.47.

Table 13.3 DMA cure times and cure rates (slopes) of synthesized resins.

F/P	Urea (%)	Cure time (min.)			Cure slope[a)]		
		90 °C	125 °C	150 °C	90 °C	125 °C	150 °C
2.16	0	33.0	20.0	12.9	21.6	29.7	55.2
	10	32.5	20.5	12.8	16.8	25.9	46.6
	18	34.3	21.4	20.5	17.3	23.6	54.0
	26	35.0	21.9	14.4	14.2	24.3	50.2
2.47	0	30.8	19.7	14.4	15.9	24.9	41.4
	10	31.1	20.0	12.7	15.7	23.8	41.4
	18	32.	21.0	13.9	13.8	22.5	37.0
	26	34.8	22.2	14.1	11.1	16.0	26.2

a) In 10^8 dynes/cm^2/minute.

resins' cure times longer and cure slopes less steep in spite of higher molecular weights and shorter gel times observed in GPC analysis and gel time tests. (iii) The higher F/P mole ratio of resins resulted in no changes in cure times, but the cure slope values decreased more with resins of higher urea levels. Thus, some deterioration of bond strength is implicated for urea-added resins and especially for higher F/P mole ratio resins.

13.2.4 Laboratory OSB Manufacturing and Testing

(i) Southern pine boles (Pinus sp., 35–40 years old) obtained in the southern Arkansas region were flaked with a CAE disc flaker into 0.025″-thick, 1″-wide, and 3″-long wood strands, screened to eliminate fines smaller than 0.25 sq. inch in size, and dried to a moisture content of 2–5% based on oven-dry wood weight. (ii) Laboratory OSB was then manufactured using the processing parameters shown in Table 13.4. In board manufacturing, the resin order, resins loading level, and hot-pressing time were randomized, but the replicate boards were made from one resin blending. A factorial design with sub-sampling was used to analyze the resulting data. (iii) Using the test specimens cut from boards as shown in Figure 13.5, IB, boil IB, MOE, MOR, and 2-hour and 24-hour water soak water absorption (WA) and thickness swell (TS) values were determined in accordance with the ASTM procedure 1037. (iv) Small sample edge-bending tests were also conducted in accordance with APA procedure S-6 using APA accelerated-aging procedure D-4. (v) Aged

Table 13.4 Laboratory OSB manufacturing parameters.

Mat formation: 55.9 cm × 61.0 cm, one-layer, no flake orientation.
Mat moisture content: 6.0–9.0%. Target density: 672.8 kg/m^3.
Press temperature: 177 °C. Press pressure schedule: 5.5–2.75–0 MPa.
Press times: 4.0 and 4.5 minutes. Post-cure: 2 hours at 90 °C.
Replication: 2 boards, from one blending (total: 128 boards)
Resin loading: 3.2% and 4.2%. Wax loading: 1.0% slack (Indrawax 210).
Board conditioning: 14 days at 21 °C in a 56% RH room (Equil. MC = 10%)

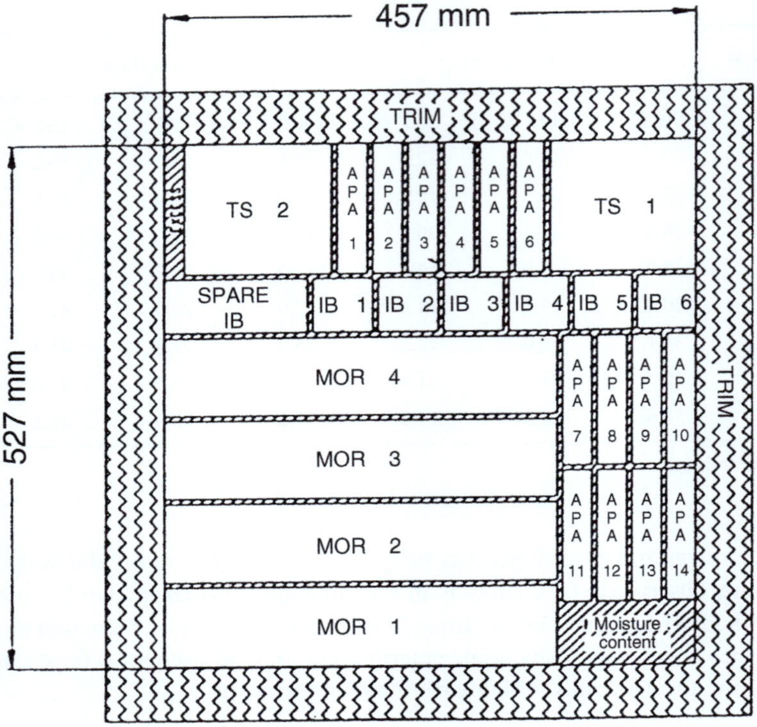

Figure 13.5 Cutting diagram of laboratory OSBs for property test samples.

MOR tests were performed after aging the specimens according to APA procedure D-4. Loads were measured using Tinius-Olsen Universal Property Testing Machine (Model 92797) and an Instron Universal Testing Machine (Model TTCML). (6) Six IB specimens of a board showed varying densities and, therefore, averaging of IB strength values was done by linear regression of strength and density values and then selecting the IB value at the target density of 672.8 kg/m^3, which is the typical value for commercial pine OSB manufactured in North America [2]. (vii) Board property results were statistically analyzed using the Statistical Analysis System program [46]. It was assumed that all panel properties adhered to the central limit theorem, the experimental units were homogeneous, and the between-treatments variance had to be greater than the total variance to be considered significant. Analysis of variance was used to generate within treatment, between treatment, and total variances. F-test was used to determine the significance of variances at less than 0.05 probability. Ficher's Protected *t*-Test (least significant difference [LSD]) was used to generate comparisons for physical properties at the 95% confidence (0.05) level [47].

13.2.5 Test Results of Manufactured OSB Panels

Test results of board performances and statistical analysis results are reported in Table 13.5. Most of all, dry and boil IB strengths and 24-hour water soak properties showed significant

Table 13.5 Physical property means and L S D letter groupings of laboratory OSB.[a)]

Density (kg/m^3)		IB (MPa)		MOR (MPa)		MOE	APA S-6			Water soak (%)			
		Dry	Boil	Dry	Agd (/dry)	(GPa)	Dry (Gpa)	1-cyc. /dry	6-cyc. /dry	2-hour TS	WA	24-hour TS	WA
Urea													
0%	740.1a	1.055a	0.627a	46.26a	0.91ab	3.30a	3.18a	0.76a	0.76a	2.7a	5.8a	12.0a	23.1a
10%	741.7a	0.986ab	0.586ab	41.02ab	0.94a	2.50a	3.12a	0.79ab	0.76a	3.0a	7.9a	13.8b	26.1b
18%	736.9a	0.958ab	0.510bc	42.95a	0.87bc	2.39a	2.79b	0.85bc	0.74a	2.7a	5.8a	14.2b	27.0b
26%	724.1a	0.986b	0.489c	39.02b	0.84c	1.68a	2.81b	0.88c	0.72a	3.7ab	8.3b	17.4c	32.0c
F/P ratio													
2.16	738.5a	1.034a	0.572a	42.06a	0.86b	2.87a	3.01a	0.83a	0.75a	2.8a	6.5a	13.7a	26.2a
2.47	735.3a	0.917b	0.537a	40.33a	0.92a	2.41a	2.94a	0.82a	0.74a	3.3b	6.9a	15.0b	27.9b
Resin load													
3.2%	735.3a	0.958a	0.482a	40.40a	0.89a	2.79a	2.90a	0.81a	0.75a	3.2a	6.7a	15.6a	28.7b
4.2%	738.5a	1.055b	0.627b	42.00a	0.89a	2.48a	3.04a	0.84a	0.75a	2.9a	6.8a	13.1b	25.4b
Press time													
4.0 min	738.5a	0.992a	0.544a	41.09a	0.89a	2.41a	2.95a	0.83a	0.75a	3.2a	7.1a	14.9a	27.7a
4.5 min	735.2a	0.958a	0.565a	41.30a	0.89a	2.87a	3.00a	0.81a	0.75a	2.9b	6.4a	13.8b	26.5b

a) L S D Letter groupings and means with same letter are not significantly different. Strength values of aged samples presented as fractions of the dry strength values. Agd: Aged. Cyc.: Cycle. TS: Thickness swell. WA: Water absorp.

decreases as the urea level increased. On average, 1% increase in the urea addition level effected decreases of dry and boil IB strength values by 6.21 and 4.83 MPa, respectively, and the water soak TS and WA values increased by 0.20% and 0.35%, respectively. Dry and aged MOR, MOE, and APA edge-bending values decreased to smaller extents. These decreasing board performances appeared like *a dilution effect by urea on the PF resin polymer* and agreed with the decreasing cure rates observed by DMA. Other resin synthesis variables also affected board performances. The *higher F/P mole ratio* used in resin synthesis resulted in poorer IB strength values and poorer water-soak properties, indicating that certain limitations exist in obtaining higher non-phenolic resin solids or faster gelation times using this approach. The lower resin-solids loading levels showed lower board performances as expected, implying that high urea addition levels would be of limited applicability for low resin-solids loadings. Thus, urea addition level, F/P mole ratio, and resin-solids loading level together showed accumulative degradation effects on board performances. The most apparent is the dry IB values, where a strength decrease from 1.055 to 0.758 MPa was observed when boards made with 26% urea addition level, 2.47 F/P mole ratio, and 3.2% resin-solids loading level were compared with boards made at 0% urea addition level, 2.16 F/P mole ratio, and 4.2% resin-solids loading level.

13.2.6 Conclusion

OSB binder PF resins synthesized with urea addition levels up to 26% based on the resin-solids weight showed increased molecular weights and drastically reduced free formaldehyde contents. But DMA experiments and evaluation of laboratory OSBs showed reduced curing rates and also gradually decreasing board performance values as the urea addition level increased. The IB strength decreases and water absorption property increases were statistically significant at the 95% confidence level. Results of this study would be useful for OSB manufacturers in making decisions on the urea addition level in their PF resin binders with respect to the resin loading level, hot-pressing time and temperature, and resulting IB and water absorption properties. At the 10% urea addition level, the board performance deteriorations were minimal and the free formaldehyde contents in the resins were extremely low.

13.3 Summary

Industry and laboratory manufacturing procedures of OSB are presented with emphases on wood drying, adhesive application, matting, hot-press curing, and post-curing steps based on open literature, invited lectures of industry experts, and experiences obtained in numerous personal plant visits. Industrially available, various PF and isocyanate resin adhesives are described regarding their methods of application, curing properties, and potential board performance differences. A publication on laboratory OSB manufacturing and testing carried out aiming at improving the board performance by experimental PF resole resins modified with post-addition of some urea is included to allow readers to obtain more realistic relationships between the resin properties and amount of materials that go into board making and the resultant board densities and other performance properties.

References

1 Sellers, T. Jr. (2001). *Forest Products Journal* 51 (6): 12.
2 Biblis, E.J. (1989). *Forest Products Journal* 39 (1): 55.
3 Carll, C.G. (1994). *Forest Products Journal* 44 (9): 26.
4 Pu, J., Tang, R.C., and Davis, W.C. (1992). *Forest Products Journal* 42 (3): 49.
5 Lowood, J. (1997). Oriented strand board and wafer board. In: *Engineered Wood Products* (ed. S. Smulski), 5–123. Madison, WI: PFS Foundation.
6 Price, E.W. and Lehmann, W.F. (1978). *Forest Products Journal* 29 (3): 29.
7 Burrows, C.H. (1961). *Forest Products Journal* 11 (1): 27.
8 Wantling, S., Zepka, B., Moberly, D. et al. (2007). *Forest Products Journal* 57 (7/8): 51.
9 Gardner, D.J., Waage, S.K., and Elder, T.J. (1990). *Forest Products Journal* 40 (7/8): 31.
10 Lehmann, W.F. (1968). *Forest Products Journal* 18 (10): 32.
11 Lehmann, W.F. (1965). *Forest Products Journal* 15 (4): 155.
12 Generalla, N.C., Biblis, E.J., and Carino, H.F. (1989). *Forest Products Journal* 39 (6): 64.
13 Wilson, J.B. (1980). *Proceedings of 14th Intern. PB/Composite Materials Symposium* (ed. T.M. Maloney), 185. Pullman, WA: WSU.
14 Galbraith, C.J. (1986). *Proceedings of 20th Intern. PB/Composite Materials Symposium* (ed. T.M. Maloney), 55. Pullman, WA: WSU.
15 Barnes, D. (2002). *Forest Products Journal* 52 (9): 31.
16 McNatt, J.D., Bach, L., and Wellwood, R.W. (1992). *Forest Products Journal* 42 (3): 45.
17 Chen, S., Du, C., and Wellwood, R. (2008). *Forest Products Journal* 58 (6): 94.
18 Wu, Q. and Piao, C. (1999). *Forest Products Journal* 49 (7/8): 50.
19 Smith, D.C. (1982). *Forest Products Journal* 32 (3): 40.
20 Wolcott, M.P., Kamke, F.A., and Dillard, D.A. (1990). *Wood and Fiber Science* 22 (4): 345.
21 Kamke, F.A. and Casey, L.J. (1988). *Forest Products Journal* 38 (6): 38.
22 Bolten, A.J., Humphrey, P.E., and Kavvouras, P.K. (1989). *Holzforschung* 43: 265.
23 Jokerst, R.W. (1990). *Forest Products Journal* 40 (2): 45.
24 Davis, C.R. (1995–2005). R & D Manager, Georgia-Pacific Corp., Data provided in invited class lectures at the Department of Forest Products of Mississippi State University.
25 Carlson, F.E., Phillips, E.K., Tenhaeff, S.C., and Detlefsen, W.D. (1995). *Forest Products Journal* 45 (3): 77.
26 Cheng, Q., Zhou, C., Jang, W. et al. (2018). *Forest Products Journal* 68 (1): 78.
27 Hse, C.Y., Geimer, R.L., and Hsu, W.E. (1995). *Forest Products Journal* 45 (1): 57.
28 Geimer, R.L. and Christiansen, A.W. (1996). *Forest Products Journal* 46 (11/12): 67.
29 Arches, S., Christian, Z., and Stapf, G. (2015). *Forest Products Journal* 65 (1/2): 60.
30 Taylor, A., Wang, S., Freitag, C., and Morrell, J.J. (2008). *Forest Products Journal* 58 (5): 77.
31 Brochmann, J., Edwardson, C., and Shmulsky, R. (2004). *Forest Products Journal* 54 (3): 51.
32 Kim, M.G., Watt, C., and Davis, C.R. (1996). *Journal of Wood Chemistry and Technology* 16 (1): 21.
33 Lee, S.M. and Kim, M.G. (2007). *Journal of Applied Polymer Science* 105: 1144.

34 Sellers, T. Jr., Kim, M.G., Kabir, F.R. et al. (1994). *Forest Products Journal* 44 (4): 63.

35 Oh, Y., Sellers, M.G. Jr., and Kim, and R. C. Strickland. (1994). *Forest Products Journal* 44 (2): 25.

36 Gramstad, T. and Sandstroem, J. (1969). *Spectrochimica Acta* 25A: 31.

37 Buschfield, A., Lattckamp, M., Ripens, G., and Shittek, H. (1991). Process for producing modified resin bonding agents and use thereof for producing particleboard. United States Patent 5,011,886.

38 De Jong, J. and de Jonge, J. (1952). *Recueil des Travaux Chimiques des Pays-Bas* 71: 3731.

39 Kim, M.G., Amos, L.W., and Barnes, E.E. (1990). *Industrial and Engineering Chemistry Research* 29: 2033.

40 Zhao, C., Pizzi, A., and Ganier, S. (1999). *Journal of Applied Polymer Science* 74: 359.

41 Pizzi, A. (1994). *Advanced Wood Adhesives Technology*, vol. 126, 192. New York: Marcel Dekker.

42 Tomita, B. (1983). *Polymer Preprints*, vol. 24, 165. Washington, DC: ACS.

43 Tomita, B. and Hse, C.-Y. (1992). *Journal of Polymer Science Part A: Polymer Chemistry* 30: 1615.

44 Kim, M.G., Nieh, W.L.-S., and Meacham, R.M. (1991). *Industrial and Engineering Chemistry Research* 29: 798.

45 Lofthouse, M.G. and Burroughs, P. (1978). *Journal of Thermal Analysis* 13: 439.

46 (1982). *Statistical Analysis User's Guide Statistical Analysis Software Institute*, 1028. Cary, NC.

47 Steel, R.G. and Torrie, J.H. (1980). *Principles and Procedures of Statistics*, 2e, 631. New York: McGraw-Hill.

14

Polyvinyl Acetate (PVAc) Emulsion Wood Adhesives

14.1 Polyvinyl Acetate (PVAc) Emulsion Wood Adhesive Resins

14.1.1 Manufacturing of PVAc Emulsion Resins

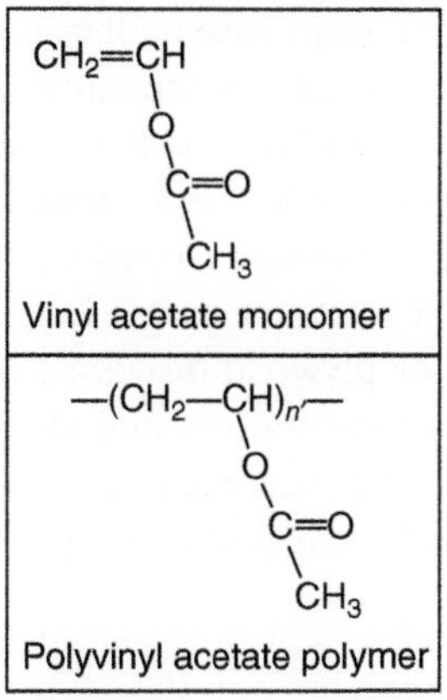

Polyvinyl acetate (PVAc) resin emulsions are manufactured in water for wood adhesive and many other binder uses such as for interior wall paints [1]. "Elmer's glue" and "Carpenter's glue" widely used in homes and furniture manufacturing and other secondary wood manufacturing processes are based on PVAc emulsion polymers. In the manufacturing of PVAc resin emulsions, vinyl acetate (monomer, shown), made from acetylene and acetic acid, is first dispersed in water by using surfactant and colloids and then vigorous stirring. Then, the free radical polymerization reaction is started by adding a catalyst (potassium persulfate) of a pre-calculated amount for the target molecular weight. The free radicals open up the vinyl acetate molecule's double bond and make bonds with another vinyl acetate monomer, and this process of addition polymerization goes on until all vinyl acetate monomer is depleted to form high molecular weight PVAc polymers. As the molecular weights increase, the polymer molecules form extremely fine solid particles that remain suspended (stabilized) in the water medium as they are formed with the help of colloids and surfactant molecules added in the beginning. The stabilization means that the particles remain suspended as fine particles, not agglomerating to form large particles that precipitate out. This stability means

Adhesives and Finishes for Wood: For Practitioners and Students, First Edition. Moon G. Kim.

an emulsion is formed and, depending on the method of preparation, PVAc emulsions can be made to be very stable, i.e. long shelf lives. If the particles agglomerate, the resultant product will become a curd or separated into a polymer-rich phase and water-rich phase, which are sometimes observed with an old bottle of Elmer's glue.

14.1.2 Properties of PVAc Emulsions and Polymers

PVAc emulsions are normally supplied with about 50% resin-solids contents and with a neutral or slightly acidic pH as a white viscous liquid that can be diluted or washed off with water if desired, as when we clean paint brushes after painting with water-based paints [2]. When isolated from the emulsion, the *PVAc* polymer is a brittle solid or powder that shows cold-flow properties with a glass transition temperature of about 28 °C or slightly higher. In order to use PVAc emulsions in film forms, by drying, as binders in surface coatings and adhesives, plasticization is often carried out by adding flexibilizers, such as 10–15% dibutyl phthalate. Thus manufactured PVAc resins are of high molecular weight, linear *thermoplastic* polymers, and therefore, as discussed in Chapter 3, do not require a further polymerization or cross-linking reactions in the application stages. This property translates into convenience and economy for the bonding process with PVAc resin-based adhesives and paints, i.e. bonding is done by drying at room or slightly elevated temperatures. On the other hand, the bond lines of PVAc resin adhesives are sensitive to heat and moisture degradation due to lack of cross-linking bonds, limiting their use to interior non-structural applications under the ordinary temperature and humidity conditions. Furthermore, because of the lack of thermosetting capability, they are not usable in the bonding of common wood composite boards such as PB, OSB, or plywood discussed in Chapters 7, 11, and 13, respectively, where high pressure, high temperature, and short bonding time are required. If PVAc resin adhesives are to be experimented in these applications, the hot-pressed board mat will simply disintegrate at the moment of hot-press opening.

14.1.3 Commercial PVAc Emulsions

One typical wood adhesives company supplies two different *PVAc* emulsion homo-polymers and another one with a modification with an un-revealed crosslinking polymer component [3]. Adhesive users can formulate adhesives themselves using these basic polymer materials, although the company also supplies various formulated PVAc wood adhesives.

One typical PVAc emulsion polymer is a low-viscosity *PVAc* homopolymer with a relatively high molecular weight, characterized by a minimal viscosity build with addition of plasticizers or solvents, so that formulators can formulate adhesives that bonds well and easily machined. The following PVAc resin properties are reported:

- The protective colloid used is polyvinyl alcohol; Viscosity: 3.0k–4.1k cps; Resin-solids content: 55.0–57.0%; pH: 4.0–6.0; Specific gravity, 1.09; Weight per gallon, 9.1 lbs; Freeze/thaw stability, excellent; Borax compatibility, no; Glass transition temperature (by DSC), 36 °C.
- Film properties – hard, glossy, translucent, good plasticizer response.

- Compounding – may be thickened using high molecular weight polyvinyl alcohol or conventional cellulosic derivative thickeners, compatible with a variety of conventional plasticizers at high loadings without excessive loss of film tensile strength properties.
- Shelf life – six months at 70 °F. To be stored in closed containers.

Another PVAc emulsion polymer offered is a relatively high molecular weight *PVAc* homopolymer, characterized with low viscosity build when formulated with plasticizers or solvents to allow formulators a wide latitude in selecting and adding modifiers so that adhesive formulations can be made that is easily machined. The major difference between this polymer and the first one appears to be a higher viscosity (4000–5500 cps) range, indicating higher molecular weights of polymers, a higher (required) minimum use temperature of 60 °F/15.7 °C, and a slightly higher glass transition temperature of 37 °C.

Another PVAc emulsion polymer offered is a *crosslinking PVAc* emulsion resin designed to be used for formulating adhesives for cold- and hot-press laminating, edge gluing, and assembly bonding applications. The nature of the crosslinker is un-revealed, but the viscosity is also higher (3000–8000 cps), and the minimum film-forming temperature {58.1 °F (14.5 °C)} and the higher glass transition temperature {107.5 °F (41.95 °C)} indicates the resultant adhesive films' increased solidity.

14.1.4 Additives Used in Formulating PVAc Emulsion Wood Adhesives

PVAc emulsions are extensively used as binders for interior paints and also as binder components in wood adhesives such as Carpenter's glue, which is formulated with a small amount of various additives without breaking the emulsion itself [1, 2]. The additives used in formulation and their positive and negative effects on the gluing of wood are as follows:

(1) *Polyvinyl alcohol (PVA)*: It increases creep resistance and machinability, i.e. clean or less gummy cuts, but it also increases the water sensitivity. *PVA* resins are thermoplastic polymers made by hydrolysis of *PVAc*, and their emulsion forms mix well into PVAc resin emulsions, resulting in harder adhesive gluelines. Modification results in only minor improvement of creep resistance, not to the extent of structural adhesives.

(2) *Starch*: Starch lowers the adhesive cost with minor positive effects on the performance. Starch is dispersible into PVAc emulsion.

(3) *PF, PRF, UF resins and animal blood*: These thermosetting components give some cross-linking capability to the PVAc emulsion resins, increasing the water resistance and creep resistance; they can be added only at low levels with an acid catalyst such as aluminum chloride that activates the curing process of additive resins. However, the extent of cross-linking is limited because the amount of additives is limited. Most commercial PVAc glues have some of these additives and are often tinted with a pale-yellow color as a marker. The acid catalyst added for curing can make the adhesives' shelf lives limited and also can cause wood degradations or a higher tool wear in machining of glued products. High moisture sensitivity of PVAc adhesives was noted early and it was reported that mixing with UF resin at a 54 : 46 ratio and with animal blood at a 83 : 17 ratio resulted in significant durability improvements [4, 5].

(4) *Plasticizers*: Plasticizers increase the flexibility of adhesive layer and lower the coalescing temperature, making the adhesive usable at colder temperatures, such as in winter.

Diethyl phthalate is the often-used plasticizer, but it increases the creep tendency of adhesives.

(5) *Co-polymerization*: External plasticizers evaporate out in the end to make the bond line brittle in long-term uses, and to overcome this limitation the emulsion polymerization of vinyl acetate is carried out with a 15–20% copolymer component such as *ethylene. Ethylene-vinyl acetate (EVA) co-polymer* resins made by co-polymerizing with ethylene at various ratios are widely used as *hot-melt adhesives* (discussed in the next chapter).

(6) *Organic solvents*: Some organic solvents can give plasticizing effects on gluelines during the bond formation by accelerating the coalescing process of PVAc particles, and because the organic solvents eventually evaporate out after curing, the creep tendency of glueline is not deteriorated, a positive for gluelines. Only small amounts of relatively high boiling point organic solvents, such as dimethyl ether of diethylene-glycol (CH_3–O–CH_2–CH_2–O–CH_2–CH_2–O–CH_3: bp 162 °C) are used.

(7) *Fillers*: Gypsum (calcium sulfate) powder, calcium carbonate (marble) powder, wood flour, etc. are used as a filler to control viscosity and consistency. Fillers also reduce adhesive penetration and improve gap-filling capability. They also increase the solids levels of adhesive and reduce adhesion problems arising from variations in moisture content and poor wood substrate quality.

14.1.5 Properties of Formulated PVAc Emulsion Wood Adhesives

Formulated PVAc emulsion adhesives with additives are widely used as ready-to-use, semi-structural, interior-use adhesives for edge- and end-jointings such as in furniture manufacturing and core composing, using spreader, dip, brush, or sprayer type applicators [1, 2]. Setting of adhesives at room temperature under 100–175 psi is relatively fast, occurring by diffusion of component water into wood substrates, followed by coalescing of PVAc polymer particles. The upper limit of use temperature of bonded products is at about 40 °C, above which the creep tendency and/or de-bonding tendency becomes appreciable due to the softening of bondline. These creep problems can appear at lower temperatures if the load is high and over a long period of time. The water resistance of PVAc adhesives is also low in comparison with thermosetting resin wood adhesives. Some limited water resistance is obtained by adding cross-linking additives, but the extent is limited because of the limited extent of cross-linking reactions possible in the system. Overall, the PVAc emulsion adhesives are useful as interior adhesives and should be used only where the load is not high and the chance of making contact with water or moist air is low.

14.1.6 Various Commercial Formulated PVAc Emulsion Wood Adhesives

Various different classes of formulated wood adhesives are also supplied by adhesive companies [3]:

- Water-resistant products for a variety of applications.
- Wide range of applications for various assembly needs.
- For radio-frequency (RF) gluing and cold- or hot-pressing.
- Higher durability, nonstructural wood bonds for furniture, cabinets, etc.

Further, several differently formulated adhesives are supplied in each class and selection guides are available for any particular wood bonding process, meeting the needs of users with necessary viscosity ranges, gluing speed, bonding temperature, glue bonds' durability, heat resistance, etc.

14.1.7 Application Methods of PVAc Emulsion Wood Adhesives

PVAc emulsion wood adhesives are used for various joinery (tabletops, etc.), laminating, core composing, overlaying [6], edge-banding, and finger-jointing in the furniture and in general wood bonding industries [4]. These gluing operations can be finished (cured) by standing the clamped wood assemblies several hours at a slightly elevated room temperature or by hot-pressing or radio-frequency heating process to speed up the drying/curing process. The adhesives are non-toxic, little odorous, water washable, and have easy-to-handle viscosity ranges to allow easy applications on small or large areas by brushing or beading applicators. They give good amounts of assembly time and have almost indefinite shelf lives if kept in a closed container at room temperature. In finger-jointing and some high-speed face gluing operations, the radio-frequency (RF) heating method is mostly used [7], but the heating is brief so as to accelerate the dissipation of water in gluelines and to initiate the coalescing process. In most cases, some heat application, such as standing the gluing assemblies in a warm room, is needed. Thus, the two important parameters in wood bonding with PVAc adhesives are the moisture content of wood substrates and the gluing room temperature. Wet wood will need a longer time to cure than dry wood.

14.1.8 Curing Mechanism of PVAc Emulsion Wood Adhesives

Coalescing of PVAc polymer particles is the main curing mechanism of gluelines [1, 2]. The temperature of the curing room should not be too low since the coalescing will be retarded if the polymer particles freeze or freeze-dried. An adequate extent of coalescing is the key to attaining strong film structures for thermoplastic polymer-based adhesives, i.e. a structurally "continuous" polymer film needs to be formed. An inadequate coalescing either by a low temperature or short bonding time will lead to poorly coalesced polymer particles with many defects in the glueline, i.e. the interstices formed between polymer particles will be too many. These defects would become the initiating points of crack formation under load. If an extremely poor coalescing had occurred, the gluelines would have "chalk-like" appearances as opposed to clear film-like layer. The cross-linking additives, added with acidic catalysts, improve the water resistance and creep resistance performances of PVAc wood adhesives to some extent [5], and although this is desirable, this is only secondary to the good coalescing that needs to occur for a good bonding. For a speedy cure, therefore, the bonding is carried out under pressure with an application of radiofrequency heating as in bonding of laminated lumber using MF or PRF resin adhesives [7]. Common problems often occur for various reasons in using PVAc adhesives for wood bonding and the troubleshooting methods are presented in Chapter 15.

14.2 Summary

Formulation and application techniques of PVAc emulsion resin wood adhesives are presented based on open literature. Properties and formulation methods of commercially available PVAc resins with various common and cross-linking additives are presented. Various grades of formulated PVAc resin wood adhesives such as water-resistant, higher durability, radiofrequency curing, and general-purpose adhesives are commercially available. The curing operations in wood laminating industry include clamping at room or slightly elevated temperatures and hot-pressing or radiofrequency heating of glued assembly. The coalescing mechanism of PVAc adhesives is presented with respect to the gluing room temperature since use of lower gluing room temperatures than the glass-transition temperature of PVAc polymers must be avoided to obtain satisfactory bonding results.

References

1 Saunders, K.J. (1988). *Organic Polymer Chemistry*, 412. New York: Chapman and Hall.

2 Geddes, K. (1994). Chapter 25, Polyvinyl and ethylene-vinyl acetates. In: *Handbook of Adhesive Technology* (ed. A. Pizzi and K.L. Mittal), 431–441. New York: Marcel Dekker.

3 Franklin International (FI). (2020). Bruck Street. Columbus, OH 43207, USA.

4 Clark, L.E. Jr. (1965). *Forest Products Journal* 15 (4): 176.

5 Carroll, M.N. and Bergin, E.G. (1967). *Forest Products Journal* 17 (11): 45.

6 Hespe, G.A. and Bergin, G. (1990). Adhesives for bonding overlays to wood-based panels. In: *Wood Adhesives* (ed. A.H. Conner, A.W. Christiansen, G.E. Myers, et al.), 166. U. S. Forest Service/Forest Products Laboratory and Forest Products Research Society.

7 Graybill, A.E. and Brown, N.R. (2005). Influence of radio-frequency time on the performance of polyvinyl acetate adhesives. In: *Wood Adhesives 2005* (ed. C.R. Frihart), 107. Forest Product Society.

15

Troubleshooting in Wood Bonding with PVAc Resin Adhesives

(Adapted with permission and with minor modifications from a paper presented by Dr. Robert Snider of Franklin Chemical International and data on their website. [1]*)*

Problems involved in gluing wood often arise from the lack of recognizing the importance of the effects of (i) moisture content of wood, (ii) water content of adhesive layer, and (iii) condition of wood surface.

15.1 Effects of Moisture Content of Wood

Figures 15.1 and 15.2 show the effect of moisture loss on the dimensional changes of wood in the tangential and radial directions. Since shrinkage in the tangential direction is almost twice that in the radial direction, moisture changes affect the wood cross-sections differently depending on their relative location in the log. Cupping and bowing are a result of the differential dimensional changes occurring with moisture changes. This also sets up severe stresses in a board or post, as shown in Figure 15.1a,b, so that in some cases the internal strength of wood is exceeded, causing a split. Table 15.1 shows the effects of moisture changes on the radial and tangential dimensions and volumes of various woods, i.e. % increases or decreases per % moisture changes. It is noted that changes in the tangential direction occur almost twice large as those in the radial direction.

Effects of radial and tangential differences in shrinkage: From the relative amounts of shrinkage shown for various woods in Table 15.1, it is noted that mahogany is one of the most stable of the common woods and beech varies considerably between the tangential and radial shrinkage. As an example, a change of 4% moisture content in a tabletop 36 in.-wide made from tangential *beech board* will change by about 0.5 in. Edge-gluing a 1-inch-thick tangential-face board to a 1-inch-thick radial-face board and changing the moisture content by 4% will cause a difference in thickness of 8 mils or a 4 mils per side. Such a difference will become readily apparent. On the other hand, mahogany would have only increased by 0.25 in. in width, and the board thickness would increase by only 1 mil per side.

Effects of moisture content differences of wood substrates: It is important to note that lumber used in panels has to have equal moisture contents. If two pieces of same kind of wood but having different moisture contents are to be bonded to a panel and sanded and

Adhesives and Finishes for Wood: For Practitioners and Students, First Edition. Moon G. Kim.

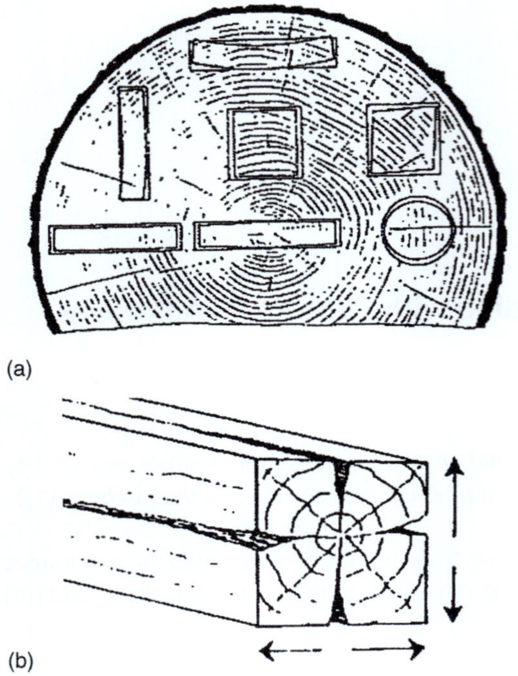

Figure 15.1 Different dimensional changes in wood by moisture content for tangential and radial directions.

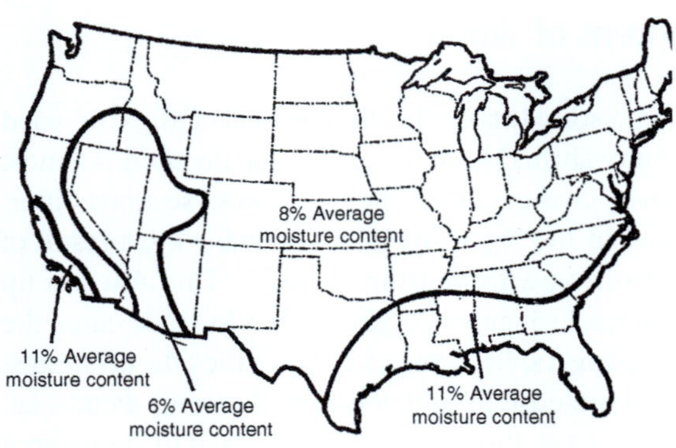

Figure 15.2 Average wood moisture contents in different regions of the United States.

then the moisture content is equalized, the thickness of the wettest board would become much less than the other.

Effects of moisture difference between manufacturing plant and location of use: The average relative moisture content varies by regions, as shown in Figure 15.2. Therefore, normally, a moisture content for wood to use for furniture manufacture is kept at 6–8%. For mobile home side walls and floors, a higher moisture content not to exceed 15% is more suitable. Also, shrinkage with moisture change in the direction of grain is comparatively very small, between 0.1% and 0.2% for most woods between green and oven-dry. Further, it is noted that (i) Boards of flat-bonded cross grains to bottom piece will result in a high-stress concentration with moisture changes and may split along the gluelines, (ii) Another example is the chair arm glued to the post, and when the arm becomes thicker by moisture increase, the joint might rupture, (iii) Another example is the face veneer bonded to lumber crossband and loss of moisture content results in splitting of face veneer, (iv) Miter joints

Table 15.1 Dimensional changes (%) in the radial and tangential directions and in volumes per % moisture change for various woods.

	% dimensional change per % moisture change		
Species	**Radial**	**Tangential**	**Volumetric**
Alder, red	0.15	0.24	0.49
Ash, white	0.16	0.26	0.45
Basswood, American	0.22	0.31	0.53
Beech, American	0.17	0.37	0.54
Birch, yellow	0.24	0.31	0.56
Cherry, black	0.12	0.24	0.38
Cherry, Pin	0.09	0.34	0.43
Cottonwood	0.13	0.31	0.47
Elm, American	0.14	0.32	0.49
Elm, Rock	0.16	0.27	0.47
Elm, Slippery	0.16	0.30	0.46
Hackberry	0.16	0.30	0.56
Hickory, pecan	0.16	0.30	0.45
Luan, red	0.11	0.27	0.39
Mahogany	0.12	0.17	0.27
Maple, Silver	0.10	0.24	0.40
Maple, Sugar	0.16	0.32	0.50
Oak, black	0.15	0.32	0.47
Oak, bur	0.15	0.29	0.42
Oak southern red	0.15	0.29	0.54
Oak, white	0.18	0.30	0.53
Sweetgum	0.17	0.33	0.50
Sycamore, American	0.17	0.25	0.47
Teak	0.08	0.14	0.23
Tupelo, black	0.15	0.26	0.46
Tupelo, water	0.14	0.25	0.42
Walnut, black	0.18	0.26	0.43
Willow, black	0.09	0.27	0.48
Willow, Pacific	0.10	0.30	0.46
Poplar, yellow	0.13	0.24	0.41

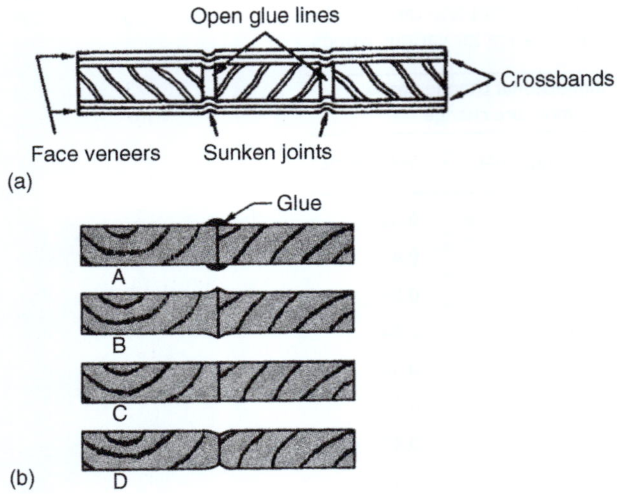

Figure 15.3 Effects of the water of water-borne adhesives on swelling/drying of wood joints in the adhesion/sanding processes.

are also highly stressed from the shrinkage of tangential or radial grains and tend to show a separation of the inner corner, (v) Bonding of wood to other materials, such as honeycomb, can result in the same separation problems when moisture relations are changed, and (vi) Bonded boards of internally stressed woods will be less strong, which arise from improper drying.

Effects of water in water-based adhesives on glue joints: Most structural glues, such as polyvinyl acetate (PVAc) emulsion adhesives, are water-based. The glue in a joint will dry and in most cases, the gap will remain the same during drying (curing) and therefore the glue layer will shrink in thickness from the loss of water, resulting in making the joint weaker and sunken as shown in Figure 15.3a. This problem will get worse when the glueline gets thicker. Further, it needs to be noted that: (i) A dowel joint should not be made so tight to push the glue to the bottom, nor should it fit too loosely for a thick glueline to form. The dowel diameter should be kept uniform such as by keeping it in a heated box to avoid the moisture content variation. Also, the dowel length should be the same as the depth of the hole to obtain the best bond strength, (ii) Thicker gluelines often occur from inadequate pressure or by improper clamping. The pressure should push both sides of the panel together with equal pressure or it will leave one side open to result in a thicker glueline. The uniformity of pressure is important, and (iii) Open or sunken joints on edge-glued panels, if present, will frequently show through the face veneer as defects of the face, as shown in Figure 15.3a, and (iv) Water-based wood adhesives such as PVAc emulsion adhesives add water to the wood in the vicinity of the bond line to result in swelling. If the panel is planed smooth immediately on removal from the press, the swelled portion will be planed to the same thickness as the drier material, and when the moisture content of the panel equalizes, there will be a valley at each glue line, as shown in Figure 15.3b.

Effects of substrate condition: Wood adhesives are commonly formulated to minimize penetration into wood in order to eliminate "starved" glue joints. Since this penetration is only a few thousand of an inch depth, this outer layer of wood is important, and almost any wood preparation method can make good or bad surface layers, such as: (i) A good saw joint can be made with a sharp saw mounted on a good equipment. Dull saw can result in a coarse surface where the surface fibers adjacent to the kerf are loosened but not removed, resulting in a weak bond layer, (ii) Abrasive planers can also result in a rough surface often when using coarser grits than usual, (iii) Burned surfaces are difficult to adhere to for conventional wood adhesives. Dowel holes can burn when the drill advances too rapidly, and the same effect can occur with edge joints, (iv) Sometimes, bleed-through of adhesive with a veneer can occur and can cause a finishing problem or difficulty in breaking down of stacked panels. A change of adhesive, reduction of spread rate, or reduction of pressure will usually solve the problem, and (v) Wood surfaces need to be cleaned before gluing. This is particularly true of flat surfaces to be laminated with veneer or high-pressure laminates. Particles left on the surface will telegraph through. A vacuum brush cleaner will eliminate most particles. Sometimes, wood chips may get into the adhesive container and cause similar problems, where a filter system is needed.

Miscellaneous causes of gluing problems: Joint failures can also be caused by a number of other errors. (i) A lack of adequate pressure can result in an ineffective transfer of adhesive in the joint, which often occurs when cold-pressing of laminates on particleboard with pressure applied only on a small area, (ii) A pre-dried joint can occur as when the glue is applied in grooves on roll coater and the thinner part dries up while the thicker part remains wet before assembly. The dried line will result in pre-dried or poor bond, (iii) Certain cured adhesives are swelled by the solvents in the finish, either the adhesive or finish needs to be changed, (iv) When the wood pores are filled with adhesives, white spots will appear under the finish; to prevent this, it is necessary either to prevent the adhesive from filling the pores or to sand off its glue spots before finishing, (v) Acidic adhesives pick up stains from contacting with iron, possibly from piping, valves on drums, or iron clamps. Small amounts of iron do not discolor the adhesive but may still stain certain woods such as oak, cedar, walnut, and mahogany, (vi) Gluing at a temperature below the chalk point with PVAc adhesives will give a weak, usually white-colored, glueline; the joint must be re-made, with the room and wood kept at temperatures above the chalk point (glass transition temperature), and (vii) The presence of finish stain in the glued joint indicates that the joint was open prior to finishing.

Quality control methods: Every woodworking operation doing gluing should institute a quality control program. This will allow corrective actions to be taken before the bonding operation gets out of control to result in many defective products. Daily charts are maintained and monitored: moisture contents from dry kiln; equilibrated moisture contents in panel department; joint failure percentage in 1000 panels of production. Joint failure values are regularly measured by breaking trim pieces along the bond line on-site using a chisel-type instrument and reading the failure values.

15.2 Troubleshooting Methods in Various Gluing Operations

PVAc resin adhesives are used in wood laminations using cold- and hot-pressing, radiofrequency edge-bonding, and edge- and face- or assembly-bonding by clamping. A troubleshooting guide [1] is presented to help readers understand the various bonding processes in the industry, and to show potential troublesome events and the methods of handling them.

15.2.1 Troubleshooting in Cold-Press Lamination

- Total delamination:
  (1) Precure of glue (no glue transfer) – increase glue spread.
  (2) Low pressure (poor contact) – increase pressure.
  (3) Short press time – increase pressure period.
- Spotty bonds:
  (1) Uneven wood surfaces – calibrate cores to uniform thickness and/or increase pressure.
  (2) Worn glue spread rolls – replace or re-groove rolls.
- Glue bleeds through veneers:
  (1) Excessive glue spread – reduce glue spread.
  (2) Excessive pressure – reduce pressure.
  (3) Wrong adhesive type – contact technical support.
- Telegraphing of core defects or banding:
  (1) Excessive pressure – reduce pressure.
  (2) Uneven or variable thickness of core – re-sand core.
  (3) Foreign materials on core wood – pre-clean wood core.
- Warpage of bonded panels:
  (1) Excessive or unequal glue spread – reduce spread and make sure equal spread.
  (2) Unbalanced construction – check grain orientation and number of plies.
  (3) Excessive moisture – 6–8% moisture content of wood is recommended.
- Brilliant white on glue squeeze-out or glueline:
  (1) Chalking caused by low temperature of bonding – raise temperature of plant, so that wood and adhesive are maintained above the chalk point of adhesive in use.

15.2.2 Troubleshooting in Hot-Press Laminating Process

- Spotty bonds:
  (1) Low spread – increase adhesive spread.
  (2) Uneven core thickness – calibrate core to uniform thickness.
  (3) Low pressure – increase pressure.
  (4) Cold spots in platen – check platen temperature with surface thermometer.
- Glue bleeds through veneers:
  (1) Excessive spread – reduce adhesive spread.
  (2) Excessive pressure – reduce pressure.
  (3) Wrong adhesive type – consult technical support.
- Steam blister:
  (1) Excessive spread – reduce adhesive spread.
  (2) High moisture content – 6–8% moisture content recommended.
  (3) High press temperature – reduce temperature.
- Warpage of panels:
  (1) Excessive spread – reduce adhesive spread.
  (2) Unequal spread – make sure that top and bottom spread are equal.
  (3) High moisture content – 6–8% moisture content recommended.
  (4) Unbalanced construction – check grain orientation.
  (5) High press temperature – reduce platen temperature.

- Checking of veneer:
  (1) Moisture content of veneer and substrate are not equal – use well-manufactured veneer.
  (2) Keep face veneer within 6–8% MC at time of pressing.
  (3) Core components and cross-band materials should be held at about the same moisture content as the face veneer.
  (4) Keep assembly time to a minimum.

15.2.3 Troubleshooting in Radiofrequency Edge-Gluing

- Under-cured joint or glueline is still wet:
  (1) Low power – increase power.
  (2) Short cure cycle – increase cycle time.
  (3) No pressure – check filler boards and pressure system.
  (4) High moisture content – 6–8% moisture content is recommended.
  (5) Improper catalyst mix – consult technical support.
  (6) RF generator malfunction – consult technical support.
- Cured joint but weak bonds:
  (1) Slightly under-cured – check power and time settings.
  (2) Incorrect pressure – check edge pressure.
  (3) Irregular or burnished surfaces – check surface preparation.
  (4) Improper moisture – 6–8% moisture content is recommended.
- Panel open on ends:
  (1) Poor stock preparation – check quality of gluing surface and end fit.
  (2) Pressure shoes overlapping on the next panel – check and correct if necessary.
  (3) Shrinkage of wood between time of surfacing and pressing – 6–8% moisture content is recommended, 30% minimum relative humidity, and gluing done within 24 hours of preparation.
- Burning in joints:
  (1) Power too high – reduce power.
  (2) Improper catalyst level – consult technical support.
  (3) Surface irregularities – check surface preparation.
  (4) Excessive spread – reduce glue spread.
  (5) Long assembly time – press within five minutes.
- Black glue lines:
  (1) Iron contamination in adhesive – eliminate the iron source of wet adhesive.

15.2.4 Troubleshooting in Edge and Face Gluing by Clamping

- Weak joints with little or no wood failure:
  (1) Poor stock machining – check for mechanical malfunction, adjust equipment, and make sure blades are sharp.
  (2) Burnished surfaces – increase feed rate on molder or jointer and sharpen or replace cutter blades.
  (3) Under-cured or frosted joints caused by low temperature – raise plant and stock temperature.

(4) Under-cured or frosted joints caused by short clamp time – increase clamp time.
(5) Under-cured or frosted joints caused by high moisture contents – 6–8% moisture content is recommended.
(6) Under-cured or frosted joints caused by thick glue lines – check part fit and/or increase clamp pressure.

- Splits and/or open joints on ends of panels:
  (1) High moisture content – 6–8% moisture content is recommended.
  (2) Dry condition in plant – 30% minimum relative humidity; and use an end sealer.
- Wood failure only over part of surface:
  (1) Uneven pressure – apply pressure equally to top and bottom.
  (2) Uneven pressure due to improper clamping spacing – apply pressure near ends and 8–12″ apart.
  (3) Uneven pressure due to poor machining – make sure that surfaces are flat and square.
  (4) Uneven glue application – glue coverage must be even over entire surface.
- Weak joints with wood failure along glue line:
  (1) Abrasive grit is too coarse – 60 grit or higher is recommended for surface preparation.
  (2) Dull saw blades – sharpen blades.
- Black glue lines:
  (1) Iron contamination – eliminate the iron source in adhesive.
- Brilliant white on glue squeeze-out or glueline:
  (1) Chalking caused by low temperature – raise temperature of plant, so that wood and adhesive are maintained above the chalk temperature of adhesive in use.

15.2.5 Troubleshooting in Assembly Gluing by Clamping

- Weak joint:
  (1) Loose part fit – part should fit snugly and not require undue assembly force.
  (2) Joint design defect or dowel too short, etc. – redesign joint.
  (3) Glue setting too quickly or too slowly – consult technical support.
  (4) Insufficient glue coverage – apply uniform coating of glue to all parts of joint.
  (5) Burnished surfaces – sharpen cutter blades.
  (6) Under-cured finish coating on gluing surface – remove finish coating before gluing.
  (7) Under-cured or frosted glue lines due to short clamp times – increase clamp time.
  (8) High moisture content – 6–8% moisture content is recommended.
- Black glue lines:
  (1) Iron contamination – remove the iron source in adhesive.
- Brilliant white on glue squeeze-out and or glue line.
  (1) Chalking by low temperature – raise temperature of plant, wood, and adhesive above the chalking temperature of adhesive in use.

Reference

1 Franklin International. (2020). Bruck Street. Columbus, OH 43207, USA. IASCS@franklininternational.com.

16

Hot-melt and Other Specialty Wood Adhesives

16.1 Introduction to Hot-melt Adhesives

Hot-melt adhesives are of recent development in wood bonding [1–9]. The major components of wood adhesive-type hot-melts are various thermoplastic materials, formulated to melt at around 120–200 °C as needed for melt application. The amount of hot-melt adhesive that is normally applied for bonding is relatively small compared with the amount of substrate surface area and, therefore, the applied molten adhesive cools quickly (a few minutes or less) to below its melting temperature (T_m) and solidifies, making bonds if the two substrates are tightly mated under pressure before solidification. This fast adhesion speed is certainly accompanied with a limited repositioning capability in comparison with other adhesives such as polyvinyl acetate (PVAc) adhesives, but hot-melt adhesives are advantageously utilized in the fast, continuous gluing processes to result in productivity gains. In these applications, current hot-melt adhesives show excellent green strength (strength right after the completion of pressure period), ease of application, low toxicity, and 100% solids (no water or solvents) without much volatile organic compounds emitted during bonding. However, hot-melt adhesives are useful only for interior, nonstructural applications due to their relatively low bond strength and low exterior durability in comparison with thermosetting-type wood adhesive resins. They are also relatively more expensive by weight basis and, therefore, used in select fast, continuous bonding processes, such as paper carton packaging and some wood gluing, including the edge-banding and overlaying of polyester-type plastic films on wood in the furniture manufacturing, and book binding, etc. In Figure 16.1, the relationship between tensile modulus and temperature range for a typical hot-melt adhesive system is shown. Hot-melts are supplied as solid, and as the temperature rises by heating, the adhesive starts to soften at T_g (glass transition temperature), then becomes leathery, then rubbery, and finally melts to a viscous liquid at T_m (melting point). The hot-melt adhesive is applied at above the melting temperature and should begin to solidify when cooled to below T_m. The modulus (stiffness) of cured hot-melt adhesive is relatively high at room temperature and below its T_g.

16.2 Requirements For Hot-melt Adhesives

(i) Figure 16.1 indicates that the temperature difference (ΔT) between T_m and T_g needs to be large as much as possible for hot-melts to allow the bond application and

Adhesives and Finishes for Wood: For Practitioners and Students, First Edition. Moon G. Kim.

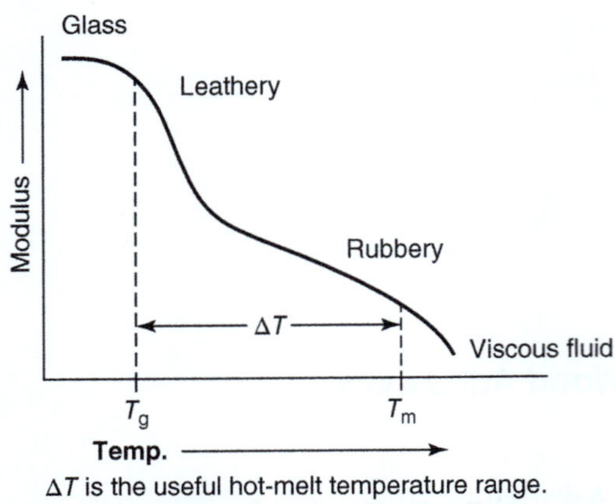

Figure 16.1 Optimum performance range of a typical hot-melt. Source: Pocius and Dahlquist [2] Reproduced with permission of the publisher American Chemical Society.

Table 16.1 Melt indices of two EVA hot-melts and their properties.

Property	High melt index HM	Low melt index HM
Bond strength at – 20 °F	High	Low
Bond strength at RT	Little difference	Little difference
Bond strength at 140 °F	Low	High
Melt viscosity	Low	High
Cohesive strength	Low	High
Molecular weight	Low	High

curing operation to be completed adequately within the given clamping time [1–9]. Most thermoplastic polymers show relatively small differences between T_g and T_m, and therefore optimum polymers for hot-melts need to be selectively manufactured and further formulated to have an adequate range between T_m and T_g. (ii) The solidified hot-melt gluelines should have a relatively high modulus at room temperature. (iii) The solidified hot-melt gluelines should have a good wetting capability (atomic attractive forces) on the substrates in both molten and solid phases. (iv) Hot-melt adhesives must be formulated to give a melt having relatively *low viscosity values* for an easy application through an orifice of a lightly pressurized gun: 20–60 k cps at 200 °C, or a melt-flow index between 2 and 20 according to the ASTM D1238 method, a viscosity measuring method for high viscosity materials at high temperatures. Higher melt index values indicate better flow speeds and low melt index values indicate lower flow speeds. Hot-melt adhesives are classified with their melt indices to guide users, for example, the commonly used *ethylene vinyl acetate* (EVA) copolymer-based hot-melt adhesives show melt indices ranges and bond strength and molecular weight properties shown in Table 16.1 [1].

Thus, it is noted that within limited ranges, the performance properties of hot-melt adhesives can be tailored by varying the base polymer material's molecular weights. Higher molecular weight polymers give low melt indices, making it necessary to use higher temperatures but result in higher cohesive bond strengths. Lower molecular weight polymers would give opposite properties. This relationship is what is expected from the polymer chemistry principles and is employed in formulating hot-melt adhesives for target applications.

16.3 Materials Used for Hot-melt Wood Adhesive Formulations

The hot-melt adhesives currently used in the wood product industry are based on one or more of the following polymers: *low-density polyethylene; EVA copolymers*; polyesters; polyamides; polypropylene; styrene-butadiene block polymers; phenoxy resins; and polyurethanes. Among these polymers, EVA copolymers are more extensively used (Figure 16.2) [1]. EVA polymers are relatively low cost and, due to their acetate groups present in the polymer structure, show relatively good wetting properties to wood. Another polymer material used in hot-melts is the *polyamides* made by condensation polymerization of long-chain dicarboxylic acids (HOOC–R–COOH) with polyamines (NH_2–R′–NH_2) as shown in Figure 16.3. The most often used dicarboxylic acid is the "dimer acids" produced by Diels–Alder-type dimerization reactions of *tall oils*, the oleic and linoleic acid mixture obtained as by-products in the Kraft pulping process of pine wood. Polyamide-based hot-melts are known to have higher heat stability. The main polymer used in hot-melts determines the hot-melt adhesive performance and application properties: viscosity and rheology, cohesive strength, adhesive strength, and flexibility.

In addition to the base polymer, other low molecular weight polymer materials are added to impart various other hot-melt adhesive properties, as follows: (i) *Tackifying agent* – it provides tack, the preliminary bond strength (grabbing) prior to the solidification of adhesive; terpene resins, poly-terpene resins, and chlorinated hydrocarbons are used. Although tack is not much related to the final bond strength, a good tack for hot-melt adhesives is an important parameter for fast bonding processes. In the fast-bonding operations, at least at the very end of the short pressuring period, the substrates should be held in position by the still hot adhesive layer. The tack is the property that gives this grabbing strength. (ii) *Plasticizer (flexibilizer)* – it improves thermal and mechanical impact resistance of cured gluelines by lowering the brittleness of the main polymer. Alkyl phthalates and polybutenes are used. Plastic overlays and edge-bonded wood composite products tend to be exposed to

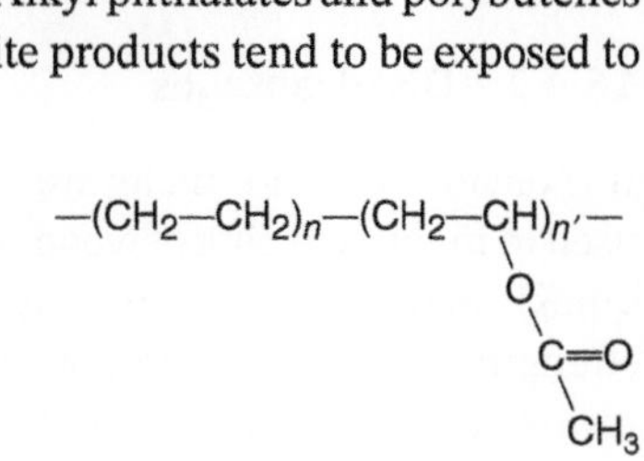

Figure 16.2 Ethylene-vinyl acetate (random) copolymer used for hot-melts.

—OC—R—**CO—NH**—R′—**NH—CO**—R—**CO—NH**—R′—NH—

Figure 16.3 Polyamides used for hot-melts: R = dicarboxylic acids; R′ = ethylene group, etc.

Table 16.2 A typical EVA-based hot-melt composition.

EVA copolymer: 39.0 parts	Hydrocarbon resin (wax): 5.5 parts
Pigment: 4.5 parts	Antioxidant: 0.2 part
Filler: 37.8 parts	Tackifier: 13.0 parts

Application temperature range: 170–200 °C [1].
Adapted from Pizzi et al. [1].

excessive mechanical impacts in use and the hotmelt adhesives used in these applications need to be low in brittleness. (iii) *Antioxidant* – it prevents the oxidation of the main polymers and other additives during the molten stage and in use. Hindered phenols (BHT, etc.) are used, similar to those we often have as preservatives in foods. Because a certain amount of hot-melt adhesive has to be kept molten (hot) in the reservoir during the bonding operation, exposed to air, an antioxidant is essential, without which hot-melt adhesives will have shorter (molten) pot-lives during the bonding operation. (iv) *Filler* – it reduces the cost, increases the opacity, or gives color and resistance to light, and modifies the flow properties. Flow properties are again an important parameter as in other adhesive applications. Finely ground clay, talc, TiO_2 (whitening), and other pigments are used, up to 50% of the adhesive volume. (v) *Wax* – it reduces the viscosity and alters the surface to a less polar one, and paraffin wax is used. A typical hot-melt adhesive formulation using EVA copolymers is shown in Table 16.2.

16.4 Advantages and Disadvantages of Hot-melt Adhesives

16.4.1 Advantages

Hot-melt adhesives are almost 100% solids adhesives, no water or solvent needs to be get rid of in the gluing process, in contrast to the water-borne PVAc emulsion adhesives or solvent-borne contact adhesives [1–9]. Hot-melt adhesives give a quick bonding because the cooling process requires only a very short time and therefore are very convenient to use in the fast, continuous wood bonding processes including overlaying and edge-banding of wood composite products and bonding of furniture base stocks.

16.4.2 Disadvantages

It is obvious that hot-melts are not capable of wood bonding in the hot-pressing processes used in manufacturing of wood composite boards since hot-melt gluelines cannot be cooled quickly in the hot-press, like any other thermoplastic adhesive materials. Also, the cohesive strengths of hot-melts are not very high in comparison to those of thermosetting polymer resins because many low molecular weight materials are added to make them work as hot-melts and also because of the relatively low strength properties of the base polymers. Also, due to the absence of cross-linking in common hot-melt adhesive polymers, the glueline is subject to creep, which increases at elevated temperatures, and this lack of cross-links also results in higher bond deterioration rates when exposed to outdoor environments and mechanical stresses.

16.4.3 Uses of Hot-melts

Hot-melt adhesives, therefore, are used mostly in nonstructural, interior applications, such as overlaying and edge-banding of vinyl sheets on particleboard-type wood products in furniture manufacturing. Overlays or edge-banding materials can be either wood or other plastic materials such as polyester, polyvinyl chloride, and melamine-formaldehyde (MF) products (Formica). The jury is still out regarding the hot-melt adhesives' long-term performance in comparison with MF resin-type adhesives used in similar wood bonding applications. Hot-melts are also used in bookbinding, shoe manufacturing, box and carton sealing, and in corrugated paperboard packaging.

16.5 Thermosetting Hot-melt Adhesives – Recent Development

16.5.1 A Typical Thermosetting Hot-melt Formulation

Major limitations of common hot-melt adhesives discussed above are the poor creep resistance, especially at elevated temperatures and the lack of exterior durability [3–9]. Although these type of deficiencies would be difficult to overcome to the extent of PF resin-type thermosetting adhesives, putting in cross-linking polymer structures into the hot-melt adhesives would boost such properties. Using pMDI resins and polyols as discussed in Chapter 12, the one-component moisture-curing polyurethane polymers were developed and incorporated at 20–60% levels into common hot-melt polymers, resulting in the reactive hot-melt (**RHM**) adhesives. In an example, the polymerization reaction of pMDI resins with polyethylene glycol-type polyols of a molecular weight of 1000 or higher was carried out at a hydroxyl: isocyanate group mole ratio between 0.1–0.7 : 1 to form polyurethane polymers having up to about 25 k cps viscosity. The excess isocyanate groups remaining in the polymer act later as moisture-curing functional groups. The second component (10–50% by weight) in the formulation is thermoplastic EVA copolymers having 50–60% vinyl acetate contents and melt index of up to about 90. The third component (20–45%) is styreneic tackifying polymer materials and antioxidant up to 5% by weight. The application temperature of this type of RHM is 110–175 °C, application viscosity of 6 k cps, and open time of 15–600 seconds. The cured glue's % elongation at 25 °C ranges 10–100%. Typical green shear strengths of wood bonds reach to about 0.2 mPa in 15 seconds and ~1.4 mPa in 120 seconds after application [9]. Thus, RHM adhesives behave like traditional hot-melts with similar melting ranges but harbor unreacted isocyanate groups to help slowly crosslink the hot-melt adhesive after application by the moisture curing mechanism. The curing step can be expedited by preheating wood substrates, especially in bonding of thick wood laminates. RHMadhesives can be used where high green strength and high tensile strength of cured bonds are needed in comparison with traditional hot melts. It is reported that this new class of hot-melt wood adhesives shows significantly improved heat resistance and better room temperature creep resistance in bonding of wood.

16.5.2 A Comparison of a Thermosetting Hot-melt Against Other Types

In Figure 16.4 the rates of strength development and final strength values of four different wood adhesives are shown: (i) Jet-weld thermosetting urethane hot-melt; (ii) PVAc

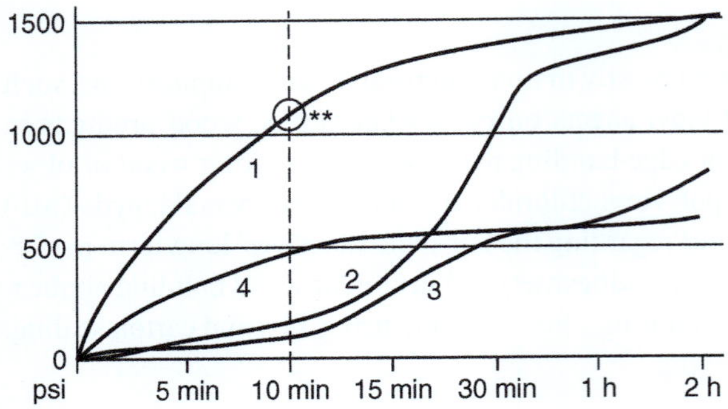

Figure 16.4 Comparison of strength development rate: **1.** Jet-Weld hot-melt. **2.** PVAc. **3.** 2-Part epoxy. **4.** Hot-melt.

emulsion adhesive; (iii) Two-part epoxy adhesive; and (iv) Regular EVA based hot-melt adhesive. [Provided by 3M Company.] The tensile strength build-up of the urethane-based hot-melt adhesive (3M Co. Jet-Weld) is fast, and the final strength is far higher than those of conventional hot-melts, approaching almost that of PVAc wood adhesive. In the bonding step, the residual isocyanate groups would come into contact with moisture in wood and cure to cross-linked polyurethane polymers. RHM adhesives are currently used in such bonding applications as metal-faced wood-based panels and floor laminates; window and door components; edge banding and vinyl wrapping of profiles; and furniture and automotive assembly. RHM adhesives must be protected similarly from the moisture from air until application, in cartridges or closed cans, and the wood substrates must have some moisture to supply moisture for curing reaction. Storage lives of these RHM adhesives would be limited and they can be higher priced than conventional hot-melts.

16.6 Key Variables in Hot-melt Edge-banding Operation

In addition to the various bonding operational variables of hot-melt adhesives discussed above, the bonding performance in high-speed bonding operations, such as in edge-banding, needs to be understood. The bonding performance in this case is closely related to the timing of adhesive solidification, in relation to the movement (transit) of the mated substrates in the pressurized zone, i.e. the passage time within the pressure zone, which tends to be short. While the specific heat of hot-melt adhesive (heat content) is nearly constant, the cooling rate of the applied adhesive layer should be at an optimum: not too fast and not too slow. It is obvious that the molten hot-melt should not cool too fast in order to allow the flow and wetting to occur on both substrates in the early part of the pressurized transit zone. The molten glue starts to cool as soon as it leaves the glue gun first by heat dissipation to air and then to the gluing substrates. In general, wood is a poor heat conductor and robs the heat slowly as long as the moisture content of wood substrate is kept low, and it is easy to see certain problems occur when the gluing speed changes from the norm. When the gluing speed slows down, the molten glue will have more time to cool before mating, resulting in poor bonding. Also, when the gluing speed becomes too fast, the mated substrates will go out of the pressurizing zone too fast before cooling down, resulting in a loosened mating. Also, the adhesive cooling will occur faster when the hot-melt spread rate becomes too thin because the glue is the only component that is hot and a less glue material means less heat on the given area of bonding. Also, when there is more depth for hot melt to penetrate into the substrates, such as pores or fuzzy fibers, the hot-melt can cool before the glue penetrates enough to give inadequate strength. Higher wood moisture content would cause similar problems, i.e. cooling the adhesive too fast and also reducing the wetting. Hot-melt adhesion failures in wood bonding can occur because of this type of fast cooling of molten glue due to the operating variables. The edge-banding or overlaying machines used in the industry carry out the bonding operation very efficiently (fast) and allow the various adjustments possible: line speed, glue spread rate, pressure control, transit time, etc. However, the fundamentals of wood gluing remain the same: clean uniform wood surface, sound wood surface structure, controlled wood moisture content, controlled constant room environment, etc., must be maintained.

16.7 Other Specialty Wood Adhesives

16.7.1 Mastic Adhesives

Mastic adhesives are made of elastomeric (thermoplastic) polymers such as natural or synthetic rubbers, and they are used in building constructions to bond wood members, such as bonding of plywood or gypsum board to wood framing members or to floor joists or bonding of staircases [10, 11]. Mastic adhesives have a paste-like consistency and are usually applied using a caulking gun or a trowel, and nails can also be used at the same time (clamping). The mastic adhesive layers solidify over several days and thereafter remain somewhat resilient for a long period of time, improving the fatigue resistance of the nailed/glued structure against repeated racking, vibrations, and impacts, due to the elastomeric nature of the polymers. Natural or reclaimed rubber, neoprene (chloroprene), styrene-butadiene (SBR), and polysulfide polymer rubbers are dissolved in organic solvents, or their emulsion (latex) forms in water are used with fillers.

Gluelines formed with mastic adhesives are generally thick and nonuniform, i.e. they are made in thick paste forms and have the gap-filling capability derived from the high levels of rubbers and fillers formulated in with low levels of organic solvents, necessary components to dissolve the rubber. The solvents evaporate out slowly to give the usual odor of mastic adhesives and this makes the gluelines somewhat rigid in the end. Mastic adhesives also contain tackifiers for the temporary "grab" property on wood surfaces, and some antioxidants are also added to prevent the rubber polymer molecules, which have some unsaturated (double) bonds, from oxidation and crack formation over time. The cure of mastic adhesives is achieved by the absorption of solvents into the wood substrates, which takes a few days in most cases. The gluelines were reported to have usually good peel strengths with low but stable shear strengths in the temperature range of −12 to 71 °C and the strength values increase markedly at lower temperatures [12]. Long-term loadings normally cause larger shear strength reduction than wood and the effect is greater at higher temperatures. Generally, use of mastic adhesives is restricted to below 150–200 °F (66–93 °C), but the creep of cured gluelines can occur even at room temperature under constant loads. Various mastic adhesives are available with different performance characteristics depending on the rubber used in the formulation.

Natural or reclaimed rubber-based mastics: solvent solutions or latex forms of natural rubber are used with excellent resilience and moisture and water resistance, but they become quite brittle with age and show poor resistance to organic solvents (to varnishes and lacquers to be applied later in the finishing). Excellent tack, good shear strength of 30–180 psi, and low cost (especially by using reclaimed rubber) are the advantages. Creep or failures occur under relatively low constant loads.

Neoprene (chloroprene) rubber-based mastics: solvent solutions or latex forms are used and they show a good resistance to water, salt spray, bio-deterioration, aliphatic solvents, and weak alkalis and acids, and give relatively good shear and peel strengths but not suitable for contact with aromatic and chlorinated solvents. They show creep tendency at shear loads over 2 psi, but are superior to other rubber-based mastic adhesives in shear strength, aging, resistance to sunlight, and other weathering factors, useful up to 200 °F (93 °C).

SBR rubber-based mastics: solvent solutions or latex forms are used and they contain plasticizers and tackifiers. SBR mastic adhesives give good heat-aging and water resistance properties, uniform appearance, and non-staining and light color. They are low cost, but the strength characteristics are poor with tendency to creep, and their bond strength is similar to those of natural rubber-based mastics.

Polysulfide rubber-based mastics: supplied as a two-part paste system to be mixed before use and they are resistant to water or organic solvents, greases, oils, salt water, aging, sunlight, and ozone, and their weathering properties are excellent and low temperature flexibility reaches down to −80 °F (−62 °C). Their high temperature resistance is poor, softening at 150–201 °F (70–94 °C) and they are not suitable for permanent load-bearing applications.

16.7.2 Contact Adhesives

Contact adhesives are also made of elastomeric polymers as main components and used in bonding, for example, Formica (MF resin-impregnated paper) sheets on wood blocks such as in manufacturing of tabletops [10]. Contact adhesives are made by dissolving various rubbers, mostly neoprene rubber, in high levels of organic solvents such as *hexane* and *methyl ethyl ketone* for low viscosity and easy spreading [13], and far less fillers are used in comparison with mastic adhesives. More uniform and thinner adhesive layers thus obtainable are suited for various overlay bonding applications, i.e. for bonding of large area surfaces. The low level of filler increases the peel strength due to greater amounts of elastomeric polymers in the glueline. The adhesive has to be applied on both pieces, and it is necessary to evaporate substantially all of the solvent in the adhesive layer before the adherends are mated, normally by standing them in open air until the glueline becomes almost dry to the touch and tacky. If solvents are incompletely removed and trapped in the glueline after mating of adherends, de-bonding problems can arise in use. After mating the adherends, a slight pressure is needed to make good contact between substrates, such as passing them through a low-cost pinch roll. One obvious advantage of using contact adhesives is the fast build-up of the bond strength, right after the mating of adherends with pressure, due to the absence of solvents and the adhesive polymers' tackiness and adhesion strength. No nailing or a long pressure period is necessary, but this adhesion process is more expensive than using casein or PVAc adhesives. Because of the nature of the adherends, i.e. a large/wide area of bonding with little chance in use of a load applied to pull them apart, bonds made with contact adhesives often last many years. However, the limited ranges of use temperature and other disadvantages are similar to those of mastic and other thermoplastic adhesives. One problem of contact adhesives has been the cost of solvent, especially the non-flammable-type solvent, and therefore, neoprene rubber-based water-borne systems have been increasing in use that need catalysts and an infrared-type heating step to expedite the drying process [14].

16.7.3 Epoxy Resin Adhesives

The principal component of many epoxy resin adhesives is the epoxy pre-polymers of diglycydyl ethers of bisphenol A (DGEBA) manufactured by reacting bisphenol A and epichlorohydrin with molar ratios of from about 1 : 1.2 to 1 : 10.0, resulting in varying decreasing "*n*' numbers in Eq. (16.1) with molecular weight in the range of from 1420 to 370 and melting points from 112 to 9 °C [15, 16]. Common clear liquid epoxy resins have a molecular weight of 400. Other epoxy resins such as novolac epoxy resins are also used. Epoxy groups are the reactive groups but they are rather stable or unreactive as is and become reactive only with the addition of curing agents. There are three (3) groups of curing agents that result in cured epoxy resin products used in different applications.

Example tertiary amine catalysts for epoxy resins and catalyzed polymerization reactions

$$\begin{aligned} &Ar - CH_2 - N = (CH_3)_2 \text{Benzyldimethylamine} \\ &(\text{BDMA})N = (CH_2CH_2OH)_3 \text{Triethanolamine} \end{aligned} \tag{16.1}$$

Various tertiary amines such as benzyldimethylamine and triethanolamine, which, with the aid of trace amounts of water or phenolic hydrogen donors, catalyze the reaction between epoxy groups (homo-polymerization), as shown in Figure 16.5, and therefore they are called as "catalytic curing agents" indicating their molecules are minimally incorporated into the cured resin structures. Thus, tertiary amines initially open up the epoxy ring to form amine (nitrogen–carbon) bonds and generate the oxide (anionic) groups, which attack another epoxy ring and make ether bonds, and this ether-bond forming reaction continues to form high molecular weight polymer molecules as shown. This curing procedure is used to obtain adhesives and coatings.

Various aliphatic and aromatic primary and secondary amines are used as hardener components and catalysts. The amine groups are reactive and, as soon as they are mixed, open up the (strained) triangular epoxy groups and form secondary or tertiary amine bonds as shown in Figure 16.6, and they are incorporated into the cured epoxy resin polymer structures and therefore change the cured resin structure and performance significantly. Various polyamines, which have more than one amino group in the molecule, such as diethylene triamine and phenylene diamine, are used. The reaction is accelerated by adding small amounts of hydrogen donors such as phenol, isopropanol, or water. The liberated hydroxyl groups from the opening of epoxy groups are not reactive under the reaction condition

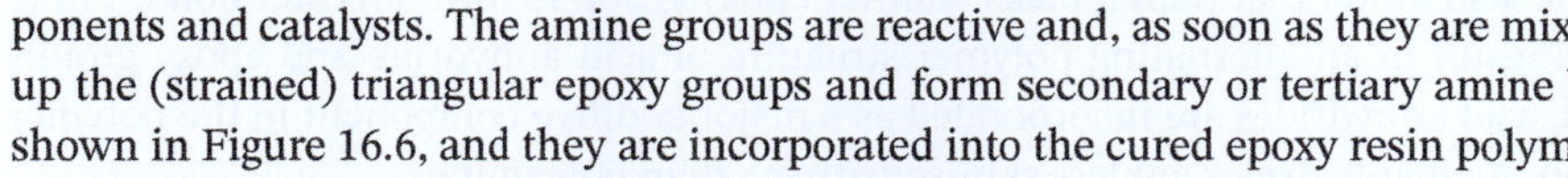

Figure 16.5 Chemical structures of diglycidyl ether of bis-phenol A. Source: Reproduced with permission from Saunders [16], © 1988, John Wiley & Sons.

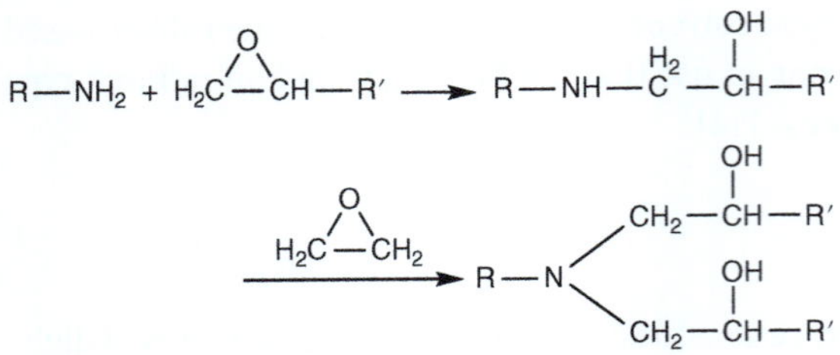

Figure 16.6 Chemical reactions involved in the curing of epoxy resin molecules with polyamines. Source: Reproduced with permission from Saunders [16], © 1988, John Wiley & Sons.

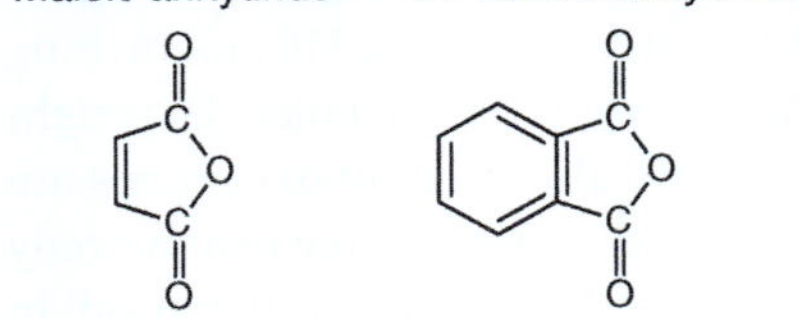

Figure 16.7 Example cyclic acid anhydrides used to cure epoxy resins.

even though some tertiary amines are generated as reaction products. Nonetheless, since polyamines have two more primary or secondary amino groups and all hydrogens on amino groups are reactive, the curing reaction at room temperature results in highly branched and crosslinked polymers through amine bonds. They are used in adhesives, castings, and laminating applications. Further polyamide materials made from diamines and dimeric fatty acids (tall oil obtained from wood pulping processes) are often used as hardeners, known to give flexible cured epoxy resin polymer adhesives, such as in bonding leather products. One deficiency of these amine procedures is the somewhat high toxic and skin-sensitizing nature of most polyamines.

Various cyclic acid anhydrides: such as maleic anhydride and *o*-phthalic anhydride (Figure 16.7), are also used as hardener components with a small amount of tertiary amine catalysts. Thus the anhydride ring opens up by tertiary amine to form tri-aralkylammonium cation and carboxylic acid anion and the latter anion react with an epoxy group to open up the ring and form an ester bond initially and results in alkoxide anion. Then, the alkoxide anion reacts with another acid anhydride molecule to form an ester bond and generate carboxylic acid anion that then attacks another epoxy group to form an ester bond. These reactions result in an alternating polymer structure of acid anhydride and epoxy groups and thus, acid anhydrides are incorporated as a major additive component in the polymer formation. Acid anhydrides are less skin-sensitizing than polyamines.

Commercial epoxy resin adhesives: Epoxy resin adhesives have been investigated as adhesives for wood laminations. Epoxy resins have excellent adhesive properties for wood and other materials due to the many hydroxyl, ester, and amino groups on the polymer structures. Thus, epoxy resin adhesives are two-part systems and cured at room temperature over a period of time or very quickly depending on the hardener component, but they are rather expensive for use in the manufacturing of common wood composite products. Epoxy resin wood adhesives would be suitable for bonding difficult-to-bond, valuable wood substrates, such as high-density woods or oily wood artifacts, or wood–concrete or wood–metal bonds. Cured epoxy resin adhesives do not have any volatile by-product and therefore cured products are rather stable and heat resistant, up to about 150 °C for amine cured and up to 200 °C for acid anhydride cured epoxy polymers.

Since many diamines and diols as well as several di-epoxy materials are commercially available, numerous epoxy resin adhesive packages are sold in the market for a variety of bonding requirements. Epoxy resin adhesives are supplied in two parts: the epoxy resin and the hardener, both of which can be liquid or semi-solid, and they should be thoroughly mixed before use. Furthermore, as with most adhesives, the commercial epoxy resin adhesives contain fillers, plasticizers, reactive diluents, solvents, elastomers, and various other additives. Epoxy resin adhesives can cure faster at elevated temperatures. The functional group ratios in the resin and hardener components are adjusted to give mixes the cross-linking thermosetting characteristics, i.e. three-dimensional, cross-linked polymer structures result when cured. In commercial products, equal volumes of resin and hardener are often used, commonly as two-syringe sets for household uses. Again, because of the higher cost and handling properties (high viscosity) and good performance characteristics such as creep resistance, the epoxy resin adhesives are used only in specialty wood bonding, such as in overlaying vinyl sheets on particleboards [17].

Are epoxy resin adhesives durable?: In the early days of wood bonding research [18, 19], various epoxy resins were investigated for bonding various wood species in comparison with resorcinol-formaldehyde resins and the conclusion was that the curing at higher temperatures resulted in better water resistance than the room temperature curing, but since there are so many different epoxy resins, hardeners, and catalysts available, the answer for the durability of wood bonding may be "yes" only selectively [20]. Some epoxies are used for boat or airplane construction and coupling agents such as the "hydroxymethylated resorcinol" discussed in Chapter 10 were reported to be useful to make epoxy wood bonds highly durable [21].

Comparison of an Epoxy Resin Adhesive with other Cold-set Adhesives in Wood Bonding: Engineered wood flooring (EWF) materials are products made by bonding a thin surface layer of various hardwood veneers (~3 mm-thick) to thick wood or wood composite board as substrates. In the reference paper [22], birch hardwood plywood, high-density fiberboard (~9 mm-thick), and OSB were used as substrates, and the surface-layers were sawn, sliced, or peeled hardwood veneers. The adhesives were cold-setting adhesives: isocyanate-modified PVAc (Hexion, XB-90 K5-LF), epoxy (Nacan, BONDMASTER E32), polyurethane hot-melt (Nacan, PUR-FECT LOK 34–777A), emulsion polyurethane (EPI, Hexion, WS799–56 L and CL-1), and moisture-curing polyurethane (Nacan, HB-22). The bonded wood products were subjected to aging up to 13 cycles, each cycle consisting of (i) 80% RH at 20 °C (1 day), (ii) 20% RH at 50 °C (1 day), (iii) 80% RH at 20 °C (3 days), (iv) 20% RH at 50 °C (2 days), and the cupping deflections were measured. The performance of substrates was in the order: hardwood plywood > high-density fiberboard > OSB, and the performance of surface-layers was in the order: peeled > sliced > sawn veneers. All adhesives performed similarly well except the epoxy adhesive, which has delaminated. The results show the relatively good performance of isocyanate-based resin adhesives as polyol emulsion forms, hot-melts, or moisture-curing systems. The variability problem for epoxy resins and the importance of choosing a right epoxy resin adhesive is noted.

References

1 Pizzi, A. (1994). Chapter 27. In: *Handbook of Adhesive Technology* (ed. A. Pizzi and K.L. Mittal), 451–457. Marcel Dekker.

2 Pocius, A. and Dahlquist, C. (1986). *Adhesion and Adhesives*, 266. Washington, DC: American Chemical Society.

3 Reischle, F., Winddhoff, J., and Bernig, W. (1986). Fusion adhesives which can comprise an isocyanate prepolymer, a thermoplastic polymer and/or a low molecular weight ketone resin. US Patent 4, 585,819. H. B. Fuller Co.
4 Markevka, V.C., Mackey, S.J, and Bunnellel, W.L. (1988). Thermally stable hot-melt moisture-cure polyurethane adhesive composition. US Patent 4, 775,719. H. B. Fuller Co.
5 Markevka, V.C., Zimmel, J.M., Messman, E., and Bunnelle, W.L. (1989). Thermally stable reactive hot-melt urethane composition having a thermoplastic polymer, a compatible curing urethane polyalkylenepolyol prepolymer and a tackifying agent. US Patent 4, 820,368. H. B. Fuller Co.
6 Gilch, H. and von Voithenberg, H., and Karl-Heinz, A. (1986). Adhesive composition. US Patent 4, 618,651. USM Corp.
7 Bowtell, M. (1987). *Adhesives Age*, 42. (9).
8 Markevka, V.C., Machay, S.J., and Bunnelle, W.L. (1990). Bonding method employing thermally stable hot-melt moisture-cure polyurethane adhesive composition. US Patent 4, 891,269. H. B. Fuller Co.
9 Higgins, E.D. (1995). *Forest Products Journal* 45 (10): 72.
10 Blomquist, R.F. (1973). Chapter 17. In: *Handbook of Adhesive Bonding* (ed. C.V. Cagle). New York: McGraw-Hill.
11 Vick, C.B. (1971). *Forest Products Journal* 21 (8): 34.
12 Hoyle, R.J. (1980). *Proceedings of 1980 Symposium on Wood Adhesives*, 133. FPL and WSU.
13 Clark, L.E. (1965). A practical guide to wood adhesives. *Forest Products Journal* 15 (4): 176.
14 Kania, W.C. (1980). *Proceedings of 1980 Symposium on Wood Adhesives*, 72. FPL and WSU.
15 Goulding, T.M. Chapter 33. In: *Handbook of Adhesive Technology* (ed. A. Pizzi and K.L. Mittal). New York: Marcel Dekker.
16 Saunders, K.J. (1988). *Organic Polymer Chemistry*, 2e, 412. London, and New York: Chapman and Hall.
17 Hespe, G.A. and Likly, C. (1990). *Wood Adhesives 1990* (ed. A.H. Conner, A.W. Christiansen, G.E. Myer, et al.), 166. FS, FPL.
18 Williamson, F.L. and Nearn, W.T. (1958). *Forest Products Journal* 8 (6): 182.
19 Olson, W.Z. and Blomquist, R.F. (1962). *Forest Products Journal* 12 (2): 74.
20 Caster, D. (1980). *Proceedings of the 1980 Symposium on Wood Adhesives*, 179. Madison, WI: FS, FPL and WSU.
21 Vick, C.B., Richter, K., and River, B.H. (1996). Hydroxymethylated resorcinol coupling agent and method of bonding wood. US Patent 5, 543, 487.
22 Blanchet, P. (2008). *Forest Products Journal* 58 (9): 37.

17

Casein, Soybean Flour, Animal Blood, and Lignin Wood Adhesives

17.1 Casein Wood Adhesives

Production and Formulation of Casein Wood Adhesives: Casein is the main protein component of cheese and is obtained from skimmed milk by precipitation with a mineral acid or an acid-producing bacterium [1, 2]. The bacterium used in cheese manufacturing converts the lactose in the milk into lactic acid, which precipitates the casein component. Casein is a high molecular weight polyamide-type polymers composed of various amino acids, and it is obtained by de-fatting cheese for industrial applications. The price of casein is directly tied to the price of food-grade cheese. Casein adhesives are often supplied as powder in sealed bags, consisting of about 60% casein, 10% fine wood flour, a crosslinking agent such as dimethyolurea, and basic catalyst components [1]. The powder adhesive is then mixed with water in the laminating plants in a ratio of 1 : 2 by weight to give a smooth white mix having a viscosity in the range of 4–8 k cps. These casein adhesive mixes are generally storage stable overnight. Once dispersed in water, casein polymer molecules swell and expand with the help of the base catalyst, sodium hydroxide, generated from reaction of the basic catalyst mixture with water. Casein polymer molecules must be opened up and expanded in water, so that the polymer molecular strands and branches to overlap each other well when laid on wood substrates and dried, before going into the curing process to form a continuous cured polymer film. It is probable that the many amide groups (—CO—NH—) in protein chains go into repositioning in such a way that hydrogen bond-type interactions (physicochemical crosslinking) occur among protein strands for the best-cured polymer structures. Small amounts of crosslinking agent such as dimethylolurea (a component of urea-formaldehyde (UF) resins) or dialdehyde starch added to casein also result in improved durability [2–5].

Application and Curing of Casein Adhesives: The adhesive mix is applied on wood substrates using a brush or beading application method. The casein adhesive mixes have long, adjustable pot-lives, are cured at room temperature or at slightly elevated temperatures [6], and have a better gap-filling capability than phenol-resorcinol-formaldehyde (PRF) resins, which means easier bonding for rough wood surfaces. The adhesive can be cured with and without heat input, but in bonding of yellow birch at 12% moisture content, the temperature of curing was found to affect the rate of curing in the range of 70–20 °F, decreasing as

Adhesives and Finishes for Wood: For Practitioners and Students, First Edition. Moon G. Kim.

temperature decreases, in which the dry strengths remain similar but the moisture resistance of glue joints decreased with the decreasing temperature of curing [5].

Uses of Casein Adhesives: These favorable adhesion/adhesive properties led to extensive uses in the industry for bonding nearly all of the interior-grade laminated lumber in 1960s [6, 7], and they were also used with a modification with soy flour in bonding of door skins to door frames [8]. Casein is still used in the wood laminated beam industry with comparable durability with PRF adhesive resins when used under dry conditions. It is known that as long as the interior condition is maintained, the strength properties of casein adhesive-bonded laminated beams are excellent for a long period of time [4]. Casein adhesives have been mostly replaced by PRF resin adhesives for exterior uses. Casein wood adhesives were also used in the past for hot-press bonding of softwood plywood.

17.2 Animal Protein-Based Wood Adhesive

Animal protein adhesives are based on high molecular weight protein materials (collagens) derived from animal skin, connective tissue, and bones [6]. The materials are separated, hydrolyzed, and isolated as dry granulated powder, cake, or jelly to be used after dissolving in hot water. They are also available in the liquid form at 35–65% solid levels with a viscosity range of 3–5 k cps for immediate applications using a spreader, brush, etc. The liquid form may contain fillers such as clay or calcium carbonate as well as fungicides and plasticizers. The warm applied animal protein adhesives in the glueline under pressure quickly form a tacky viscous film, which upon cooling a few degrees to room temperature passes to a firm gelled state, providing an immediate, moderately strong initial bonds. Subsequent natural or forced cooling results in strong permanent bonds with wood substrates, effective for use under dry, ambient temperature conditions, but they are prone to get enzymatic and bacterial degradations and not durable under severe moisture exposure conditions such as >85% relative humidity for extended period of time. Use of animal protein adhesives in bonding of large-scale wood products such as in furniture manufacture has been mostly replaced by polyvinyl acetate resin adhesives, but they are still used in bonding a wide range of small items, including toys, sporting goods, musical instruments, and other household goods.

17.3 Soybean Meal and Soybean Protein Wood Adhesives

Production and Formulation of Soybean Meal Wood Adhesives: Soybean meals are obtained as solid residues from the oil extraction process of soybean, consisting of 45–55% proteins by weight with the balance consisting of starch-type carbohydrates that do not have much durable bond-forming functionality [1, 6]. The general scheme of formulating soybean meal wood adhesives is similar to that of casein adhesives, i.e. dispersion in water with the help of sodium hydroxide. But the polymer structure and properties of soybean proteins are quite different because the constituting amino acids are different in kind and ratio and also the presence of a large portion of starch-type carbohydrate components. The soybean meals for wood adhesives must be processed at a relatively low temperature, below about 70 °C, and ground to a fine powder (325 mesh; 46 μm) because of the fast dispersion rates in water needed in the glue preparation stage. Thus, the soybean meal is first dispersed in water in the presence of a strong alkali such as sodium hydroxide to open up and expand the protein polymer molecules by breaking up internal hydrogen bond structures. But the alkali eventually destroys the protein structures, so that the pot-life of the mixed adhesive normally ranges only 6–12 hours. Sodium silicate is often added to improve the water resistance. Fungicide is normally added.

Uses of soybean meal wood adhesives: they were widely used in 1940–1960s as main adhesive components for bonding of interior-grade softwood plywood but were replaced by phenol-formaldehyde (PF) resin adhesives because of their improved water resistance properties and consistent bonding qualities.

Recent Research Activities on Soybean Meal Wood Adhesives: (i) Soybean meal adhesives have been investigated aiming at making the alkali-treated water dispersions more stable for use in bonding of particleboard or hardwood plywood for replacement of the formaldehyde-emitting UF resin binders. One key has been assumed to be finding the right treatment procedures for dispersing the soybean meal proteins in alkaline water: optimum high shear rates of mixing with stepwise increases of temperature and caustic, followed by optimum cooling to room temperature while stirring to result in stable, reactive mix. But the objective appears to have been met with difficulty in achieving reproducible products, especially on large scales [9]. The resultant adhesive needs to be stable for at least two to three weeks in order to be useful in industrial particleboard production. (ii) Various modification efforts on soybean meal or soybean protein isolates were reported using additives such as formaldehyde or using them as additives to alkaline PF resole resins [10–13] by adjusting the pH of the PF resin to the acidic side to obtain dispersion-type acid-curing PF resin adhesives [12]. In most cases, attaining the viscosity stability and retaining the reactivity of modified resins appears to be met with problems and there is presently no big need for modifying PF resole resins in industry. (iii) Addition of soy flour to resorcinol-formaldehyde (RF) finger-jointing adhesives was reported with no positive results [14]. (4) Industrial soybean protein isolates were tested as binders for fiberboards to result in low strengths of boards with low durability [15]. Industrial soybean protein isolates, soy flour, or casein were also tested as additives to acid-curing UF resins up to 30% with no reduction of formaldehyde emission from the cured resins [16]. The acid-curing condition of UF resins appears to be unfavorable to opening up of the protein polymer structures. (v) A study on the addition of soy flour to phenylene methylene-di-isocyanates (pMDI) resins for bonding oriented strand board (OSB) reported that 10–20% of soy flour can be added to pMDI, which gave similar bonding results except the higher edge-swelling values. Further, the increased viscosity of adhesive mix needed to be lowered by heating the adhesive mix to 40 °C or higher for application [17]. (vi) The addition of various protein dispersing agents such as urea, surfactants, guanidine hydrochloride, dicyanamide, and lithium perchlorate to soy powder formulations with and without 5% addition of polyamidoamine epichlorohydrin (PAE) condensation product was reported to help wood adhesion only minimally [18]. (vii) On the other hand, PAE has been reported to be a key material for formulating soy powder-based adhesives for bonding hardwood plywood with no formaldehyde emission potential, from Ashland Chemical Company, but the details were not reported [19]. (viii) Apparently, soy-modified polyvinyl acetate resin adhesives are available from Franklin International (Columbus, OH) and Momentive Specialty Chemicals (Columbus, OH) [20].

17.4 Animal Blood-based Wood Adhesives

Beef and hog bloods were used as wood adhesives in the past. The main component is albumin with molecular weights of about 69 000 Da [1, 6]]. Adhesive-grade animal bloods were supplied as spray-dried powders that have water solubilities in the range of 20–95%. The optimum water solubility range is 20–40 % for wood adhesive uses. Because of the relatively good water solubility in comparison to other protein adhesives discussed above, blood products can be dispersed in water with a milder alkali such as ammonium hydroxide. A small amount of para-formaldehyde or PF resole resin is normally added to denature and help cross-link the protein structures in the curing step. Blood adhesives were the most water-resistant adhesives available until the emergence of PF resins. Blood adhesives are also excellent adhesives for interior-grade softwood plywood, cured either by hot or cold pressing. They were used extensively in the past with the advantage of very short hot-pressing times. However, animal bloods are not much used currently due to various problems arising from variable qualities, fluctuating supplies, and inconvenience of collecting and spray-drying operations of animal bloods. A small volume of 80%-soluble spray-dried bloods is used as foaming agent [21], where the blood is added to the PF resin plywood adhesive mix at about 3% levels based on glue mix weight and put through a fast-stirring/air-injection system, resulting in a fine foamed glue mix that has a specific gravity of about 0.2. The foamed, so-called air-extended glue is then extruded onto the passing veneers, followed by assembling and hot-pressing to give exterior-grade softwood plywood. This method is currently used in some parts of the softwood plywood industry with reported adhesive cost savings of about 25%. Also, adding about 5% blood to PF resin adhesives was reported to improve the overall bonding performances of southern pine plywood [22].

17.5 Various Lignins and Uses in Wood Adhesives

Lignosulfonates are currently available in large quantities at about 25 cents per pound in the United States, a by-product of the sulfite pulping process of wood [23, 24]. The sulfite pulping process runs in such a way that the lignin by-products cannot be burnt for energy production, and the industry has to find other uses or dispose of them at a cost due to the sulfur content. The burden of disposal is one of the major reasons for the declining number of sulfite pulping mills. On the other hand, as of now, the lignosulfonate lignin is the only commercial lignin available in large quantities at a low cost in the United States and in the world, and they are currently used in many application areas in small quantities. Lignosulfonates consist of phenolic-type polymers having sulfonate ($-OSO_2^-Na^+$) groups derived from the pulping process and, therefore, they are available either as an ammonium, sodium, or calcium salt and readily soluble in water, which means they can have some hygroscopic property when added to wood adhesives. Currently, lignosulfonates are available from two pulping plants in North America (Georgia-Pacific/Bellingham, WA, and Tembec/Temiscuming, Quebec).

Kraft lignin is currently available from kraft-pulping processes possibly at about 35 cents per pound, similar price with phenol [25]. The high cost is due to the high cost of extraction of lignin from the black liquor of the kraft pulping process, achieved only by some expensive procedures. Kraft lignin polymers have no hygroscopic sulfonate groups and, therefore, they are soluble in water only with the addition of alkali such as sodium hydroxide because the phenolic hydroxyl groups present in lignin need to be ionized to make the polymer soluble. The unit price comparable with that of phenol discourages its use in PF adhesive resins.

Organosolv lignin was available in pilot-scale quantities from a company engaged in exploring the organosolv pulping process of wood [26, 27]. This pulping process uses ethyl alcohol and mild acidic catalyst as the pulping medium. In the course of recovering the solvent in the pulping process, the lignin is separated out as brown powder without any inorganic components (like kraft lignin) or sulfonate groups (like lignosulfonate) in it. The lignin yield is about 15–21% of the weight of wood used. The organosolv lignins have no ionic groups and are soluble in water only when an alkali is added, but the properties are different from those of kraft lignins because of the different pulping processes used. A large-scale commercial availability of organosolv lignin is not forthcoming, however. The organosolv lignins have the following characteristics: number average molecular weight of about 1000 with a poly-dispersity of about 3; it has a glass transition temperature of around 130 °C. Preliminary laboratory and pilot plant evaluation results indicated that about 30% of PF resin solids can be replaced with Organosolv lignin in bonding of OSB, particleboard (PB), and softwood plywood.

Formaldehyde Reactivity of Lignins: Lignins react with formaldehyde to the extent of about 0.4 (12 g) mole per 100 g of lignin, a level of about 1/3 of phenol. The reaction is similar to the

Reaction of formaldehyde with lignin [28]

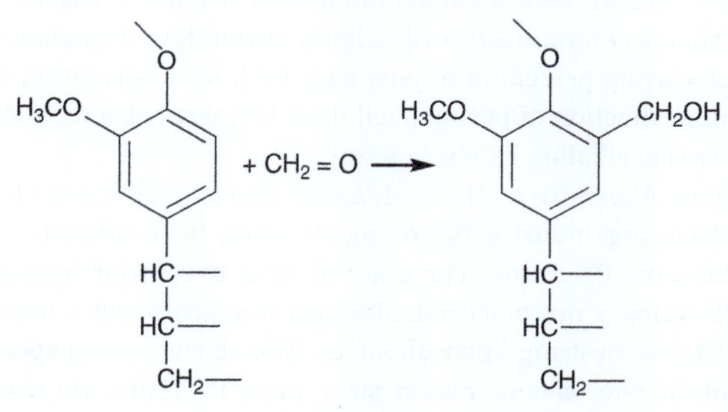

methylolation reaction of phenol in the PF resins synthesis as shown in Eq. (17.1) [29, 30]. This formaldehyde reactivity is the key to potential uses of lignin as additives in PF resins, although the amount of reactive methylol groups thus formed will be significantly lower than for the un-modified PF resole resins.

Research Efforts on Using Various Lignins in Alkaline PF Resole Resins: Much research has been done on possible uses of various lignins as an additive component of alkaline PF resole resins, focused on the methylolation of lignosulfonates or kraft lignins with formaldehyde and mixing the resulting products with alkaline PF resole resins, a sound approach based on utilizing the formaldehyde reactivity of lignins. However, because of the limited amount of formaldehyde reactivity, the methylolated lignins can be added in favorable cases only up to about 40% to PF resins [28, 30]. There is also an indication that these lignin-modified PF resins cure somewhat slower than unmodified resins. The following mill-trial experimental reports are noted. (i) *Organosolv lignins*, after methylolation with formaldehyde, were reported to be useful with powder PF resole resins added up to 38% replacement with a benefit of reducing the formaldehyde emission levels from the hot press [27]. (ii)*Sodium lignosulfonates*, 10–17%, were cooked in alkaline PF resole resins and spray-dried and the resultant powder resins were used in bonding of 6.4 mm-thick single-layer OSB and 11.1 mm-thick 3-layer OSB in actual OSB production lines for one month without any problem and the resultant bond strength values passed the Canadian standards (CSA) for OSB panels but were somewhat lower than control PF powder resins [31]. (iii) *Kraft lignins* were investigated by methylolation in alkaline pH, resulting in satisfactory bonding of OSB when the wood flakes were pre-treated with 3N sodium hydroxide solution [32]. (iv) *Kraft lignins* were also investigated by methylolation in alkaline pH followed by acidification to pH 5 and mixing the lignin product with acid-curing PF resole resins [33], which resulted in satisfactory bonding of OSB. (v) *Calcium lignosulfonates* (Lignosol Co., Québec) were also extensively investigated. After acidification with sulfuric acid followed by removal of calcium sulfate, the lignin product is added with more sulfuric acid and spray-dried to a powder lignin product and tested as a sole binder for bonding of particleboards [34]. At 10% binder loading levels, an acid level of 0.9–1.0% weight based on binder weight was found necessary to cure to yield satisfactory boards [34]. (vi) *Ammonium lignosulfonates* (Temfibre Co., Québec) were extensively investigated: an acidic dispersion mixture made by mixing 30% lignin with an alkaline PF resole resin followed by acidifying the adhesive mixture to pH 4 resulted in satisfactory bonding of waferboards, but the strength values decreased rapidly at higher pH values [35]. Resin adhesives that need to be cured by acidic catalyst would be difficult to be accepted in industry without a long-term durability assurance. (vii) *Other lignin or tannin materials*: Lignin/tannin-type degradation materials obtained from an autoclaving procedure of pine bark [36], from separation of wood pyrolysis oil [37], and from extraction of peanut shell flour [38] were also reported to be useful as partial replacements of alkaline PF resole resins.

Current Status of Using Lignin Materials in Wood Adhesive Resins [39]: Currently, however, the PF resin manufacturing industry is not much using lignosulfonate or Kraft lignins in most wood adhesive PF resins. There are at least two major reasons why lignin incorporation in PF resins is not practiced, although it appears that a lower adhesive cost can be realized in case of using lignosulfonates. The slower curing speed of lignin-incorporated PF resins is one obvious reason since most PF resins are used in manufacturing of commodity-type wood composite products, where the production volume greatly affects profit margins of the plants. Secondly, the available lignosulfonates lignin at a reasonable price is relatively inconsistent in quality: the formaldehyde reactivity or the degree of polymerization is not consistent from one lot to the next, resulting in variability of resultant resin properties. This variability would add to the manufacturing cost of resin, and the incorporation of lignin at about 40% substitution level does not bring down the total manufacturing cost much at all.

Use of Lignins in Wood as Thermoplastic Binder: A new method development in wood bonding is the mechanical fusion method, i.e. rubbing wood pieces against each other under pressure till the surfaces melt, followed by letting the glue line set to a solid bond line [40]. Thus, *beech wood* pieces were bonded using the following parameters: wood dimensions of 150 × 20 × 15 mm; wood moisture content: 12%; frequency of vibration: 100 Hz; welding pressure: 0.4–2.3 MPa; vibrational amplitude: 3 mm; contact holding time: 2–5 seconds; contact holding pressure: 1.33–2.7 MPa; bond line maximum temperature attained: ~170 °C; followed by equilibration of bond at 65% RH for a week. Bond strength obtained: 7.61 vs. 10.0 MPa required. Water resistance of the bond was relatively poor as of the year 2004. The authors observed that the bond line is formed by melting and setting principally of the lignin and hemicellulose components with the cellulose fibers not directly participating. Solid-state ^{13}C NMR studies of the bond line materials indicated that some of the lignin and hemicellulose components do appear to make new crosslinking bonds, indicating that it is possible to develop the method further for stronger, durable bonds.

References

1 Lambuth, A. (1994). Chapter 13. Adhesives for wood. In: *Handbook of Adhesive Technology* (ed. A. Pizzi and K.L. Mittal), 259–282. Marcel Dekker.
2 Weakly, F.P., Ashby, M.L., and Mehltretter, C.L. (1963). *Forest Products Journal* 13: 51.
3 Blomquist, R.F. (1963). *Forest Products Journal* 13 (2): 50.
4 Bergin, E.G. (1965). *Forest Products Journal* 15 (4): 152.
5 Raknes, E. (1968). Long-term durability of structural adhesives for wood. The Norwegian Institute of wood-working and wood technology. Meddelelse NR. 31:1-16.
6 Blomquist, R.F. (1958). Developments in glues and gluing – 1957. *Forest Products Journal* 8 (2): 45.
7 Clark, J. (1965). *Forest Products Journal* 15 (4): 176.
8 Sellers, T. Jr., (2001). *Forest Products Journal* 51 (6): 12.
9 Thames, S.F., Cook, R.C., and Mendon, S.K. (2004). Soy protein-based adhesives containing a vegetable oil derivative. US Patent 6, 790,271. September 14.
10 Hse, C.-Y., Fu, F., and Bryant, B.S. (2001). Development of phenolic resins co-reacted with soy flour hydrolyzate. In: *Extended Abstracts of Wood Adhesive 2000 Conference*, 3. Madison, WI: FPS.
11 Kuo, M., Myers, D., Heemstra, H. et al. (2001). Soybean-based adhesive resins and composite products utilizing such adhesives. US Patent 6, 306,997.
12 Westcott, J.M., Traska, A., Hart, C.R., and Lorenz, L. (2005). Durable soy-based adhesive dispersions. In: *Wood Adhesives 2005*, 263. FS: FPL.
13 Lorenz, L., Frihart, C.R., and Wescott, J.M. (2005). Analysis of soy flour/phenol-forrnaldehyde adhesives for bonding wood. In: *Wood Adhesives 2005*, 501. FS: FPL.
14 Vrazel, M. and Sellers, T. (2004). *Forest Products Journal* 54 (3): 66.
15 Kuo, M., Adams, D., Myers, D. et al. (1998). *Forest Products Journal* 48 (2): 71.
16 Lorenz, L.F., Connor, A.H., and Christiansen, A.W. (1999). *Forest Products Journal* 49 (3): 73.
17 Cheng, Q., Essien, C., Via, B., and Banerjeo, S. (2019). *Forest Products Journal* 69 (2): 154.
18 Frihart, C.R. and Lorenz, L. (2013). *Forest Products Journal* 63 (3/4): 138.
19 Li, K. (2007). Formaldhyde-free lignocellulosic adhesives and composites made from the adhesives. US Patent 7, 252,735.
20 Frihart, C.R. (2011). *Forest Products Journal* 61 (1): 4.
21 Cone, C. (1969). *Forest Products Journal* 19 (11): 14.
22 Fischer, C. and Bensend, D.W. (1969). *Forest Products Journal* 19 (5): 32.
23 Glennie, D.W. (1971). Chapter 15. In: *Lignins, Occurrence, Formation, and Structure and Reactions* (ed. K.V. Sarkanen and C.H. Ludwig), 507. New York: Wiley.

24 Shen, K.C. and Calvé, L. (1980). Exterior adhesive from sulfite mill effluent. In: *Proceedings of 1980 Symposium on Wood Adhesives* (ed. W.E. Johns and R.H. Gillespie), 54. FPL and WSU.

25 Marton, J. (1985). Chapter 16. In: *Lignins, Occurrence, Formation, Structure, and Reactions* (ed. K.V. Sarkanen and C.H. Ludwig), 639. New York: Wiley.

26 Lora, J.H. and Aziz, S. (1985). *Tappi Journal* 68 (8): 94.

27 Senyo, W.C., Creamer, A.W., Wu, C.F., and Lora, J.H. (1996). *Forest Products Journal* 46 (6): 73–77.

28 Sellers, T., McGinnis, G.D., Ruffin, T.M., and Janiga, E.R. (2004). *Forest Products Journal* 54 (9): 45.

29 Gratzl, J.S. (1979). Potential of technical lignins as extenders in phenolic resins: an assessment of reactive sites. Weyerhaeuser Science Symposium II, Weyerhaeuser Co., Tacoma, WA. 291.

30 Wooten, A.L., Sellers, T., and Paridah, T. (1988). *Forest Products Journal* 38 (6): 45.

31 Calve, L., Shields, J.A., and Sudan, K.K. (1991). *Forest Products Journal* 41 (11/12): 36.

32 Krzysik, A. and Young, R.A. (1986). *Forest Products Journal* 36 (11/12): 39.

33 Dolenko, A.J. and Clarke, M.R. (1978). *Forest Products Journal* 28 (8): 41.

34 Shen, K.C. and Fung, D.P.C. (1979). *Forest Products Journal* 29 (3): 34.

35 Calve, L.R., Shields, J.A., Blanchette, L., and Frechet, J.M. (1988). *Forest Products Journal* 38 (5): 15.

36 Zhao, Y., Yan, N., and Feng, M. (2016). *Forest Products Journal* 66 (1/2): 19.

37 Mao, A., Xu, W., Xi, E. et al. (2018). *Forest Products Journal* 68 (2): 113.

38 Chen, C.M. (1982). *Forest Products Journal* 32 (11): 14.

39 Sellers, T. Chapter 37. Adhesives in the wood industry. In: *Handbook of Adhesive Technology* (ed. A. Pizzi and K.L. Mittal), 605. Marcel Dekker.

40 Gfeller, B., Lehmann, M., Properzi, M. et al. *Forest Products Journal* 54 (7/8): 72.

18

Theory and Practices of Adhesive Bonding for Wood

18.1 Formation of Interphase in Wood Adhesive Bonds and Failure Modes

18.1.1 Interphase Present Between Wood and Adhesive Layer in Adhesive Bond

The adhesive layer (glueline) between two solid (wood) surfaces holds the bonded product against the internal and external pressures, load of use, intrusion of water, etc. In the overall scheme of adhesive bonding of wood or any other material, it is a given fact that the adhesive layer's *cohesive strength* and *adhesive strength to the substrates* must be as strong as, or stronger than, the strength of substrates they hold. Thus far, we have discussed on how the wood adhesive molecules grow, cross-link, and cure to a solid cohesive structure of glue layers. Though, it has been also understood that the adhesive layer, over the course of curing, has to create its surface that has the necessary characteristics for a good adhesion to occur with the surface of wood substrate. This adhesion between the glue layer and wood surface arises from complex physical and chemical attraction forces involving the atomic contacts between them as well as various macroscopic entanglement factors. This layer where these various attractive forces occur can be defined as an interphase (note that it is not interface) composed of adhesive and substrate molecules intermixed. Details of these attractive forces will be discussed in Chapter 19.

In this lecture, we will first concern ourselves with the wood adhesion phenomena at the macroscopic and practical levels. Thus, there are three phases in a wood adhesive joint: *wood phase*, *interphase*, and *adhesive phase*, as presented by Marra [1]. A phase means a uniform region of material composition, separated by a boundary from other types of regions. Phase separations come to exist due to the different materials structures. A simple example will be the ice water in a glass, where the ice phase and the liquid water phase can be clearly differentiated. In wood adhesive bonds, we can readily see wood and adhesive layer phases, but the interphase region is often very thin and does not show a clear picture of composition, but a uniform appearance and structure can be imagined for the interphase. Although this interphase cannot be readily identified by visual inspection, a bond failure that occurred neither in the adhesive layer nor in the wood layer can be concluded to have occurred in the interphase.

Adhesives and Finishes for Wood: For Practitioners and Students, First Edition. Moon G. Kim.

18.1.2 Failure Modes in Mechanical Testing of Wood Bonds

In practice, when a wood bond is stressed to break along the glueline, the breakage can occur through all three phases: interphase, adhesive phase, or wood phase, and the failure mode can be defined as: *Interphase region – adhesion failure*; *Adhesive layer phase – adhesive or cohesive failure*; and *Wood phase* failure – *wood failure*. The breakage that occurs only through the wood phase will satisfy the glue supplier and glue operator because then the glueline made is judged to be stronger than the wood substrates. This statement will be justified more readily if the wood substrates used have no more defects than the average wood the manufacturer uses in the production. On the other hand, more often than not, the breakage can occur in the interphase, which indicates that some unfavorable physical and chemical processes have occurred because of an ineffective glue or a poor gluing/curing procedure. Also, the failure occurring in the glue layer itself indicates that the blame has to go to the glue supplier and glue applicator. In practice, however, the adhesion and adhesive (cohesive) failures are often difficult to differentiate, and therefore they are lumped together as the *adhesive/adhesion failure* vs. the *wood failure*, which can be measured by eyeballing the area of wood failure on broken samples and reported in percentages based on the total glued area.

18.1.3 Causes of Adhesive-layer or Interphase Failures in Wood Adhesive Bonds

We discussed in Chapters 3 and 4 the thermosetting/thermoplastic *glueline's curing problems:* under-cure, over-cure, dry-out, starvation, and pre-cure. These problems often lead to adhesive/adhesion failures. Under-cures often occur when the curing temperature and/or the curing time are inadequate and over-cures occur under the opposite conditions. Furthermore, under-cures or over-cures can be caused by catalysts and extra acidity, and high buffer capacity of wood for UF or PF resins. Adhesive glueline's dry-outs are due to the evaporation of too much water from gluelines during the assembly times because of the gluing environment being too dry or too hot, or the assembling time being too long, which makes the adhesive layer dry out or advance before the full compaction of substrates. All these will result in an inadequate flow in the hot-pressing and weak gluelines. Glueline starvation occurs when the adhesives penetrate into wood pore structures too much or exude too much in pressing, leaving inadequate amounts of adhesive solids in the glueline. Wet wood, short assembly times, low molecular weight adhesives, and adhesives that cure too slowly can result in squeeze-outs. Porous wood structures can also cause glueline starvation from an excessive absorption of adhesive solids. Most thermosetting adhesives are composed of molecules of varying molecular weights; therefore, they can be stratified in the course of penetration into wood (low molecular components penetrate more severely) and set in that manner, forming gluelines that can lack in some essential curing components and result in a weak link. In the case of thermoplastic adhesives, an inadequate coalescing of polymer particles can lead to adhesive failures.

Interphase failures occur often due to the same reasons as those associated with adhesive glueline failures mentioned above. On the other hand, the pre-cure and dry-out problems can be often visually identified as failures of the interphase that did not form completely due to the advancing or drying of the adhesive polymers before making full contact with substrates. Furthermore, the inadequate adhesive bonding or curing phenomena due to wood acids or wood extractives can occur more often in the interphase because the extraneous chemical species originating from wood first diffuse out into the interphase ahead of or during the assembly and curing period.

18.1.4 Possible Further Divisions of Interphase in Wood Adhesive Bonds

As implied above, the interphase in wood adhesive bonds may not be uniform, i.e. consists of two or more similar but slightly differing phases (layers). Changes of composition and molecular structure in the interphase can occur to form various compositionally different mini-layers with different strengths [1]. For one wood species, when the adhesive penetrates into wood surface structures in forming of interphase, a fractionation of the adhesive chemical components can occur: water and inorganic and small resin molecules can preferentially penetrate, leaving the larger resin molecules behind. Setting of the adhesive bond will then lead to form these varying mini-interphase layers within the interphase layer. Thus, the interphase link we assumed to be a uniform layer to begin the discussion can be expanded into several different mini-interphase layers depending on our point of interest in looking into the failure modes and causes. For example, if the wood acid effect mentioned above comes into play within the interphase, one possible scenario is that a thin top layer of interphase is affected the most and the next layers of interphase are affected less and less. Also, wood extractive chemicals often move to wood surfaces during drying or hot-pressing, resulting in making contact and reacting with adhesive components. These extractive chemicals, such as fatty oil, wax, and terpenes, existing on wood surfaces would act as extraneous materials participating in a destructive manner in the formation of interphase, causing discontinuities between the wood surface and the glue layer proper. Extraneous materials derived from wood preparation procedures can be similarly incorporated into the interphase as nonparticipating adhesive constituents; dirt, wood fines, cutting fluids, burned surfaces, deactivated surfaces, loose fibers in weakened wood structures due to dull knives or sanding, low wood (composite) density areas, etc. You now see why the interphase can be, at least conceptually, differentiated further into various mini-layers. Although the various distinctive mini-interphases are difficult to examine visually or analytically, this differentiation can greatly help operators/researchers in their efforts to find the causes of failed wood adhesive bonds. At least, three (3) mini-layers in the interphase have been proposed to be necessary for differentiating interphase failures in wood bonds [1].

18.1.5 Adhesive Failure vs. Wood Failure in Testing of Adhesive Bonds

In practice, the adhesive layer and interphase failures are lumped together as the adhesive/adhesion failure vs. the wood failure, observed in the broken adhesive-bonded samples under both dry and wet or boiled conditions. Commonly, the *wood failure* values (%) are reported with the balance being the adhesive/adhesion failure values. The breaking strength values are also reported. High wood failure values indicate that the adhesive and interphase layers are properly made, and wood failure values obtained under wet conditions indicate a measure of the water resistance of the glueline. In a set of bonded samples from same panels or lots, the breaking strength and wood failure values often vary and, therefore, average values of multiple samples should be used as the representative values.

A commercial, bonded wood product must show wood failure and strength values above the minimum required values, for example, 85% and 110 psi, etc. A confusion can occur when one of these two values is slightly lower than the required value and the other value is significantly higher than the minimum value. Assuming that the composition of the supplied adhesive is constant, the wood failure value can fluctuate daily by the various adhesive application process parameters. If the strength of wood substrates is assumed to be unchanging, it is expected that as the wood failure value increases, the bond strength value would increase, up to the strength of the wood substrate. This is what is commonly observed, i.e. one cannot obtain a good wood failure value and an exceptionally low strength value at the same time. If such a trend is not observed, the changing quality of wood substrate is implicated, i.e. weak wood surface layers giving shallow breaks with low strength values and high wood failure values. On the other hand, the wood substrate could have an unusually high strength and then, even when the bond is made well, the wood failure value tends to go lower while the breaking strength goes higher. This trend can occur in bonding of strong wood species, such as oak. On the other hand, in softwood plywood tests, veneers often contain lathe checks and other disturbed surfaces and tend to be weaker than well-made solid wood surfaces. Therefore, the wood failure values in plywood tests tend to go higher and be more consistent, and the breaking strength values are somewhat lower than those of solid wood planks or laminated beams. Thus, wood failure values and strength values should be considered together because of this relationship: weak woods tend to give high wood failure and low strength values, while strong woods tend to give lower wood failure values and high strength values.

A practical note regarding the wood/adhesive failure values is that the damage done to a victim by a wood bond failure would be the same as whether it occurred from a weak strength of wood or a glueline failure, but the particular failure mode can often be identified and traced back to the cause of failure by inspecting the failed wood products. The results would be useful in deciding who would be liable for the damage and also in improving the adhesive/bonding process technique. Since the damage can often be costly, the wood product manufacturer and adhesive supplier try to pinpoint the cause of failure, whether it is the adhesive or something else. Overall, the manufacturer of bonded wood products must use sound wood in the first place, and the next is a sound understanding of the adhesive and its adhesion process.

18.1.6 Monitoring is a Must: Adhesive Quality, Adhesion Operation, and Bonded Products

The adhesive and interphase layers in bonded wood products are almost inseparable. They are both very thin and their volumes amount to only a small fraction of the bonded product. If the interphase region is to be isolated for analyses, chemical and physical alterations are likely to occur. Also, the chemical analytical techniques available to inspect the interphase *as is* are currently inadequate. The general knowledge discussed above is deduced from past basic research and experiences. In the past, adhesion failures and mechanisms have been studied for various adhesive/substrate pairs, and several bonding mechanisms were proposed based on the physical and chemical phenomena. Each proposition tends to explain one type of bonding system better than other types, and these adhesion mechanisms will be discussed in Chapter 19. Here, it should be recalled that most current industrial wood adhesive bonds are made based on the principles proven to give satisfactory adhesive bonds only under certain restricted or optimum gluing conditions. The adhesion procedures, however, can deviate often without due cause, and one of the main deviations has been found to be human errors, such as mixing inappropriate amounts of adhesive components or simply missing or applying less than the adequate amount of adhesive in a continuing process due to malfunction of machine or operator errors. Once an error is made somewhere in the adhesive/adhesion process and not corrected on the spot, it is extremely difficult to find it later by the operator or by anyone else. The author observed one most apparent reason for failure in a court case: no adhesive was found in the glueline. Most adhesive-bonded wood products need to be suspected until they are proven to be good, which is a difficult thing to do without breaking the products, since there is no sure way to evaluate the glueline and interphase directly. However, the past history tells us that a continuous monitoring of the adhesive and gluing operations and measuring of strength and wood failure values of samples taken often and comparing them against standard values can assure the adhesion quality of the products. That is, a good quality control program is essential. Also, wood product users, especially when the bonded products carry a heavy loads, such as laminated beams and I-beams, due attention should be paid to any changes occurring in the bond lines over the long years of use.

18.2 Wettability of Solid Surface, Contact Angle, and Surface/Interface Tensions

It is now apparent that the interphase holds the key to a good bond strength and good adhesion and that the adhesion or adhesive bonding arises between two distinctively different materials because the atomic attractive forces between them are satisfied, as discussed in Chapter 19. In this chapter, we will first look at these questions of mutual attraction from the classical, phenomenological point of view by looking at the *interphase* as an *interface*, where the wood and adhesive surfaces are forced to interact either in an attractive or repulsive way.

Good wetting and contact angle: It has been known that a close contact attained between two surfaces is essential to have a good bonding as can be surmised from the fact that most glues are liquid and gluing operations need a pressuring of the two component pieces. In classical terminology, a *close contact* between a liquid and solid material occurs when the liquid *wets* the solid surface. Good wetting often occurs when a drop of water is put on a piece of wood, or more spectacularly when the lead is melted between two copper pieces with the help of soldering acid. This concept of wetting, first of all, presupposes that one of the materials must be a liquid, i.e. solid materials cannot change shapes to wet another solid material. This is the reason why adhesives should be a liquid. But can then any liquid wet any solid material well? No. The extent of wetting is different for different pairs of liquid and solid material and this relationship is described by the wetting (or contact) angle concept. In Figure 18.1, a drop of a liquid, such as water, is put on a piece of (impenetrating) solid material [2]. The drop of water on the wood surface will not flatten as shown, at least for a few minutes before absorption. On the other hand, a molten paraffin drop on wood surface would behave like a ball. This different phenomenon is explained in terms of surface tensions of liquid and solid materials and their interactions.

Surface Tensions of Liquid and Solid Materials: Surface tensions (force or energy) of liquid materials, such as a water drop, arise in its surface layers because the top-most surface

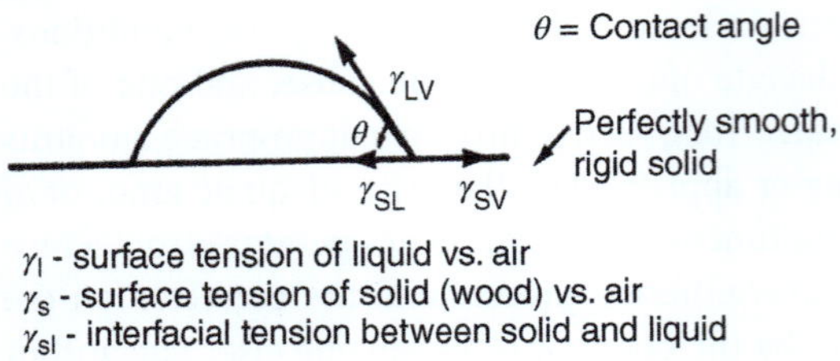

Figure 18.1 Schematic diagram of a liquid droplet on flat solid substrate (Letter V for air is omitted in titles for simplicity). Source: Pocius and Dahlquist [2] Reproduced with permission of the American Chemical Society.

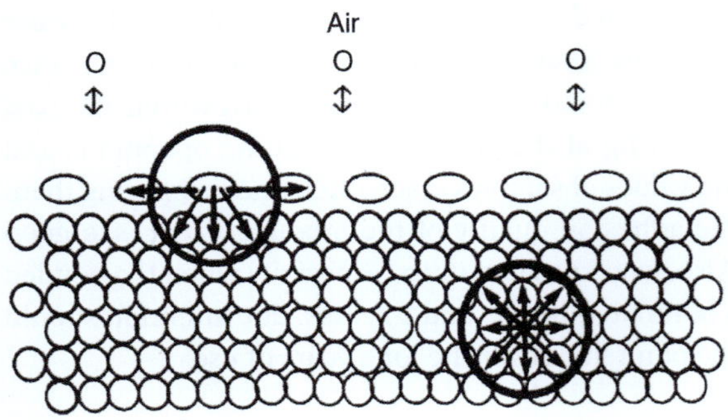

Figure 18.2 Schematic diagram showing the presence of surface tension: Surface molecules lack in the surrounding stabilizing attraction forces compared with the ones in the middle of liquid. Source: Pocius and Dahlquist [3] Reproduced with permission of the American Chemical Society.

molecules lack in the stabilizing attraction forces among themselves (with stretched bonds between molecules) compared with others existing in the middle of liquid, as shown in Figure 18.2 [3]. The air molecules only partly satisfy the attraction forces. The imbalance of attracting (stabilizing) forces on molecules in the top-most surface layer finds its way to exert extra-attracting forces on the neighboring surface molecules to result in surface tension forces, resulting in helping the surface area unbroken toward forming a ball as in the case of a water drop thrown into the air. Solid materials cannot change shape, but the surface tensions nonetheless exist because the molecules in the outermost surface layers are only partly stabilized in comparison with molecules in the deeper layers. (The surface tension energies arise from the hydrogen bonding, dipole–dipole interaction, and dispersion interaction forces discussed in the next Chapter 19.) For example, the surface tension (force) of water is 72.2 mN (milli-newton), wood surfaces is in the range of 40 – 55 mN [4], that of paraffin oils and common organic solids is in the range of 20 ~ 35 mN [5], and that of mercury is ~485 mN (https://en.wikipedia.org/wiki/Surface-tension_values). Surface tension forces can be measured or calculated. The interest of adhesive scientists is on the interaction between the surface tensions of a solid (wood) and a liquid (adhesive) arising when they are put together, i.e. whether the interaction results in an attraction or repellation, and then how much, etc.

Relationship between the contact angle (θ) and the interfacial tension (γ_{LV}) and the solid surface tension (γ_{SV})

$$\gamma_{SL} = \gamma_{SV} - \gamma_{LV} \times \text{Cos}\,\theta \tag{18.1}$$

$$\gamma_{SV} - \gamma_{SL} = \gamma_{LV} \times \text{Cos}\,\theta \tag{18.2}$$

$$\text{Cos}\,\theta = (\gamma_{SV} - \gamma_{SL})/\gamma_{LV} \tag{18.3}$$

Example calculation:

if Cos $\theta \sim (\gamma_{SV} - \gamma_{SL})/\gamma_{LV} \sim 0.00$; then $\theta \sim 90°$ (γ_{SL} is as large as γ_S)

if Cos $\theta = (\gamma_{SV} - \gamma_{SL})/\gamma_{LV} = 0.25$; then, $\theta = 75.5°$

if Cos $\theta = (\gamma_{SV} - \gamma_{SL})/\gamma_{LV} = 0.90$; then, $\theta = 25.8°$

if Cos $\theta \sim (\gamma_{SV} - \gamma_{SL})/\gamma_{LV} = 1.00$; then, $\theta \sim 0°$ (γ_{SL} is small and γ_{LV} equals to γ_{SV})

Interfacial Tensions Between a Liquid and Solid: [2] It is easily recognized that when two such tensioned surface layers of adhesive and substrate molecules of Figure 18.2 are brought together in a bonding, the molecules meet new neighbors, interact with their respective intermolecular forces, and come to a term. Attractive interactions will cause the liquid adhesive drop to flatten (low contact angles), and repelling interactions will result in curling-up of the liquid drop (high contact angles). The result of this attractive/repelling interaction comes out as *the interfacial tension* (γ_{SL}), which cannot be measured directly, but measured indirectly by measuring the wetting or contact angle (θ) with a goniometer, as shown in Figure 18.1 and Eq. (18.1). Thus, a contact angle established for a liquid drop put on a solid surface is a compromising result of attractive and repelling forces among the molecules in their upper-most surface layers, and the resultant value is a measure of the repelling force, i.e. a higher value means poor adhesion or vice versa. The air also interacts with the solid and liquid drops. In other words, the resin adhesive and solid material under air have their own surface tensions, and when they are brought to meet (by gravitation and external pressuring), the molecules of solid with a surface tension (γ_{SV}) and those of liquid with a surface tension (γ_{LV}) interact to reach to an equilibrium wherein each other's attractive/repelling forces are mutually satisfied or compromised by rearranging their relative molecular positions as much as possible, i.e. reaching to *the state of least interfacial tension* (γ_{SL}). Naturally, the magnitude of interfacial tension is mathematically defined via the contact angle by the Young's equation [2] shown in Eq. (18.1) and in rearranged forms, in Eqs. (18.2) and (18.3).

These equations indicate that interfacial tension values between solid and liquid materials of known surface tensions are obtained *through* a measurement of the contact angle. Thus, as shown in example calculations, when the contact angle is large, i.e. $\theta \sim 90°$or Cos $\theta \sim 0$, the interfacial tension (γ_{SL}) becomes as large as the solid surface tension (γ_{SV}) from Eq. (18.1), which gets worse as the liquid surface tension (denominator in equations) becomes larger from Eq. (18.3), for example, for a drop of mercury, a case where the molecules are in a fully repelling state. On the other hand, when the contact angle is small, an ideal case of good wetting, it means that the liquid's surface tension (γ_{LV}) is close to the solid's surface tension (γ_{SV}) and the interfacial tension (γ_{SL}) becomes small, which gets better as liquid surface tension (denominator) gets larger. Thus, since surface tensions of wood range from 40–55 mN [2], the surface tensions of wood adhesives should be in the similar range to have the interfacial tension in the low range in order to have a good wetting and low contact angles. Further, it is noted that materials that have very high or very low surface tensions, such as mercury or paraffins, will not be making small contact angles and small interfacial tensions with various woods and wood adhesives. Thus, contact angles and interfacial tensions describe the attraction and repulsion forces between the molecules of the solid and liquid materials, i.e. high attractions result in low contact angles or *low interfacial energy or good bonding*, and this occurs only between materials of similar surface tensions.

Contact Angles of Wood Surfaces vs. Good Bonding: For example, a water drop on wood surface would spread well, making a small contact angle, to indicate that the interfacial tension between the water drop and wood surface is small. On the other hand, a paraffin oil drop would not spread well, making a large contact angle, to indicate a large interfacial tension. For PF resole resins put on wood, it was reported that as the sodium hydroxide content increases, the contact angle decreases [6], and also the latewood of southern pine with PF adhesives shows higher contact angles than that of earlywood. The delamination values of bonds between latewood–latewood were higher than those between earlywood–earlywood probably due to the fact that early wood surfaces were rougher than latewood surfaces [7]. Therefore, not only the chemical composition but also the surface character or roughness affects the contact angle. For many liquid/solid combinations, often, the contact angle is also a time-dependent value, decreasing from the value at the starting point due to the absorption (penetration) of liquid into the solid surface and also due to other factors. Therefore, the contact angles should be measured over a period of time, and the rate of change thus obtained is also a useful parameter in evaluating surface properties of wood materials. Overall, the surface tension and contact angle phenomena are the results of interaction of atomic attractive/repelling forces between two materials: *like materials attract each other and dislike materials repel each other,* according to an old adage.

Are the interfacial tension values: obtained from liquid adhesive drops good enough to predict the bond quality? A low interfacial tension possibly means a good adhesion between the liquid and solid materials and these measurements have been utilized in the adhesive/adhesion research of wood materials. However, it is important to recall that the contact angle measurement is possible only for liquid materials or adhesives, and in bonded products, the adhesive layer becomes a solid, and the contact angle between two solids cannot be measured, even though there could exist a significant solid–solid interfacial tension to reflect on whether the bonding is good (low tension) or bad (high tension). Then, a good wetting of a liquid adhesive on a substrate is desirable for an adequate adhesion, but it has to be remembered that a good wetting in the uncured adhesive stage does not guarantee a good adhesion in the resultant solid bond, although it could be a good predictor. On the other hand, a bad wetting in the liquid stage of adhesive would rarely result in a good wetting after hardening. In general, wood surfaces are polar and, therefore, adhesives of polar molecules, such as current wood adhesive resins, would lead to good attractive forces with most wood surfaces. Highly nonpolar materials, such as paraffin or polyethylene, would not establish such attractive forces with wood.

Effects of Wood Qualities, Extractives, Waxes, and Wood Moisture Contents: Often, abnormal factors can interfere with the wetting process to cause bond failures. (i) A high viscosity of (usually old) adhesive will prevent a good flow, which has to precede the wetting process, and therefore a proper wetting may not be obtained. (ii) Inactivated wood surfaces, such as over-dried wood veneers, will have a reduced polarity and result in increasing the contact angle with wood adhesives, and the curing could result in poor bonding. (iii) Wax is normally applied in the particleboard, MDF, and OSB manufacturing up to about 1.5% based on dry wood weight to increase the water repellency of boards, but if it is to be used at higher levels, the wood composite boards will have an increased wood surface area covered with wax molecules and adhesive drops on these wax film spots will not be able to wet well, resulting in poor bonds. (iv) Extractives that moved to wood surface by aging of wood similarly affect the surface characteristics [8]. A bonding study of southern pine plywood made with and without extraction of extractives of veneers reported that acetone-extracted wood bonds yielded the highest bond strength, followed by unextracted wood bonds and hot-water extracted wood bonds. The authors reasoned that the acetone extraction effectively removed the resin acids (pine's major extractive compounds) to result in better wetting and also better penetration of adhesive molecules into wood. (vi) Another factor is the moisture content of wood because if it is higher than normal, for example in OSB manufacture, the liquid PF resin will wet well but can also penetrate too much, and it has been noted early on that the best bond strengths are obtained when the resin wets but not over-penetrates [9].

18.3 Work of Adhesion

Work of adhesion of an adhesive bond in theory (W_A) is the breaking strength of the bond, wherein a solid surface of substrate with surface energy γ_{sv} and another solid surface of adhesive layer with surface energy γ_{LV} are created with help of the pre-existent interfacial tension (repelling) force (γ_{SL}):

$$W_A = \gamma_{SV} + \gamma_{LV} - \gamma_{SL}$$

Thus, the work of adhesion of an adhesive bond cannot be directly measured but can be calculated if the surface tension energies of both materials and their interfacial tension forces are known. For most solids, the surface tension energies can be measured or derived, and the resultant interfacial tension energies, if known, can be incorporated in calculating the work of adhesion. But for adhesive bonds between solid substrates and solid adhesives, which comprise most practical adhesive bonds, the interfacial tension cannot be directly measured, and various theoretical calculation methods have been developed especially for bonding of organic polymer materials [2].

References

1 Marra, G.G. (1980). The role of adhesion and adhesives in the wood product industry. In: *Proceedings of 1980 Symposium of Wood Adhesives* (ed. W.E. Johns and R.H. Gillespie), 1. USDA Forest Service/Forest Products Laboratory.

2 Pocius, A. and Dahlquist, C. (1986). *Adhesion and Adhesives*, 32. Washington, DC: American Chemical Society, p. 31.

3 Pocius, A. and Dahlquist, C. (1986). *Adhesion and Adhesives*, 27. Washington, DC: American Chemical Society, p. 27.

4 de Meijer, M., Haemers, S., Cobber, W., and Militz, H. (2000). *Langmuir* 16 (24): 9352.

5 Anonymous (2009). Surface energy data for paraffins, CAS # 8002-74-2, Diversified Enterprises.

6 Hse, C.Y. (1972). *Forest Products Journal* 22 (1): 51.

7 Hse, C.Y. (1968). *Forest Products Journal* 18 (12): 32.

8 Perry, D.A. and Choong, E.T. (1968). *Forest Products Journal* 67 (12): 112.

9 Bryant, B.S. (1968). *Forest Products Journal* 18 (6): 57.

19

Physical and Chemical Mechanisms of Adhesive Bonding for Wood

(This Chapter is in part an adaptation with permission of the publisher and minor revisions from the short course presentation given by. Pocius, A. and C. Dahlquist. 1986. Adhesion and Adhesives. American Chemical Society, Washington, D. C. [1])

Physical and chemical attractive/repelling forces are playing in all materials including the surface tension forces discussed in Chapter 18 and also in the interphase of adhesive bonds. Various different physical and chemical adhesion bond theories based on the contact angle/wetting and some other concepts have been advanced. Depending on the nature of adhesives and substrates, one theory can explain the particular adhesions better than others. These molecular adhesion mechanisms and theories are discussed and related with wood adhesions.

19.1 Adsorption (Secondary Bond Forces) Theory of Adhesion

This theory is related to the good wetting/low contact angle requirements needed for a successful adhesion discussed in Chapter 18 [1]. The good wetting or attractive interaction forces between the solid glue layer and wood surface are the basis of the adsorption (forces) theory, i.e. when applying and curing a liquid adhesive on a clean solid substrate (assuming there are no chemical reactions to form covalent bonds), atoms of the adhesive and substrate surface layers need to be brought into within atomic contact distances (about 5 Å or less) if they are to be atomically attractive. Pressure is often applied to expedite this process and also extraneous materials have to be expelled from the bonding interface. The fact that these atomic distances are attained means the atoms' attractive forces are dominant forces over the various repelling forces present. On the other hand, poor wetting means that the atomic forces between surfaces of the adhesive and substrate are repelling, and molecules do not come close to each other within the atomic distances.

Molecular Basis of the Adsorption Theory: Three different atomic forces are at play as shown with dotted lines in Figure 19.1 between surfaces of adhesives and adherends (wood): (*1*) *hydrogen bonds* (b), (*2*) *van der Waals bonds* (c), and (*3*) *dispersion forces* (d). Each force type is effective with its own varying distance ranges as attractive or repulsive

Adhesives and Finishes for Wood: For Practitioners and Students, First Edition. Moon G. Kim.

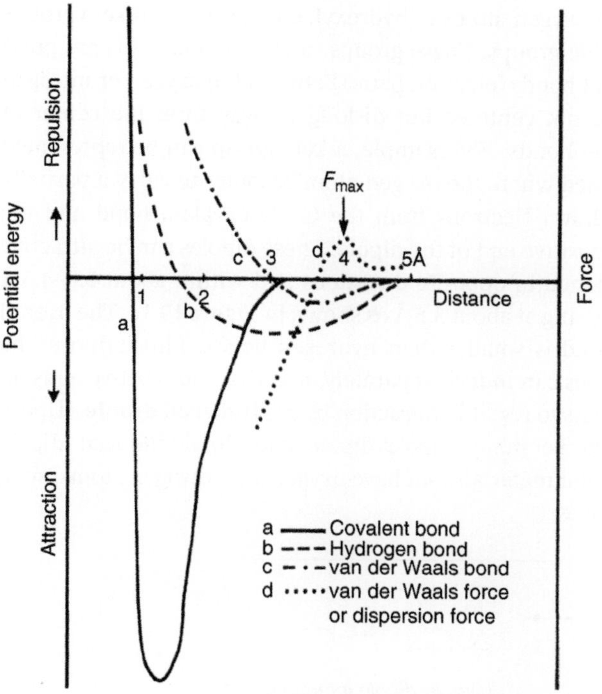

Figure 19.1 Potential energy curves for various chemical bond forces. Source: Pocius and Dahlquist [1] Reproduced with permission from the American Chemical Society Education Division. © 1986.

forces within about 1—5 Å distance ranges. Atoms separated by more than about 5 Å do not attract nor repel each other and of course no bonding occurs. These different atomic forces are further explained below:

Hydrogen Bonds can occur among and between hydroxyl (—O—H), amine/amide (=N—H), and keto (=C=O) groups of amides, amines, and hydroxy organic compounds, by forming bonds between their hydrogens and oxygen or nitrogen atoms when they are positioned within 1.8–4.5 Å with the maximum attraction force occurring when they are distanced at about 3 Å. These hydrogen bonds do not occur (breakdown) when the atoms are pushed closer to less than about 1.5 Å. Hydrogen bond forces occur due to the special attracting bond forces arising between electropositive hydrogen atoms and electronegative oxygen or nitrogen atoms, which are present plenty in both wood and thermosetting wood adhesive molecules. As shown in Figure 19.1b hydrogen bonds are stronger and have longer effective distance ranges than the van der Waals bonding forces.

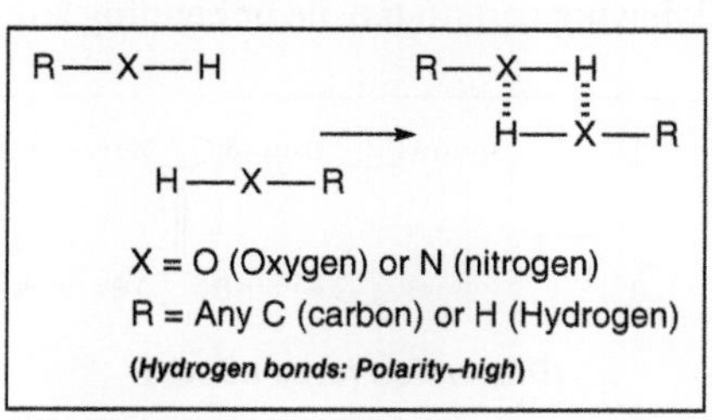

van der Waals Bonds occur with oxygen atoms of hydroxyl, ether, ester, and keto groups and nitrogen atoms of amine/amide groups. These groups in organic materials are polar groups, indicating that the covalent bonds (electron pairs) between the oxygen or nitrogen atoms and the carbon atoms are not centered but dislodged away from the center of carbon–oxygen or carbon–nitrogen bonds. For example, a keto group can be represented as $R_2{=}C^{+}{=}O^{-}$, i.e. a dipole is formed where the oxygen atom becomes to carry a partially negative charge because it withdraws electrons from the C—O covalent bond and the carbon atom becomes a partially positive end of the dipole. These dipoles can be attractive to other dipoles if they are positioned in opposite directions and within about 3.0–4.3 Å distances with the maximum occurring at about 3.5 Å as shown in Figure 19.1c. The overall **dipole–dipole attraction** per bond is smaller than hydrogen bonds. Furthermore, the positive and negative ends of dipoles can *induce*, separately, negative and positive ends on other neighboring, non-dipole atoms to result in attraction by the **induced dipole–dipole** mechanism. Overall, there could be many dipole–dipole and dipole-induced dipole interactions in common organic solid materials that have oxygen and nitrogen atoms, such as wood and thermosetting adhesives.

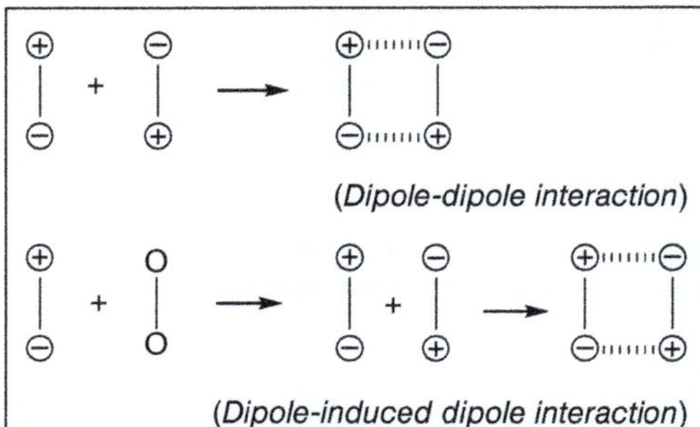

(Van der Waal's bonds: Polarity – high but lower than hydrogen bonds.)

Dispersion forces can occur between carbon, hydrogen, oxygen, and nitrogen atoms in any organic molecules when they are positioned within a narrow range of 2.7–3.6 Å with the maximum attraction forces occurring at about 2.7 Å, decreasing rapidly at 3.5 Å, and then becoming repulsive as the distance increases, reaching to its maximum repulsion at about 4 Å as shown in Figure 19.1d. Dispersion forces arise between any adjacent atoms (protons) because their electrons are circling around their respective nuclei in such a manner that the electron clouds of the two atoms, when positioned within the narrow distances, are pushed and pulled concertedly to become charged in opposite ways as shown, resulting in attraction to each other all the time. This dispersion force, for example, explains the fact that a candlestick is a stick having certain tensile or bending strengths.

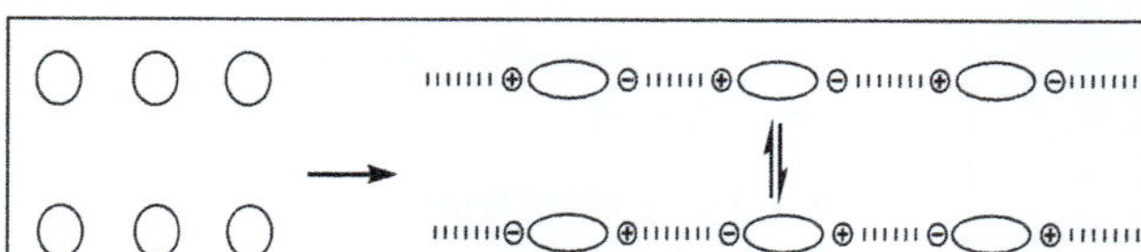

(Dispersion force bonds)

Candles consist of linear and slightly branched hydrocarbons of 18–45 carbons and each molecule is highly hydrophobic, but they exist together as a solid stick because of the dispersion forces arising between the neighboring hydrogen-hydrogen atoms throughout all the hydrogen atoms. Dispersion bond forces can play a significant role in bonding of UF and PF resins on wood surfaces because most molecules in them would come into this atomic distance range to wood surface molecules due to the hydrogen bonds and/or van der Waals forces present. However, for paraffin waxes added as water repellants to wood particles, the molecules have no oxygen or nitrogen atoms, and hydrogen bonds and van der Waals bonds are not possible with the wood surface molecules. Therefore, they probably cannot approach close enough to position themselves in this close distance ranges of dispersion forces. Normally, mechanical pressures alone are unlikely to push molecules to come into that close ranges. This means that paraffin wax added to wood composite boards would not be bonded to any significant extent by any of these atomic secondary forces bonding. The water repellency effect gained from paraffin wax added to particleboard or OSB mostly disappears when boards are soaked in water for about 24 hours.

Adsorption Theory in Wood Bonding: *The atomic distances between atoms* in solid materials or in gluelines are established during the bonding process, i.e. in curing of adhesive layers between two pieces of wood by the external pressure and various internal atomic attracting/repelling forces. Good materials selection and good bonding methods are essential to corral these forces into reaching to the optimum distances. Overall, from the secondary bond forces graphs in Figure 19.1, the optimum distance between the adhesive and adherend surface molecules would be within 1.5–3.5 Å to avoid the occurrence of the repulsion from dispersion forces. In bonding with wood, it is important to recognize that wood and UF and PF resin adhesives have many hydroxyl and amino groups that can result in forming hydrogen bonds and van der Waals forces in addition to the more universal dispersion forces. It should also be noted that the adhesive bonds become ineffective at longer distances, and this is why adhesives should be a liquid to make close contacts with wood surfaces in the first place. Two solid materials cannot be pushed so closely by mechanical pressures alone. *The adsorption force theory of bonding thus calls for these secondary chemical attracting forces to play the major role in the bonding with adhesives.* A good molecular contact (good wetting or low contact angle) means that a large number of atoms are contacting each other through these secondary attractive forces. In many practical wood-adhesive bonds, both polymeric adhesives and wood surfaces are polar and therefore this adsorption theory appears to be the major contributor. The fact that certain two adhesive systems are different in performance means that each has its own proportion of secondary chemical bonding forces. For example, PF resin adhesives would have high proportions of van der Waals forces and some hydrogen bond forces while a rubber cement (contact adhesive), albeit it is not a structural adhesive, has a high proportion of dispersion forces with no hydrogen bond or van der Waals bond forces available. Therefore, the former adhesives would bond well with polar substrates such as wood, and the latter adhesives bond well with nonpolar substrates such as reinforced plastics or rubber. In general, poor adhesive bonds often obtained with PF or UF resins mean that the secondary bond forces are not effectively positioned due to some unfavorable bonding process factors.

19.2 Mechanical Interlocking Theory of Adhesion

Lock and key effect in bonding: The mechanical interlocking adhesion theory emphasizes the mechanical imbedding of solidified adhesive into the cavities and other irregular geometries present in the surface of substrates as shown in Figure 19.2. This so-called anchoring, or lock and key effect in adhesion, is self-evident and explains the difficulty of achieving good adhesive bonds for microscopically smooth surfaces such as of glass or metals. In fact, in all adhesive bonding operations and especially with the smooth surface materials, one key to successful bonding lies in the surface preparation that invariably involves a *roughening step* of surface.

Greater Bonding Surfaces from Rough Surfaces: Another apparent fact is that a roughened surface provides a greater surface area in comparison to flat surfaces for the given area of bonding, i.e. more bonding forces are established between the adhesive and adherend on roughened surfaces.

Wood Adhesion and Wood Surfaces: Planed wood surfaces have craggy shapes and many cut lumens (4–25 μm) and they are generally well suited for adhesive bonds to be made according to this key and lock mechanism. On the other hand, even for same knife-planed surfaces the latewood or knots of coniferous wood (higher density and lower porosity) are more difficult to make good bonds than early wood having higher porosities [2] and this means that the anchoring mechanism is at play more effectively for early wood surfaces. Similarly, it is also known that wood adhesive bonds made on tangential surfaces (less porosity) are often less effective than those made on the radial surfaces (higher porosity). This tangential/radial surface effect is aptly shown in a recent report on the bonding of quarter-sawn and flat-sawn lumbers into composite lumbers using PRF and pMDI (PU) resin adhesives [2]. In the 12-Cycle Accelerated Aging (ASTM D 2559) and block shear strength tests of quarter-sawn samples, the PRF resins were more resistant than polyurethane resins, and for PRF resin-bonded samples, the quarter-sawn samples were more resistant than the flat-sawn samples [3]. On the other hand, the end-grains of

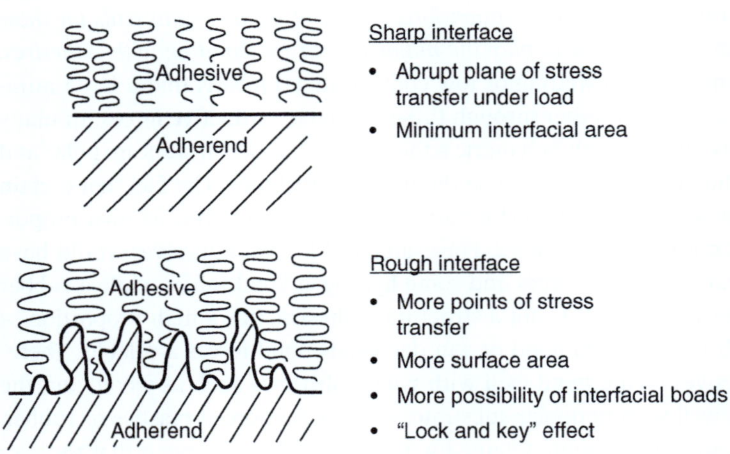

Figure 19.2 Mechanical inter-locking theory. Source: Pocius and Dahlquist [1] Reproduced with permission from the American Chemical Society Education Division. © 1986.

wood have the porosity okay, but the pores are too deep and big, causing the adhesive to penetrate into them too much to result in starved gluelines. In cross-cut surfaces of wood, for example in butt joints, the surfaces are damaged at least to a depth of 0.1–0.2 mm when sawing or planing of spruce [4] to result in poor bonds that can be remedied possibly by a heavier spread of adhesive. Thus, it is clear that this anchoring adhesion mechanism is operating in the adhesive bonding of wood at varying levels, which must be utilized as much as possible for maximum bond strengths. However, it should be recognized that this anchoring mechanism would be a secondary contributor to the adsorption mechanism.

19.3 Diffusion Theory of Adhesion

The diffusion theory of bonding emphasizes the "bridging" effect in adhesion by (long) polymeric adhesive molecules straggling between two substrates, polymer material #1 and polymer material #2, as shown in Figure 19.3, where polymer #1 can be wood and polymer #2 a hardened adhesive layer. This theory proposes that a good surface adsorption between adhesive and substrate layers is necessary but it alone will not be sufficient for a durable adhesive bonding to occur. The ideal adhesive interphase envisioned in the diffusion theory is an alloy interphase formed of substrate and adhesive molecules. This type of inter-diffusion of polymer molecular strands can occur at elevated temperatures below their melting points, for example, between chemically similar and compatible materials, but the process may need some length of time to occur. This mechanism would play a major role in the adhesion of, for example, polyethylene pieces with melted polyethylene. This theory is not readily acceptable between, for example, a metal surface and polymeric adhesive, because the two materials are so different in polarity and structure that the diffusion of polymer molecules into metal structures is less likely.

In case of wood bonding, this theory has been proposed as an important contributing factor in the formation of *durable adhesive bonds*, i.e. the diffusion of adhesive molecules into **cut** cell wall structures has been considered necessary to obtain durable bonds. Wood cell walls are composed of crystalline cellulose structures imbedded in the matrix of polymeric hemicellulose and lignin molecules. Undamaged cells or cut lumen walls have pores of 0.8–3.6 nm (or 8–36 Å) sizes [5] and can allow water or small resin molecules

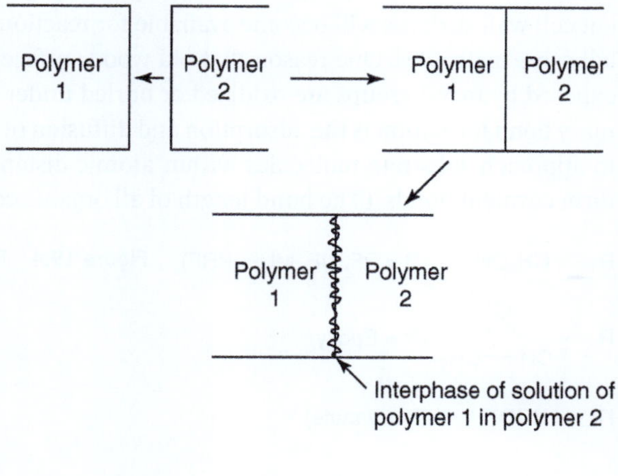

Figure 19.3 Diffusion theory of bonding. Source: Pocius and Dahlquist [1] Reproduced with permission from the American Chemical Society Education Division. © 1986.

to diffuse into these cellulose/lignin/hemicellulose structures. Also, in the presence of alkalis such as in PF resole or PRF resins, the cell wall structures can open up further for increased diffusion of adhesive molecules. It is known that these resins have some fractions of molecules with sizes smaller than those pore sizes so that they would diffuse into the cell wall structures. Larger adhesive molecules would not diffuse into them and, therefore, advancing resins in manufacture to an optimum extent appears to be very necessary. Overall, the diffusion by adhesive molecules appears to occur to some useful extents for alkaline PF and PRF type resins especially by smaller adhesive molecules considering these type resins' excellent durability of cured bonds, although the theory is a debated area in wood adhesion. This diffusion theory in wood bonding must be differentiated from the penetration of adhesives occurring through broken cells and cut cell lumens discussed in the mechanical adhesion theory.

19.4 Primary Chemical Bond (Covalent Bond) Theory

This theory emphasizes primary chemical bonds, or covalent bonds, to be formed between the surfaces of adhesive layer and substrate as a major contributor to the adhesive bonding. Thermosetting wood adhesive resins are formulated to have the chemical functional (reactive) groups shown in Figure 19.4: hydroxymethyl, epoxy, or isocyanate groups. Based on the known organic chemical reactions occurring in solutions, these groups can affect the cure of adhesive resins but also have the potential to diffuse into cut-cell walls, come into atomic contacts, and react with hydroxyl groups of wood to form covalent bonds of ether (R—CH_2—O—cellulose) or urethane (R—NH—CO—O—cellulose) bonds. Such covalent bonds would result in strong or durable adhesive bonds because covalent chemical bonds formed within 1.5 Å atomic distances are far stronger, as shown in Figure 19.1a, and durable in comparison with secondary chemical forces. Though, these types of reactions are likely to occur only to an extremely limited extent in wood adhesive bonds and, actually, the occurrence of such covalent bonds has been exceedingly difficult to prove by analytical methods. One of the reasons for analytical difficulty is the low number of such bonds to arise in the interphase. There are many hydroxyl groups possibly available for such reactions on wood, but the molecules in wood cell walls exist in well-ordered solid phases (crystalline cellulose phase, for example) and most of the hydroxyl groups are tied up in internal hydrogen bonds and buried under the surface. Only some hydroxyl groups newly exposed on the cut cell wall surfaces will become available for reaction to form covalent bonds with active adhesive molecules. One reason that old wood surfaces are difficult to bond would be the exposed hydroxyl groups are oxidized or buried under extractives. The prerequisite to primary bond formation is the adsorption and diffusion of adhesive molecules since they need to approach substrate molecules within atomic distances of about 1.0–1.5 Å to react and form covalent bonds. (The bond length of all organic covalent bonds is within this range.)

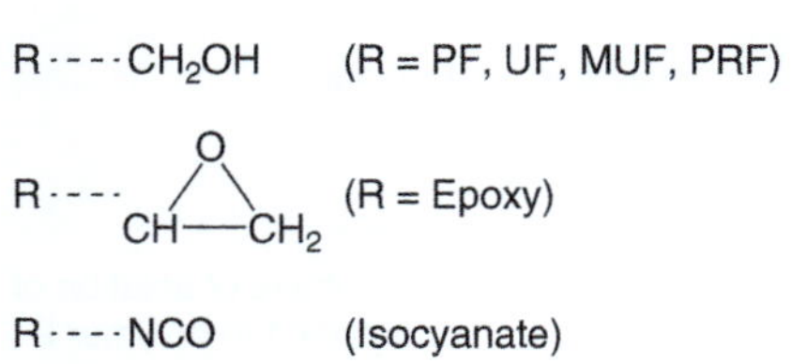

Figure 19.4 Primary chemical bond theory.

Wood bonding with isocyanate (—N=C=O) resins have been known to give higher bond strengths in dry and wet tests than PF resins in OSB, PB, and fiberboard manufacture, [4] corroborating with the primary bond theory. Further, PF and UF resins have hydroxymethyl (—CH_2—OH) functional groups and, although they are known to be less reactive than the isocyanate groups, the possibility of forming primary bonds is there since they give good dry bond strength values reflecting their molecules' adsorption/diffusion capabilities to reach to atomic distances. However, nobody thus far has proven the existence of covalent bonds between wood and cured thermosetting adhesive resin layer, since isolating these adhesive fractions intact for chemical or physical analyses and testing is extremely difficult. Overall, the author considers the contribution of primary bonds to wood adhesion to be significant for all current thermosetting wood adhesive resins.

19.5 Summary of Adhesion Mechanisms

The various adhesion mechanisms discussed above are operating together in wood adhesive bonds and in general, various adhesion mechanisms contribute to the bond strength with differing extents depending on the kind of substrate and adhesive. This means that different adhesives are needed for different substrates in order to make satisfactory adhesive bonds. It also reminds us that all bonding mechanisms should be utilized as much as possible to maximize the bond strength values for a given pair of substrate and adhesive, as summarized in Table 19.1.

19.6 Glueline Layer Thickness and Bond Strengths

Another important theory on adhesive bonds is related to the thickness of gluelines. Adhesive bonds made between loosely fitting gaps do not perform well and all-purpose gap-filling structural adhesives are rarely available. Also, due to practical limitations, often, the glueline thickness can vary along the bonding area with locations having thicker

Table 19.1 Summary of adhesion mechanisms.

1. Good interfacial contact and good interfacial Interactions:
    a. Maximum adsorption
    b. Maximum spreading
    c. Good wetting by proper glue viscosity
    d. Maximize covalent bonding
2. Good inter-diffusion:
    a. Match solubility parameters-polar/non polar relationship
    b. Make diffuse interphase
3. Good mechanical interlocking:
    a. Spread stresses to a wide area
    b. Increase the surface bonding area
    c. Appropriate surface roughness is needed

gluelines than other locations, especially for long and wide substrates. Or, sometimes, a whole glueline of one specimen can be thicker than that of another sample. These thick gluelines were found to give weaker strengths than normal (thin) gluelines. Adhesive bonds, first of all, mean *thin gluelines* formed between two adherends. The reason for the dependence of adhesive strengths on the glueline thickness has been explained by *the defect theory*. In all (organic) structural materials manufactured or naturally obtained including cured adhesive layers, their internal structures never go to perfection, i.e. they harbor defects, or weak points (structural imperfection) due to various reasons such as impurities (air, water, extraneous materials, etc.) or improper curing reactions. These defects are distributed uniformly more or less throughout the entire volume of specimens. When a specimen is stressed to break, in a strength testing for example, any one of these defects becomes the initiating point of rupture. In tests of adhesive bond strengths, it is seen that as the thickness of glueline becomes thinner and thinner, the total number of defects in the test piece decreases accordingly. This decrease in the number of defects means a decreased probability of breakage initiation. That is, the decreasing number of defects will cause the breakage to occur at higher stress values for thinner gluelines than thicker gluelines. In Figure 19.5, the relationship between the bond strength and glueline thickness of butt joints is shown [5]. The increase in glueline thickness causes rapid decreases of breaking strength until about 0.05 cm thickness and then more gradually till about 0.2 cm thickness. Thus, the optimum glueline thickness in this adhesion system appears to be less than about 0.05 cm. However, gluelines that are too thin would show poor strengths due to the lack of glue, thus gluelines should be neither too thin nor too thick and should be within the optimum ranges.

For wood adhesive bonds, the glueline thickness has been long recognized for its importance in the lumber lamination industry. PRF resin adhesives perform unsatisfactorily when the glueline becomes too thick or too thin. Long lumber pieces are often difficult to manufacture with *perfectly un-wavy and smooth* surfaces throughout the whole mating area and, therefore, the gluelines tend to have thin and thick areas or spots. Therefore, PRF resin-type adhesive formulations have been developed to have some gap-filling capability to compensate for roughness and thickness variations. It has been known in the wood

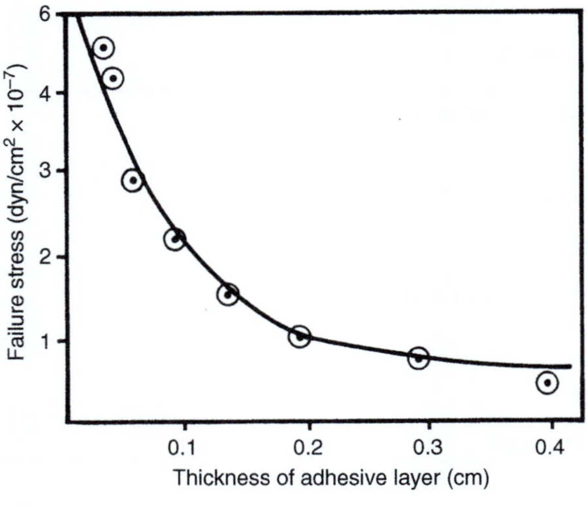

Figure 19.5 Adhesive layer thickness and strength relationship of butt joints. Source: Nordström and Johnson [6] Reproduced with permission. © 1995 of Springer-Verlag.

laminating industry that well-planed wood surfaces are satisfactory while sawn wood surfaces are not in the lamination with PRF resin adhesives, and the thickness of cured gluelines is required to be inspected and be within 0.006 in. (0.0152 cm) plus or minus 0.004 in. (0.0101 cm) for no-gap-filling adhesives [7]. Further, softwood plywood veneers have rough surfaces and this roughness has been compensated by the gap-filling PF resin adhesive formulations using fillers and thickeners and high viscosity values, an essential characteristic needed for the gap-filling capability.

There is another aspect to look at the thick gluelines from the fact that they incur *higher extents of contraction of adhesive layers upon curing* as discussed in Chapter 15. The adhesive contraction occurs more where it received the lowest pressure during pressing to result in thicker gluelines. Most adhesive bonding practices involve pressuring during the solidification period, i.e. during the liquid, gel, rubberly, and cured stages. What does the continued pressuring do? It keeps the glueline thinner by compensating (filling up) for the contraction (voids) of gluelines occurring from curing and loss of water or solvent molecules. Thus, thicker gluelines will harbor more voids or defects. On the other extreme, gluelines that are too thin, or starved gluelines, due to lower amount applied glue or an excessive pressure application would be lacking in the glue mass needed to bridge the gap completely and the entailing voids will become defects. These thin glueline layers will be formed also when the amount of applied glue is okay, but the mating surfaces have waviness or receive uneven mating (clamping) pressures.

19.7 Summary

Various adhesion and adhesive bonding theories were presented based on the past seminar presentation given by A. Pocius and C. Dahlquist with emphasis on theories related to wood adhesive bonds. Thus, adsorption theory based on various secondary atomic bond forces, mechanical interlocking theory by imbedding of adhesive materials into (micro) pores of substrate, diffusion theory of adhesive molecules penetrating into substrate molecular structures, primary chemical bond theory of forming chemical covalent bonds between polymer molecules of adhesive and substrate in the adhesion interphase, and optimum glueline thickness theory are discussed.

References

1 Pocius, A. and Dahlquist, C. (1986). *Adhesion and Adhesives*. Washington, DC: American Chemical Society, p. 23; 56; 128; 286.

2 Hse, C.Y. (1968). *Forest Products Journal* 18 (12): 32.

3 Wan, H., Dahlen, J., Mao, A. et al. (2017). *Forest Products Journal* 67 (1/2): 112.

4 Sun, B.C.H., Hawke, R.N., and Gale, M.R. (1994). *Forest Products Journal* 44 (3): 34.

5 Stamm, A.J. (1964). *Wood and Cellulose Science*, 333. New York: The Ronald Publishing Co.

6 Nordström, J.-E.P. and Johnson, I. (1995). *Holz als Roh-und Werkstoff* 53 (1): 38.

7 American National Standard for wood products—Structural Glued Laminated Timber, AITC, Section 5.8. (h).

20

Evaluation of Wood Adhesive Bonds, Quality Control, and Bond Durability

Wood adhesive bond layers do not lend themselves isolated for examination, and they are therefore commonly examined as part of bonded wood products. This leads to the question regarding the quality of wood substrates, and it is tacitly assumed that the wood quality is normal (good). However, when wood failure (WF) values are high, the wood quality factor must be considered in the interpretation of test results, as discussed in Chapter 18. In this chapter, we will emphasize the methods and procedures of testing of bonded wood products as practiced in routine quality control and product performance evaluations, including the long-term durability tests practiced by the wood product industry.

20.1 Mechanical Testing Modes and Methods for Measuring the Adhesive Bond Strength

There are four basic modes of stress application in testing of bonded wood samples: shear, tensile, cleavage, and peel. In bending tests, shear and tensile as well as compression modes of stress are involved.

20.1.1 Shear Strength Test

Block shear test for face-laminated lumber: The force in the shear test is applied parallel to the glueline, and therefore the entire glueline area (all adhesive components) in the specimen is stressed in each moment of testing. The elements in the glueline volume, especially those that lie along the breakage line, are moving sideways against each other. The shear modulus is measured as the coefficient of the force vs. shear angle, and the shear force can be applied until a total failure occurs instantly to give the breaking shear strength. The breaking strength value depends much on the rate of stress application or the speed of the instrument's test arm's movement. Generally, use of a fast speed increases the breaking strength value, but this also increases the variability of resultant values, while use of slow speed decreases the breaking strength value and lowers the variability. Because tests should be done within a reasonably short time with a reasonably narrow range of variability of results, the standard test methods prescribe the instrument arm movement speed at a relatively low but finite value for each type of test method. The block shear test is commonly used for face-laminated wood products as shown in Figure 20.1 (ASTM D 905-98, block shear), where the breaking is designed to occur along the bond line.

Adhesives and Finishes for Wood: For Practitioners and Students, First Edition. Moon G. Kim.

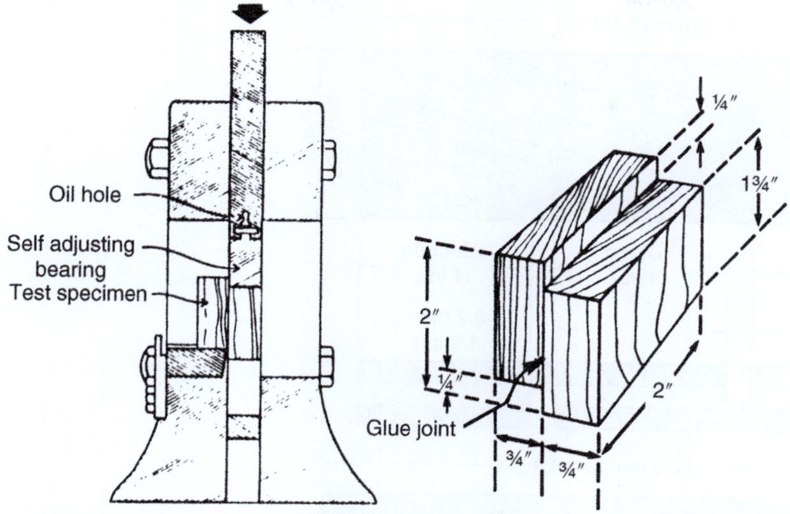

Figure 20.1 Laminated lumber block-shear test specimen (ASTM D905-49) and shearing instrument. Source: Reproduced with permission of ASTM and the American Chemical Society Education Division.

Lap shear test for plywood: The lap shear test is used for softwood and hardwood plywood as shown in Figure 20.2 (PS1-19, APA), where the specimen is kerfed one-third of the length from each end to provide a 2.54 mm^2 test area. The breaking by pulling is designed to occur in the middle-layers, and the sample cutting can be done in two different ways, so that in the first one, the breaking occurs with lathe checks being closed in the face layer while in the other with lathe checks being opened to eliminate the lathe check effects.

Commonly, in block shear and lap shear tests, an equal number of specimens is tested, and an average is obtained. The unit of test strength is force per unit area or pound-force per square inch (**psi**), and the WF value is commonly read by eyeballing, in percentages (%) based on total broken test area. Also, image analysis methods were claimed to be as accurate as readings from experienced technicians [1]. The tests can be done with dry, wet, or aged samples.

Torsion shear tests for wood composite boards: In this test, the torsion force is applied to the rectangular (commonly 1″ square) test specimens, forcing the break to occur in the middle layer of wood composite board [2]. Oriented strandboard, particleboard, and medium density fiberboard can be tested for shear strengths using this method with greater ease and speed than the common internal bond (IB) (tensile) strength test discussed below. The torsion shear and IB tensile test values were reported to be linearly related [3], suggesting the validity of this method for routine uses. The test can be done using a 1″ socket wrench equipped with a torsion gauge.

20.1.2 Tensile Strength Tests

Tensile tests for internal bond (IB) strength of wood composite boards: This test is commonly used for measuring the IB (breaking) strength of wood composite boards, such as OSB, PB, and fiberboard, as shown in Figure 20.3. The stress in this tensile test is applied in the direction perpendicular to, and away from, the plane of bond line, and the results are the Young's

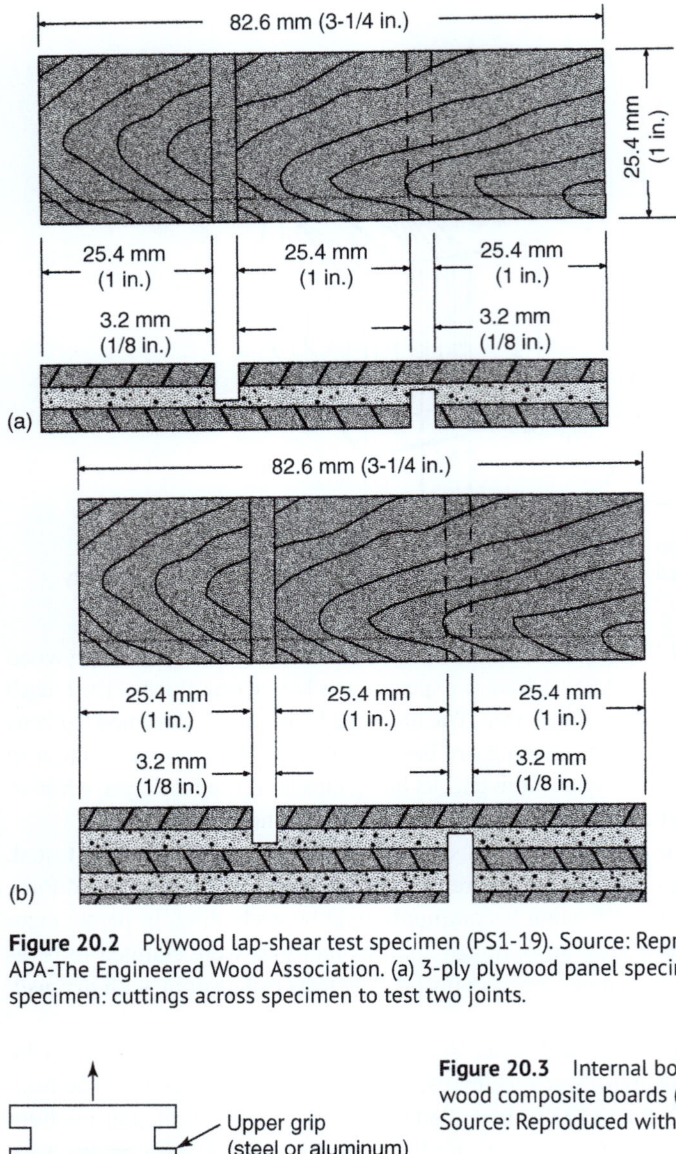

Figure 20.2 Plywood lap-shear test specimen (PS1-19). Source: Reproduced with permission of APA-The Engineered Wood Association. (a) 3-ply plywood panel specimen. (b) 5-ply plywood specimen: cuttings across specimen to test two joints.

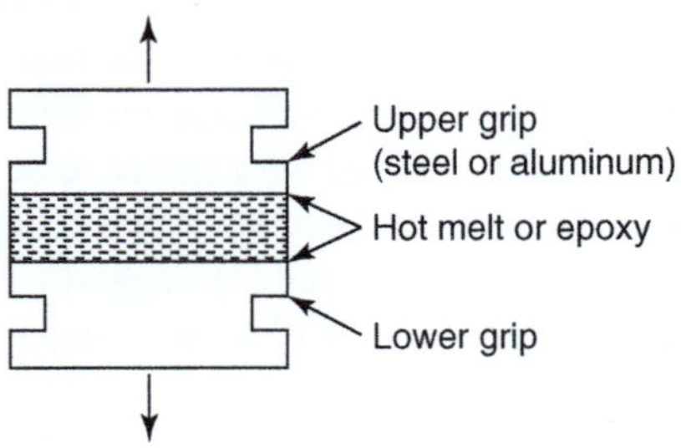

Figure 20.3 Internal bond strength test method for wood composite boards (ASTM Procedure D 1037). Source: Reproduced with permission of ASTM.

modulus (stretching). As in the shear test, the pulling stress is applied until a total failure occurs to give the breaking strength values. The entire bond line area in the test specimen is equally stressed (stretched) in each moment of testing. All elements in the test specimen volume, especially those existing along the breaking line, are moving away from each other. Although the glueline is not well defined and can exist throughout the entire sample

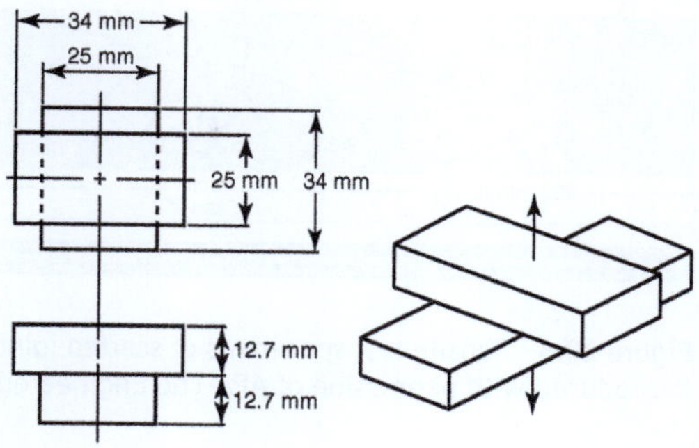

Figure 20.4 Tensile test specimens of cross-laminated lumber (ASTM D 897-78). Source: Reproduced with permission of ASTM.

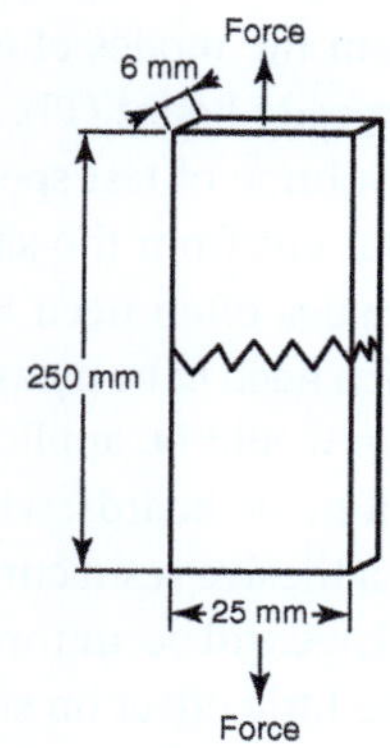

Figure 20.5 Tensile test specimen of finger joint (ASTM D-4688). Source: Reproduced with permission of ASTM.

volume, the breakage normally occurs through the middle layer, the weakest bond layer formed in manufacturing of boards. If needed, however, IB strengths (bond quality) of surface layers or other intermediate layers can be measured separately by slicing the layers and testing them individually. For IB strength tests, 2″ × 2″ board specimens are generally used with the two surfaces bonded with a hot-melt to an aluminum adapter (grip) for loading on the testing machine (Instron). As in the shear test, the speed of pulling is standardized for each product type. (Of course, the bond strength of hot-melt needs to be greater than bond strength.)

Tensile test for face-laminated lumber: The test is done with parallel or cross-laminated lumber as shown in Figure 20.4. In these cases, a neck can be carved out around the bond line to induce the breakage to occur through the bond line rather than wood.

Tensile test for end-jointed lumber: Lumber end-joints, including butt, scarf, and finger joints, are tested by tensile test as shown in Figure 20.5 (ASTM D-4688). In the finger and but joint tests, the direction of stress is not perpendicular but diagonal to the glueline and, therefore, the stress exerts in *tensile* as well as *shear* modes. Scarf and finger joints are more common in lumber manufacturing using 2 × 4, 2 × 6, etc. In order to make them testable by using small-capacity testing machines or to obtain more consistent values, the bonded samples can be sliced into longitudinal, thin strips (1/4″ thick) and tested them individually and obtained the average values.

The unit of test results is also in psi, commonly calculated based on the cross-sectional area, not on the glueline area.

Tensile test for scarfed end-jointed plywood: The test is done with sample specimens, which are cut to have the entire scarf-line in the breaking region, as shown in Figure 20.6.

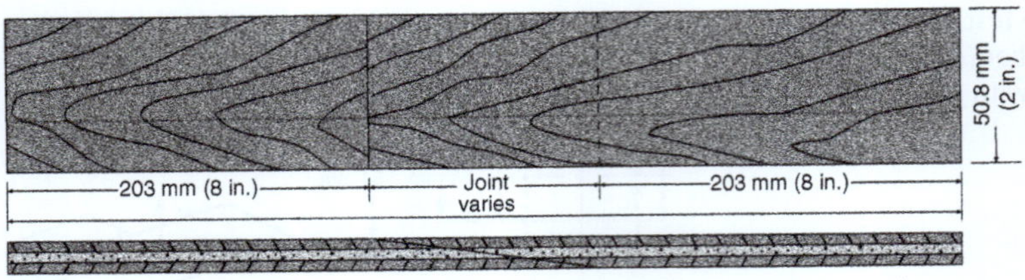

Figure 20.6 Tensile test specimens of scarfed jointed plywood panel (PS1-19). Source: Reproduced with permission of APA-The Engineered Wood Association.

Effects of board density of wood composite boards on tensile strength: One consequence derived from the modes of elemental movements in tensile tests is that as the density of wood composite board (PB, MDF, and OSB) increases, there are more bonds to break in the given volume of test specimen, i.e. *higher values for higher density boards.* Also, since test samples cut from the same board tend to have varying densities, the average tensile strength values often need to be reported along with the average density values of specimens, which need to be measured before the testing (breaking). This *tensile strength/density relationship* would be applicable mainly when other board-making parameters are equal, i.e. for well-made boards, where the adhesive resin curing is attained at an optimum level. When the adhesive resin curing involved certain amounts of pre-curing or over-curing, the bond breakage will occur through the defects of adhesive layers, and the board density value would have little effect on strength values. Good tensile strength/density relationships for wood composite boards are often observed in laboratory-made boards, and a regression analysis of strength values against density values usually results in a linear relationship, from which the average value at the target density value can be obtained. Industrial boards also show this tensile strength/density relationship.

20.1.3 Cleavage and Peel Strength Tests

Cleavage tests: The cleavage test exerts the "prying open" mode of force on the specimen's glueline, often used in testing of laminated wood beams, as shown in Figure 20.7. The applied force in the breaking zone is similar to that of the tensile test in that the elements are moving away from each other. Such cleavage tests, often called *fracture mechanics tests*, give the breaking strength (fracture toughness) value of bond line that is similar in meaning to the IB strength values with breaking occurring mostly along the bond line but only from one end of the test specimen. The advantage of cleavage tests in comparison to common tensile or shear tests is that the rate of breakage propagation along the glueline can be significantly slowed, or accurately controlled to occur in small increases, by adjusting the applied load continuously, causing the breakage to propagate slowly over an extended period of time [5]. In general, the use of slow breakage rates allows to obtain more accurate bond strength values or to predict long-term failure loads of wood beams that might contain stress concentrations such as knots, notches, splits, and other discontinuities [5]. The instant breakage occurring through the entire bond line in common tensile and shear test procedures emphasizes too much on the few defects that catastrophically initiate the breakage process due to the excessive loading necessary to break the specimen in a short time. These rapid-breaking test values are not the typical

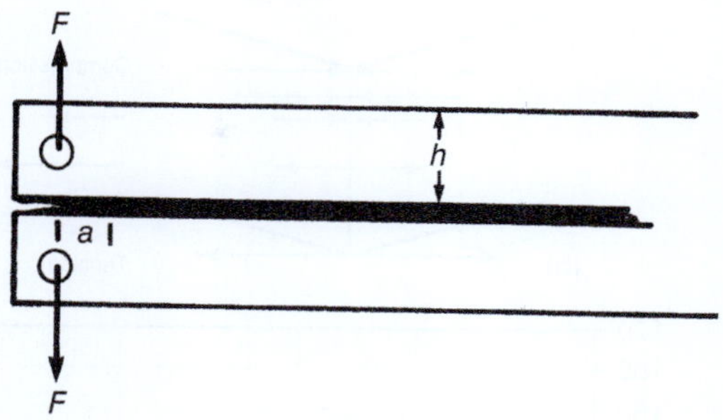

Figure 20.7 Simple cleavage (fracture) test: double cantilever beam [4]. Source: Reproduced with permission of the American Chemical Society.

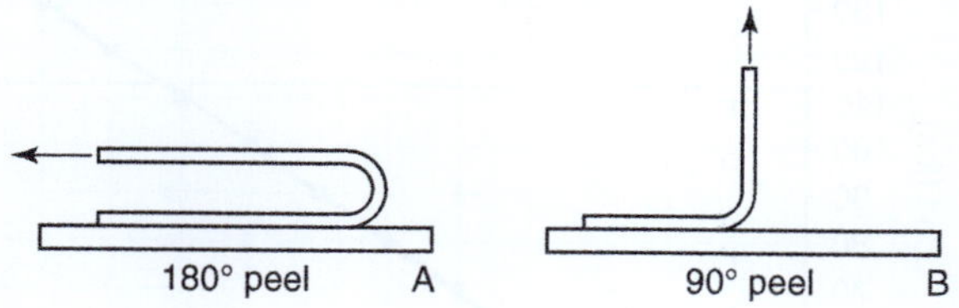

Figure 20.8 Various peel test modes: *T* peel, 180° peel, 90°, and drum peel [4]. Source: Reproduced with permission of the American Chemical Society.

stress modes in practical failures. Many practical stress modes of wood composite products are slow stresses. The fracture mechanics cleavage tests thus give useful data in designing large structures, such as laminated beams, as described in ASTM D 3433-75, but the test takes long time to finish and therefore is not used in daily manufacturing for quality control purposes.

Peel tests are used when flexible substrates (films) are bonded to wood, which is common in the overlay wood product industry (Figure 20.8). This test is used in evaluating the adhesion strengths of overlays and coatings applied on wood composite boards. For example, in the APA (American Plywood Association) test method F-2, the adhesion strength of a color stain is compared to the standard stain by measuring the peel strengths of cotton cloth strips imbedded with the stain and cured on the surface of wood panels.

20.1.4 Bending Strength Tests

Bending strength test method: Many structural wood composite products are used where bending-type stresses are applied. In the bending test, the force is applied by a blunt-edged arm at mid-span on a piece of beam (boards) sitting on a pair of fulcrums at a given distance as, shown in Figure 20.9a. In this test mode, the stress becomes a tensile type on the bottom layer and a compression type on the top layer of boards as well as the shear-type stresses mixed-in in the inner layers of board throughout the whole thickness of specimen (Figure 20.9b).

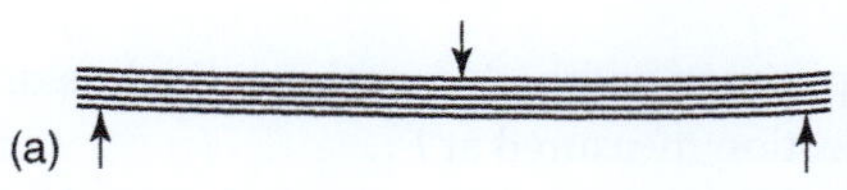

Figure 20.9 (a) Bending test of a beam: the force (*P*) is applied on the midpoint incrementally and measure the deflection in inches until sample's breakage. (b) Distribution of compression, shear, and tension stresses in the bending test of a beam. (c) A load (lbs) vs. deflection (in″) graph obtained in the bending test of a beam. Tests continue till the breaking point.

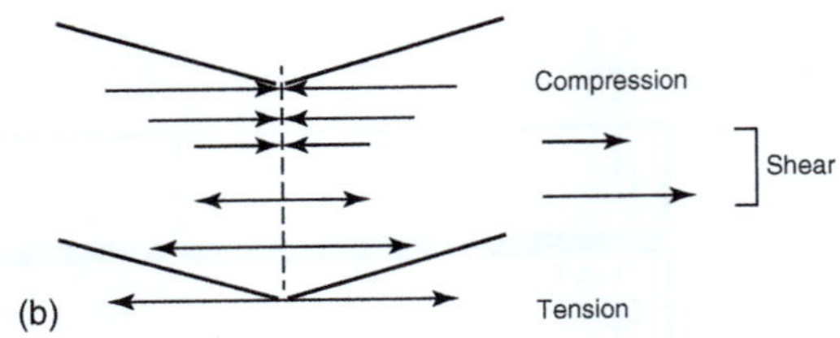

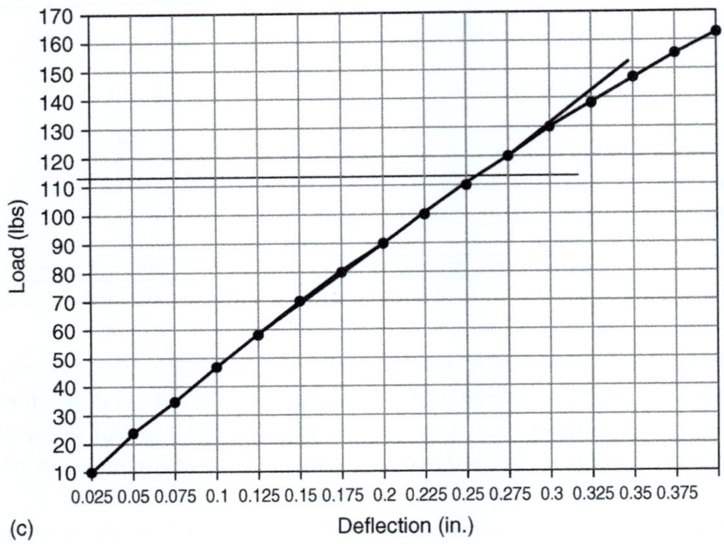

Figure 20.9 *(Continued)*

In this test, the load is increased slowly at a prescribed rate while the deflection in inches is measured at the midspan, resulting in a graph as shown in Figure 20.9c. The loads at the proportional limit (P_1) and at the breaking point (P), as well as the deflection inches (y) are obtained, and the results and sample dimensions are put into the *modulus of rupture* (**MOR**, breaking strength) and *modulus of elasticity* (**MOE**, stiffness) equations as shown in Eq. (20.1). MOE value represents the load necessary to bend a unit distance at mid-span within the elastic limit and MOR the breaking strength. Bending test values are practically useful information; for example, different grades of OSB, PB, and MDF boards are specified in industry based on MOR/MOE values.

Calculation of MOR and MOE from the bending test

$$\mathrm{MOR} = P \times \frac{(3L)}{2bd^2}$$

$$\mathrm{MOE} = \frac{P_1(L^3)}{y(4bd^3)} \tag{20.1}$$

P = breaking load, L = specimen span length, b = specimen width, d = specimen thickness, P_1 = load at proportional limit, y = distance of deflection measured at P_1.

Bending strength and board density: Tensile, shear, and compression forces are involved in bending tests, where board elements move against each other or undergo a compaction. The breaking area in bending tests is not as large as in IB tests and, therefore, MOR and MOE values are affected less by the specimen's density values compared with IB test values. Wood composite boards manufactured in a plant would experience minor changes in the average density values unless intentionally made, but some limited variations are inevitable due to the matting and other operational variables. A lower or higher overall density affects the IB test values accordingly, and the bending test values will vary too to a less degree if the adhesive curing process has gone normally. If the density values are the same but the bending strength is found to be unusually poor, the adhesion must be questioned, especially on *the face layers of boards*. Also, wood composite boards manufactured by hot-pressing processes have certain *density gradients* through the board thickness, i.e. the face-layers have higher and the core-layer lower densities due to the compaction and resin curing occurring progressively through the thickness of board. The higher density of face-layers greatly helps the overall wood composite's bending strength, as well as increases the surface finishing properties, although the resultant lower density core-layers tend to give weaker structures potentially causing weaknesses in screw holding strength. On the other hand, for boards made with the steam-injection pressing method, the density difference between the face and core layers is less and the advantages and disadvantages are less pronounced.

20.2 Quality Control, Certification Tests, and Adhesive Bond Durability

20.2.1 Objectives and Scope of Quality Control Processes

In wood composite board manufacturing plants, manufactured products are routinely tested to assure their performance against the standards of certification that the plant maintains to satisfy their customers. This quality control program involves selecting samples and testing them at regular intervals. This day-to-day testing is to maintain the product's performance qualities with respect to minor manufacturing parameter variations that are often experienced in daily operations, such as mat moisture content, furnish quality, matting quality, hot-press operation parameters, weather changes, and adhesive resin quality and dosage. More severe manufacturing parameter variations would be encountered when an intentional change is made, for example, in wood species, adhesive resin suppliers or kinds, hot-pressing parameters, etc. In these latter cases, the situation is handled as *an experimental trial,* and it precedes a good amount of laboratory work to back up its feasibility ahead of the process changes. The day-to-day quality control program includes various mechanical strength tests for dry samples as well as some aged samples obtained by using various accelerated aging treatments. The kind of aging procedures to use depends on the product type, for example, exterior-use or interior wall panels, etc. This will be discussed below.

20.2.2 Durability Evaluation of Wood Composite Boards and Certification

One major objective in the evaluation of exterior-grade bonded wood products is to predict their water resistance and long-term durability. Obviously, the direct *outdoor exposure* of products under the actual exterior use condition is one method, but this takes long time and there are also many varying exposure conditions, such as different locations having different humidity and temperature parameters. Such outdoor exposure experiments have been conducted in the past for major wood adhesive resin systems: PF, PRF, MF, UF, etc., and the conclusions obtained from those tests, in general, have been assumed to be similarly applicable to current adhesives. This is why PF and PRF resins are exclusively used for products designed for exterior applications, while MF and UF resins are used in products for interior applications. Based on the general knowledge of adhesive performances, various accelerated aging tests were developed, and they are now used in evaluation of products using shorter testing times (accelerated aging procedures) against the benchmark values established from the past research results. In accelerated aging tests, the strength and WF values from the aged samples are compared with the results of the dry, un-aged samples, resulting in *the extent (%) of strength losses or WF value decreases*. At the same time, the dimensional and weight changes can be tested.

The reason for doing the accelerated aging and other quality control tests on exterior-use wood composite boards despite using the known "exterior durable adhesive resins" is that the bond quality of PF or PRF resin adhesives (as well as others) greatly depends not only on the quality of adhesives but also on how adequately the bonding procedure has been performed. Furthermore, one of the tenets used in the wood bond evaluation is that any deficiency in adhesive bonds is accentuated by aging procedures, i.e. a small difference in the dry bond test results, if the underlying cause is an improper cure or something related to adhesion procedural deficiencies, the performance difference would get amplified by the aging test procedure. Thus, one objective of quality control tests is to catch the causes of improper adhesion as early as possible before the product is distributed widely and used for long durations.

The quality control programs and standards for exterior-grade wood products are voluntary, i.e. wood composite manufacturers follow them to be certified by an independent institution of their own choice. Large-volume, commodity-type, exterior-grade softwood plywood, and OSB panels are mostly thus certified. Small volume, specialty products can be manufactured and sold without such a certification if buyers agree with whatever specifications are supplied by the manufacturer. Various accelerated aging methods are used in testing of exterior-grade wood composites: water, water and heat, water and heat and dry, various aging durations, biological agents, and physical stresses, such as load or vibration, and expansion/contraction by moisture or heat. These various aging methods are used selectively according to the product's intended uses. The effects of wood-only degradations are already accounted for in the standards, but any excess degradation due to deficiency of adhesive layers is presented as deviations from the standards. Therefore, in accelerated aging tests of adhesively bonded products, the WF values are reported to be the more consistent predictor than strength test values [6].

20.2.3 Certification of Exterior-Use Wood Composite Products

The National Institute of Standards and Technology, the Department of Commerce, has established the product standard for structural (softwood plywood and overlaid) panels: *PS1-19*, Construction and Industrial Plywood, and *PS2-92*, Performance Standard for Wood-based Structural-use Panels (OSB, hardboards). The general class of durability, including overlaid softwood plywood panels, is defined as follows with the necessary test procedures:

- Exterior: can be used for permanent exposure to weather or moisture.

- Exposure 1: can be used for partial exposure to weather or moisture, such as against the effects of moisture due to construction delays, high humidity, and water leakage.

Plywood (PS1-19): with various aging test methods depending on the grades, the adhesive bond requirements are as follows:

(1) *Exposure* 1: For this and following Exterior classes tests, eight plywood shear specimens (82.6 mm × 25.4 mm) from one panel are treated either by Method (A) or Method (B), below, and shear tested. The WF value for Exposure 1 class shall not be less than 80%. When more than one panel is tested, (i) at least 90% of the panels shall have 60% WF or better, and (ii) at least 95% of the panels shall have 30% WF or better.
   (A) *Vacuum pressure test:* Specimens are submerged in cold tap water. A vacuum of 83.5–98.5 kPa (24.7–29.0 in.) is applied for 30 minutes, followed immediately by pressuring at 450–480 kPa (65–70 psi) for 30 minutes. Specimens are strength tested wet, followed by drying the tested pieces and reading the WF values. Bonds between overlaid sheets of overlaid panels are treated similarly with separately prepared specimens kerfed only through the layer of overlay, and the overlay layer is broken by inserting a sharp, thin blade to completely separate, not peeling nor cutting into overlay, followed by reading WF values.
   (B) *Boiling test:* Specimens are boiled in tap water for 4 hours and dried for 20 hours at 63 + 3 °C to a maximum moisture content of 8%. The specimens are boiled again for four hours, cooled in water, and tested wet, followed by drying of test pieces and reading the WF values.

(2) *Exterior*: The same aging procedure of (A) or (B), above, is carried out independently and separately, and the WF value for exterior use classes must be greater than 85%. When more than one panel is tested, (i) at least 75% of the panels shall have 80% WF or better, (ii) at least 90% of the panels shall have 60% WF or better, and (iii) at least 95% of the panels shall have 30% WF or better.

(3) *Other tests are also prescribed* and carried out as needed as follows:
   (C) *Heat stability test:* Bunsen burner, 800–900 °C for 10 minutes – "no delamination" of veneers is necessary to pass.
   (D) *Mold test:* Specimens are subjected to mold condition and tested according to (A) vacuum pressure test and (B) boiling test described above.
   (E) *Bacteria test:* Specimens are subjected to bacterial condition and tested according to (A) or (B) above.
   (F) E*nd-joints*: If they are involved in panels, they are tested separately.

20.2.4 Certification of Exterior-Use Wood-Based Structural Panels (PS2-92)

This standard covers the wood-based panels made with veneer and other types of wood composites, such as OSB, which is classified as Exposure 1, not as Exterior class. The following aging procedures are used.

- Exterior: Same as that of plywood (PS1)
- Exposure 1: Same as PS1 except for wood composite boards, such as OSB, for which a residual strength (RS) of greater than 50% is needed in comparison with the un-aged controls from *the six-cycle test treatment (below).*
- Exposure 2: Same as that of plywood (PS1-95)

Six-cycle test treatment: The testing is done by the small-specimen bending test method: 25 mm × 125 mm (1″ × 5″ specimens (10 samples) of any thickness, 5 specimens for each direction if panels have directional properties. Also, side-by-side matching control samples (untreated samples) are needed. A set of specimens for five consecutive tests held in a rack is submerged in water at 66 °C (150 °F), 50.6 kPa (15 in.) vacuum is applied for 30 minutes, the specimens soak in the same water for 30 minutes without heating, and are dried for six hours at 82 °C (180 °F). This cycle is repeated once more. Then, the final drying is carried out for 15 hours and the first test specimens are tested. With the rest of four sets of specimens, this vacuum soak cycle is repeated and tested after each cycle. Each test is done as a bending test with a 100 mm (4″) clear span with the 25 mm specimen dimensions acting as the depth of beam; load is applied at mid-span with a speed less than 25 mm (1″) per min. The breaking load is measured, and the average is obtained and compared with the strengths of un-aged controls to obtain the RS values, which should be greater than 50% to pass for Exposure Class 1.

20.2.5 Cyclic Delamination Test Procedure (AITC Test 110)

This bond durability test is done for wet use face and end joint bonds of laminated beam products with a sample block cut to be thicker than 6″ and 3″ in length and, after the vacuum soak and pressure and drying treatments, the bondlines are evaluated for delamination on the sawn (end) faces of the samples as described in Section 10.7.2.

20.2.6 Other Common Laboratory Aging Test Methods Used in the Industry

(1) *Water-soak (2-hour, 24-hour, etc.)*: Keep samples underwater at 20 °C; the dimensional (thickness) increase and water absorption values are measured. An important criterion for temporary water resistance performance of PB and MDF.
(2) *Boil in water:* Two-hour boil in water, dry and measure the IB strengths, used for OSB for adhesive bond's durability test.
(3) *Boil in water and dry cycle*: ASTM D 3434-75 (Automatic Boil Test [ABT]): [7] A cycle of 10 minutes in boiling water, 4 minutes in 23 °C forced air circulation, and 57 minutes in 107 °C forced air circulation. This cycle is repeated as many times as needed with lap shear or block shear test samples. The results are used for long-term exterior durability tests for laminated beams and softwood plywood.
(4) *12 Wet-dry cycle and delamination test (ASTM D2559):* Laminated beam test sample is cut into three sets of six end-matched block shear test samples having a 1.12 square inch glued area, and they are subjected to (i) control, (ii) 12 wet–dry cycles, and (iii) 12 wet–freeze–dry cycles. Each wet cycle consists of 12 hours of de-chlorinated water spray, which brought the MC to 30%, and each dry cycle consists of drying at 70 °C for 24 hours, which brought the MC to about 10%. This cycle is repeated 12 times. The wet–freeze cycle test includes a step, after the wet cycle, of freezing samples at−18 °C for 24 hours, followed by the usual drying cycle. The dry samples are tested for block shear strength and WF values.
(5) *Outdoor regional test methods:* Humid climate (US SE); cold winter/hot summer (US MW); hot and dry/all year (US SW); mild, wet winter/dry cool summer (US NW).

20.2.7 Testing and Certification Organizations Involved

Certification and associated tests' procedures for plywood and wood composite panels (OSB) are described in the APA – The Engineered Wood Association (APA) booklet APRP-108, Performance Standards and Policies for Structural-Use Panels, and in Product Standard (PS)1-19 and PS2-92 issued by the National Institute of Standards and Testing (NIST). Similar procedures for laminated beams are described in the American Institute of Timber Construction (AITC) manuals (ANSI/AITC A190.1.1992). Separately, some of these testing procedures are described by ASTM (American Standards of Testing and Materials) methods, which are often imbedded into the APA or AITC procedures. For PB and MDF panels, the Composite Panel Association (CPA) spearheaded the testing procedures for interior use boards, such as formaldehyde emission, water repellency, and mechanical strength values. There are also several other contractual testing organizations that do the necessary tests regularly to issue the grade certifications according to the respective standards.

20.3 Comparison of Various Wood Adhesive Bonds by Accelerated Aging or Exterior Exposure Tests

ABT and exterior exposure Test results: Wood, when subjected to only natural (chemical) effects of aging, has been known to last thousands of years with only small changes occurring in the mechanical properties, chemical composition, and morphology. The durability of adhesives has been compared in the past with that of wood substrates (control) by accelerated aging procedures discussed in Section 20.2.6. The past research results indicated that the kind of adhesive, wood species, method of adhesive bonding, and exposure environment (moisture, pH, and temperature) play predominant roles in determining the durability of adhesive-bonded wood products [7, 8]. One typical procedure developed and used extensively is the *ABT* (ASTM 3434-75) method [8]. Typical test results of various adhesive-bonded solid wood products (laminated beam and softwood plywood) obtained from the ABT and *Outdoor Exposure Tests (Exterior)* done for 16 years are reported by Dick Caster of Weyerhaeuser Company [9] as shown in Figure 20.10 with shear strength values retained (%) at the indicated number of aging cycles or years of exposure. Conclusively, the tested adhesives were classified into three performance groups: (i) Top performing – Epoxy (SCE and FCE), PF, PRF, and Urethane #2; (ii) Middle-performing – MUF, MF, and Casein; (iii) Poor performing – Urethane #1 and UF. Also, southern pine did not perform as well as Douglas Fir in outdoor exposure, data of which were not included in the exterior exposure graph. It is noted that urethane resins can perform well or not as well depending on the formulation.

Also, the results of various wood adhesives-bonded birch plywood exposed to the exterior condition for nine years or longer are reported by E.G. Bergin. and V. Godin, which are shown in Table 20.1 [10]. Similarly, with the results of the ABT, the PF and PRF resin adhesives are good, MF or MUF resins are intermediate, and UF and other resins, irrespective of the curing methods, showed relatively fast degradation of bond strengths.

Effects of aging of wood before gluing: Another factor affecting the durability of adhesive-bonded wood products was found to be the aging history of the wood used for bonding and the presence of extractives. The effects of aging of veneers and extractive amounts on bonding of southern pine with PF resin adhesives were studied and the results

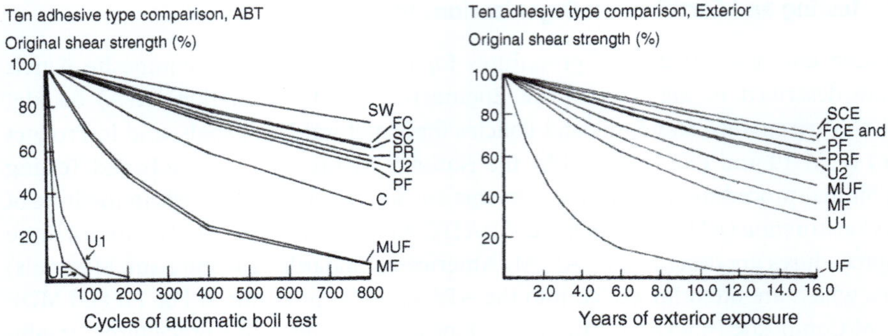

Figure 20.10 *ABT:* Original shear strengths of various wood adhesives bonds retained in the *Automatic Boil Test* results using southern pine and Douglas fir and; *Exterior*: the exterior exposure test results using Douglas fir at three US locations (average): Seattle, WA (mild and wet), Modesto, CA (hot and dry); Electric Mills, MS (hot and humid). Abbreviations: *SW*: solid wood; *SCE*: slow-cure epoxy; *FCE*: fast-cure epoxy; *PF*: phenol-formaldehyde, heat-cured; *PRF*: phenol-resorcinol formaldehyde; *U2*: urethane #2; *MF*: melamine-formaldehyde; *MUF*: melamine-urea-formaldehyde (60/40 M/U ratio); *C*: Casein; *U1*: urethane #1; *UF*: urea-formaldehyde; *SCE*: slow-cure epoxy. Test samples were made using lumber (6 mm-thick) cut in parallel, 2-ply and cross-grain 3-ply bonds, rather than veneer, same dimensions (25.4 mm × 24.5 mm bonding area) similar to plywood shear specimens [9].

Table 20.1 Summary of results after 15 years of outdoor exposure to birch plywood bonded with various wood adhesives.

Adhesive type		Time required to produce 10% delamination	Time required to produce 100% delamination
Phenol	(H.T.)	—	—
Phenol	(R.T.)	—	—
Resorcinol	(R.T.)	—	—
Resorcinol-Phenol	(R.T.)	—	—
Melamine	(H.T.)	6.5 yr	9 yr
Mel–Urea (50 : 50)	(H.T.)	6.5 yr	9 yr
Urea	(H.T.)	2 yr	5 yr
Urea	(R.T.)	2 yr	5 yr
Urea-S.A.	(R.T.)	10 mo	3 yr
Urea-Ext.	(H.T.)	1 yr	2.5 yr
Casein		10 mo	32 mo
Soybean		10 mo	26 mo
Vegetable		1 mo	18 mo
Polyvinyl "B"		1 mo	24 mo
Polyvinyl "A"		0.5 mo	3 mo
Animal		1 mo	3 mo

Abbreviations: R.T. = Room-Temperature-Curing, H.T. = High-Temperature-Curing, Ext. = +100% Wheat flour, S.A. = Separate-Application-Catalyst.
Source: Bergin and Godin [10]/Newfoundland and Labrador.

showed that the bond strength decreased faster with days of aging of veneers, and the chemically extracted veneers retained their bond quality even after four months of aging [11], indicating the extractives on the surface of wood are detrimental for durability of adhesive bonds. Also, the influence of surface aging prior to gluing on bond strength of Douglas fir and redwood was studied with PRF and casein adhesives by examining samples at 15- and 30-days intervals over a period of three to five months [12]. Migration of extractive materials to the wood surface, temporal chemical changes in wood that resulted in weaker wood fibers at the surface, and absorption of airborne contaminants, all were found to be detrimental to wood bond durability.

20.4 Nondestructive Testing of Wood and Wood Adhesive Bonds

Nondestructive testing methods are useful as quality control methods in the production line. For example, in the finger-jointing plant, it is accomplished by applying a load to induce an allowable stress for the piece. A limit switch system discontinues the load application if there is any tendency for excessive deflection. The sequential loading/deflection data are also analyzed to eliminate any damaged pieces in the proof loading process [13, 14]. Also, inspection of bonded products in irretrievable structures needs to be done nondestructively, and some are simple methods and others need special instruments [15, 16].

Visual or simple measuring: Voids, cracks, color, etc., can be inspected visually, and thickness of boards is measured to show blows or poor bonds formed due to thickness expansion (spring-back). Also, the on-floor testing of trims using a crude hammer, knife, or chisel for WF values is widely used in plywood and laminating plants.

Mechanical proof loading: A certain passing weight is loaded on test boards or adhesive joints [13] in the production line to detect and eliminate the weak boards while not damaging sound boards in bending, tension, or compression modes. Also, probing or coring tests are carried out.

Physical tests: Acoustic emission [17], X-ray wave propagation, vibrational properties, dielectric properties, electrical properties, and near-infrared (NIR) spectral tests are used for test of defects, decays, voids, termite attacks, or any density variations.

Chemical tests: Chemical composition tests for the presence of preservatives and fire retardants are carried out.

20.5 *In-situ* Adhesion Testing

In adhesive evaluation research on bonding of wood with various adhesives, the Adhesive Bonding Evaluation System (ABES) has been widely used and described in ASTM D 7998-15 [18]. This instrument can measure lap shear values of adhesion between two solid wood pieces (20 × 20 mm) anytime during or after the bonding procedure with a controlled bonding time, temperature, and pressure. Measurements of cure extents and development of strength values over cure times can be investigated, which is a very useful instrument in screening many different adhesive formulations without making boards.

References

1 Zink-Sharp, A.G. and Kartunova, E. (1998). *Forest Products Journal* 48 (4): 69.
2 Shen, K.C. and Carroll, M.N. (1969). *Forest Products Journal* 19 (8): 17.
3 Shen, K.C. and Carroll, M.N. (1971). *Forest Products Journal* 21 (6): 40.
4 Pocius, A. and Dahlquist, C. (1986). *Adhesion and Adhesives*, 146–163. Washington, D. C.: American Chemical Society.
5 Patton-Mallory, M. and Cramer, S.M. (1987). *Forest Products Journal* 37 (7/8): 39.
6 Northcott, P.L. and Colbeck, H.G.M. (1960). *Forest Products Journal* 10 (8): 403.
7 Freeman, H.G. and Kreibich, R.E. (1968). *Forest Products Journal* 18 (7): 39.
8 Gillespie, R.H. (1980). In: *Proceedings of the 1980 Symposium on Wood Adhesives, FS, FPL and WSU*. Madison, WI, p. 168.
9 Caster, D. (1980). In: *Proceedings of the 1980 Symposium on Wood Adhesives, FS, FPL and WSU*. Madison, WI, p. 179.
10 Bergin, E.G. and Godin, V. (1971). Department of Fishery and Forestry, Canada. Publication No. 1296, pp. 1–22.
11 Perry, D.A. and Choong, E.T. (1968). Louisiana State University, Wood Utilization Note No. 13, p. 1.
12 Stumbo, D.A. (1964). *Forest Products Journal* 14 (5): 582.
13 Eby, R.E. (1981). *Forest Products Journal* 31 (1): 37.
14 Bolger, R.J. and Rasmussen, C.A. (1962). *Forest Products Journal* 12 (9): 422.
15 Ross, R.J., Brashaw, B.K., and Pellerin, R.F. (1998). *Forest Products Journal* 48 (1): 14.
16 Brashaw, B.K., Bucur, V., Divos, F. et al. (2009). *Forest Products Journal*. 59 (3): 7.
17 Birks, A.S. (1972). *Forest Products Journal* 22 (6): 23.
18 Adhesive Evaluating Systems, Inc. 55 SW Madison Avenue, Corvallis, OR 97333.

21 Introduction to Coatings Technology for Wood

21.1 Three Components of Coatings

Coating materials include paint, enamel, varnish, lacquer, and stains. These materials consist of *binder polymer, pigment and filler, and carrier solvent* as major components. When the applied coating is dried (cured), the binder and pigment components become the two major components of the film formed. The binder in coatings is often called a "vehicle" because it carries the pigments, i.e. without the binder, the pigments (and fillers) will fly away. Vehicles are mostly colorless (transparent) organic polymers, and pigments and fillers are very fine solid particles with colors originated from organic or inorganic solid materials. The total amount of pigments that give the color is relatively small (a few percentages); a greater amount of lower-cost fillers is used to augment the main color pigment and to bulk the film, which also improves certain other film properties. Carrier solvents keep the other components in solution, suspension, or in emulsion forms. Small amounts of additive materials, such as for increasing the viscosity, flow control, wetting properties, and fungal protection, are also added.

21.2 Pigment Volume Concentration (PVC)

$$\text{PVC} = V_{\text{f+p}}/(V_{\text{f+p}} + V_{\text{v}})$$

$V_{\text{f+p}}$: Combined volume of filler and pigment.

V_{v}: Combined volume of vehicle polymers.

The volume proportion of fillers and pigments vs. vehicle polymers in a piece of dried coating film is an important parameter [1, 2]. Fillers are relatively low-cost, and vehicles (polymers) are more expensive materials. Color pigments are somewhat expensive, but the amount is small and classified as parts of fillers. The ratio of these two components is related to the integrity of coating films and also determines other film properties, such as porosity, durability, gloss, and adhesion to the substrates. In Figure 21.1, a dried coating film is shown schematically that consist of fillers, vehicles, and voids, occupying the whole space of the film volume. The carrier (solvent) does not remain in the

Adhesives and Finishes for Wood: For Practitioners and Students, First Edition. Moon G. Kim.

Figure 21.1 Schematic representation of a dried coating film: dark = vehicle polymer; white hatched = filler particles; white = unfilled voids.

dried film. If comparing a series of coatings composed of the same kinds of fillers and vehicles, the volume ratio of fillers (V_{f+p}) vs. fillers (V_{f+p}) plus vehicles (V_v) defines the film characteristics and is termed as the *pigment volume concentration* (PVC) of a particular coating system. A high PVC value means a high amount of pigments/fillers and a low PVC value means a low amount of filler/pigments. Thus, if the PVC value is incrementally increased from zero by adding fillers/pigments, the coating system will go through the low PVC regime first, where much of the dried coating film is composed of polymer vehicles. The dried film would be possibly impermeable to a liquid such as water or water vapor since all interstices of filler/pigment particles are filled up with the impermeable vehicle polymer. Then, it will come to a point where the film starts to show some permeability because the amount of polymer vehicle becomes insufficient to fill all the interstices between the filler/pigment particles. This point is called the *critical pigment volume concentration* (cPVC) for a particular coating system. In summary, coating films made with a PVC value above cPVC value will be permeable, while those films made with a PVC value below the cPVC will be impermeable to water, water vapor, or air.

Chemical structures of common fatty oils

$$\begin{array}{l} CH_2\text{-}O\text{-}C_{16}\text{-}H_n\text{-}CH_3 \\ | \\ CH_2\text{-}O\text{-}C_{16}\text{-}H_n\text{-}CH_3 \\ | \\ CH_2\text{-}O\text{-}C_{16}\text{-}H_n\text{-}CH_3 \end{array} \tag{21.1}$$

n = 26, 28, 30, 32

This measure of PVC is often used in formulating coatings for wood and dry walls because some porosity of coating film is sometimes needed to allow the water vapor in the room to diffuse in and out. For example, interior wall/ceiling paints are commonly designed to be permeable to water vapor in order to make moisture changes possible between the room and the wall and also with the spaces behind the wall. In addition to film porosity, the PVC value also affects the appearance of film and the gloss. Normally, the PVC values of enamels are 10–12%, for exterior paints it is 30–40%, and for interior paints it is 30–50%. These gradations of PVC values are also in line with the decreasing order of the richness and solidity of coating films: enamels have the highest luster and solidity, the exterior paints are the next, and the interior paints at times can appear like thinly bound chalk powders. The sheen (gloss, semi-gloss, and mat) of paint films also goes with the PVC values to some extent. The required PVC values are different for different polymer/filler combinations and need to be determined by experiments for each coating system.

Common fatty oils used in coatings with the position of double bonds indicated starting from the carboxylic acid carbon as No.1.

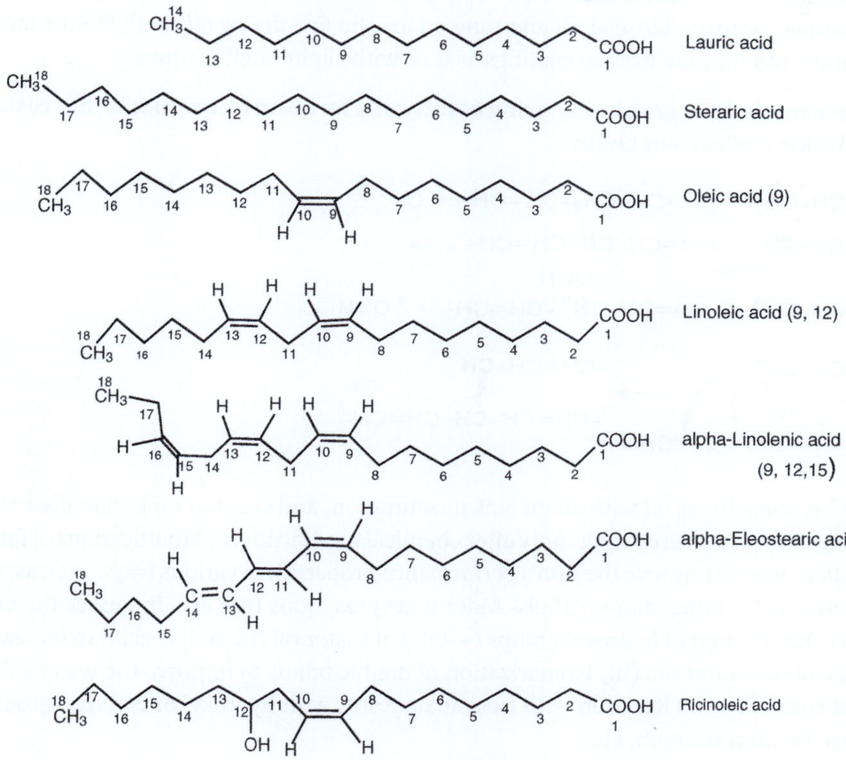

21.3 Various Kinds of Vehicle Polymers

21.3.1 Fatty Oils and Modified Fatty Oils

Natural oil-based coatings are the oldest type of coating vehicles. Unsaturated oils, such as linseed oil, tung oil, soya oil, castor oil, tall oil, and other vegetable oils, as well as saturated oils, such as coconut oil, have been used in coatings [1–3]. *All fatty oils* are glycerol esters of long-chain carboxylic acids of 12 carbons or higher, as shown (Eq. 21.1), and each carboxylic acid can harbor up to three (3) conjugated or unconjugated *carbon–carbon* double bonds (Eq. 21.2), the content of which is defined as *the extent of unsaturation*, which determines the curing reactivity of coating films with atmospheric oxygen. The curing reaction is often accelerated by metal salt catalysts, called *dryers*. The fatty oils are monomers, not yet polymers. In the formation of polymers from fatty oil monomers, the methylene carbons (—CH_2—) next to the carbon–carbon double bonds and those in between two carbon–carbon double bonds are activated to react with air oxygen (O_2) to form peroxides {=CH—(OOH)}, which decompose to reactive free radicals (=CH$^\cdot$) (Eq. 21.3) that react to other carbon radicals to form carbon–carbon bonds resulting in branched polymer chains and eventually cure to three-dimensional, cross-linked polymers. Therefore, the extent of unsaturation of fatty oils determines the curing speed of film, often called the "drying speed" in the coatings industry. Linseed oil and tung oil are the *fast-drying oils* with 60% or more unsaturation and they are used in coatings as is or with slight modifications.

Reactive methylene groups in curing of fatty oils forming free radicals that result in crosslinking of carbon chains

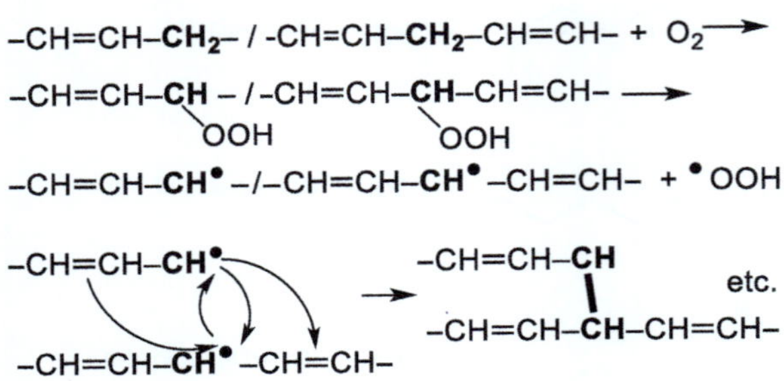

Soya oil is *semi-drying oil* with about 50% unsaturation, and coconut oil is classified as a *non-drying oil* and used in coatings only after chemical modifications. Modifications of fatty oils are often done to improve the film's performance properties in various ways, such as: (i) Heat-treatment for dimerization (Diels–Alder types) reactions that also increases the viscosity. (ii) Dehydration of hydroxyl groups (—OH) of castor oil (ricinoleic acid) to increase the degree of unsaturation. (iii) Isomerization of double bonds to improve the water solubility and curability. (iv) Reaction with isocyanate resins with hydroxyl or carboxyl groups to improve the film strength, etc.

A schematic representation of alkyd polymer formation from partially hydrolyzed fatty acid and di-acid.

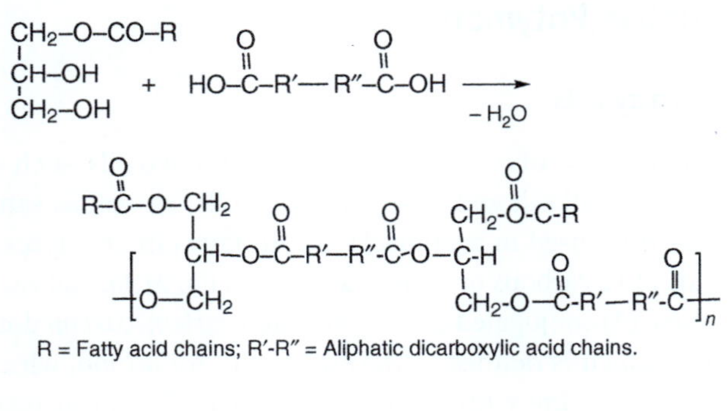

R = Fatty acid chains; R′-R″ = Aliphatic dicarboxylic acid chains.

21.3.2 Alkyd Resins

Alkyd resins were developed by condensation of fatty oils with diacids (HOOC—R′—R″—COOH) in the early days when the current high-performing polymers were unavailable, but they are still widely used today. Alkyd resins are made by partially hydrolyzing the glycerol ester bonds of fatty oils, adding some additional glycerol, and then reacting with a dicarboxylic acid, such as phthalic or maleic acid, to form *ester bonds*, resulting in longer chain prepolymer molecules that still have carbon–carbon double bonds in the fatty oil chains (Eq. 21.4). The resultant alkyd polymers show improvements in the mechanical property, drying speed, durability, etc., in comparison with unmodified or lightly modified fatty oils. There are many variations in the formulation of alkyd resin polymers depending on the *oil type* and *oil length*. The oil length is how much oil is used relative to the glycerol and dicarboxylic acid components added in the formulation, and this value is a barometer for the alkyd polymer's flexibility and solubility in organic solvents and also the drying speed when drying oil is used: short, medium, and long (high) oil-length values are available. For example, the long oil-length alkyd resins are only slightly modified fatty oils, and short oil-length ones are highly modified materials. Alkyd resins are often further modified for improvement of various film properties, most of all, to toughen the film for durability. Various modifiers are added to give certain advantages: (i) *Rosins*, in enamels for fast drying and increased hardness. (ii) *PF resins*, in primer coatings for increased hardness and chemical resistance, (iii) *Acrylics*, for fast drying and better color and gloss retention, (iv) *Styrenics*, in aerosol paints for fast drying, (v) *Epoxy resins*, for fast drying and film toughness, (vi) *Urethane polymers*, in floor finishes for increased abrasion resistance, (vii) *Silicone resins*, in exterior coatings for excellent gloss retention.

The solvent (carrier) in alkyd resins is usually xylene, an aromatic solvent that has a high solvent power. Naturally, alkyd paints are *oil-based paints* and are currently used mostly as exterior paints to give the wholesome (high build) appearance of well-painted houses in comparison to much of current exterior paints based on acrylic polymers. Alkyd resins are low-cost and versatile in formulating for various different performance properties with a wide-ranging compatibility with other film-forming resins.

21.3.3 Polyester – Unsaturated and Saturated Polyesters

Unsaturated polyester resins

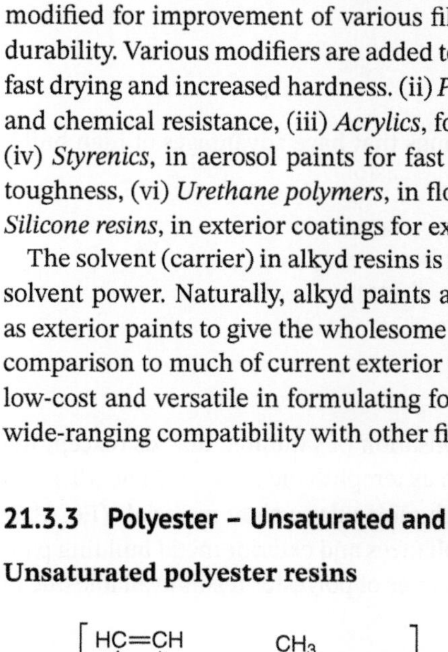

Unsaturated polyesters: There are two types of polyesters used in coatings: *unsaturated and saturated polyesters*, made by condensation of one or more di-acids and a polyol. Both are used as oil (solvent)-based coating polymers. *Unsaturated polyesters* are made from unsaturated diacids such as maleic acid (HOOC—CH=CH—COOH) and diols (HO—R—OH) for condensation. An example is shown here, composed of maleic acid and 1,2-propanediol (Eq. 21.5). These components are polymerized by esterification to some low extent of polymerization to give low-viscosity unsaturated polyester polymers, which are then mixed with styrene monomer (low viscosity), and the double bonds in the polyester polymers (Eq. 21.6) and styrene are all copolymerized using a peroxide catalyst (free radical) to cure the resin. Curing of these binder vehicle polymers together with fiberglass mats results in reinforced plastic structures used for building of boat, car, bathroom fixtures, etc. Also, curing of these resins with UV light or electron beam is carried out in the manufacturing (finishing)

Curing of unsaturated polyesters with styrene crosslinker

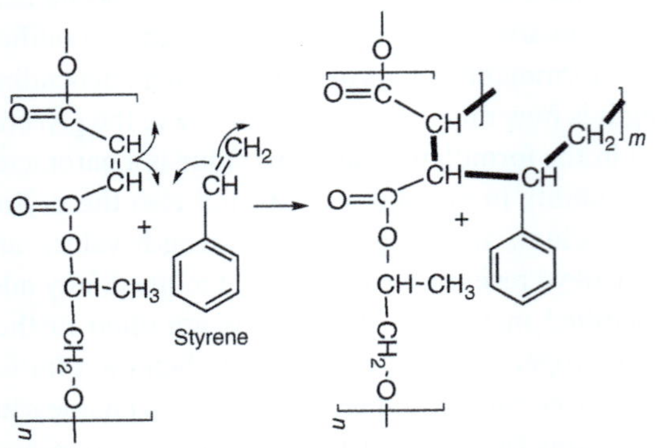

of decorative wood wall panels to give finish coatings that have advantages of high build, chemical resistance, and good hardness.

A typical saturated polyester resin

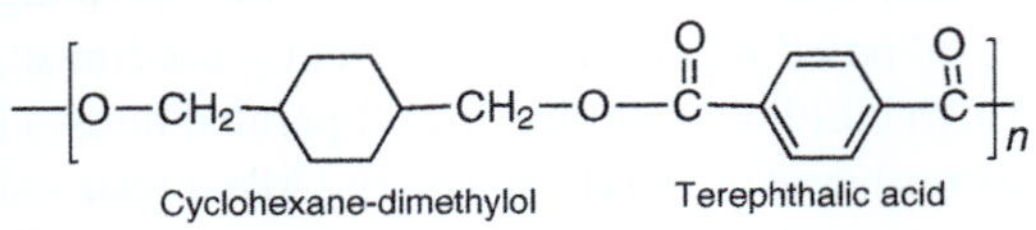

Saturated polyesters are manufactured by condensation of saturated diacids (except the unreactive double bonds in the benzene ring) such as terephthalic or isophthalic (di-) acid and a diol such as cyclohexylene dimethylol to high molecular weight materials (Eq. 21.7) and used as solvent-borne lacquer coatings for appliances and exterior metal building panels, and also for wood finishes. There are many varieties of polyester resins available due to the many different diols available in the industry.

21.3.4 Cellulosics

Cellulose esters are widely used as film formers in the coatings industry. The major cellulose esters currently used for coatings are the reaction products of cellulose with nitric acid, acetic acid, and acetic/butyric acids, as discussed in Section 1.5. In these derivatization reactions, cellulose molecular weights values are unchanged. Therefore, these high molecular weight cellulose esters are brittle and used as film formers only by blending with other resins and plasticizers that improve the flexibility, gloss, build, etc. These hard-coating materials based on cellulosics are usually called lacquers and characterized by the high drying speed and hard film properties. They do not need long drying times or heating for curing of film.

Nitrocelluloses: Nitrocellulose-based lacquers are currently used in large volumes for wood finishes. It was the main automobile coating material (black Fords), which was later replaced by other polymers due to the low exterior durability. They still hold a strong position in furniture finishes because of the ease of application, good drying speed at ambient temperature, easy removal for refinishing and spot patching, and especially in finishing of high-quality furniture (piano, etc.) because of the richness and depth of color that can be attained by rubbing operations. Nitrocelluloses with high nitrate group contents are flammable (gun powder).

Cellulose acetates (CA): are used as solvent-borne, grease-resistant coatings for paper, wire, and cotton fabrics, and also as lacquers for electrical insulation, release coatings for pressure-sensitive tapes, and coatings for various plastic products. Good resistance to weather exposure and sunlight. It requires very strong solvents for solubilization.

Cellulose acetate butyrates (CAB): have better solubility and wider compatibility with various resins and plasticizers than cellulose acetate (CA) and are used as clear and pigmented coatings for metal, wire, paper, plastics, wood, and cotton fabrics. They have good resistance to UV light and also show high heat stability, high moisture resistance, and good dimensional stability. They have good exterior durability, excellent color retention, and resistance to flame spread. On the other hand, compared to nitrocellulose, it has a slow solvent release rate and less resistance to alcohol and hydrocarbons.

21.3.5 Acrylic Resins

Various acrylic resins are produced by addition polymerizations of acrylic acid (***i***), methacrylic acid (***iv***), or their ester/amide derivatives (***ii***, ***iii***, ***v***) to high molecular weight thermoplastic polymer materials. The acrylic polymers commonly used in coatings are the copolymers of methyl or ethyl acrylate (***ii***, $R = —CH_3—$ *or* $—CH_2CH_3$) (for flexibility) and esters of methyl methacrylic acid (***v***, $R = —CH_3$) (for hardness) produced in emulsion (latex) or solution forms. They are used as lacquers and paints for automobiles, appliances, and general metal products, and exterior and interior finishes (acrylic paints) for wood, concrete, paper, leather, and textiles. They are excellent in color and gloss retention and have good exterior durability, good chemical resistance, and resistance to stains and other environmental agents. Various performance characteristics are available through compositional variations. Currently, most exterior house paints used in the United States are based on these acrylic emulsion polymers.

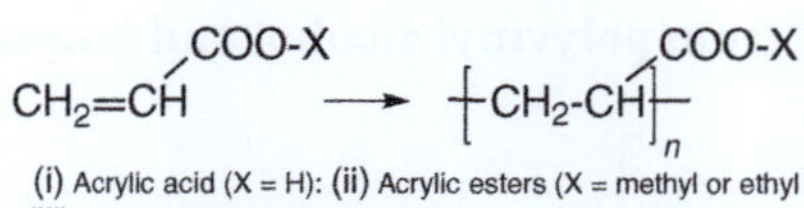

(i) Acrylic acid (X = H): (ii) Acrylic esters (X = methyl or ethyl groups): (iii) Acrylamide (X = NH_2 group).

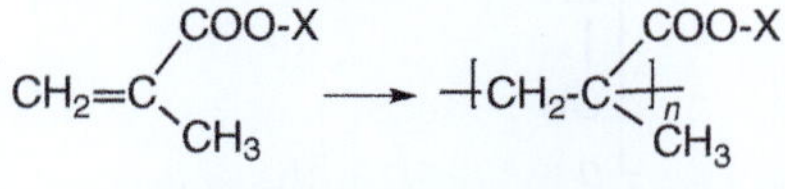

(iv) Methacrylic acid (X = H): (v) Methacrylic esters (X = methyl or ethyl groups

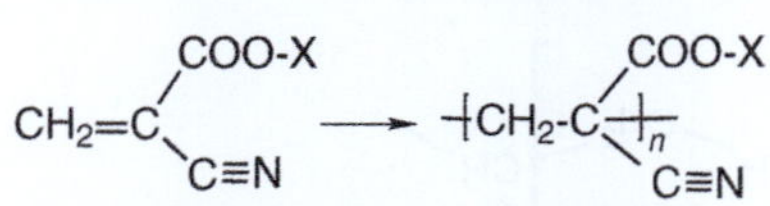

(vi) Cyanoacrylic esters (X = methyl or ethyl groups

Thermosetting Acrylic Resins: Another class of acrylics based on the *methyl methacrylic acid esters* (***v***) is obtained by combining the hydroxyethyl methacrylate (***v***, $X = —CH_2—CH_2—OH$) resin and butylated melamine-formaldehyde (MF) resins. The hydroxyethyl groups and MF resin molecules react to form crosslinks by acid catalysts and heat to result in *thermosetting coating vehicles* used for finishing autos, appliances, and other metals products, and also for wood finishes for hardboard-based door panels, renowned for their exceptional durability and color retention.

Methyl or ethyl cyanoacrylates (***vi***) are also a special class of adhesive polymers of the so-called "super glues." The presence of the cyano (—C=N) and carboxylic acid groups on carbon #2 in the molecule, both of which are strong electron-withdrawing groups, strongly polarizes the double bond and makes the monomer highly reactive to anionic polymerization, so that weak bases, such as water (moisture in wood or air), act as catalyst to bring about a rapid polymerization of the monomeric glue, resulting in fast curing and bonding.

21.3.6 Vinyl Resins

Polyvinyl aetate resins

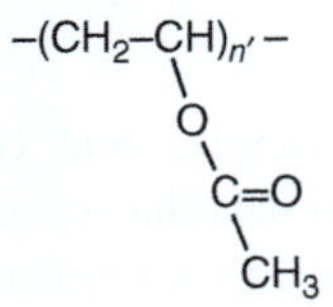

Polyvinyl Acetate (PVAc): As discussed in Chapter 14, addition polymerization of vinyl acetate results in PVAc polymers commonly in emulsion form (latex). The glass transition temperature of commercial solid PVAc polymers is about 28 °C, and they are modified with about 10–15% of plasticizer (dibutyl phthalate) for use as binders in coatings. They are extensively used as interior use wood adhesives and as vehicle polymers of interior flat wall paints and water-based sealers. The excellent adhesive characteristics of adhesives and paint vehicles are due to the acetate groups present in comparison to polyethylenes or polypropylenes.

Synthesis polyvinyl butyral from polyvinyl alcohol and butyraldehyde

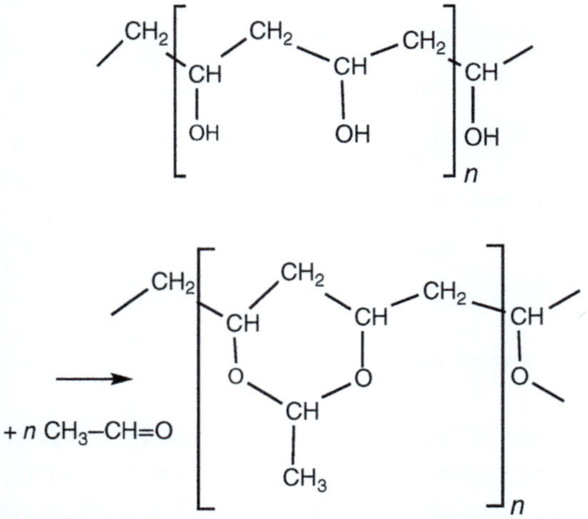

Polyvinyl Butyral: Polyvinyl alcohol is first produced by hydrolysis of *PVAc* and then reaction with n-butyraldehyde results in polyvinyl butyral. The polymer is used widely as metal wash primers and as sealers and wash coats in wood product finishing. It shows excellent adhesion to metal and wood and increases the flexibility, toughness, resistance to sunlight when used as a modifier to other polymers in coating resins.

Polyvinyl chloride

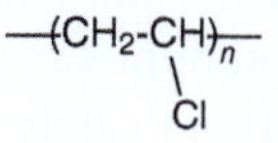

Polyvinyl Chloride: Obtained by the addition polymerization of vinyl chloride. It is the major binder in organosol and plastisol coatings. It is low-cost and gives water-/chemical-resistant coatings. An external plasticizer is needed. The coating shows poor heat and light resistance. For organosol coatings, the polymers and plasticizers are suspended as fine particles in a solvent system, a suspension, and not as a solution. After the coating is applied and dried, heat is applied to melt the polymer particles, resulting in their bonding to each other and to substrates. Use of the expensive and often toxic solvents is required.

Styrene-butadiene Resins: Addition *copolymerization* of styrene and butadiene results in various useful products such as styrene butadiene rubber (SBR), as discussed in Chapters 1 and 16, and used as binders of latex paints and mastics. They offer chemical-resistant coatings, masonry paints, concrete floor coatings, barrier coatings, and traffic marking paints. *Excellent* film stability, acid and alkali resistance, good resistance to oil, grease, and water. Poor color and gloss retention and limited compatibility with other type of polymer vehicles. High molecular weight SB resins modified with polyacrylonitriles are known as acrylonitrile-butadiene-styrene (ABS) polymers, which are extensively used as engineering plastics and coating vehicles for various appliances and office machines.

Formation of styrene butadiene copolymers

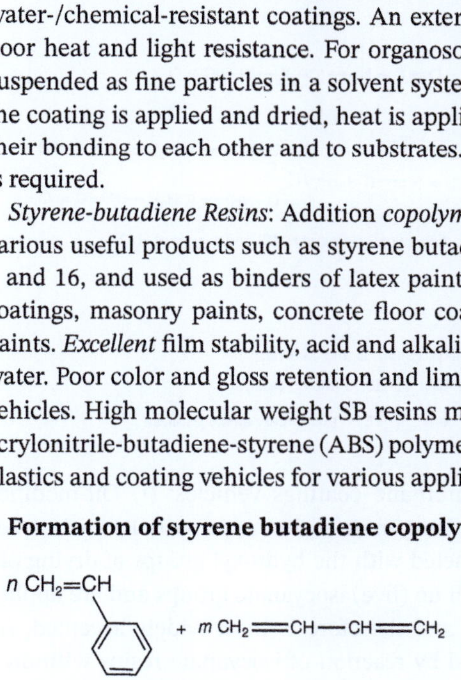

21.3.7 Epoxy Resins

Epoxy resins were discussed in Section 16.6.3. Most common epoxy resins are based on reaction products of epi-chlorohydrin and bis-phenol-A or phenol-formaldehyde novolac resins, resulting in oligomeric epoxy resins having two or more terminal epoxy groups. Principal uses of epoxy resins are for coatings (paints) where exceptional chemical and alkali resistance are needed, including industrial primers, exterior wall coatings, and marine coatings that require resistance to saltwater exposure.

21.3.8 Isocyanate Resins (urethane resins)

Isocyanate resins based on phenylene methylene-di-isocyanates (pMDI) were discussed in Chapter 12. As shown in the equations below, there are other isocyanate materials used in coatings: tolylene di-isocyanates (TDI), isophorone di-isocyanates (IPDI), hexamethylene di-isocyanates (HMDI), and adducts of HMDI and water. They give more flexible and therefore more durable coating films such as desired in floor varnishes. In coatings industry, isocyanate resin-based coatings are called *urethane resins*, and the hardener (B) components are polyols, polyamines, and polyphenolics, and their optimum curing conditions range from room to slightly elevated temperatures with or without catalysts depending on the isocyanate resins. The HMDI adduct has an higher molecular weight than HMDI, and it is therefore possible to have less personnel exposure to the toxic resin vapors in the coatings application stage.

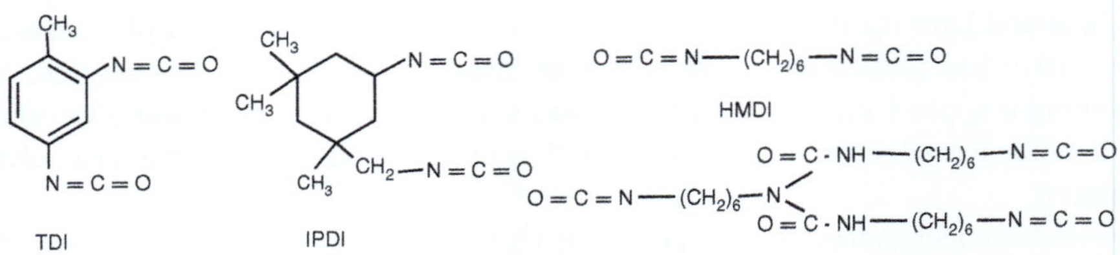

TDI IPDI

Adduct of 3 moles of HMDI and 1 mole of H_2O.

Typical reactions and reaction condition of isocyantes

R-N = C = O + Alkyl OH → R-NH-CO-O-Alkyl, 25–50 °C and catalyst.

R-N = C = O + Alkyl NH_2 → R-NH-CO-NH-Alkyl, 10–35 °C and no catalyst.

R-N = C = O + Phenyl OH → R-NH-CO-O-Phenyl, 50–75 °C and tertiary amine catalyst.

There are three approaches in making urethane coatings vehicles: (i) Oil-modified urethane resins are obtained by reaction of isocyanate resins with partially hydrolyzed drying oils until all isocyanate groups are reacted with the hydroxyl groups of drying oils to form urethane bonds. These coatings entail no (live) isocyanate groups and are applied similarly to common one-package systems, and therefore is most widely accepted. (ii) Moisture-curing urethane resins are obtained by reaction of isocyanate resins with diols and triols in such group mole ratios that the resultant polymer products still contain an adequate amount of reactive isocyanate groups, and when the coating is applied, the moisture from atmosphere reacts with the isocyanate groups to complete the cure of coating systems. (iii) Two-package systems: the first package is obtained by reacting isocyanate resins with polyols in such a mole ratio that relatively low molecular weight polymer resins, that still harbor a good amount of isocyanate groups, are obtained. The second package consists of diols and triols and catalysts, and the two packages are mixed prior to application of coatings on substrates. Two package systems, once mixed, have limited pot-lives. Urethane coatings are widely used as floor varnishes, seamless flooring and marine finishes, and other industrial maintenance and corrosion-resistant finishes because of their unique properties of toughness, flexibility, and abrasion resistance. However, the isocyanate resins and catalysts are toxic and require special handling procedures. Urethane coatings based on aliphatic backbones (HMDI and IPDI) show better color resistance to sunlight than those based on aromatic backbones (TDI or pMDI).

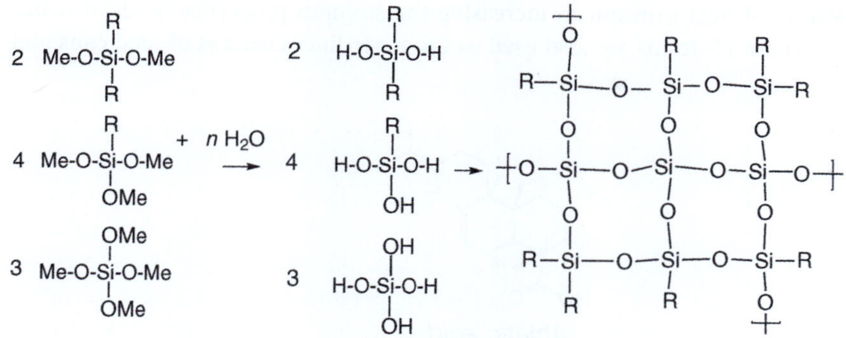

Figure 21.2 Schematic representation of silicone resins and their polymerization and crosslinking reactions initiated by water molecules (P = polymer extension).

21.3.9 Silicone Resins

Silicone resins are somewhat expensive materials due to the elemental silicon to be obtained from silica, followed by reactions to form mono- and di-methyl (R= —CH_3)) and/or di-, tri-, and tetra-methoxy (—OCH_3) silicon monomers, as shown in Figure 21.2. They are then polymerized in an appropriate ratio to low molecular weight silicone resins and processed in the presence of a limited amount of water, so that the molecular weight can grow to some intermediate stages with some methoxy groups remaining for curing reactions. The methyl or phenyl (—R) groups are stable substituents to the silicon atom and remain in the polymer, but the methoxy groups are reactive with water to hydrolyze to hydroxyl groups (—Si—OH) that polymerize among themselves to form silicon–oxygen–silicon (Si—O—Si) bonds of linear and branched silicone polymers and with aliphatic hydroxyl groups to form silicon ether (—Si—O—CH_2—) bonds in cases of using as modifiers of aliphatic resins. In the curing stage of the resins, the remaining methoxy groups react similarly with moisture from the air to further polymerize and crosslink among themselves or to make bonds with other hydroxyl group-containing coatings polymers (**P**) and modify them, resulting in imparting water repellency and stability for exterior exposure. Silicone resins are used as modifiers of various other coatings polymers as well as by themselves as sealants (caulking), rubbers, and adhesives.

21.3.10 PF, MF, UF Resin Derivatives

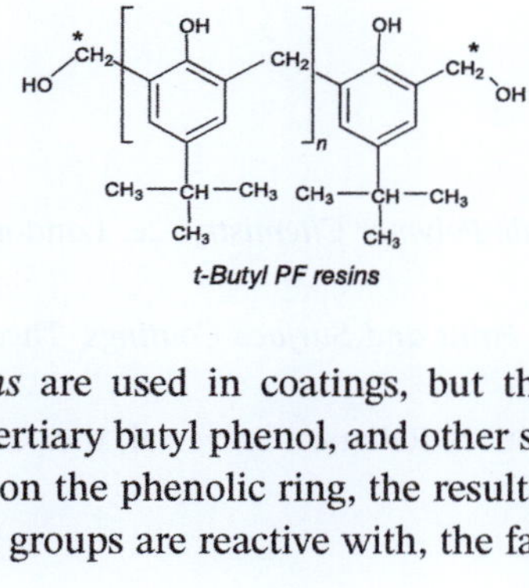

t-Butyl PF resins

Phenol-formaldehyde resins are used in coatings, but the phenols used are phenol-derivatives such as cresols, tertiary butyl phenol, and other substituted phenols as shown. Because of the alkyl groups on the phenolic ring, the resultant PF resins become soluble in, and their hydroxymethyl groups are reactive with, the fatty-oil based coating vehicles under application of heat to result in increasing the strength properties of alkyd resins or drying oils. These PF resins are also used as the inner liner coatings of food cans and drums.

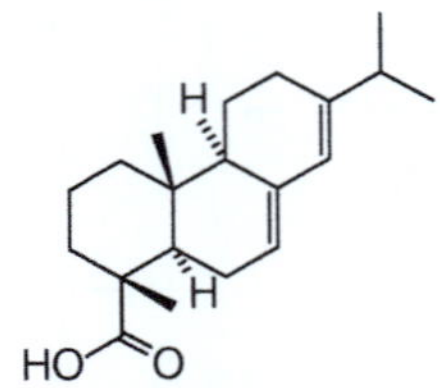

Abietic acid

Rosin modified phenolic resins are also used in coatings. Rosin is composed of polycyclic aromatic carboxylic acid materials obtained from conifer trees during pulping. The major component is the abietic acid and they are derivatized by Diels-Alder type reactions on their double bonds with PF resole resins in quinone methide forms and they are esterified with polyhydric alcohols for use as coatings and modifiers, especially for alkyd types.

Butylated urea-formaldehyde (UF) and melamine-formaldehyde (MF) resins are used as cross-linkers for thermosetting acrylics coatings vehicles for increasing the hardness and durability as discussed in Section 21.3.5. Introduction of butyl (ether) groups to UF and MF resins make them more stable and miscible with acrylics, and during the thermosetting curing reactions the butyl groups are broken and evaporated out, but the main amino methyl (*) resin groups are imbedded into the cured acrylic polymer structures.

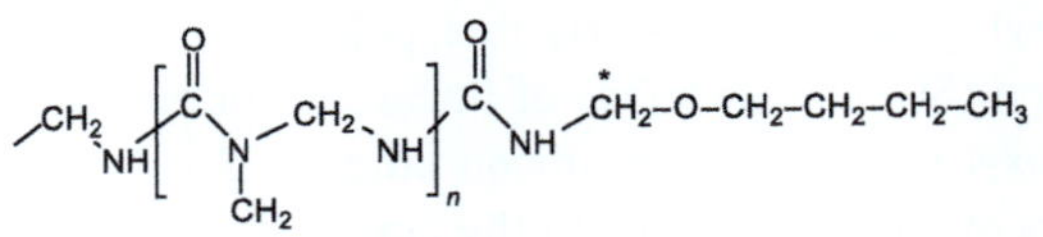

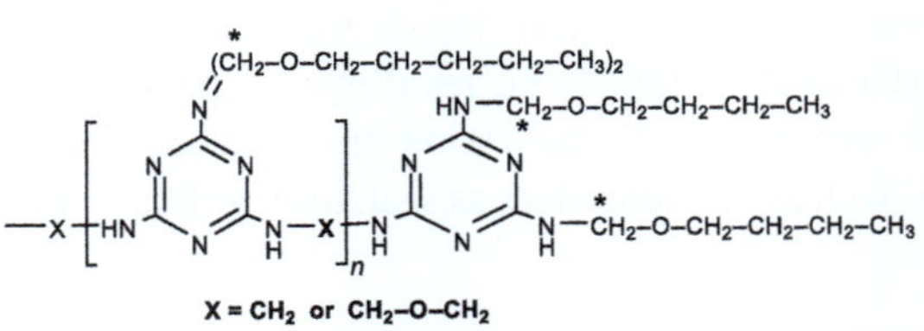

Butylated MF resins

References

1 Saunders, K.J. (1988). *Organic Polymer Chemistry*, 2e. London, and New York: Chapman and Hall.

2 Lambourne, R. (ed.) (1987). *Paint and Surface Coatings, Theory and Practice*. West Sussex, England: Ellis Horwood Ltd.

3 *Useful Facts & Figures*, 3 Edition, Reliance Universal, Inc., Louisville, KY, USA.

22

Introduction to Coatings Technology for Wood. II

22.1 Pigments and Fillers

Pigments and fillers are extremely fine solid particles to be suspended in the carrier and vehicle polymers and, when the coating is dried (cured), they are uniformly imbedded throughout the vehicle polymer film to give the coatings various performance properties, as shown in Table 22.1 [1]. It should be noted that the coatings as we see on a painted wall are mostly from the characteristics of the fillers and pigments, bound in place by transparent vehicle polymers. The vehicle polymers are not readily discernible except that they control much of the overall gloss and texture.

22.2 Manufacturing and Kinds of Pigments and Fillers

Pigments and fillers are manufactured from organic or inorganic materials, synthetic or natural, by grinding them to extremely fine particles as shown in Table 22.2 [1]. Especially, color pigments are ground (ball mill) to nanometer sizes in the presence of stabilizing colloids, resulting in thick pastes of pigments. The subsequent dispersion into the coating system (emulsion) can be done in a low shear-rate (slow) stirring process. Pigments can be classified into inorganic and organic pigments. Inorganic pigments are generally of dull color, high opacity, low color strength, and low cost. Organic pigments are of bright color, low opacity, high color strength, and high cost. Fillers are mostly inorganic materials and have white colors with low color strengths. They extend and modify the white color of the more expensive white pigment, titanium dioxide, as well as bulking the coating films for the various performance characteristics discussed above. The following white fillers are used: calcium carbonate (chalk), calcium sulfate (gypsum), aluminum silicate (clay and kaolin), magnesium silicate (talc), barium sulfate (barytes), and silica (celite and mica).

22.3 Color Control Methods

The visible white light is a combination of lights of varying wavelengths [1]. When the white light is beamed on a coating, lights with certain wavelengths are absorbed by fillers and

Adhesives and Finishes for Wood: For Practitioners and Students, First Edition. Moon G. Kim.

Table 22.1 Functions of pigments and fillers.

- Color – pigments
- Opacity – pigment and fillers
- Weather and corrosion resistance – fillers
- Photostability – fillers and pigments
- Film reinforcement – fillers
- Hardness – fillers
- Texture and gloss – fillers and pigments
- Body and consistency – fillers
- Leveling and brushing/roller qualities – fillers
- Fire retardancy – filler
- Non-skid – fillers

Table 22.2 Color pigments.

- White – Titanium dioxide (TiO_2), lead carbonate, zinc oxide, zinc sulfide
- Black – Carbon black, iron oxide, aniline black
- Brown – Iron oxide, organic browns
- Blue – Iron blue, cobalt blue, phthalocyanine blue
- Green – Chromium oxide, phthalocyanine green
- Yellow – Lead chromate, azo yellows
- Red – Azo reds, cadmium red
- Others – Aluminum flakes, pearlescent pigment, fluorescent pigment, zinc dust (for corrosion resistance)

pigments and the energy is dissipated into the surrounding materials as heat. Those lights that are not absorbed are reflected and reach our eyes to give us the colors of color pigments. There are many color pigments available in industry that can be used as is or mixed in various ratios with other color pigments to get desired colors. The various colors present in this world have been described in terms of color coordinates and model color solids as shown below:

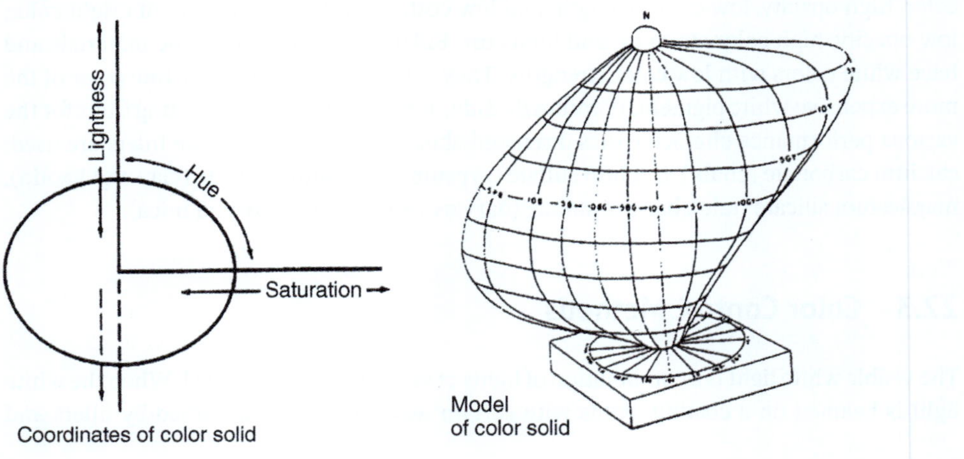

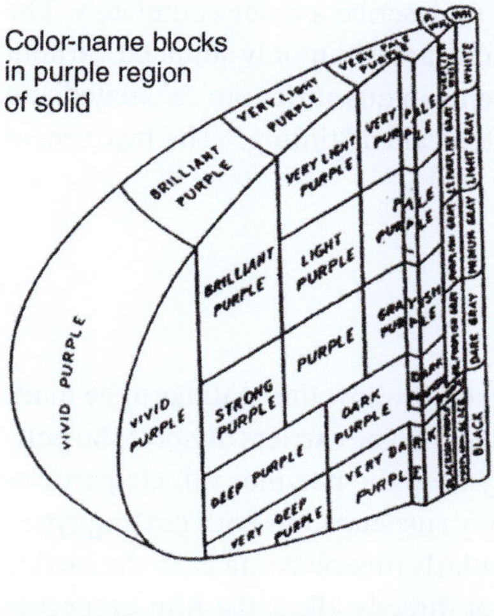

Reproduced from: The American Heritage Dictionary of the English Language. 1975. W. Morris, Editor. American Heritage Pub. Co, New York.

In principle, the three primary color pigments (yellow, blue, and red) can be mixed in various proportions to make various colors. To a color pigment thus obtained, the black or white color pigments can be added to vary the *lightness or color value*, which increases from black to white. On the other hand, the same color pigment can be added to the coating formulation in increasing amounts to get from weak to strong colors, i.e. increasing the saturation or the chroma values. *Hue* means the differences between a color and another similar color, i.e. yellow from yellow pale, etc. White, gray, and black colors are regarded as having no hues. Thus, the color-name blocks in the purple region of the color solid (shown above) show the various purple colors, indicating the chroma and color values. *Opacity* is often desired in coatings, i.e. masking the substrates. High chroma values or high color values do not guarantee opacity, and it is controlled by white pigments and fillers. The presence of white color pigments in base coatings (primers) means reflecting all visible wavelengths, resulting in blocking the view of substrates. Also, color pigments that have higher refractive indices relative to the refractive indices of polymer vehicles are selected to obtain increased opacity values: higher refractive index materials can lead to higher opacities.

22.4 Color Scales – Hunter *L*, *a*, *b*, and CIE 1976 *L*a*b** (CIELAB) Color Scales

Both the Hunter and CIE color scales consist of three parameters (scales) for defining colors, and color measuring instruments developed using these scales were based on the opponent-color theory, which assumes that receptors in human eyes perceive a color as the following three pairs of opposite colors: *L scale*: light vs. dark, where a low number (0–50) indicates dark, and a high number (51–100) indicates light. *a Scale*: red vs. green, where a positive number indicates red, and a negative number indicates green. *b Scale*: yellow vs. blue, where a positive number indicates yellow, and a negative number indicates blue [1].

Thus, *L value* indicates the level of light or dark, *a value* redness or greenness, and *b value* yellowness or blueness. All three values are required to describe a color completely. The color matches done in paint stores use these scales. Paints are commonly supplied without color pigments to paint stores, and desired color pigments in quantities are calculated and added with a good shaking before sale. This step is often called "tinting." The instrument can also read any (old) paint colors and match them.

22.5 Carriers of Coatings

Carriers (solvents) of coating formulations are used firstly to allow the coatings to be made in a liquid form so that they can be applied on substrates [1]. The carriers dissolve the polymer vehicles in *solvent-borne coatings*, while they only keep the polymer vehicle particles dispersed in *emulsion coatings*. Pigments and fillers stay suspended in both coating types. Sometimes, coatings contain dyes, which are in (molecularly) dissolved states in the carrier. Carriers in coatings evaporate upon curing and do not directly affect the film properties of coatings. However, carriers can affect the film formation characteristics: hardness, film formation speed, film adhesion on substrate, etc. The selection of carrier solvents for a particular coating system is made depending on the kinds of polymer vehicles, pigments and fillers, other additives for solvency, evaporation rate, odor, toxicity, flammability, environmental effect, cost, etc. Water-borne (emulsion, latex) systems are preferable in many cases and, if possible, powder and radiation coating methods are more desirable because of little volatile organic components (VOCs) that are produced in curing. Thus, carriers of coatings are organic solvents for oil-based paints, enamels, and varnishes and water for latex paints and partially water-borne coatings or water-reducible coatings. No carriers in powder coatings and electron-beam (EB) or ultraviolet (UV) curing coatings.

22.6 Additives to Coatings

Antifoam, anti-settling, leveling, antiskinning, dispersion aid, preservative, deodorant, optical whitener, UV absorber, and curing catalyst [1].

22.7 Manufacturing Procedures of Coatings

In manufacturing of coatings, the total solids level, pigment volume concentration (PVC), polymers, colors and pigment, and solvents, and order of putting them together must be first worked out. The dried film properties of finished coatings must be measured for intended applications. The color pigments are normally obtained as semi-dispersed pastes of agglomerated particles from suppliers, and they must be first dispersed completely to individual particles using small amounts of carrier solvents and a high-shear stirring procedure. Additives are normally needed to aid the dispersion process. Then, the polymer vehicles, fillers, balance of carrier solvents, and other additives are added under a relatively low-shearing condition (slow stirring). This low-shearing condition is important, especially in using emulsion forms of polymer vehicles. A uniform mix and other coatings' target physical properties (viscosity) need to be attained at the end of mixing. On storage of coatings, some separation of materials normally occurs, but it should be recouped by re-stirring before use. Separated coatings, if used without a good re-dispersion, will give different colors and different hiding qualities. Reactive coatings, which are mostly reactive with moisture, must be packed under an anhydrous condition, and once opened, all must be used soon.

22.8 Film Formation Mechanisms

When a *solvent-borne or emulsion coating* is applied on substrates and the drying step begins, the vehicle polymers start to lose solvent (supporting carrier) molecules and try to form solid particles from the solution or emulsion [2]. The curing process proceeds to form a continuous film that embeds the pigment and filler particles and also adheres to substrates. There are several paths to form the continuous adhering film depending on the kind of polymer vehicles and coating types used, as follows:

Cure by air oxidation aided by metal salt catalysts: fatty oils and alkyd-type vehicles; need catalysts to speed up the curing rates, normally called a "dryer"; carbons next to the double bonds in the fatty acid chains are activated and form peroxide groups, which decompose to free radicals to initiate the formation of cross-linked polymer networks out of double bonds in the fatty acid molecules.

Cure by reaction by heat: thermosetting-type vehicles, such as acrylic-melamine-formaldehyde resin mixtures used for automobile coatings and door skin manufacturing, are cured by heating to form cross-linked polymer structures.

Cure by reaction without heating: Isocyanates-based two-part systems, cured by catalysts.

Cure by reaction with moisture: Isocyanates-based one-part systems.

Cure by electron beam or UV/infrared (IR) lights: Unsaturated polyester-type vehicles containing styrene monomer. The double bonds are opened up to cross-link the polymer system by the free radicals produced from initiators by application of the lights.

Cure by drying: Varnishes and lacquers are made in organic solvents as carriers and drying off of carrier solvents usually gives the desired quality of films. High molecular weight polymers are used, and no cross-linking reaction occurs in the drying (curing) process. However, the high molecular weight polymers in solution are normally swelled and well stretched out until the drying step, and the molecules overlap with each other well to form a continuous film simply by drying.

Cure by fusion: Polyvinylchloride polymer resin particles used in organosol type coatings are only suspended (not dissolved) in the solvent and, after evaporation of solvent, they need to be fused by heating to form continuous films. No cross-links are formed between polyvinyl chloride polymers. Also, all powder coatings, which consist of fine polymer particles, are sprayed on heated substrates, and those landed on the substrates are melted and form films either by just melting or by heat-curing mechanisms.

Cure by coalescing: All latex paints form films cured by coalescing. No chemical cross-linking occurs, but coalescing agents are often used in small proportions. They are solvents with somewhat higher boiling points than the main carrier of water. After the water carrier is evaporated off, the coalescing agents remain in the drying film for a while to partially help dissolve or soften the polymer particles to allow engagement with each other for film formation. Later, the coalescing agent molecules eventually dissipate from the film to the environment. Thus, even latex paints are not emission-free, although the amount is exceedingly small. This is one of the reasons why water-borne latex paints smell when they are drying. Another reason for the smell of latex paints is the small amounts of monomers that remain unreacted in the initial polymerization process, but this source of odor has decreased greatly in recent years due to measures taken by polymer manufacturers because of the health concerns of various monomers.

22.9 Water-Borne Coatings and Coatings for Less VOC Emissions

Physical properties	Emulsion/water	Colloidal dispersion	Water-reducible	Solvent-borne (conventional)
Appearance	Translucent to opaque	Translucent to opaque	Transparent	Transparent
Molecular weight	Very (V) high, 1 M or higher	Up to 200 000	Up to 20 000	Up to 30 000
Viscosity	Low	Low to intermediate	High	
Need for auxiliary for pigment dispersion	Needed	Dependent of pigments	Not needed	
Need for film-forming aids	Needed	Usually needed	Not needed	
Foaming tendency	High	Intermediate	Intermediate	Least
pH Adjustment	Needed	Needed	Needed	N/A
Flow and leveling	Poor to good	Fair to good	Good to excellent	
Application solids	Highest	Intermediate	Lowest	Intermediate to lowest
Ability to wet substrates	Poor to good	Fair to good	Good to excellent	
Early water resistance	Very good	Very good	Fair to very good	Very good to excellent
Film toughness	Highest	Intermediate	Intermediate to lowest	
60° Gloss	Lowest	Intermediate	Highest	
Exterior durability, flexibility, hardness, corrosion and humidity resistance	Poor to excellent depending on the coating chemistry and cure goodness			

Development of water-borne coatings has been pushed in recent years by new regulations for lowering the emissions of organic solvents from traditional solvent-borne coatings [2]. Accordingly, aside from the latex-based coatings, some polymers are chemically modified to make them soluble in or at least compatible with water, either by derivatizing with certain functional groups or controlling the molecular weights.

Thus, water-reducible and colloidal dispersion coatings were developed as shown in the above table of a comparison chart of *the emulsion-based coatings, colloidal dispersion coatings, and water-reducible coatings vs. the traditional solvent-borne coatings* and their handling and performance properties. Appearances, molecular weights, and film toughness show clear differences, and different properties, such as exterior durability, hardness, and corrosion/humidity resistance, can vary depending on the kind of polymers and how well the curing process is carried out.

Comparison of various coatings for emission potential and energy usage

	Conventional >50% solids	**High solids 60–80% solids**	**Water-borne 30–50% solids**	**Radiation >95% solids**
Solvent usage	High	Moderate	Low	Very low
Pollution potential	High	Moderate	Low	Very low
Energy usage in curing	High	High	High	Low

High solids coatings have been similarly advanced for lower VOC emissions from solvent-borne coatings. The high molecular weight polymers cannot be readily formulated to have higher solids contents because of the high viscosity. Therefore, the polymers for this purpose are initially made to have lower molecular weights (about 5000) and, since these low molecular weight polymers do not perform as well, thermosetting groups are attached to them to cure in the film-forming steps to high molecular weight cross-linked polymers. These polymers are often called *oligomeric polymers* or *telechelic polymers*, the latter definition referring to low molecular weight polymers with functional groups on both ends.

Radiation curing coatings were discussed in Section 21.3.3 for unsaturated polyester/styrene coatings. This type of coating is made possible by the relatively low molecular weights of the polyesters and monomeric styrene, resulting in low-viscosity mixes applicable as thin coating films on substrates. They have been used for wood, paper, and plastic substrates, especially in the finishing operations of *hardboard and hardwood plywood wall panels*. Radiation curing includes EB, UV, and IR light heating methods. UV and IR methods, respectively, need photo-initiators and thermal-initiators formulated in the coatings to cure by free-radical reactions. Unsaturated polyesters, epoxy resins, and polyurethane resins are used this way by adding monomers to appropriately polymerized polymer components. EB curing equipment can be more expensive and also use higher levels of energy for curing than UV or IR light, but the latter methods are not useful for colored coatings because of their optical reflective properties.

Powder coatings polymers are made by evaporation of carrier solvents or water from the thermosetting polymer coatings (by spray drying, etc.) and then ground into fine powders. Heated metal substrates go through the powder spray chamber and catch the coating particles, which are then cured by thermosetting reaction mechanisms. Epoxy resins with amino or phenolic crosslinkers and acrylics with melamine cross-linkers are widely used this way for coating of appliances, toys, and metal furniture parts. Especially, the *electro-spray process* uses a highspeed rotating disc or an electron gun to spray the powder particles with negative charges directed to the heated coating substrates that are positively charged, resulting in more effective applications. This method produces not much VOC, at least in the coatings process, and the loss of coating material is minimal in comparison to liquid coating operations using compressed-air or pressurized gun sprayers.

22.10 Exterior Coatings for Wood

Stains, varnish, and paints are the coatings used for finishing wood for exterior uses [3]. They are used to protect wood from degradation but differ in appearance, durability, and ease of application. The demarcation between stains and paints can often be ambiguous.

22.10.1 Stains

The pigments used in stains are commonly red or yellow iron oxides, giving the "natural wood" shades, and although transparent or semi-transparent stains show the UV screening effect needed to protect wood from sunlight (UV) degradation. Low solids, penetrating oil-borne stains containing fungicides and preservatives are used on fencing, garden post, (sawn) cladding, rough-sawn wood, and softwood plywood. Stains are easier to apply and reapply than paints or varnishes. Stains are in general formulated to have less than ~20% solids (binder resin and pigment) and low binder resin loadings, which means the dried coatings do not form a continuous film and the binder polymers only bond the pigment particles on to the wood surface. Therefore, dried stains do degrade in time not by flaking or crack formation but mostly by erosion. Better penetration is attained with solvent-borne stains than with water-borne stains. Stains are not a wood preservative and therefore are best applied on preservative-treated wood. For solvent-borne stains, alkyds and oils are often the binder polymers. Water-borne stains are based on aqueous latex polymers, but they do not penetrate wood well. Stains are also modified with silicone resins, waxes, etc. to give some water repellency. Also, there are oil-based, or latex-based commercial finish products known as *heavy-bodied solid color or opaque stains*, but they are more similar to paints because of their film-forming characteristics, and they are used widely on textured panel products such as hardboard.

22.10.2 Varnishes

Varnishes are based on solvent-borne systems, have good flow properties in application, and result in transparent, clear glassy, attractive coat on wood without covering up the surface features. Varnishes are used extensively in finishing interior use woods, such as furniture and flooring. Exterior use varnishes have been decreasing because of their flaking and discoloration tendencies upon sunlight exposure and weathering and therefore high maintenance costs are involved, but there still remains some demand for good quality varnishes to set off the aesthetic qualities of wood in semi-exterior uses such as front doors. The degradation of varnish coatings arises from the film's low permeability and brittleness that gets severe with aging as well as their transparency that allows sunlight to penetrate to wood surfaces and gradually degrade the wood surface and the bonds with binder polymers. Traditionally, matrix resins of varnish have been tung and linseed oils modified with PF resins because of their good UV absorbing (blocking) characteristics. UV absorbers and iron oxides are also added to increase the UV absorbing (dissipating) capacity. The resistance to flaking and crack formation of varnish films can be improved by applying thinner films, but a yearly maintenance is often needed in case of severe exposure to sunlight. The urethane and alkyd-modified urethanes are good vehicles for interior use varnishes but they do not perform as well in exterior uses.

22.10.3 Paints

Compared with stains and varnishes, paints provide a better protection against erosion of wood from weathering and sunlight, but they normally are not formulated with preservatives and therefore do not protect wood against decay caused by fungal growth. Paints are commonly divided into solvent-borne or oil-borne and latex or water-borne systems. The traditional high-build full-gloss paint systems for exterior woodwork are commonly comprised of a primer coat, undercoat, and topcoat.

Primer coats provide bonding between the substrate and the subsequent coating layer. In old days, pigments like white lead or red lead dispersed in linseed oil were the major primer coats, but the toxicity of lead led to the use of other common pigments, dispersed in various oils or alkyd resins. The PVC values of current primer coatings reach to about 30–40%, by which some flexibility of the film is attained. Primer coatings are usually formulated to contain some fungicides.

Undercoats contribute to the build and opacity of the paint system. Use of undercoats in repainting jobs helps attain better adhesion with the old glossy paints and on sharp edges. They are formulated as solvent-borne or water-borne systems with relatively brittle alkyd-type polymers at high PVC values to make the sanding easier in (future) repainting, although their durability would become low. If the undercoat is replaced with an extra topcoat, the durability of the film would be improved, but the usual high-build, full-gloss appearance may not be obtained.

Topcoats contribute most to the protection of wood against erosion by sunlight and weathering and also determine the color and sheen. Flexibility and durability of topcoat vehicle polymers are highly desired for erosion resistance, and inclusion of an appropriate fungicide is necessary. Various alkyd resins have been used to obtain the traditional high-build, full gloss appearance in exterior paints, but they tend to lose flexibility and often require repainting in four to five years to maintain the full gloss appearance and also to repair flakes and cracks that appear as the exterior exposure time goes. In recent years, water-borne acrylic topcoats have become the well-established products and are often available as primer/topcoat combinations without the undercoat system. Acrylic vehicles are more flexible and more durable, and their high gloss lasts far longer than alkyd vehicle polymers. Most exterior topcoat paints used on houses and buildings nowadays are based on these water-borne acrylic latex polymers. Topcoats are best to be applied within two or three days after the primer coat to avoid poor bondings between the two layers. Paints are generally best applied on warm days without direct sunlight exposure to allow an ample time for the film to form gradually or for polymer particles to coalesce adequately, without drying up.

22.11 Summary

Pigments and fillers and color pigments in coatings are discussed with the production methods of materials and color control principles. Film-forming mechanisms of various polymer binder materials for coatings are also presented. Water-borne and reduced emission coatings are emphasized, and stains, varnishes, and paints as well as primer coats, undercoats, and topcoats of paints are discussed, all based on cited open literature.

References

1 Lambourne, R. (ed.) (1987). *Paint and Surface Coatings, Theory and Practice*. West Sussex, England: Ellis Horwood Ltd.
2 *Useful Facts & Figures*, 3e. Louisville, KY, USA: Reliance Universal, Inc.
3 Feist, W.C. (1989). Protective finishes and coatings. In: *Wood and Wood-Based Materials*, 230. Cambridge, MA: The MIT Press.

23

Industrial Coating Application Processes

(This Chapter is wholly adapted from Useful Facts & Figures, 3rd Ed., Courtesy of Reliance Universal, Inc., Louisville, KY, [1]).

23.1 Application Methods of Coatings

Airless spray guns: Since the early days of wood finish operations by brushing, wiping, dipping, and hand spraying, there have been significant advances in the speed and material efficiency attained with spray guns. Air pressure sprayers with siphon feeds of coatings have been the more common methods, but the *airless spray guns developed*, where the coatings feed is hydraulically pressured, resulted in nearly 15% savings on coating materials because of less overspray, fewer rejections, more easily controlled film thickness, improved adhesion, and uniformity of film (Figure 23.1). Where heavy films are desired, a single coat with airless spray can often give results that would require two coatings by air spray systems. Further, the *heated airless spray guns* later introduced led to finer finishes possible at lower hydraulic pressures by using higher viscosity coatings as lower viscosity materials. This method also results in faster buildup of film thickness due to fast evaporation of solvent, resulting in time savings when thicker coatings are necessary. Also, airless spray guns were further developed to have a mechanism of mixing resin with catalysts in desired proportions, driven by air pressure, and became useful in such applications as urethane foams, fiber-reinforced plastics, gel coats, epoxy adhesives, and just about any multi-component systems.

Electrostatic spray methods using spray gun or high-speed rotating disc sprayer: This method uses high voltage electric power (50k–150k V) within an enclosed chamber, where individual coating particles get charged when they pass through the spray gun or high-spinning disk sprayer (Figure 23.2) that works as one terminal, while the substrate objects are hung and oppositely charged and pass through the chamber, resulting in painting on all exposed areas of the object articles. In this process, sharply angled corners are not painted well and need touch-up paints. Metal objects, being electrical conductors, readily accept electrical charges to become ready acceptors of coatings. This process works for wood products when the moisture content (MC) of wood is kept in the range of 10–20%, the average wood MC values in some European countries. Since the average MC in the

Adhesives and Finishes for Wood: For Practitioners and Students, First Edition. Moon G. Kim.

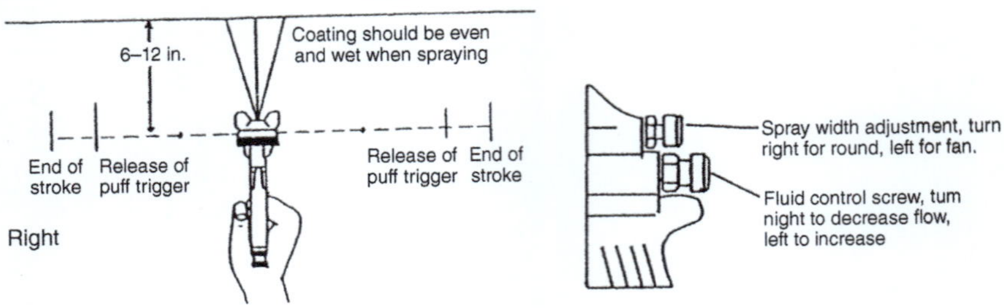

Figure 23.1 Optimum spray gun operation methods and control valves.

Figure 23.2 A high-speed rotating disc sprayer in motion.

U.S. runs in the 6–8%, the coating method is inadequate and therefore, the wood needs to be made conductors by applying thin electrical conducting sealers or chemicals in simple dip-and-dry treatments. Overall, this process results in significantly higher application efficiency than with conventional spray guns.

Roll coating method: The direct roll coating (DRC) method continues to be the high-speed means of applying coatings such as lacquers, enamels, sealers, primers, and waxes to all flat work substrates such as plywood, hardboard, and other composition boards (Figure 23.3a). There are top, bottom, or top and bottom roll coaters as shown in Figure 23.3b. Curved or routed surfaces cannot be coated. Commercial heavy-duty roll coaters feature rugged construction and rigid framing to maintain high quality, high volume, trouble-free production, and performance.

Reverse roll coater: In the early days of finishing of *hardboard*, the rough surfaces resulted in the "hungry look" of finishes due to the "soak-in" tendency of boards, and this was greatly improved by the reversing the direction of the panels going into the roll coating process. This forcing of panels against the pressure of coatings on rollers resulted in pressure-filling the pores on board surfaces with coating primers and sealers, resulting

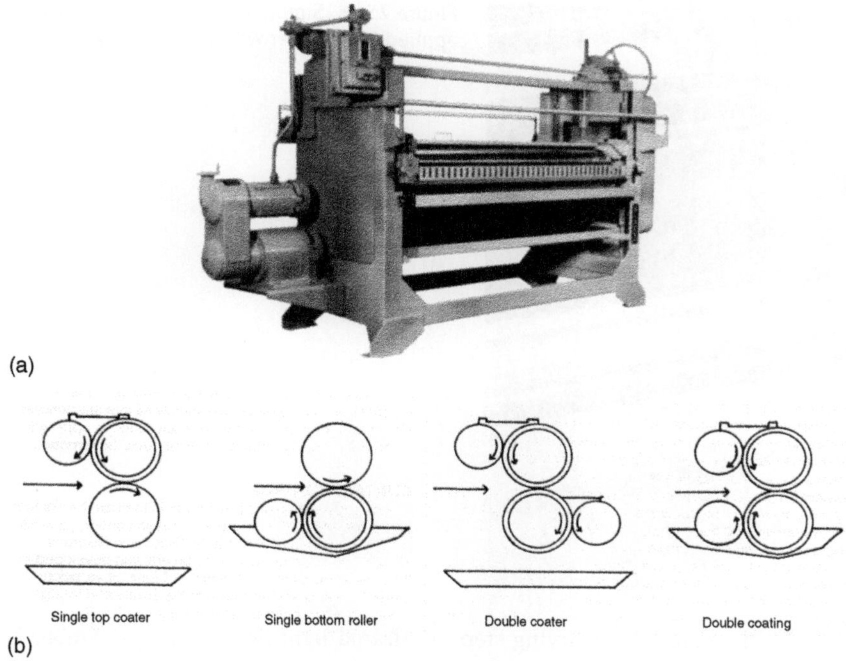

Figure 23.3 (a) High speed roll coating machine. (b) Various roll coaters: double ringed drums carry coatings; straight arrows indicate panel directions.

in better, nonporous surfaces for subsequent finishing processes. This process has been further refined to become useful for finishing hardwood veneer and wood composite board panels, filling imperfections, pore structures, cracks, and indentations with sealers.

Precision roll coaters: Precision roll coater uses engraved rollers that apply stains, toners, and toner inks onto well-prepared hardboard or hardwood surfaces to produce patterns such as natural wood grains as shown in Figure 23.4. This method involves the printing of desired patterns, and the substrates must be well-prepared with no voids or blemishes. This method is extensively used in finishing wood wall panels and flat furniture stock pieces.

Curtain coating: This method is a high-speed production process for applying smooth films of lacquers, paints, varnishes, adhesives, waxes, etc., to substrates that are flat or moderately curved or routed. The substrates pass under a falling, continuous stream (curtain) of a coating material. The production rate is extremely high, for example, 150 m per minute, made possible by passing the coating materials through a pressure head or a weir-head. The excess coating material that is not deposited on the substrates is collected at the bottom and recycled. The applied coating could be thin in one pass, and two passes may be needed; also, the sides and edges of substrates may not be coated as well, and it would be necessary to reprocess or use some other methods to overcome this.

Knife coating: This method is akin to the reverse roll coating, i.e. pushing fillers or sealers into voids on plywood or any solid wood surfaces, which are fibrous, porous, and less resilient than hardboard or other composition boards, which are best treated with the reverse roll coaters but decided to be knife-coated for some other reasons. In the knife coating process, an excess sealer or filler is applied, and a doctor's blade follows to scrape

Figure 23.4 Simulated woodgrain pattern applied on a printer with an engraved cylinder.

off the excess coating materials. A drying step in infrared light or air-drying chamber is followed. A high-speed coating is possible.

Flow coating: This coating procedure includes a coating chamber where the articles hung on overhead conveyors are moving in and out while several vertical coating streams are shot from the bottom to fully drench the articles. At the end point of this chamber, the coated articles just look like the dip-treated articles. After waiting some time, the drying or baking step is followed to cure the coatings. The utilization efficiency of coating materials is high, reaching 90–95% in comparison to about 70–80% for dipping processes and 50% for manual or automatic spraying processes.

23.2 Sanding Processes

Proper and thorough sanding is needed to carry out finishing jobs successfully for protecting or preserving the wood qualities and enhancing the beauty or accentuating the grain patterns of wood [2]. There are various sanding machines, depending on the sizes and shapes of the articles.

Drum sander: is used for flat and large pieces just after planer or veneering procedure, to remove roughness and to bring to desired dimensions (thicknesses). The abrasive to use with respect to its backer, adhesive, mineral, grade, form, and flex depends on the mechanics and severity of the operation and materials to be sanded. In very heavy sanding operations, where the abrasive drum wrap is subjected to high stresses, fiber-backed abrasives are recommended. In light sanding operations, "E" weight paperback abrasives are adequate and more economical.

Wide-belt sander: uses an abrasive belt of 12 in. and wider, has straight-through feeding device, and it is often used in place of drum sander. The feed speed can be high, 30–200 ft per minute. The sanders are available for single- or double-side sanding, with one or more sanding heads per side, and abrasives constructed of resin over resin-bonded aluminum oxide on X-weight cloth belts are suitable.

Stroke sander: This method of sanding normally means a touch-up sanding after the drum or wide belt sanding operations, and it is also the common procedure done just before the coating material is applied.

Mold sander: This method is used for articles of simple curvatures such as found in interior house trims as well as intricately shaped edges of a furniture top. Through-feed mold sanders are equipped with series of abrasive wheels of different grades, starting with coarser and ending with finer grades. Curved moldings that cannot be sanded on belt abrasives are sanded *on spindle sanders* with spool discs or solid abrasive wheels. The spindle is motorized and set to protrude from the horizontal tabletop, and a spool made of a hard felt or rubber is mounted on the spindle and shaped to conform to the molding. Aluminum oxide bonded on cloth backing is tightly wrapped on the spool. Usually, 80 or 100 grits are used. Also, solid, resilient sanding wheels have been developed that are shaped and used on a spindle sander-like spool discs, such as nylon plastics spray-bonded with abrasive grains and polyurethane plastics impregnated with 40- or 60-grit garnet grains.

Turning sander: is used to cross-grain sanding of turnings. A brush forces a strip of coated abrasive, which can be slashed or not slashed depending on the contour of the turning, against rotating workpiece, such as a table leg, to achieve an extremely smooth finish. Slashed rolls are used for sanding beaded turnings and un-slashed rolls for tapered turnings. Grits 80 through 180 are commonly used on turning sanders.

Edge sander: The abrasive belts run vertically against edges of workpieces that are lying and moving on a table. The *belt is supported by a flat platen, deadhead, or contact roll.*

Pneumatic drum sander: This sanding equipment is comprised of an air-inflated contact roll, around which the abrasive is fitted tightly when it is fully inflated. The pneumatic drum may also be used as a conventional contact roll with long belts and a backstand idler. It is used for sanding mildly contoured areas or for breaking sharp edges. Air pressure may be varied for different degrees of resiliency.

Backstand sander: This sanding equipment is comprised of a contact roll mounted on the drive spindle, a long abrasive belt, and an adjustable backstand idler for applying tension on the belt and providing guidance. Compared to open drum or conventional pneumatic sander, the advantage of this setup is the more economical use of abrasives. The contact roll can be a pneumatic drum, multiple sewn cloth sections, or a metal hub covered with rubber of selected hardness. The rubber-covered hub can be flat-faced or serrated and can be slightly contoured for different operations.

Brush back Vonnegut wheel sander: This sanding machine is used to sand intricate shapes or areas inaccessible to other abrasive forms. It consists of a hub into which broom-like brush assemblies are fastened. These brushes serve as sanding supports for the abrasive loadings made of regular or slashed abrasive cloth rolled up in the hub and fed out to the work area at the periphery of the wheel as the abrasive wears. Generally speaking, the more intricate the shape of the workpiece, the narrower the slash pattern. Aluminum oxide on waterproof X-weight cloth and 80–150 grit are recommended.

Open drum sander: The sander consists of a steel drum approximately 18 in. in diameter and two-feet long turning that normally runs at 400–600 rpm. The drum is usually covered with a cushion of 1/4″-thick felt, and the abrasive is held by a clamp in a slot across the face of the drum. This machine is usually employed for sanding simple contoured furniture parts, such as chair parts.

Drawer sander: This setup is designed to sand the dove-tail jointed and glued area where the side of a drawer is fitted to the front. The purpose is to remove the extruded glue and to trim the joint. Commonly, edge sanders and portable sanders will be employed for assembling whitewood pieces in furniture plants for the purposes of blending, fitting, repair, and overall touch-up, where considerable stock removal is required, such as in fitting of drawers. Also, the following hand-held equipments are often encountered.

Portable belt sander: Resin over resin aluminum oxide and open coat on X-weight cloth are recommended.

Orbital and reciprocating pad sander: This setup is used to remove coarse surfaces and where considerable durability is required, such as in working over curves, corners, and edges. Resin over resin aluminum oxide and closed coat on flexible X-weight cloth sheets are recommended.

Orbital disc sander: The spindle rotates in a small orbit at a high rpm (6k–8k). The orbital action helps to remove unwanted swirl marks that the disc sander might has left. Silicon carbide sheets of A-weight paper and resin over glue aluminum oxide on E-weight paper are recommended.

Hand-held block sander: The same recommendations as for orbital and reciprocating sander apply except that lighter weight backings may be used since the abrasives are subjected to less strain than with machine sanding.

23.3 Sanding Abrasives, Construction, and Flexing

Construction, materials, and grades of commercial abrasives are summarized in Figure 23.5. Briefly, an abrasive consists of hard mineral abrading grits bonded on a backer with a binder glue and then filled with size coat. *Backer materials* are papers of various weights (A, C, D, and E), cloths of different thread counts (J and X), and vulcanized rubbers of different thicknesses. They can be flexed in various ways to better conform to the shapes of sanding surfaces. *Bonding adhesives* are hide glues, urea-formaldehyde resins, and phenol-formaldehyde resins. *Size coats* are either the same glues or some kind of (natural) resins. *Mineral abrading materials* are normally natural grains of garnet, synthetic aluminum oxide, and silicon carbide with mesh sizes of 80–400.

23.4 Typical Furniture Finishing Procedures and Materials

The following procedures describe various wood finishing procedures roughly in the order of application for highly finished products, such as pianos and cabinets. In other cases, these procedures are selectively used.

1. *Size coating and/or bleaching*: Sizing is an application of a thin mixture of water-soluble binder to the surfaces of furniture parts between sanding operations. The increased MC raises loose and ruptured wood fibers and the sizing binder, once hardened, holds them stiff in the raised position, to be cut off in the subsequent sanding operation to give

Abrasive technical data

Basic construction of coated abrasives

The three basic components of coated abrasives are:

1. A backing – to support the coating.
2. An abrading agent (mineral).
3. An adhesive coat (commonly known as "bond") – to hold the abrading agent in place.

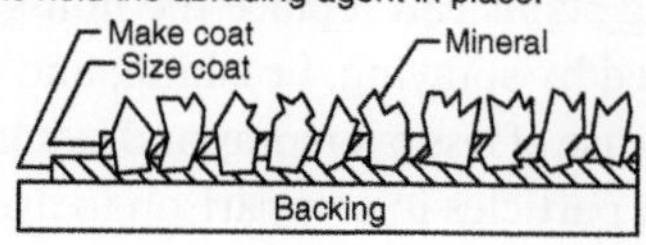

Backing materials

Papers

Letter indicator	Weight (480 sheets 26″ × 36″)	End product usage
A	40 lb.	Sheet goods
C	70 lb.	Sheet goods
D	90 lb.	Sheet goods
	110 lb.	In combination with cloth
E	130 lb.	Rolls - Belts

Cloths

	Thread count	
J (Jeans)	96 × 64	Belts, Rolls, Discs, Sheets
X (Drills)	76 × 48	Belts, Rolls, Discs

(Note: All are twill weave)

Vulcanized fibres

Thickness	
20 mils	Rolls 24″ × 36″
30 mils	Disc, 7″ × 9″ Dia.

Combination

100 lb. Paper and Print Cloth — 28″ × 36″ Rolls

Bonding adhesives

Principal bonding Adhesives	Product adhesive combinations Make coat	Sandsize coat
1. Hide glue	1. Glue	Glue
2. Urea resins	2. Glue	Resin
3. Phenolic resins	3. Glue	Resin

Mineral grading There are two methods of identifying mineral grades. One is the use of symbols. The other is mesh numbers. Symbols, the first system used, were arbitrary means of identification.

Mesh No.	Symbol	Mesh No.	Symbol
400	10/0	60	½
320	9/0	50	1
280	8/0	40	1½
240	7/0	36	2
220	6/0	30	2½
180	5/0	24	3
150	4/0	20	3½
120	3/0	16	4
100	2/0	12	4½
80	1/0		

Coated abrasive minerals

Commercial name	Origin	Specific gravity	Hardness scales MOHS	KNOOP	Grain-shape
Garnet	Natural	3.4-4.3	7.5-8.5		Light wedges
Aluminum oxide	Syn.	3.96	9.4	2050	Heavy wedges
Silicon carbide	Syn.	3.2	9.6	2480	Sharp wedges

Flexing

Flexing is a controlled mechanical cracking of the adhesive bond to render the proper degree of flexibility.

Standard flexes

Single flex – flexed at right angle (90°) to the material, making the product flexible in the length-wise direction.

Biflexed – flexed at two opposite 45° angles to the length of the material. Since this is done after single flex, the product is more flexible in all directions.

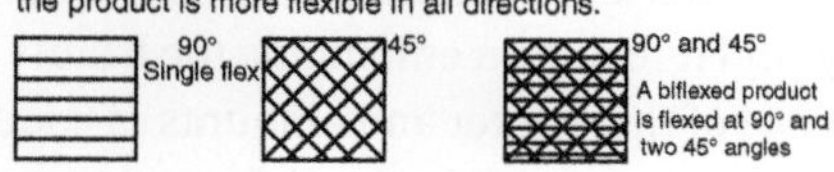

Figure 23.5 Construction and flexure of backers of abrasive sheets used in sanding of wood. Source: Reproduced with courtesy of Reliance Universal, Inc. Louisville, KY.

smooth level surfaces for finishing. Also, sizing results in making the wood surfaces absorb stains more uniformly than unsized wood surfaces. *Bleaching* is sometimes applied to equalize color variations between various veneer flitches. Bleaching is a costly and hazardous process and also can result in destroying the wood substrates, so it needs to be done minimally and the subsequent drying step be promptly completed for removal of residue bleach materials, or poor finish or stain application could result.

2. *Stain or pigmented lacquer toner to achieve color*: The commonly used non-grain-raising stains are applied by a roll coater or pouring machine, followed by drying times of 15–30 minutes. These stains and toners contain methanol or ethanol and *acid dyes*. There are also other types of stains that contain pigments/binders or pigments/binders/dyes made in organic solvents.
3. *Wash-coat of lacquer to smooth the wood surface*: A wash coat is commonly applied between staining and filling steps and it consists of thin coating material, usually clear lacquer, or vinyl sealer of low solids content, in organic solvents. The process results in stiffening the small wood fibers that are raised in the staining operation, so that they are cut off sharply in the subsequent fine sanding operation. The process also seals the stain and helps ease the wiping and cleaning up in the subsequent filler process.

4. *Wood filler application*: A filler is applied to fill pores and grain structures to make the surface smooth and to accentuate the wood grain patterns. Fillers are purchased as pastes, diluted with solvent at time of use, and applied by spraying, brushing, or dipping, followed by thorough drying.
5. *Pigment wiping stains*: Application of pigment wiping stains can replace the non-grain raising stains on close grain woods and can be applied by spraying, brushing, and dipping. The dipping method is the most economical because of less overspray and less labor involved. This stain consists of finely ground pigment particles in comparison to dyes of non-grain raising stains, but it also contains some binder materials. This step can be used on non-grain raising stained substrates for improved staining effects.
6. *Glazes*: Glazes are used mostly in manufacturing better-grade furniture, increasing the depth and richness of grain patterns. They do not change the color but enhance the three-dimensional effects. Also, the highlighting process is done to further accentuate the grain patterns after the pigment wiping stain is applied by rubbing with various fine-grade sandpapers or steel wool.
7. *Sealers*: The purpose of the sealer is to seal the wood and color coats. Color-coated wood surfaces tend to harbor broken fibers that can be the source of a rough appearance later on. Sealer application results in raising the fibers, and a following sanding step removes the fibers and excess sealer materials with minimal damage to the preceding color coats.
8. *Padding stains*: Padding stains are made of dyes similar to those used in non-grain-raising stains but also contain a vehicle polymer and are applied by hand with a rag moistened with the stain in discretionary manner to accentuate the figures or grain patterns of wood. The areas are generally the opposites of those areas that were highlighted previously, and it results in enhancing the beauty and depth of the finish. Depending on the experience of the worker and amounts of padding done, these operations can contribute greatly to the beauty of final finish.
9. *Topcoats*: are the final clear coat applied, the type of which is selected based on the desired physical characteristic, application characteristics, and cost. Physical characteristics include resistance to checking and cracking, toughness and adhesion, resistance to mars and scratches, permanency of color, resistance to alcohol, moisture, hot dishes, and solvents. Characteristics related to production and finishing are the ease of application, drying specifications, reparability, amount of build per coat, and resistance to printing. Such properties as polishing and rubbing, clarity, gloss, and depth are related to customer appeal and salability. Lacquers account for the major volume by far. Lacquers are film-forming materials that dry by evaporation of solvents, and nitrocellulose and cellulose acetate butyrate (CAB) finishes are the major polymer materials. Also, some *catalyzed* vinyl compositions are used as topcoats in furniture finishes. Small quantities of melamine- and urea-modified alkyds are also used.

23.5 Flat Line Finishing Procedures of Wood Composite Boards for Furniture Production

Large quantities of hardboard, particleboard, and medium-density fiberboard are used, after printing and application of some flat line finishes, as flat boards in furniture manufacturing. Flat finishing lines are used to produce printed panels that closely resemble quality hardwoods (Figure 23.4). After the panels are filled, base coated, and printed, they may be cut and used in manufacturing of furniture, followed by finish applications using the same procedures as those used on regular wood furniture. The flat line finishing procedure is as follows: (i) Panels or cut-to-size parts are prepared by sanding with 150-grit paper. (ii) Voids of panels are filled with a filler, which is normally composed of 100% solid acrylic monomers/prepolymer mixtures with heavy pigment loadings. The pigments selected need to be transparent to ultraviolet light, so that a thorough conversion of monomers and prepolymers is accomplished in the subsequent UV light curing step. Also, the filler application is normally done with the reverse roll coater to accomplish maximum penetration. (iii) The filler-coated panel is then passed through a UV chamber, for a few seconds, to cure the coating. (iv) The cured panel is then sanded with 320-grit paper, and the base coat is applied by a curtain coating, roll coating, or precision roll coating procedure. Base coated vehicles consist of catalyzed butylated urea-formaldehyde or vinyl-modified nitrocellulose materials. The vehicle polymer must have a good bonding property with the applied filler and subsequent coats since the UV-cured surface is very hard and often difficult to bond. (v) Then, printing of patterns on the panel is done on rotor-gravure presses that can allow the use of six different colors of ink depending on the application. The inks used are likely formulated with nitrocellulose vehicle in a very slow drying solvent such as ethylene glycol mono-butyl ether with finely dispersed pigments. (vi) Low solids lacquers are sometimes applied to protect the inks from smudging and marking during handling of panels in the furniture assembly and finish application steps.

23.6 Kitchen Cabinet Finishing Procedures

Finished kitchen cabinets are a special segment of the housing construction market, and various lacquers such as catalyzed urea-alkyd conversion coatings, oil and resin-modified alkyds, and epoxy varnishes are used as topcoats. Vinyl lacquers and vinyl-modified conversion coatings are also used for prefinished kitchen cabinets. The coating application process includes: (i) Sanding and wood-sizing application, (ii) Pigment stain application, (iii) Sealer application, and (iv) Topcoat application. Also, this industry expanded into functional furniture, where the stains and coating systems became more sophisticated, closer to those of conventional furniture manufacturing.

For kitchen cabinets, several qualification tests for finishes are commonly required: (i) Shrinkage and heat resistance. The sample is subjected for 24 hours in a hot box maintained at 115–125 °F and 65–75% relative humidity and then equilibration and visual inspection are conducted for no discoloration, no blistering, no checks, or no other film failures. (ii) Hot and cold check resistance. The sample is subjected for one hour in a hot box at 115–125 °F and 65–75% relative humidity, followed by placing it in a cold box at −5 °F for one hour. Repeat this cycle for a total of five cycles and examine for any film failure. (iii) Chemical resistance. The test is done by exposing the sample for 24 hours to vinegar, lemon and grape juices, tomato catsup, coffee (115 °F), olive oil, and 100 proof alcohol, followed by washing with water and drying and examination for any film failure. (iv) Detergent and water resistance. The test is done on 1–1/2″ thick cellulose sponges cut to the size of test piece or, for door frames, two inches shorter than the shoulder length of the door frame. Detergent is dissolved in water at 1% concentration and poured into the container to fill ½″ below the top of the sponge. Place the test piece on the sponge face down for 24 hours, followed by washing with water and inspecting for any film failure. No delamination, swelling, discoloration, blistering, checking, or whitening.

23.7 Hardwood Plywood Panel Prefinishing

Hardwood plywood panels are finished either as "natural" or "print" prefinishing. *Natural prefinishing* keeps the color, grain pattern, and texture of the natural veneer face, and the coating is essentially clear coats or includes one or more color coats. Sometimes, embossing or overprinting is done but the essential wood surface characters remain. Common species of hardwood used in panel production include birch, pine, ash, oak, elm, cedar, pecan, maple, cherry, hickory, and walnut. *Printed panel finishing* of hardwood plywood panels was developed in the United States and Canada by printing natural or unnatural wood grain patterns on lower-cost substrates, such as lauan plywood imported from Southeast Asia.

Natural prefinishing procedure: (i) Preheat to about 70 °F by infrared lamp to remove the surface moisture to reduce variation in finish results. (ii) Groover – Grooves are cut to provide a "planked" effect, and they are spaced so that they coincide with construction studs spaced on 16″ centers so that fixing nails are concealed in the grooves. (iii) Sanding – wide belt sanders are used. (iv) Brush and exhaust – the sander dust is removed. (v) Embossing – some manufacturers use embossing for accentuating the grain texture, by lightly pressing with an embossed plate. (vi) Stripe or flood – a coating of contrasting color is applied on groove lines by "gang-sprayers" or "flood and picking up" procedure. (vii) Dry the coatings in the grooves. (viii) Toner – Principal color is applied on a direct roll coater, followed by drying. (ix) Toner coating – is applied, followed by drying. (x) Sanding – Roughness is removed by 400-grit wide belt sander. (xi) Mismatch – A very thin color coat is applied to alternate groove-outlined sections to create a flitch effect, and the small amount of solvent is often dried off in transit to the next station. (xii) Topcoat – A clear protective finish coat is applied by direct roll coater. A second coat may be applied. (xiii) Drying – Topcoat is dried under infrared lamp and/or in hot-air oven at 170 °F. (xiv) Brush – any debris and roughness remaining in the topcoat are removed by a rotary brush. (xv) Cool, grade, and stack.

Print prefinishing procedure: (i) Preheat. (ii) Sanding. (iii) Sealer – A clear or lightly colored sealer is sometimes applied by direct or reverse roll coater. (iv) Dry at surface temperature of 120 °F. (v) Fill – A high viscosity, highly pigmented coating is applied by reverse roll coater to fill the pores. (vi). Dry – At surface temperature of 150 °F. (vii) Sanding with a belt sander till the filler is only left in the pores. (viii) Fill – Filler is applied once more to fill the open part of the first filler coat after drying/shrinking. (ix) Dry. (x) Sanding – A thin film of filler is left on the surface after this sanding to ensure a flush surface. (xi) Grooves are cut. (xii) Flood – Coating is applied on groove lines often by the flooding and picking up method. (xiii) Drying – Groove coating is dried at 140 °F. (xiv) Basecoat (or ground coat) – An opaque coat is applied by a direct roll coater or occasionally by a curtain coater. A second coat may be applied. The basecoat is the background color of the final composite panels. (xv) Dry – The basecoat is dried in an infrared or warm air oven at surface temperature of 150 °F. (xvi) Burnish – Dried base coat is lightly sanded by a wide belt sander or rotary brush to remove debris for a smooth, uniform surface. (xvii) Embossing. (xviii) Valley print – Ink is applied to embossed depressions in panels by printing with inks only on the raised points of embossed roll. (xix) Print – One to four rotogravure printers are used in register with each other, from which with the help of embossing and valley printing, more authentic wood grain patterns are obtained. The solvents are normally flashed off in transit. (xx) Topcoat. (xxi). Dry/Cool/grade/stack.

Quality control: Finished panels are compared with the standard color control panels set up at the end of the coloring station. Physical properties are checked at regular intervals for cure and hardness, adhesion, gloss, film clarity, mar resistance, smoothness, etc. *Current concerns of panel producing industry* are: *Governmental rules* and regulations that lead to using more water-borne coatings for less air emissions. *Current lower-quality wood substrates* available require more of the high solids water-borne coatings due to splits, torn and brash grain, vine marks, wormholes, etc.; *reducing the flame spread rates* for all panels is important. Methods of eliminating *toxic odors and fumes* are being developed, such as after-burners, in addition to using more of the above mentioned electron beam, UV, and free radical curing coating compositions.

23.8 Hardboard Panel Finishing Procedures

In the beginning of the industry, hardboards were sold primarily as "brown boards" as basic construction materials, but they were developed into various specialty engineered products principally by the application of coating materials, as discussed in Chapter 9. Hardboards are produced with PF resole resin binders to have smooth, tight surfaces needed for application of coating/finishing materials and are particularly adapted to wood grain reproduction processes or to lamination with decorative overlays. When brown boards are used in finishing procedures, sanding or washing with a mild soap on finishing surfaces is needed. Sanding needs to be done with slightly heavier pressures and faster cutting abrasives than for solid wood or plywood. Sanding tends to fuzz up, which can be controlled by applying a size or wash coat and drying/re-sanding. Silicon carbide abrasives are preferred because of their fast-cutting properties. Some of the largest coating lines in the world are operating to manufacture finished hardboard panels, and the positive factor for future growth is the industry's unique ability to tailor panels to suit various end uses. The following key points are noted:

1. Hardboards require a good primer coat of heavily pigmented primer–sealer, and for use in high humidity or moisture conditions, the back and edges of panels also need to be protected by an oil-based paint, primer, or a good water-repellant material. In finishing hardboard panel faces, tinted semi-gloss enamel paints and colored flat paints are often used.
2. Hardboard panel furniture components are also finished in the conventional manner with non-grain raising stains to produce the body or ground color, followed by finishing with clear primer and sealer and topcoats of lacquer, varnish, or synthetic resin materials. Many current topcoats provide resistance to alcohol, fruit acids, household alkalis, nail polish, and heat marring.
3. Enameled hardboards such as door panels are popular where a hard-wearing, hygienic, smooth, dust-free surface is required, and they are available in many attractive patterns and range of colors. Door skins and brick texture panels are manufactured not by using the printing process but by using deeply textured hot-press plates in the board manufacturing step, followed by enamel finishing operations designed for high durability, etc.

Enameled finishes are usually cured by baking at about 300 °F in a forced-air drying or infrared heating oven.

4. *Class I hardboard coating finishes* are usually obtained by curtain coatings with solvent-borne thermosetting acrylic resins, followed by curing at temperatures in excess of 300 °F preferably using factory-tempered hardboards; suitable for use in stores, restaurants, bathrooms, kitchens, and other areas where extreme durability is needed.
5. *Class II hardboard coating finishes* are usually applied by direct roll coaters intended for use as interior wall paneling.

23.9 Summary

Various industrial wood coatings and application methods and procedures are presented based on open literature. Spray gun, electrostatic sprayer, high-speed rotating sprayer, roll coater, precision roll coater, curtain coater, knife coater, and flow coater are described. Also, sanding materials and various sanding methods and processes are presented. Furniture finishing procedures, including the higher-grade furniture, flat line finishing procedure of wood composite boards, kitchen cabinet finishing procedure, natural and print prefinishing procedures of hardwood panels, and hardboard panel finishing procedure are discussed.

References

1 *Useful Facts & Figures*, 3e. Louisville, KY, USA: Reliance Universal, Inc.

2 Richter, K.W., Feist, C., and Knaebe, M.T. (1995). The effect of surface roughness on performance of finishes. Part I. Roughness characterization and stain performance. *Forest Products Journal* 45 (7/8): 91.

24

Advanced Reading Materials on UF Wood Adhesive Resins

24.1 Introduction to the ^{13}C NMR Spectroscopic Analysis Method

Principle of the ^{13}C NMR spectroscopy [1]: ^{13}C nuclear magnetic resonance (NMR) analysis method became a useful tool for all organic compounds, including urea-formaldehyde (UF) and PF resins since 1970s. The schematic diagram of an early NMR instrument is shown in Figure 24.1. The strong magnet is the main component, and the sample tube of organic materials in solution is put in between the two poles. Around the sample tube are placed the radiofrequency (RF) transmitter and detector. RF transmitter radiates RF excitation pulses, to which the carbon nuclear spins respond and give back certain RF waves that are detected and analyzed. A brief introduction to the principle of ^{13}C NMR Fourier-transform spectroscopic method of analysis is presented in this chapter. Carbons consist of ^{13}C isotope (6 protons and 7 neutrons in the nucleus) at about 1.1% of their natural abundance with the balance being ^{12}C isotope. Only the ^{13}C nuclei due to their odd proton/neutron number in the nucleus have a spin angular, and therefore magnetic (dipole) moment (often called as "spin"), i.e. the ^{13}C nucleus itself is a tiny magnet that is continuously undergoing molecular Brownian motions at ordinary temperatures, and thus the magnets are unoriented in nature.

^{13}C magnetic spin population established under a strong magnetic field [2]: When an organic carbon sample in solution is put in the strong magnetic field in NMR experiment, the ^{13}C magnetic spins go into precessing movements at a **Larmor frequency** around the axis of the applied magnetic field (Ho) with an angle (θ), as shown in Figure 24.2a. The Larmor frequency value is proportional to the applied magnetic field strength, and at 23.5 kG (kilogauss) magnetic field strength, which was common in the early days, the Larmor frequency value of ^{13}C nuclear spins goes up to 25.2 MHz, and nowadays with higher power magnets it goes up further, all in the RF region. In these levels of high magnetic fields, the precessing ^{13}C magnetic spins in a sample are populated in two nuclear magnetic energy levels, magnetic quantum numbers, +1/2 (lower energy) and −1/2 (higher energy), i.e. the lower-level spins are aligned to, and the higher-level spins are aligned against the applied magnetic field axis with the same angle θ, as shown in Figure 24.2b. At ordinary temperatures under the strong magnetic field, the spin population in the lower level is slightly higher than in the upper level, and therefore a **net magnetic moment (spin)** is established in the sample, as shown in Figure 24.3d. Thus, the sample becomes an ensemble of

Adhesives and Finishes for Wood: For Practitioners and Students, First Edition. Moon G. Kim.

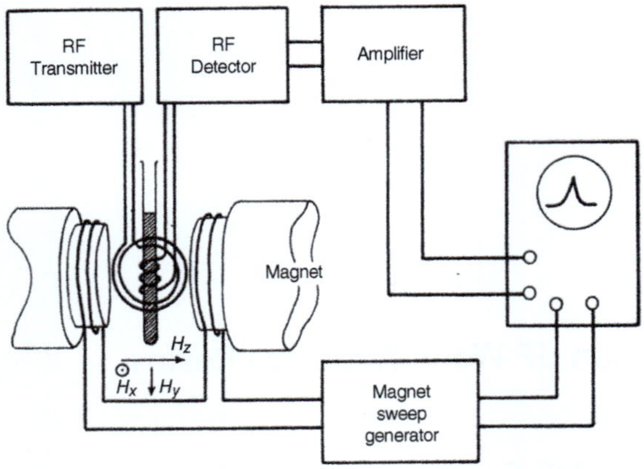

Figure 24.1 A schematic of an NMR spectrometer, Courtesy of the Ronald Publishing Company [1].

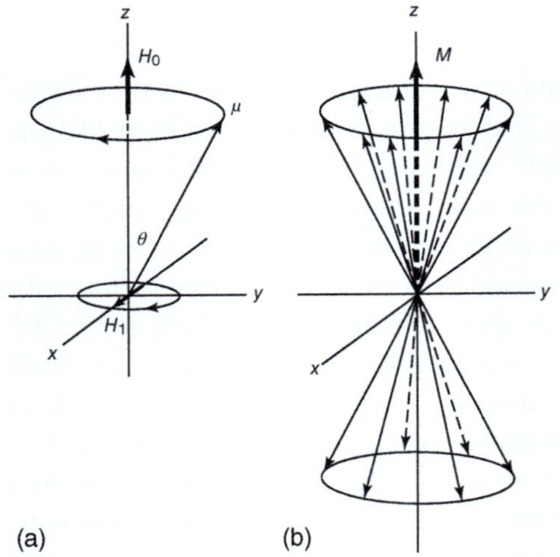

Figure 24.2 (a) Magnetic spin precessions around the main magnetic field (H_o).
(b) Precession of an ensemble of identical spins of nuclei in the lower (−1/2) (upper portion) and higher (+1/2) (lower portion) spin energy levels. Source: [2] Courtesy of Academic Press.

^{13}C magnetic spins acting as a small magnet that can be measured and manipulated. Since the magnetic spins are rotating at frequencies in the RF region, applying (irradiating) on the sample with a pulse of same (rotating) radio-frequencies brings on energy absorptions by ^{13}C spins (resonance), resulting in some spins in the lower level moving up to the upper level (**excitation**). After the pulse, the resultant excess spins in the upper level relax to the lower level by emitting energies of same radio-frequencies (**relaxation**), which are detected in the receiver coil in the ^{13}C NMR spectroscopy. The strength of main magnet is a fixed value in a spectrometer, and using higher power magnets (more expensive NMR machines) means the precession frequency becomes higher and the spin energy level differences also become larger to allow a more accurate measurement possible.

Different shieldings of ^{13}C spins from the applied magnetic field strength and chemical shifts [1, 2]: Various different carbons in organic molecular structures were found to have (very slightly) different Larmor frequency values because each carbon's nuclei have their own umbrellas of different thicknesses of surrounding atoms that shield them from the onslaught of applied magnetic field strength. For 100 MHz magnets, the extent of shieldings of various organic carbons spans about 5 kHz within 25.2 MHz of carbon's Larmor frequency, i.e. in a range of about **200 ppm** (parts per million). For example, carbons in organic molecules are often bonded (surrounded) with protons and other carbons that shield them like an umbrella from the magnetic field strength, and therefore the carbons of methyl groups ($-CH_3$) are shielded more than those of methylene groups ($-CH_2-$), etc. Successful measurements of these shielding differences in ppm (*chemical shifts*) and also relative quantification of their spin populations lead to identification of various carbons of the molecular structure. The instrument is complex, and although resultant spectral data were found to be extremely useful for various organic and organic polymeric materials, it entails some quantification limitations that need to be kept in mind in analyses of UF resins or many other organic polymer materials, some of which are discussed below.

Excitation and relaxation processes [2, 3]: During a Fourier-transform NMR experiment, the sample, i.e. the aligned net magnetic moments composed of spins of varying Larmor frequencies, shown in Figure 24.3d, are irradiated (pulsed) for a short time, commonly with a 90° pulse, with a (rotating) RF wave of a broad bandwidth (H_1 in Figure 24.2a) that covers the Larmor frequencies of all different carbons. Because of a short irradiation, only some of the magnetic spins in the lower energy level (−1/2) pick up the RF waves of same frequencies, i.e. go into the nuclear magnetic resonance (**NMR**), i.e. they jump into the upper energy (−1/2) level (**Excitation**). The short pulse-time used is to excite only some of the magnetic spins, so that the net magnetic moments' direction is turned just by a 90° angle into the X–Y plane with respect to the applied magnetic field direction (Z-axis) as shown in Figure 24.3a. Since the pulse is so short that all spins retain the coherence of rotation at their Larmor frequencies, even at the 90° angle position. Proton decoupling irradiation is also done during the pulse time (see below). Then, the RF signal receiver is turned on to receive and measure the energy (RF waves) given off from the 90°-excited spins while they return to the lower level to re-equilibrium (**relaxation**), which occurs through two simultaneous but different paths with different rates: (i) by reverting spin direction to z-axis, i.e. to alignment with the applied magnetic field (***T_1*** relaxation time) by spin–lattice interactions as shown in Figure 24.3c and (ii) by losing the coherence of spin rotation within the X–Y plane mainly by the inherent inhomogeneity of applied magnetic field strength present in the X–Y plane (***T_2*** relaxation time) as shown in Figure 24.3b, and finally to the equilibrium state as shown in Figure 24.3d, discussed further below.

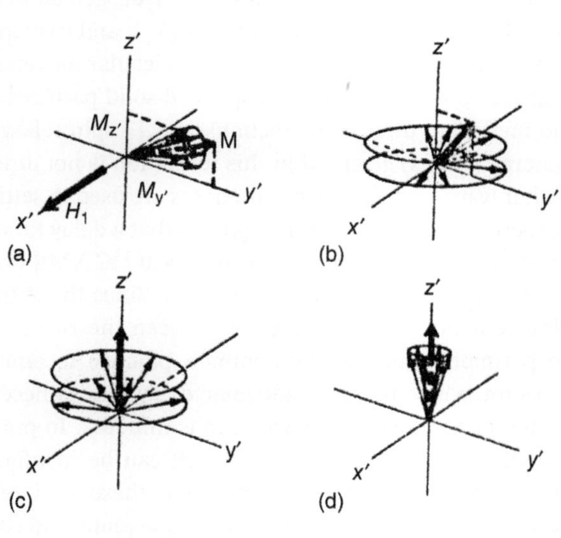

Figure 24.3 (a) Tipping of magnetic spins and macroscopic magnetization through an angle (90°) and establishment of $M_{y'}$; (b) Dephasing of nuclear moments by spin–spin (T_2) interaction. (c) Reduction of $M_{y'}$ to ~0. (d) Reestablishment of $M_{z'}$ at equilibrium value M_o. Source: [2] Courtesy of Academic Press.

T_2 relaxation time [2, 3]: Both relaxation processes give off RF waves to be detected by the receiver, but due to the configuration limitations around the sample tube in between two magnetic poles, the receiver coil sits to detect waves in the X–Y plane, i.e. the RF waves given off from the relaxation path ***2*** (Figure 24.3b) are detected in the form of *free induction decays* (FID), which are sine waves that diminish in amplitude within the time span of up to about 10 seconds (defined as relaxation time $\boldsymbol{T_2}$) [2]. One FID wave entails decreasing amplitudes and unchanging frequency values vs. elapsed time. The frequency values of FID waves reflect the distances in Hz from the center frequency value of excitation RF pulse applied [3], for example, if an FID wave shows a 23 Hz frequency, it means that the magnetic spin is precessing at a Larmor frequency value that equals to the center frequency value of applied RF pulse wave *minus* 23 Hz if the center frequency value of excitation pulse was set at the far right-end of the spectral range, a common method. The methyl group carbon of tetramethylsilane (TMS) is used as a reference with its FID frequency value set at 0 Hz and other carbons' frequency (Hz) values are re-calculated using their frequencies of FIDs and dividing them with the center RF frequency value of excitation RF pulse (for example, 23 Hz/25.2 MHz) to obtain the *chemical shift* values in parts per million (ppm). Thus, one broad-band pulse excitation results in as many FIDs as the number of different Larmor frequencies of carbons present in the sample. The FID data are picked up by the RF detector, and FID frequencies and amplitudes are converted using the Fourier transform (FT) mathematical method into the frequency (Hz or ppm) and intensity (peak height) formats of ^{13}C NMR spectra.

T_1 relaxation time [2, 3]: Further, in addition to losing the coherence, the excited spins start to fall back to the lower level by path ***1***, i.e. to align with the applied magnetic field. This process occurs by dipolar (through space) exchanging of spin energy with the *surrounding atoms' magnetic spins* (the lattice) that have the same frequencies as the carbon spins' Larmor frequencies. The dipolar magnetic spins in the lattice arise at ordinary temperatures because of the incessant vibrational movements of atoms having magnetic spins. These spin–lattice relaxations occur over a period of time and T_1 is defined as the half-lives, normally a time span of a few seconds to many seconds depending on the availability of magnetic spin interactions with surrounding atoms and they are usually longer than T_2. For example, carbons bonded with a hydrogen atom (every hydrogen atom has a magnetic dipole) have short relaxation times (T_1), and carbons having no bonded hydrogen atoms have longer T_1 values. Also, when molecular movements are restricted in samples, such as carbons and hydrogens in suspended solid particles or viscose (polymeric) materials, due to inefficient dipolar interactions, the T_1 spin relaxation times tend to become long. The energy (waves) given off in this relaxation is not directly detected but can be measured in other ways, and these measurements are used in setting up the delay times to use in repeat experiments. It has been recognized that a delay time of $5 \times T_1$ is required to reach the full re-equilibrium of carbon spins in repeat ^{13}C NMR experiments.

Repeat scans and pulse-delay [2, 3]: Since the abundance of ^{13}C isotope in nature is so low and the energy difference between the two magnetic energy levels is small, repeat experiments became the common practice to enhance the signal/noise (S/N) ratio of spectra, where using an adequate delay time is necessary since the spin system has to be fully relaxed before a repeat scan is initiated. In practice, the delay time between repeats is set at about 5–10 seconds, which can be insufficient and shorter than $5 \times T_1$ values for some slow-relaxing carbons. For these slow-relaxing carbons, the spin population excess in the lower level becomes inadequate, and comparatively less spin is excited in the subsequent pulse to result in lower spectral peak intensities vs. other carbons that have shorter, adequate T_1 values. The number of repeat experiments is often large for polymer solutions, four hundred or more for modern 360 MHz machines and 1000 scans or more for older 100 MHz machines. Thus, resultant *averaged* spectra can harbor spectral intensity inequalities arising from inadequate delay times used, which is amplified with increasing number of scans. For example, the peak intensity ratio value of combined CH_2 groups and C = O groups in spectra of UF resins often turns out to be higher than the material charge ratio values used in resin syntheses because of longer T_1 values of carbonyl groups in comparison with methylene groups of shorter T_1 values. Long delay times would be desirable, but it means longer analysis times.

Proton-decoupling and nOe [2, 3]: Another handicap present in ^{13}C NMR is the magnetic (through chemical bonds) spin–spin coupling of carbon spins with those of neighboring hydrogens. Hydrogen (^{1}H) atoms also have magnetic spins and are similarly populated in the upper and lower levels. Without applying the proton spin-decoupling procedure, the NMR spectrum of a carbon having one bonded hydrogen, for example, would show up as two equal half-strength peaks separated by about two times the carbon-hydrogen spin-coupling constant (~125 Hz) values from the carbon peak (ppm) obtainable with the application of decoupling method. This spin-coupling also occurs with hydrogens at 2–4 chemical bonds away from the carbon in the molecule with smaller spin-coupling constants. Therefore, complex organic molecules result in exceedingly complex spectra that are difficult to delineate all spin–spin couplings. The spin-decoupling procedure results in collapsing the spin-coupled multiplets into single higher-intensity peaks. In ^{13}C NMR experiments, all proton nuclear spins are populated slightly more in the lower than in the upper level and also similarly undergo the proton (^{1}H) Larmor precessions with frequencies far away from ^{13}C carbon Larmor frequencies and not normally detected. In the current decoupling procedure, the NMR sample is irradiated not only for ^{13}C spins for excitation but also for ^{1}H spins for disturbing the spin population with broad-band RF waves that span the Larmor frequencies of all protons. This proton irradiation results in disorienting the proton spins and thus losing protons' capability of spin–spin coupling with carbon spins. In this procedure, however, the carbons' spectral peaks increase in intensity not only from the collapse of (spin-coupling) multiplets into one peak but also about 3 times more by an effective magnetic spin–spin interaction mechanism. The excess proton spin population attained in the upper level by irradiation interacts with carbon spins in the upper level to make the carbon spins relax faster, and therefore it results in increasing the carbon spin population in the lower level, making the repeat pulse excitations more effective for carbons. This effect is called the *nuclear Overhouse effect* (nOe). Thus, adoption of the proton-decoupling procedure improves the ^{13}C NMR spectroscopy, but the nOe effect is more effective for carbons having dipole–dipole spin relaxation (T_1) mechanisms available, such as carbons having bonded hydrogens as the major relaxation route, and carbons without bonded hydrogens or fast-moving small symmetrical molecules tend to have weaker nOe effects to result in lower peak intensities. Thus, the proton-decoupling procedure is necessary for simplification of ^{13}C NMR spectra, but it can result in distorted carbon intensities or quantitation results, and in recent years, a method has been used to suppress the nOe effect for minimization of spectral intensity deviations, by turning on the decoupling irradiation only during the data acquisition times and keeping it off during the delay times. This procedure was applied in obtaining all spectra of UF and PF resins reported in this book.

Precautions in reading ^{13}C NMR spectra: Usually, one ^{13}C NMR experiment includes several hundred repeat scans using a delay time of 5–10 seconds or longer, and this leads to an experimental time of about 30 minutes to an hour or more per sample, i.e. expensive. (Also, the machine is expensive.) The result is a spectrum consisting of peaks for all different carbons (with different chemical shift values in ppm) and their relative population (peak heights). For example, the spectrum of pure ethanol will show two peaks, one for methyl carbon at around 30 ppm and the other for methylene carbon at around 60 ppm with about equal intensities. In the case of mixtures or polymers, similar carbons occur at similar chemical shift values and their intensities would show the mole ratios of the mixture components. It needs to be remembered that quantitative data distortion is possible due to using delay times that are shorter than $5 \times T_1$ relaxation times and nOe effects that are not fully nullified for some carbons. This deviation of carbon populations (concentrations) can be minimal when intensities of same kinds of carbons, i.e. only protonated carbons or only non-protonated carbons, are compared among themselves in a spectrum. Thus, analyses of UF resins discussed in this book are based on this premise since carbons of UF resins have only formaldehyde-derived (two protons) groups, and urea carbonyl (no bonded hydrogen) groups and comparing various methyleneic groups quantitatively among themselves would result in reasonably good values while comparing these groups with urea carbonyl groups would result in some deviations depending on the NMR experimental conditions. Another limitation of ^{13}C NMR method is that the sample needs to be dissolved in a suitable solvent, such as D_2O or other deuterated organic solvents. Further, the organic sample concentration needs to be high, and the solution viscosity must be low to allow molecules to undergo molecular/atomic (Brownian) motions freely. Insoluble or partially soluble fractions of UF resins or of any solid polymer materials suspended in the sample would not be detected in the liquid ^{13}C NMR experiments. With these various precautions, the NMR method has been reported to be giving reliable analytical data for UF and PF resins [4] and many other polymer materials.

Summary: Introductory principles of ^{13}C NMR spectroscopic analysis method of organic materials, including wood adhesive resins, are presented. Chemical shift and intensity values of carbons in organic compounds or polymer materials are defined, and relaxation times, scan delay times, proton-decoupling, nOe effects, and repeat spectral scans and averaging are explained to be affecting the quantitative accuracy of ^{13}C NMR spectroscopy results. These parameters need to be optimized, although it is noted that the deviations can be minimal when protonated carbons or non-protonated carbons are compared among themselves in a spectrum. Another limitation to note is that the sample needs to be dissolved in a solvent such as D_2O or acetone-d_6 since insoluble fractions of polymer materials present suspended in the sample would not be detected. Overall, ^{13}C NMR methods have been giving quite useful analytical results for wood adhesive resins and other polymeric materials.

24.2 Introduction to ^{13}C NMR Analysis Methods of UF and UMF Wood Adhesive Resins

History: Ebdon and Heaton [5] Tomita and Hatono [6] de Breet, et al. [7] and Slonim, et al. [8] pioneered the use of ^{13}C NMR analysis method for UF resins, identifying the chemical shift values of various carbon groups as shown in Figure 24.4. Later, Rammon et al. [10]

advanced the branching theory of UF resin polymers using this analysis method, and Kim and Amos [11] applied this method to analysis of various UF resin intermediate samples taken during UF resin synthesis and showed the growth of UF resin polymers and other behaviors of UF resin systems. Later, Kim and co-workers [9, 12–15] published a systematic study on various UF and urea-melamine-formaldehyde (UMF) resin systems using this method, and these publications are presented with some modifications as advanced reading materials in this chapter. In the past, the functional group contents of UF resins have been analyzed using wet chemical methods [16] and ^{13}C NMR results obtained since 1980 indicated the two different analysis methods agreed relatively well.

Current ^{13}C NMR analysis results of UF resins' chemical structures: Greater details of UF resin structures were obtained from ^{13}C NMR. All possible carbon groups in UF resins are identified as shown in Figure 24.4a–o with carbon peak positions (*chemical shift values in ppm*) shown in parentheses. Example of ^{13}C NMR spectra of a UF resin intermediate with an F/U_1 mole ratio of 2.10 is shown in Figure 24.5a and that of the finished resin after adding the second urea to an F/U mole ratio of 1.15 in Figure 24.5b, where peaks were identified with the same lower case letters shown in Figure 24.4. Chemical shift values range from zero from the right end to about 200 ppm, and the peak heights (or areas obtained by integration) represent the *relative population* of the carbon groups. TMS is commonly added in the test sample as a reference compound that gives a sharp peak of the methyl groups (not shown) defined to be at 0.00 ppm [3], from which the carbon groups of UF resins occur in the down-field region in the following order: [5–7] methylene groups (47 – 60 ppm), hydroxymethyl groups (65 – 72 ppm), methylene-ether groups (69 – 79 ppm), hemi-formal and formaldehyde oligomers (80 – 91 ppm), and finally, urea carbonyl groups (154 – 164 ppm).

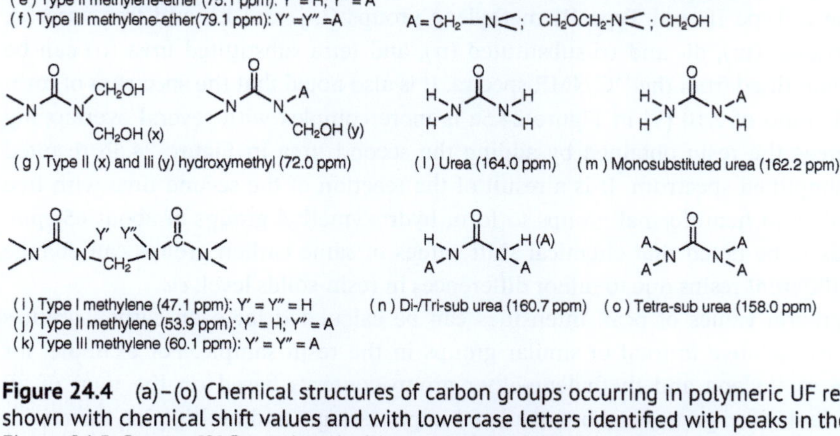

Figure 24.4 (a)–(o) Chemical structures of carbon groups occurring in polymeric UF resins are shown with chemical shift values and with lowercase letters identified with peaks in the spectra in Figure 24.5. Source: [9] Reproduced with permission of the American Chemical Society.

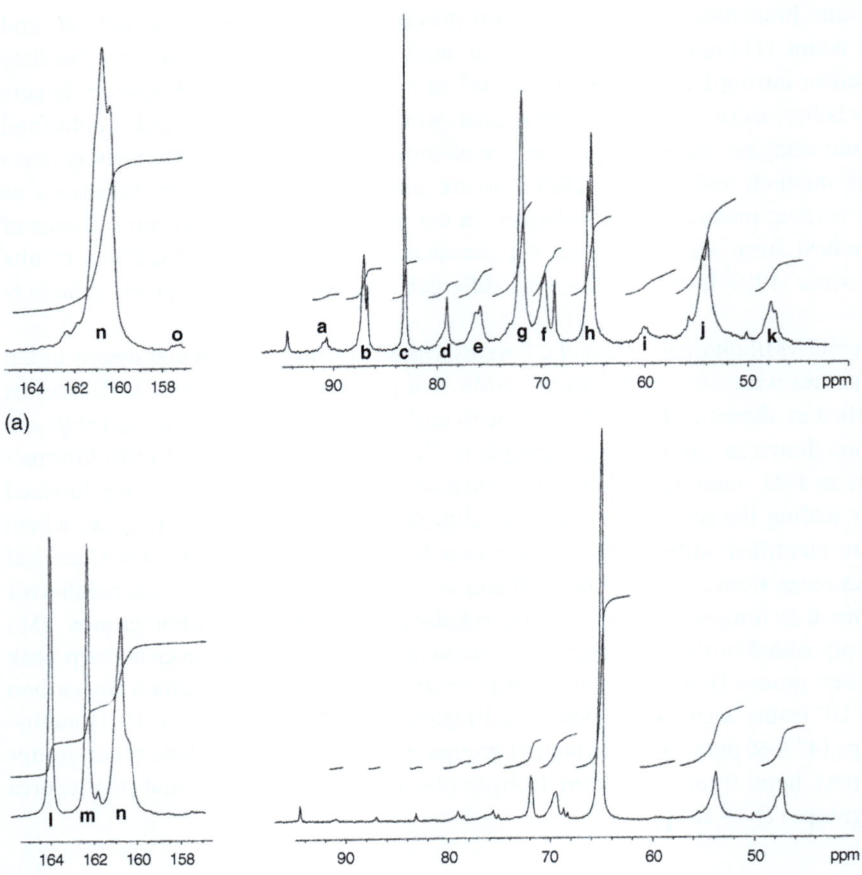

Figure 24.5 (a) ^{13}C NMR Spectra of UF resin samples (4Bb, F/U = 2.1) (top) and (b) (4BU, F/U = 1.15) (bottom) [9].

The free and oligomeric formaldehyde species (a, b, and c), Type I and Type II hydroxymethyl groups (h and g), Type I, Type II, and Type III methylene-ether groups (d, e, and f), Type I, Type II, and Type III methylene groups (i, j, and k), and free urea (l), mono-substituted (m), di- and tri-substituted (n), and tetra-substituted urea (o) can be separately quantified from the ^{13}C NMR spectra. It is also noted that the spectrum of resin with an F/U ratio of 2.10 [9] in Figure 24.5a is more complex with several overlapping peaks, whereas the resin obtained by adding the second urea in Figure 24.5b resulted in a quite simplified spectrum. It is a result of the reaction of the second urea with free formaldehyde and hemi-formal groups to form hydroxymethyl groups at about 65 ppm. It also needs to be noted that chemical shift values of same carbon groups can change slightly for different resins due to minor differences in resin-solids level, *etc.*

The integration values of peak intensities can be calculated for each carbon group's population (%) relative to total of similar groups in the resin sample. For example, for a UF resin, methylene and methylene-ether group contents based on the total of all formaldehyde-derived groups can be calculated in percentages and multiplying them with moles of formaldehyde used in resin synthesis (2.10, for example, per mole of urea), results in moles of these groups per 1.00 mol of urea used. Then, the degree of polymerization (n) values can be calculated by using the total methylene and methylene-ether group contents. Further, the F/U mole ratio values of (unknown) UF resins can be calculated by using the total of formaldehyde-derived group values divided by all urea carbonyl group values, or the ratio value determined can be compared with the charged F/U ratio values used in resin synthesis for checking the accuracy of the ^{13}C NMR experimental procedure or the resin synthesis procedure, although the ratio values need to be scrutinized for possible measurement deviations as discussed in the previous section (Section 24.1).

^{13}C NMR experimental procedure: All spectra reported in this book were taken using adequate delay times and suppressing the nOe. [3] For the two UF resin samples in Figure 24.5, two grams of 60% resin-solids resin were dissolved in 1.0 g of D_2O (internal lock and diluent) with resulting samples' viscosity ranging 60 – 90 cps. Spectra were obtained on a *CFT-20* ^{13}C NMR spectrometer for the first paper shown in the following section, and then on *Techmag 360* MHz ^{13}C NMR spectrometer using 22-microsecond pulse-with of an 80° angle and 10-second pulse-delay times. Proton decoupling was kept on during the data acquisition and off during the delay times to suppress the nOe effect. The relaxation times (T_1) measured on resin Sample A_{12} (Section 24.3), obtained on a CFT-20^{13}C NMR instrument using the inversion recovery method were 1.4–6.2 seconds for urea carbonyls, 1.1 seconds for methanediol, 5.8 seconds for methanol present as an impurity, and 0.16 seconds or less for all methyleneic carbons of UF resin molecules, and therefore the 10 second pulse-delay times used appeared adequate for most carbons of UF resins. About four hundred scans were accumulated to result in these and other spectra presented in this section but results sometimes showed deviations and this possibility needs to be kept in mind whenever ^{13}C NMR numerical results are interpreted.

Another factor that affects ^{13}C NMR spectra of UF resins is the reversibility of hydroxymethyl groups that is progressing slowly over days after resin manufacture because hydroxymethyl groups are bonded in excess to the first urea (U_1) than to the second urea (U_2), and the final cooling stage of resin manufacturing is often too short, not allowing the reverse reaction of hydroxymethyl groups to reach to the equilibrium stage. Depending on the thermal/physical history of resin samples, this re-distribution affects the spectra. For example, for types II and III methylene or methylene-ether groups with hydroxymethyl side groups (Y' and Y'' in Figure 24.4), the movement of one or both these side groups will result in downgrading the methylene or methylene-ether groups to lower type groups, resulting in increasing the intensities of the lower type groups, as discussed in detail in the next several sections.

Summary: Past research of using the ^{13}C NMR analysis method on UF resins is reviewed, and the various chemical/polymeric resin structures obtained from various UF resin model compounds are summarized in terms of chemical shift values of carbons, and they are classified as formaldehyde oligomers, hemi-formal groups, Type I and II hydroxymethyl groups, Type I, II, III methylene and methylene-ether groups, and urea carbonyl groups. Also, quantitative ^{13}C NMR analysis experimental procedures are presented for laboratory synthesized UF resin samples: dissolved in D_2O with resultant samples' viscosity of 60–90 cps, using adequate delay times, suppressing of nOe [3], and proton decoupling, on *CFT-20 (100 MHz)* and higher power T*echmag 360 MHz* ^{13}C NMR spectrometers.

24.3 ^{13}C NMR Analysis Results of Reaction Intermediates Taken in UF Resin Syntheses

(Adapted with permission from the publisher and minor revisions from Kim, M. G. and Amos, L. W. 1990. Quantitative ^{13}C NMR study of UF resins in relation to the formaldehyde emission level. Ind. and Eng. Chem. Resch. 29, 208–212. [10]*)*

In this paper, intermediate UF resin samples were taken during the synthesis of typical UF resins as described in Table 24.1, and ^{13}C NMR spectra were obtained on CFT-20 NMR spectrometer, peak assignments and integration values of peaks are reported in percentages, and the calculated ratio values are reported in Table 24.2 on the next page. The ^{13}C NMR spectrum of Sample A_3, taken at the end of the alkaline first step of resin synthesis is shown in Figure 24.6.

First step of UF resin synthesis: From the data for *Samples* $A_1 - A_3$ in Table 24.2 it is seen that the hydroxymethylation reaction at an F/U_1 *mole ratio of 2.10* proceeds to completion in about 20 minutes (*Sample* A_2). As shown in Eq. (24.1), the apparent second-order reaction rate calculated from the NMR data is 2.23×10^{-3} l/mol s. (90 °C), which is approximately twice as large as that predicted by de Jong and de Jonge [17–21]. *Sample* A_3, the last sample of the first alkaline reaction step whose spectra shown in Figure 24.6 shows that: (i) Type I **hydroxymethyl groups** amount to 1.449 groups/mol of urea and Type II **gem-dihydroxy methyl** groups amount to 0.416 group/mol of urea; (ii) **Methylene-ether** groups are also formed from the formation of hemi-formals on hydroxymethyl groups (R–NR′–$\mathbf{CH_2}$–**O**–CH_2–OH) as well as of (real) **methylene-ether** (f) groups formed from the reaction of two hydroxymethyl groups to result in oligomeric polymers of methylene-ether hydroxymethylureas. The (real) methylene-ether groups reach about 17% of charged formaldehyde, but for some reasons, they do not increase much beyond this level; (iii) **Methylene** bonds at about 48 ppm appear only in very small amounts. Minor levels of hemi-formals and free formaldehyde oligomers (peaks a–c) are present with the total reaching about 7% reduced from 100% of theory at the start of

Table 24.1 Resin samples taken in synthesis of UF resins [11].

1. A weighed amount of 50% formaldehyde solution was taken in a reactor equipped with a stirrer, thermometer, and cooling coil, and adjusted pH to 7.8 with 8.0% sodium hydroxide solution and heated to 70 °C.
2. A sufficient amount of urea (first urea) was added in small portions to bring the $F/U_1 = 2.10$ at 70 °C, and the reaction exotherm and heating control were used to raise the temperature to 90 °C, which was maintained for 30 min, while the pH was also maintained at 7.5. *Samples* A_1, A_2, and A_3 were taken at 10, 20, and 30 min of reaction time.
3. Then, the pH was adjusted to 5.1 using 8% sulfuric acid solution, and raised the temperature was raised to 95 °C and maintained for 110 min, during which the resin increased in viscosity to 500 cps and *Samples* A_4–A_9, and A_{10} were taken at 15-min intervals. The samples were neutralized to pH 7.0 and stored at 4 °C until analysis.
4. The resin batch was also neutralized, divided into 3 portions, and added the second urea to $F/(U_1 + U_2) = 1.50$ for *Sample* A_{11}, = 1.00 for *Sample* A_{12}, and = 0.80 for *Sample* A_{13}.
   - *Resin Samples B* and *C* were similarly prepared with an F/U_1 ratio of 2.40 and 2.80, respectively.
   - *Resin Samples D* and *E* were similarly prepared with an F/U_1 of 1.75 and 1.50, respectively, and added the second urea to F/U mole ratio of 1.20.
   - ^{13}C NMR analyses were done on CFT-20.

Table 24.2 ^{13}CNMR spectroscopic integrals (%) of various carbons of urea-formaldehyde resin samples[a),b)].

Resin	Charge F/U ratio	urea[f)]	MU[c)]	DU/TU[1]	TeU[d)]	NCH_2N	$=N^+CH_2N$	$=N+CH_2N^+=$	Total CH_2	NCH_2^+OH	$=N^+CH_2OH$	$NCH_2O\text{-}CH_2OH$[b)]	Total CH_2OH	$CH_2=O$	Totalnet ether CH_2[c)]	XCH_2X/urea[d)]	XCH_2X/Total CO[e)]
		163.3	161.7	160.2	157.8	47.3, type 1	54.0, type(II)	60, type(III)		65.1, type(I)	71.8, type(II)	87.1		82–95			
A1	2.10	<1	24	75	1	<1	<1	<1	<3	61	15	3	79	7	11	2.02	2.20
A2	2.10	<1	21	78	1	<1	<1	<1	<3	55	15	3	73	6	13	1.93	2.05
A3	2.10	<1	24	74	1	1	1	1	3	42	17	3	72	7	17	1.98	1.99
A4	2.10	<1	10	89	1	7	8	2	17	34	18	3	55	7	21	2.31	2.15
A5	2.10	<1	<5	91	3	8	15	4	27	24	19	3	46	7	21	2.54	1.90
A6	2.10	<1	<5	89	5	8	17	4	29	19	17	3	39	9	17	2.40	1.84
A7	2.10	<1	<1	93	5	9	20	5	34	18	17	3	38	10	16	2.56	1.78
A8	2.10	<1	<1	92	6	9	20	5	34	17	18	3	38	13	13	2.50	1.89
A10	2.10	<1	<3	92	8	9	20	4	33	19	16	4	39	13	16	2.54	1.70
A11	1.50	10	26	92	2	13	19	4	37	30	13	0	43	0	15		1.36
A12	1.00	33	26	39	2	20	21	4	44	28	13	0	41	0	14		0.88
A13	0.80	40	24	36	<1	18	19	5	42	24	13	0	37	0	18		0.81
B	2.40	<1	<1	96	4	6	15	5	26	14	16	4	34	4	15	2.42	1.96
C	2.80	<2	<3	91	4	5	10	4	19	16	17	2	35	2	23	2.69	2.95
D[e)]	1.75/1.2	20–	22	56	2	15	21	7	43	27	13	0	40	0	14		1.07
E[e)]	1.50/1.2	15	22	63	1	24	33	2	59	24	10	0	34	0	7		1.17

a) Degree of polymerization (DP) = 1/[1-(total methylene + [1/2] net ether)].
b) +sign (superscript) denotes two substituted groups on the nitrogen.
c) MU = mono-substituted urea species.
d) TeU = tetra-substituted urea species.
e) Charged F/U ratios were 1.75 and 1.50; data were obtained with F/U = 1.2.
f) Determined from urea CO carbon ratios (157.8–163.3 ppm).
g) Methylol groups protected as hemi-formal.
h) Ether group minus hemi-formal and other minor ethers.
i) (Methylene X 2+ methylol + ether)(charged F/U ratio): total substitution of urea.
j) Total XCH_2X groups/total CO groups = determined F/U ratio. X = oxygen or nitrogen.
k) DU/TU = di- or tri-substituted urea species.

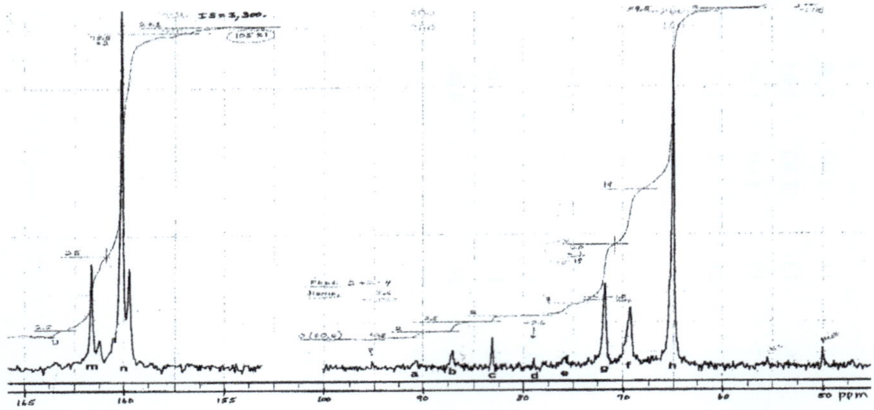

Figure 24.6 ^{13}C NMR spectra of Sample A_3 ($F/U_1 = 2.10$) obtained on CFT-20 (100 MHz).

reaction. The results indicated that the reacted F/U mole ratio reached 1.98 in the first step of resin synthesis in comparison with the charged mole ratio of 2.10.

$$\begin{aligned}
&NH_2CO - NH_2 + 2.10CH_2 = O \rightarrow \\
&\quad \rightarrow NH_2 - CO - NH - CH_2 - OH \text{ (mono-hydroxymethylurea, 24.0\%) :} \qquad K_1 = 170 \text{ l/mol} \\
&\quad \rightarrow NH_2 - CO - NH - (CH_2 - OH)_2 \quad \text{(1,1-di-hydroxymethylurea, 6.3\%) :} \qquad K_2 = 17 \text{ l/mol} \\
&\quad \rightarrow HO - CH_2NH - CO - NH - CH_2 - OH\text{(1,3-di-hydroxymethylurea, 53.2\%) :} \\
&\quad \rightarrow HO - CH_2NH - CO - NH - (CH_2 - OH)_2\text{(tri-hydroxymethylurea, 14.5\%) :} \quad K_3 = 1.25 \text{ l/mol}
\end{aligned} \tag{24.1}$$

In *Sample A_3*, as shown in Eq. (24.1), urea carbonyl groups consist of about 24.0% of **mono-substituted urea** (m) and 75.0% of **di- and tri-substituted urea** (n), and free urea and tetra-hydroxymethyl urea (~158 ppm) are present only in small amounts. Considering the methylene-ether groups as hydroxymethyl group equivalents, and combining hydroxymethyl groups contents and urea carbonyl groups contents resulted in a distribution of mono-, 1,1-di-, 1,3-di-, and tri- hydroxymethyl-urea components, as shown in Eq. (24.1). The unsymmetrical 1,1-dihydroxymethylurea shown in this analysis has been suggested to occur in the first step of UF resin synthesis as a minor component [17–19]. Further, the formation constants K_1, K_2, and K_3 for mono-hydroxymethyl-, di-hydroxymethyl, and tri-hydroxymethyl-ureas calculated for sample A_3 were about one order greater than the reported values measured at 59 and 35 °C [17–19]. The greater extent of reactions observed can be ascribed to factors such as higher temperature, higher concentration, and formation of ether and hemi-formal groups.

Second step of resin synthesis – methylene bond formation: ^{13}C NMR data of *Samples A_4–A_8* in Table 24.2 are noted for the following features of polymerization reaction: (i) continuous increases in methylenes, (ii) rapid decreases in hydroxymethyls, (iii) slow decreases in methylene-ethers, and (iv) slow increases in free formaldehyde content. The extent of methylene bond formation reached 15% in the first 15 minutes, resulting in mostly type I methylenes, and more slowly thereafter, resulting in type II and type III methylenes. The principal groups that decreased were type I hydroxymethyl groups. Application of the second-order kinetic equation to the decreases of hydroxymethyl groups resulted in the apparent reaction rates shown in Eq. (24.2). The rate of reaction between free urea amide

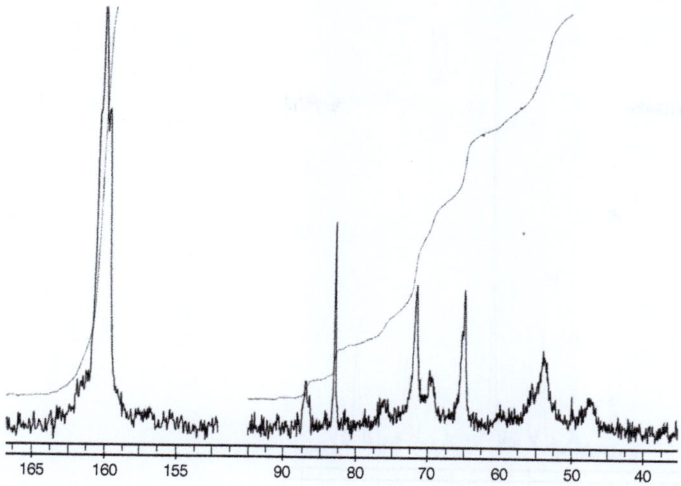

Figure 24.7 ^{13}C NMR spectra of Sample A_8. ($F/U_1 = 2.10$) obtained on CFT-20.

group and Type I hydroxymethyl group was reported to be $1.78 - 10 \times 10^{-4}$ l/mol s at pH 5.0 and 90 °C (extrapolated value), and the rate of reaction between two sym-dihydroxymethyl urea molecules to be 3.0×10^{-6} l/mol s [20, 21]. The fact that our data stand in between these values and the increasing free formaldehyde content from Sample A_4 to Sample A_8 suggests that the reaction in the later stages entails a rate-determining step, in which free amide groups are formed first by the reverse reaction of hydroxymethyl groups and the formed free amide groups go into reaction with hydroxymethyl groups of other molecules to form methylene bonds. ^{13}C NMR spectra of sample A_8 ($F/U_1 = 2.10$) are shown in Figure 24.7, in which Type I and Type II methylene groups are prominent at about 48 and 54 ppm, and Type III methylene groups appear as a small peak at about 60 ppm.

Type I and II hydroxymethyl groups are also prominent at about 65 and 72 ppm, respectively, as well as methylene-ethers at about 69 ppm and formaldehyde species at about 83 ppm. Urea carbonyl peaks are all those of di- and tri-substituted urea groups at about 160 – 162 ppm. Mostly all peaks are broad, indicating the minor differences existing in the structures of each carbon species.

$$\begin{aligned}
&\text{Sample } A_3 \rightarrow \text{Sample } A_4 \; K_2 = 3.97 \times 10^{-5} \text{ l/mol s}\\
&\text{Sample } A_4 \rightarrow \text{Sample } A_5 \; K_2 = 3.23 \times 10^{-5} \text{ l/mol s}\\
&\text{Sample } A_5 \rightarrow \text{Sample } A_6 \; K_2 = 3.88 \times 10^{-5} \text{ l/mol s}\\
&\text{Sample } A_6 \rightarrow \text{Sample } A_7 \; K_2 = 4.55 \times 10^{-6} \text{ l/mol s}\\
&\text{Sample } A_7 \rightarrow \text{Sample } A_8 \; K_2 = \text{too slow to calculate. (At 90°C, pH 5.0, and concentration of 8.5 mol/l.)}
\end{aligned} \tag{24.2}$$

Addition of the second urea and resultant polymer structures: The second urea is normally added when the polymerization is completed and after the acid catalyst is neutralized. As shown by Tomita and Hatono [6], the addition of second urea lowers the free formaldehyde species level, as shown in ^{13}C NMR spectral data of *Samples* A_{11} ($F/U = 1.50$) to Sample A_{13} ($F/U = 0.8$) in Table 24.2, resulting in formation of hydroxymethylureas. The spectra of *Sample* A_{12} ($F/U = 1.00$) are shown in Figure 24.8. Also, it is noted from data in Table 24.2 that methylene group integrals increased dramatically from *Samples* A_8 to A_{12}, i.e. from

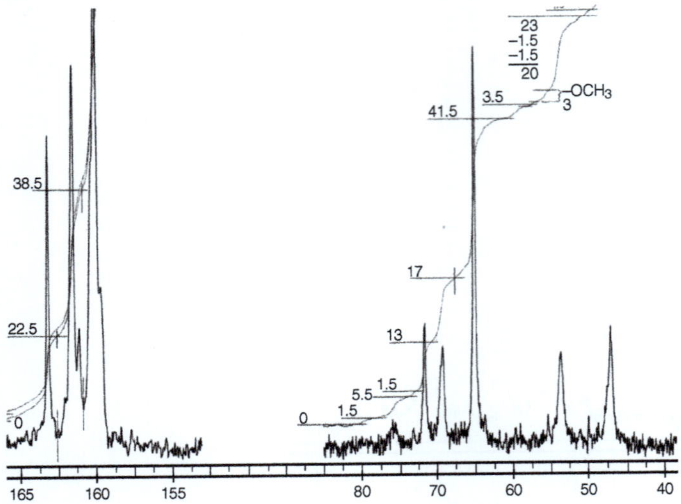

Figure 24.8 ^{13}C NMR spectra of Sample A_{12} (F/U = 1.00) obtained on CFT-20.

34% to 44%, especially for Type I methylene groups, which have increased from 9% to 20%. In spite of a similar report [6], the increases were not explainable in terms of actual bond formation since the second urea addition was conducted in a mildly alkaline pH and at a low temperature, where methylene groups are not known to form. A reasonable explanation is that the ^{13}C NMR sensitivity of methylenes was depressed in *Sample A_8*. Since the ^{13}C NMR pulse-delay time used in the experiment exceeded all UF resins' methyleneic carbons' relaxation times by more than 5 times [22], the signal depression appears to arise from restricted molecular motions like those in partly cross-linked polymers [23]. Such a molecular restriction appears quite possible in certain UF resins like *Sample A_8*, in which the molecular steric hindrance is expected to be high, as shown by the fact that the substitution number per urea molecule for *Sample A_5* reaches already to 2.50 and did not increase till the final *Sample A_8* (column d in Table 24.2). The sensitivity depression was also present in the calculated F/U_1 mole ratio values (column d in Table 24.1), where the F/U_1 mole ratio values are appreciably less than the charged ratio of 2.10 for *Samples A_4–A_8*. When the total methyleneic values are adjusted to 2.10, the degree of polymerization (DP) for *Sample A_8* was calculated to be 20, and other group content values were also obtained as reported in Table 24.3.

Methylene bond formation at F/U_1 = 2.4 and 2.8: From the data in Table 24.2, *Resin B* synthesized with an F/U_1 mole ratio of 2.4 was similar to *Sample A_8*, but the presence of

Table 24.3 Group contents of *Sample A_8*.

Type I hydroxymethyl group:0.320/urea.
Type II hydroxymethyl groups:0.325/urea.
Type I methylene groups: 0.251/urea.
Type II methylene groups: 0.559/urea.
Type III methylene groups:0.149/urea.
(Methylene-ethers are included in methylenes after multiplying by 0.5 for convenience.)

an increased amount of Type IIi hydroxymethyl groups and lower extents of methylene group formation was evident. Furthermore, in case of *Resin C* synthesized with an F/U_1 ratio of 2.8, the viscosity did not increase beyond about 50 cps despite a prolonged reflux at 100 °C and pH 5.0. This suppression of polymerization has been known in industry [24] and is reflected in the low extents of methylene bond formation (methylene/urea ratio of 0.53–0.62) and high free formaldehyde contents. Excess hydroxymethyl groups are present and the total substitution number reached about 2.70 per urea, which is higher than *Sample* A_8.

Methylene bond formation at $F/U_1 = 1.75$ *and 1.50*: Using lower F/U_1 mole ratio values in resin synthesis was considered as a useful method of controlling the polymer chain branching. *Resin D* synthesized with an F/U_1 mole ratio of 1.75 polymerized slightly faster in viscosity advancement than those of higher F/U_1 mole ratio resins and resulted in a turbid resin. *Resin E* synthesized with and F/U_1 mole ratio of 1.50 polymerized even faster and the viscosity advancement was difficult to control. Increasing the pH from 5.0 to 6.0, lowering temperature from 90 to 70 °C, and adding 2-aminoethansulfonic acid helped the viscosity control to a limited extent. The resulting resin was very turbid with a short storage life and non-Newtonian viscosity behavior. The turbidity appears to arise from the formation of insolubles, and the ^{13}C NMR data of the resins therefore would represent only the soluble portions. From what was observed in the spectra, these resins were similar in general to those made with higher F/U_1 mole ratios, but their methylene contents were higher and some reduction in methylene-ethers was also noticeable. Since both of these resin properties were expected to affect the formaldehyde emission properties positively, a board evaluation was conducted despite these handling difficulties.

Wood particleboard properties: Particleboards (PBs) were prepared using D. fir wood particles with a 6.0% resin-solids loading level. Boards of 405 × 405 × 19 mm in dimensions, a target density of 800 kg/m^3 were hot-pressed at 150 °C for 5 and 6 minutes. The internal bond (IB) strength and free formaldehyde emission data of boards prepared with resins A_{12}, D, and E, all at a final F/U mole ratio of 1.0 are reported in Table 24.4. The formaldehyde emission values by the 2-Hour Desiccator method [25] were slightly lower for resins D and E, which were validated in repeat experiments, but the IB strength reduction was apparent, partly due to the inefficiency in the spraying operation of these resins.

Summary: Quantitative ^{13}C NMR analysis using CFT-20 resulted in details on the synthesis chemistry and polymer structures of UF resins. The first step of UF resin synthesis in the alkaline pH at F/U_1 mole ratio of 2.10 resulted in various hydroxymethylureas and small amounts of methylene-ether bonds. The second acidic polymerization reaction step

Table 24.4 Douglas fir particleboard data.

Resin	F/U	Mat moisture (%)	Thickness (cm)	Density (g/cm^3)	IB[a] (kPa)	Emission[b] (μg/ml)
A_{12}	1.0	8.5	1.753	0.804	1176	0.87
D	1.0	8.7	1.753	0.815	1117	0.80
E	1.0	8.7	1.753	0.816	909	0.74

a) Internal bond strength according to ASTM test method (1989).
b) 2-h Desiccator formaldehyde emission value according to NPA.

progressed gradually by forming linear and branched chains of methylene bonds to result in typical polymeric UF components. The F/U_1 mole ratio was shown to significantly affect the polymerization reactions: (i) F/U_1 mole ratios of 2.50 or higher resulted in only a limited amount of the methylene bonds, (ii) F/U_1 mole ratios of about 1.75 or lower resulted in forming methylene bonds fast but resulted in turbidity and unpredictable viscosity behaviors of resins, and (iii) F/U_1 mole ratios of about 2.10 resulted in linear and branched chains of methylene bonds of typical polymeric UF components. Further, the second urea (U_2) addition resulted in forming monomeric hydroxymethylureas by reaction with the remaining formaldehyde and also with the formaldehyde formed by dissociation of hydroxymethyl groups of polymeric UF resin components. The ^{13}C NMR analysis results affirmed the typical polymeric UF resin structures reported from wet chemical analysis methods, including branched UF polymer chain structures via methylene bonds.

24.4 ^{13}C NMR Analysis Results of Reaction Intermediates Taken in UF Resin Syntheses with a Higher Power Instrument

(Adapted with permission from the publisher and minor revisions from Kim, M. G. 1999. Examination of selected synthesis parameters for typical wood adhesive-type UF resins by ^{13}C NMR spectroscopy. I. J. Polym. Sci.: Part A: Polym. Chem. 37, 995–1007. [11])

Resin synthesis and general spectral features observed with a higher power ^{13}C NMR: A typical UF resin was synthesized using the similar procedure as those used in the previous section (Section 24.3), and various time samples were taken in the acidic polymerization steps, followed by addition of second urea to F/U ratio of 1.15 as described in Table 24.5. ^{13}C NMR spectra of *Samples 4Bb* and *4BU* are shown in Figure 24.5a,b, and group percentages and calculated group ratio values for selected samples are reported in Table 24.6. Each formaldehyde-derived group (—CH_2—) value is calculated in percentages in relation to the total values of methyleneic groups, and each urea carbonyl carbon's (—CO—) group value is calculated in percentages based on total of urea carbonyl groups. Data from the previous

Table 24.5 Samples taken in the syntheses of a typical UF resin at $F/U_1 = 2.10$.

1. 1260 g 50% formaldehyde solution was taken, adjusted the pH to 7.8 with 8.0% sodium hydroxide solution and heated to 70 °C.
2. 600 g of urea was added to bring the $F/U_1 = 2.10$, and the reaction exotherm and heating control were used to keep the temperature at 90 °C, which was maintained for 30 min while the pH was maintained at 7.8.
3. The temperature was then lowered to 70 °C, pH adjusted to 4.0 using 8% sulfuric acid solution, and raised to 95 °C and kept the temperature for the entire reaction period. During the resin synthesis, samples of indicated viscosity values were taken at indicated times, neutralizing to pH 8.0, and adding the second urea to $F/(U_1 + U_2) = 1.15$, resulting in B series samples:
   - Sample 1BU at 45 min, C; Sample 2BU at 60 min, F.
   - Sample 3BU at 85 min, M.; Sample 4BU at 105 min, T.
   - Sample 4Bb was taken before the second urea addition to Sample 4BU.
   - Sample 4BUHT was taken by subjecting Sample 4BU to stirring and heating at 60 °C for 170 min with viscosity decreasing from N to F.
   - ^{13}C NMR analyses were done on all samples on Techmag 360 MHz NMR.

Table 24.6 Carbon group content (%) changes in the alkaline and acidic reactions in UF resin syntheses by ^{13}C NMR.[a)]

pH of reaction		pH 7.0–8.0[b)]		pH 4.0					
Sample Name		A2	A3	4Bb	1BU	2BU	3BU	4BU	4BUHt[c)]
Rx time at 95 °C, min. & Vis.		20; A_1	30; A_1	105; T	45; C	60; F	85; M	105; T	105; F
Mole ratio		$F/U_1 = 2.10$			$F/(U_1 + U_2) = 1.15$				
Free Form (%)	91.0 ppm	9	10	1.10	0.00	0.41	0.25	0.39	0.16
	87.0 ppm			6.18	0.46	0.33	0.33	0.35	0.29
	83.1 ppm			7.46	0.51	0.40	0.36	0.34	0.34
	Total	*9*	*10*	*14.73*	*(0.97)*	*(1.14)*	*(0.94)*	*(1.08)*	*(0.79)*
Methylol (%)	Type II 72.0 ppm	15	17	18.99	5.14	7.18	5.67	7.48	5.18
	Type I 65.2 ppm	55	42	16.09	(54.31)	(47.62)	(50.99)	(47.74)	(50.16)
	Total	*60*	*59*	*35.08*					
Methylene-ether (%)	Type III 79.1 ppm	13	17	3.18	2.17	1.65	1.91	1.84	1.79
	Type II 75.1 ppm			5.78	2.93	3.78	3.11	3.17	2.81
	Type I 69.5 ppm			10.13	14.50	10.60	9.92	8.96	10.25
	Total	*13*	*17*	*19.09*	*19.60*	*16.03*	*14.94*	*13.97*	*14.85*
Methylene (%)	Type III 60.1 ppm	<1	1	5.02	1.08	1.70	1.90	1.54	1.91
	Type II 53.9 ppm	<1	1	19.53	9.81	15.92	14.86	17.21	15.40
	Type I 47.4 ppm	<1	1	6.54	9.06	10.42	10.71	10.95	11.70
	Total	*<1*	*<1*	*31.09*	*19.95*	*28.40*	*27.47*	*29.70*	*29.01*
Urea, free (%); 164.0 ppm		<1	<1	0	22.3	25.7	25.2	25.5	18.7
Urea, mono-sub (%) 162.2 ppm		21	24	0	37.6	26.9	29.5	28.7	35.5
Urea, di- & tri-sub (%); 160.7 ppm		78	74	99	40.1	47.4	45.3	45.9	45.8
Urea, tetra-sub (%); 157.8 ppm		<1	<1	0	0	0	0	0	0
—CH_2—/CO (Urea)		1.93	0.98	2.49	1.48	1.43	1.49	1.50	1.45

a) Formaldehyde-derived (CH_2) groups and urea carbonyl groups are separately grouped and % values are calculated. CH_2/CO ratios are direct spectral integration values.
b) Reference 11 (previous paper).
c) heated Sample 4BU at 60 °C for 170 min. pH of analysis 7.8.

section (Section 24.3) of *Samples A_2* and A_3 are also included for comparison in Table 24.6. In comparison with spectra obtained with (100 MHz) CFT-20 NMR spectrometer, the spectra obtained with (360 MHz) *Techmag 360 NMR* spectrometer offered better resolution of peaks, but quantitative aspects were not known. The chemical shift values of methyleneic and carbonyl carbons measured in this work are referenced to the small peak at 50.4 ppm of methanol present in most UF resin samples [6]. For convenience of discussion on polymer structures, various methylenes and methylene-ethers are defined similarly as in the previous section (Section 24.3) [11] into Type I, II, and III classes, depending on whether 0, 1, or 2 methylol, methylene, or methylene-ether groups are bonded to the two urea nitrogen atoms involved. Further, Type I hydroxymethyls are single hydroxymethyl groups, Type IIt hydroxymethyl groups are *gem*-dihydroxymethyl groups bonded to the same urea nitrogens on UF polymer chain ends, and Type IIi hydroxymethyl groups are single hydroxymethyl groups bonded to urea nitrogens in the middle of UF polymer chains.

Chemical shift assignments of various chemical structures: (i) Methylene groups of Type I, Types II, and Types III occur at 47.4, 53.9, and 60.1 ppm, respectively, all of which are somewhat broad. (ii) Methylene-ether groups of Type I, Type II, and Type III occur, respectively, at 69.5, 75.7, and 79.1 ppm, although the last group's peak was weak, and the assignment remains speculative because of the lack of appropriate model compounds. (iii) Type II methylene-ether peaks at 75.7 ppm are in the same chemical shift range as the methylene-ethers of uron-type ether structures [5, 6], but uron carbonyl peaks at about 157 ppm are almost absent in common UF resins, and it is more appropriate to assign the peak to ordinary Types II methylene-ethers. (iv) Several small sharp peaks often occur overlapping with various methylene-ethers, which were included in methylene-ethers in deriving polymer structures.

Reaction paths to various functional groups in UF resins: Figure 24.9 shows a schematic representation of various paths to polymeric UF components and hemi-formal groups. The following observation is noted: (i) In UF resins that contain some free formaldehyde, i.e. samples taken before adding the second urea, all types of hydroxymethyl groups go to partially form methylene-ether type hemi-formals in both acidic and basic pH ranges {(Figure 24.9 ***(1)–(14)***} and they overlap with respective methylene-ether peaks. This

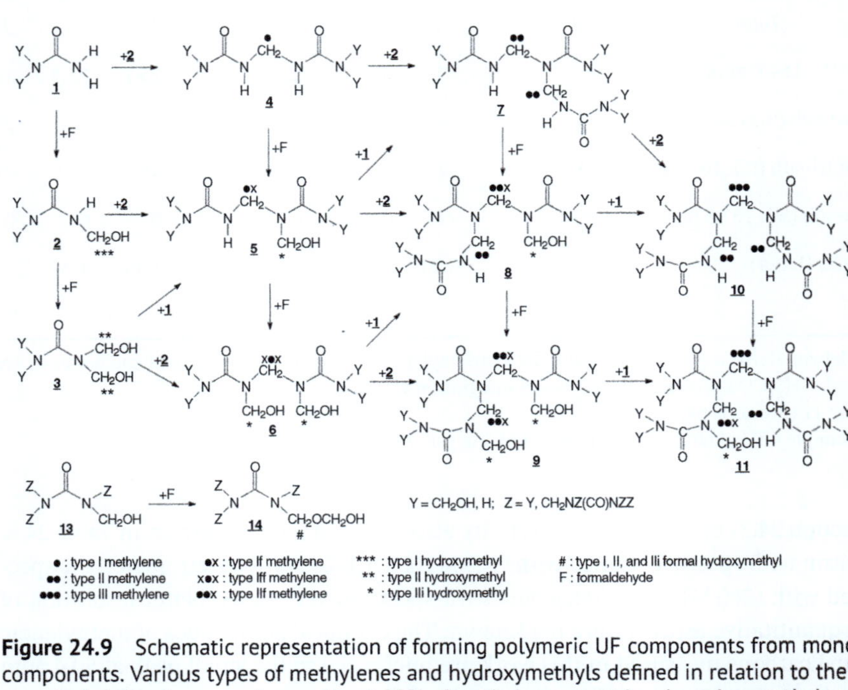

Figure 24.9 Schematic representation of forming polymeric UF components from monomeric UF components. Various types of methylenes and hydroxymethyls defined in relation to their ^{13}C NMR chemical shift values are shown. About 20% of methylenes must be viewed as methylene-ethers for Sample 4Bb.

prevents direct measurements of methylene-ether groups until an enough amount of second urea is added to unzip the formal groups; (ii) Type I and IIt/IIi hydroxymethyl groups occur as relatively sharp peaks at 65.2 and 72.0 ppm, respectively, and the peaks of Type IIt/IIi groups tightly overlap preventing their separate measurements. Type IIi hydroxymethyl groups exert both β- and γ-substituent effects on the chemical shift values of neighboring methylene and methylene-ethers [3], elevating them to the levels of higher branched structures; [6, 7]; (iii) Type IIt/IIi hydroxymethyl groups can be labile by the reversibility [17–19] and are not always considered to create polymer chain branches on curing, and their concentration is more appropriate to be subtracted from Type II or III methylene and methylene-ether values to obtain un-incumbered polymer chain branching structures; (iv) Other methyleneic peaks are methanediol (formaldehyde) at 83.1 ppm and its oligomers at 91.0 ppm and formals at 87.0 ppm. These free formaldehyde species are low in resins that have second urea added to F/U mole ratio values of common UF resins and almost disappear when more urea is added, for example, in *Sample 4BUU* with an F/U ratio of 0.75 (spectrum not reported); (v) The carbonyl peak at 164.0 ppm is of free urea, mono-substituted ureas occur at 162.2 ppm, and di- and tri-substituted ureas occur at 160.7 ppm [6]. The carbonyls of tetra-substituted ureas and urons occur in the 155 to 157 ppm range [5–8] observed at trace levels in resin samples studied.

24.4.1 Effects of the Addition of Second Urea and Mild Heat Treatments

Free formaldehyde species and formal groups diminish: The second urea was added in a weakly alkaline pH and during the cooling period without any external heating unless indicated otherwise. ^{13}C NMR data in Table 24.6 indicate that: (i) the addition of second urea results in transforming most of free formaldehyde species and hemi-formals into hydroxymethyl groups. No methylene or methylene-ether bonds would form or break apart by the second urea and, therefore, *Sample 4BU*, typical of finished UF resins, can be considered to have a composition consisting of *unreacted urea and monomeric hydroxymethylureas in the resin derived from the second urea (44%) and the polymeric UF components derived from the first urea (56%)*. (ii) The decrease in hemi-formal groups (5.83%) results in decreases in various methylene-ether peaks (5.12%) by reverting hemi-formal methylene-ethers to hydroxymethyl groups (Table 24.6, Figure 24.9 *14* → *13*), where the difference of the two values (0.71%) is relatively small. (iii) The decrease in Type IIi hydroxymethyls occurring with the second urea addition also brings about changes in methylenes: an increase in Type I (4.41%) and decreases in Types II and III methylenes (5.80%, Table 24.6, Figure 24.9 *6* → *5* → *4*), where the difference (1.39%) is significant and can come from similar methylene-ether group changes, which cannot be separated out from decreases occurred from the removal of hemi-formals. (iv) The total changes in methylene-ethers appear to be 2 – 3% increase in Type I methylene-ethers. Theoretically, the increases and decreases, when combined for methylenes and methylene-ethers, should cancel out. (v) The combined increase in Type I methylenes and methylene-ethers attained is calculated to be about 6.9%, and since it is from the removal of Type IIi hydroxymethyl groups from the polymeric UF components, it is a minimum estimate value of the groups' level before the second urea addition. (vi) The significant amount of increase in Type I methylene groups observed here is the same phenomenon reported for samples obtained with the second urea addition in earlier reports [6, 11], which was ascribed to the variation of ^{13}C NMR relaxation parameters [11]. The earlier workers, however, were unable to accurately detect the concurrent decreases in Type II and III methylene groups, probably due to the instrumental limitations related to the lower magnetic strengths available then and to ^{13}C relaxation parameters for these somewhat crowded structures of carbons with limited mobilities. Some variations in results due to the ^{13}C NMR relaxation parameters may still be present in this work done with a higher-power NMR machine.

Type IIt/IIi hydroxymethyl groups decrease: Another main spectral change due to the second urea addition and mild heat treatments revealed is the large extent of decrease of Types IIt/IIi hydroxymethyl groups, from 22.49% to 7.48%, including those generated from the removal of hemi-formal groups. The result indicates that these groups are transferred from the polymeric UF resin components to form Type I hydroxymethyl groups on the second urea. This movement of hydroxymethyl groups is understood based on the reversibility of hydroxymethylation reactions between formaldehyde and urea occurring in alkaline pH [17–19]. A similar movement of Type I hydroxymethyl groups should have occurred from the polymeric UF components, although it is not measurable. It was found that the hydroxymethyl group transfer depends on the temperature and duration of mixing of second urea. When the synthesized resin sample was cooled rapidly in ice water bath and stored at about 4 °C (*Sample 4BUc*, spectrum not reported), Type II/IIi hydroxymethyl groups were not reduced at all. On the other hand, when a third urea is added to *Sample 4BU* at room temperature to attain an F/U of 0.75 (*Sample 4BUU*, spectrum not reported), only a very minor intensity change occurred further for Type IIt/IIi hydroxymethyls, indicating that the equilibrium was almost reached in Sample 4BU. It should be also noted that the amount of free formaldehyde and hydroxymethyl groups involved in the reaction with second urea is only about 1.0 mol per mole of second urea and, therefore, the Type II/IIi hydroxymethyl groups formed on second urea would be quite low, and those Type IIt/IIi groups observed in spectra would be mainly those already bonded to the polymeric UF components formed with first urea.

Estimate of type IIt hydroxymethyl groups: Another interesting observation in comparing spectra of *Samples 4Bb* and *4BU* is that the estimate of Type IIi hydroxymethyl groups, about 6.9% derived from combined increases of methylene and methylene-ethers, is roughly one-half of the total amount of reduction in Type IIt/IIi hydroxymethyl groups. The other half of decrease could be the movement of Type IIt hydroxymethyl groups.

24.4.2 Polymerization in the Acidic Polymerization Step Observed with Second Urea Addition

Methylene group increases: The progression of polymerization in the second acidic step of resin synthesis is clearly seen from *Sample A_3* to *Samples 1BU*, *2BU*, *3BU*, *4Bb*, and *4BU*, where *Type I, II, and III methylene groups gradually increased from about 3% to 19.95% and then to about 29.01%*. The values for various methylene and methylene-ether groups would not be distorted much by the (low) levels of Type IIi hydroxymethyl groups or hemi-formals. *Type II methylene groups grew more (1.00% → 9.81% → 15.4%) than Types I (1.00% → 9.06% → 11.70%) and Type III (1.00 → 1.08% → 1.91%) groups, indicating the formation of branched polymer chain structures*. The methylene group increases were accompanied by decreases in Type I hydroxymethyl groups from 42% to 16%. Type IIt/IIi hydroxymethyl groups changed only a little and methylene-ether groups also did not change much. If comparing *Samples A_3* and *4Bb*, i.e. samples before the second urea additions, the free formaldehyde species increased from 7% to 14%, mono-substituted urea species decreased nearly to 0%, and di- and tri-substituted urea species increased to 99%.

Reaction mechanism of methylene group formation : ^{13}C NMR results indicate that Type I hydroxymethyl groups reacted predominately with free amide groups to form Type I methylene groups but also moderately with imide groups to form Types II and III methylene groups, respectively. It is also apparent that *some hydroxymethyl groups are broken off as formaldehyde.* De Jong and de Jonge [17, 19] considered that the reaction rate difference of hydroxymethyl groups between amide and imide nitrogens of urea molecule is so large that the polymerization in the early stages would go through a two-step reaction, i.e. de-hydroxymethylation (by the reversibility) of Types I and IIt hydroxymethyl groups to free amide groups, which then react with Type I hydroxymethyl groups of another molecule to form methylene bonds. In this regard, the high levels of Type IIt/IIi hydroxymethyl groups in pre-urea addition samples with F/U_1 mole ratio greater than 2.50 in the previous section (Section 24.3) are noteworthy: the emergence of polymer molecules is prevented as a result of the slow reverse reactions of hydroxymethyl groups. While this theory appears plausible and explains the increase in formaldehyde content during the polymerization, it also seems to predict UF polymers formed in the end would be mostly of linear polymer structures through formation of Type I methylene bonds. Thus, the relatively fast polymerization reaction rate in the beginning appears to be due to the ready availability of free amide groups from mono-hydroxymethylurea and *unsym*-dihydroxymethylurea present at a level of nearly 34% of first urea used. However, this appears to be only a part of the story. General schematic reaction paths leading to Types II and III methylene group structures are shown in Figure 24.9 and increases in methylene group contents observed in the time samples from *Samples 1BU* to *4BU* shown in Table 24.6 indicate that the polymerization reaction proceeds relatively fast in the beginning with more than 60% of methylene groups forming in *Sample 1BU* (45 minutes), mostly Type I and II methylenes in about equal proportions, indicating the former is an intermediate to the latter.

Branched polymer structures are formed: In the later samples, Type II methylene groups increased further and then Type III methylenes also emerged. Type I methylenes remained at about the same level until the later part of the reaction, indicating that they are continuously replenished. In order to form Types II and III methylenes, the imide nitrogens of Type I methylenes must react with hydroxymethyl groups of another molecule (Figure 24.9 ***4–7***), or Type IIi hydroxymethyl groups must react with amide nitrogens. The combined Type II/III methylene and methylene-ether bond contents were significantly higher than Type II/IIi methylol bond contents, indicating that the branch point methylene bonds that appear as such by Type IIi methylol groups would be small. Since methylene groups are nearly **irreversible groups,** these branched structures will remain to exist even in cured resins and can thus significantly affect the curing rheology and bonding performance of resins.

Role of methylene-ether bonds: Most of the methylene-ethers are formed in the initial alkaline condensation step [5–8, 10, 26] as indicated by their high concentrations in *Sample 1BU*. Type I methylene-ethers, normally the highest among methylene-ethers, started out in *Samples 1BU* at a relatively high level and decreased slowly as the polymerization reaction progressed. On the other hand, Type II methylene-ethers increased slowly from *Samples 1BU* to *Samples 3BU* and then decreased slowly thereafter, and Type III methylene-ethers were low in *Samples 1BU* and showed small increases thereafter. Overall, methylene-ethers are slightly decreased in the acid-catalyzed polymerization reaction period. This observation indicates that methylene-ethers in UF resins could be a transient species that would eventually break up in the acidic curing process to emit formaldehyde [26]. On the other hand, methylene-ethers, although lower in concentration than methylenes, partly constitute the polymer chain structure of polymeric UF resin components and therefore can modulate various rheological and curing characteristics of UF resins until they break up.

Hydroxymethyl group changes during polymerization reaction and by second urea addition: Type I hydroxymethyl groups showed an overall decrease as the polymerization progressed, reflecting their transformation into methylenes. Type I hydroxymethyl groups bonded to polymeric and monomeric UF components occur together at 65.2 ppm in ^{13}C NMR spectra and are not differentiated. On the other hand, Type IIt/IIi hydroxymethyls show an overall increase as the polymerization reaction progresses, bonded mostly to the polymeric UF components, and a major part of them is removed by second urea addition, and the residual Type IIt/IIi hydroxymethyl groups would be low and it appears that they are mostly of Type IIt hydroxymethyl groups. Formaldehyde and its oligomers reached to low levels because of the second urea addition, and they are reduced further in Sample 4BUU (F/U ratio of 0.75, spectrum not reported), showing that the free formaldehyde level in resins is dictated by final $F/(U_1 + U_2)$ ratios.

24.4.3 Polymerization In the Acidic Reaction Step Observed Without the Second Urea Addition

Increases in free formaldehyde species: The free formaldehyde species in *Sample 4Bb*, present as methanediol and its oligomers and formals, are about 0.30 mol per mole of first urea, an increase from about 0.15 mol present at the beginning of the polymerization step (Table 24.6) [9]. The reacted F/U_1 ratio reached to about 1.80, a consequence of transformation of hydroxymethyl groups into methylene groups that require two urea nitrogen sites for making a bond. Some hydroxymethyl groups are expelled via de-hydroxymethylation reaction, and the urea nitrogen sites thus freed become available for methylene bond formation (Figure 24.9 ***8*** → ***7*** → ***10***). The rate of de-hydroxymethylation reaction under the acidic reaction condition and at an F/U_1 ratio of 2.10 appears to be just fast enough to keep the polymerization reaction go at a controllable speed at common resin synthesis temperatures of about 95 °C.

Role of formaldehyde species in the polymerization: In the previous section (Section 24.3) [11], it was noted that as the F/U_1 mole ratio used in resin synthesis is lowered to 1.80 or lower, the viscosity of reaction mixture increases faster and its turbidity also begins to arise and become severe, resulting in resins having a low storage stability and poor wood bonding performance. In these cases, the urea amide nitrogens are too abundant, and the emerging UF resin polymers lack hydroxymethyl groups to result in destabilizing the UF polymers in solution. On the other hand, when the F/U_1 mole ratio is increased in resin synthesis, the polymerization reaction slows down, and at an F/U ratio of about 2.60 or higher, the reaction proceeds only to limited extent even when the pH is lowered to near 1.0 [11, 27]. Overall, the availability of urea amide nitrogens during the polymerization reaction is controlled by the amount of formaldehyde or F/U_1 mole ratio in the resin, a key ingredient parameter for synthesizing useful UF resins. Certain specialty wood adhesive-type UF resins are known to be manufactured by conducting both the hydroxymethylation and polymerization reactions simultaneously at pH between 4 to 5, but the procedure requires an F/U ratio of about 2.50 or higher, and the resulting resins tend to become turbid and show lower wood binding performance [28]. It is probable that under acidic pH, the hydroxymethylation and methylene bond-forming reactions occur concurrently even at high F/U_1 mole ratios, resulting in the formation of sparingly soluble methylene-di-urea type precipitates to make UF resins turbid.

Bond restriction of urea molecule observed: Another observation made on *Sample 4Bb* (F/U_1 = 2.10) is that Type I hydroxymethyl groups are present at a high level despite the excess free formaldehyde present without their being converted into Type IIt hydroxymethyl groups on polymer chains. The levels of Type IIt/IIi hydroxymethyl groups are attained in the acidic polymerization and unaltered by the weakly alkaline pH of the resin sample. Also, heating of *Sample 4Bb*, under an optimum condition for hydroxymethylation reaction (pH 8.0, 90 °C for 20 minutes), did not increase Types I or IIt/IIi hydroxymethyl groups (Sample 4BbHT, spectrum not reported). Thus, the imide nitrogens, including those to which Type I hydroxymethyl groups are bonded, appear to be prevented from reacting with formaldehyde that is present, probably indicating the fact that one urea molecule can form only three methyleneic bonds at most [26]. Urea units having one Type I hydroxymethyl group on one nitrogen would be allowed to have *two* methyleneic groups on the other nitrogen, and urea units having one Type IIt hydroxymethyl group on one nitrogen would be allowed to have only *one* methyleneic group on the other nitrogen. Once such structures are attained, further reactions with formaldehyde would be restricted. Thus, the molecular restriction appears intertwined with reversibility of hydroxymethyl groups to play out in the polymerization reactions.

24.4.4 Summary

Quantitative analysis of UF resin intermediates taken during the acidic polymerization in resin synthesis at an F/U_1 ratio of 2.10, followed by second urea addition to $F/(U_1 + U_2)$ = 1.15, was re-examined on a higher power ^{13}C NMR. Various monomeric hydroxymethylureas and some methylene-ethers of hydroxymethylureas are formed in the initial alkaline reaction step. In the subsequent acidic polymerization reactions, various methylene bonds increased accompanied with viscosity increases to result in forming poly(methylene methylene-ether hydroxymethylureas). The methylene-ethers formed in the alkaline step remained little changed and are present as minor polymer components. In the polymerization reaction, some hydroxymethyl groups are reverted to formaldehyde, resulting in a reacted F/U ratio of about 1.80 with the remainder present as (excess) formaldehyde. The de-hydroxymethylation reaction appears to be the rate-determining methylene bond forming process in the later part of polymerization reaction step. The addition of second urea to resin intermediate samples resulted in its reaction with free formaldehyde present and also with those formed from dissociation of especially Type IIi hydroxymethyl groups from polymeric hydroxymethylureas, resulting in decreases in Type II/III and increases in Type I methylene groups, revealing the linear and also branched UF resin polymer structures, the values of which would entail only small deviations due to presence of Type IIi hydroxymethyl groups. Type IIi hydroxymethyl group contents are low in UF resins, although their presence would still make some Type I and II methylene/methylene- ether groups appear as Type II and III groups. Polymeric UF resin structures are thus concluded to be branched via Type II and III methylene groups with polymer chain-end groups capped with Type I and Type IIt/IIi hydroxymethyl groups as well as free urea amide groups.

24.5 Chemical Changes Occurring in UF Resins on heat/stirring and Room-temperature Storage Treatments by ^{13}C NMR and Formaldehyde Emission Tests of Particleboard

(Adapted with permission from the publisher and minor revisions from Kim, M. G. 2000. Examination of selected synthesis parameters for typical wood adhesive-type UF resins by ^{13}C NMR spectroscopy. II. J. Applied Polym. Sci.: Part A: Polym. Chem. 75, 1243–1254. [12]*)*

Storage stability of UF resins and variation of bond performance over the transportation and storage period are of interest in industry. The lower the temperature, the slower the resin property changes and the longer the storage lives. At elevated temperatures, the storage time is reduced significantly, and in warm climates liquid UF resins are advised to be stored in a cool place. This paper is a continuation of the previous section (Section 24.4), and one typical UF resin $\{F/(U_1 + U_2) = 1.15)\}$ is similarly synthesized, cooled rapidly in ice bath, added the second urea, and subjected to varying treatments: (i) stirring at room temperature (*Sample* A_1) followed by stirring/heat treatments at 40 – 60 °C (*Samples* $A_2 \rightarrow A_5$) and (ii) stirring at 70 °C (*Sample* A_6) for various lengths of time as described in Table 24.7. ^{13}C NMR analysis results of these resin samples are reported in Table 24.8. (iii) Also, the untreated resin samples were stored at room temperature without stirring for up to 30 days as shown in Table 24.9 with the ^{13}C NMR results reported in Table 24.10.

24.5.1 Effects of Mild and Intermediate Level Heat/Stirring Treatments (Samples A_2–A_6)

Free formaldehyde and hemi-formal groups: ^{13}C NMR data in Table 24.8 indicate that in *Samples* A_2 *(room-temperature treatment)* and A_3 most free formaldehyde and hemi-formals present in Sample A_1 have reacted with second urea to form hydroxymethyl groups

Table 24.7 Various heat treatments of UF resins.

- *Resin synthesis*: 3650 g batch, $F/U_1 = 2.10$, reacted 30 min at 95 °C and pH 7.8, then reacted for 105 min at 95 °C and pH 4.0, W viscosity, adjusted to pH 7.8, and cooled to 4 °C.
- *Sample* A_1 – Sample A_1 (2.0 g) is taken,
- Added second urea (1239 g), F/U = 1.15, cooled to 4 °C, stirred for 90 min; warmed to 20 °C, and *Sample* A_2 *(800 g)* is taken, viscosity S.
- Residual, heated/stirred to 40 °C, held for 25 min, and *Sample* A_3 *(800 g)* is taken, viscosity RS.
- Residual, stirred for 45 min at 40 °C, viscosity unchanged at RS. Heated to 50 °C and stirred 15 min, *Sample* A_4 is taken, viscosity QR.
- Residual, stirred for 30 min at 50 °C, viscosity QR, heated to 60 °C and stirred for 30 min, *Sample* A_5 *(800 g)* is taken, viscosity P.
- Residual, stirred for 30 min at 60 °C, viscosity N, heated to 70 °C and stirred for 30 min, and *Sample* A_6 is taken, viscosity K, light turbidity.
- Residual, stirring for 30 min at 70 °C, turbidity increased, viscosity lowered, but samples were not taken. pH values were 7.7–8.0 during heat treatments. All samples were stored at 4 °C for one week. ^{13}C NMR was obtained and PBs were made and tested.

Table 24.8 Percentage values of various methyleneic carbons and carbonyls of UF resin Samples A1–A6 determined by ^{13}C NMR, sample viscosity, and 2-h Desiccator formaldehyde emission values (Added 2020).[a)]

Groups	ppm	A_1	A_2	(A_2-A_1)	A_3	A_4	A_5	A_6	(A_5-A_2)	(A_6-A_2)
Free formaldehyde (%)	91.0	1.42	0.72	−0.70	0.70	0.82	0.40	0.59	−0.32	−0.13
	87.0	5.40	0.00	−5.40	0.00	0.00	0.00	0.00	0.00	0.00
	83.1	5.88	0.67	−5.21	0.00	0.37	0.00	0.33	−0.67	−0.34
	Total	*12.70*	*1.39*	*−11.31*	*0.70*	*1.19*	*0.40*	*0.92*	*−0.99*	*−0.47*
Methylene-ether (%)	79.1	2.30	2.30	0.00	2.15	2.15	2.01	1.91	−0.29	−0.39
	75.1	7.28	5.40	−1.88	5.61	5.17	4.55	4.40	−0.85	−1.00
	69.5	9.56	8.31	−1.25	8.31	8.61	8.69	10.68	+0.38	+2.37
	Total	*19.14*	*16.01*	*−3.13*	*16.07*	*15.81*	*15.25*	*16.99*	*−0.76*	*+0.98*
Hydroxymethyl (%)	72.0	19.80	17.06	−2.74	18.60	15.95	13.72	7.70	−3.34	−9.76
	65.2	15.36	31.81	+16.45	31.03	31.91	37,50	39.06	+5.69	+7.25
	Total	*35.16*	*48.87*	*+13.88*	*49.63*	*47.86*	*51.22*	*46.76*	*+2.35*	*−2.11*
Methylene (%)	60.1	5.50	3.42	−2.08	3.21	3.00	2.28	2.57	−1.14	−0.85
	53.9	20.96	22.21	+1.25	22.55	23.11	22.27	19.70	−0.06	−2.51
	47.4	6.53	8.13	+1.60	7.48	9.03	10.37	13.06	+2.19	+4.93
	Total	*32.99*	*33.76*	*+0.77*	*33.60*	*35.44*	*34.92*	*35.33*	*+0.99*	*+1.57*
Urea, free (%)	*164.0*	0.00	32.91	+32.91	33.97	31.42	28.07	21.73	−11.80	−11.18
U-mono sub (%)	*162.2*	4.50	20.86	+16.36	20.01	21.46	24.80	31.24	+3.94	10.38
U, di–/tri-sub (%)	*160.7*[b)]	95.50	46.22	−49.28	46.02	46.96	47.12	47.03	+0.90	+0.81
CH_2O emission[d)]		—	0.70	—	0.73	0.84	0.82	0.94	—	—
Viscosity values[c)]		W	S	—	RS	QR	P	K	—	—

a) *Chemical structures are defined as before in ppm. Values in parentheses are the differences between samples.*
b) *Group value includes 2–3% of tetra-substituted urea carbonyls.*
c) *Viscosity of resin sample at time of test in Gardner-Holdt scale.*
d) 2-H Desiccator Formaldehyde Emission (ASTM D 5582) Values of PB in μg/l.

Migration of types IIt/IIi hydroxymethyl groups (72.0 ppm) in mild stirring and heat treatments: ^{13}C NMR data in Table 24.8 indicate stepwise migrations of Types IIt/IIi hydroxymethyl groups to second urea in stages, gradually from *Sample A_1* to *Sample A_2 (−2.74%)*, to Sample A_3 (−1.20%). Samples A_2 and A_3 are similar in ^{13}C NMR and viscosity values, indicating that the room-temperature stirring and heat treatment at 40 °C used to obtain *Samples A_2 and A_3* had only a minor effect. Furthermore, the viscosity of *Sample A_3* did not change in the next 45 minutes of heating at 40 °C, indicating that the hydroxymethyl group migration rate is relatively slow at 40 °C, which can be considered quite high compared with the temperatures encountered in the storage and transportation of industrial UF resins. In industry, wood PB adhesive-type UF resins are transported and stored at cool temperatures and normally used within a week.

Migration of types IIt/IIi hydroxymethyl groups (72.0 ppm) in intermediate level stirring and heat treatments: Type IIt/IIi hydroxymethyl groups migrated faster: *Sample A_4*

Table 24.9 Various room-temperature storage treatments of UF resin samples.

- Resin synthesis: 3650 g batch, $F/U_1 = 2.10$, reacted for 30 min at 95 °C and pH 7.8, then reacted for 105 min at 95 °C and pH 4.0, W viscosity, adjusted to pH 7.8, cooled to 4 °C in ice bath.
- Sample B_1 – Sample B_1 (2.0 g) is taken, vis. V.
- Residual, divided into 6 portions (785 g each); $U_2 = 207.0$ g each.
- All samples are stored at 4 °C.
- Sample B_6 taken, added U_2, $F/(U_1 + U_2) = 1.15$, stirred at room temp., vis. P, stored at room temp. For 30 days, vis. K at 12 days and then increased to N with turbidity.
- Sample B_5, same treatment with U_2, stored for 22 days, viscosity M and light turbidity.
- Sample B_4, same treatments with U_2, stored for 16 days, vis. L.
- Sample B_3, same treatments with U_2, stored for 7 days, vis. L.
- Sample B_2, same treatment with U_2, stored for 1 day, vis. OP
- ^{13}C NMR were obtained, and PBs were made.

(−3.85%), *Sample A_5* (−6.08%), and *Sample A_6* (−12.10%). *Sample A_4* (additional heating for 30 minutes at 50 °C) showed a slightly reduced viscosity of QR and the viscosity remained unchanged for the next 30 minutes of heating at 50 °C, indicating a similarly good stability at this increased temperature, indicating that the migration of hydroxymethyl groups occurring during transportation and storage of UF resins in the industrial practice would be relatively slow. In *Sample A_5* (additional heating at 60 °C for 40 minutes), the viscosity was reduced to P and when the heat treatment at 60 °C was continued for the next 30 minutes, the viscosity decreased relatively fast to N. For resin *Sample A_6*, the temperature was raised to 70 °C and held for 30 minutes, resulting in a rapid decrease in viscosity to K with a light turbidity arising in the resin sample. The viscosity reduction and turbidity increase became relatively faster when the resin sample was heated further. The viscosity decreases of resin samples in these heat treatments can be ascribed to decreases in the molecular weight of first-urea resin components with the migrated hydroxymethyl groups appearing as Type I hydroxymethyl groups on second urea. In industry, the second urea addition is normally conducted while cooling down, starting at about 70 °C, and therefore the time/temperature history of the initial cooling period would make some large differences in the extent of hydroxymethyl group migration. The migration rate of hydroxymethyl groups will decrease as batch temperature decreases and, once the resin batch is cooled to room temperature, the hydroxymethyl groups would migrate only very slowly. Therefore, some variability in the hydroxymethyl group distribution between the polymeric and monomeric UF resin components is expected to exist in industrial UF resins, depending much on the batch cooling rates.

Methylene and methylene-ether group changes due to migration of hydroxymethyl groups: ^{13}C NMR data in Table 24.8 indicate that the migration of Types IIt/IIi hydroxymethyl groups discussed above is reflected in decreases of Type II and III methylene-ether groups totaling −1.39% and also decreases of Type II and III methylene groups totaling −3.36% to a combined total of −4.75%, which can be considered to be equal to the value of Type IIi hydroxymethyl groups that have migrated. This value becomes 5.15% when counted from Sample A_1, and the amount of Type IIt hydroxymethyl groups that migrated to second urea would be the difference, 6.95%.

Polymeric structure changes in heat/stirring treatments: Methylene and methylene-ether bonds occur in a ratio of 81% to 19%, and the Type IIi hydroxymethyl group content of the resin assumed to be 3.85% is relatively small and can be ignored as an approximation in calculation of an approximate polymer chain structure based on methylene and methylene-ether bond contents. The polymer structure of *Sample* A_6 can be calculated by combining methylene and methylene-ether (×0.5) bonds as one bond group as in the following bond-type values per mole of first urea: Type I, 0.386 bond (18.4 % × 2.1), Type II, 0.460 bond (21.9 % × 2.1), and Type III, 0.074 bond (3.5 % × 2.1) with a total of 0.920 bond-group. This group value indicates a number-averaged 12.5-mer {n = (1−0.92)}, [29],which possesses a total of 5.81 urea-chain branches $(2+0.46\times12.5\times0.5+0.074\times12.5\times1)$ for the polymeric UF resin molecules. Since UF resin polymer chain structures based on methylene and methylene-ether bonds would change little by mild heat treatments conducted in the weak alkaline pH, it is useful to compare the number of urea polymer chain branches obtained (5.81) with the number of Type I and IIt hydroxymethyl terminal groups for *Sample* A_1, where little of them migrated to second urea. The number of Type I hydroxymethyl groups $\{4.36 = (0.154+0.013)\times2.1\times12.5\}$ and that of one-half of Type IIt hydroxymethyl groups $\{1.42 = (0.198+0.019)/2\times2.1\times12.5/2)\}$, assuming Type IIt hydroxymethyl groups constitute about one-half of the Type IIt/IIi groups, amounts to 5.78, which compares well with the number of urea chain branches obtained. Some free amide nitrogens were indicated to be present for Sample A_1 type resins in the previous section (Section 24.4), [9] but *Sample* A_1 used in this work indicates very few free amide terminal groups. Uncertainty exists in the 1 : 1 proportioning of Type IIt/IIi hydroxymethyl groups, but the result indicates that the acidic polymerization reactions in later parts of resin synthesis would occur mostly by de-hydroxymethylation reactions, as suggested in Section 24.3.

Type IIi hydroxymethyl groups remaining in Sample A_6 are unknown and if they are assumed to be one-half of the residual Type IIt/IIi hydroxymethyls (3.85%), the total number of polymer branches obtained based on methylene and methylene-ether bonds need to be revised down by 0.51, a refinement of the polymer chain structure. Overall, the polymer structure of *Sample* A_1 obtained above appears to be the closest one obtainable and it is in accord with the general concept of branched, resole-like structures, the long-held theory in UF resin chemistry. Thus, *Sample* A_1 can be represented as a 12.5-mer having 5.81 polymer chain branches capped with Type I or IIt hydroxymethyl groups, and Type IIi hydroxymethyl groups are present bonded to some internal urea nitrogens in polymer chains. The overall reacted F/U mole ratio is approximately 1.70 with an average functionality of 2.39 for (first) urea. The amounts of Type IIt/IIi hydroxymethyl groups formed on the second urea during heat treatments appear to be small. For *Sample* A_6, for example, the total hydroxymethyl groups that migrated to the second urea amounts to 0.365 mol per mole of second urea, at which mole ratio the Types IIt/IIi hydroxymethyl groups formed on the second urea will be small.

Reaction of second urea and an indication of polymerization occurring in stirring/heating treatments: Overall, the migrated formaldehyde reacted mostly with the free urea (164.0 ppm), the level of which is reduced by 11.8%, resulting in mono-substituted urea, more likely hydroxymethylurea (162.2 ppm), which increased by 10.38% based on total urea in the resin. This change is also reflected in Type I hydroxymethyl groups that grew by 7.25% in *Sample* A_6 from *Sample* A_2, and the difference (2.51%) from the Type IIt/IIi group decrease value of 9.76%, or the disappearance of formaldehyde, is difficult to ignore, and a part of it is considered to be a telltale sign of polymerization, i.e. the increase in methylene bond contents by about 3.24% observed, which is relatively small in contrast to the room-temperature storage results discussed in the next section.

Stepwise viscosity decreases in heat/stirring treatments?: The somewhat stepwise manner of viscosity decreases observed during the heat/stirring treatments is an interesting phenomenon. All hydroxymethyl groups in *Sample* A_1 type UF resins are likely to be bonded to di- and tri-substituted urea units and their migration to second urea entails de-hydroxymethylation reactions that could be responsible for the stepwise manner of viscosity decreases observed. It is probable that hydroxymethyl groups bonded to di-substituted urea units are slower to dissociate than those bonded to tri-substituted urea units, and this appears to make some barriers to de-hydroxymethylation reactions, making UF resins used in the way we currently manufacture, transport, store, etc. Hydroxymethyl groups are stabilizing groups for polymeric UF resin molecules, i.e. help keep them swollen and suspended in the aqueous medium. An excessive loss of hydroxymethyl groups from polymeric UF components would result in agglomeration of polymer molecules to form solid-like particles and cause turbidity to arise, as partly seen in *Sample* A_6 and beyond. However, totals of methylene and methylene-ether bonds did not change much within the temperature treatment regime used, indicating that the final cooling rate in UF resin manufacture might be only a small factor as long as the second urea is promptly added with vigorous stirring at about 70 °C.

24.5.2 Effects of Room-temperature Storage Treatments of UF Resins

Ripening times of UF resins: One characteristic of UF resins is the various slow physical changes going on in manufactured resins during the transportation and storage periods: (i) viscosity decreases initially and increases later on, (ii) appearance of turbidity, and (iii) variations in bonding performance and formaldehyde emission values of boards. The causes of these changes could be related firstly to the reversibility of hydroxymethyl groups. These changes could be minimal when stored at cool temperatures (not freezing), but they could become significant if the resin is unused for more than about 2 weeks or longer, even at room temperature. The question: is there an optimum storage or ripening time for a UF resin in terms of storage time and temperature for maximum adhesive performance?

A study on ripening times of UF resins: The same typical UF resin used in the heat treatment study above was synthesized till the end of the acidic polymerization step, cooled in ice water bath before addition of second urea, and divided it into 6 samples, which are taken out one by one and warmed up to room temperature mixed with second urea, and stored at room temperature for up to 30 days as described in Table 24.9 (*Samples* B_1–B_6), and ^{13}C NMR analysis results of the samples are reported in Table 24.10. ^{13}C NMR samples (2.0 g) were taken from resin samples at the end of storage period and stored at 4 °C until sending them overnight for analysis. All resin samples were used directly in PB manufacture on the 30th day of storage for *Sample* B_6. This resin sampling sequence was based on a reasonable assumption that *Sample* B_1 $(F/U_1 = 2.10)$ would change little on storing at 4 °C, at which temperature it only increased in viscosity a little, from V to W, while the resin sample reached to YZ viscosity when stored at room temperature. The sampling sequence was designed to minimize resin synthesis duplication errors expected when resin batches are made at various times. The pH values of resin samples did not change much during the storage periods. *Sample* B_1, made by using the same procedure and formula as Sample A_1 discussed in the previous section (Section 24.4), turned out to be slightly lower in viscosity and slightly different in structure: increased substitution on urea units, as seen by higher

Table 24.10 Percentage values of various methyleneic and carbonyl carbons of UF resin Samples B1–B6 determined by ^{13}C NMR, sample viscosity, and 2-h Desiccator formaldehyde emission values.[a)]

Groups	ppm	B_1	B_2	(B_2-B_1)	B_3	B_4	B_5	B_6	(B_5-B_2)	(B_6-B_2)
Formaldehyde, free	91.0	1.14	0.55	−0.59	0.65	0.58	0.66	0.54	+0.11	−0.01
	87.0	5.55	0.00	−5.55	0.11	0.00	0.00	0.00	0.00	0.00
	83.1	6.77	0.17	−5.60	0.20	0.19	0.31	0.23	+0.14	+0.06
	Total	*13.46*	*0.72*	*−11.74*	*0.96*	*0.77*	*0.97*	*0.77*	*+0.25*	*+0.05*
Methylene-ether	79.1	1.41	1.05	−0.40	0.76	1.00	1.12	1.26	+0.07	+0.21
	75.1	7.18	4.73	−2.45	4.64	5.21	4.95	4.94	+0.22	+0.21
	69.5	8.66	7.20	−1.46	8.00	9.35	9.89	11.34	+2.69	+4.14
	Total	*12.75*	*12.98*	*−4.31*	*13.40*	*15.56*	*16.06*	*17.54*	*+2.98*	*+4.56*
Hydroxymethyl	72.0	19.71	20.65	+0.95	17.02	14.73	13.94	12.93	−6.71	−7.72
	65.2	14.76	30.41	+15.60	30.76	29.52	28.86	22.82	−1.55	−7.59
	Total	*34.47*	*51.06*	*+16.60*	*47.78*	*44.25*	*42.80*	*35.75*	*−8.26*	*−15.31*
Methylene	60.1	6.54	5.18	−1.36	4.22	5.33	5.32	6.02	+0.14	+0.84
	53.9	20.84	22.71	+1.87	24.43	22.62	22.41	22.20	−0.30	−0.51
	47.4	7.44	7.35	−0.09	9.22	11.25	12.52	17.72	+5.17	+10.37
	Total	*34.82*	*35.24*	*+0.42*	*37.87*	*39.20*	*40.25*	*45.94*	*+5.01*	*+10.70*
Urea, free	164.0	0.00	26.73	+26.73	22.75	18.60	16.88	12.71	−9.85	−14.02
Urea, mono-sub	162.2	4.83	17.83	+13.00	23.31	27.31	29.97	33.15	+12.14	+15.32
Urea, di–/tri-sub	160.7[b)]	94.17	55.44	−38.77	53.94	54.08	53.15	54.15	−1.29	−1.29
Formaldehyde[c)]		ND	1.20	—	1.23	1.36	1.50	1.26	—	
Storage days[d)]		0	1d	—	7d	16d	22d	30d	—	
Viscosity[e)]		V	P	—	L	L	M	N	—	

a) Chemical structures are defined as before in ppm. Values in parentheses are the differences between samples.
b) Group value includes 2–3% of tetra-substituted urea carbonyls.
c) 2-Hour desiccator form. Emission (ASTM D 5582) values of PB in μg/l.
d) Number of days of storage at room temperature.
e) Viscosity of resin samples at time of test in GH scale.

values of Type II/IIi hydroxymethyl and Type III methylene groups. These differences are probably due to some resin synthesis parameters being inadequately attended, but overall, they were similar.

Comparison of room storage temperature treatments vs. heat/stirring treatments: In general, the effects of storage treatments done at room temperature in Table 24.10 were similar to those of mild heat/stirring treatments discussed in the previous section (Section 24.4). The second urea reacts with most of the free formaldehyde species to form Type I hydroxymethyl groups, and Types IIt/IIi hydroxymethyl groups on the polymeric UF molecules migrated to second urea during the storage time. However, some differences were noted. In the initial period of room-temperature storage, the viscosity decreased from P reaching L on the fifth day and it remained at L until the 7th day (*Sample* B_3) and decreased to K on the 12th day. ^{13}C NMR data and viscosity changes indicated that the seven- to twelve-day storage samples are comparable to *Sample* A_4 obtained by heat treatments carried out up to 15 minutes at 50 °C. In comparison, the viscosity changes and migration of Type II/IIi hydroxymethyl groups were negligible when stored at 4 °C for 22 days for Sample B_5c (NMR data not reported). However, some unexpected behaviors were observed in the later period of room-temperature storage. After about 12th day, the viscosity of resin samples started to gradually increase; a phenomenon not observed in the heat/stirring treatments. The turbidity of resin samples also arose between 7 and 14 days (*Samples* B_4–B_5) and it increased gradually further, although it was still relatively light even for 30 day-stored *Sample* B_6. Reasons for the viscosity increases observed were not clear initially, but ^{13}C NMR results indicated that, although the migration of Type II/IIi hydroxymethyl groups continued, Type I hydroxymethyl groups also decreased, and Type I methylenes and Type I methylene-ethers increased in *Sample* B_6.

Formation of methylene-di-urea type groups (polymerization): Methylene-ether groups would be formed to some extent in the alkaline pH, but the significant increases in Type I methylenes in *Samples* B_5 (+5.17%) and B_6 (+10.37%) without much decreases in Type II or III methylene groups were not expected. The increases were also accompanied by *decreases* in hydroxymethyl groups in contrast to the heat-treated samples, and the fact that mono-substituted urea species have increased significantly indicates that the entailing reaction product is likely *methylene-di-urea*, not mono-hydroxymethylurea. Di-methylene-tri-urea-type products do not appear to be forming much since di- and tri-substituted urea species (160.7 ppm) did not increase but decreased slightly. Thus, the formation of methylene-di-urea explains the significant increase in Type I methylene groups during the later part of the room-temperature storage, in contrast to a smaller increase occurring in the heat-treated *Sample* A_6. Methylene groups have been known to form only in the acidic pH within the common reaction time regime used in manufacturing of UF resins [11, 20, 21, 29–32]. In addition, the rising turbidity of *Samples* B_5 and B_6 can be explained to be the precipitation of some methylene-di-urea and this precipitation raises some uncertainty about the accuracy of ^{13}C NMR results, but it appears to be a minor factor. Thus, the increases in Type I methylene groups in *Samples* B_5 and B_6 are accompanied by significant decreases in the overall hydroxymethyl group content as well as decreases in the free urea content, as observed in the ^{13}C NMR carbonyl group data. The author earlier observed a similar increase in Type I methylene group content for UF resin samples stored for an extended period of time (6 months or more) at about 4 °C. This was not a well-recognized phenomenon until now.

Transport and storage practices in UF resin industry: UF resins in weak alkaline pH are observed to turn into flimsy gels in about 30 minutes at 100 °C, and on storage for extended times at room temperature, they increase in viscosity and turbidity and eventually form a water-in-oil type gel. Therefore, certain polymerization reactions are progressing in UF resins in the weak alkaline pH especially on standing for extended times. In this regard, it is noted here that a suggestion was made previously that methylene bonds would form in UF resins even in the weak alkaline pH if an ample amount of free amide groups is available [29]. Also, a trace amount of methylene groups was observed to be forming in samples taken after the first weak alkaline condensation step of UF resin synthesis (30 minutes at 90 °C), although it was assumed to be insignificant and unexplained in the previous section (Section 24.4) [9]. The reaction appears to occur by general acid-catalysis by unionized acidic components present. Overall, extended resin storage at room temperature results in increasing the extent of hydroxymethyl group migration as well as forming of methylene groups, including those of methylene-di-urea-type poorly soluble products.

24.5.3 Formaldehyde Emission Test Results of Particleboard (PB)

Particleboards bonded with heat-treated UF resin samples: The IB and bending strength values of boards manufactured with heat-treated resins were in the range of typical laboratory PBs made with common UF resins, indicating that the board manufacturing parameters used were adequate to allow the comparison of the formaldehyde emission values. 2-Hour Desiccator method [25] test results for *Sample A* series (Table 24.9) indicate that the formaldehyde emission was lowest in *Sample* A_2 and increased gradually until *Sample* A_6, from 0.70 to 0.94 μg/ml. Thus, the formaldehyde emission values showed increases as the amount of hydroxymethyl groups migrated from the polymeric to monomeric UF resin components increased. The reason behind this phenomenon is unclear, but this formaldehyde emission trend observed from *Samples* A_1 to A_6, of up to 35%, appears to be large enough to deserve a detailed study.

Particleboards bonded with room temperature stored UF resin samples: For *Sample B series* (Table 24.11), the formaldehyde emission level increased gradually from 1.20 μg/ml for Sample B_2 to 1.50 μg/ml for *Sample* B_5 and then decreased to 1.26 μg/ml for *Sample* B_6. First of all, the overall formaldehyde emission level is appreciably higher than those of *Sample A series*, which is apparently due to some differences in the board manufacturing procedure including the season of the year and wood raw materials used. Nevertheless, the increasing trend of formaldehyde emission level with respect to the extent of the hydroxymethyl group migration up to *Sample* B_5 is similar to the trend observed in *Sample A series*. The results indicate that shorter storage times would show less formaldehyde emissions for UF resins made by adding the second urea at low temperatures. About a week's resin storage time at room temperature hardly increases the formaldehyde emission. The lowered formaldehyde

Table 24.11 Syntheses of UF resins with different F/U_1 mole ratios.

3150 g 50% formaldehyde; 1500 g U_1 and 1237 g U_2 were used, $F/U_1 = 2.10$, $F/(U_1 + U_2) = 1.15$.
• 30 min at 95 °C & pH 7.8; then, cooked for 120 min at 95 °C and pH 4.5 to V viscosity.
• Cooled to 4 °C and 10 g *Sample 2.10a* was taken.
• Added U_2, stirred/warmed up 1.0 h to room temperature and *Sample 2.10b* (400 g) was taken.
• Residual was stirred/heated for 40 min to reach to 60 °C and continued stirring for 90 min and 400 g *Sample 2.10c* were taken.
• Residual was stirred at 60 °C for 60 min and 400 g *Sample 2.10d* were taken.
• Similarly, *Samples 2.40a, 2.40b, 2.40c, & 2.40d* were made. Acidic cook was made at pH 4.10 and 95 °C for 140 min to X viscosity, $U_1 = 1313$ g; $U_2 = 1426$ g.
• Samples 1,80a, 1.80b. 180c & 1.80d were made with a minor modification as described in the text. Acidic cook was made at pH 4.5 and 95 °C for 50 min to V viscosity, $U_1 = 1750$ g; $U_2 = 989$ g.
• All samples were analyzed by ^{13}C NMR.
• b-series samples were also used to make PB and tested for formaldehyde emission.

Source: Adapted with permission of the publisher and minor revisions from Kim, M. G. (2001). Examination of selected synthesis parameters for typical wood adhesive-type UF resins by ^{13}C NMR spectroscopy. III. *J. Applied Polym. Sci.*, 80, 2800–2814 [13].

emission of *Sample* B_6 is interesting since it coincides with, and may be traced to, the increase in Type I methylene group content (formation of methylene-di-urea) and decrease in hydroxymethyl group content. Overall, the results indicate that typical industrial UF resins manufactured by adding second urea at 60 – 70 °C and thus having a considerable amount of hydroxymethyl groups migrated to second urea at the end of resin manufacture would show initially (1–2 weeks) lower formaldehyde emission values, which may increase when the resin storage time is lengthened or the cooling time in the last step of resin manufacture is delayed too much. This observation thus is in accord with the optimum aging time concept used in the UF resin industry.

24.5.4 Summary

UF resin samples synthesized similarly as in previous sections (Section 24.4) were subjected to various lengths of heating/stirring treatments at elevated temperatures up to about 70 °C and, in the second part of study, the fresh resin samples made by adding second urea at room temperature were subjected to storage treatments at room temperature for up to 30 days. Resin samples were analyzed using ^{13}C NMR and evaluated as PB binders. The migration of hydroxymethyl groups from the polymeric components occurred slowly and in somewhat stepwise manner, resulting in decreases in the viscosity of resins and changes in Type I, II, and III methylene and methylene-ether bond group contents. Thus, commonly synthesized UF resins showed an ample room temperature stability as practiced in the current UF resin industry. However, in the later part of treatments, especially in the room-temperature storage treatments, some other reactions occur with increases in viscosity and turbidity arising from the formation of methylene-di-urea-type insoluble products. The formaldehyde emission properties of PBs bonded with resin samples also varied, indicating that industrial formaldehyde emission performances would vary depending on the time/temperature parameters used in the second urea addition step of resin manufacturing and in the transport and storage periods.

24.6 Effects of Mild Heating/Stirring Treatments on UF Resins Synthesized with Different F/U_1 Mole Ratios by ^{13}C NMR

24.6.1 Syntheses of UF Resins with Varying F/U_1 Mole Ratios

F/U_1 mole ratios in UF resin synthesis affect various resin synthesis parameters and result in resins having different storage and performance properties due to different polymer structures. To investigate the mole ratio effects in detail, three typical F/U_1 mole ratios, 1.80, 2.10, and 2.40 were chosen, and the resin syntheses were finished by adding second urea at room temperature to an $F/(U_1 + U_2)$ mole ratio of 1.15. The resin samples were then subjected to various lengths of stirring/heat treatments as described in Table 24.11. One minor synthesis modification was applied for resin *Samples 1.80* series to ameliorate the problem of turbidity arising in the acidic polymerization step: the first urea was divided into two parts and the resin was cooked first with an F/U_{11} mole ratio of 1.85 until "B" viscosity, followed by neutralization to pH 7.8, adding urea to an $F/(U_{11} + U_{12})$ ratio of 1.80, and reacting for 20 minutes, followed by acidification to pH 4.5 and advancing to "V" viscosity.

24.6.2 Heat/Stirring Treatments of Synthesized UF Resins

As shown in Table 24.11, for the three F/U_1 mole ratio resins, the *-a series* represents the samples taken before the addition of the second urea, *–b series* are the samples taken after the addition of the second urea at 4 °C and stirred until it reached to room temperature, *–c series* are the samples taken after heat/stirring for 40 minutes to reach to 60 °C, followed by heat/stirring for 90 minutes, and *-d series* are the samples taken after an additional heat/stirring treatment for 60 minutes at 60 °C. ^{13}C NMR data of all UF resins obtained using the same procedure as in the previous section (Section 24.5) are reported in Table 24.12.

24.6.3 Viscosity and Turbidity Changes in Heat Treatments

The heat-treated resin 2.40c and 2.10c series (1.5-hours treatments) and d-series (2.5-hours treatments) did not show turbidity, but Sample 1.80d series showed minor turbidity. The viscosity decreases observed in heat treatments (Figure 24.10) were similar to the

Table 24.12 Percentage values of various methyleneic carbons and carbonyl carbons of UF resin samples determined by ^{13}C NMR and 2-h Desiccator formaldehyde emission values.[a)]

	Synthesized/Heat-treated UF Resin Samples											
Group (ppm)	**2.40a**	**2.40b**	**2.40c**	**2.40d**	**2.10a**	**2.10b**	**2.10c**	**2.10d**	**1.80a**	**1.80b**	**1.80c**	**1.80d**
91.0	2.31	0.79	0.79	0.85	1.23	0.52	0.83	0.65	0.98	0.36	0.41	0.40
87.0	8.00	0.24	0.27	0.25	4.84	0.19	0.46	0.37	3.13	0.10	0.20	0.20
83.1	10.49	0.28	0.20	0.62	7.39	0.29	0.29	0.39	4.13	0.28	0.37	0.26
Total	*20.80*	*1.31*	*1.26*	*1.32*	*13.46*	*1.00*	*1.58*	*1.41*	*8.24*	*0.74*	*0.98*	*0.86*
79.1	3.20	2.66	2.43	1.87	1.90	2.22	2.22	2.12	2.74	2.15	2.28	2.05
75.1	7.15	4.98	3.76	3.97	6.69	4.82	4.82	4.07	5.40	4.50	4.58	4.21
69.5	8.52	6.79	7.67	8.12	9.18	7.27	7.27	9.32	10.57	8.92	9.50	9.58
Total	*18.87*	*14.43*	*13.86*	*13.96*	*17.77*	*14.31*	*15.51*	*15.51*	*18.71*	*15.57*	*16.36*	*15.84*
72.0	17.04	17.91	11.89	10.51	18.24	18.52	12.83	11.97	16.50	15.19	11.68	11.35
65.2	11.87	35.01	40.20	40.36	15.39	29.66	34.26	35.11	18.91	26.70	29.40	30.56
Total	*28.91*	*52.92*	*52.09*	*50.87*	*33.63*	*48.18*	*47.09*	*47.08*	*35.41*	*41.89*	*41.08*	*41.91*
60.1	6.52	4.64	4.84	4.23	5.81	4.48	2.71	4.56	5.39	4.29	4.88	2.75
53.9	19.50	21.06	19.33	19.53	21.40	23.79	23.26	20.59	22.28	2581	23.50	24.70
47.4	5.40	5.64	8.65	9.69	7.93	8.24	9.85	10.85	9.97	11.70	13.20	13.94
Total	*31.42*	*31.34*	*32.82*	*33.45*	*35.14*	*36.51*	*35.82*	*36.00*	*37.59*	*41.80*	*41.58*	*41.39*
164.0	0.00	28.01	23.32	16.70	0.00	26.39	23.36	21.61	0.00	25.08	21.81	20.76
162.2	0.95	23.58	29.94	37.35	1.83	18.44	25.60	27.62	5.11	15.91	23.50	24.75
160.7	94.29	45.86	45.47	44.02	94.63	54.03	49.51	49.21	90.77	56.78	53.02	52.88
158.0[b)]	4.76	2.55	2.26	1.93	3.54	1.14	1.53	1.56	4.12	2.23	1.67	1.61
CH_2O[c)]	—	0.60	—	—	—	0.70	—	—	—	0.72	—	—

a) Chemical structures are defined as before in ppm.
b) 158.0 ppm is tetra-substituted urea carbonyls.
c) 2-Hour Desiccator formaldehyde emission (ASTM D 5582) values of PB in μg/l.

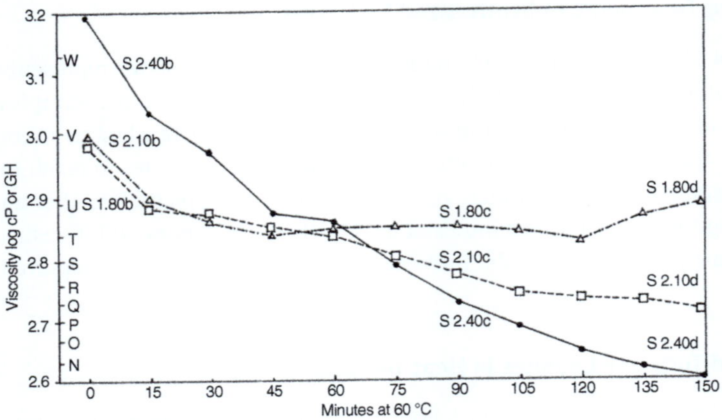

Figure 24.10 Viscosity changes of Resin Samples 1.80b, 2.10b, and 2.40b during stirring/heating treatments at 60 °C for 2 hours.

general downward trend observed in the previous sections (Sections 24.4 and 24.5) [9, 12], ascribed to the migration of Type IIt/IIi hydroxymethyl groups from polymeric UF molecules to second urea. The viscosity decreases were found to be different for different F/U_1 mole ratios, an indication that different polymer chain structures resulted in resin syntheses. First of all, the low viscosity values and the low or no or minor turbidity appearance of the resin samples indicate that the ^{13}C NMR experiments, obtained with a long pulse delay time, would not incur too much quantitative inaccuracies from suspended solid particles or molecular movement limitations. Heat treatments at 60 °C for longer than 150 minutes were observed to increase the turbidity significantly for all resin types, which were not studied.

24.6.4 F/U_1 Mole Ratio Effects on UF Resin Structures in Heat Treatments Observed by ^{13}C NMR

Effects of F/U_1 mole ratios on functional groups and changes in heat treatments (Table 24.12): In b-series samples the formaldehyde and hemi-formal species decreased to low levels before any heat treatment. Also, the destruction of ether-type bonds of hydroxymethyl formal groups of a-series samples is shown by the significant decreases in the total methylene-ether group levels. In heat-treated c- and d- series resin samples, Types IIt/IIi hydroxymethyl groups (72.0 ppm) bonded to polymeric UF molecules decreased. In comparing -b and -d series samples, the decreases of Type IIt/IIi hydroxymethyl groups by migration to second urea to form Type I hydroxymethyl groups were *7.40% for Resin 2.40*, *6.55% for Resin 2.10*, and *3.84% for Resin1.80*. A similar migration of Type I hydroxymethyl groups from the polymeric UF polymers would have occurred, although they are not measurable in ^{13}C NMR spectra. Type IIi hydroxymethyl groups can be viewed as polymer chain branches, but their level can change or decrease to low values during resins' lifetimes due to their tendency to migrate to the second urea. It is interesting to estimate Type IIi hydroxymethyl groups after heat-treating of resin samples, *i.e.* after much of them have migrated to the second urea. The heating/stirring treatment method at 60 °C for resin samples was chosen to minimize side-reactions observed in the previous section

(Section 24.5), [11] while allowing hydroxymethyl group migrations to go completion within a reasonably short time. These c-series heat-treated resins were expected to be similar to freshly made industrial resins since a relatively long batch-cooling time is usually taken in industry. Further, it is noted that the migration of Type IIt/IIi hydroxymethyl groups in *Samples c-series* reached relatively high levels, and the migration proceeded only slightly further in *Samples d-series* with additional 1.0 hour heat treatments, indicating that, first of all, no significant polymerization or de-polymerization reaction occurred during the heating/stirring treatments. On the other hand, the migration of Type IIi/IIt hydroxymethyl groups resulted in decreases in Type II and III methylene and methylene-ether groups in d-series in comparison with resin Samples b-series, with values reaching to *4.56% for resin Sample 2.40d, 4.66% for resin Sample 2.10d*, and *2.97% for resin Sample 1.80d*. Also, as a result, Type I methylene-ether and methylene groups increased in comparison with -b series, with values reaching *5.38% for Sample 2.40d*; *4.60% for Sample 2.10d*; and *2.90% for Sample 1.80d*. These methylene-ether and methylene group level changes can be directly ascribable to the migration of Type IIi hydroxymethyl groups to the second urea, and therefore, the values are their (minimum) level estimates in b-series resin samples. Although there is no good reason to suggest for a preferential migration of Type IIt or IIi hydroxymethyl groups in heat treatments, it is reasonable to assume that the observed residual levels of Type IIt/IIi hydroxymethyl groups, 10.51 – 11.97% in Sample d-series, appear to harbor only small amounts of Type IIi hydroxymethyl groups, about 1/3 of the values, as discussed below further. The viscosity changes that occurred in heat treatments are ascribed to these functional group changes.

Degree of polymerizations of resins calculated with data obtained with heat treatments: Some calculated ^{13}C NMR data are reported in Table 24.13. First of all, the *methylene-ether: methylene* group ratio values obtained for all resin samples were constant, 17 – 18 % vs. 83 – 82%, and not much affected by F/U_1 mole ratios. The degree of polymerization (**n**) calculated on the basis of total methylene and methylene-ether bond contents per mole of first urea (U_1) was *37.0 for Sample 2.40d, 12.4 for Sample 2.10d*; and *8.90 for Sample 1.80d*. The cause of these differences is traceable to the similar viscosity end-points used in resin syntheses. For all three resin formulas, the same 50% formaldehyde solution was used without any added water and only the amount of first urea was varied. Resins with higher F/U_1 ratios resulted in more dilute systems to result in higher degrees of polymerization in resin syntheses. On the other hand, the amount of second urea added, which decreases the viscosity of resulting resins, is greater with higher F/U_1 ratio resins, and this compensation results in similar final viscosities and overall average molecular weights for the three different resin samples. (The final viscosity was slightly higher for *Sample 2.40b* than for *Samples 2.10b* or *1.80b*, but this cannot be used to directly explain the molecular weight differences.) Overall, the degree of polymerization values based on first urea observed appear to be within the range where polymer structure comparisons can be made with appropriate adjustments.

Polymerization reaction times, functionality of first urea, and F/U_1 mole ratio: Thus, the polymerization reaction time, measured by end viscosity, was longest for *Sample 2.40* (140 minutes) and shortest for *Sample 1.80* (50 minutes). Although the dilution factor mentioned above could be considered for the differences, the major reason for the slower reaction rates of higher F/U_1 mole ratio resins appears to be the decreasing availability

Table 24.13 Polymer structure calculation results of Resins 2.40, 2.10, and 1.80 for the number of polymer chain branches (dB) per mole of first Urea (U_1) before (Sample b-series) and after (Sample d-series) stirring/heating treatments at 60 °C for 2.5 h.[a)]

	Sample 2.40			Sample 2.10			Sample 1.80		
	d	dB	b	d	dB	b	d	dB	b
Type III M/ME	0.125	4.63		0.119	1.47			0.61	
Type II M/ME	0.517	9.56		0.473	2.92			2.16	
Type I M/ME	0.331	2.00		0.327	2.00			2.00	
Total	0.973	16.19		0.919	6.39			4.77	
DP	37.0			12.32					
Total branch/U_1	(0.438)			(0.517)					
Side branch/U_1	0.384			0.355					
Type I HM/U_1	0.285		0.285	0.323		0.323	0.341		0.341
Type II(/II i) HM/U_1	0.126		0.126	0.126		0.126	0.102		0.102
Free amide group/U_1	0.027		−0.007	0.068		0.046	0.091		0.071
Total branch/U_1	0.438			0.517			0.534		
Type IIi HM, mig.			0.109			0.089			0.053
Type II HM, mig.			0.034			0.022			0.020
Total branch/U_1			(0.547)			0.606			(0.587)
Side branch/U_1			0.493			0.444			0.363
Functionality U_1			2.65			2.56			2.33
M/ME group ratio	82.18			83.17			83.17		

a) Types of groups are as defined in the text and Figure 24.5. M: methylene bonds. ME: methylene-ether bonds. HM: hydroxymethyl groups. DP: degree f polymerization. dB: number of polymer branches per molecule. U1: one mole of first urea. Mig.: migrated Type IIt/IIi HM groups in heat treatments. b and d: resin sample b- and d-series. Side branch: number of polymer chain branches excluding the number of linear portion branches. Calculation procedures were discussed in the text.

of urea amide/imide nitrogens for hydroxymethyl groups to form methylene bonds. The maximum functionality of first urea (U_1) under the resin synthesis condition has been known to be near 2.50. The attained functionality of first urea shown in the syntheses of resins, calculated from resin Samples a-series including the hemi-formal factor derived from *Samples b-series*, shows it decreases as the F/U_1 ratio decreases: *2.65 for Sample 2.40*, *2.56 for Sample 2.10*, and *2.33 for Sample 1.80*. Thus, the accessibility of urea nitrogens for reaction increases as the F/U_1 ratio increases, to limited extents, but it should be noted that when the F/U_1 mole ratio increases to about 2.60 or higher, only a limited extent of polymerization occurs, indicating that the functionality of 2.65 observed is over the maximum value attainable in syntheses of typical wood adhesive-type UF resins. In UF resins made with exceptionally high F/U_1 mole ratios, the reaction is known to occur only when acidifying the reaction mixture to about pH 1.0, where uron-type methylene-ether bonds are formed with increased functionality for the first urea. UF resins containing low levels of uron groups have been known to form and their ^{13}C NMR data were reported recently [33]. In this regard, it should be noted that all three UF resin types of this study showed small levels of urea carbonyl peaks at near 155.0–158.0 ppm, corresponding to tetra-substituted, possibly, uron structures. These uron-type groups and ordinary methylene-ether groups in resins showed little changes in the heat treatments, i.e. no breaking or forming, raising an intriguing question regarding their formaldehyde emission potential. A source of more emissions?

Polymer chain branches increasing with F/U_1 mole ratios: Molecular data calculated based on heat-treated, d-series (heat-treated) resin samples are reported in Table 24.13 (top section). The number of polymer branches per mole of the first urea calculated using methylene and methylene-ether bond contents showed unexpectedly an increasing trend as the F/U_1 mole was decreased: *0.438 for Sample 2.40, 0.517 for Sample 2.10*, and *0.534 for Sample 1.80*. A direct comparison of these values is unfair, however, because the degree of polymerization factor is included in assigning two chain ends to Type I methylene and methylene-ether bond values in the calculation. Excluding this factor results in the values for the number of side-chain branches: *0.384 for Sample 2.40, 0.355 for Sample 2.10*, and *0.310 for Sample 1.80*, i.e. lower number of side-chain branches for lower F/U_1 mole ratio resins. The maximum side-chain value would be 1.0 for polymer chains with every urea unit having one side branch, and the minimum value would be zero with all urea units having no branch, i.e. linear polymer chain structures, which were suggested to be formed in the early days of UF resin development [26].

Comparison of polymer chain branch numbers with terminal hydroxymethyl group numbers: For heat-treated resin *Samples 2.40d, 2.10d*, and *1.80d*, Type I hydroxymethyl groups bonded to polymeric UF components are not measurable, but it is reasonable to first assume that the maximum values would be the same as those of corresponding a-series resin Samples, i.e. values obtained before adding the second urea. The sum value of Type I, IIi, and one-half of Type IIt hydroxymethyl group values reflect the number of polymer chain ends, and its shortage against the number of polymer branches is assignable to free amide terminal groups. The free amide group terminal branch values thus obtained were: 0.027 for Sample 2.40d, 0.068 for Sample 2.10d, and 0.091 for Sample 1.80d, increasing, as expected, as the F/U_1 ratio decreased (Table 24.13, middle section). These values should be viewed as the minimum values due to the assumption made that the Type I hydroxymethyl group values are the same as those of a-series. Thus, the results show that freshly made industrial UF resins with relatively long cooling times after the addition of second urea would have free amide groups as polymer chain branch-ends at very low levels depending on the F/U_1 ratio as well as the cooling temperature/time history as indicated in an earlier section (Section 24.5) [11].

Polymer chain structures of the untreated sample -b series: Furthermore, more definite polymer structures of untreated b-series resin samples can be obtained by comparing them with d-series resin samples. First, the level of Type IIt/IIi hydroxymethyl groups that migrated to second urea in heat treatments was obtained, and the values were proportioned into Type IIt and IIi hydroxymethyl group values by assigning the averages of the decreasing/increasing values of methylene/methylene-ether group levels among different types in heat treatments to the Type IIi hydroxymethyl group levels. The balance amount was assigned to Type I hydroxymethyl groups. These Type IIt and IIi hydroxymethyl group values were then combined with Type I and IIt/IIi hydroxymethyl group and free amide group values of d-series samples to obtain the total number of polymer branches of b-series samples. In order to avoid a double counting, however, the summation was made after the free amide group value of d-series samples was adjusted down with the migrated Type II hydroxymethyl group values. The obtained total number of side-chain branches, which includes Type IIi hydroxymethyl groups, similarly decreased as the F/U_1 mole ratio decreased: *0.493 for resin Sample 2.40b*, *0.444 for resin Sample 2.10b*, and *0.363 for resin Sample 1.80b* (Table 24.13, lower section). The latter two values included some free amide terminal group values, indicating that freshly made UF resins, even if the second urea is added at room temperature, could harbor free amide groups on the polymeric resin molecules when lower F/U_1 mole ratios were used. It is also interesting to see that resin *Sample 1.80b* showed the lowest level of Type II/IIi hydroxymethyl groups and the highest level of free amide groups. Although originated from the low F/U_1 ratio used in resin synthesis, these polymer molecular structural differences appear to be responsible for the unusual rheological and storage behaviors for UF resins synthesized at F/U_1 ratios below about 1.80, as reported in the previous section (Section 24.5).

24.6.5 Summary

UF resins were synthesized using three different F/U_1 mole ratios (180, 2.10, and 2.40) till the end of acid polymerization step and cooled to 4 °C (–a series samples) and the second urea was added to $F/(U_2 + U_2) = 1.15$, stirred at room temperature for one hour (–b series samples), heated/stirred to 60° C in 40 minutes and stirred for 90 minutes at the same temperature (–c series samples), and heated/stirred for additional 60 minutes (–d series samples). The migration of hydroxymethyl groups to second urea (U_2) occurred in heat/stirring treatments allowed a more accurate measurement of methylene and methylene-ether group contents by ^{13}C NMR, resulting in the following degree of condensation values (n) of the polymeric UF resin components based on the first urea: *Sample 2.40* ($n = 37.0$), *Samples 2.10* ($n = 12.4$) and *1.80* ($n = 8.9$). Also, the extent of branching of polymeric UF resin components calculated based on the first urea was *0.384* for *Resin 2.40b*, *0.355* for *Resin 2.10b*, and *0.310* for *Resin 180b*. The second urea reacted with formaldehyde and reduced the viscosity of resins with the reduction being greater for higher F/U_1 mole ratio resins, thus evening out the molecular weight difference effects to some extent. The degree of polymerization (**n**) calculated based on the combined amount of first and second urea was *1.80* for *Resin 2.4b*, *2.01* for *Resin 2.10b*, and *2.33* for *Resin 1.80b*. The polymer chain branching differences occurred from using different F/U_1 mole ratios appear to result in varying viscosity behaviors in resin syntheses and posttreatments.

24.7 Effects of Room-temperature Storage Treatments on UF Resins Synthesized with Various F/U_1 Mole Ratios by ^{13}C NMR and Formaldehyde Emission Tests of Particleboard

(Adapted with permission of the publisher and minor revisions from Kim, M. G., H. Wan, No, B. Y., and W. L. Nieh. 2001. Examination of selected synthesis parameters for typical wood adhesive-type UF resins by ^{13}C NMR spectroscopy. IV. J. Applied Polymer Science 82, 1155–1169. [14]*)*

A detailed study on longer room-temperature storage treatment effects of UF resins is reported in the previous section (Section 24.5). As shown in Table 24.14, the syntheses of

Table 24.14 Syntheses of UF resins with different F/U_1 mole ratios, room-temperature storage, & PB preparation and testing.

- 3150 g 50% formaldehyde; 1500 g U_1. (F/U_1 = 2.10)
  1. 30 min at 95 °C & pH 7.8; then, cooked for 120 min at 95 °C and at pH 4.5 to X viscosity, and adjusted pH to 7.8.
  2. Cooled to RT and divided into *six* portions, 775 g each, numbered *2.10b, 2.10e, 2.10f, 2.10 g, 2.10 h, and 2.10i*, and all were stored at 4 °C. The amount of second urea to be added is 206.2 g each to F/(U_1 + U_2) = 1.15.
- Similarly, *Samples 2.40b, 2.40e. 2.40f, 2.40 g, 2.40 h, & 2.40i* were made. 3150 g 50% formaldehyde; U_1 = 1313 g (F/U_1 = 2.40). Acidic cook at pH 4.10 and 95 °C for 140 min to X viscosity, and adjusted pH to 7.8. The amount of the second urea to be added is 237.7 g to F/(U_1 + U_2) =1.15.
- Similarly, *Samples 1.80b, 1.80e, 180f, 1.80 g, 1.80 h, 1.80i* were made. 3150 g 50% formaldehyde;, U_1 = 1750 g. (F/U_1 = 1.80). Acidic cook at pH 4.5 and 90 °C for 40 min to V viscosity with a minor modification as in the previous Section (Section 24.5), and adjusted pH to 7.8. The amount of the second urea to be added is 164.8 g to F/(U_1 + U_2) =1.15.
- Resin samples were warmed up to air-conditioned room temperature, and added with U_2 and stirred for 3 hours, and then stood on a bench for given number of days, as follows: –b series (2 days); –e series (7 days); –f series (15 days); –g series (30 days); –h series (45 days); and –i series (50 days). Then, ^{13}C NMR and PB preparation and tests were done with the aged samples.

resin samples were similar to those used in the previous study with the second urea added to F/(U_1 + U_2) = 1.15. In this study, the resin samples were stood on the bench and checked for viscosity changes for *2 days (–b), 7 days (–e), 15 days (–f), 30 days -g), 45 days (–h),* and *50 days (–i).* Resin samples were also analyzed by ^{13}C NMR and used for PB preparation and testing for strength properties and formaldehyde emission values. ^{13}C NMR data of selected resin samples are summarized in Table 24.15.

24.7.1 F/U_1 Mole Ratio Effects on Viscosity/Turbidity Changes in Room-temperature Storage Treatments

As shown in Figure 24.11, the viscosity values of resins decreased initially to some lower values with the extent being the highest for *Sample 2.40* and lowest for *Sample 1.80*, reaching to the low points at about 15 days of storage time. This trend of initial viscosity reductions is ascribable to the migration of hydroxymethyl groups from the polymeric UF components to monomeric second urea components shown in Figure 24.12a, similarly as observed in the heat/stirring experiments. The viscosity of resin samples then began to increase from the 15th day and reached to some high values at the end of the experiment, with the rate of increase being lowest for *Sample 2.40*, faster for *Sample 2.10*, and fastest for *Sample 1.80*. *Sample 1.80i* (50-day storage) was barely mobile, and the *Resin 1.80 hour* (45 days) in the bag gelled, preventing board preparation with it.

These viscosity increases are in pace with the increases in methylene and methylene-ether group contents as shown in Figure 24.12b and c till 30th day of storage and then, the viscosity kept increasing while the methylene and methylene-ether groups stopped increasing, as further discussed below. Since UF resin molecules are associative polymers in the aqueous medium and at the high resin solids regime, the molecular weight increases appear to enhance the associative power of resin molecules. According to the dynamics theory of associative polymer systems [34–37], molecular weight increases of polymers push the resin

Table 24.15 Percentage values of various methyleneic carbons and carbonyl carbons of UF resin samples determined by ^{13}C NMR method.

	Synthesized/room-temperature-treated UF Resin Samples												
Group (ppm)	**2.40b**	**2.40 g**	**2.40 h**	**2.40i**	**2.10b**	**2.10e**	**2.10f**	**2.10 g**	**2.10 h**	**2.10i**	**1.80b**	**1.80 g**	**1.80 h**
91.0	0.79	0.00	0.00	0.00	0.31	0.31	0.25	0.00	0.00	0.00	0.36	0.00	0.00
87.0	0.24	0.00	0.00	0.00	0.19	0.06	0.05	0.00	0.00	0.00	0.10	0.00	0.00
83.1	0.28	0.18	0.14	0.18	0.29	0.10	0.10	0.22	0.12	0.14	0.28	0.23	0.08
Total	*1.31*	*0.18*	*0.14*	*0.18*	*1.00*	*0.47*	*0.40*	*0.22*	*0.12*	*0.14*	*0.74*	*0.23*	*0.08*
79.1	2.66	2.12	2.17	2.13	2.22	2.12	2.10	1.86	1.64	1.37	2.15	2.07	1.78
75.1	4.98	4.51	4.79	4.38	4.82	4.80	4.83	4.82	4.52	4.14	4.50	4.31	4.33
69.5	6.79	11.60	12.12	12.70	7.27	7.38	8.63	11.52	12.06	12.51	8.92	11.24	10.36
Total	*14.43*	*18.23*	*19.08*	*19.21*	*14.31*	*14.30*	*15.56*	*18.20*	*18.22*	*18.02*	*15.57*	*17.62*	*16.47*
72.0	17.91	11.34	10.63	9.13	18.52	17.82	15.65	11.69	10.59	9.19	15.19	9.46	0.92
65.2	35.01	26.19	25.55	26.50	29.66	29.76	29.42	24.40	24.55	26.17	26.70	27.46	27.62
Total	*52.92*	*37.53*	*36.18*	*35.63*	*48.18*	*47.58*	*45.08*	*36.09*	*35.14*	*35.36*	*41.89*	*36.92*	*38.54*
60.1	4.62	3.20	3.09	3.13	4.48	4.43	4.15	3.25	2.99	2.77	4.29	3.10	3.12
53.9	21.06	22.63	22.56	22.29	23.79	23.90	23.62	24.03	23.71	22.71	25.81	23.86	25.27
47.4	5.64	18.24	18.96	19.55	8.24	9.32	11.20	18.22	19.81	20.98	11.70	18.27	16.51
Total	*31.32*	*44.07*	*44.61*	*44.97*	*36.51*	*37.65*	*38.97*	*45.50*	*46.51*	*46.46*	*41.80*	*45.23*	*44.90*
164.0	28.01	13.56	12.68	11.76	26.39	22.76	19.10	14.32	11.76	11.50	25.08	12.65	15.62
162.2	23.58	37.37	38.42	38.47	18.44	23.39	28.10	35.81	35.90	37.69	15.91	34.34	30.36
160.7	45.86	47.32	47.17	47.32	54.03	52.50	51.33	48.21	50.31	49.38	56.78	51.33	52.65
158.0	2.55	1.75	1.73	2.45	1.14	1.35	1.47	1.75	2.03	1.42	2.23	1.68	1.37

Sample numbers are as defined in the text: –b series (2 d); –e series (7 d); –f series (15 d); –g series (30 d); –h series (45 d); and –i series (50 d).

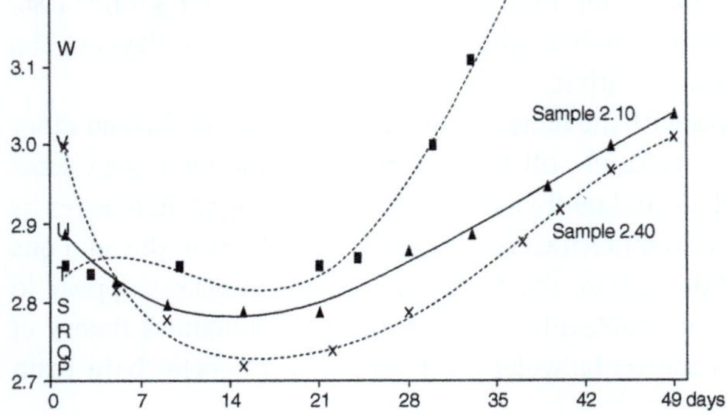

Figure 24.11 Viscosity changes of Resin Samples 1.80b, 2.10b, and 2.40b during room-temperature storage treatments for 50 days: 1.80 (■); 2.10 (▲); 2.40 (x).

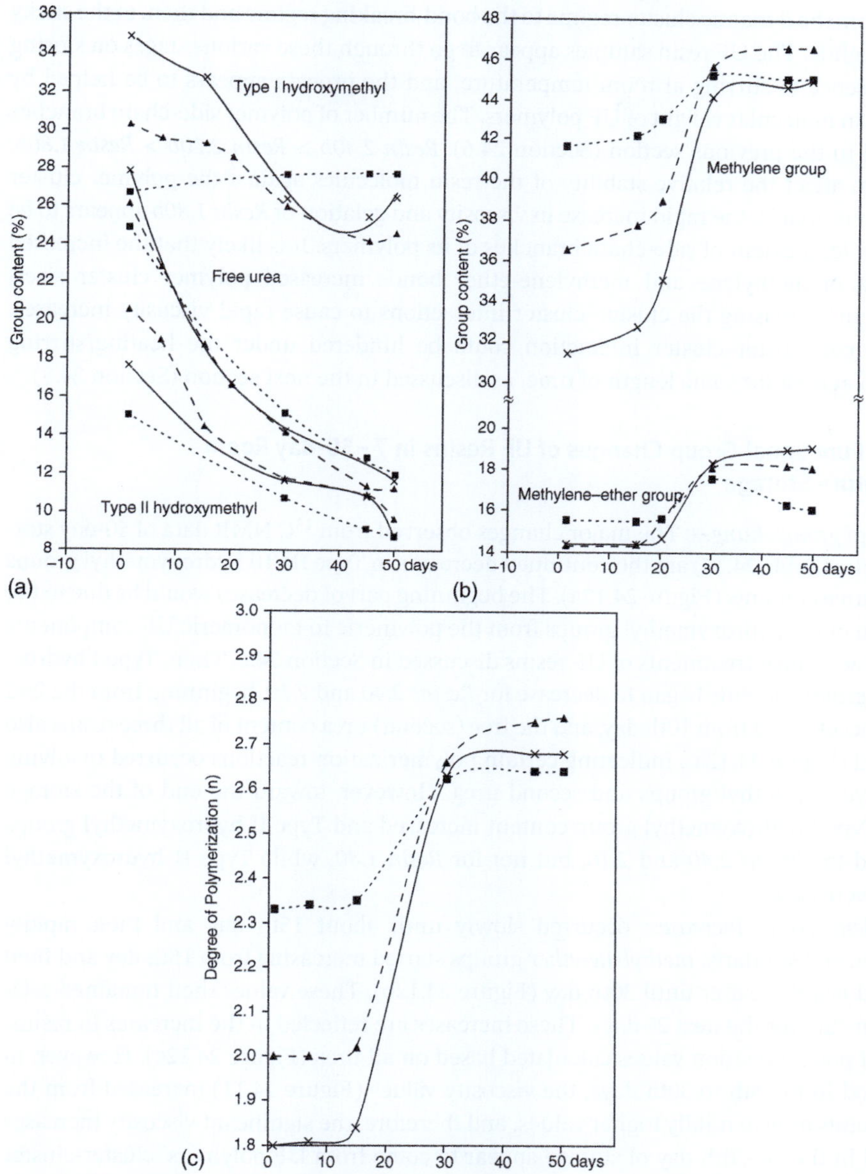

Figure 24.12 (a) Changes in ^{13}C NMR values of total Type I and Type II/III hydroxymethyl groups and free urea contents of UF resins stored at room temperature for 50 days:1.80 (■); 2.10 (▲); 2.40 (x). (b) Changes in ^{13}C NMR values of total methylene and methylene-ether group contents of UF resins stored at room temperature for 50 days: 1.80 (■); 2.10 (▲); 2.40 (x). (c) Changes in the degree of polymerization based on all urea of UF resins stored at room temperature for 50 days: 1.80 (■); 2.10 (▲); 2.40 (x).

system from the dynamic cluster regime to the bond-breaking regime and then, to the sticky motion regime. The UF resin samples appear to go through these various stages on storing in the absence of stirring at room temperature, and the process appears to be helped by increases in molecular weight of UF polymers. The number of polymer side-chain branches discussed in the previous section (Section 24.6), *Resin 2.40b* > *Resin 2.10b* > *Resin 1.80b*, appears to affect the relative stability of the resin molecules against the polymer cluster associations, that is, the rapid increase in viscosity and gelation of *Resin 1.80b* appears to be due to the least extent of side-chain branches of its polymers. It is likely that the increased formation of methylene and methylene-ether bonds increased polymer cluster sizes, resulting in increasing the cluster–cluster interactions to cause rapid viscosity increases. The polymer cluster–cluster interaction could be hindered under the heating/stirring condition at least for some length of time, as discussed in the next section (Section 24.8).

24.7.2 Functional Group Changes of UF Resins in 7–50-day Room Temperature Storage

Functional group changes: The major changes observed from ^{13}C NMR data of 50-day storage samples (Table 24.15) are the continued decreases in Type IIt/IIi hydroxymethyl groups and free urea contents (Figure 24.12a). The beginning part of decreases would be due to the migration of the hydroxymethyl groups from the polymeric to monomeric UF components as observed in heat treatments of UF resins discussed in Section 24.6. Then, Type I hydroxymethyl group contents began to decrease for *Resins 2.40* and *2.10*, beginning from the 2nd day and accelerated from 10th day, and the free (second) urea content of all three resins also decreased (Figure 24.12a), indicating certain polymerization reactions occurred involving Type I hydroxymethyl groups and second urea. However, toward the end of the storage period, Type I hydroxymethyl group content increased and Type II hydroxymethyl groups decreased for *Resins 2.40* and *2.10*, but not for *Resin 1.80*, while Type II hydroxymethyl groups decreased.

Methylene group increases: occurred slowly until about 15th day and then rapidly thereafter, and similarly, *methylene-ether* groups started increasing from 15th day and then increased fast thereafter until 30th day (Figure 24.12b). These values then remained relatively constant for the next 20 days. These increases are reflected in the increases in resins' degree of polymerization values calculated based on all urea (Figure 24.12c). However, in this period from 15th to 30th days, the viscosity values (Figure 24.11) increased from the lowest points only to mildly higher values, and therefore, the significant viscosity increases observed in the 30–50th day of storage appear to come from UF polymers' cluster-cluster interactions initiated by increases in the methylene and methylene-ether bond contents.

Formation of methylene and methylene-ether bonds in UF resins in alkaline pH: The increases in degree of polymerization and viscosity values observed are in line with the fact that most UF resins maintained in weak alkaline pH, on storing for extended periods, become highly viscous and finally become immobile. UF resins in weak alkaline pH also turn into weak gels at about 30 minutes or longer when they are heated/stirred slowly at 100 °C with minimal evaporation of content water. In this study, methylene groups increased more than methylene-ether groups. Although the time length used in this storage experiment was long in comparison with practices in resin-using industries, *the sharp increases in mostly Type I methylene and methylene-ether bond contents in alkaline pH at room temperature have been documented for first time in this work.* This methylene bond-forming reaction in alkaline pH appears not much accelerated at higher temperatures, and the reaction mechanism is therefore puzzling, although the occurrence of such a reaction was suggested earlier [29]. Also, agglomeration of certain molecular weight fractions into solid particles occurring in resins as seen by turbidity increases during storage is noteworthy. The turbidity could have distorted ^{13}C NMR determinations, but this would not explain the significant increases in the methylene group content. When resin samples were stirred and heated at 60 °C for several hours as described in Section 24.6, such a high extent of polymerization reaction was not observed, and therefore the reaction appears to be promoted by undefined factors related to storage times.

24.7.3 Test Results of Particleboards Bonded with Various UF Resins

24.7.3.1 Internal Bond and Bending Strengths of Particleboards (PBs)

Both resin storage time and F/U_1 mole ratio of resin synthesis significantly affected the IB strength values of boards. Industrial PBs manufactured with density values close to the value used in this study normally show IB strengths of 100 – 160 psi, bending modulus of elasticity (MOE) value of 300 – 600 kpsi, and bending modulus of rupture (MOR) values of 1800–3000 psi [38]. *(i)* At 2-day storage time, the board IB strength values (Figure 24.13a–c)

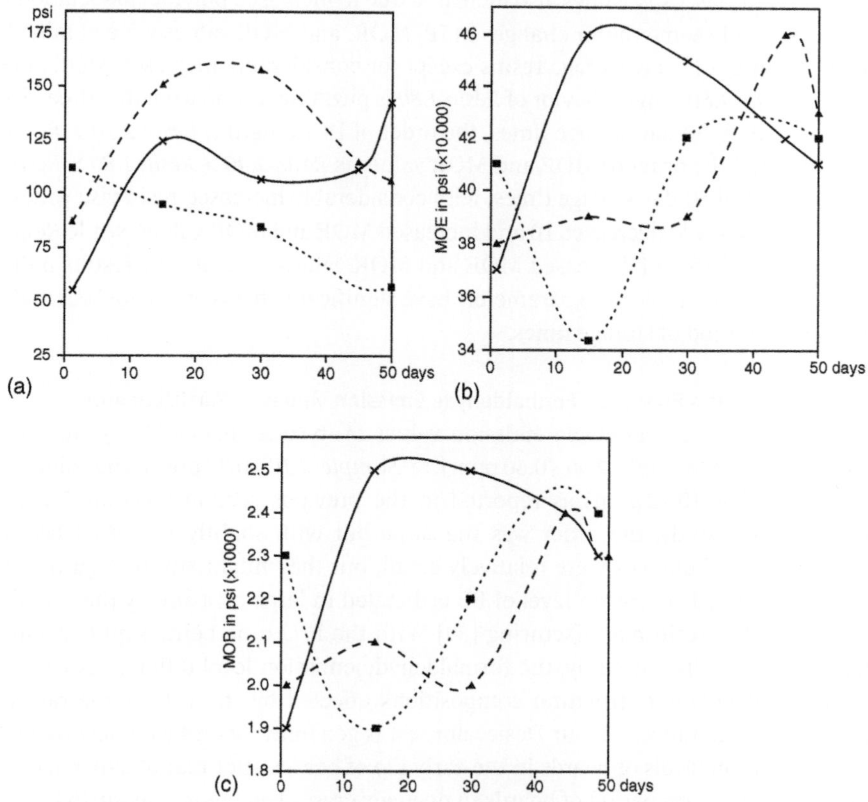

Figure 24.13 (a) IB strength values of particleboard bonded with UF resins stored at room temperature for 50 days: 1.80 (■); 2.10 (▲); 2.40 (x). (b) MOE values of particleboard bonded with UF resins stored at room temperature for 50 days: 1.80 (■); 2.10 (▲); 2.40 (x). (c) MOR values of particleboard bonded with UF resins stored at room temperature for 50 days:■); *2.10 (▲); 2.40 (x).*

indicate that *Resin 1.80* showed relatively good IB, MOE, and MOR values, while *Resins 2.10* and *2.40*, in that order, showed significantly lower values. This order in performance is in accord with the average degree of polymerization of UF resin polymers. *(ii)* At 15-day storage time, however, *Resin 1.80* showed significantly lowered IB, MOE, and MOR strength values, while *Resins 2.10* and *2.40* showed appreciably increased strength values. ^{13}C NMR data indicated that this storage period involves the migration of hydroxymethyl groups from the polymeric to monomeric UF resin components to result in a more uniform distribution of functional groups on all urea molecules. This migration of functional groups appears to be the reason behind the so-called optimum ripening times used in the UF resin industry.

The deteriorating performance of Resin 1.80 at 15-day storage time is puzzling, but it appears to be due to an increased pre-curing tendency of resins in the hot-pressing step of boards. Further, the IB values of *Resin 2.10* were appreciably higher than those of *Resin 2.40* while the MOE and MOR values of *Resin 2.40* were higher than those of *Resin 2.10* (Figure 24.13b–c), indicating that *Resin 2.10* resulted in a faster curing speed, making it more suitable as a core layer binder, while *Resin 2.40* is more suitable as a face-layer binder in the normal industrial three-layer board mat construction. The higher polymer chain branching extent of *Resin 2.40* in comparison with *Resin 2.10* may have contributed to this difference in curing speed. *(iii)* At 30-day storage times, although the degree of polymerization increased and the viscosity values also increased due to increased polymer associations for all three resins, only some minor changes in IB, MOR, and MOE values were obtained in comparison with the 15-day storage resins except for considerably increased MOE and MOR values for *Resin 1.80*. This behavior of *Resin 1.80* is puzzling. Comparing the three different types of resins at 30-day storage times, the order of IB strength is *Resin 2.10* > *Resin 2.40* > *Resin 1.80*, and the order of MOE and MOR values is *Resin 2.40* > *Resin 1.80* > *Resin 2.10*. *(iv)* At 45-day and 50-day storage times, with considerably increased resin viscosities, *Resins 2.10* and *1.80* showed decreased IB and increased MOE and MOR values, while *Resin 2.40* showed increased IB and decreased MOE and MOR values. Overall, the results indicate that the various F/U_1 mole ratio parameters have significant effects on boards' strength properties over the period of storage times.

24.7.3.2 F/U_1 Mole Ratio Effects on Formaldehyde Emission Values of Particleboards

The 2-hours desiccator formaldehyde emission values of -b-series resins (2-day storage) showed increases from *Sample 2.40* (0.60 ppm) to *Sample 2.10* (0.70 ppm) and slightly further to *Sample 1.80* (0.72 ppm) as reported in the previous section (Section 24.6.4) (Table 24.15). In this study, the order was the same but with slightly increased levels (Figure 24.14a). The differences were relatively small, but they might still be significant in practice. Formaldehyde emission level of PB is dictated in large measure by the overall final F/U ratio used in resin manufacturing [39] With the F/U ratio being kept constant for all three resin types in this study, the formaldehyde emission level differences would reflect the different polymer structural compositions effected by the F/U_1 mole ratios, procedures, and storage times. 2-Hour Desiccator test is generally viewed as reflecting the formaldehyde emission levels of boards in the early life of boards after manufacturing due to the two-day equilibration period of boards in open air prescribed before measuring. It is probable that free formaldehyde species trapped in boards are the sources.

Further, in the *2-Hour Desiccator* emission data of boards (Figure 24.14a, top), the emission level of *Resin 2.40* was the lowest in the beginning, decreased a little at 15 days,

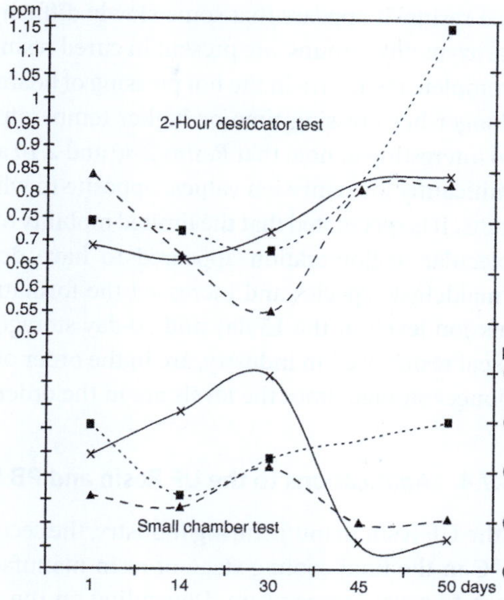

Figure 24.14 (a) Two-hour desiccator (ASTM D 5582) and (b) Small-chamber formaldehyde emission (ASTM D 6007) values of particleboards bonded with UF resins stored at room temperature for 50 days: 1.80 (■; 2.10 (▲); 2.40 (x).

but increased at 30 days and further increased at 45 and 50 days. *Resins 2.10* showed the highest value in the beginning but decreased rapidly at 15 and 30 days and then increased to significantly higher values at 45 and 50 days, to the same level as those of *Resin 2.40*. *Resin 1.80* showed intermediate emission values until 30 days but increased to a very high value at 50 days. In overall, the 2-Hour Desiccator emission levels at the 15-day and 30-day storage times, which would correspond to the values of typical UF resins used in the industry, are in the order of *Resin 2.40* > *Resin 1.80* ≫ *Resin 2.10*. The migration of hydroxymethyl groups that occurred predominantly until the 15-day storage samples appeared to lower the formaldehyde emission values for all three resins, while the increased degree of polymerization observed predominantly in the 30-day storage samples lowered the formaldehyde emission levels for *Resins 2.10* and *1.80* and slightly increased the emission for *Resin 2.40*. Thus, the 2-Hour Desiccator values reached minimum levels between 15 and 30 days of storage time, where both the hydroxymethyl group migration and the degree of polymerization increased. In the 45-day storage samples, where mostly the polymer agglomeration continuously increased, the 2-Hour Desiccator values significantly increased for all three resins, and the emission levels remained the same at 50 days for *Resins 2.40* and *2.10* while further increased for *Resin 1.80*.

The small-chamber (SC) formaldehyde emission tests were carried out after the prescribed 8 days of equilibration in open air, and therefore the more slowly defusing formaldehyde species or those in deeper layers are measured (Figure 24.14b, lower curves). The emission value of *Resin 2.40* was higher than that of *Resin 2.10* at 2-day storage, and it significantly increased at 15-day and 30-day storage times and then decreased to low levels at 45- and 50-day storage times. The emission value of *Resin 2.10* was relatively low until 15 days, increased a little at 30 days, and decreased to low values at 45 and 50 days. The emission value of *Resin 1.80* was initially high but decreased at 15 days and then continuously increased thereafter, reaching the highest value among the three resin types at 50 days. It is interesting to note that *Resin 2.40* at 30 days reached the highest value among the three resins. It appears that some slowly diffusing formaldehyde species, such as residual hydroxymethyl groups, are present in cured resins to slowly breakdown and diffuse out. An incomplete resin cure in the hot pressing of boards is suspected, especially in the core layer. A longer hot-pressing time or higher temperature might be needed with this resin. It is also interesting to note that *Resins 2.40* and *2.10* at 45 or 50 days of storage times resulted in significantly low emission values, opposite to what was observed in the 2-Hour Desiccator results. It is speculated that the limited mobility of polymer resin molecules due to increased molecular agglomeration appeared to have decreased the formation of slow-diffusing formaldehyde species and increased the formation of fast-diffusing ones. Overall, the SC emission levels at the 15-day and 30-day storage times, which would be comparable with typical resins used in industry, are in the order of *Resin 2.40* ≫ *Resin 2.10–Resin 1.80*, while at longer storage times the levels are in the order of *Resin 1.80* ≫ *Resin 2.0–Resin 2.40*.

24.7.4 Applications to the UF Resin and PB Manufacturing Industry

In the UF resin manufacturing industry, the second urea addition step would start at about 70 °C in the final cooling stage of resin manufacturing, and it takes at least two hours to reach to room temperature. Depending on the efficiency of this mixing and cooling step and also on the subsequent transportation and storage times until use, boards bonded with UF resins would show varying board strengths and formaldehyde emission values. The F/U_1 mole ratio value used in the first step of resin synthesis would also play an important role, although this parameter is usually a constant for a given resin but not provided to resin users. The following is noted. *(i)* The two-day storage resin samples are materials likely to be rarely encountered in industry because of the drastic batch cooling method used before the addition of the second urea in this research. *(ii)* The 15-day storage resin samples would be more like the freshly delivered industrial resins and, here, *Resin 1.80* appears to be the least suitable because of the poor IB and MOE/MOR values of boards, and *Resin 2.40* be similarly so due to the high formaldehyde emission value in the SC emission test in spite of good MOE/MOR values. *Resin 2.10* would be the better resin type overall because of the good IB and decent MOE/MOR values as well as low 2-Hour Desiccator and SC formaldehyde emission values. *(iii)* The 30-day storage samples would be like about two week-aged industrial resins, which is commonly the maximum storage time recommended in the industry. Here, *Resin 2.10* is again the better resin type overall due to the similar reasons as for the 15-day storage resins. The high MOE/MOR values of *Resin 2.40* at intermediate storage days appear to indicate that this resin cures adequately in the surface layers but does not fully cure in the core layers of boards, giving mediocre IB and high SC formaldehyde emission values probably due to residual hydroxymethyl groups in the cured resin. If it is used only in face layers of boards with other resins, such as *Resin 2.10* used in the core layer, boards of good MOE/MOR and lower formaldehyde emission values might be obtained. *(iv) Resin 1.80* is also interesting due to its relatively high IB value at 2-day storage and high MOE/MOR values at 30-day storage, and it might be useful as face-layer resins. *(v)* For 45-day and 50-day storage resins, which would be less frequently encountered in UF resin industry and quite inconvenient to produce purposedly, the results are interesting for *Resins 2.10* and *2.40* that gave relatively good IB and MOE/MOR and significantly low SC emission values, although their 2-Hour Desiccator emission values were high. Since the SC emission values of boards are viewed as a better indicator in addressing the public health concerns of boards, these low SC emission values effected by long storage times indicate a practical interest. *(vi)* Formaldehyde emitted from UF resin-bonded boards is thought to be mostly traced to the hydroxymethyl groups that break off during hot-pressing and those that remain as hydroxymethyl groups in cured resins [9, 12, 13]. In this formaldehyde emission mechanism, the increased agglomeration of resin molecules due to long storage times appears to have effected certain molecular rearrangements for polymers where more hydroxymethyl groups, possibly due to restricted access to urea amide groups, break off in hot pressing as formaldehyde and are trapped in boards to result in high 2-Hour Desiccator values. One result of this increased breakage of hydroxymethyl groups in curing would be a cured resin in boards that contain less residual hydroxymethyl groups, to give lower SC values. Therefore, for boards made with resins with 45-day or 50-day storage times, an airing would be a good method to obtain low formaldehyde emission values in both 2-Hour Desiccator and SC emission tests. The results discussed in this work were obtained in January and February, and the same experiment was repeated in April and May. The results showed an accelerated aging of resins by about 20% in time due to the slightly increased room temperature in May, but the overall trend of data from ^{13}C NMR and PB tests was similar.

24.7.5 Summary

UF resins synthesized with three different F/U_1 mole ratios, Resin 1.80b, 2.10b, and 2.40b, all at $F/(U_1+U_2)$ ratio of 1.15 and in weakly alkaline pH, were subjected to room-temperature storage treatments over a period of 50 days and resin samples were analyzed using quantitative ^{13}C NMR methods. The UF resins were found to undergo polymer structural changes arising from the migration of hydroxymethyl groups from the polymeric to monomeric UF resin components and then from forming of mostly Type I methylene and methylene-ether bonds involving the second urea, resulting in increasing the degree of polymerization of resins to higher values. Toward the later period of storage days, associative interactions of polymer molecules appeared to occur to result in forming of bigger polymer clusters and precipitates. The viscosity values of UF resins decreased initially and then increased and reached high values. The polymer structure variations and changes that occurred during the storage treatment days varied depending on the different F/U_1 mole ratios, resulting in different increases and decreases in UF resins' wood bonding performances and formaldehyde emission values. The resin synthesis and room-temperature storage parameters obtained in this work in conjunction with the parameters obtained in the heating/stirring treatments of UF resins obtained in, Sections 24.5 and 24.6 would be useful in optimizing UF resin manufacturing and PB bonding processes.

24.8 Effects of Mild Heat/Stirring and Room-temperature Storage Treatments of UF Resins Synthesized with Various F/U_1 Mole Ratios by ^{13}C NMR and Formaldehyde Emission Tests

(Adapted with permission of the publisher and minor revisions from Kim, M. G., No, B. Y., Lee, S. M., and Nieh, W. L.-S. 2003. Examination of selected synthesis parameters for typical wood adhesive-type UF resins by ^{13}C NMR spectroscopy. V. J. Applied Polym. Sci. 89 1 896–1917. [15]*)*

24.8.1 Resin Synthesis and Treatment Procedures

Extended heat/stirring treatments at 60 °C of UF resins and room-temperature storage treatments of 2.5-hour heat-treated UF resins were investigated. The syntheses of UF resins *2.40b*, *2.10b*, and *1.80b* were conducted similarly as in Section 24.7. UF resins *1.60b* and *1.40b* were also made with minor modifications similar to those used in the synthesis of UF resin 1.80b. All resins' final F/U mole ratios were 1.15. As shown in Table 24.16, the synthesized resin samples were subjected to heat treatments with stirring at 60 °C for 0.17, 2.5, 5.0, 7.5, 8.75, and 13.5hours for 2.40 series resins, but for other resins the heat/stirring treatment was cut short when unstable resins resulted. Also, the 2.5-hours treatment samples were stood on the bench for 7 days, 15 days, and 27 days to simulate industrial situations. All room-temperature storage treatment samples were about 900 g each in a polyethylene bag and evaluated by PB preparation and testing for strength and formaldehyde emissions.

24.8.2 Collection of Data

^{13}C NMR data of selected (most) resins obtained in this study are summarized in Table 24.17. All resins behaved normally in heat/stirring treatments or room-temperate storage treatments except *Resins 1.60b* and *1.40b*, which were stable only for several days and allowed only ^{13}C NMR spectra to be obtained. ^{13}C NMR data were combined with data reported in Sections 24.6 and 24.7 for resins treated at 60 °C for up to 2.5 hours [13] and the resultant changes occurring in both treatments are graphically shown in Figure 24.15 for hydroxymethyl groups, in Figure 24.16 for methylene and methylene-ether groups, Figure 24.17 for degree of polymerization values, and Figure 24.18 for urea carbonyl groups. The degree of polymerization values were obtained from methylene and methylene-ether group contents as described in Section 24.5. The viscosity changes of resins that occurred in heat treatments are reported in Figure 24.18.

Table 24.16 UF resins obtained by heat-treating synthesized resins (F/U = 1.15) at 60 °C for given times (h) and then storing at room temperature for two days or indicated numbers of days (d).

UF resin	0.17 h	2.5 h	5.0 h	7.5 h	8.75 h	13.5 h
2.40b	2.40b′	2.40j (7 d)	2.40m	2.40n	2.40o	2.40p
		2.40 k (15 d)	—	—	—	—
		2.40 l (27 d)	—			—
2.10b	2.10b′	2.10j (7 d)	2.10m	2.10n	2.10o	—
		2.10 k (15 d)	—	—	—	—
		2.1 l0 (27 d)	—	—	—	—
1.80b	1.80b′	1.80j (7 d)	1.80m	1.80n	—	—
		1.80 k (15 d)	—	—	—	—
		1.80 l (27 d)				—
1.60b	1.60b′	—	—	—	—	—
1.40b	1.40b′	—	—	—	—	—

Table 24.17 Percentage values of various methyleneic carbons and carbonyl carbons of UF resin Samples determined by ^{13}C NMR method.

Group (ppm)	Synthesized/room-temperature-treated UF resin samples																
	2.40b	**2.40j**	**2.40 k**	**2.40 l**	**2.40 m**	**2.40o**	**2.40p**	**2.10b**	**2.10j**	**2.10 k**	**2.10 m**	**2.10o**	**1.80b**	**1.80j**	**1.80k**	**1.60b**	**1.40b**
91.0	0.79	0.69	0.73	0.66	0.60	0.82	0.58	0.52	0.57	0.51	0.67	0.55	0.36	0.27	0.28	0.00	0.00
87.0	0.24	0.06	0.16	0.00	0.00	0.00	0.00	0.19	0.00	0.06	0.00	0.00	0.10	0.03	0.00	0.28	0.24
83.1	0.28	0.41	0.51	0.50	0.26	0.20	0.27		0.25	0.18	0.18	0.22	0.28	0.09	0.20	0.00	0.00
Total	*1.31*	*1.16*	*1.40*	*1.16*	*0.86*	*1.02*	*0.85*	*1.00*	*0.82*	*0.75*	*0.85*	*0.77*	*0.74*	*0.39*	*0.48*	*0.28*	*0.24*
79.1	2.66	1.95	2.26	2.26	1.82	1.77	1.73	2.22	1.66	1.67	1.36	1.38	2.15	1.52	1.26	1.80	2.58
75.1	4.98	4.43	4.12	4.35	3.75	3.54	3.33	4.82	4.32	4.24	4.95	4.59	4.50	3.17	4.06	4.06	4.14
69.5	6.79	10.03	10.27	10.90	9.40	11.06	12.15	7.27	9.70	10.46	10.80	11.30	8.92	10.12	10.39	8.50	10.13
Total	*14.43*	*16.41*	*16.65*	*17.51*	*14.97*	*16.37*	*17.21*	*14.31*	*15.68*	*16.37*	*17.11*	*17.27*	*15.57*	*14.81*	*15.71*	*14.36*	*16.85*
72.0	17.91	13.88	12.23	12.83	10.47	7.90	5.97	18.52	12.93	11.12	12.32	10.34	15.19	11.01	10.43	11.53	10.27
65.2	35.01	30.31	30.24	25.07	36.53	36.79	36.82	29.66	30.91	30.94	27.20	28.11	26.70	29.34	27.76	27.56	24.01
Total	*52.92*	*44.19*	*42.7*	*37.90*	*47.00*	*44.69*	*42.79*	*48.18*	*43.84*	*42.06*	*39.52*	*38.45*	*41.89*	*40.35*	*38.19*	*39.09*	*34.28*
60.1	4.62	4.58	4.70	5.20	1.53	1.05	1.15	4.48	3.89	4.60	4.12	4.26	4.29	4.43	4.24	4.07	5.57
53.9	21.06	21.17	20.60	20.88	22.32	21.18	20.47	23.79	22.84	21.83	22.98	22.30	25.81	24.06	24.43	25.85	24.68
47.4	5.64	12.67	14.18	17.35	13.31	15.69	17.48	8.24	12.88	14.35	15.42	16.95	11.70	16.05	16.95	16.35	18.38
Total	*31.32*	*38.42*	*39.48*	*43.43*	*37.16*	*37.92*	*39.10*	*36.51*	*39.61*	*40.78*	*42.52*	*43.51*	*41.80*	*44.54*	*45.62*	*46.27*	*48.63*
164.0	28.01	18.32	14.71	15.30	15.74	12.83	11.17	26.39	18.05	16.51	14.18	12.75	25.08	16.87	15.61	18.17	15.89
162.2	23.58	31.42	32.72	34.97	30.81	33.53	35.77	18.44	28.14	30.49	30.35	32.20	15.91	26.47	29.10	20.73	23.66
160.7	45.86	47.84	49.60	46.90	51.43	51.50	50.83	54.03	50.70	50.18	53.86	52.93	56.78	53.64	52.28	58.59	56.87
158.0	2.55	2.43	2.97	2.83	2.01	2.14	2.22	1.14	3.11	2.81	1.61	2.02	2.23	3.02	3.01	1.51	3.58

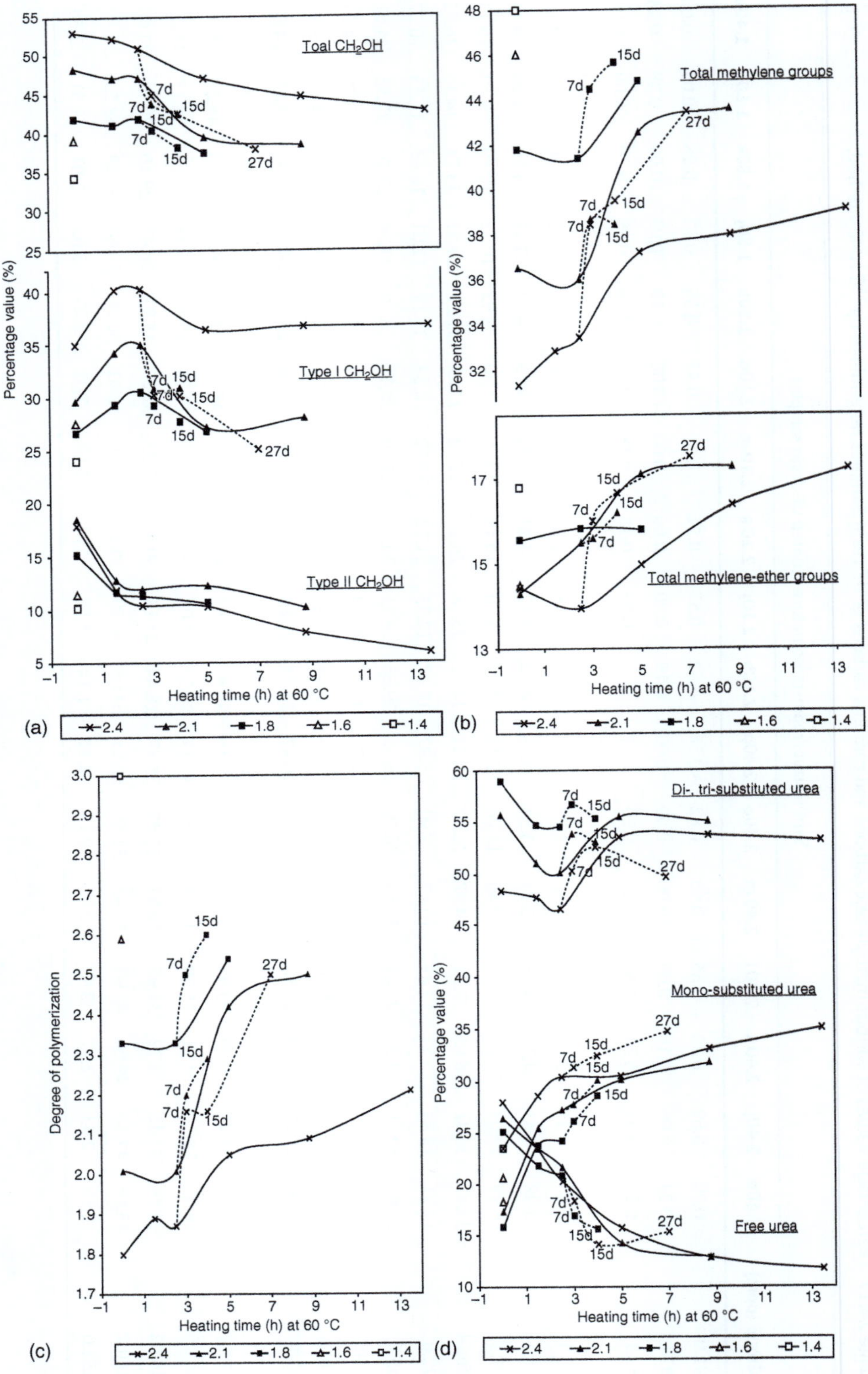

Toal CH_2OH
Type I CH_2OH
Type II CH_2OH
Percentage value (%)
Heating time (h) at 60 °C
(a)
2.4 2.1 1.8 1.6 1.4
Total methylene groups
Total methylene-ether groups
(b)
Degree of polymerization
(c)
Di-, tri-substituted urea
Mono-substituted urea
Free urea
(d)
7d
15d
27d

Figure 24.15 (a) Changes in ^{13}C NMR values of total and types I and IIt/IIi hydroxymethyl groups based on all urea of UF resins 1.40–2.40 heated at 60 °C for up to 13.5 hours (solid line) and heat-treated at 60 °C for 2.5 hours followed by storing at room temperature for 7, 15, 27 days (dotted line). (b) Changes in ^{13}C NMR values of total methylene and methylene-ether groups in UF resins 1.40–2.40 heat-treated at 60 °C for up to 13.5 hours (solid line) and heat-treated at 60 °C for 2.5 hours followed by storing at room temperature for 7, 15, and 27 days (dotted line). (c) Changes in the degree of polymerization based on all urea of UF resins 1.40–2.40 (solid line) heated at 60 °C for up to 13.5 hours and heat-treated at 60 °C for 2.5 hours followed by stored at room temperature for 7, 15, and 27 days (dotted line). (d) Changes in ^{13}C NMR values of urea carbonyl carbons in *UF resins 1.40–2.40* heated at 60 °C for up to 13.5 hours (solid line) and heat-treated at 60 °C followed by storage at room temperature for 7, 15, and 27 days (dotted line).

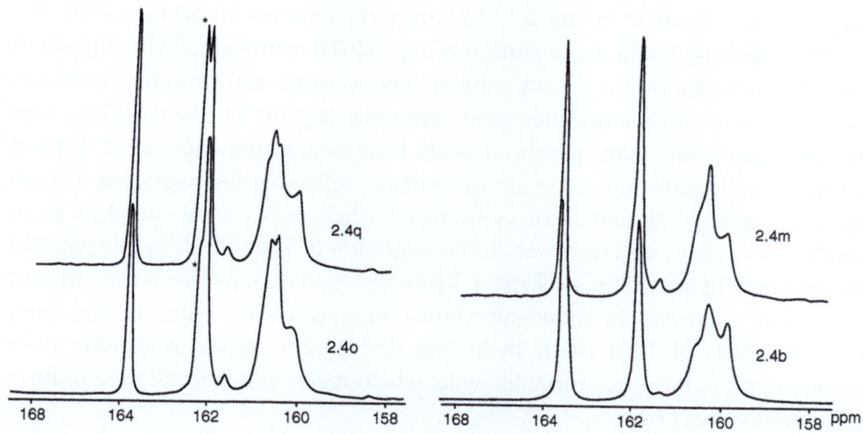

Figure 24.16 ^{13}C NMR spectra of urea carbonyl carbons of Resins 2.40b, 2.40 m (5.0 hours 60 °C), 2.40o (7.75 hours 60 °C), and 2.40q (2.5 hours at 60 °C and 25 days at room temperature) showing (*) the growth of the mono-substituted urea species peak at ~162.1 ppm (2.40q = 2.40 l).

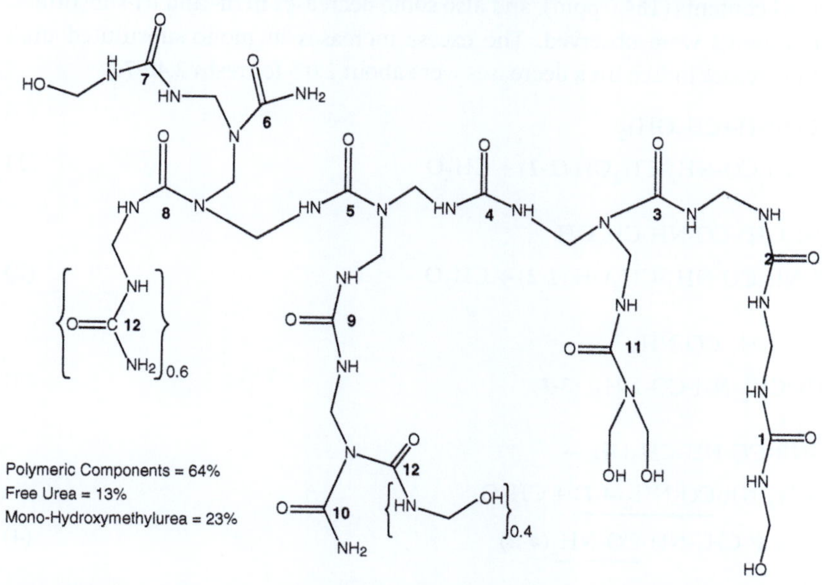

Figure 24.17 A schematic polymer structure for polymeric components of resin 2.4o.

24.8.3 Heat Treatment Effects of UF Resins Made with Various F/U_1 Mole Ratios

The ^{13}C NMR data of *-b series* resins in Table 24.17 indicate that resins synthesized with decreasing F/U_1 mole ratios, from 2.40 to 1.40, have increasing methylene group contents, from 31.32% to 48.63% and decreasing hydroxymethyl group contents, from 52.92% to 34.28%. It is the result of making resins using 50% formaldehyde solution and 100% urea without adjusting the water level in the first and second steps of manufacturing and ending the second step of polymerization step at about same viscosity points for all resins. These differences need to be considered in resin syntheses and data analyses.

Re-examination of heat treatment effects at 60 °C for 2.5 hours: Data for heat treatments with stirring of resins at 60 °C for up to 2.5 hours were reported in Section 24.7. The viscosity of resins decreased, and the migration of Type IIt/IIi hydroxymethyl groups from polymeric (U_1) to monomeric (U_2) resin components occurred with minimal increases in total methylene and methylene-ether group contents (*Figure* 24.15*a–d*). Also, Type I methylene and methylene-ether group contents increased at the expense of Type II and III methylene and methylene-ether group contents, reflecting the migration of Type IIi hydroxymethyl groups. Similar changes occurred when resins were stored at room temperature for up to about 20 days. Overall, the migration of Type IIt/IIi hydroxymethyl groups also resulted in increases in Type I hydroxymethyl groups, decreases in free urea contents, and increases in mono-substituted urea contents. Thus, in the early treatment period, parts of Type IIt/IIi hydroxymethyl groups on the polymeric resin components break off to form free formaldehyde, which reacts with second urea to form mono-hydroxymethylurea (***3-1***) (Eq. 1–3).

Re-examination of spectral data indicated that other reactions also began to occur in the heat treatment as shown in graphs in Figure 24.15a–d. *(i)* Firstly, the increases in mono-substituted urea carbonyl contents (162.2 ppm) were greater than the decreases in free urea carbonyl contents (164.0 ppm), and also some decreases in di- and tri-substituted urea carbonyl contents were observed. The excess increases in mono-substituted urea carbonyls over decreases in free urea decreases were about 2.0% for resin 2.40-7d,

$$\mathbf{P}\text{-NH-CO-NH-}(\text{CH}_2\text{OH})_2 \rightarrow \text{P-NH-CO-NH}_2\text{-CH}_2\text{OH}\ (\textbf{\textit{1-1}}) + \text{CH}_2\text{O} \tag{1}$$

$$\mathbf{P}\text{-N(CH}_2\text{OH)-CO-NH-CH}_2\text{OH} \rightarrow \text{P-NH-CO-NH}_2\text{-CH}_2\text{OH}\ (\textbf{\textit{1-2}}) + \text{CH}_2\text{O} \tag{2}$$

$$\text{CH}_2\text{O} + \text{NH}_2\text{-CO-NH}_2 \rightarrow \text{HOCH}_2\text{-NH-CO-NH}_2\ \ (\textbf{\textit{3-1}}) \tag{3}$$

$$\text{P-CH}_2\text{-NR-CO-NH-CH}_2\text{OH} \rightarrow \text{P-CH}_2\text{-NR-}\underline{\text{CO-NH}_2}(\textbf{\textit{4-1}}) + \text{CH}_2\text{O} \rightarrow \mathbf{P}\text{-CH}_2\text{-NH-}\underline{\text{CO-NH}_2}(\textbf{\textit{4-2}}) \tag{4}$$

$$(\mathbf{3\text{-}1}) + NH_2\text{-CO-NH}_2 \rightarrow \underline{H_2N\text{-CO-NH}}\text{-CH}_2\text{-}\underline{NH\text{-CO-NH}_2}(\mathbf{5\text{-}1}) \quad (\mathbf{P} = \text{UF polymer}) \tag{5}$$

$$(\mathbf{3\text{-}1}) + \mathbf{P}\text{-NR-CO-NH}_2(\mathbf{4\text{-}1}) \rightarrow \text{P-NR-CO-NH-CH}_2\text{-NH-CO-NH}_2(\mathbf{6\text{-}1}) \tag{6}$$

$$\mathbf{U} + \mathbf{P}\text{-NR-CO-NH-CH}_2\text{OH} \rightarrow \text{P-NR-CO-NH-CH}_2\text{-NH-CO-NH}_2(\mathbf{7\text{-}1}) \tag{7}$$

$$(\mathbf{3\text{-}1}) + (\mathbf{3\text{-}1}) \rightarrow NH_2\text{-CO-NH-CH}_2\text{-NH-CO-NH-CH}_2\text{OH}(\mathbf{8\text{-}1}) \tag{8}$$

$$(\mathbf{3\text{-}1}) + \text{P-NR-CO-NH-CH}_2\text{OH} \rightarrow \text{P-NR-CO-NH-CH}_2\text{-NH-CO-NH-CH}_2\text{OH}(\mathbf{9\text{-}1}) \tag{9}$$

3.9% for Resin 2.10-7d, and 9.4% for Resin 1.80-7d. One explanation for this secondary reaction is the migration of Type I hydroxymethyl groups from the polymeric resin components leaving behind free amide-end groups on UF resin polymers (***4-1***) (Eq. 4). When the other neighboring nitrogen is bonded to UF polymers only through one methylene group (R = H), it will be mono-substituted urea-groups bonded to UF polymers (***4-2***). This reaction path will increase mono-substituted urea group contents and decrease di-substituted urea group contents; *(ii)* Secondly, increases in Type I methylene/methylene-ether group contents were slightly higher than decreases in Type II/III methylene/methylene-ether group contents, indicating that total methylene/methylene-ether group contents also began to increase (*Figure* 24.15*b*). A plausible explanation for this increase is the reaction of mono-hydroxymethylurea (***3-1***) with free urea to form methylene-di-urea (***5-1***) (Eq. 5), which is a mono-substituted urea, which will increase mono-substituted urea group contents as well as Type I methylene group content. Data presented below for extended storage treatment samples indicate that ^{13}C NMR peaks at about 162.2 ppm for mono-substituted urea carbonyls in these resins could represent not only mono-hydroxymethylurea (***3-1***) but also methylene-di-urea (***5-1***) and singly bonded urea-groups (***4-2***).

Methylene and methylene-ether bond formation in heat treatments of UF resins at 60 °C for longer than 2.5 hours: ^{13}C NMR data of resins treated at 60 °C for longer than 2.5 hours indicated a continued migration of hydroxymethyl groups (Table 24.17 and Figure 24.15 a–d), which was shown by decreases in both Type IIt/IIi hydroxymethyl groups and free urea contents and increases in mono-substituted urea species content. The other significant changes were increases in the total and Type I methylene and methylene-ether group and di- and tri-substituted urea species contents and decreases in Type I hydroxymethyl group contents. All these changes indicated that the resins' degree of polymerization values increased significantly, principally between 2.5 and 5.0 hours of heat treatments, as shown in Figure 24.15c. The polymerization reactions then slowed down. Similar changes in resins were observed to occur between 15-day and 30-day storage treatments at room temperature in Section 24.7, indicating that the underlying chemical reactions were accelerated by the longer heat treatments. Decreases in free urea contents paralleled the increases in di-substituted urea species, and decreases in hydroxymethyl group contents were about one-half of these amounts. Mono-substituted urea species increased modestly. Normally, peaks for di- and tri-substituted urea carbonyls are only slightly separated, but a close inspection of these peaks indicated that most increases were for di- rather than tri-substituted urea carbonyls. These functional group changes can be summed up as a result of reactions of Type I hydroxymethyl groups of UF polymers (***1-1***) and mono-hydroxymethylurea (***3-1***) with the amide groups of mono-hydroxymethylurea (***3-1***) or free urea or urea-groups bonded to UF polymers (***4-1/5-1***). The resultant species would be methylene-di-urea (Eq. 5) and urea end-groups that are singly bonded to UF polymers (Eqs. 6 and 7) as well as their hydroxymethyl group derivatives (Eqs. 8 and 9). ^{13}C NMR data in Table 24.17 also indicated that Type I methylene and methylene-ether groups were the primary bonds formed in this heat treatment period, with negligible increases in Type II/III methylene and methylene-ether groups.

Do methylene bonds really form in the weak alkaline pH?: Unreported until now, it appears possible for long treatment times in that a slow but steady reaction mechanism is playing out, and details on some of these reactions were directly observed in ^{13}C NMR spectra of *Resin 2.40* series as shown in *Figure* 24.16, where the mono-substituted urea carbonyl peak (161.9 ppm) of untreated -b series resins agrees with the reported value for mono-hydroxymethylurea (***3-1***) when the carbonyl peak of free urea in this work (164.0 ppm) is adjusted to the literature value (163.60 ppm) [6]. The mono-substituted urea carbonyl peak is normally somewhat broad at the base indicating a hidden peak, which did not show up as a separate peak in *Resins 2.40b* (no heat treatment) or *2.40m* (5-hours heat treatment). However, the hidden peak began to show up as a shoulder in *Resin 2.40o*, (7.75-hours heat treatment) and then as a separate peak at 162.01 ppm (asterisk) in *Resins 2.40q (=2.40 l)* and *2.40i (in Section* 24.7*)*, separately from the main peak at 161.9 ppm of mono-hydroxymethylurea Figure 24.16. The chemical shift value of the shoulder peak at 162.01 ppm is very close to that of methylene-di-urea (***5-1***) peak reported at 162.1 ppm measured at 60 °C for better solubility [6] This peak can also be assigned to urea-groups singly bonded to UF polymers (***6-1***, ***7-1***). Thus, the extended heat treatments of resins gave rise to a peak that is assignable to methylene-di-urea and/or singly bonded urea-groups shown in *Eqs.* (*5–7*). The combined level of methylene-di-urea and/or singly bonded urea-groups in *Resin 2.40o* is approximately 11.1% of total urea. Since di- (and tri-) substituted urea species increased by 5.2% in this resin compared with *Resin 2.40b*, and this increase can be assigned mostly to the singly bonded urea-groups, and the amount of methylene-di-urea could be balanced at 5.9%. Overall, these results indicated that (free) second urea reacted with mono-hydroxymethylurea to form methylene-di-urea and also reacted with UF polymers to become incorporated as singly bonded urea-groups between 2.5–8.75 hours of heat treatment periods.

P-N-CO-NH_2 (***10-1***)
|
P-NR-CO-NH-CH_2 (10)

(P = UF Polymer)

An Illustration of UF Resin Structures Attained in Heat Treatments at 60 °C - The composition of Resin 2.40o (8.75 hour heat treatment) consists of free urea (12.8%), mono-hydroxymethylurea (22.4%), methylene-di-urea (5.9%), and polymeric UF components (58.1%). The polymeric UF components of Resin 2.40o, including methylene-di-urea, showed a number average degree of polymerization of about 12, and each molecule has about 2.4 Type I hydroxymethyl groups, 1.1 Type IIt hydroxymethyl groups, and 0.6 singly bonded urea- group, for a total of 3.6 end-groups. Type IIi hydroxymethyl groups were assumed to be negligible. Another polymer end-group possible in UF resins is the gem-di-substituted urea-group (**10-1**) located in the middle of polymer chains in Eq. 10. An approximate value can be calculated from values of Types I, II, and III methylene/methylene-ether groups for a UF polymer, [12] and a calculation showed that the polymeric resin components of Resin 2.40o would have 5.7 polymer end-groups. Therefore, there should be about 2.1 gem-di-substituted urea-end groups. A schematic structure of this UF polymer thus obtained is shown in Figure 24.17.

24.8.4 Heat Treatments at 60 °C for 2.5 hours Followed by Storage at Room Temperature for up to 27 days

The following observation is noted: *(i)* As shown overlaid with dotted lines in *Figure 24.15a–d*, the room-temperature storage-treated resins showed functional group changes that are quite different from the continuous heat-treated resins. The methylene/methylene-ether group content changes were most prominent for Resin 2.40, its value increasing rapidly in 7 days and then slowly afterwards until 27 days of storage time (*Figure 24.15b*), significantly exceeding the value of 13.5 hours heat-treated resins; *(ii)* Other major differences observed in the storage treatments of Resin 2.40 were that free urea content decreased and mono-substituted urea carbonyl species at 162.2 ppm, i.e. methylene-di-urea (***5-1***), and singly bonded urea-groups (***7-1***) increased significantly while di–/tri-substituted urea species decreased only minimally. Also, Type I hydroxymethyl group, *i.e.* hydroxymethylurea, decreased drastically, which suggests that free urea reacted mainly with mono-hydroxymethylurea to form methylene-di-urea (***5-1***) and only minimally with polymeric resin components to form the urea-groups singly bonded to UF polymers (***7-1***) in storage experiments; *(iii)* Resins 2.40 series showed these increases more prominently due to the larger amount of free urea available and thus *Resin 2.40 l* (27 days) was estimated to have about 35.0% of total urea as mono-substituted urea and half of that, 17.5%, as methylene-di-urea (***5-1***) and singly bonded urea (***7-1***) groups. However, since di- and tri-substituted urea species, which represent singly bonded urea-groups, increased only by about 1.3 – 3.7%, *the methylene-di-urea would amount to the difference*, 13.8 – 16.2%, a high value compared with 5.9% for *Resin 2.40o* (8.75 hours at 60 °C) discussed above; *(iv)* In room storage treatments, the formation of methylene-di-urea is favored over that of forming urea-groups singly bonded to UF polymers (Eq. 6 and 7). Stirring of resins during storage treatments would bring a more uniform reaction among the various monomeric and polymeric resin components. On the other hand, storing reactive resins without stirring thus appears to favor reactions among monomeric or low molecular weight components, such as mono-hydroxymethylurea and urea, to react among themselves rather than reacting with polymeric UF molecules. Thus, methylene-di-urea components thus formed at a higher level are expected to make the resin composition different and result in differences in curing of resins; *(v)* Resin 1.80 showed only minor increases of methylene and methylene-ether groups, and Resin 2.10 showed only about one-half of increases in methylene and methylene-ether groups in comparison with the continuous heat treatment; and *(vi)* Similar changes occurred between 15–30 d of storage treatment at room temperature without prior heat treatments [12, 13].

24.8.5 Reaction Mechanisms of Methylene and Methylene-ether Group Forming Ractions

In Mild Acidic pH: In mildly acidic aqueous media, UF resins and similar melamine-urea-formaldehyde (MUF) resins undergo facile methylene bond-forming reactions by the (specific) hydrogen ion (H^+) catalysis process (Eq. 11), where Type I hydroxymethyl groups are activated in the form of a hydrogen ion complex (***11-1***), [17] which then breaks up to form the cationic methylene intermediate (***11-2***) by expelling a water molecule in the rate-determining step. The cationic methylene intermediate is then resonance-stabilized with the methylene imidinium ion (***11-3***), making the activation process occur at a lower energy level. These reactions go readily at mildly acidic pH 4–5 at elevated temperatures as observed in resin synthesis of UF resins. The cationic methylene intermediate, a strong electrophile, would attack the amide nitrogen nucleophiles preferentially to form methylene bonds. When the activated hydroxymethyl group is Type I, the cationic methylene imidinium ion (***11-3, R = H***) may equilibrate further to methylene imine (***11-4***) to lower the reaction energy barrier further. On the other hand, for Type IIt hydroxymethyl groups, which create branched polymer structures in resin synthesis or in curing, the formation of imidinium ions (***11-3, R = –CH₂–OH***) is a quaternization reaction on the nitrogen, which has normally a higher reaction barrier. Thus the first reaction (***11-3, R = H***) would go faster than the second reaction (***11-3, R = –CH₂–OH***) in curing, and Type IIi hydroxymethyl groups (***11-1, R′ = CH₂-NR″-CO-***) would cure similarly but at a slower rate. Particleboard binder-type UF resins synthesized with an F/U_1 mole ratios of about 2.0 with post-urea addition to F/U mole ratios of 1.15 would harbor significant and varying levels of Type IIt/IIi hydroxymethyl groups. That these UF resins give well-cured, crosslinked wood adhesives only in the presence of an acid catalyst indicates that both reactions would readily occur, but some differentiating effects would linger till the end of curing period, and some Type IIt/IIi hydroxyl groups would be preferentially left unreacted to become a source of formaldehyde emission.

$$\text{P}{\sim\sim}\text{CH}_2\text{-NR-CO-NR}'\text{-CH}_2\text{OH}\ (\textbf{11-1}) + \text{H}^+ \longrightarrow \text{P}{\sim\sim}\text{CH}_2\text{-NR-CO-NR}'\text{-CH}_2\text{-OH}_2^+$$
$$\xrightarrow[-\text{H}_2\text{O}]{} \text{P}{\sim\sim}\text{CH}_2\text{-NR-CO-NR}'\text{—}\mathbf{CH_2}^+\ (\textbf{11-2}) \rightleftharpoons \text{P}{\sim\sim}\text{CH}_2\text{-NR-CO-N}^+\text{R}'{=}\mathbf{CH_2}\ (\textbf{11-3}) + \text{H}_2\text{O}$$
$$\longrightarrow \text{P}{\sim\sim}\text{CH}_2\text{-NR-CO-N}{=}\mathbf{CH_2}\ (\textbf{11-4}) + \text{H}^+ \quad (\text{R}' = \text{H}) \tag{11}$$

$$\text{P}{\sim\sim}\text{CH}_2\text{-NR-CO-NR}'\text{-CH}_2\text{OH} + \text{H-X} \longrightarrow \text{P}{\sim\sim}\text{CH}_2\text{-NR-CO-NR}'\text{-CH}_2\text{-OH}_2\text{-X}\ (\textbf{12-1})$$
$$\text{P}{\sim\sim}\text{CH}_2\text{-NR-CO-N(H}^\oplus\text{)—CH}_2\text{-OH}_2\text{-X}\ (\textbf{12-2})\ (\text{R} = \text{H}) \xrightarrow[\text{OH}^\ominus]{} \text{P}{\sim\sim}\text{CH}_2\text{-NR-CO-N}{=}\text{CH}_2 + \text{H}_2\text{O} + \text{HX}\ (\textbf{12-3}) \tag{12}$$

In mildly alkaline pH: The methylene group forming reactions in alkaline pH were observed to occur as well as the methylene-ether bond forming reactions. It is known that UF resins in alkaline pH can transform into a solid mass when stored for an extended period of time at room temperature or heated at elevated temperatures, but the solidified resins did not show the kind of cure required as wood adhesives. In the similar melamine-formaldehyde resins, the methylene/methylene-ether bond-forming reactions in neutral or mildly alkaline pH proceed more readily during resin synthesis, but the good cure of these resins is achieved similarly with strong acid catalysts. The transformation of UF resins into solids in alkaline pH would entail the formation of some methylene/methylene-ether groups that also accelerates physical agglomeration of resin molecules. In the previous Section 24.7, the methylene/methylene-ether group forming reactions in alkaline pH were suggested to entail the general acid catalysis process, *i.e.* general acids, such as water, resin's hydroxymethyl groups, and undissociated acid species would be the catalysts for the activation of hydroxymethyl groups. In the *general acid catalysis mechanism (Eq. 12)* [38], an undissociated acid species (H-X) forms an acidic complex-type intermediate on the hydroxymethyl group (***12-1***), which would be expelling a water molecule to form cationic methylene intermediate only very slowly. The slow advancement of UF resins in weak alkaline pH corroborates with this weak general acid complex theory, where a mechanism of some activation appears available only for *Type I hydroxymethyl groups (**12-1**, **R′ = H**)* in the alkaline pH, through the abstraction of the weakly acidic hydrogen on the imide nitrogen by a hydroxide ion (OH^-) (Eq. 12). Initially, the imide anion (***12-2***) will be formed, and the negative charge dislocates to form methylene imine species (***12-3***) and a water molecule is eliminated. The methylene imine is a cationic methylene equivalent and can react concertedly with amide nitrogens (–HN–H) or hydroxyl (–O–H) groups of another urea molecule to form methylene or methylene-ether bonds. The reaction rates would increase with pH increases since the imide anion-forming step (***12-2***) is likely to be the rate-determining step. Also, the observation that methylene bond forms only at trace level while methylene-ether bond forms at about 15% level in the first alkaline reaction step of UF resin synthesis appears to be due to this mechanism, forming mostly Type I methylene-ether bonds but little methylene bonds are formed due to lack of imide nitrogens (–HN–H) in resins with F/U mole ratios of about 2.0 or higher. Also, this path would be not very effective for Type IIt/IIi hydroxymethyl groups since the corresponding imide anion (***12-2***) could not be formed due to lack of hydrogen atoms for abstraction by hydroxide ions. This might also be the reason Type IIt/IIi hydroxymethyl groups are less reactive and remain at relatively high concentrations, as observed in the highly heat- or storage-treated samples in alkaline pH. This reaction mechanism is a speculation but agrees with the fact that Type II/III methylene/methylene-ether groups do not form much in the posttreatment experiments as well as that the polymerization reactions in alkaline pH normally do not go to give completely cured products even if hot-pressed for a long time, *i.e.* UF polymers appear to grow only through the activation of Type I hydroxymethyl groups in alkaline pH and the relatively high level of Type IIt/IIi hydroxymethyl groups brake the resin system at an incompletely cured stage of resin.

Does the type IIt/IIi hydroxymethyl group migration stop in the end of treatments?: One puzzling question regarding the reactions occurring in alkaline pH is the observation that the free formaldehyde species are low, and the free urea level and Type IIt/IIi hydroxymethyl group contents are relatively high in the longest heat treatment or 50-day stored resin samples. For example, the free formaldehyde species are 1.02%, the free urea level is about 11%, and Type IIt/IIi hydroxymethyl group level is about 6.0% in *Resin 2.40o*, indicating that the migration of hydroxymethyl groups from polymeric resin components to free urea almost came to a standstill. Similarly, the formation of methylene/methylene-ether groups significantly slowed down after the initial rapid pickup during the heating/storing treatments, even though Type I hydroxymethyl group and free urea contents are relatively high. It is not clear why these reactions slow down, but it may be that UF polymers in posttreatments at an elevated temperature attain in time certain tertiary molecular structures or agglomerates via strong intra-molecular hydrogen bond interactions to result in preventing Type IIt/IIi hydroxymethyl groups from dissociation reactions. As the treatment temperature is increased, this tertiary structure would be steadily overcome and the methylene and methylene-ether group contents would increase to eventually reach to the gelation stage as observed in gel tests carried out at 100 °C, which usually take about 30 minutes or longer. Though, Type IIt/IIi hydroxymethyl groups appear to remain till the last minutes to prevent the resin system from reaching the fully cured state. Unfortunately, ^{13}C NMR tends to give less reliable results for resins in this later stage of treatments due to the molecular agglomeration and solid-like materials formed in resins.

24.8.6 Viscosity and Phase Changes of UF Resins During Heating/Storage Treatments

Normal viscosity behaviors of resins 2.40 and 2.10: The initial viscosity decreases in resins observed on heating/stirring at 60 °C in the first 2.5 hours were reported in Section 24.6 (Figure 24.10). In the extended heating/stirring treatments of resins carried out in this study, as shown in Figure 24.18, the viscosity of *Resin 2.40* decreased the most, while that

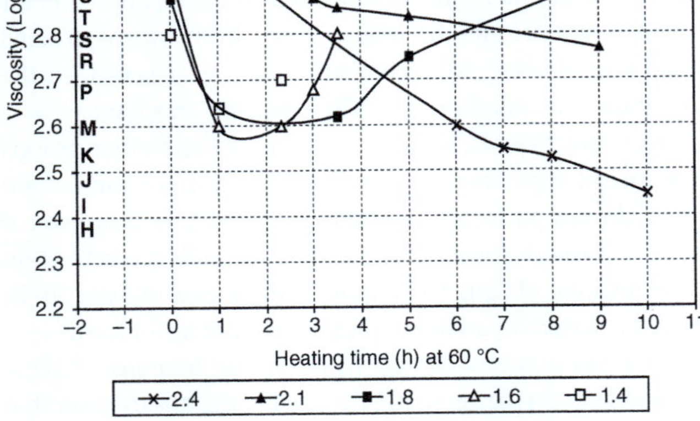

Figure 24.18 Changes in viscosity of UF resins 1.40–2.40 stirred/heat treated at 60 °C for up to 10 hours.

of Resin 2.10 decreased modestly until 7.5 hours and then remained stable until 13.5 hours of heat treatment. ^{13}C NMR data showed, as in the room-temperature storage experiments, that a gradual migration of hydroxymethyl groups initially occurred from the polymeric to monomeric resin components in the early heat treatment period and these resulted in viscosity decreases. On the other hand, the methylene and methylene-ether group contents increased sharply between 3 – 5 hours heat treatments (Figure 24.15b) but the viscosity of Resin 2.40 and 2.10 kept decreasing while the viscosity of Resin 1.80 increased slightly. Thus, it appears that the viscosity values of finished resins do not necessarily increase much as the molecular weight increases as long as the resins are stirred at mild temperatures. However, both *Resins 2.40* and *2.10* became slightly turbid at the end of the heating periods, as expected from the formation of methylene-di-urea-type insoluble products.

Abnormal viscosity behaviors of resins 1.80, 1.60, and 1.40: On the other hand, Resin 1.80, slightly turbid as made, showed a slight decrease in viscosity in 2.5-hours heat treatment and then the viscosity continuously increased until 8.5 hours of heat treatment with the turbidity also increasing. The viscosity-increasing effect, due to increased molecular association and increasing methylene/methylene-ether group contents, appears to outweigh the viscosity-decreasing effect of hydroxymethyl group migrations. Furthermore, *Resins 1.60* and *1.40* were already somewhat turbid as made although their starting viscosity values were comparable to or lower relative to higher F/U_1 mole ratio resins. These low F/U_1 mole ratio resins, when second urea was added at about 4 °C, were stable for about 5 days at room temperature and their ^{13}C NMR spectra were obtained for the first time in this laboratory (Table 24.17 and Figure 24.19a/b). When heated at 60 °C, the two resins in alkaline pH increased in viscosity and turned into semi-solid mass in about 1.0 hours for *Resin 1.6* and 0.5 hours for *Resin 1.40*. For these resins, the amount of the second urea was lower, and the proportion of polymeric resin components and the average degree of polymerization was higher. All of these factors and the lower levels of hydroxymethyl groups would have contributed to the short lifetimes of the heat treatments.

Lesser branching, greater tendency of agglomeration: The increased molecular association/agglomeration tendency of lower F/U_1 mole ratio resins appeared to have played a large role in the rapid viscosity increases. For example, the methylene/methylene-ether group contents of *Resins 1.60b* or *1.40b* with no heat treatments were comparable with those of *Resin 2.40o* heat-treated at 60 °C for 8.75-hours. But *Resin 2.40o* remained at the low viscosity attained by heat treatment for several weeks, while *Resins 1.60b* and *1.40b* gelled within about 5 days at room temperature. It was also observed that the viscosity increases of resins generally lagged behind the methylene/methylene-ether group increases. Thus, although the viscosity increases would be helped by increases in the degree of polymerization or methylene-di-urea type precipitates, the increased agglomeration tendency of resin molecules appears to be playing a major role arising because of the lower extents of polymer chain branching [12, 13].

24.8.7 F/U_1 Mole Ratio and Posttreatment Effects on Resin Polymeric Structures and Compositions

The F/U_1 mole ratio and posttreatment parameters together affect UF resins to advance and reach to some uniquely different polymer compositions. First of all, the methylene/methylene-ether group content, or the average degree of polymerization,

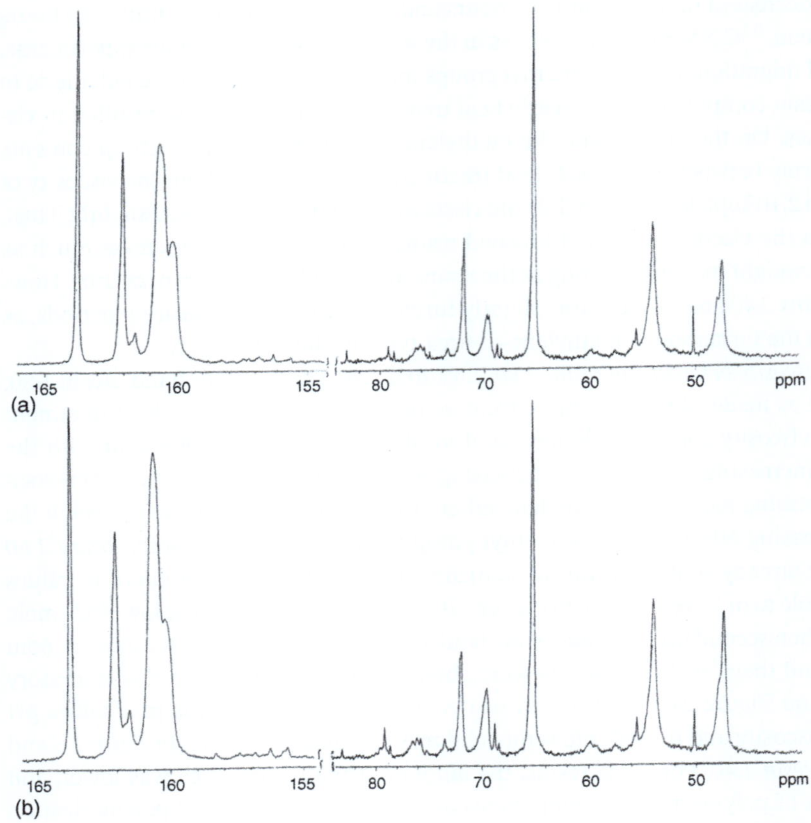

Figure 24.19 (a) ^{13}C NMR spectra of UF resin 1.60(b) and (b) UF resin 1.40(b) made with F/U_1 mole ratios of 1.60 and 1.40, respectively, and the second urea added to an F/U mole ratio of 1.15.

for untreated resins were in the order of *Resin 1.40b > 1.60b > 1.80b > 2.10b > 2.40b* (*Figure* 24.12). Then, resins made with higher F/U_1 mole ratios have greater amounts of second urea to be added and both types of posttreatment result in mostly methylene-di-urea and UF polymers having singly-bonded urea-groups, while resins with lower F/U_1 mole ratios result in Type I methylene/methylene-ether groups that are mostly part of the polymeric components. Further, the methylene-di-urea contents would be higher in resins obtained from non-stirring, storage treatments than from heat treatments. Therefore, *one way to increase the posttreatment effects* could be to synthesize resins using F/U_1 mole ratios at higher levels at about 2.60–3.30 and at pH 1.0 to carry out the polymerization reaction, [27, 33] resulting in resins with polymer structures having some intra-molecular ether bonds, *i.e.* uron rings, and since these resins need larger amounts of second urea, more room for posttreatments would be possible to bring higher extents of posttreatments.

Another way to increase the posttreatment effects could be to synthesize resins at lower F/U_1 ratios and with lower extents of polymerization. Because the varying U_1/U_2 mole ratios used in the resin syntheses in this work, the ratio of polymeric components vs. monomeric components increases as the F/U_1 mole of resin synthesis is decreased: 47%,

55%, 64%, 71%, and 75%, even though similar viscosity values were attained at the end of the acidic polymerization step. The higher proportions of polymeric components for lower F/U_1 mole ratio resins appear to have led to shorter short storage lives, and therefore, stopping the polymerization reaction earlier than those used in this study appears to give resins with longer storage lives and lengthen the room for post-heat treatments. UF resins with more linear UF resin polymer structures might be obtained. Overall, one limitation in this approach would be the time-consuming nature of posttreatments by industry standards; however, these various UF resin polymer structures and compositions are difficult to obtain from ordinary resin synthesis procedures, and thus the scope of the UF resin system might be expanded possibly leading to resins with better bonding and formaldehyde emission performances.

24.8.8 F/U_1 Mole Ratio and Heat Treatment Effects on Particleboard Strength Properties

Industrial PB manufactured with densities close to the value used in this study is the highest-grade materials and normally show IB strengths of 100–160 psi, bending MOE values of 300–600 kpsi, and bending MOR values of 1800–3000 psi [39]. The resin samples used in bonding of PBs were heat-treated at 60 °C for 0.0, 0.17, 2.5, 5.0, and 7.5 hours, which were only slightly different in treatment times than for samples used in ^{13}C NMR analyses. IB, MOE, and MOR values of PBs obtained were in the normal ranges but showed some differentiating effects of heat treatment and F/U_1 mole ratio parameters (Figure 24.20a–d) discussed below.

Physical strength values of boards bonded with resins with short heat treatment times: First, the board strength values of untreated resin samples were unusually low for all F/U_1 mole ratio resins compared with treated resins. The migration of hydroxymethyl groups did not occur in these resins and, therefore, the poor performances could be traced to the second urea components that are deficient of functional groups for curing. A 10-minutes heat treatment resulted in improved strength values for all resins. The extent of hydroxymethyl group migration occurring in 10-minutes heat treatments would be relatively small but appears to have sufficiently increased the functionality of second urea components. In curing of UF resins, a uniform distribution of hydroxymethyl functional groups among urea molecules appears to be needed. The higher formaldehyde emission values of boards bonded with these resins, especially for *Resin 2.40b* (Figure 24.20d) can be explained this way. Therefore, the extent of hydroxymethyl group migration in UF resins could be a useful industrial parameter, which has been vaguely known, up to now, only as ripening times of resins, which have been roughly met in most cases in the industry. Boards bonded with *Resin 1.80* and particularly with *Resins 1.40b* and *1.60b*, performed somewhat poorly. The latter two resins, evaluated only with 10-minutes heat treatments due to their limited storage stability, showed IB values lower than the other resins, although MOE and MOR values were comparable or slightly higher. Past experience indicates that thermosetting resins like UF or phenol-formaldehyde resins cure poorly when some of the resin components have a limited miscibility and remain so in the vitrification stage. These low F/U_1 mole ratio resins appear to have resulted in an incomplete curing and low bond strengths because the temperature of core layers (~127 °C) was too low to unscramble (or melt) the agglomerated

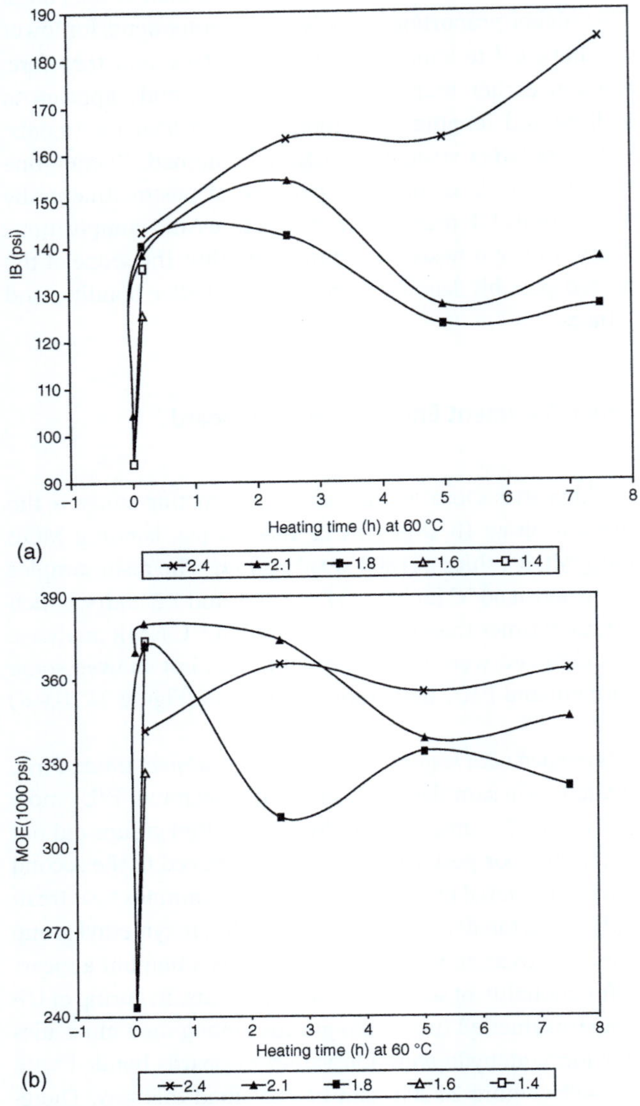

Figure 24.20 (a) IB strength values of particleboards bonded with UF resins 1.4–2.4 heat treated at 60 °C for 0.0, 0.17, 2.5, 5.0, and 7.5 hours. (b) MOE values of particleboards bonded with UF resins 1.4–2.4 heat treated at 60 °C for 0.0, 0.17, 2.5, 5.0, and 7.5 hours. (c) MOE values of particleboards bonded with UF resins 1.4–2.4 heat treated at 60 °C for 0.0, 0.17, 2.5, 5.0, and 7.5 hours. (d) **Top**: PE and **Bottom:** SC formaldehyde values of particleboards bonded with UF resins 1.4–2.4 heat treated at 60 °C for 0.0, 0.17, 2.5, 5.0, and 7.5 hours.

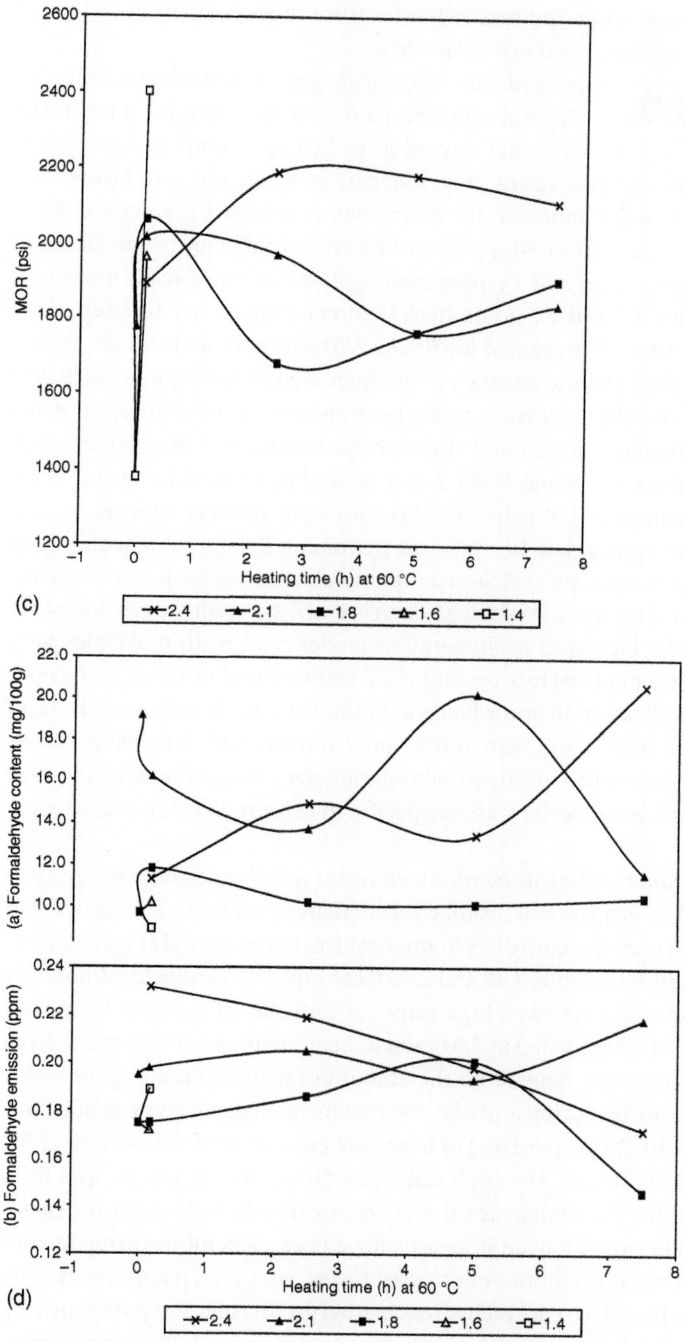

Figure 24.20 *(Continued)*

resin molecules during curing, while the higher temperature in surface layers (~160 °C) adequately assimilated the agglomerated resin fractions.

Physical strength values of boards bonded with resins with long heat treatment times: At the 2.5 hours heat treatment stage, all resins showed increased IB strength values in the order of *Resin 2.40 > 2.10 > 1.80*. Thus, the migration of hydroxymethyl groups in this heat treatment period made all three resins cure faster or better in the core layers. The MOR/MOE values of boards were improved for *Resin 2.40*, remained the same for *Resin 2.10*, but decreased for *Resin 1.80*. *Resin 1.80* appears to be adversely affected by the extended migration of hydroxymethyl groups and by increased agglomeration of resin molecules in the surface layers. At the 5.0 and 7.5 hours heat treatment times, the IB strength of boards increased for *Resin 2.40* but decreased for *Resins 2.10* and *1.80*, with similar trends observed for MOE/MOR values (Figure 24.20a–c). The increased IB strength of *Resin 2.40* is interesting and indicates that the increase in methylene/methylene-ether group contents with the formation of methylene-di-urea and urea-groups bonded to UF polymers during these long heat treatments has made *Resin 2.40* a better binder in core layers. Thus, the optimum resin composition of UF resins for PB core layer binders appears to be a high F/U_1 mole ratio and an appropriate heat/storage treatment. Resins made with higher F/U_1 mole ratios would, in general, provide broader parameter ranges for posttreatments. Boards bonded with *Resins 2.10* and *1.80* showed decreased IB strength values, but since this appears to be due to the increased agglomeration tendency of resin molecules with increased molecular weights, and therefore optimum UF resins would be obtained by optimizing the extent of polymerization in accordance with the F/U_1 mole ratio used in resin synthesis. Resins other than *Resin 2.40* made in this study, if made with lowered polymerization end-points, would have wider posttreatment parameter ranges available and may be made to exhibit optimum bond performances specific to certain board manufacturing conditions.

Formaldehyde emission values of boards bonded with resins with heat treatments: *(a) perforator extraction (PE)* [40] values, or total residual formaldehyde values in boards, were in general lower for *Resin 1.80* over the entire 7.5-hours heat treatment period (Figure 24.20d *Top*). The formaldehyde values of *Resins 2.40* and *2.10* were higher and affected differently by heat treatment times. *Resin 2.10* showed high values at zero and 5.0 hours of heat treatment, and *Resin 2.40* showed a high value at 7.5 hours of heat treatment. PE formaldehyde contents of boards would primarily depend on the amount of hydroxymethyl groups that break off as formaldehyde during curing and also the free formaldehyde content of resins. PE formaldehyde values of 10–20 mg per 100 g of board are equivalent to 0.13–0.26% of the formaldehyde used in making resins. Although some free formaldehyde gas escapes from boards during hot-pressing, the formaldehyde values in resins are relatively small and therefore, the sources of residual formaldehyde in boards have been difficult to correlate with specific polymer structures or compositions of uncured resins. However, it appears that the low values of *Resin 1.80* are traced to reduced hydroxymethyl group contents, partly proving that the amount of hydroxymethyl groups that break off to free formaldehyde during curing is proportional to the residual formaldehyde in boards. The question remains whether the poorer bond performance of low F/U_1 mole ratio resins can be improved without increasing the residual formaldehyde values. Furthermore, the variable PE formaldehyde values observed of *Resins 2.40* and *2.10* series indicate that the posttreatments of resins could play a significant role.

(b) SC formaldehyde emission (ASTM D6007) values measured after airing for 8 days (from the same board pieces used in PE formaldehyde tests) showed varying effects of F/U_1 mole ratios and heat treatment times {Figure 24.20d *(bottom)*}. The SC emission values of boards showed minor changes between 2.5 and 5.0 hours heat-treated resins where the degree of polymerization increased the most. The SC method is the preferred formaldehyde emission measurement method used in industry. The high formaldehyde emission value of *Resin 2.40* at 10 minutes heat treatment is indicative of the high (over-crowded) proportion of hydroxymethyl groups in the high molecular weight components. Heat treatments continuously decreased SC emission values for *Resin 2.40*. On the other hand, *Resin 2.10* showed only minor changes until 5.0-hours treatments but an increase at 7.5-hours treatment opposite to the low PE formaldehyde value. *Resin 1.80* showed slight increases initially but gave a low value at the 7.5-hours treatment, agreeing with the low PE formaldehyde values. Both PE and SC formaldehyde emission results are somewhat different from the results obtained from the 50-d storage experiment reported in Section 24.7.3.2. SC formaldehyde emission values can be considered to be related to the residual formaldehyde content and porosity of boards, and therefore SC values would follow PE values when boards' porosity characteristics are similar. However, the SC and PE formaldehyde values showed several conflicting values, indicating that the porosity characteristics would be somewhat different for those boards. Porosity of boards is related to the density and density profiles of boards. The density and density profiles of all boards showed the usual varying density profiles, higher in surface layers than in core layer, but they are all similar as expected from using the same press-closing speed and same target board density values. In this case, the porosity of boards for formaldehyde diffusion would be differentiated by the quality of adhesive layers in boards. Resins that flow well in curing would result in better bonding and better continuity in adhesive layers and lower SC formaldehyde values. These trends are not well identified except perhaps for *Resin 2.40*, which resulted in board strength values that generally increased while SC formaldehyde values decreased as the heat treatment time was extended. Overall, both PE and SC formaldehyde values obtained were in the normal ranges as laboratory PBs and suggest further studies are needed to lower the formaldehyde emissions.

24.8.9 Particleboards Bonded with Resins of 2.5-hours Heat and Followed Room-temperature Storage Treatments

The boards were made with resins that were heat-treated for 2.5 hours, followed by room-temperature storage treatments for up to 27 days. Most methylene/methylene-ether groups were formed within 7 d of storage time. In overall, boards bonded with *Resin 1.80* performed poorly in IB strength values in comparison with those bonded with *Resins 2.40* and *2.10* for all storage days (Figure 24.21a–c), similarly with the results from the continuous heat treatments discussed above in Section 24.8.8. In comparison with boards made with resins with only 2.5 hours heat treatments, boards bonded with 7-d stored *Resins 2.40* and *2.10* showed only small changes in IB but their MOE and MOR values significantly increased and also showed decreased SC formaldehyde emission values, indicating a higher bond quality and, therefore, an optimum posttreatment parameter. This trend was different than those observed in the continuous heat treatments. These board performance improvements coincided with the formation of more methylene-di-urea and less increases in molecular weights for the polymeric resin components in comparison with changes incurred in continuous heat treatments. Boards bonded with 14-d and 27-d stored *Resins 2.40* and *2.10* showed some increases in IB strength, some decreases in MOE and MOR values, and increases in SC formaldehyde emission values compared with boards made with 7-d stored resins. The increased levels of methylene-di-urea type components in resins appear to have increased the IB values but adversely affected the surface layers, lowered MOE/MOR values, and increased SC formaldehyde emission values. A short heat treatment followed by room-temperature storage appears to have some merit for controlling the resin compositions, especially for core layer binders. A more detailed study is needed.

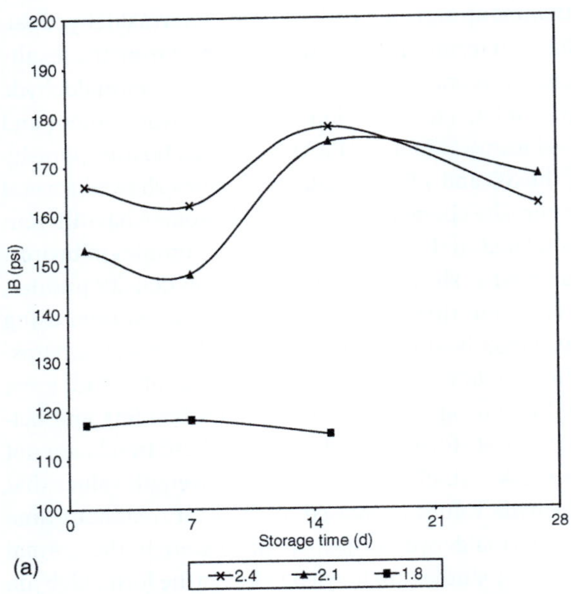

(a)

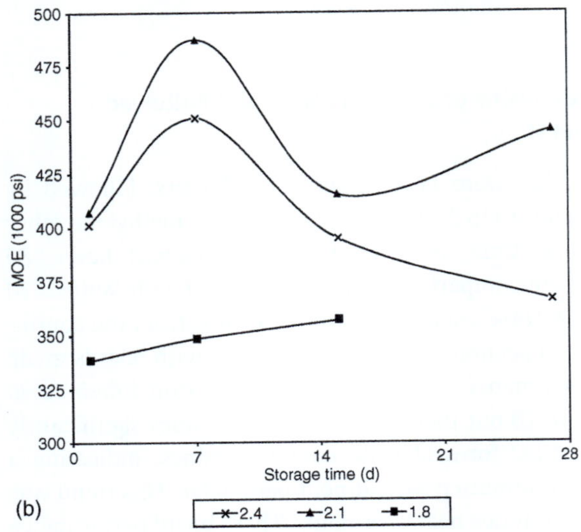

(b)

Figure 24.21 (a) IB strength values of particleboards bonded with UF resins 1.8–2.4 treated at 60 °C for 2.5 hours and stored at room temperature for up to 27 days. (b) MOE values of particleboards bonded with UF resins 1.8–2.4 treated at 60 °C for 2.5 hours and stored at room temperature for up to 27 days. (c) MOR values of particleboards bonded with UF resins 1.8–2.4 treated at 60 °C for 2.5 hours and stored at room temperature for up to 27 days. (d) SC formaldehyde emission (ASTM D 6007) values of particleboards bonded with UF resins 1.8–2.4 treated at 60 °C for 2.5 hours and stored at room temperature for up to 27 days.

Figure 24.21 (*Continued*)

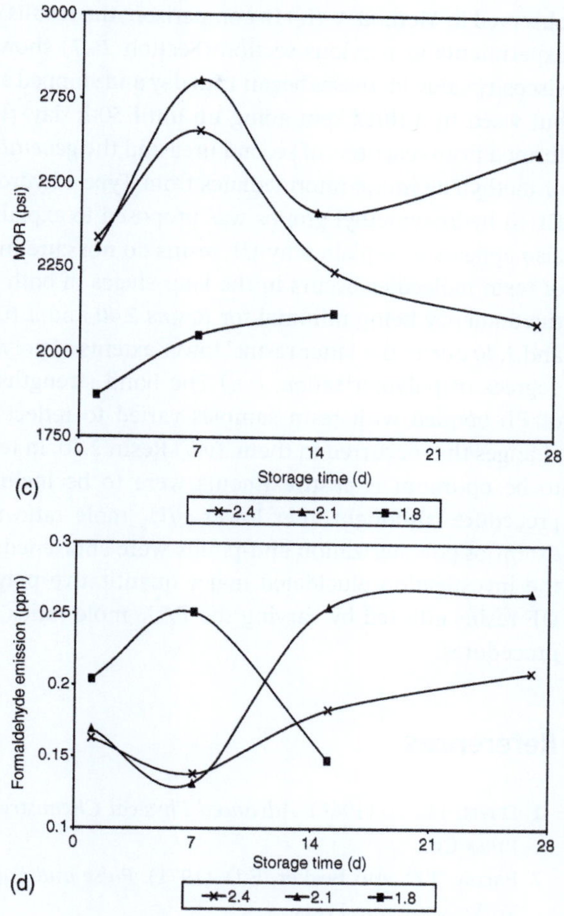

24.8.10 Summary

Typical wood adhesive-type UF resins synthesized with F/U_1 mole ratios of *1.40*, *1.60*, *1.80*, *2.10*, and *2.40*, respectively, with the second urea additions to an overall $F/(U_1 + U_2)$ mole ratio of 1.15 in alkaline pH, were subjected to two different posttreatment procedures: (Method A) Heating/stirring treatments at 60 °C up to ~12.5 hours and, (Method B) Heating/stirring treatments for 2.5 hours at 60 °C followed by room-temperature storage treatments for 27 days. The results are: *(i)* in both methods, in early periods some hydroxymethyl groups migrated from the polymeric resin components to second urea to form mono-hydroxymethylurea, accompanied with decreases in resin viscosity more prominently for higher F/U_1 mole ratio resins; *(ii)* Then, in method A, methylene/methylene-ether (M/ME) group contents increased sharply in the 2.5–5.0 hours heating period but the resin viscosity values kept decreasing for Resin 2.40 and 2.10, while it kept increasing for lower F/U_1 ratio resins leading to earlier gelation; *(iii)* In method B, the M/ME group contents increased sharply by 7th day of storage and kept increasing till 27th day for Resin 2.40 with its extent being the highest among resins exceeding the value observed in Method A. *(iv)* In comparison, the results of straight room-temperature storage experiments in previous section (Section 24.7) showed both M/ME group contents and viscosity value increases began 15th day and stopped at 30th day for M/ME group increases, but viscosity values kept going up until 50th day. *(v)* Mainly Type I M/ME bonds were formed from reactions of second urea and the *general acid catalysis* mechanism of forming of methylene imine intermediates from Type I hydroxymethyl groups and not from Type IIt/IIi hydroxymethyl groups was proposed to explain the M/ME bond formation, which also appears to explain why UF resins do not cure in the alkaline pH. *(vi)* Agglomeration of resin molecules occurs in the later stages of both heating and storage treatments with the tendency being minimal for *Resins 2.40* and *2.10* but significant for *Resins 1.80, 1.60,* and *1.40* due to the latter resins' lower extents of polymer side-chain branching and higher degrees of polymerization. *(vii)* The bond strengths and formaldehyde emission values of PB bonded with resin samples varied to reflect the resin composition and physical changes that occurred in them. *(viii)* Resin 2.40, in terms of IB strength of boards, appears to be optimum if posttreatments were to be included as part of resin manufacturing procedure, although other lower F/U_1 mole ratio resins would behave similarly if the extent of polymerization end-points were shortened in resin syntheses. *(ix)* Conclusively, the investigation elucidated major quantitative polymer structures and compositions of UF resins effected by varying the F/U_1 mole ratios in resin synthesis and posttreatment procedures.

References

1 Davis, J.C. Jr. (1965). *Advanced Physical Chemistry*, 479–513. New York: The Ronald Press Co.
2 Farrar, T.C. and Becker, E.D. (1971). *Pulse and Fourier Transform NMR*, 6–12. New York: Academic Press.
3 Levy, G.C. and Nelson, G.L. (1972). *C-13 NMR for Organic Chemists.* New York: Wiley Interscience.
4 Valdez, D. (1995). *Wood Adhesive 1995* (ed. A. Christiansen and A. Corner), 193. FS: FPL.
5 Ebdon, J.R. and Heaton, P.E. (1977). *Polymer* 18: 971.
6 Tomita, B. and Hatono, S. (1978). *Journal of Polymer Science Part A: Polymer Chemistry* 16: 2509.
7 de Breet, A.J.J., Dankelman, W., Huysman, W.G.B., and de Wit, B. (1977). *Die Angewandte makromolekulare Chemie* 62: 7.
8 Slonim, I.Y. et al. (1979). *Journal of Polymer Science* 20: 2569.
9 Kim, M.G. (1999). *Journal of Polymer Science Part A: Polymer Chemistry* 37: 995.
10 Ramon, R.M., Johns, W.E., Magnuson, J., and Dunker, A.K. (1986). *Journal of Adhesion* 19: 115.
11 Kim, M.G. and Amos, L.W. (1990). *Industrial & Engineering Chemistry Research* 29: 208.
12 Kim, M.G. (2000). *Journal of Applied Polymer Science* 75: 1243.
13 Kim, M.G. (2001). *Journal of Applied Polymer Science* 80: 2800.
14 Kim, M.G., Wan, H., No, B.Y., and Nieh, W.L.-S. (2001). *Journal of Applied Polymer Science* 82: 1155.

15 Kim, M.G., No, B.Y., Lee, S.M., and Nieh, W.L. (2003). *Journal of Applied Polymer Science* 89: 1896.

16 Graves, G. (1993). *Proceedings of the 27th International PB and Composite Materials Symposium*, 221. Pullman, WA: WSU.

17 de Jong, J.I. and de Jonge, J.J. (1952). *Recueil des Travaux Chimiques des Pays-Bas* 1952 (71): 643.

18 de Jong, J.I. and de Jonge, J.J. (1952). *Recueil des Travaux Chimiques des Pays-Bas* 71: 661.

19 de Jong, J.I. and de Jonge, J.J. (1952). *Recueil des Travaux Chimiques des Pays-Bas* 71: 891.

20 de Jong, J.I. and de Jonge, J.J. (1953). *Recueil des Travaux Chimiques des Pays-Bas* 72: 139.

21 de Jong, J.I. and de Jonge, J.J. (1953). *Recueil des Travaux Chimiques des Pays-Bas* 72: 1027.

22 Levy, G.C. and Craik, D.J. (1984). *Topics in Carbon-13 NMR Spectroscopy*, vol. 4 (ed. G.C. Levy). New York: Wiley Interscience.

23 Ford, W.T. and Balakrishnan, T. (1983). *Polymer Characterization* (ed. C.P. Craver), 477. Washington, DC: ACS.

24 Williams, J. H. (1982). *Presentation, Forest Products Soc. 36th Annual Meeting (June).*

25 2-Hour Desiccator Method, FTM 1–1983, CPA, Gaithersburg, MD.

26 Wirpsza, Z. and Brezezinski, J. (1973). *Aminoplasty*, 96. Moscow: USSR.

27 Anilin, B. and Soda-Fabrik, A.G. (1972). German Patent No. 2,207, 921.

28 Gu, J., Higuchi, M., Morita, M., and Hse, C.Y. (1995). *Mokuzai Gakkaishi* 41: 1115.

29 Wegler, R. (ed. E. Mueller) *Methoden der Organischen Chemie*, 4the, vol. 14½, 332. Stuttgart: Georg Thieme Verlag.

30 Crowe, G.A. and Lynch, C.C. (1948). *Journal of the American Chemical Society* 70: 3795.

31 Crowe, G.A. and Lynch, C.C. (1949). *Journal of the American Chemical Society* 71: 3731.

32 Smythe, L.E. (1951). *Journal of the American Chemical Society* 1951 (73): 2735.

33 Soulard, C., Kamoun, C., and Pizzi, A. (1999). *Journal of Applied Polymer Science* 72: 277.

34 Rubinstein, M., Colby, R.H., and Gilmor, J.R. (1989). *Space-Time Organizations in Macro Molecular Fluids* (ed. F. Tanaka, M. Doi, and T. Ohta). Berlin: Springer-Verlag.

35 Cates, M.E., McLeish, T.C., and Rubinstein, M. (1990). *Journal of Physics: Condensed Matter* 2: 749.

36 Baxandal, L.G. and Edwards, S.F. (1988). *Macromolecules* 21: 1763.

37 Tanaka, F. and Edwards, S.F. (1992). *Macromolecules* 25: 1516.

38 March, J. (1968). *Advanced Organic Chemistry*, 226. New York: McGraw-Hill.

39 American National Standards, ANSI/A208.1–1989. CPA, Gaithersburg, MD.

40 DIN EN 120–92 (1992). Deutsches Institute fuer Normung: Berlin, Germany.

25

Advanced Reading Materials on UMF Resins Modified with 6–12% Melamine

25.1 Introduction and Synthesis and Analysis Results of UF and UMF Resins

(Adapted with permission of the publisher and minor revisions from No, B. Y. and M. G. Kim, 2004. (2004). Syntheses and properties of low-level melamine-modified UMF resins. J. of Applied Polym. Sci. 93, 2559–2569.)

25.1.1 Background and Known Resin Synthesis Methods of UMF Resins

Lowering the F/U mole ratio of wood adhesive urea-formaldehyde (UF) resins decreases the formaldehyde emission values of wood composite boards, but it also decreases the bond strength and water resistance properties due to lowered functionality of urea components, limiting the extent of crosslinking of cured UF polymers. Melamine, with a functionality of 3 or higher [1], has been considered as a ***low-level*** **copolymer component** of particleboard binder-type UF resins. With their triple functionality and reaction chemistry that is similar to that of urea, the **urea-melamine-formaldehyde (UMF**) resins have been claimed to enhance the water resistance and physical strength properties and reduce the formaldehyde emission levels of wood composite boards [2–8].

Reactions of melamine with formaldehyde (MF), often carried out at pH between 6 and 9, initially result in a mixture of various hydroxymethylmelamines, dimers, trimers, etc., and most dimers and trimers tend to precipitate out easily from the aqueous phase (Figure 25.1) and therefore, the common melamine-urea-formaldehyde (**MUF) resins** made with high melamine levels are often spray-dried soon after resin manufacturing. The melamine-formaldehyde (M-F) reactions are relatively fast at pH 6 and somewhat stable, clear liquid MF resin products are obtained at pH 8.3–8.6 [1] because the formation rate of dimers and trimers is slow at this mild alkaline pH range. In **UMF** resins, the melamine level is low anywhere from about 2% to 10% or slightly higher but it is limited, first of all, due to its high cost. The first reported method of synthesis of UMF resins is adding small amounts of melamine (M) with the first urea (U_1) to the formaldehyde (F_1) solution at pH 8–9 and reacting while the reaction mixture's pH is allowed to drift down to about 6.2, where the reaction rate (viscosity increase) speeds up [2, 4]. With this method, it is not certain whether the reaction is ended at the full extent of polymerization for the major UF components, since UF components would advance only minimally at this pH range while MF components are most likely to advance readily to bring the resin system to the

Adhesives and Finishes for Wood: For Practitioners and Students, First Edition. Moon G. Kim.

target viscosity. In another method, the UF component is first advanced to the high extents commonly used in synthesis of UF resins, and then the melamine is added with or without some more formaldehyde (F_2) and reacted further, followed by the addition of the second urea [2, 8]. In this latter method, MF components would advance minimally. In both procedures, the initial $(F_1 + F_2)/(U_1 + M)$ mole ratio value is normally kept at 2–3 during the polymerization reaction, followed by carrying out the reaction until a predetermined viscosity or water dilutability is obtained. The water dilutability is determined mostly by the MF components. The second urea is then added to lower the $(F_1 + F_2)/(U_1 + M + U_2)$ mole ratio to 1.0–1.3.

25.1.2 Reaction Chemistry of MF and UMF Resins

Reactions of Formaldehyde with Melamine: Similarly as with urea, melamine reacts to form initially hydroxymethyl-melamines and their dimers, trimers, etc. by forming methylene or methylene-ether bonds as shown in Figure 25.1. For MUF or UMF resins, some melamine-urea copolymers have been assumed to be formed through the methylene or methylene-ether bonds as shown in Figure 25.2, but it has been generally difficult to analyze MUF resins because of poor solubility. In past research, a reaction product of urea, melamine, and formaldehyde at pH 6.5 was shown to have methylene groups formed between urea and melamine by ^{13}C-NMR analysis, a result of copolymerization [9], and more recently solid-state NMR analysis of cured MUF resins revealed the formation of methylene bonds appearing at 45.5–46.5 ppm and methylene-ether bonds appearing at 67.5–68.5 ppm [10].

On the other hand, UMF resins synthesized with an F/(U + M) mole ratio of 3.0 and at 4–8% melamine levels in the polymerization step using the first procedure discussed above (Section 25.1.1) showed turbidity in 30 minutes of cooking at pH 6.5 [11]. The resultant UMF resins showed lower formaldehyde emission values than control UF resins, but the internal bond (IB) strength values of boards remained about the same, not much improved. UMF resins synthesized using the second procedure discussed above (Section 25.1.1), resulted in lower formaldehyde emission values of boards, but the IB and water soak thickness (TS) values sometimes deteriorated as the melamine level was increased from 9.5% to 34.4% [12]. Thus, both resin synthesis and curing procedures appear to be less than

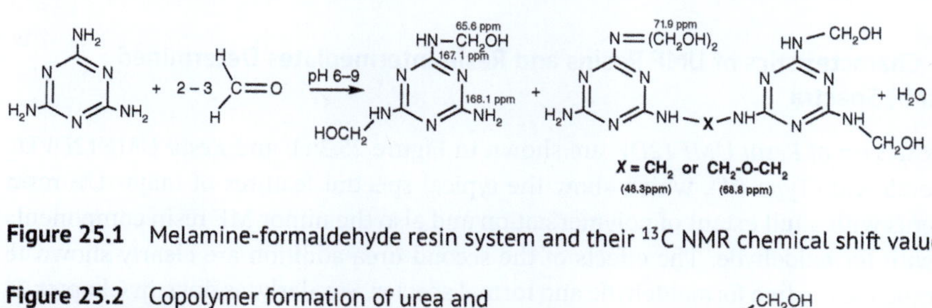

Figure 25.1 Melamine-formaldehyde resin system and their ^{13}C NMR chemical shift values.

Figure 25.2 Copolymer formation of urea and melamine in UMF resins.

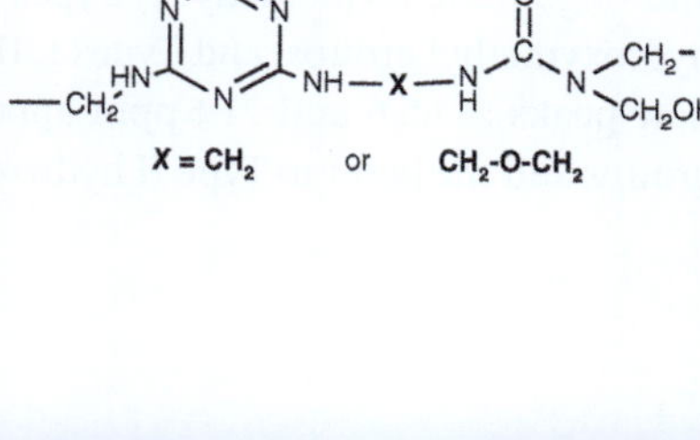

satisfactory in showing the advantage of incorporating low levels of melamine in UMF resins, apparently because appropriate experimental procedures were not realized with respect to the degrees of advancement of the UF and MF resin components, respectively, which were explored in this Chapter.

Stronger Acid Catalysts are Needed for UMF Resins Because of the Melamine: It is also known that the curing of UMF resins requires a stronger catalyst and higher press temperatures or longer press times compared to UF resins because of the lower reactivity of hydroxymethyl groups on the melamine molecule [1]. Another possibility for varied board performances of boards bonded with UMF resins could be the different extents of migration of hydroxymethyl groups of the polymeric UMF resin components to the second urea during the cooling and storage periods, as discussed in Chapter 24.

25.1.3 Resin Syntheses of UF and UMF Resins

The bonding and formaldehyde emission improvements that parallel the melamine levels used in UMF resins (higher costs) appear to be realizable by optimizing the extents of polymerization for both UF and MF components in resins as well as by understanding the optimum curing parameters of resins. For UMF resins, it appears important to attain an appropriately high extent of polymerization for the major UF resin components since, under the prevailing condition of pH 6–9, the minor MF components tend to advance faster to bring the resin viscosity to the synthesis endpoint before UF components advance to an adequate extent, resulting in unbalanced degrees of polymerization. To uncover the effects of this possible mismatch of polymerization, three resin synthesis methods were devised where the melamine was added to UF resin components in the beginning (Type #1), or in the middle viscosity point of D or K (Type #3a and Type #3b) of UF resin synthesis, or in the high viscosity point of V (Type #4), as shown in Table 25.1 for 6% and 12% melamine addition levels. Melamine level was also varied over 2–12% based on the liquid resin weight (or 4.0–25.0% of urea by weight), and the F/(U + M) mole ratio was kept at 2.10 during the polymerization step and at 1.15 after the second urea addition. Several other resin intermediates were also synthesized, as shown in Table 25.2. Synthesized resin intermediates and finished resins were analyzed by ^{13}C NMR and examined for storage stability and other resin characteristics using the general laboratory testing methods.

25.1.4 Characteristics of UMF Resins and Resin Intermediates Determined by ^{13}C NMR Spectra

^{13}C NMR spectra of Resin UMF12DV are shown in Figure 25.3a,b and *Resin UMF12DVU$_2$* in Figure 25.3c,d (Type #4), which show the typical spectral features of major UF resin components with a full extent of polymerization and also the minor MF resin components reacted with formaldehyde. The effects of the second urea addition are clearly shown to mob up much of the free formaldehyde and formal species. Similarly, as done in Chapter 24 for UF resins the methyleneic groups are classified as shown in Table 25.3 (page 425) as free and oligomeric formaldehyde; Types I, II, and III methylene-ether groups; Types I and II hydroxymethyl groups; and Types I, II, and III methylene groups. For the MF components, new peaks at 65.6 and 71.6 ppm appeared, the former assigned to Type I hydroxymethyl groups and the latter to Type II hydroxymethyl groups bonded to melamine molecules from

Table 25.1 Syntheses of UF and UMF resins.

Resins	First step, mild alkaline pH	Second step	Third step
UFA (Control) *UFA$_1$* ($F/U_1 = 2.10$)	*Control UF*: 1697 g 50% form. pH 7–8, 70 °C, added 807 g urea ($F/U_1 = 2.10$), 90 °C, 30 min	pH 4.6, 120m min at 90 °C, viscosity X, pH 8.0. Sample *UFA$_1$* taken at 30 min of reaction time	Cool to 60 °C, and add 668 g urea ($F/(U_1 + U_2) = 1.15$ & also 1.05 and 1.25
UMF12 A$_2$VU$_2$ ***(Type #1)***	1974 g 50% form, pH 7.0–8.0, 80 °C, Added 700 g urea and 484 g melamine {$F/(U_1 + M) = 2.10$); 85 °C, 30 min	Adjusted to pH 6.7, vis. A_2; 65 min reaction at 85 °C, *pH 6.3*, vis. V	Adjusted to pH 8.3. *Sample UMF12A$_2$V*, cool to 60 °C and add 775 g urea $F/(U_1 + M + U_2) = 1.15$
UMF12VVU$_2$ ***(Type #4)***	*UF base*: 1450 g 50% form. pH 7–8, 70 °C, added 690 g urea ($F/U_1 = 2.10$), 90 °C, 30 min. (***Step 1***)	Adjusted to pH 4.6, 90 °C, advanced about 120 min to vis. V; (***Step 2***)	
	Adjusted to pH 8.0, added 524 g 50% form, pH 8.0, and 484 g melamine, $\{(F_1 + F_2)/(U_1 + M) = 2.10$, 85 °C, 20 min, vis. J. (***Step 3***)	Adjusted to pH 6.7, 20 min at 85 °C, vis. V. (***Step 4***)	Adjusted to pH 8.0, cooled, and mixed with 795 g urea $\{(F_1 + F_2)/(U_1 + M + U_2) = 1.15$
UMF12DVU$_2$ **(Type #3a)**	*UF base*: 1365 g 50% form. pH 8.0, 70 °C, added 649 g urea ($F/U_1 = 2.10$), 90 °C, 30 min. (***Step 1***)	Adjusted to pH 4.6, 90 °C, advanced to vis. D (#3a) or K (#3b). (***Step 2***)	
UMF12KVU$_2$ **(Type #3b)**	Adjusted to pH 8.0, added 540 g 50% form. and 458 g melamine, $\{(F_1 + F_2)/(U_1 + M) = 2.15$, 85 °C, 20 min, vis. J. (***Step 3***)	Adj. to pH 6.7, 20 min at 85 °C, vis. V. (***Step 4***)	Adj. to pH 8.0, cooled, and mixed with 795 g + urea. (***Step 5***) $\{(F_1 + F_2)/(U_1 + M + U_2) = 1.25$ $\{(F_1 + F_2)/(U_1 + M + U_2) = 1.15$ $\{(F_1 + F_2)/(U_1 + M + U_2) = 1.05$
UMFXXDVU$_2$	Same procedure of UMF12DVU$_2$, but melamine levels (XX) were varied to 2%, 4%, 6%, 8%, and 10%		
UMF6DVU2 1.05 and 1.25	Same procedure of UMF12DVU$_2$, but the melamine level at 6% based on total resin weight and $\{(F_1 + F_2)/(U_1 + M + U_2) = 1.05$ &1.25		
UMF6KVU$_2$ 1.05 and 1.25	Same procedure of UMF12KVU$_2$, but the melamine level at 6% based on total resin weight and $\{(F_1 + F_2)/(U_1 + M + U_2) = 1.05$ &1.25		

Table 25.2 Syntheses of UF and UMF resin intermediates.

Resins	First step	Second step	Third step
UFD	*UF cook at pH 6.3*: 126 g 50% form. pH 8.0; 70 °C, added 60.0 g of urea ($F/U_1 = 2.10$), 90 °C, 30 min; vis. A_2	Adjusted to pH 6.3, reacted 6 h at 90 °C, vis. B	Adjusted to pH 8.0. Sample UFD obtained $F/U_1 = 2.10$
UFA_2BU_2			Added 49.6 g urea (U_2) to UFD (above) & cooled down ($F/U_1 + U_2$) = 1.15
$MFaU_2$	*MF intermediate*: 150 g 50% form. and 20 g water, pH 7–8 and 80 °C; 150 g melamine and 80–90 °C (F/M = 2.10)	~70 min at 90 °C and pH 8.3–8.6, dilutability 1 : 2 resin to water	Cooled to RT, vis. O. (Sample gelled in 2 days); Added 59.0 g U_2 F/(M + U) = 1.15
UMF12DN	Sample taken in synthesis of $UMF12DVU_2$ at viscosity N		
UMF12DV	Sample taken in synthesis of $UMF12DVU_2$ at viscosity V before adding U_2		

similarities with the chemical shift values of UF resin components at 65.2 and 71.9 ppm, respectively. They were close together and calculated together in this study. The methylene (48.3 ppm) and methylene-ether (68.8 ppm) groups on MF components appeared as broad peaks and were not differentiated from the broad Type I methylene (47.4 ppm) and methylene-ether groups (69.5 ppm) of UF components. It was difficult to conclude whether any copolymerization had occurred between urea and melamine units. Further, melamine ring carbons occurred as broad peaks at 167.7, 167.0, and 166.7 ppm, assigned to melamine ring carbons with amino groups having none, one-, and two-substitution groups [13], respectively, away from the urea carbonyl carbons occurring at about 164–158 ppm [14]. Integration values of melamine carbons can be compared with urea carbonyl groups to calculate M/U charged mole ratios. Further details are discussed below.

UF Resin Intermediate (UFD) and Resin UFA_2BU_2 Synthesized at pH 6.3: The UF resin intermediate made by polymerizing UF resin components at pH ~ 6.3 and 90 °C for 6 hours with an F/U_1 mole ratio of 2.1 in order to investigate the extent of polymerization of UF components in Type #1 UMF resin synthesis method became cloudy in the first hour of reaction and then reached to only B viscosity at the end of reaction time (*Resin intermediate UFD*), indicating that UF resin components do advance but not much in viscosity at a mild acidic pH 6.3. After the addition of the second urea [$F/(U_1 + U_2) = 1.15$], the viscosity decreased to A_1 (*Resin UFA_2BU_2*) and its ^{13}C NMR results (Table 25.3) showed a (methylene + 1/2 methylene-ether) bond content of 0.722, corresponding to a low degree of polymerization of 3.6 for the *polymeric UF resin* components. Also, the methylene-ether: methylene group ratio was about *40 : 60*, a higher methylene-ether group content compared to the ratio of *23 : 77* found for a common Resin UFA_1 synthesized at pH 7.8–4.6. These results indicate that the polymerization reaction of UF components occurs at pH 6.3 but it is very slow, and more methylene-ether groups are formed than at pH of 4–5

Table 25.3 Percentage values of various methyleneic and carbonyl carbons of urea portions of UF and UMF resins determined by ^{13}C NMR spectroscopic method.

Carbon groups	(ppm)	Synthesized UF and UMF resin samples								
		UMF12A_2V	UMF12A_2VU2	UMF12DN	UMF12DV	UMF12 DVU_2	UMF12 VVU_2	UFD	UFA_2BU2	MFaU_2
Free formaldehyde and formal	91.0	0.59	0.41	0.61	0.42	0.34	0.33	0.56	0.65	0
	87.0	1.04	0	1.56	1.72	0.15	0.10	2.72	0	0.33
	83.1	1.56	0	2.74	3.52	0.17	0.19	2.98	0	0.25
	Total	*3.19*	*0.41*	*4.91*	*5.66*	*0.66*	*0.62*	*6.26*	*0.65*	*0.58*
Types III, II, and I methylene-ethers	79.1	0.93	0	1.56	1.99	1.24	1.85	1.02	2.64	0
	74.8*/75.1	4.26	3.93	7.03	7.63	5.66	4.43	5.04	6.62	3.43
	69.5/68.8[a]	19.15	20.50	12.14	12.10	11.85	9.76	8.39	18.20	16.14
	Total	*24.34*	*24.43*	*20.73*	*21.72*	*18.25*	*16.04*	*14.45*	*27.46*	*19.57*
Types IIt/IIi and I hydroxymethyls	71.9/71.6[a]	16.60	12.53	26.40	24.69	19.83	16.18	14.20	16.03	10.05
	65.2/65.6[a]	44.75	51.62	22.90	22.48	34.87	38.90	13.13	35.10	64.78
	Total	*61.35*	*64.15*	*49.30*	*47.17*	*54.70*	*55.08*	*27.33*	*51.13*	*74.83*
Types III, II, and I methylenes	60.1	0	0	2.19	2.84	1.82	0.94	3.13	2.03	0
	55.0*/53.9	7.17	6.66	16.02	15.69	15.79	17.16	14.64	10.48	1.87
	47.4/ 48.3[a]	3.96	4.35	6.86	6.87	8.29	10.16	5.83	8.25	5.02
	Total	*11.13*	*11.01*	*25.07*	*25.40*	*25.90*	*28.26*	*23.60*	*20.76*	*6.89*
Urea CO	164.0	2.86	38.80	0	0	34.74	34.59	0	38.11	46.84
Mono-sub. CO	162.2	30.02	30.21	9.67	11.03	21.36	23.06	6.07	19.84	44.59
Di-/tri-sub CO	160.7	65.68	30.75	87.72	84.81	42.06	39.64	91.85	39.21	8.57
Tetra-sub. CO	158.0	1.45	0.24	2.60	4.16	1.84	2.70	2.08	2.84	0

Sample numbers were defined in the text, and the chemical shift values are related to the chemical structures described in previous papers.

a) Chemical shift values are assigned to similar groups bonded to melamine. The amount of formaldehyde in Resin UFD is smaller than that in others and its methyleneic group percentage values were proportioned to make them comparable. For Resin UMF12 series, to convert the values into moles based on the first urea, multiply by 2.935. Chemical shift values of methyleneic groups bonded to melamine are marked with asterisks. Melamine ring carbon peaks appeared at 167.1–168.1 ppm and are not included in this table.

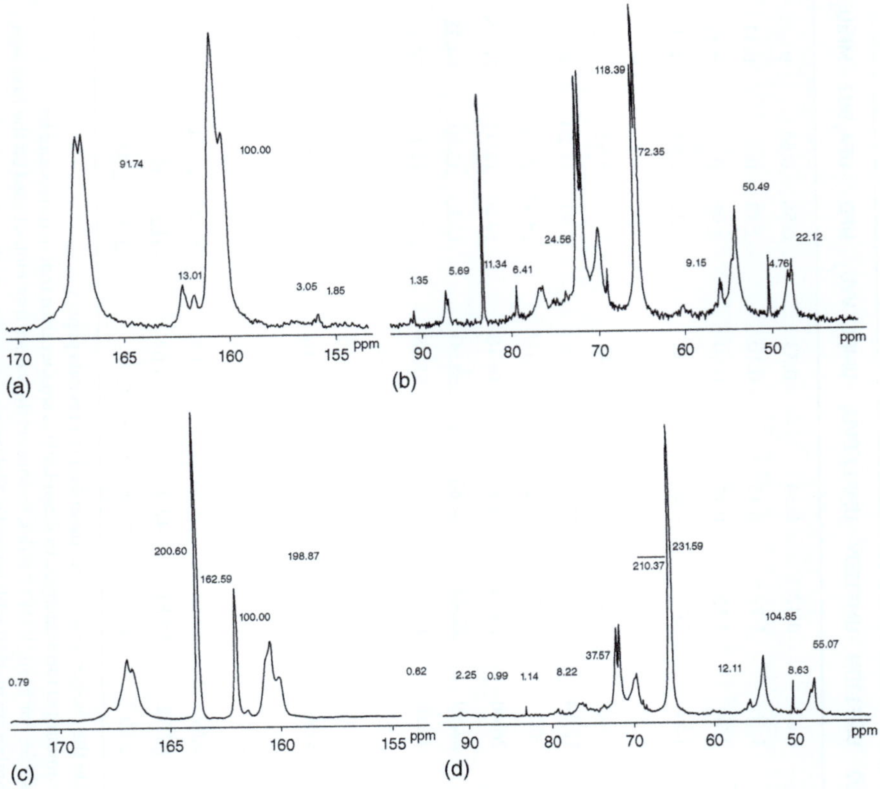

Figure 25.3 (a) ^{13}C NMR spectrum of *Resin UMF12DV* – Urea carbonyl (l, m, n) and melamine (p′, q′) ring region. (b) ^{13}C NMR spectrum of *Resin UMF12DV* – Aliphatic carbon region. (c) ^{13}C NMR spectrum of *Resin UMF12DVU$_2$* – Urea carbonyl (l, m, n) and melamine (p′, q′) ring region. (d) ^{13}C NMR spectrum of *Resin UMF12DVU$_2$* – Aliphatic carbon region.

used in typical UF resin synthesis. Two different reaction mechanisms for methylene and methylene-ether bonds appear to be effective for UF resins in these pH regions (Eqs. 25.5 and 25.6) discussed in the next section).

$$\text{Mel-NH-CH}_2\text{-OH} + \text{H-X} \longrightarrow \text{Mel-NH-CH}_2\text{-OH---H-X} \quad (25.1)$$

$$\text{Mel-N(H}^{\oplus}\text{)-CH}_2\text{-OH---H-X} \xrightarrow{\text{OH}^{\ominus}} \text{Mel-}\overset{\ominus}{\text{N}}\text{-CH}_2\text{-OH---H-X} + \text{H}_2\text{O} \quad (25.2)$$

$$\text{Mel-}\overset{\ominus}{\text{N}}\text{-CH}_2\text{-OH---H-X} \longrightarrow \left[\text{Mel-}\overset{\ominus}{\text{N}}\text{-CH}_2^{\oplus}\right] + \text{H}_2\text{O} + \text{X}^- \quad (25.3)$$

$$\longrightarrow \left[\underset{\text{(Ionic Form)}}{\text{Mel-}\overset{\ominus}{\text{N}}\text{-CH}_2^{\oplus}} \rightleftharpoons \underset{\text{(Imine form)}}{\text{Mel-N=CH}_2}\right] \quad (25.4)$$

$$\text{(Imine form)} + \text{Mel-NH-CH}_2\text{-OH} \longrightarrow \underset{\text{(Metheylene ether bond)}}{\text{Mel-NH-}\mathbf{CH_2\text{-}O\text{-}CH_2}\text{-NH-Mel}} \quad (25.5)$$

$$\text{(Ionic form)} + \text{Mel-NH}_2 \longrightarrow \text{Mel-NH-}\mathbf{CH_2}\text{-NH-Mel (Methylene bond)} \quad (25.6)$$

Melamine-formaldehyde Reaction Products (Resin MFaU$_2$): MF resins synthesized with an F/M ratio of 2.50 at pH 6.3 quickly became cloudy and partly insoluble. On the other hand, in the synthesis of *Resin MFaU$_2$* at F/M mole ration of 2.10 and pH 7–8 for 70 minutes at 90 °C, the initial reaction mixture remained clear and its water dilutability decreased slowly to 1 : 2 (resin:water) at the end of reaction, at which point the sample was taken as *Resin MFa* (F/M = 2.10) with O viscosity, which gelled on storage at 4 °C overnight indicating an appreciable reactivity of melamine hydroxymethyl groups at pH 8.4, but *Resin MFaU$_2$*, obtained by adding second urea to [F/(M + U) = 1.15], showed a G viscosity and remained clear at room temperature for at least 10 days, with the ending viscosity increasing to P. ^{13}C NMR spectra of Resin MFaU$_2$ (Table 25.3, *Figure 25.4a,b*) indicated an overall (methylene + 1/2 methylene–ether) bond content of 0.415, mostly of Type I bonds, per mole of melamine. Because little polymerization would have occurred for the second urea component in this synthesis reaction sequence at alkaline pH, this bond content value would be mostly of the melamine components, corresponding to a low extent of polymerization of about 1.71. Furthermore, the methylene–ether: methylene group ratio of *Resin MFU$_2$* was 74 : 26, a high methylene-ether group content than those of UF resins (40 : 60) synthesized

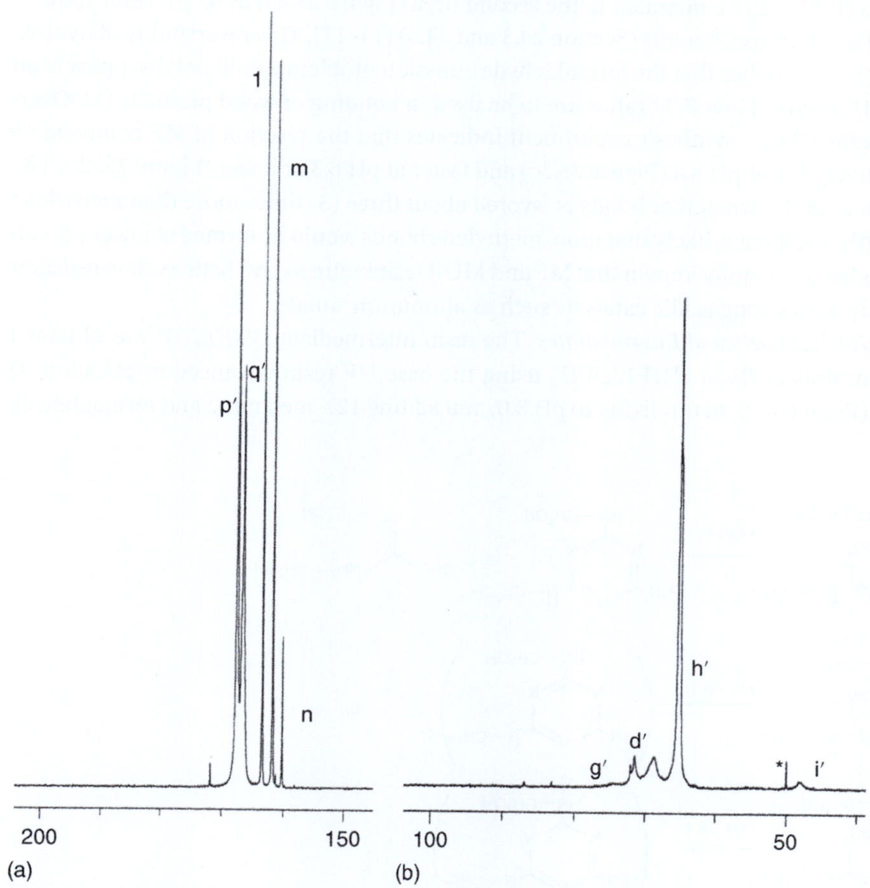

Figure 25.4 (a) ^{13}C NMR spectrum of *Resin MFaU$_2$* – Urea carbonyl (l, m, n) and melamine carbon (p′, q′) region. (b) ^{13}C NMR spectrum of *Resin MFaU$_2$* – Aliphatic region. * Methanol peak.

at pH ~6.3 or those of UF resins (22 : 77) synthesized at pH 4.5, a characteristic of MF or MUF resins generally known in industry [1]. The ^{13}C NMR results further indicated that the second urea added had reacted to form mono- and di-hydroxymethylureas, consuming about 0.64 out of 2.10 moles of formaldehyde used in resin synthesis, leaving the balance 1.46 of formaldehyde per mole of melamine to be bonded. It also appears that MF resin dimers, etc. are formed at pH ~6.3 by the general acid-catalysis (Eqs. 25.1–25.3) through the formation of methylene-imine intermediates (Eq. 25.4) as similarly discussed in the previous Chapter (Section 24.8.5) on the polymerization of UF resin components in alkaline pH. Eqs. (25.2/3/4) would be more favorable for the melamine (pKb = 9.0) system than the urea (pKb = 13.9) system, and the resultant methylene-imine intermediate preferably reacts with a hydroxymethyl groups to result in formation of methylene-ether bonds (Eq. 25.5), the major pathway, while the ionic intermediate (Mel-NH-CH_2^+) preferably reacts with amino groups to result in formation of methylene bonds (Eq. 25.7), the minor pathway.

Reverse Hydroxymethylation Reactions of Melamine: Although a direct comparison is not possible because of the premature gelation of Resin MFa, the ^{13}C NMR results of Resin MFaU$_2$, thus indicate that some of the hydroxymethyl groups, initially bonded to melamine molecules (F/M = 2.10), migrated to the second urea (Figure 25.5a) as in UF resin systems discussed in previous Chapters (Section 24.3 and 24.4) [14–17]. This reversibility of hydroxymethyl groups implies that the formaldehyde emission problem would not disappear even if 100% MF resins of low F/M ratios are to be used in bonding of wood products [1]. Overall, the *Resin MFaU$_2$* synthesis experiment indicates that the reaction of MF components goes relatively fast at pH 8.4 (Figure 25.5c) and faster at pH 6.3 or lower (Figure 25.5b). The formation of methylene-ether bonds is favored about three (3) times more than methylene bonds at pH 8.4, but it is likely that more methylene bonds would be formed at lower pH values. It has been generally known that MF and MUF resins cure to give better water-resistant bonds only with strong acidic catalysts such as aluminum sulfate.

Resin UMF12D Series of Intermediates: The resin intermediate *UMF12DV* was obtained in the synthesis of *Resin UMF12DVU$_2$* using the base UF resin advanced at pH 4.6 to D viscosity (*Resin UFD*), neutralizing to pH 8.0, and adding 12% melamine and formaldehyde

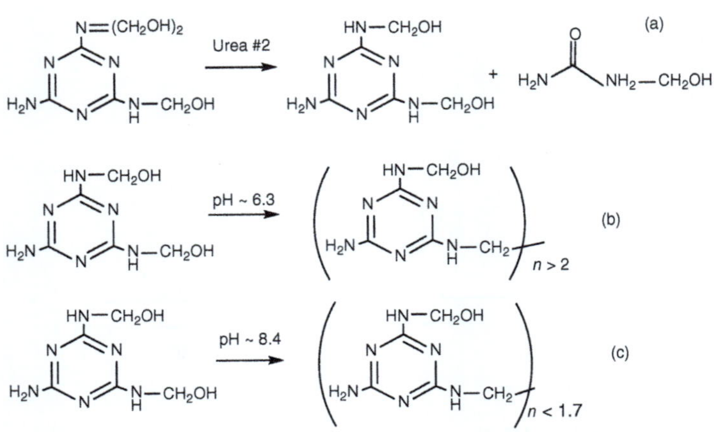

Figure 25.5 (a)–(c) Reaction schemes of melamine and formaldehyde resins.

to $(F_1+F_2)/(U_1+M)=2.15$, and reacting to V viscosity at pH 6.7. The ^{13}C NMR spectra of the resin are shown in Figure 25.3a, b, and addition of the second urea resulted in the final resin sample (Figure 25.3c, d). ^{13}C NMR data of these resin samples in Table 25.3 indicate the various reaction characteristics of UF and MF resin components, summarized in the following: (i) The methylene + methylene-ether group content, adjusted for the second formaldehyde added with the melamine, gradually increased to: 0.146 groups per mole of first urea for Resin UFD and to 0.774 per mole of first urea for Resin *UMF12DVU*$_2$. Because the polymerization of UF resin components would have progressed little after the melamine addition, this increase of methylene + methylene-ether group contents in comparison with Resin UFD would be mostly coming from the MF components, which translate into an extent of polymerization 1.58 for MF components. Further, the methylene-ether groups increased more (7.27%) than the methylene groups (1.80%) as expected from the pH value used; (ii) That the free formaldehyde level was relatively high until the second urea addition (Table 25.3) indicates that the formaldehyde mole ratio used, $(F_1+F_2)/(U_1+M)=2.15$, was in excess for the MF resin components, not fully reacting in this polymerization reaction period; (iii) Type I hydroxymethyl groups (~65 ppm) increased from 13 to 23% when the second formaldehyde and melamine were added, and it then further increased to 35% with the addition of second urea $[(F_1+F_2)/(U_1+M+U_2)=1.15]$ similarly as observed with Resin $MFaU_2$ discussed above (Section 25.1.4). This migration of hydroxymethyl groups also resulted in decreasing Type II and Type III methylene and methylene–ether group contents similarly as in UF resins [14–17]; (iv) In ^{13}C NMR spectra of Figure 25.3a,c, the melamine ring carbons appeared as a broad peak at 168.1–167.1 ppm, composed of three peaks, the first weak peak (highest ppm value) being assigned to the melamine ring carbons having no substitution and last two peaks assigned to the ring carbons bonded to amino groups having mono- and di-substitutions.

25.1.5 General Resin Characteristics of Synthesized UMF Resins

Resin UMF12A$_2$VU$_2$ Made Without Pre-advancement of UF Base Resin: With this Type #1 UMF resin synthesis procedure, often used in early research [2–8, 18], the reaction was started out with a 12% melamine, first urea, and formaldehyde at an $F/(U_1+M)$ mole ratio of 2.10 at *pH 6.3*. The reaction mixture reached to the target viscosity VW in about 60 minutes. ^{13}C NMR results of the final resin, the sample taken before the addition of second urea, *UMF12A$_2$V* in Table 25.3 indicated a (methylene + 1/2 methylene–ether) bond content of 0.465 groups per mole of first urea and melamine combined, a DP value of 1.87, in spite of the high viscosity (VW) attained. Further, the methylene-ether: methylene bond ratio was also high at 69 : 34. This synthesis procedure thus resulted in resins having very low molecular weight values compared to the DP values of 10 or higher for common UF resins' DP values of polymeric components [14–17], The finished resin with second urea addition was cloudy and the viscosity values measured at room temperature for 50 days increased faster for resins with higher melamine contents as shown in Figure 25.6. With these UMF resins, since the UF components would have experienced only a small increase in the methylene + methylene-ether bond content (0.12 group), the balance of about 0.345 bond per 0.336 mol of melamine for the MF components indicates a high degree of polymerization value. This over-polymerization of MF components in *Resin UMF12A$_2$VU$_2$* is evidenced by the turbid appearance. Also, the UMF resin synthesized with a 2% melamine

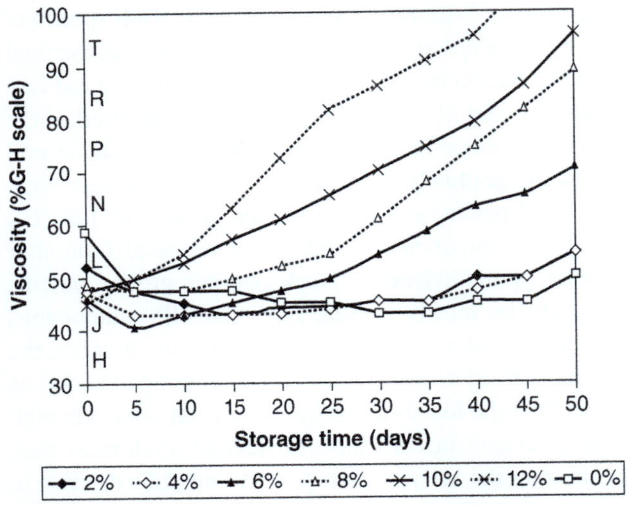

Figure 25.6 Viscosity changes of UMF resins synthesized in Table 25.1 (2–12% melamine) and control UF resin observed in storing at room temperature.

level according to this procedure did not even reach to the target viscosity in 5.5 hours of reaction, apparently because of the lack of the reactive MF component materials and UF resin components' lack of reactivity at pH 6.3. UMF resins made with 2% and 4% melamine contents, albeit the presence of turbidity, and UF control resin were stable for more than 30 days, as shown in Figure 25.6. With these UMF resins, the MF components appear to have polymerized to the point where they are no longer soluble, forming solid particles that remain suspended among the low-molecular weight UF components. On the other hand, UMF resins with higher melamine contents, the major MF components remained still reactive and caused to increase the viscosity during the storage periods.

Resin UMF12VVU$_2$ Made with a UF Base Resin Advanced to Full Extent: With this Type #4 method of UMF resin synthesis procedure used in early research [2–8], 12% melamine and second formaldehyde were added to the UF components, which was pre-advanced to V viscosity. The reaction time to the target viscosity of V at pH 6.3, after the addition of melamine, was short and it resulted in a clear resin as expected. The resin sample taken before the addition of second urea gelled overnight at pH 8.0, indicating that the MF components were low molecular weight, reactive components. The ^{13}C NMR results of *Resin UMF12VVU$_2$*, compared with those of typical UF resins [14–17], showed that the methylene-ether: methylene groups ratio is at 36 : 64 and the overall extent of polymerization attained by the melamine components was very low high in comparison with *Resin UMF12A$_2$VU$_2$* (Table 25.3).

UMF resins made with UF base resins advanced to intermediate extents and with various melamine levels: In these experiments, the UF base resin was polymerized at pH 4.6 to D, K, and S viscosities and then reacted at pH 6.3 with 2–12% levels of melamine and second formaldehyde at $[(F_1 + F_2)/(U_1 + M) = 2.1]$. Selected resin property data of these UMF resins are reported in Table 25.4. The reaction time taken after the melamine addition at pH 6.3 until the target viscosity of V was long but became shorter as the melamine level was decreased, as shown in Figure 25.7, indicating that the major reacting components in this stage are MF components. Also, the reaction time was slightly shorter as the UF

Table 25.4 Physical properties of UF and UMF resins synthesized with F/U + M = 1.15.

Resins	pH	Gel times (s)	Specific gravity	Viscosity (G–H scale)	Solids content (%)
UF Control	8.5	88	1.26	N	61.3
$UMF6DVU_2$	8.5	135	1.27	JK	62.5
$UMF12DVU_2$	8.5	129	1.29	J	64.5
$UMF12KVU_2$	8.6	126	1.29	KL	64.4

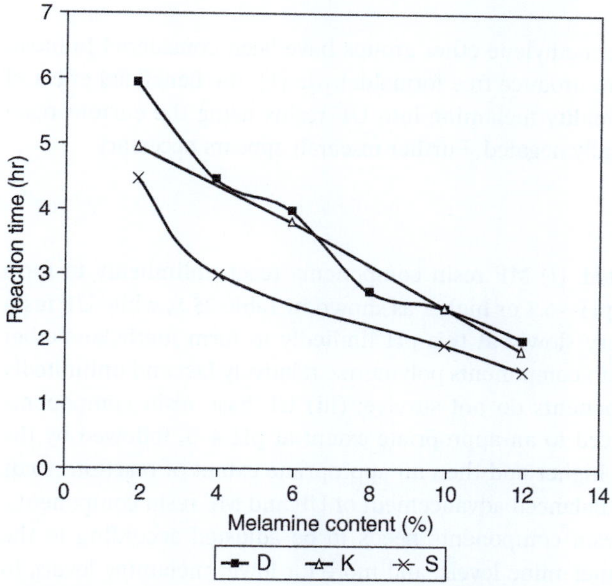

Figure 25.7 Reaction times taken in syntheses of UMF resins depending on the starting UF base resin viscosity values and melamine levels (Table 25.1).

base resin viscosity increased from D to S at the same melamine levels. The UF base resin components would have incurred only a small extent of polymerization during the melamine reaction time and therefore, the cloudiness of UMF resins appears to arise from melamine components. Resins made with a 12% melamine level were typical, changing gradually from opaque to translucent and to clear with UF base resin viscosity increases from A_2 to V in pre-advancement, reflecting the MF components' decreasing extents of polymerization. Also, for a given UF base resin the resultant resins' cloudiness tended to be greater for lower melamine levels because of the higher extents of polymerization incurred by MF components. Similarly, the viscosity increases in the room-temperature storage of resins were faster for resins made with more advanced UF base resins.

General Resin Properties of UMF Resins: Synthesized UMF resins generally showed longer gel times than the control UF resin with ammonium sulfate catalyst (Table 25.4). This trend of gel time variations agrees with the generally known fact that MF resins cure slower than UF resins [1, 18], Another interesting note to make is the increasing ratios of methylene-ether vs. methylene groups in these UMF resins synthesized at about pH 6.3,

Table 25.5 Methylene-ether: methylene group ratio.

UF	23:77
UF, pH 6.2	40:60
UMF12VV + U_2	36:64
UMF12DV + U_2	41:59
UMF12A2V + U_2	69:31
MF + U, pH 8.4	74:26

summarized in Table 25.5. Since methylene-ether groups have been considered to break up in the curing stage of resins to produce free formaldehyde [1], the beneficial effect of incorporating the higher functionality melamine into UF resins using the current resin synthesis methods could be partially negated. Further research appears necessary.

25.1.6 Conclusion

The following conclusion is noted: (i) MF resin components react unlimitedly to form mostly *methylene-ether bonds* at pH ~6.3 or higher as shown in Table 25.5, while UF resin components polymerize only very slowly at this pH limitedly to form methylene-ether and methylene bonds; (ii) UF resin components polymerize relatively fast and unlimitedly at pH 4–5 where the MF components do not survive; (iii) UF base resin components therefore need to be first advanced to an appropriate extent at pH 4–5, followed by the melamine addition at pH 6.3 or higher and then an appropriate extent of reaction needs to attained; (iv) In order to have balanced advancement of UF and MF resin components, the advancement of UF base resin components needs to be adjusted according to the melamine level, less for higher melamine levels, and more for lower melamine levels, to avoid an over-polymerization of MF resin components; (v) Turbid UMF resins result from excessively advanced MF resin components and they show limited storage lives; (vi) The storage stability of clear UMF resins made by minimally advancing MF components could be somewhat shorter, especially for UMF resins made with higher melamine levels; (vii) Overall, pre-advancing the UF base resin components to an appropriate extent, followed by an appropriate extent of advancing the MF resin components appear to be the key parameters in synthesizing various low-level melamine-modified, storage-stable UMF resins; (viii) Gel times of UMF resins measured at 100 °C are significantly longer than UF resins with ammonium sulfate catalysts and they varied in minor ways by the extent of polymerization of base UF polymers.

25.2 DMA Curing Properties of UF and UMF Resins

25.2.1 Past Research Using Dynamic Mechanical Analysis (DMA) Method on Wood Adhesive Resins

The DMA method based on TA Model 983 DMA has been used to monitor the curing of thermosetting wood composite binder resins, as discussed in Lecture 6. This method

Table 25.6 Various UF and UMF resins synthesized in the previous section (Section 25.1) evaluated with DMA.

Resin $UFA_{1.05}$	Resin $UFA_{1.15}$	Resin $UFA_{1.25}$
Resin $UMF6D_{1.05}$	Resin $UMF6D_{1.15}$	Resin $UMF6D_{1.25}$
Resin $UMF6K_{1.15}$	Resin $UMF6K_{1.15}$	Resin $UMF6K_{1.25}$
Resin $UMF12D_{1.05}$	Resin $UMF12D_{1.15}$	Resin $UMF12D_{1.25}$
Resin $UMF12K_{1.05}$	Resin $UMF12K_{1.15}$	Resin $UMF12K_{1.25}$

UFA – Control UF resins. **UMF** – Resins with numbers 6 and 12 indicate melamine levels based on total resin weights and letters D and K indicate viscosity points of UF base resins attained in acidic pH 4.5 and $F/U_1 = 2.10$. Then, pH was adjusted to ~8.0, and melamine and second formaldehyde were added to $(F_1 + F_2)/(U_1 + M) = 2.10$ and reacted at 85 °C at pH 6.7–6.3 until V viscosity, pH is then adjusted to 8.0 and second urea is added to final mole ratios of $(F_1 + F_2)/(U_1 + M + U_2) = 1.05$, 1.15, or 1.25.

continuously measures the storage and loss moduli and tan δ (the ratio of loss modulus to storage modulus) values of a resin sample as the sample temperature is increased and maintained at a fixed temperature [19, 20]. Follensbee, et al. found that PF wood composite resins' storage moduli decreased as the extent of pre-staging of resin was increased [21]. Kim, et al. reported that the curing rate of PF plywood binder resins increased with curing temperatures, extents of polymerization, and F/P mole ratios of resins [21]. Kim, et al. also found that the urea commonly added to PF resins lowered the curing speed of resins but resulted in minimal deleterious effects on the bond strength of boards [22, 23]. Ebewele claimed that certain modified UF resins cure to give an improved polymer structure based on DMA results [24]. Overall, the groundwork on the general curing characteristics of UF or UMF resins, such as gel time, cure time, and curing rate using the DMA method has not been reported. In this study, selected UMF resins and a control UF resin discussed in the previous paper, shown in Table 25.6, were studied on DMA, including the effects of catalyst levels, catalyst mixing temperatures and mixing time, and curing temperatures.

25.2.2 DMA Sample Preparation and Experiments

A resin mix with catalyst was made in 25 °C water bath by stirring for 15 minutes, and approximately 25 mg of mixed resin was evenly spread on the fiberglass braid (16.15 × 10.95 × 0.12 mm in dimension), which was then tested on DMA Model 983 [25], of which mechanism and principles were discussed in Section 6.8.2. The DMA test results are the rigidity (storage modulus, G′), loss modulus (G″), and tan delta (δ, G″/G′) curves. For UF control and all UMF resins, the catalyst level was set at 0.5% of the ammonium sulfate catalyst solution, and the resin mix time and temperature were kept constant at 15 minutes at 25 °C.

25.2.3 Three Curing Segments of DMA Curing of Resin $UFA_{1.15}$

General Curing Pattern on DMA: DMA results of *Resin* $UFA_{1.15}$ obtained at 120 °C with 0.5% ammonium sulfate catalyst and pre-treatment for 15 minutes at 25 °C are shown in Figure 25.8a,b. In *the (expanded) first segment* shown in Figure 25.8b, the rigidity curve (solid line) started at a small value due to drying of resin and initially decreased to a lower value due to thermal softening. The rigidity curve then begins to increase rapidly at about 60 s during the temperature ramping-up period to reach to a small value while the tan delta curve was rapidly decreasing, indicating that the gelation of resin occurred from the formation of an incipient infinite polymer network [20]. This time to reach to the rigidity increasing point was defined as the *gel time*. The rigidity increase of gelled resin then continued until about 130 s, and the rate of increase in this first segment is defined as *Cure rate I*. The curing temperature reached 120 °C at about 110 s and, therefore, the rigidity increase entailed the rigidity decreasing effect of increasing temperature. In this early period of resin cure, the loss modulus also increased due to the infinite polymer network being expanded, and the tan delta curve decreased rapidly due to the more rapid increase in the rigidity than the loss modulus. The loss modulus increase reflects the increasing amounts of mechanical energy lost, by the infinite polymer network being formed, to the uncured resin components that are thermally coupled to it. In another word, the amount of uncured resin components coupled to the infinite polymer network would increase as the polymer network grows, to be reflected as increasing of the loss modulus. The rigidity value attained in the first segment is small compared to the final rigidity value and the significance of *Cure slope I* to the overall curing quality is unknown.

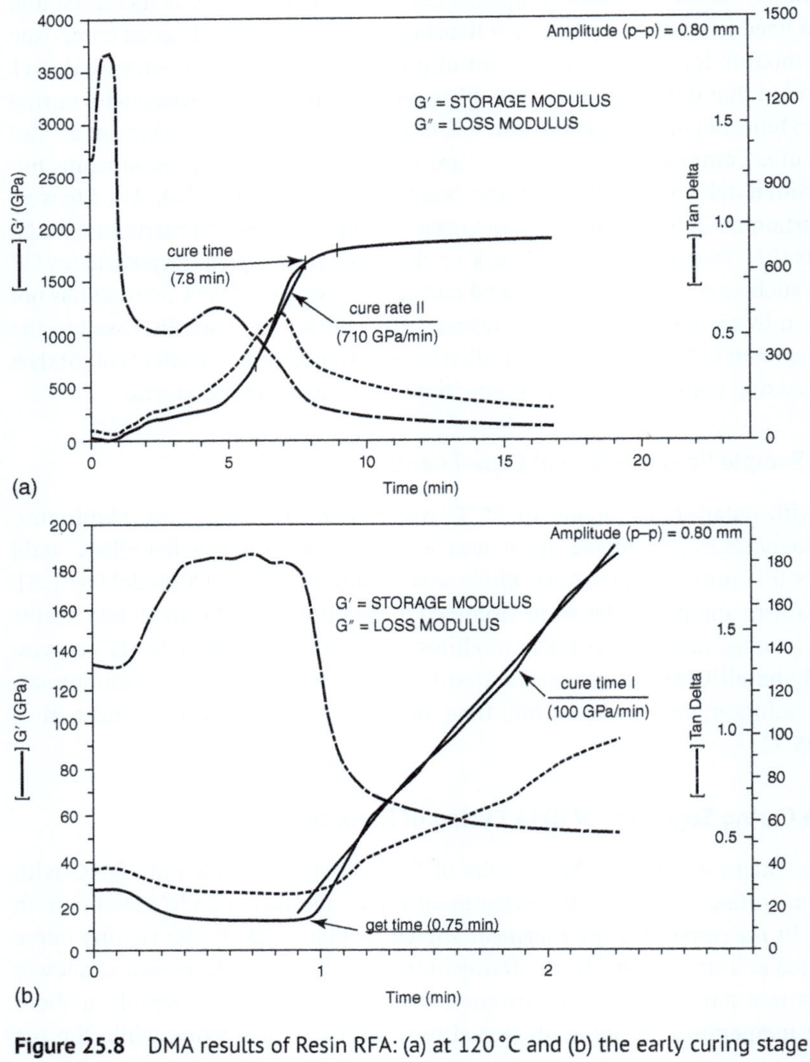

Figure 25.8 DMA results of Resin RFA: (a) at 120 °C and (b) the early curing stage expanded.

The second segment of cure, in Figure 25.8a, began and continued until about 260 s, a relatively long time, where both the rigidity and loss modulus increases were relatively small, indicating a minimal expansion of the infinite polymer network occurring. The tan delta curve ranged between 0.6 and 0.7, reflecting the rubbery stage of sample at the measuring temperature.

The third segment of cure, in Figure 25.8a, then began with the rigidity curve rapidly increasing and tan delta rapidly decreasing, and it was completed at about 480 s, passing through the vitrification stage and further advancing to a highly elastic state. The rigidity and loss modulus curves reached their final values and the tan delta curve reached a low plateau value, indicating that most low-molecular weight resin components were incorporated into cured polymer networks. The rate of rigidity increase in this final segment was defined as *Cure rate II*. Both the rigidity and loss modulus increased in the first part of this segment and then the loss modulus reached the maximum at about 420 s and then began to decrease. The loss modulus decrease reflects the decreasing levels of uncured resin components remaining in the sample. At the maximum of loss modulus curve, the rigidity value reached 1600 GPa and the tan delta value decreased to 0.4, indicating that the vitrification point of resin at the measuring temperature is passing [26]. The resin curing continued, and the rigidity reached a high plateau value (~2800 GPa), and the loss modulus a low plateau value (~250 GPa) with the tan delta decreasing to about 0.09 at the measuring temperature. The tan delta value was less than 0.05 when the cured sample was cooled and a DMA re-run was made at room temperature, indicating that a highly glassy state of cured resin was attained. The *Cure time* of resin was defined in this work as the time when the rigidity rise ended in this segment, as shown in Figure 25.8a. Furthermore, it is noted that the rigidity curve remained stable for the next 15 minutes, indicating that no degradation of UF polymer network occurred at 120 °C.

25.2.4 Curing Chemistry of Resin $UFA_{1.15}$

Curing Chemistry in the First Segment with an Early Gelation: The DMA curing profile of Resin $UFA_{1.15}$ (Figure 25.8a) is generally similar to those of other UF and UMF resins studied in this work, represented by the three segments of rigidity development. First of all, the early gelation appears to arise from the presence of certain fast-reacting components. In curing of thermoplastic polymers to cross-linked products, such as rubber products, it has been reported that cross-linking extent at about 1~2% monomer levels, or one out of 50~100 monomers, would give the typical rubbery products [27].

$$6\,CH_2O + 2\,(NH_4)_2SO_4 \rightarrow 4\,H^+ + 2\,SO_4^{--} + (CH_2)_6(NH)_4 + 6\,H_2O \qquad (25.7)$$

$$\text{P-NR'-CO-NR-}CH_2\text{-OH} + H^+ \rightarrow [\text{P-NR'-CO-NR-}CH_2\text{-}OH_2^+] \rightarrow \qquad (25.8)$$

(2-1)

$$[\text{P-NR'-CO-NR-}CH_2^+ \leftrightarrow \text{P-NR'-CO-}N^+R{=}CH_2] + H_2O$$

(2-2) (2-3)

$$\rightarrow [\text{P-NR'-CO-}\mathbf{N{=}CH_2}] + H^+ \quad (R{=}H)$$

(2-4)

$$(\underline{2\text{-}2});\ (\underline{2\text{-}3});\ (\underline{2\text{-}4}) + \mathbf{NH_2}\text{-CO-NH}^{\sim\sim} \rightarrow \text{P-NR'-CO-}\mathbf{NR\text{-}CH_2\text{-}NH}\text{-CO-NH-}CH_2^{\sim\sim} + H^+$$

(2-5) *(P=Polymer)*

As discussed in Chapter 6 and Section 25.1 above, in curing of UF or UMF resins, the formaldehyde present reacts with ammonium salts to generate free acid (Eq. 25.7) and weakly basic hexamethylenetetramine (hexa), and the acid starts a series of curing reactions described in Eq. (25.8) [1]. Thus, the free acid activates hydroxymethyl groups of UF polymers by forming an acid complex (***2-1***), which dissociates into the cationic methylene intermediate (***2-2***), which is resonance-stabilized to an amidinium ion (***2-3***). This intermediate is convertible to an imine group (***2-4***) when the substituent (R) on the nitrogen is a hydrogen and a mild base is available to abstract the hydrogen to result in a further lowering of the activation energy of ionization [17], similarly with the melamine system discussed in the previous section (Section 25.1). The cationic reaction intermediates (***2-2; 2-3***) would react with urea amide groups to form methylene bonds (***2-5***) between urea molecules, i.e. dimers, etc. are formed. It appears that intermediate (***2-4***) plays only a minor role in curing of UF resins, since UF resins are usefully curable only in the relatively strong acidic pH.

The early gelation characteristics of UF and UMF resins appear to originate from thus far not well-identified polymeric UF resin components and structures. Polymeric UF resin components are branched at every three or four urea units via methylene and methylene-ether groups and harbor the functional groups shown in Eq. (25.9) [14–17] that can bring the resin system to an early gelation. It is especially noted that UF polymer chains harbor *gem*-di-substituted urea units (***3-4***) in the middle of polymer chains with their concentration increasing as the resin ages due to the reverse reaction of hydroxymethyl groups. The free amide groups of monomeric UF resin components (***3-5***) are also present in high concentrations due to the second urea addition [16]. It was shown that Type I hydroxymethyl groups in acidic pH of resin synthesis react fast with free urea amide groups (***3-4*** and ***3-5***), but they react slowly with imide groups [16].

Functional groups involved in cuing of UF resins

$$\begin{aligned}
&\mathrm{P-NH-CO-NH-\underline{\mathbf{CH_2OH}}(Type\ I)}\ (\textit{\textbf{3-1}})\\
&\mathrm{P-NH-CO-N\underline{(-\mathbf{CH_2OH})_2}(Type\ IIt)}\ (\textit{\textbf{3-2}})\\
&\mathrm{P-N\underline{(\mathbf{CH_2OH})}-CO-NHR(Type\ IIi)}\ (\textit{\textbf{3-3}})\\
&\mathrm{P_2=N-\underline{\mathbf{CO-NH_2}}}\ (\textit{\textbf{3-4}})\\
&\mathrm{NHR-\underline{\mathbf{CO-NH_2}}}\ (\textit{\textbf{3-5}})\\
&(P = UF\ Polymer; R = H, P, or\ CH_2 - OH)
\end{aligned} \tag{25.9}$$

On the other hand, Type IIt (***3-2***) and Type IIi (***3-3***) hydroxymethyl groups are relatively less reactive since the cationic methylene intermediate to be formed would not be as much resonance-stabilized compared to Type I hydroxymethyl groups because the amidinium ion intermediate (***2-3***, R=CH_2OH) would not be formed as readily, nor converted to the imine intermediate (***2-4***). In the first stage of curing, Type I hydroxymethyl groups bonded to high molecular weight components would react not only with monomeric urea amide groups (***3-5***) to extend polymer structures but also with the unsubstituted amide groups of *gem*-di-substituted urea units (***3-4***). This latter type reactions will result in fast molecular weight increases and possibly cross-linking to bring the early gelation of UF resins observed. The concentration of *gem*-di-substituted urea groups (***3-3***) in fresh resins can be low because amide groups of polymeric resin components are mostly bonded to hydroxymethyl groups, but it would increase by the reverse hydroxymethylation reactions as the curing reaction progress. The reverse reactions occur relatively slowly at room temperature but faster at elevated temperatures and, for example, one UF resin that was post-treated at 60 °C in alkaline pH for several hours was found to have the *gem*-di-substituted urea group increased to 0.17 group per mole of the first urea (U_1), or one out of about 6 urea units [16]. These groups might be one of the causes of early gelation and pre-curing problems often occurring in PB industry.

-NH-CO-NH-**CH_2-O-CH_2**-NH-CO-NH-P →
P-NH-CO-NH-**CH_2-O^-** + **$^+CH_2$**-NH-CO-NH-P

+H_2O
→ P-NH-CO-NH-**CH_2-O-H** + **HO-CH_2**-NH-CO-NH-P

(P=Polymer) (25.10)

Curing Chemistry of Resin $UFA_{1.15}$ in the Second Segment with a Long Rubbery Stage: In the second segment of curing (rubbery stage), both the rigidity and loss modulus increased quite slowly in comparison to the first or last segment. The underlying mechanism for this segment may be that the small resin molecules react among themselves or with larger polymers only to extend the linear polymer structures. Alternatively, the reaction that occurs in the first segment may have slowed down because of depletion of the more reactive groups discussed above. It is also possible that polymers are forming while other polymer structures break down. UF polymer chain structures consist of methylene (~80%) and methylene-ether (~20%) bonds, and the latter groups are known to break down at elevated temperatures, such as in curing (Equation 25.10) [1]. Such breakdown of polymer chains would lengthen this rubbery stage or delay the development of rigidity. Also, UF resins are composed of first urea-derived polymeric components (~55%) and second urea-derived monomeric components (~45%) [14–17] with the relatively large proportion of the latter components leading to delayed growth of cured polymer structures.

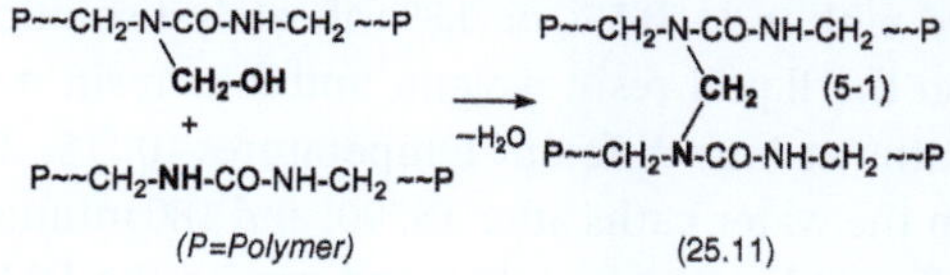

(P=Polymer) (25.11)

Curing Chemistry of Resin $UFA_{1.15}$ in the Third Segment: The rigidity rises rapidly to a high value, and the loss modulus also increases but then starts to decrease and reaches to a low plateau value, passing through the vitrification stage and then reaching to the highly glassy state. This characteristic is similar to those of other highly cross-linking thermosetting resins, such as PF resins [22, 23], and this suggests that cross-linking methylene (***25-11***) bonds occur extensively between polymer chains. The reaction between Type IIi hydroxymethyl groups and imide groups appears to occur primarily in this final curing stage.

The current F/U mole ratio, about 1.15, can be reasonably equated to the methylene group content in cured resins, to a cross-linking level of about 30%, or one cross-linking methylene bond for every 3.3 urea units. This level of cross-linking is a relatively high value even as for a thermosetting resin [26] and the characteristics of rapid rise to a high rigidity value shown in the third segment of curing of *Resin $UFA_{1.15}$* appear to be in accord with this

Table 25.7 DMA results of Resin $UFA_{1.15}$ obtained with ammonium sulfate catalyst (0.50%) and various catalyst mixing parameters.

Catalyst mixing parameters			DMA results at 120 °C		
Mix. temp (°C)	Mix. time (min)	Gel time (min)	Cure rate I (GPa/min)	Cure rate II (GPa/min)	Cure time (min)
0	15	0.75	150	787	5.8
	90	0.75	150	784	5.9
	180	0.75	67	520	5.8
15	15	0.76	290	667	7.5
	90	0.76	150	427	7.8
	180	0.73	180	400	7.9
25	15	0.74	100	710	6.9
	90	0.72	120	674	7.0
	180	0.7	140	867	7.8
35	15	0.59	200	767	7.4
	90	0.65	250	467	8.0

conclusion. Some hydroxymethyl groups would remain unreacted to cause formaldehyde emission problems although the amount of formaldehyde measured in particleboard tests constitutes only a very small fraction of the total formaldehyde used in resin manufacture [28, 29].

25.2.5 Effects of Catalyst Level and Mix Time and Temperature on DMA Curing of Resin $UFA_{1.15}$

Resin $UFA_{1.15}$, a typical UF resin, was mixed with 0.3%, 0.5%, or 0.8% of catalyst solution (25% ammonium sulfate in water) based on the liquid resin weight, and four resin mix samples were made and placed in water baths at four different temperatures (0, 15, 25, and 35 °C). The resin mixes were taken from the water baths after 15, 90, and 180 minutes of mixing time, and approximately 25 mg of mixed resin was taken and run on the DMA at 120 °C. Gel times, Cure slopes I and II, and Cure times obtained at 0.5% ammonium sulfate catalyst level with mix times of 15, 90, and 180 minutes at 0, 15, 25, and 35 °C are reported in Table 25.7. Similar data obtained at 0.3% and 0.8% catalyst levels (not reported) indicated that 0.3% level was short, but 0.8% level did not show any improvements. *Gel times* were generally shortened with increasing of mix time except at 0 °C but did not change significantly at different mixing temperatures except at 35 °C. All resin mixes made with 180-min mix time at 35 °C gelled before DMA runs.

Cure rate I increased with mix times and catalyst levels, except at mix temperature 0 °C, similarly with the shortening of gel times. It may be that the polymer components that caused the early gelation continue to react in the first segment of curing and therefore, the pre-curing tendency of UF resins may be predicted from gel time and Cure rate I.

On the other hand, *Cure rate II values* decreased as the resin mix time increased from 15 to 90 minutes at all temperatures except at 0 °C, indicating that the resin advancement

Table 25.8 DMA cure time, cure rate II, and gel time of resins obtained at various temperatures with 10 minutes mix times at room temperature.

		UF and UMF Resins				
	Cure temp. (°C)	UFA	UMF6D	UMF6K	UMF12D	UMF12K
Cure time (min)	120	8.5	13.7	12.4	18.3	16.2
	150	5.7	10.8	8.5	10.8	9.2
	170	4.5	8.5	7.5	6.5	6.3
Cure rate II (GPa/min)	120	258	191	188	89	123
	150	525	253	291	256	295
	170	180	206	240	266	330
Gel time (min)	120	0.67	1.12	1.05	1.17	1.15

occurring in resin mixes hampers the rigidity development in the final curing segment. Higher catalyst levels generally resulted in higher Cure rate II values. *Cure time values* were shortest for the resin mixes made at 0 °C; a result of higher Cure rate II values, ascribable to the little advancement incurred in resin mixes. Wood composites binder resins that cure fast in the final curing stage would be desired for productivity reasons. Overall, the results indicate that UF resins tend to advance in resin mixes, engendering pre-cure problems and hampering the proper curing in the final segment. Thus, low mix temperatures and short mixing times preferred in industry thus appear proper.

25.2.6 Comparison of Resins $UFA_{1.15}$ and $UMF_{1.15}$ by DMA at Three Different Temperatures

DMA of Resins $UFA_{1.15}$ *and* $UMF_{1.15}$ were isothermally conducted at 120, 150, and 170 °C with resins mixed with 0.5% ammonium sulfate catalyst at room temperature for 10 min, and the results are summarized in Table 25.8 and Figure 25.9a and b.

Cure rate II values of UF and UMF resins increased as the temperature increased from 120 to 170 °C, but the *UMF resins'* Cure rate II values were slower than UF resins at all temperatures'

The Cure rate II values of UMF12K and 12D were lower than those of UMF6K and 6D, indicating that the curing rate is slowed with increasing melamine levels from 6 to 12%. This melamine level effects were most pronounced at 120 °C, indicating that higher temperatures are needed to cure UMF resins. The reason for melamine's cure-retarding effect has been ascribed to the higher basicity of melamine than urea, tying up the acid catalyst generated in the curing process [19].

Resins UMF6K and 12K showed higher *Cure rate II values* than *Resins UMF6D* and UMF*12D*, respectively, reflecting the base UF resins' higher extent of polymerization attained at K viscosity point in comparison with the D viscosity point with the effect reflected in the cure rates.

Cure time data of these resins also led to similar conclusions regarding the effects of melamine levels and extents of pre-advancements of UF base resins on cure rates. This

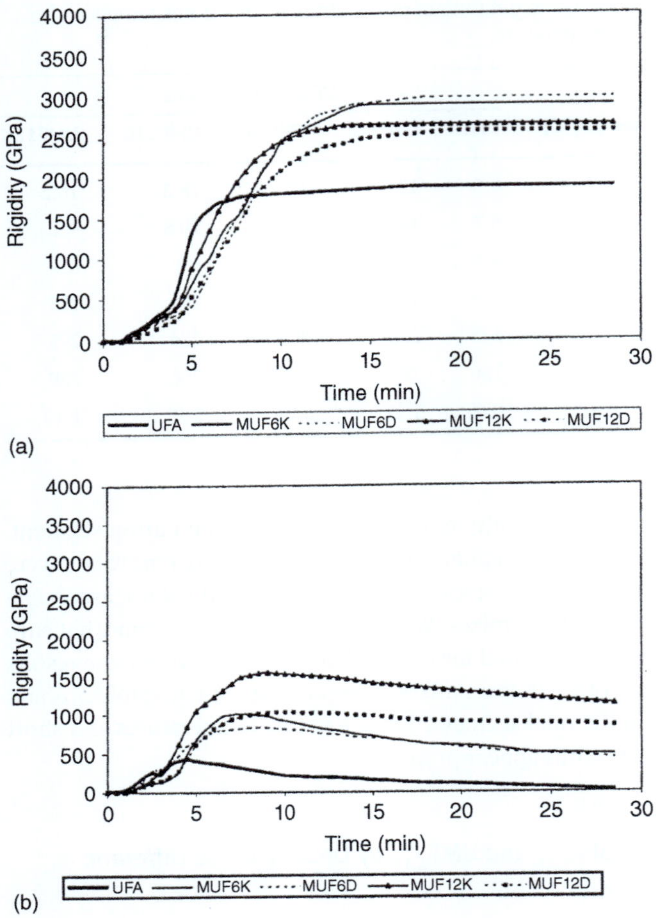

Figure 25.9 (a) Comparison of rigidity curves obtained at 150 °C. (b) Comparison of rigidity curves obtained at 170 °C.

observation may be useful in synthesizing UMF resins, especially at relatively low levels of melamine usage.

Final rigidity values of resins attained at 150 °C as shown in Figure 25.9a show that UF and UMF resins cured well to give high, stable values indicating the optimum curing temperature range used. UMF resins showed higher final rigidity values than the UF resin, and UMF6K and 6D resins appear curing better than UMF12D and 12K resins at this temperature to give slightly better rigidity values. Also, the rigidity value for UMF12K is slightly higher than UMF12D, possibly reflecting the fact that the base UF resin synthesis extended beyond D viscosity till K viscosity is preferable.

UMF and UF resins cured but showed degradation of rigidity values at 170 °C as shown in Figure 25.9b with the resistance to degradation of rigidity being higher for UMF12K and 12D resins than UMF 6K and 6D resins, and UF resin showed a very poor resistance. Further, UMF12K showed better degradation resistance than UMF12D, strengthening the proposal that the UF base resin synthesis needs to be advanced to a higher or optimum level.

Table 25.9 DMA cure rate II and cure time of resins at different F/(U + M) mole ratios at 150 °C.

	Cure rate II (GPa/min)			Cure time (min)		
F/(U+M)	UFA	UMF6D	UMF12D	UFA	UMF6D	UMF12D
1.05	340	270	150	5.7	9.2	10.9
1.15	500	320	260	4.5	8.7	9.5
1.25	900	520	510	4.3	5.7	8.0

Thus, the MF components appear to help lessen the over-curing tendency of UF resin components as well as better performances for resins with higher or optimum advancements of base UF resins.

25.2.7 Effects of Three Different F/(U + M) Mole Ratios on Curing Rates at 150 °C

Resins UFA, *UMF6D*, and *UMF12D* made at F/(U + M) ratios of 1.05, 1.15, and 1.25, respectively, were cured at 150 °C to investigate the effects of resin F/(U + M) mole ratios on curing and the results are summarized in Table 25.9 and in Figure 25.10a and b. The results indicate that *Resins UMF6D* and *12D* have longer cure times than *Resin UFA* for all mole ratios, indicating the melamine's cure-retarding effect is not overcome by the mole ratio increases. However, the *cure times* of all resins became slightly shorter as the mole ratio increased from 1.05 to 1.25 and the cure times for *UMF resins* shortened somewhat more than *Resin UFA*. Similarly, the Cure rate II values of all resins increased as the mole ratio increased from 1.05 to 1.25, as expected by the higher functionality. Similarly, the *Cure rate II* values of *Resin UMF6D* and *12D* were lower than that of *Resin UFA* at all mole ratios and the *Cure rate II* value of *UMF12D* was lower than that of *UMF6D* at all mole ratios. These melamine level effects agree with the discussion made above (Section 25.2.7) with mole ratio of 1.15 and that the Cure rate II value differences were smaller at higher mole ratios indicating that the delay of resin cure due to melamine is partly compensated by higher mole ratios. The final rigidity values for all resins were higher at higher mole ratios and the rigidity values of *Resin UMF6D* and *12D* were higher than that of *Resin UFA* at all mole ratios. Overall, the final rigidity values of UMF resins at mole ratio of 1.05 are comparable with those of *Resin UFA* with a mole ratio of 1.15. Thus, UMF resins would be useful at lower F/(U + M) mole ratios than UF resins to give particleboard with similar strength with reduced formaldehyde emission potentials [7].

25.2.8 Summary

DMA gel-time, cure-time, and cure-rate values of UF and UMF resins measured at 120, 150, and 170 °C were found to be useful in optimizing resin formulation, catalyst level and mix time, and curing temperature and time. The DMA curing curves of UF and UMF resins distinctively showed three segments of curing: (i) gelation with a small increase in rigidity; (ii) rubbery stage over a relatively long segment of time; and (iii) rapid increase in rigidity due to vitrification and further curing to reach the high modulus stage. Higher

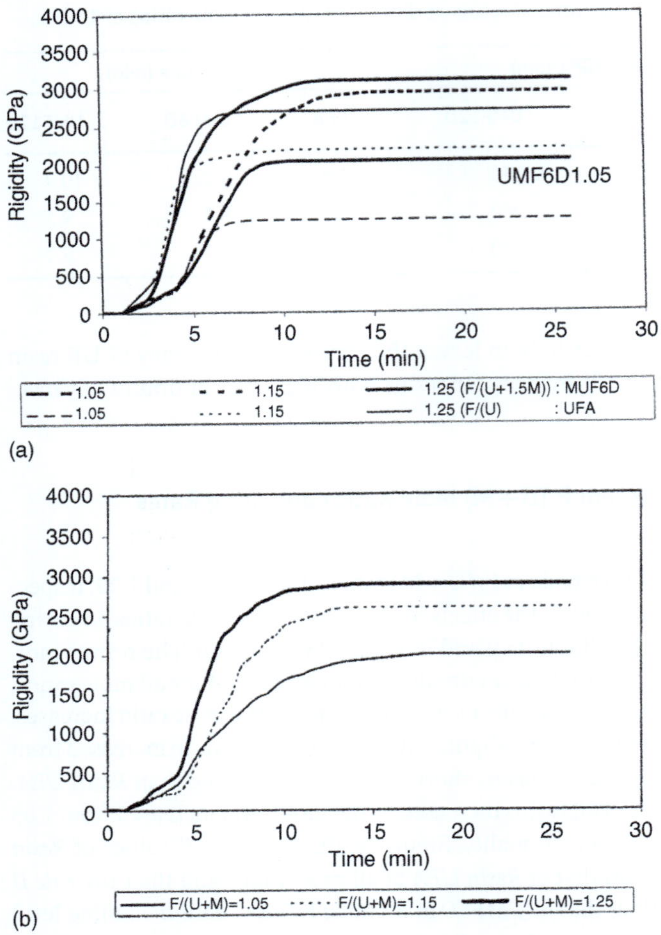

Figure 25.10 (a) DMA rigidity curves of Resins UFA and UMF6D at three different F/(U + M) mole ratios measured at 150 °C. (b) DMA rigidity curves of Resins UMF 12D at three different F/(U + M) mole rations measured at 150 °C.

curing temperatures shortened the overall cure times of resins. Optimum catalyst levels were between 0.5% and 0.8% of 25% ammonium sulfate solution based on the liquid resin weight. With this level of catalyst, the gel time shortened but Cure rate II slowed as the catalyst mix time increased, indicating that the advancement of resins incurred in the resin mix could degrade the final modulus development. *For UMF resins, higher advancements of UF base resin increased the Cure rate II and shortened the Cure time.* Cure times of UMF resins were longer than UF resins and get longer with increases in melamine levels, suggesting that higher temperatures or stronger catalysts are needed to cure UMF resins in practice. The final rigidity and thermal stability of cured UMF resins were higher than cured UF resins and increased with the increasing melamine level and, therefore, the thermal degradation often observed in UF resin-bonded particleboard can be lessened with UMF resins. The resultant higher rigidity values of UMF resins would allow the

F/(U + M) ratio to be lowered, for example, to reduce the formaldehyde emission levels of particleboard compared to UF resins, without lowering the strength properties.

25.3 Bond Performance of UMF Resins as Particleboard Binders

(Adapted with permission of the publisher and minor revisions from No B. Y. and Moon G. Kim. 2007. Evaluation of UMF resins as particleboard binders. J. of Applied Polym. Sci. 106, 4148–4156.)

25.3.1 Preparation and Tests of Laboratory Particleboards

Single-layer boards were made with UF and UMF resins catalyzed with 0.5% ammonium sulfate (25%) solution based on the liquid resin weight with resin-solids loading levels of 8% based on oven-dry wood weight. *No wax was added.* The press closing rate was initially 1.27 cm/s to a mat thickness of 2.53 cm and then 0.08 cm/s to reach the target thickness. Panels (860 × 860 × 12.65 mm) were made for each resin with a target density of 801 kg/m^3 (50 lbs/ft^3). Press temperatures were 325 and 365 °F and press times were 4.0 and 5.0 min for boards bonded with Resins UFA and UMF at an F/(U + M) mole ratio of 1.15. Boards were conditioned and trimmed to 660 mm × 660 mm in size, sanded, and cut for testing of IB, modulus of rupture (MOR) and modulus of elasticity (MOE), thickness swelling (TS) and water absorption (WA), and formaldehyde extraction (PE) and small-scale chamber (SC) emissions.

Tests of Internal Bond, Bending Strength, and Dimensional Stability of Boards: All physical tests of boards were carried out according to the ASTM test methods [30, 31]. Cut test specimens were equilibrated for 3 weeks in a constant humidity chamber to about 7% moisture content. Wet IB samples were obtained by immersing samples in water *at room temperature* for 24, 48, and 72 hours and then oven-drying at 158 °F for 24 hours to a moisture content of about 8%. An Instron testing machine was used for both IB and static bending tests. TS and WA samples were immersed in water *at room temperature* for 2 and 24 hours and then measured for TS and WA.

Formaldehyde Emission Tests – Small Chamber (SC) [32] method was carried out at Georgia-Pacific Resins laboratory (Decatur, GA) [33]. In this test, the hot-pressed boards were allowed to stand for 24 hours in the laboratory and three test specimens (379.5 mm × 199.2 mm) were cut from each board and the edges were sealed with aluminum adhesive tape giving a total exposed board's face area of 0.4536 m^2 per board. The test samples were aired for 7 days at 23.9 °C at a relative humidity of 50%. *Perforator Extraction* (PE) procedure [34] was carried out using one-day aired boards cut to 150 mm × 150 mm dimensions, which were kept in a sealed plastic bag until the tests, all tests done within one week.

25.3.2 Test results of Particleboards Bonded with Resins at an F/(U + M) Ratio of 1.15

Dry Internal Bond Strengths: PB bonded with control Resin UFA and UMF resins at an F/(U + M) ratio of 1.15 and press times of 4 and 5 min at press temperatures of 325 and 365°F

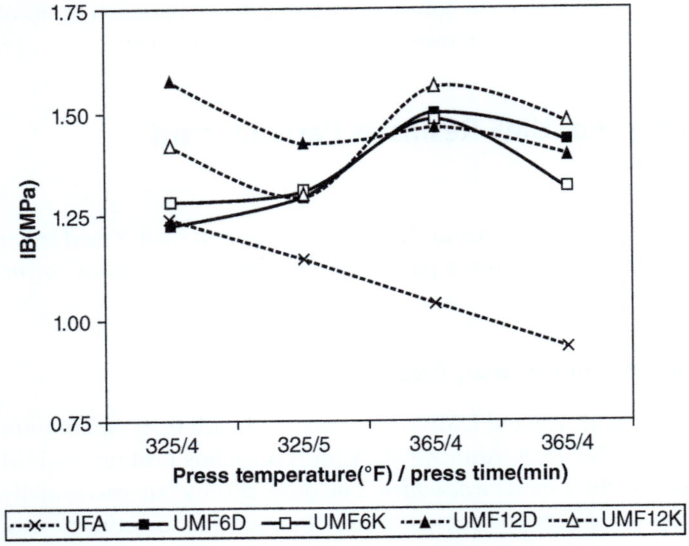

Figure 25.11 Internal bond (IB) strength values of particleboards bonded with Resins UFA and UMF at an F/(UM) mole ratio of 1.15.

showed IB strength (IB) values shown in Figure 25.11, summarized as follows: (i) *UMF resins* exhibited higher IB values than *Resin UFA* throughout most conditions except the boards made with UMF resins with 6% melamine level pressed at the lower press temperature and shorter press time, giving strength values similar to that of *Resin UFA*. (ii) *Resin UFA* showed low IB values at the higher press temperatures probably because of a greater tendency of over-curing/degradation at high temperatures and longer press times while UMF resins exhibited better bond strength values due to the slow curing characteristics and better thermal stability as revealed from the DMA curing study discussed above. (iii) There were no significant differences in IB strength values of boards among UMF resins between "D" and "K" series indicating that these resins are in the

$$IB = 213.44\text{–}8.82X_2 + 4.61X_3 (\text{for } D \text{ resins}).$$

$$IB = 382 : 61 + 0 : 33X_1 - 11.68X_2 + 4.64X_3 - 5.48X_4 (\text{for } K \text{ resins}). \quad (25.12)$$

optimum range of advancement for UF base resin components. (iv) UMF resins with 12% melamine level had higher IB values than those resins with 6% melamine level for all hot-pressing conditions. It is believed that the higher melamine content enhanced the bond strength of boards because of the higher functionality and rigid ring structure of melamine molecules. (v) Higher press temperatures also gave higher bond strength of boards bonded with UMF resins probably because of the higher temperature that compensated for the slower curing rates of UMF resins and also from the retarded degradation. (vi) For all conditions, boards bonded with UMF resins exceeded the minimum strength requirements for IB required in the US ANSI H-3 [32]. (vii) Statistical analyses using SAS program [35] showed that IB strength values were highly correlated with melamine contents. The Pearson correlation coefficients between IB strength values and melamine contents were 0.83 for #D and 0.79 for #K resin series. Regression analyses resulted in Equation 25.12, where X_1 is press temperature, X_2 is press time, X_3 is melamine level, and X_4 is board density. For D series resins, X_1 and X_4 were the least significant factors and therefore removed from the equation. The melamine content (X_3) was the most significant factor affecting IB values of boards for UMF resins.

MOR and MOE strength values and thickness swell values showed the same trends as dry IB strength values of boards for *Resin UFA* and *UMF Resins*. Further, the effects of F/(U + M) mole ratios of UF and UMF resins on particleboard strengths and formaldehyde emission values were studied but details are omitted in this lecture since the melamine levels used in this work were considered to be on the high side.

25.3.3 Summary

PBs bonded with control Resin UFA and various UMF resins at an F/(U + M) ratio of 1.15 and press times of 4 and 5 minutes at press temperatures of 325 and 365°F showed adequate IB strength values in that 6% and 12% melamine-modified *UMF resins* exhibited higher IB values than *Resin UFA* except the boards made with UMF resins with 6% melamine level pressed at the lower press temperature and shorter press time, giving strength values similar to that of *Resin UFA*. *Resin UFA* showed lower IB values at the higher press temperatures because of their higher tendency of over-curing/degradation while UMF resins exhibited better bond strength values due to the slower curing characteristics and better thermal stability in accord with the results observed from the DMA curing study. There were no significant differences in IB strength values of boards among UMF resins between 'D' and 'K' series to indicate that these resins are in an optimum range of advancement for UF base resin components. A further board study is needed with UF base resins with "A" and "V" viscosity points of melamine addition, as pursued in the following Chapters. UMF resins with 12% melamine level had higher IB values than those resins with 6% melamine level for all hot-pressing conditions. The bond strength variations observed in past research thus appear to be related also to the hot-pressing temperatures and times.

References

1 Wirpza, Z. and Brezezinsky, J. (1973). *Aminoplasty*, Kimya (Chemistry). Moscow, 71–76.

2 Shiau, D.W. and Smith, E. (1985). Low formaldehyde emission urea-formaldehyde resins containing a melamine additive. US Patent 4, 536,245. Borden Inc. Columbus, OH.

3 Graves, G. (1993). *Proceedings, 27th International PB and Wood Composite Materials Symposium* (ed. T.M. Maloney), 221. Pullman, WA: WSU.

4 Maylor, R. (1995). *Proceedings of Wood Adhesives Symposium*, 115. Madison, WI: FPS.

5 Breyer, R.A., Hollis, S.G., and Jural, J.J. (1997). Low mole ratio melamine-urea formaldehyde resin. US Patent 5, 681,917. Georgia-Pacific Resins, Inc. GA.

6 Ramon, R.M. (1977). *Proceedings of 31st International PB and Wood Composite Materials Symposium* (ed. T.M. Maloney), 177. Pullman, WA: WSU.

7 Parker, R. and Crews, G.M. (1977). *Proceedings of 33rd International PB and Wood Composite Materials Symposium* (ed. T.M. Maloney), 57. Pullman, WA: WSU.

8 Druet, B. and Rochet, G. (1991). US Patent 5, 008,365. Orkem Group, Paris LaDefense, France.

9 Tomita, B. and Hse, C.Y. (1995). *Mokuzai Gakkaishi* 41: 349.

10 Philbrook, A., Dunlop, N., Easton, C.J., and Keniry, M.A. (2005). *Wood Adhesive 2005* (ed. C.R. Frihart), 143. FS, FPL.

11 Tomita, B. and Hse, C.Y. (1995). *Mokuzai Gakkaishi* 41: 490.

12 Hse, C.Y. and He, Z. (1990). *Proceedings of Wood Adhesives Symposium*, 155. Madison, WI: FPS.

13 Christensen, G. (1980). *Prog. Org. Coatings* 8: 211.

14 Kim, M.G. and Amos, L.A. (1990). *Ind. Eng. Chem. Res.* 29: 208.

15 Kim, M.G. (2001). *J. Appl. Polymer Sci.* 80: 2800.

16 Kim, M.G., Wan, H., No, B.Y., and Nieh, W.L. (2001). *J. Appl. Polym. Sci.* 82: 1155.

17 Kim, M.G., No, B.Y., Lee, S.M., and Nieh, W.L. (2003). *J. Polym. Sci.* 89: 1896.

18 Oh, Y.S. (1999). *Forest Products Journal* 49 (11/12): 31.

19 Higuchi, M., Shimokawa, H., and Sakada, I. (1979). *Mokuzai Gakkaishi.* 25: 630.

20 Babayevsky, P.C. and Gillham, J.K. (1973). *J. Appl. Polym. Sci.* 17: 2067.

21 Follensbee, R.A., Koutsky, J.A., Christiansen, A.W. et al. (1993). *J. Appl. Polym. Sci.* 47: 1481.

22 Kim, M.G., Nieh, W.S., and Mecham, R.M. (1991). *Ind. Eng. Chem. Res.* 29: 798.

23 Kim, M.G., Watt, C., and Davis, C.R. (1996). *J. Wood Chem. Tech.* 16 (1): 21.

24 Ebewele, R.O. (1995). *J. Appl. Polym. Sci.* 58: 1689.

25 Lofthouse, M.G. and Burrough, P. (1978). *J. Therm. Anal.* 13: 439.

26 Ferry, J.D. (1980). *Viscoelastic Properties of Polymers*, 3e. New York: Wiley, Chapter 15.

27 Fahrenhorst, H. (1955). *Kunststoff* 45: 219.

28 Broline, B.M., Holloway, T.C., Moriarty, C.J., and C. J. (1995). *Wood Adhesives 1995* (ed. A.W. Christiansen and A.H. Conner), 97. Madison, WI: FPS.

29 Jiang, T., Gardner, D.J., and Baumann, M.G.D. (2002). *Forest Products Journal* 52 (11/12): 66.

30 American Society for Testing and materials. ASTM D 1037-93.

31 American Society for Testing and materials. ASTM D 6007-96.

32 Composite Panel Association (1999). *ANSI A 208.1-1999*. New York: American National Standards Institute.

33 Nieh, W. L-S. (2006). Georgia-Pacific Resin, Inc. generously provided the analysis service.

34 (1992) DIN EN 120-92; Deutsches Institute fuer Normung: Berlin, Germany.

35 SAS Institute (1998). Release 7.00 Software. SAS Institute: Cary, NC.

26

Advanced Reading Materials on UMF Resins Modified with 2.5% and 5.0% Melamine

26.1 UMF Resins Synthesized with 2.5% and 5.0% Melamine Levels

(Adapted with permission from An Mao, El Barbary Hassan, and Moon G. Kim. 2013. Investigation of low mole ratio UF and UMF resins aimed at lowering the formaldehyde emission potential of wood composite boards. BioResources 8(2), 2453–2469.)

One current challenge for the UF wood adhesive resin industry is the high formaldehyde emission (FE) levels of boards. Formaldehyde is a human carcinogen, and its exposure has been known to cause various health problems [1]. The emission levels are regulated by new state and federal laws to less than 0.09 ppm for particleboards (PB) measured by the large-chamber test method [2, 3], which is lower by one-half of the recent regulation values. Further, the long-term amount of formaldehyde releasable from wood composite boards calculated is large and considered to be a significant burden on the environment [2]. In the previous papers, UMF resins at melamine levels of 6–12% based on liquid resin weight were studied, but the melamine levels were considered to be on the high side for industrial practice. In this study, control UF resins and UMF resins having 2.5% and 5.0% melamine addition levels based on the liquid resin weight were studied using the resin synthesis *Methods #4* at F/(U + M) mole ratios of 1.05. The resins were analyzed and evaluated by making PBs and testing for physical properties and FE levels. A part of this paper was presented in Chapter 5.

26.1.1 Experimental Procedures

Resin Synthesis: UF control resins with an F/U mole ratio of 1.05, 1.15, and 1.25 and experimental UMF resins were synthesized, as described in Table 26.1 according to the Method #4 described in previous chapter (Section 25.1.3), i.e. the melamine was added in the last step of UF base resin synthesis procedure. Synthesized resins showed pH values of 8.0, specific gravity values of 1.262 to 1.271, viscosity values of G-H to I-J on the G-H scale, and solids content values of 62.29–63.78%. The test values obtained were in accord with expected values from the synthesis procedures used and were also within the range of industrial values.

Gel Time and Curing Time Measurements: Gel time was measured according to the conventional method at 100 °C. Also, gel times and cure times of catalyzed resins were measured in duplicates on an oscillatory rheometer (AR1500ex, TA Instruments Corp., DE) with an 8 mm diameter probe at test temperatures of 90, 120, 135, and 145 °C. In this

Adhesives and Finishes for Wood: For Practitioners and Students, First Edition. Moon G. Kim.

Table 26.1 Syntheses of UF and UMF resins (Method #4).

Urea-formaldehyde concentrate (UFC) was made: 50% formaldehyde solution (1229.3 g, 50.0%) was reacted at pH 8.0 and 70 °C with urea (278.8 g) for 30 min and then reacted for 30 min at 90 °C (F/U = 4.50).

- *Control UF resins*: *(i)* UFC (1508.1 g) and first urea (348.4 g) were mixed at pH 8.0 and 70 °C, reacted ($F/U_1 = 2.00$) and heated to 90 °C, and maintained for 30 min at pH 8.0. *(ii)* pH was adjusted to 4.75 and reacted to viscosity of P-Q (1 h and 25 min.). *(iii)* pH was adjusted to 8.0 and the second urea (567.4 g) was added $\{F/(U_1 + U_2)\} = 1.05$, giving *Resin UF1.05* with viscosity of *I–J* on the G-H scale. Two other UF resins were similarly synthesized and added the second urea weighing 464.3 and 376.9 g, respectively, to have an F/U mole ratio of 1.15 for *Resin UF1.15* and an F/U ratio of 1.25 for *Resin UF1.25*, with viscosities of HI and IJ, respectively.
- *UMF resins according to Method #4*: Control Resin UF 1.05 described above was made until the end of the second step. *(i)* At pH 8.0, for 2.5% melamine level, melamine (61.4 g) was added and reacted at 90 °C for 60 min and then, the second urea (537.9 g) was added and cooled to RT, resulting in Resin 2.5%UMF1.05 with viscosity of GH. *(ii)* At pH 8.0, for 5.0% melamine level, the same procedure was followed, using melamine (124.5 g) and second urea (507.5 g), resulting in Resin 5.0%UMF1.05 with viscosity of GH.

test, the resin sample was allowed to equilibrate at 20 °C for 10 seconds, heated to the target temperature in 50 seconds, and time sweep was performed at a frequency of 1.0 Hz with a strain of 1.0%. Two different catalysts were used in this study: *Catalyst A* is the common 25% ammonium sulfate solution in water and *Catalyst B* consists of 25% ammonium sulfate and 5% sulfuric acid in water solution.

Laboratory Particleboard Manufacturing: PBs were made similarly as in the previous studies using the blending parameters shown in Table 26.6. Wood particles were dried to 5.0% moisture content and put in a rotating blender, and wax and catalyzed resin were successively sprayed in using an air-atomizing nozzle with a blending time of approximately 15 minutes. Face particles and core particles were blended separately and they were weighed and laid on a steel plate in a box to form three-layer mats with a 1 : 2 : 1 weight ratio for top, core, and bottom layers. Two mats were made, one was pressed for 3.0 minutes and the other was pressed for 3.5 minutes. The press closing rate was initially 0.5 in./s to a mat thickness of 1.0 in. and then 0.03 in./s to reach the target board thickness of 0.5 in. Boards were cooled to room temperature for one day. Other target parameters were as follows: resin solids loading level of 9.0%, wax content of 1.0%, mat moisture content of 8.0% all based on oven-dried wood weight; board dimensions of 24 in. × 22 in. × 0.5 in., board density of 50.0 pcf, and hot-pressing temperature of 350 °F. Boards were cut and evaluated for internal bond (IB) strength (eight test pieces) and bending strengths (MOR and MOE) (four test pieces) on an Instron machine (Norwood, MA). Water-soak thickness swelling (TS) and water absorption (WA) values (two test pieces) were measured at 20 °C in a water bath according to ASTM D 1037-06a.

Formaldehyde Content (FC) Measurements: The tests were conducted according to the European standard method (EN120 2001) [4]. Board samples had moisture contents of about 3.0% and the resultant FC data were not corrected, although the standards are based on 6.5% moisture content of boards. In this study, tests were done after about three months of hot-pressing of boards due to an instrument breakdown. FC test samples with dimensions of 6 in. × 6 in. were cut from each board after one-day of airing of the boards after hot-pressing. Because of this unexpected delay, each test sample was sealed at the

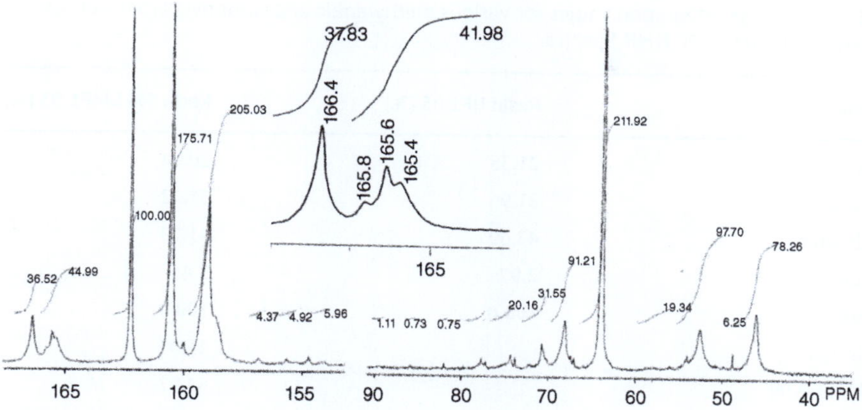

Figure 26.1 ^{13}C NMR spectra of *Resin 5.0%UMF1.05* in water (D_2O).

edges with duct tape, wrapped in saran film, and put in a sealed plastic bag and all samples were kept in a refrigerator at 4 °C until testing. The formaldehyde content (FC) decreases of boards from the long storage time appeared minimal in comparison with reported data of UF resin-bonded boards tested in one to two weeks of board preparation in the previous paper.

^{13}C NMR spectra: Quantitative ^{13}C NMR spectra were obtained with a Techmag 360-MHz NMR spectrometer by Spectral Data Services, Inc., Champaign, IL. The spectra were analyzed similarly as in the previous publications [5, 6].

26.1.2 Resin Chemical Structures Determined by ^{13}C-NMR

Chemical structures of resin UF1.05 and resin 5.0%UMF1.05 from ^{13}C NMR are summarized in Table 26.2 and the spectrum of the latter resin is shown in Figure 26.1 with an inset of expanded melamine carbon peaks. Chemical shift values were referenced to the free urea peak at 164.0 ppm. The *methyleneic groups* expected of melamine-formaldehyde components were not well differentiated from those of UF resin components. The calculated F/(U + M) mole ratios (total CH_2 groups/total urea and melamine carbons) were only slightly lower than the charged F/U mole ratio of 1.05. The degrees of polymerization of both resins were low at about 1.92, partly due to the low F/U mole ratios used with large amounts of second urea portion needed in the syntheses.

Several new findings from the ^{13}C NMR data were noted. *(i) Resin 5%UMF1.05* showed four peaks for melamine carbons (165.4–165.8 ppm), in that 55.0% of melamine is shown to have reacted with formaldehyde and *45.0% melamine* (166.4 ppm) remained unreacted, indicating the low concentration of formaldehyde available at the point of melamine addition in the resin synthesis Method #4, where no second formaldehyde addition was included, differently than in the previous chapter (Section 25.1.4), where melamine molecules were all in the reacted form. Un-reacted melamine in resins tends to come out as precipitates to shorten the storage times, since melamine itself has a low solubility in (cold) water. The optimum efficiency or resin stability would be obtained when melamine reacts with two to three formaldehyde molecules. Since the melamine addition point selected in this study led to about 45% un-reacted melamine, the full utilization of the

Table 26.2 Percentage integration values for various methyleneic and carbonyl carbons of resin samples determined from ^{13}C NMR Spectra.

Carbon groups	Resin UF1.05 (%)	Resin 5% UMF1.05 (%)
Free urea	21.38	20.04
Mono-sub. urea	31.96	35.22
Di, tri-sub. urea	43.69	41.29
Cyclic urea	2.97	3.45
Total urea	**100.0**	**100.0**
Free melamine		45.03
Mono-, di-sub Mel.		54.97
Total melamine		**100.0**
Free formaldehyde	0.60	0.47
Total hydroxymethyl	**43.30**	**44.59**
Type I	34.46	38.81
Type II	8.84	5.78
Total methylene-ether	**17.92**	**17.82**
Type I	11.71	12.20
Type II	4.59	3.68
Type III	1.62	1.92
Total methylene	**38.18**	**37.12**
Type I	14.95	15.69
Type II	19.96	17.89
Type III	3.28	3.54
Total CH_2	**100.0**	**100.0**
CH_2/CO	**1.023**	**1.038**
Degree of polymer	1.931	1.915

Notes: CH_2/CO refers to methyleneic carbons/carbonyl ratios calculated from the integration values; Degree of polymerization according to: DP = 1/[1−(methylene + 0.5 × methylene-ether)/urea]; Chemical structures of carbon groups in the table, refer to the cited Ref. [7].

higher-cost melamine may not be realized. *(ii) Another finding from the ^{13}C NMR data* was the fact that *the level of methylene-ether groups* was the same as in UF resins [6] because of the melamine addition done in the third step of resin synthesis.

(iii) Another interesting result is that Resin *5.0%UMF1.05* had slightly higher hydroxylmethyl group content than *Resin UF1.05*, wherein the Type I hydroxylmethyl group increased from 34.46% to 38.81%, and the Type II hydroxylmethyl group decreased from 8.84% to 5.78%. These changes indicate that melamine reacted not only with free formaldehyde present but also with formaldehyde formed from de-hydroxymethylation reaction of Types I and II hydroxylmethyl groups of polymeric UF resin components. The increased content of the mono-substituted urea from 31.96% to 35.22% would also be the result of this transfer of hydroxylmethyl groups.

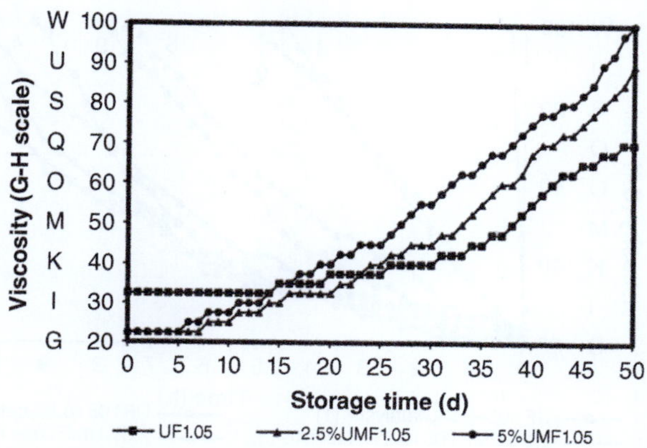

Figure 26.2 Viscosity increases of UF and UMF resins during storage at 30 °C (storage lives).

(iv) During the resin synthesis procedure, it was also observed that the addition of melamine lowered the viscosity from P-Q to N-O on the G-H scale after 1.0 hour of reaction, in agreement with the migration of hydroxymethyl groups from the polymeric UF resin components to melamine. Overall, the ^{13}C NMR results indicated that UMF resins made with low melamine levels in this work are similar to those of typical UF resins with only minor variations occurring due to melamine addition.

26.1.3 Storage Stability of UF and UMF Resins

The viscosity changes that occurred during the resin storage at 30 °C for 50 days are shown in Figure 26.2. *Resins 2.5%UMF1.05* and *5.0% UMF1.05* remained clear for a few days and then turned cloudy, indicating the separation/precipitation of free melamine or melamine-formaldehyde reaction products, which would affect the viscosity and storage stability negatively. *Resin UF1.05* showed the highest storage stability, while Resin 2.5%UMF1.05 showed a higher stability than Resin 5.0%UMF1.05, indicating that storage stability decreased with increasing levels of melamine. If the viscosity K on the G-H scale is considered to be the maximum value for spraying in board manufacturing, *Resin UF1.05* has a storage life of 31 days, Resin *2.5%UMF1.05* has 24 days, and Resin *5.0%UMF1.05* has 19 days of storage lives. Thus, the choice of melamine level for use in UMF resins synthesized according to Method #4 could partly hinge on the storage life required for a particular board manufacturing plant.

26.1.4 Properties of Catalyzed Resins

Pot-lives: Pot-lives of catalyzed resins measured at 30 °C are shown in Figure 26.3. Assuming that a viscosity of S on the G-H scale is the target level, UMF resins with a 2% level of Catalyst B showed pot-lives of about 3.5 hours and with a 0.5% level of the same catalyst about 10.5 hours. *Resin UF1.05* gelled in 15 minutes with 0.5% catalyst B, indicates that Catalyst B is not appropriate for UF resins. In industrial board manufacturing plants, the resin is catalyzed and then soon sprayed onto wood particles. The mat is then promptly formed and hot-pressed within about 30 minutes, and therefore, the applied resin should not be advancing too quickly before the hot-pressing. It appears that both *UMF resins* catalyzed at 1.0% to 1.5% levels of Catalyst B would have comparable pot-lives with *Resin UF1.05* catalyzed

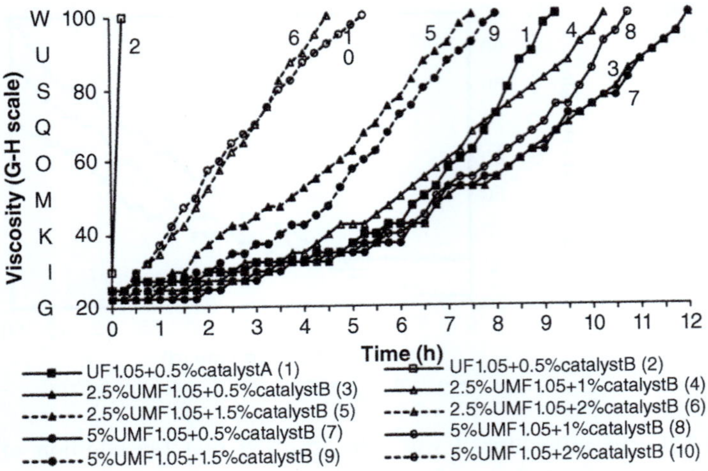

Figure 26.3 Pot-lives of catalyzed UF and UMF resins with different catalysts and contents.

Table 26.3 Gel times (s) of resins with Catalysts A and B at various levels at 100 °C.

Catalyst level	0.5%		1.0%		1.5%		2.0%		2.5%		3.0%
Catalyst	A	B	A	B	A	B	A	B	A	B	B
UF1.05	131	105	103	80	85	65	86	64	—	—	—
2.5%UMF1.05	219	187	201	172	179	162	169	146	167	138	139
5.0%UMF1.05	223	204	197	179	184	168	169	162	168	157	158

at a 0.5% level of Catalyst A. Thus, no particular problem is expected for the UMF resins regarding the pot-lives.

Gel times of catalyzed resins measured at 100 °C are shown in Table 26.3. Gel times were shorter with increasing catalyst levels and were also shorter with Catalyst B than Catalyst A, reflecting the free sulfuric acid content in Catalyst B. Resin 2.5%UMF1.05, has shorter gel times than *Resin 5.0%UMF1.05*, indicating that melamine increases the gel time even with the stronger Catalyst B. Thus, UMF resins even with Catalyst B are significantly slower than UF resins, indicating that melamine ties up the acid catalyst and delays the onset of curing reactions.

Gel and Cure Times of Catalyzed Resins Measured with Rheometer: Curing data obtained in the dynamic mechanical analyzer (DMA) mode of the rheometer are reported in Table 26.4. In this test, the resin sample was heated from room temperature to the target isothermal temperatures of 90 °C, etc. in about 50 seconds and then held for 25 minutes. The moisture in the resin sample would be mostly gone when it reached the curing temperature, and this fact makes this test different from the gel time results discussed above and more likely to reflect the resin curing times in the hot-pressing of boards. In typical runs, storage modulus, loss modulus, and tan δ curves were obtained vs. elapsed times. The gel time was taken as the time needed to reach to the maximum value of tan δ, and the cure time is defined as the time when the storage modulus curve in the maximum slope region meets the tangent

Table 26.4 Gel and cure times obtained under isothermal conditions at different catalyst levels from rheometry.

Resin type	Catalyst type	Catalyst level (%)	Gel time (s)				Cure time (s)			
			90 °C	120 °C	135 °C	145 °C	90 °C	120 °C	135 °C	145 °C
UF 1.05	A	0.5	186	152	128	109	400	265	180	—
	B	0.5	60	—	—	—	137	—	—	—
2.5% UMF1.05	B	0.5	181	117	102	95	525	210	165	130
		1.0	123	116	88	87	333	185	133	140
		1.5	122	116	87	81	333	203	152	140
		2.0	87	81	81	74	253	145	123	117
5.0% UMF1.05	B	0.5	176	158	102	102	483	300	242	215
		1.0	123	119	95	88	325	227	187	157
		1.5	115	112	96	88	325	183	207	120
		2.0	96	94	91	81	275	180	147	135

drawn from its maximum value at the end of the run [7–9]. All resin samples showed tan δ values of about 0.10 at the end of curing, indicating that typical extents of vitrification and complete curing were achieved in tests [7]. The gel times and cure times in Table 26.4 were shortened with increasing curing temperatures and also with higher catalyst levels, as expected. *Resin 2.5%UMF1.05* showed shorter gel and cure times than *Resin 5.0%UMF1.05*, indicating the slower curing of melamine components as observed in pot-life and gel times discussed above. In industrial hot-pressing of PBs, the core-layer temperature normally reaches the range of 140 to 150 °C within about 150 to 180 seconds of pressing times for a typical 0.5-inch-thick board [10]. The gel and cure time values of ~109 and ~180 seconds for *Resin UF1.05* with typical Catalyst A measured at 145 °C appear to indicate that the UF resin could bond boards well at hot-pressing times of about three minutes. Similar values are obtained for *Resins 2.5%UMF1.05* and *5.0%UMF resin1.05* with Catalyst B at 1.0% to 2.0% levels, but Catalyst B at 0.5% levels appeared to be somewhat weaker.

26.1.5 Particleboard Test Results

Control UF Resin in the Face- and Core-layers: The test results are shown in Table 26.5, arranged with the board numbers and preparation parameters as shown in Table 26.6. PBs made with Resin UF1.05 in the face-layer and Resins UF1.05, 1.15, and 1.25 in the core-layers with Catalyst A (Boards 1–3) represent the current range of UF resins used in the industry, of which the last resin combination, Resin UF1.25 in the core-layer, being the more common one used for short hot-pressing times and good physical properties of boards [11]. However, the formaldehyde content (FC) value by the perforator extraction (PE) test method with this mole ratio was 16.8 mg/100 g board, which equaled the typical values of boards manufactured in the industry before the enactment of the new FE laws [3]. This FC

Table 26.5 Formaldehyde content (FC) [4] and physical property test results of particleboards.

Bd no.	CH_2O mg/100 g Bd Press time (min) 3.0	 3.5	Physical properties of particleboards. Averages of 3.0- and 3.5-min press times Density (lbs/ft³)	 IB (psi)	 MOR (psi)	 MOE (kpsi)	 24-h TS (%)	 24-h WA (%)
1	7.9	8.0	48.6	58.0	940	141	29.1	76.6
2	13.6	12.8	49.7	74.4	1295	180	20.9	58.5
3	16.8	15.9	50.8	90.4	1495	202	18.5	51.2
4	7.8	7.3	50.6	74.4	1355	182	28.7	61.0
5	6.7	6.1	50.8	96.8	1530	215	25.1	61.9
6	6.2	6.0	50.5	78.1	1365	183	27.3	64.9
7	5.7	5.4	50.4	71.8	1285	185	27.0	71.3
8	8.0	7.1	51.2	90.9	1640	227	26.8	68.6
9	7.3	6.0	51.5	124.3	1720	235	23.6	65.3
10	6.2	5.8	51.3	113.0	1895	250	22.3	74.7
11	5.6	5.2	50.8	103.2	1610	222	22.1	73.6
12	6.7	6.1	51.7	116.3	1815	262	16.8	49.6
13	6.6	5.9	50.9	116.3	1635	225	18.7	52.2
14	7.4	6.8	51.3	127.8	1925	273	16.7	41.3
15	7.0	5.9	52.0	134.3	2160	299	16.1	42.0

value corresponds to a FE value of about 0.20 ppm in the large-scale chamber test method [12], which is significantly higher than the new emission limit of 0.09 ppm. *Resins UF 1.05* and *1.15* used in the core-layers lowered the FC values to 8–13 mg/100 g ranges, but the boards showed lower IB and lower bending strength values due to an insufficient extent of cross-linking in cured core-layer UF resins.

Control Resin UF1.05 in the Face-layer and Resin 2.5%UMF1.05 in the Core-layer: PBs bonded with this set of resins at various levels of Catalyst B (Boards 4 to 7) showed lower FC values of 6–8 mg/100 g boards with acceptable IB values when using Catalyst B at a 1.0% level in the core-layer. These PE FC values correspond to the E1 class of European Standards. The longer press time of 3.5 minutes showed slightly lower PE values, as expected. The UMF resin required the stronger catalyst B, of which the free acid content needs to be weighed in with respect to its long-term wood degradation potential [13, 14]. The UMF resins' curing rates were found to be too slow when using the common *Catalyst A* under the laboratory hot-pressing conditions. Use of higher levels (1.5% and 2.0%) of Catalyst B tended to show lower IB values and poorer water-soak test values, indicating the over-catalyzing effects. The approach of using a UF resin in the face-layer and UMF resins in the core-layer would lower the overall resin cost. Using UMF resins in the core-layer fits more to capture the free formaldehyde that is accumulating during the hot-pressing of boards because of melamine's higher reaction capacity for formaldehyde.

Table 26.6 Particleboard preparation: board number & parameters.

	Face-layer		Core-layer	
Bd no.	**Resin/mole ratio**	**Catalyst level (%)**	**Resin/mole ratio**	**Catalyst level (%)**
1	UF 1.05	A 0.5	UF 1.05	A 0.5
2			UF 1.15	
3			UF 1.25	
4	UF 1.05	A 0.5	2.5% UMF1.05	B 0.5
5				B 1.0
6				B 1.5
7				B 2.0
8	UF 1.05	A 0.5	5.0% UMF1.05	B 0.5
9				B 1.0
10				B 1.5
11				B 2.0
12	2.5% UMF1.05	B 0.5	5.0% UMF1.05	B 1.0
13		B 1.0		
14	5.0% UMF1.05	B 0.5	5.0% UMF 1.05	B 1.0
15		B 1.0		

Control Resin UF1.05 in the Face-layer and Resin 5.0%UMF1.05 in the Core-layer: Boards bonded with this set of resins at various levels of Catalyst B (Boards 8–11) showed PE values of 5.2–8.0 mg/100 g boards with acceptable IB values only with 1.0% to 1.5% levels of Catalyst B. These FC values are of slight improvements over Resin *2.5%UMF1.05* used as core-layer resin discussed above but similarly corresponded to E1 Class of European Standards. IB and bending strength values and water-soak test values were also improved.

Resin 2.5%UMF1.05 in the Face-layers and 5.0%UMF1.05 in the Core-layer: Boards bonded with this set of resins at the 1.0% B catalyst level (Boards 12–13) showed similar PE FC values of 5.9–6.7 mg/100 g boards with acceptable IB values. The PE values were slight improvements over using Resin UF1.05 in the face-layer (Boards 8–11) and similarly correspond to the E1 class of European Standards. The main advantage of using the UMF resin in the face-layer is the improved water-soak thickness-swell and WA values. On the other hand, IB and bending test values indicated that the catalyst level in the face-layer with *Resin 2.5%UMF1.05* showed higher values with 0.5% catalyst level than at 1.0% catalyst level, indicating that the catalyst levels of UMF resins also need to be adjusted according to the layer of board.

Resin 5%UMF1.05 in Both Face- and Core-layers: Boards bonded with this set of resins at 0.5–1.0% levels of Catalyst B (Boards 14–15) showed similar PE values of 5.9–7.4 mg/100 g boards with relatively good IB, bending strength, and water-soak test values. These PE FC values did not improve over the values of using Resin 2.5%UMF1.05 in the face discussed above (Boards 12–13). However, IB, bending strength values, and water-soak test values of boards were improved further due to the extra melamine in the face-layer UMF resin.

The 1.0% catalyst content in the face-layer appears to perform better than 0.5% level, indicating again the need of adjusting the catalyst content according to the melamine level of UMF resins and board layers. The good physical properties of boards resulting from this set of resins appear to indicate that the mole ratio could be lowered further to 1.00 or below to further lower PE FC values of boards, although the catalyst strength level and the hot-pressing time may have to be increased. Overall, boards made with the longer press time of 3.5 minutes showed slightly lower PE FC values than 3.0 minutes press time, indicating that formaldehyde values are dependent on the expulsion of steam during hot pressing. It should be also noted that industrial boards made with the same resins and same hot-pressing temperature and time would show lower IB strength and higher PE values than the boards made in the laboratory because of bigger sizes of boards manufactured in the industry. The moisture in the mat, which entails the free formaldehyde generated from resins during hot-pressing, takes longer time to get expelled from bigger boards.

26.1.6 Summary

UMF resins made with Method #4 of adding melamine at the beginning of the third step of UF resin synthesis with 2.5% and 5.0% levels of melamine at a mole ratio of 1.05 were found to be adequate as core-layer binders for PB. The PE FC values of boards reached the required value of the E1 Class by the European Standards. UMF resins needed to be catalyzed with stronger, free acid-containing catalysts at appropriate levels, depending on the melamine levels and also on the layers of boards to be applied. Curing rates of UMF resins were still slower than typical UF resins and needed a stronger catalyst. Use of the UMF resins in both face- and core-layers of PBs resulted in improving the strength and water-soak test values of boards significantly. The chemical structures of the UMF resins analyzed by ^{13}C NMR were similar to those of typical UF resins with similar methylene and methylene-ether group contents in agreement with the mode of melamine addition used. About 55% of added melamine reacted with formaldehyde in the resins. The room temperature storage stability of the UMF resins was slightly poorer than the control UF resin.

26.2 UMF Resins Synthesized by Adding Melamine at Different Points

(Adapted with permission and minor revisions from: An Mao, El Barbary Hassan, and Moon G. Kim. 2013. The effects of adding melamine at different resin synthesis points of low mole ratio urea-melamine-formaldehyde (UMF) resins. BioResources 8(4), 5733–5748.)

In addition to Method #4 discussed above (Section 26.1), in the past, melamine was used at higher levels and the addition was often carried out at the beginning or in the middle of UF resin synthesis so that the melamine components were forced to go through the usual acidic second step of reaction as shown in the previous paper and other reports [15–20]. It is known that dimeric or higher melamine-formaldehyde reaction products are quickly formed in the acidic reaction, and they precipitate out of the water phase [21]. The precipitated MF polymers would shorten the storage lives of resins and affect performance properties and although the effects could be less severe at low melamine addition levels. In this study, adding melamine at three different points in the early part of UF base resin synthesis procedure was investigated.

26.2.1 Syntheses of UMF Resins

In *Method #1 (UMF-1)* the melamine addition was done with the first urea; in *Method #2 (Resin UMF-2)* at the mid-point in the reaction of the first urea; and in *Method #3 (Resin UMF-3)* at the end of reaction of first urea as shown, respectively, in Tables 26.7a–c. No second formaldehyde addition was used. Then, the acidic polymerization step was conducted at pH 5.25 till to P-Q viscosity, followed by the second urea addition in alkaline pH as usual. The resulting resins were investigated and compared with the results of the previous report (Section 26.1), where the melamine addition was done in the last step just before the addition of second urea (*Method #4 and Resin UMF-4*). Test procedures and evaluation methods were all kept the same as in the previous study [22]. PBs were made with *2.5%* and *5.0% Resins UMF-1* and evaluated. Test results for *control UF resins* having F/U mole ratios of 1.25, 1.15, and 1.05 and *Resins 2.5%UMF-4* and *5.0%UMF-4* were taken from the previous paper for comparison.

Table 26.7a Syntheses of Resin UMF-1 according to Method #1.

UFC made in the previous reports was used.

(1) UFC (1508.1 g) was taken and mixed at pH 8.0 and 70 °C with melamine (61.4 g) and then the temperature was raised to 90 °C and maintained for 30 min at pH 8.0. The first urea (319.2 g) was then added, reaching an F/(U + M) mole ratio of 2.0, and the temperature was maintained at 90 °C for another 30 min at pH 8.0.
(2) Then, the pH was adjusted to 5.25, a slightly higher pH than the typical values used for UF resins, which was found necessary to slow the reaction rate within a controllable range. It was then reacted to *P–Q* viscosity (1 h and 25 min).
(3) Then, the pH was adjusted to 8.0 and the third urea (567.1 g) was added, reaching to an F/(U + M) mole ratio of 1.05, giving ***Resin 2.5%UMF-1*** with a viscosity of G-H.
(4) For the 5.0% melamine level, the same procedure was followed, using the UFC and first urea (289.1), melamine (124.5 g), and the second urea (566.9 g), resulting in ***Resin 5.0%UMF-1*** with a final viscosity of H-I.

Table 26.7b Syntheses of Resin UMF-2 according to Method #2.

UFC made in the previous reports was used.

(1) UFC (1508.1 g) was taken and reacted at pH 8.0 and 70 °C with the first half of the first urea (159.6 g) and then the temperature was raised to 90 °C and maintained for 15 min; then, melamine (61.4 g) was added and reacted at pH 8.0 and 90 °C for 30 min. Then, the second half of the first urea (159.6 g) was added to an F/(U + M) mole ratio of 2.0, and the temperature was maintained at 90 °C for another 30 min at pH 8.0.
(2) Then, the pH was adjusted to 5.25 and reacted until *P–Q* viscosity (1 h and 25 min).
(3) Then, the pH was adjusted to 8.0 and the third urea (567.1 g) was added, reaching to an F/(U + M) mole ratio of 1.05, giving ***Resin 2.5%UMF-2*** with a final viscosity of *H–I*.
(4) For the 5.0% melamine level, the same procedure was followed, using the first urea portion (144.6 g/144.6 g), melamine (124.5 g), and the second urea portion (566.9 g), resulting in ***Resin 5.0%UMF-2*** with a final viscosity of *H–I*.

Table 26.7c Syntheses of UMF resins according to Method #3.

UFC made in the previous reports was used.

(1) UFC (1508.1 g) was taken and reacted at pH 8.0 and 70 °C with the first urea portion (319.2 g), then the temperature was raised to 90 °C and maintained for 30 min at pH 8.0. Then, melamine (61.4 g) was added, reaching to an F/(U + M) mole ratio of 2.0, followed by reacting at pH 8.0 and 90 °C for 30 min.
(2) Then, the pH was lowered to 5.25 and reacted to *P–Q* viscosity (1 h and 15 min).
(3) Then, the pH was adjusted to 8.0 and the second urea (567.1 g) was added, reaching an F/(U + M) mole ratio of 1.05, giving ***Resin 2.5%UMF-3*** with *G–H* viscosity.
(4) For the 5.0% melamine level, the same procedure was followed, using the first urea (289.1 g), melamine (124.5 g), and the second urea (566.9.5 g), resulting in ***Resin 5.0% UMF-3*** with a final viscosity of *H–I*.

26.2.2 Physical Properties of Synthesized UMF Resins

The synthesized UMF resins showed pH values of 8.0, specific gravity values from 1.262 to 1.281, viscosity values from G-H to I-J on G-H scale, and solids content values ranging from 62.3% to 64.0%. The obtained values were within the expected ranges from the synthesis methods and procedures used and also within the range of current industrial values. However, all six synthesized UMF resins in this work were turbid.

26.2.3 Resins' Chemical Structures From ^{13}C NMR Results

All chemical shift values were referenced to the free urea peak set at 164.0 ppm. The turbidity of the three synthesized UMF resins (5.0%UMF-1, 5.0%UMF-2, and 5.0%UMF-3) could have resulted in some distortion of the ^{13}C NMR quantification results due to the (liquid-state) NMR method's inability to detect solid components in samples. Calculated chemical structures from the ^{13}C-NMR spectra are summarized in Table 26.8 and the spectra of the three synthesized UMF resins are shown in Figure 26.4a–c, and the different features are noted in the following: *(i)* MF components had fully reacted with formaldehyde in all three resins due to the high concentrations of formaldehyde available at the addition points of melamine, while only 55.0% of melamine reacted in Resin 5.0%UMF-4. *(ii)* The total methylene-ether group contents of the three resins were increased in comparison with UF control or UMF-4, from forming of dimers and trimers, etc. of MF resin components in the mild acidic step at pH 5.25. *(iii)* For resin 5.0%UMF-4, the melamine components did not go through the acidic step, and its total methylene-ether group contents were lower and similar to those of the typical resin UF1.05. Thus, the present resins' turbidity and some possible deviations of NMR results mentioned above agree with the increased methylene-ether group contents of MF resin components. *(iv)* Another interesting result is that the three resins showed higher hydroxymethyl group contents in comparison with control resins, from 34.46% to about 40.0% for Type I, and from 8.84% to about 12.0% for Type II hydroxymethyl groups, indicating that the UF components have not fully polymerized at the milder acidic reaction condition used (pH 5.25), which was used to keep the viscosity advancement rate under control. *(v)* Thus, although all resins were synthesized to have similar end-viscosity values of about GH, a considerable portion of the viscosity values appears to have arisen from the polymerization of MF components at the

Table 26.8 Percentage integration values for various methyleneic and carbonyl carbons of resins determined from ^{13}C NMR spectra.

Bond groups	UF 1.05 (%)	5.0% UMF-1 (%)	5.0% UMF-2 (%)	5.0% UMF-3 (%)	5.0% UMF-4 (%)
Free urea	21.38	26.18	26.48	27.3	20.04
Mono-sub urea	31.96	34.58	33.10	34.61	35.22
Di- & tri-sub urea	43.69	32.84	34.01	32.79	41.29
Cyclic urea	2.97	6.39	6.41	5.30	3.45
Total urea	**100.0**	**100.0**	**100.0**	**100.0**	**100.0**
Free melamine	--	0	0	0	45.03
Mono-/di-sub Me.	--	100.0	100.0	100.0	54.97
Total melamine	--	**(100.0)**	**(100.0)**	**(100.0)**	**100.0**
Free formaldehyde	0.6	0.17	0.17	0.16	0.47
Total hydroxymethyl	**43.3**	**51.54**	**50.33**	**51.98**	**44.59**
Type I	34.46	42.33	40.78	42.94	38.81
Type II	8.84	8.21	9.55	9.03	5.78
Total methylene-ether	**17.92**	**27.91**	**28.01**	**27.45**	**17.82**
Type I	11.71	16.44	16.01	16.81	12.2
Type II	4.59	6.92	7.25	6.73	3.68
Type III	1.62	4.54	4.75	3.91	1.92
Total methylene	**38.18**	**21.09**	**21.48**	**20.41**	**37.12**
Type I	14.95	10.68	10.61	10.23	15.69
Type II	19.96	10.40	10.87	10.18	17.89
Type III	3.28	0	0	0	3.54
Total CH_2	**100.0**	**100.0**	**100.0**	**100.0**	**100.0**
CH_2/(CO+Mel)	**1.023**	**0.980**	**1.026**	**0.986**	**1.038**
Degree of Polymer	1.931	1.553	1.606	1.533	1.915

Notes: CH_2/CO ratio refers to methyleneic carbons/carbonyl ratios calculated from the integrated values. Degree of polymerization was calculated using DP = 1/[1-(methylene + 0.5 × methylene-ether)/urea]. For chemical structures and names of functional groups in the table, refer to the previous papers.

expense of UF resin components. This result also agrees with the lower methylene group contents in the UMF-1, UMF-2, and UMF-3 resins, which also indicates their significantly lower degree of polymerization, 1.57 vs. 1.92.

26.2.4 Storage Stability of Un-catalyzed Resins

Viscosity increases that occurred during the 50-day storage period at 30 °C are shown in Figure 26.5. Resins 2.5%UMF-1, 5.0%UMF-1, 2.5%UMF-2, 5.0%UMF-2, 2.5%UMF-3, and 5.0%UMF-3 were cloudy at room temperature from the beginning. Resins 2.5%UMF-4 and 5.0%UMF-4 of Section 26.1 remained clear for a few days and then turned cloudy,

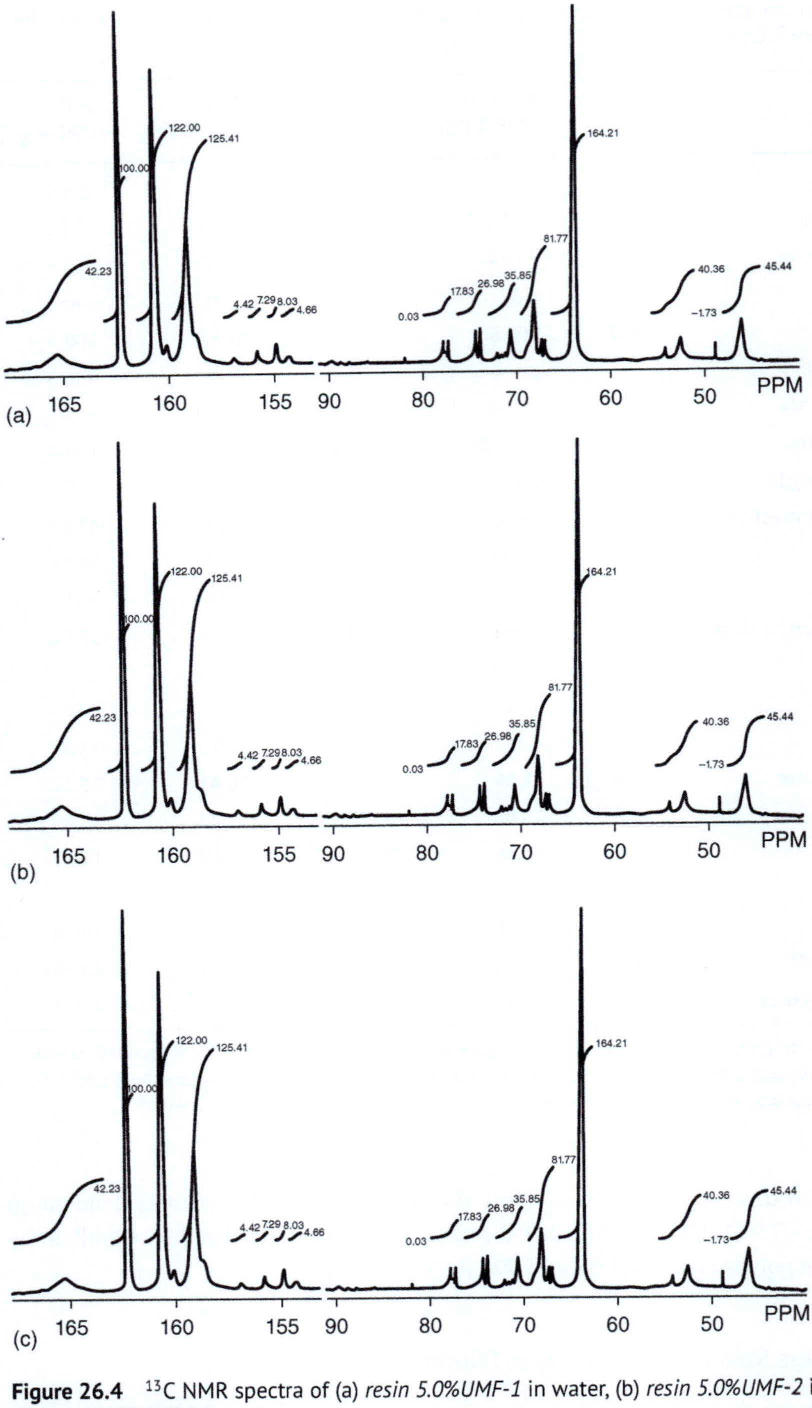

Figure 26.4 ^{13}C NMR spectra of (a) *resin 5.0%UMF-1* in water, (b) *resin 5.0%UMF-2* in water, and (c) *resin 5.0%UMF-3* in water.

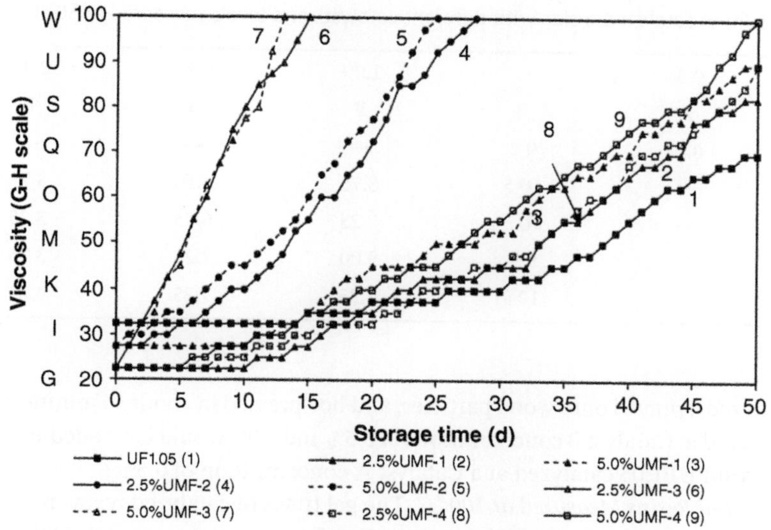

Figure 26.5 Viscosity increases of UF and UMF resins observed during storage at 30 °C.

indicating a longer delay in the precipitation of solid particles of MF reaction products. Turbidity of these resins may indicate a lower viscosity storage stability. Resin UF1.05 showed the longest storage stability, and moreover, the storage stability of the UMF resins decreased with increasing levels of melamine content. If viscosity K on the G-H scale is considered to be the maximum value allowed in PB manufacturing, then resin UF1.05 has a storage life of 31 days, resin 2.5%UMF-1 has a storage life of 22 days, resin 5.0%UMF-1 has a storage life **of 17 days**, resin 2.5%UMF-2 has a storage life of 9 days, resin 5.0%UMF-2 has a storage life of 7 days, resins 2.5%UMF-3 and 5.0%UMF-3 have storage lives of 3 days, resin 2.5%UMF-4 has a storage life of 24 days, and resin 5.0%UMF-4 has a storage life of 19 days. Currently, U.S. PB manufacturing plants use resins within 14–21 days of delivery. Thus, Resins UMF-2 and UMF-3 types were judged to be unacceptable on the basis of their storage lives and were excluded from further testing in this study. The short storage lives of Resins UMF-2 and UMF-3 types can be explained in terms of the oligomeric MF components formed in the acidic polymerization step, which have a higher tendency of precipitation because they have less than full substitutions with hydroxymethyl groups (3 or higher) due to the melamine addition points used, where the amount of available formaldehyde is limited in comparison with resin UMF-1 types. On the other hand, for UMF-4 type resins, the MF components did not go through the acidic polymerization step and therefore less dimers or trimers were formed.

26.2.5 Properties of Catalyzed UMF Resin

Pot-lives of catalyzed resins measured at 30 °C are shown in Table 26.9 by assuming viscosity S on the G-H scale is the target value. As will be discussed later, UMF resins used as PB binders need to be cured with the stronger catalyst B; Resin UF1.05 gelled in 15 minutes with 0.5% Catalyst B at room temperature, indicating that Catalyst B is too strong for UF resins. Further, it is apparent that different UMF resin types have different pot-lives due to the melamine addition points used in resin synthesis. However, in PB manufacturing

Table 26.9 Pot-lives (h) of catalyzed resins with different catalysts and levels measured at 30 °C.

Catalyst level	0.5%		1.0%	1.5%	2.0%
Catalyst kind	A	B	B	B	B
UF1.05	8.25	0.2	--	--	--
2.5%UMF-1	--	10.5	8.75	6.50	3.75
5.0%UMF-1	--	10.5	9.25	6.75	3.75
2.5%UMF-4	--	12+	9.00	7.25	3.75
5.0%UMF-4	--	12+	9.25	7.25	4.25

plants, resin is catalyzed, sprayed onto wood particles, and hot-pressed in about 30 minutes; all UMF resins catalyzed at Catalyst B concentrations of 0.5% and 1.0% would have adequate pot-lives, as would resin UF1.05 catalyzed at a Catalyst A concentration of 0.5%.

Gel Times of Catalyzed Resins Measured at 100 °C: The gel times of catalyzed resins measured at 100 °C are shown in Table 26.10. Gel times became shorter with increasing catalyst levels and also with catalyst B. Resin UF1.05 showed shorter gel times than UMF resins. Furthermore, UMF-1 resins showed shorter gel times than UMF-4 resins, as indicated by (negative) difference values. Gel times are an indication of the curing speed of resins, and all UMF resins' gel times were longer and became comparable with those of resin UF 1.05 only with 2.0% level of Catalyst B, the stronger catalyst. The shorter gel times of UMF-1 resin compared to Resins UMF-4, although the differences were small, appear to be related to the higher levels of solid particles of oligomeric MF resin components in the former resins.

Gel and Cure Times of Catalyzed Resins Measured on Rheometer: The gel and cure times of catalyzed resins measured on the rheometer in the DMA mode are reported in Table 26.11. The resin sample was heated from room temperature to the target isothermal temperatures, 120 or 145 °C, in about 50 seconds and then held for 25 minutes. The moisture in the resin sample was mostly evaporated out by the time it reached the curing temperature, making the test results different from the (liquid) gel time results discussed above. The gel time and cure times were significantly longer for UMF resins than the control UF resins, and for all resins, they were shorter with increasing curing temperatures and also with higher catalyst levels, as expected. Further, UMF resins with 2.5% melamine levels showed shorter gel and cure times than those with the 5.0% melamine level, indicating the effects of slow-curing melamine components as observed in the pot-lives and gel times of catalyzed resins discussed above. Further, Resins UMF-1 showed slightly longer gel times but slightly shorter cure times in comparison with Resins UMF-4 types, indicating a faster curing of UMF-1 resins because the resins' MF components are more fully substituted with hydroxymethyl groups than in Resins UMF-4.

26.2.6 Particleboard (PB) Test Results

PB test results of UMF-1 resins shown in Table 26.12 are arranged according to the board preparation parameters shown in Table 26.13. Additionally, the increase and decrease (−) of test values calculated in comparison with the previous results of resins of *Method-#4* (UMF-4) are reported with a slash (/).

Table 26.10 Gel times (s) of resins with catalysts A and B at various levels measured at 100 °C.

Catalyst level	0.5%		1.0%		1.5%		2.0%		2.5%		3.0%	
Catalyst kind	A	B	A	B	A	B	A	B	A	B	A	B
Resin UF1.05	131	105	103	80	85	65	86	64	—	—	—	—
Resin 2.5%UMF-1	212/−7	175/−12	183/−18	180/8	175/−4	152/−10	146/−25	134/−12	145/−22	124/−14	212/−7	175/−12
Resin 5.0%UMF-1	242/19	188/−16	201/4	165/−14	171/−7	152/−16	157/−13	136/−26	162/−6	131/−26	242/19	188/−16

Notes: Values separated by a slash (/) indicate the increases or decreases in comparison with data from Resin UMF-4 reported in the previous Section 26.1.

Table 26.11 Gel times and cure times (s) obtained under isothermal conditions at different catalyst levels by rheometry.

		Gel time (s)		Cure time (s)	
Resin	**Catalyst level (%)**	**120 °C**	**145 °C**	**120 °C**	**145 °C**
UF1.05	A/0.5	152	109	265	--
	B/1	--	--	--	--
2.5% UMF-1	B/0.5	132/15	101/6	207/−3	132/−2
	B/1.0	117/1	90/3	174/−11	120/−20
	B/1.5	96/20	82/1	175/−28	120/−20
	B/2.0	88/7	74/0	139/−6	101/−16
5.0% UMF-1	B/0.5	175/17	116/14	226/−74	166/−49
	B/1.0	166/47	118/30	207/−20	133/−24
	B/1.5	127/15	110/22	206/23	121/−1
	B/2.0	105/11	96/15	182/2	123/−14

Notes: Values separated by a slash (/) indicate the increases or decreases (−) in comparison with data from resin UMF-4 reported in the previous Section 26.1.

Table 26.12 Formaldehyde content (FC) and physical property test results of particleboards and difference values from resins of Method #1.

CH_2O mg /100 g board [4]		**Physical properties of particleboards. Averages of 3.0- and 3.5-min press times**				
Press time (min)		IB (psi)	MOR (psi)	MOE (kpsi)	24-h TS (%)	24-h WA (%)
3.0	3.5					
7.9	8.0	58.0	940	141	29.1	76.6
13.6	12.8	74.4	1295	180	20.9	58.5
16.8	15.9	90.4	1495	202	18.5	51.2
8.0 /0.2	7.7/0.4	82.2/7.8	1327/−8	186/4	26.1/−2.6	57.2/−3.8
7.2/0.5	6.5/0.4	92.5/−4.3	1438/−92	219/4	22.0/−3.1	56.8/−5.1
6.4/0.2	6.1/0.1	85.1/7.0	1338/−27	202/19	24.7/−2.6	58.9/−6.0
6.0/0.3	5.5/0.1	78.7/6.9	1223/−62	191/6	26.5/−0.5	63.7/−7.6
8.2/0.2	7.4/0.3	108.5/17.6	1740/100	245/18	24.8/−2.0	62.6/−6.0
7.2/0.1	6.3/0.3	114.9/−9.4	1713/−7	230/−5	22.1/−1.5	57.4/−7.9
6.4/0.2	6.0/0.2	111.8/−1.2	1770/−125	240/−10	20.1/−2.2	63.9/−10.8
5.5/−0.1	5.3/1	105.5/2.3	1670/60	233/11	22.3/−0.2	66.6/−7.0
6.7/0.0	6.2/0.1	115.8/−0.5	1832/17	256/−6	14.5/−2.3	45.6/−4.0
6.5/−0.1	6.1/0.2	119.3/3.0	1759/124	228/3	15.1/−3.6	47.3/−4.9
7.5/0.1	7.0/0.2	122.9/−4.9	2038/113	272/−1	14.2/−2.5	40.0/−1.3
7.2/0.2	6.4/0.5	127.9/−6.4	2087/−73	286/−13	14.5/−1.6	39.7/−2.3

Table 26.13 Particleboard preparation: board number and parameters.

Bd no.	Face-layer		Core-layer	
	Resin/mole ratio	Catalyst level (%)	Resin/mole ratio	Catalyst level (%)
1	UF 1.05	A/0.5	UF 1.05	A 0.5
2			UF 1.15	
3			UF 1.25	
4	UF 1.05	A/0.5	2.5% UMF-1	B 0.5
5				B 1.0
6				B 1.5
7				B. 2.0
8	UF 1.05	A/0.5	5.0% UMF-1	B 0.5
9				B 1.0
10				B 1.5
11				B 2.0
12	2.5% UMF-1	B/0.5	5.0% UMF-1	B 1.0
13		B/1.0		
14	5.0% UMF-1	B/0.5	5.0% UMF-1	B 1.0

PBs Bonded with Resin UF1.05 in the Face-layer and Resins UF1.05, 1.15, and 1.25 in the Core-layer (Board 1–3): Test results of PBs made with this set of resins, taken from the previous work (Section 26.1), represent the current range of UF resin mole ratios. Using the high F/U mole ratio of 1.25 in. the core-layer had been common in the board manufacturing industry until recently, but it results in high PE FC values, 16.8 mg/100 g dry board, which corresponds to a formaldehyde emission (FE) value of more than 0.20 ppm in the large-scale chamber test method, which is significantly higher than the new emission limit of 0.09 ppm and far higher than E1 class European Standards [12].

PBs bonded with Resin UF1.05 in the face-layer and Resins 2.5%UMF-1 and 5.0%UMF-1 in the core-layer with various levels of Catalyst B (Boards 4–11) showed lower PE FC values of 5.3–8.2 mg/100 g dry boards, decreasing with increasing levels of melamine and catalyst and also with longer hot-press times. Most of these FC values of UMF resins correspond to the E1 class of European Standards. Resin 2.5%UMF-1 used in the core-layer showed acceptable IB values with a Catalyst B level of 1.0%. Higher catalyst levels of 1.5% and 2.0% tended to show lower IB values and poorer water-soak test values, an effect of over-catalyzing. The IB, MOR/MOE, and water-soak values of PBs with resin 5.0%UMF-1 used in the core-layers showed significant improvements over those using resin 2.5%UMF-1. In comparison with boards bonded with *Resins of UMF-4*, the differences were small in physical strength values, but the FC values were slightly worse (higher). The water-soak thickness swell and WA values slightly improved, indicating that the resin synthesis method of Method #1 might find some applications in the industry.

PBs Bonded with Resin 2.5%UMF-1 in the Face-layer and 5.0%UMF-1 in the Core-layer with Various Levels of Catalyst B (Boards 12–13): PBs made with this set of resins with 0.5–1.0% catalyst levels in the face-layer and at 1.0% catalyst levels in the core-layers showed similar

FC values of 6.1–6.7 mg/100 g dry boards, and IB, MOR/MOE, water-soak thickness swell, and water-absorption values were improved due to the use of resin 2.5%UMF-1 instead of resin UF1.05 in the face-layer, but only small improvements in FC values were obtained. In comparison with boards bonded with Resins of UMF-4 at the same melamine levels, the results were generally similar, i.e. FC values were slightly worse (higher) and the physical strength, thickness swell, and WA values were slightly better.

PBs made with Resin 5.0%UMF-1 both in the Face and Core-layers and in the Core-layers at the 1.0% Catalyst B Level (Boards 14–15): PBs made with this set of resins at 0.5–1.0% Catalyst B level in the face-layers and at 1.0% Catalyst B level in the core-layer showed similar FC values of 6.4–7.5 mg/100 g dry boards with higher IB, bending and water-soak test values due to the higher melamine contents in both layers of boards. The bending test values indicated that the catalyst level of face-layer resin 5.0%UMF1.05 had higher values at the 0.5% catalyst level than at the 1.0% catalyst level, indicating that the catalyst levels of UMF resins should be adjusted not only for different melamine levels in the resin but also for which layer of board it is to be used. In comparison with boards bonded with resins of UMF-4 at the same melamine levels, the results are similar, and the FC values were slightly worse (higher), but the physical strength values were slightly lower, and the thickness swell and WA values improved slightly.

Overall, the PB test results indicate that *Resins UMF-1* and *UMF-4* performed with some differences in FC values and water-soak test values, which could be traced to the differences in resins' physical and structural differences, resulting from the different synthesis procedures. The formaldehyde released during hot-pressing may be re-captured more effectively by the free melamine present in *Resin UMF-4*, while the fully-reacted melamine in *Resin UMF-1* is more efficient as a bonding agent. On the other hand, since Resins UMF-1 were turbid as synthesized and Resin UMF-4 were clear as synthesized, there may arise problems related to storage and spraying operations for the former type of resins.

26.2.7 Summary

UMF resin synthesis methods were modified by adding melamine in the beginning with the first urea (Resin UMF-1), in the middle of first urea by dividing the first urea into two portions (Resin UMF-2), and at the end of reaction of the first urea (Resin UMF-3) with an F/(U + M) mole ratio of 1.05%, and 2.5% and 5.0% melamine levels. The polymerization reactions were all carried out at pH 5.25 to keep the cooking times in similar ranges with those of current UF resins. All three UMF resins were cloudy with higher hydroxymethyl and methylene-ether group contents and lower methylene group contents compared with Resin UMF-4 made by adding melamine in the last alkaline step in the previous paper. Resins UMF-1 and Resins UMF-4 showed adequate storage lives of two to three weeks, similarly to the control UF resin. Resins UMF-2 and -3 showed poor storage lives and therefore did not offer any practical usefulness. UMF resins were found to require stronger, free acid-containing catalysts at varying levels depending on the melamine levels and layers of boards. Boards bonded with Resin UMF-1 showed slightly higher FC values and slightly improved water-soak test values than boards bonded with Resin UMF-4, but they are considered equally useful, although the cloudiness of Resin UMF-1 may become problematic in large-scale resin productions. It is concluded that the approach of adding low levels of melamine to UF resins for FE reduction is constrained by the need for a stronger acid catalyst and/or longer hot-pressing time, which can get worse at higher melamine levels.

References

1 Internatinal Agency for Research on Cancer (2004). IRC classifies formaldehyde as carcinogen to humans, Press release No. 153, 15. WHO, Lyon, Cedex, France.
2 California Air Resources Board (2008). California Code of Regulations, Sections 93120-93120.12, Title 17.
3 U. S. Senate Bill 1660 (2011). American Jobs Act of 2011. ASTM E 1333 Large chamber form emission test.
4 DIN EN 120-92 (1992). Deutsches Institute fuer Normung. Berlin, Germany.
5 Kim, M.G. and Amos, L.A. (1990). *Industrial & Engineering Chemistry Research* 29: 208.
6 Kim, M.G. (2001). *Journal of Applied Polymer Science* 80: 2800.
7 No, B.Y. and Kim, M.G. (2005). *Journal of Applied Polymer Science* 97: 377.
8 Gillham, J.K., Benci, J.A., and Noshay, A.J. (1974). *Journal of Applied Polymer Science* 18: 951.
9 Laza, J.M., Julian, C.A., Larrauri, E. et al. (1999). *Polymer* 40: 35.
10 Wang, W., Gardener, D.J., and Baumann, M.G.D. (2003). *Forest Products Journal* 53 (3): 65.
11 Go, A. T. (1991). *Proceedings of 25th Internship PB and Composite Materials Symposium*, WSU, Pullman, WA, USA, p. 285.
12 Schwab, H., Marutzky, R., and Meyer, B. (2012). *European Regulation for Formaldehyde*. Braunschweig, Germany: Fraunhofer Institute for Wood Research, Wilhelm-Klauditz-Institut.
13 Myers, G.E. (1984). *Forest Products Journal* 34 (5): 35.
14 Myers, G.E. (1982). *Wood Science* 15 (2): 127.
15 Oh, Y.S. (1999). *Forest Products Journal* 49 (11/12): 31.
16 Park, B.D., Lee, S.M., and Roh, J.K. (2009). *European Journal of Wood Products* 67 (1): 121.
17 Park, B.D., Kang, E.C., and Park, J.Y. (2006). *Journal of Applied Polymer Science* 101: 1787.
18 Dunky, M. (1998). *International Journal of Adhesion and Adhesives* 18 (2): 95.
19 Sun, Q.N., Hse, C.-Y., and Shupe, T.F. (2011). *Journal of Applied Polymer Science* 119: 3538.
20 Tohmura, S., A. Inoue, and, H. Sahari. 2001. *Journal of Wood Science* 47, 451.
21 Higuch, M., Shimokawa, H., and Sakada, I. (1979). *Mokuzai Gakkaishi* 25: 630.
22 Mao, A.E., Hassan, B., and Kim, M.G. (2013). *BioResources* 8: 2453.

27

Advanced Reading Materials on Diethylene Tricarbamide-Formaldehyde Resins

(Adapted with permission of the publisher and minor revisions from: Kim, M. G., S.-Y. Lee, and B. Sukhbaatar, Syntheses of diethylene tricarbamide-formaldehyde (DF) and copolymer resins and their low formaldehyde-emitting binder performance of wood composite boards, Forest Products J. 68(3):226–235. 2019; and from M. G. Kim. 2019. Synthesis process development of diethylene tricarbamide as a starting material for wood composite binder DF resins. Forest Products J. 68(3):236–255. 2019.)

27.1 Introduction

Wood composite boards bonded with UF resins entail the formaldehyde emission problem, unresolved or only partially resolved until recently [1–4]. The emission values in the recent past ranged from about 0.15–0.20 ppm by the standard Small Chamber test method as discussed in Chapters 6 and 26, far higher than current U. S. federal and state formaldehyde emission limits [5, 6]. In UF resin syntheses, the initial formation reaction of hydroxymethyl groups is slightly reversible and a small amount of formaldehyde remains unreacted and the formaldehyde generated during curing in boards due to the reversibility is the cause of the formaldehyde emission problems [7, 8]. Suppression of the reverse reaction of hydroxymethyl groups is possible only by increasing the urea amide group content of resins by, for example, adding more urea, which results in lowering the formaldehyde/urea (F/U) mole ratio of the resins. This method has been used to the extreme extent in industry, but the emission values are still significant, the hot-pressing speed of boards decreased, and the board strength values deteriorated due to the reduced cross-linking capacity of the resins [1, 2, 9]. Addition of the tri-functional melamine (M) at low levels in UF resins in the form of UMF resins can result in some improvements as discussed in Chapters 25 and 26, but this method is also limited due to the slower curing speed and the fact that high levels of melamine are difficult to incorporate into liquid UMF resins due to the limited storage stability of resins and the need of stronger acid catalysts.

Adhesives and Finishes for Wood: For Practitioners and Students, First Edition. Moon G. Kim.

27.2 A Higher Functionality Urea Analogue – Diethylene Tricarbamide

Synthesis of diethylene tricarbamide

$$\underset{\text{Diethylene triamine}}{HN-(CH_2CH_2-NH_2)_2} + \underset{\text{urea}}{3NH_2-CO-NH_2} \underset{\text{(1st-Step)}}{\rightarrow} \text{Mono-carbamides}$$

$$\underset{\text{(2nd-Step)}}{\rightarrow \text{Di-carbamides}} \underset{\text{(3rd-Step)}}{\rightarrow} \underset{\text{Diethylene tricarbamide (D)}}{NH_2-CO-N-(CH_2CH_2-NH-CO-NH_2)_2} + \underset{\text{Ammonia}}{3NH_3} \quad (27.1)$$

In theory, the F/(U+M) mole ratio of UMF resins needs to be greater than 1.0 to result in useful cross-linked thermosetting network polymers with higher mole ratios increasing the polymer strength values [2]. UMF resins with an F/(U+M) mole ratio of 1.05 or lower in the core-layer of boards may produce boards that could meet the current formaldehyde emission regulation levels as discussed in Chapter 26, but the curing speed is significantly slower and the melamine level in resins is difficult to increase to further lower the formaldehyde emission values. This dilemma encountered in raising the urea amide group content of UF and UMF resins was considered to be resolvable by using a new monomer, *diethylene tricarbamide (D),* which has three carbamide groups in the molecule [10]. D has been a little-known material in literature [10, 11], synthesized by reacting diethylene triamine (DA) with urea in a three-step reaction process using water as reaction medium, as shown in Eq. (27.1) [10]. The high current selling price of DA (~\$1.75/lb) compared to urea (~\$0.30/lb) appeared to be a major roadblock to a large-scale production of D, but a higher volume production of DA could narrow this price gap since it is currently manufactured from low-cost starting materials: ethylene and ammonia. On this premise, an efficient synthesis procedure of D from DA and urea was first sought since the reported synthesis procedure using water as a reaction medium gives poor yields of about 56% with a long reaction time and a long work-up procedure [10]. Then, D-formaldehyde (DF) and several copolymer resins were investigated.

27.3 Chemical and Physical Properties of Diethylene Tricarbamide

D synthesized in this study was found to be a stable, colorless (white), non-hygroscopic, and microcrystalline powder with a melting point of 218–222 °C, which was slightly higher than the reported value of 210 °C [10]. It is sparingly soluble in water (~2.0/100.0 g) at room temperature, and the solubility increases to about 30/100.0 g water at 90 °C. It is nearly insoluble in methanol and slightly soluble in warm dimethylsulfoxide (DMSO). CHN analysis for $C_7H_{16}N_6O_3$ (232.24): Calculated (%), C 36.20, H 6.95, N 36.19; Found (%), C 36.12, H 7.13, N 35.72. ^{13}C NMR (DMSO-d_6, ppm): 38.53 (2), 47.44 (2), 158.7 (1), 159.26 (2). FTIR (KCl, cm^{-1}): 1,560, 1,660, 2,950, 3,220, 3,370, and 3,420. Thus, D is qualified to become a chemical raw material for commercial uses that does not need any special handling procedure similar to urea. D was also found to react with formaldehyde to give DF resins.

27.4 An Efficient Synthesis Method of Diethylene Tricarbamide

The three-step carbamylation reaction of DA with urea was investigated in pyridine, 1-butanol, pentanols, and also with polar cosolvents additions to find an efficient synthesis procedure of D, aiming at a procedure that can be scaled up for industrial production. One typical method is based on using n-butanol as the reaction medium, as in the following: Into a stirred 1,000-ml reactor equipped with a condenser, 450.0 g of (undried) n-butanol and 190.0 g of urea were added, and separately, 78.0 g of DA dissolved in 100.0 g of *n-butanol* mixture were set up in an addition funnel, and about one-eighth of the DA/*n*-butanol mixture was added to the reactor. The reaction mixture was then heated to and maintained at 110 °C. Dropwise the addition of DA solution was then initiated and completed in 1.6 hours. Then, at about 2.5 hours of reaction time, white voluminous precipitate particles of di-substitution products began to appear, and then hard compact fine particles of final product soon began to appear. Both types of particles increased in amount over time, causing a stirring problem, but it could be handled by increasing the stirring speed. The latter type of particles soon increased in amount and became the dominant particles in the reaction mixture. After 13.5 hours of total reaction time, the reaction mixture was allowed to cool to about 90 °C, and the precipitates were filtered hot at 90 °C, washed with warm ethanol, and dried to yield 166.0 g of D (94.5%) with a melting point of 213–215 °C. Relatively high yields in the range of 94.5–95.7% were obtained using various solvent systems.

27.5 Synthesis Chemistry and Procedure and Properties of D-formaldehyde (DF) and Copolymer Resins

Synthesis chemistry of DF resins is similar to those of UF resins: adding D to the formaldehyde solution at a mildly alkaline pH and elevated temperatures, as shown in Eq. 27.2. One major difference is that the synthesis procedure can be done in one step, i.e. there is no need to have a step of adjusting the pH to 4–5 for polymerization reactions. Also the F/D mole ratio can have any target value from the beginning since the polymerization reaction proceeds well at pH 6–7. Addition of D to formaldehyde solution can also be done at one time or in a stepwise manner. The polymerization step can be done at a lower pH and at lower temperatures.

Resin synthesis scheme of DF resins

$$\mathbf{NH_2CO - N -} (\underset{(D)}{CH_2CH_2} \mathbf{- NH - CO - NH_2})_2 + \mathbf{n\,} \underset{(F)}{\mathbf{CH_2O}} \rightarrow$$

$$\underline{HOCH_2} - NH - CO - \underset{\text{(Tri-hydroxymethyl D)}}{N} - (CH_2CH_2 - NH - \underline{CH_2OH})_2 \rightarrow$$

$$\mathbf{(HOCH_2 - NH - CO)_2 - N -} CH_2CH_2 - NH - \underset{\text{(Dimeric DF resins)}}{\underline{CH_2}} - NH - CO - N - (CH_2CH_2 \mathbf{- NH - CH_2OH})_2$$

$$\rightarrow \textbf{Oligomeric DF resins} \tag{27.2}$$

Synthesis Reaction Characteristics of DF Resins vs. UF and UMF Resins: (i) D is soluble in water at room temperature but becomes increasingly more soluble at higher temperatures and readily dissolves in the 50% formaldehyde solution at above 60 °C under mild alkaline pH of 7.0 or higher, indicating its ready reaction with formaldehyde to form hydroxymethyl groups. (ii) The pH value of reaction mixture slowly decreases from the initial set value of about 7.5–8.5 down to about 6.2 by itself during resin synthesis as in the first alkaline stage cook of UF resins without pH adjustment; and the reaction mixture increases in viscosity relatively rapidly in this very mild acidic pH range, indicating a ready polymer formation through the major methylene and minor methylene-ether bonds as shown in Eq. (27.2). (iii) The F/D mole ratio in the polymerization step can be set to any value above 1.10 in contrast to UF resin systems, and thus, DF resins can be synthesized by either one-step or multi-step addition of D to the formaldehyde solution in mild alkaline/acidic pH ranges to reach to the target F/D mole ratio values of 1.10–1.25. (iv) The pH of reaction mixture that has gone down to about 6.2 by itself during the resin synthesis reaction comes back to close to 7.0 when the finished resin is cooled to room temperature, which can be adjusted further by adding sodium hydroxide solution. (v) The dissolution speed of D decreases as the resin F/D mole ratio is decreased similarly as in the addition of melamine in UMF resin syntheses, but the phase disturbances observable in UF or UMF resins [9] at low mole ratio values in the polymerization step were not observed for all mole ratios.

The resultant (aqueous) DF resins were found to behave well, clear, and easily washable with water unlike UF or UMF resins, of which dimers or trimers are often precipitating and difficult to wash off with water.

Copolymer DUF Resins Synthesis: D-urea-formaldehyde (DUF) copolymer resins, which could provide lower cost alternatives, were synthesized at 10–20% urea levels by weight by advancing the DF resin components with an appropriate intermediate F/D mole ratio to an appropriate degree of polymerization, followed by adding urea and further reaction. The starting F/D mole ratio was pre-calculated based on the amount of urea to be added to reach to the target F/(D + U) mole ratio values with about 10% and 20% addition levels of urea by weight. In this type of urea addition method, the resins are expected to have all urea components in monomeric forms and DF resin components in polymeric forms.

Copolymer DMF resins syntheses: Since the pH range required in the polymerization step of DF resins is only between about 6.0 and 7.0, melamine can be added at any point in syntheses of D-melamine-formaldehyde (DMF) copolymer resins with melamine levels of between 7.5% and 50.0% relative to the weight of D. Melamine is only sparingly soluble in water, but DMF resins came out clear and were generally more stable than UMF resins in similar mole ratio ranges albeit the instability of DMF copolymer resins became apparent at 30–50% levels of melamine in resins. One difficulty in syntheses of urea-melamine formaldehyde (UMF) resins with low formaldehyde mole ratio values has been the limited storage stability of resultant UMF resins due to the tendency of forming insoluble precipitates of oligomeric MF components as discussed in Chapter 26. However, DMF resins were more stable than UMF resins, as shown in Table 27.2.

27.6 Synthesis Formulations of DF Resins vs. UF Resins

In Table 27.1 are shown the resin synthesis formulations based on obtaining 100 g of DF and UF resins, having the same final mole ratio of 1.15 and same resin-solids levels of about 59%. It is readily recognized in the table that the amount of formaldehyde needed for DF resin is only 31% of that for UF resin due to the fact that the molecular weight of D (*232*) is 3.87 times greater than urea (*60*). This fact also indicates that cured DF resins' life-time

Table 27.1 Comparison of formulations for syntheses of 100 grams of DF and UF resins.

Mole ratio	Resins	D or urea weight (g)	50% Form solution (g)	Water (g)	Resin-solids content (%)	Total resin weight (g)
1.15	DF	55.16	16.42	28.42	59.0	100.0
	UF	46.51	53.49	0		

Table 27.2 Storage stability of resins at room temperature (~22 °C) measured by viscosity changes and the formaldehyde contents of selected resins in the beginning.

Days	0	3	4	5	6	7	8	10	12	14	17	20	21	23	24	CH_2O Content
UF1.05	G				H		I	J			K		P			0.12%
DF1.15	F			G						G	G	GH				0.02%
15%UMF1.05	I	J		J				K		M			P			0.10%
7.5%DMF1.25	I													N		0.03%
15.0%DMF1.25	E													I		—
80%DUF1.20	HI					HI		HI	I	IJ		J	K		KL	—

formaldehyde emission potentials are that much lower. The starting thesis of this work of choosing a material of higher functionality (3) resulted in this formulation differences due to the presence of two ethylene groups in the molecule. Use of less formaldehyde in resin synthesis would result in less formaldehyde emission.

27.7 Room Temperature Storage Properties of DF and Copolymer Resins

Synthesized DF, DUF, and DMF copolymer resins at low F/D ratio values with resin-solids levels in range of 50–65% and viscosity values anywhere between D–X by G–H scale were found to be clear and adequately stable, better than current UF or UMF resins in most cases as shown in Table 27.2, as viscosity increases of resins were observed at room temperature over a period of three weeks. Also, the free formaldehyde contents of DF resins determined were less than 20% of the common values of UF resins at similar mole ratio values. DMF resins synthesized with about 25% or higher melamine levels were found to have shortened storage lives, but they might still be useful.

27.8 Pot-lives and Curing Characteristics of DF and Copolymer Resins

DF, DUF, and DMF copolymer resins were found to need the ammonium sulfate-type acid catalysts for curing similarly as UF resins. Gel times measured at 100 °C, and

Table 27.3 Pot-lives observed by viscosity changes at 30 °C and gel times at 100 °C of resins catalyzed with 1.0% ammonium sulfate solution (25%).

Hours	0	2.25	6.0	10.5	17.5	24.0	30.0	Gel time (sec.)
UF1.10	E	F	gel					88
DF1.10	D	E	EF	FG	I	J	J	91
DF1.15	D	E	F	GH	J	K	MN	75
DF1.25	E	F	GH	HI	P	TU	V	77
80%DUF1.20	E	F	G	H	O	Q	T	84
8.8%DMF1.10	D	E	G	J	N	RS	U	80

pot-lives measured at 30 °C of catalyzed DF and copolymer resins are shown in Table 27.3, comparable with those of UF resins at similar mole ratio values.

27.9 Preparation of Laboratory Particleboards and Testing

Board Preparation: Boards were made with selected synthesized resins along with some commercial UF (CUF) and UMF (CUMF) resins obtained from suppliers, according to the same procedure used in previous studies. Wood particles were dried to 4.0–5.0%; 1.0% wax solids and 8.0% catalyzed resin solids were applied; face-layer and core-layer, three-layer mats with a 1 : 2 : 1 weight ratio; 3.0 and 3.5 min hot-press times at 350 °F; *board dimensions of 22″ × 22″ is × 0.5″*;board density of 50.0 pcf; physical test samples were allowed to equilibrate in a 60% RH room for a week.

Formaldehyde test samples (6″ × 6″) were wrapped in saran wrap, stored in a refrigerator, and evaluated within three days according to perforator extraction (PE) method [12]. Cutting boards into small test pieces were done about 0.5 hours prior to test. Board samples evaluated had moisture contents of about 3.0% and the resultant FC data were not corrected.

Physical and Mechanical Testing of Particleboards: Equilibrated boards were evaluated for internal bond (IB) strength (8 test pieces) and bending strength (MOR and MOE) (4 test pieces) on an Instron machine (Norwood, MA). Board densities were measured by measuring dimensions and weights of bending test samples. Water-soak thickness swelling (TS) and water absorption (WA) values (2 test pieces) were measured at 20 °C in a water bath according to ASTM D 1037-93. Exploratory 6″ × 6″ boards were also made on the small press and they were cut into two 2″ × 2″ samples and tested for PE formaldehyde content values and 24 hours water-soak properties at 20 °C in a water bath, but data are omitted in this lecture.

27.10 Test Results of Prepared Laboratory Particleboards

Physical strength and water soak test values: Test results of boards are shown in Table 27.4. All boards have shown similar board density values to allow board properties to be compared directly among themselves. Boards prepared for comparison with laboratory

Table 27.4 Test results of DF resins alone or in combination with UF resins in core and face layers of 22″ × 22″ × 0.5″ laboratory boards: board density, internal bond (IB), 24 hours water absorption (WA) and thickness swell (TS), and PE formaldehyde contents.

Face layer	Core layer	Board density	IB (psi)	MOE (kpsi)	MOR (psi)	24 h WA (%)	24 h TS (%)	PE (mg/100g)
UF 1.05	UF 1.10	0.77	123	280	1559	54	19	12.8
	UF 1.20	0.79	119	224	1550	40	18	14.4
CUF	CUF	0.77	109	248	1556	45	16	24.5
CUF	CUF	0.78	63	187	905	44	24	13.6
CUMF	CUMF	0.78	76	143	781	29	10	11.1
CUF	CUMF	0.79	72	135	932	40	18	8.5
DF1.30	DF 1.40	0.78	104	256	1377	39	20	3.0
DF1.25	DF 1.25	0.80	160	297	2196	42	15	1.8
DF1.20	DF 1.20	0.78	94	225	1259	43	17	1.6
UF1.05	DF 1.10	0.77	100	210	1056	48	20	2.3
	DF 1.20	0.79	94	226	1188	46	18	2.8
	DF 1.30	0.79	139	142	1362	45	18	3.9
	DF 1.35a[a)]	0.76	139	261	1881	44	20	4.0
	DF 1.35b[a)]	0.78	148	246	1781	47	22	4.4
	DF 1.35c[a)]	0.78	147	235	1460	50	26	2.9
	DF 1.35d[a)]	0.76	142	250	1777	54	20	3.9

a) a, b, c, and d indicate different batches of DF resin preparation. Boards were pressed at 350 °F for 3.0 and 3.5 minutes and averages obtained. CUF and CUMF resins are industrial UF or UMF resins that were going through mill trials in board manufacturing plants in Mississippi in 2012.

synthesized UF resins gave satisfactory IB, modulus of elasticity (MOE), modulus of rupture (MOR), and 24-h water-soak weight gain and thickness swell (TS) values as expected. Boards made with DF resins in both core and face layers and also those made with DF resins in core and UF resins in face layers showed some scatter in strength data, but they are judged to be comparable with boards bonded with UF resins.

PE Formaldehyde Content Values: Boards bonded with DF resins in core layer only or in core or face layers showed PE values reduced by 75–80% in comparison with same mole ratio UF resins: DF resin mole ratio values between 1.15–1.35 show PE content values of 2.3–4.4 mg/100 g of boards, significantly below the new emission limits of about 8.0 mg/100 g of wood [13]. The fact that DF resins used in core layer only or in both face and core layers show similarly low FC values indicates that DF resins, which are expected to be higher in price than UF resins, can be used only in core layers to save costs. Four sets of test results of commercial CUF or CUMF resins obtained as trial resin samples from nearby particleboard plants over a period of time, evaluated in this laboratory, showed low strength values and high formaldehyde content values, reflecting the hectic transition period being experienced by the particleboard industry since the new formaldehyde emission regulations were enacted [5, 6].

27.11 Summary

D was conceived as a monomer material that can form polymers like UF resins with one more amide group in the molecule, and since its hydroxymethylation reaction with formaldehyde would go forward more forcibly forward than urea, a lower formaldehyde emission potential of boards was expected. Synthesis procedure of D from DA and urea was shown to proceed at high yields and synthesis procedure of DF resins using D and formaldehyde was found to be relatively simple because of no need of an acidic pH adjustment step during the cook, and the resin synthesis reactions mostly resulted in clear stable resins, indicating that current resin manufacturing industries can conduct DF resin manufacturing without modification of resin reactors. DF resins showed adequate storage lives at room temperature and also have curing parameters that are similarly as fast as UF resins, indicating that current board manufacturing industries also would not need any modification to their hot-pressing lines. Particleboards bonded with DF and DMF resins used in the core layer only or in both core and face layers showed similarly good physical properties as UF resins in the same mole ratio ranges, and the formaldehyde content values of boards were lower than those of UF resins by about 75–80%. The PE formaldehyde content values of 2.3–4.4 mg per 100 g boards obtained with DF resins correspond to emission levels significantly below the current US regulation limits. Thus, the reversibility of hydroxymethyl groups in DF resins is suppressed proportionately to its functionality, proving the thesis that a higher functionality monomer can reduce the formaldehyde emission potentials of wood composite boards.

References

1 Graves, G. (1993). *Proceedings of 27th International PB and Wood Composite Materials Symposium* (ed. T.M. Maloney), 221. Pullman, WA: WSU.
2 Go, A.T. (1991). *Proceedings of 25th International PB and Composite Materials Symposium* (ed. T.M. Maloney), 285. Pullman, WA: WSU.
3 Dunky, M. (1998). *International Journal of Adhesion and Adhesives* 18 (2): 95.
4 Sigvartsen, T. and Dunky, M. (2006). *Wood Adhesives 2005* (ed. C.R. Frihart), 23. Madison, WI: FPS.
5 California Air Resources Board (2008). California Code of Regulations, Sections 93120-93120.12, Title 17.
6 U. S. Senate Bill 1660 (2011). American Jobs Act of 2011.
7 de Jong, J.I. and de Jonge, J.J. (1952). *Recueil des Travaux Chimiques des Pays-Bas* 71: 661.
8 Myers, G.E. (1982). *Wood Science* 15 (2): 127.
9 Kim, M.G. and Amos, L.A. (1990). *Industrial & Engineering Chemistry Research* 29: 208.
10 Bentoniere, N.R. and Rowland, S.P. (1979). *Journal of Applied Polymer Science* 23: 2567.
11 Higuchi, M., Shimokawa, H., and Sakata, I. (1979). *Mokuzai Gakkaishi* 25: 630.
12 (1992). DIN EN 120-92; Deutsches Institute fuer Normung: Berlin, Germany, .
13 Athanassiadou E., Tsiantzi, S., and Markessini, C. (2009). Wood Adhesive Symposium, Lake Tahoe, NV. Sept. 29.

28

Advanced Reading Materials on PF Resole Wood Adhesive Resins

28.1 Chemical Structures of PF Resole and Novolac Resins by ^{13}C NMR Spectroscopy

(Adapted with permission and minor revisions from Kim, M. G., Amos, L. W., and Tiedeman, G. T. 1979. Carbon-13 NMR study of phenol-formaldehyde resins. Weyerhaeuser Company Science Symposium II: 263–289. Weyerhaeuser Company, Tacoma, WA.)

^{13}C NMR spectroscopy was used for the first time in 1970s for studying PF resins with the substituent and solvent effects of various simple phenolic molecules elucidated [1, 2]. The applicability of the additivity rules [3] to predict chemical shift values of various carbon peaks of organic compounds and also on PF resin molecules were reported [3–5]. Also, ^{13}C NMR studies on ortho-condensed novolac oligomers were reported [6]. Despite the lack of a high degree of quantitative accuracy as described in the previous Chapter (Section 24.1), ^{13}C NMR method became useful for comparing various PF resins. It was then known that ^{13}C NMR chemical shift values of phenolic carbons of PF resins vary in small but significant extents when the basicity of solvent is changed, or in different solvents of determination, [2] to make it somewhat confusing to compare a PF resin to another PF resin using this method.

28.1.1 PF Resin Model Compounds

The assignments of chemical shift values of various model compounds of PF resins are shown in Table 28.1, obtained with increasing levels of caustic in the range of 0–64 mol% per phenolic ring in water solutions. With these solvent changes, C_1 carbons vary from 155 → 165 ppm; unsubstituted C_2 carbons vary from 115 → 120 ppm; unsubstituted C_4 carbons vary from 120 → 116 ppm; substituted C_2 carbons vary from 127 → 131 ppm; and substituted C_4 carbons vary from 133 → 128. These chemical shift value changes occur due to the increasing extents of ionization of the phenolic hydroxy group (de-shielding) and thereby delocalization of the anionic charges onto C_2 and C_4 carbons (shielding). C_3 carbons at about 129 ppm are affected little. Further, in the absence of any extra formaldehyde, ortho-hydroxymethyl groups vary from about 59 → 62 ppm, *p*-hydroxymethyl groups

Adhesives and Finishes for Wood: For Practitioners and Students, First Edition. Moon G. Kim.

Table 28.1 Changes of ^{13}C NMR chemical shift (ppm, TMS) values of PF resin model compounds by changes of the NMR solvents and basicity [10].[a)]

C	So.	C_1	C_2	C_3	C_4	C_5	C_6	C_7	C_8	C_9
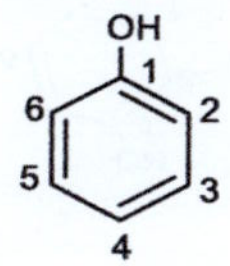	1	157.63	115.56	129.62	119.10					
	2	157.85	116.21	130.21	120.12					
	3	159.74	116.97	130.34	118.92					
	4	161.80	117.71	130.29	117.71					
	F	157.21	120.50	130.34	116.12					
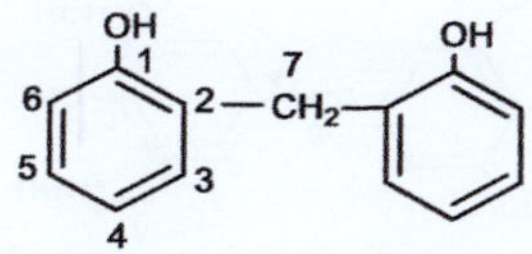	1	155.30	127.29	130.46	119.21	127.14	115.16	29.61		
	2	155.60	127.92	131.16	120.17	127.92	115.87	30.29		
	3	157.57	129.20	131.04	119.16	128.11	116.94	32.08		
	4	159.51	130.88	130.48	118.10	128.21	117.94	33.81		
	F	154.87	127.82	131.07	120.71	128.10	116.03	30.36		
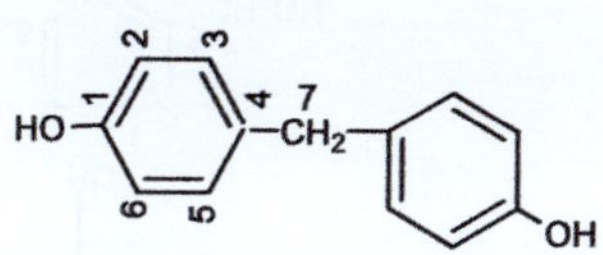	1	155.70	115.44	129.75	132.48			39.89		
	2	155.02	116.15	130.47	133.50			40.50		
	3	157.76	116.83	130.39	132.42			40.50		
	4	159.59	117.50	130.32	131.41			40.87		
	F	155.25	116.03	130.38	133.98			40.25		
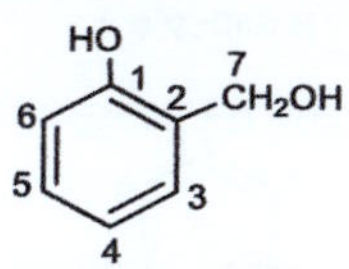	1	154.76	128.75	128.01	119.26	128.01	115.14	39.08		
	2	155.10	128.89	128.89	120.17	128.89	115.80	59.75		
	3	157.20	129.04	128.88	119.09	128.88	116.67	60.51		
	4	159.41	129.40	129.11	117.88	128.93	117.51	61.25		
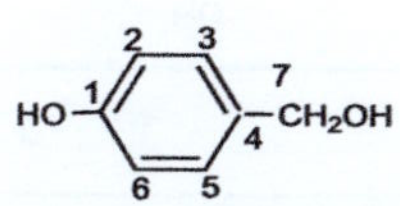	1	155.70	115.38	128.69	133.09			63.42		
	2	156.98	116.02	129.53	133.38			63.99		
	3	158.76	116.68	129.66	132.20			64.23		
	4	160.75	117.37	129.78	130.89			64.45		

(Continued)

Table 28.1 (Continued)

C	So.	C_1	C_2	C_3	C_4	C_5	C_6	C_7	C_8	C_9		
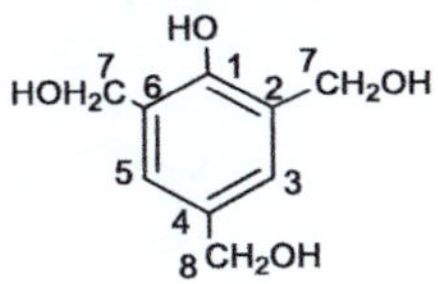	1	150.79	128.02	125.50	133.06							
	1'	151.50	128.23	126.91	135.17			63.68	60.27	~40.0		
	2	151.33	128.35	126.69	133.46							
	2'	152.32	128.51	127.87	133.74			64.23	60.87			
+ 4,4′-dimer	3	152.84	128.83	126.98	133.00							
+ 4,4′-dimer	3'	154.78	128.83	128.03	132.00			64.47	61.38			
	4	154.55		127.25	132.14			61.89				
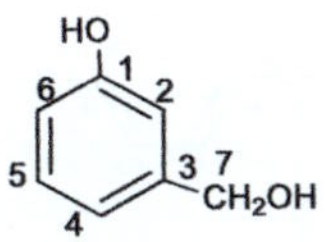	1	157.73	114.16	144.42	117.60	129.56	113.85	63.47				
	2	157.89	114.94	144.42	118.66	130.36	114.52	64.08				
	3	159.55	115.64	144.20	117.75	130.32	115.28	64.34				
	4	161.11	116.81	143.93	116.27	130.21	115.93	64.53				
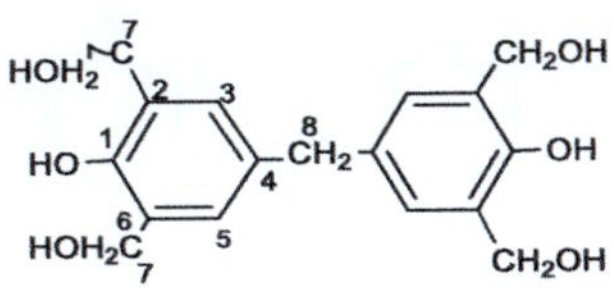	1	150.84	127.03	128,32	132.98			60.32				
	2	151.25	127.87	128.63	133.79			60.81	41.25			
	3	152.22	128.05	128.89	133.42			60.09	41.33			
	4	153.48	128.11	129.06	132.78			61.37	41.20			
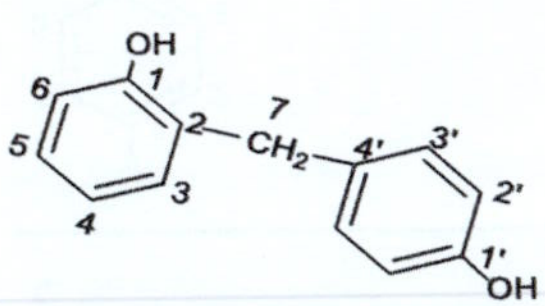		C_1	C_2	C_3	C_4	C_5	C_6	C_7	$C_{1'}$	$C_{2'}$	$C_{3'}$	$C_{4'}$
	1	155.19	128.50	130.32	119.29	127.26	115.34	34.48	155.61	115.34	129.94	131.79
	2	155.51	129.03	131.21	120.17	128.02	115.98	35.36	155.79	115.98	130.96	132.71
	3	156.37	129.50	131.16	119.72	128.03	116.51	35.51	157.07	116.51	130.65	132.06

Solvents (So.): **1**- About 0.40 g of sample dissolved in 1.40 g of DMSO and 0.40 g of DMSO-d_6 (internal lock); **2**-Added 0.40 g water; **3**-Added 25% NaOH solution to 16.0 mol% of solutes; **4**-Added the NaOH solution to 32.0 mol% of solutes. F: Formamide is used as sole solvent.

a) Trihydroxymethylphenol and its dimer were prepared according to the known methods [11, 12].

vary from about 63 → 65 ppm, *o–o*-methylene peaks vary from about 30 → 34 ppm, *o–p* methylene groups vary only a little at about 35 ppm, and *p–p* methylene groups varied from about 40 → 41 ppm (The latter peak overlaps with DMSO solvent peak.) Thus, comparing different PF resins by ^{13}C NMR needs to be done in conjunction with the changes in sodium hydroxide contents and solvent kinds and levels. It is noted that chemical shift values of some phenolic ring carbons occur in a narrow range and often overlap.

28.1.2 An Example Spectrum of a PF Resole Resin

The ^{13}C NMR spectrum of a green PF resole resin, Resin #1 type described in Chapter 9, is shown in Figure 28.1. The resin was prepared at about 50% resin solids content in water according to a US patent [7], using 1.0% sodium hydroxide as catalyst based on total resin weight, followed by neutralization with citric acid. This PF resole resin results in a complex spectrum due to the extra formaldehyde present as hemi-formal groups. The "a" peaks are *o–o*, *o–p*, and *p–p* methylenes, "b" peaks are *o*- and *p*-hydroxymethyl groups, and "c" peaks are the oxymethylene ether carbons of hemi-formal groups, and "d" peaks are those of formal methylenes or hemi-formal groups. The "e" peaks are hydrated formaldehyde oligomers. Formal groups mostly disappear when the resin was treated with acidic hydroxylamine reagent [8], which is accompanied by increases in the intensity of hydroxymethyl groups (b's). This reveals the facile reactions between hydroxymethyl groups and formaldehyde monomers to form hemi-formal groups. Further, the f′ peaks are unsubstituted C_2 carbons and the "f_4" peak is C_2 carbons of (pure) phenol itself. The "f_5" peaks are unsubstituted C_4 carbons. The "g" peaks are prominent in various resins that contain benzyl ether of hemi-formal groups. The "h", "i", and "j" peaks are of C_3 and C_5 and substituted C_2 and C_6 carbons. The "k" peaks are assigned to substituted C_4 carbons. The "l_4" peak is identical to C_1 peak of phenol itself by a spectral comparison. The "l_3" "l_2", and "l_1" peaks are roughly grouped into mono-, di-, and tri- substituted phenolic C_1 carbons. This type of neutralized PF resole resins was studied in the past as acid-curing PF resole resin wood adhesives, which are currently not much in use because of the acid curing catalysts' degrading effects of wood as discussed in Chapter 9.

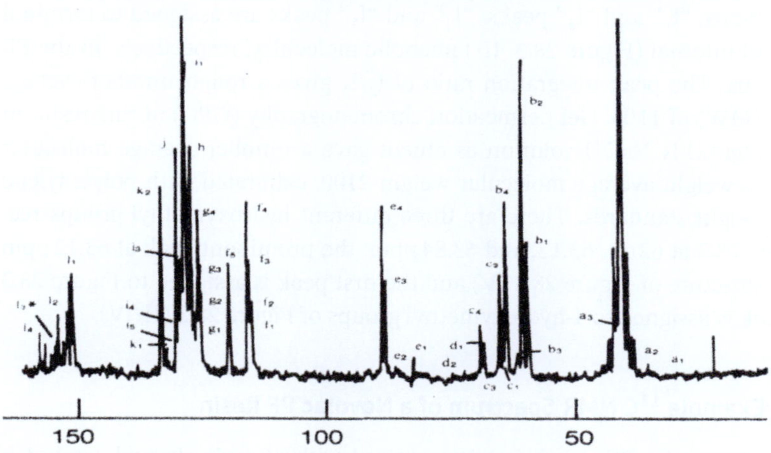

Figure 28.1 A green PF resole resin synthesized with an F/P mole ratio of 2.5 using sodium hydroxide catalyst and neutralized with citric acid [10].

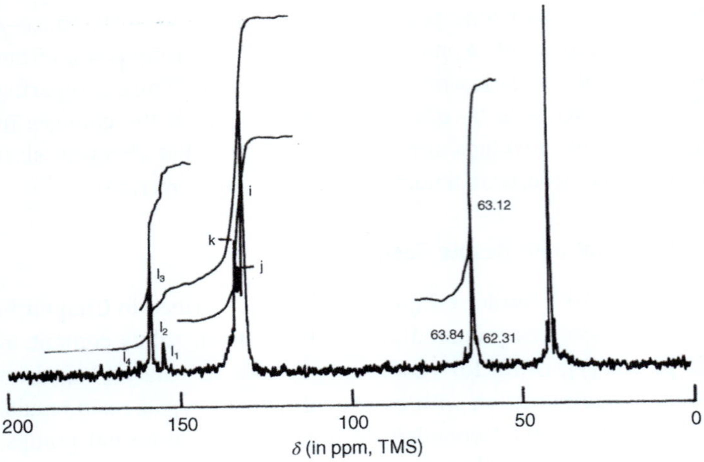

Figure 28.2 ^{13}C NMR spectrum of an F/P mole ratio of 1.80 PF resole resin in DMSO (at 40 ppm) with NaOH level of Resin #4 type described in Chapter 9. Integration curves of peaks are shown and DMSO solvent peak at about 40 ppm (cut), obtained on CFT-20 [10].

28.1.3 A Second Example Spectrum of a PF Resole Resin

Another example ^{13}C NMR spectrum of PF resin is shown in Figure 28.2, a more advanced resin with higher sodium hydroxide content like Resins #3 or #4 types described in Chapter 9. First, in the aliphatic region the methylene peaks in the 30–40 ppm region are recognizable but the integration is difficult due to the strong solvent (DMSO) peak at about 40 ppm. Ortho-hydroxymethyl groups at 63.12 ppm are strong and para-hydroxymethyl groups at about 63.84 ppm are weak. In the aromatic region, a nearly unrecognizable hump near 118.5 ppm can be assigned to unsubstituted C_2 and C_4 carbons, and the peaks around 130 ppm assigned are for substituted $C_{2,6}$, $C_{3,5,}$ and C_4 carbons, which are bunched up and unresolved, which is true whenever the condensation reaction of resin is advanced far. There are two prominent groups of peaks in the C_1 carbon region, "l_2" and "l_3" and other weak absorptions, "l_1" and "l_4" peaks. "l_2" and "l_3" peaks are assigned to terminal (Figure 28.3, V) and internal (Figure 28.3, III) phenolic molecules, respectively, in the PF resin polymer chains. The peak integration ratio of l_2/l_3 gives a rough number-average molecular weight (MW) of 1100. Gel permeation chromatography (GPC) of this resin on Sephacryl S200 using 0.1 N NaOH solution as eluent gave a number-average molecular weight of 720 and a weight average molecular weight 2100, calibrated with polyethylene glycol molecular weight standards. There are three different hydroxymethyl groups recognizable in Figure 28.2 at 62.31, 63.12, and 63.84 ppm: the prominent peak at 63.12 ppm is assigned to the structure of Figure 28.3 (V) and the first peak is assigned to Figure 28.3 (I/IV). The last peak is assigned to 4-hydroxymethyl groups of Figure 28.3 (II/IV).

28.1.4 A Third Example ^{13}C NMR Spectrum of a Novolac PF Resin

The ^{13}C NMR spectrum of a PF novolac resin prepared with phosphoric acid catalyst is shown in Figure 28.4 in formamide solvent (~163 ppm), showing the presence of methylene group carbons at 30–40 ppm and phenolic carbons at 115–155 ppm and no hydroxymethyl

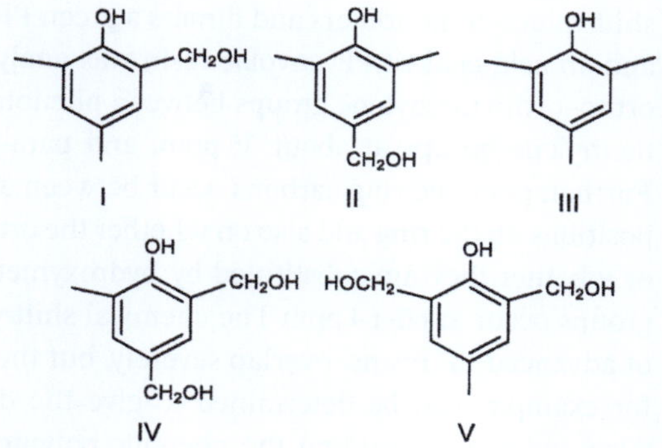

Figure 28.3 Possible PF resole molecular structures with full substitution of *o*- and *p*-carbons.

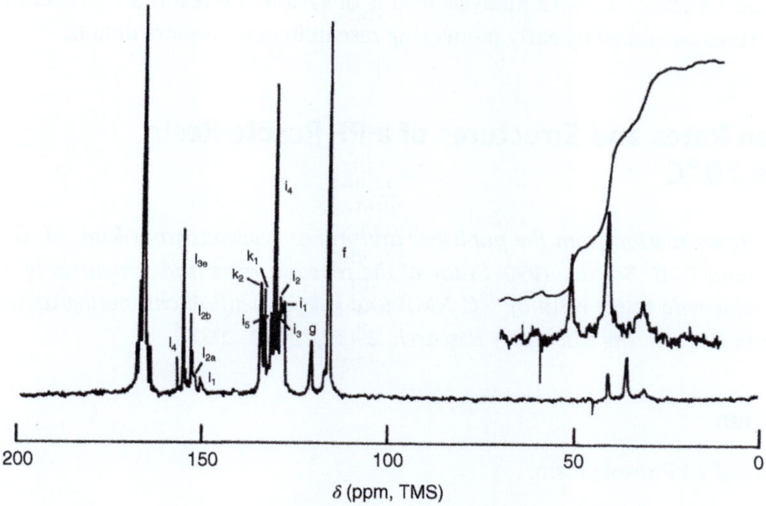

Figure 28.4 ^{13}C NMR spectrum of a novolac PF resin with an F/P mole ratio of 0.76 in $HCONH_2$ (Formamide at ~157 ppm) with no water, no NaOH. Integration curves of methylene peaks are shown at about 30–40 ppm (expanded), obtained on CFT-20 [10].

or formaldehyde species at 50–115 ppm The phenolic carbons' chemical shift values were assigned similarly to those of PF resole resins marked with lower case letters from **f** to $\mathbf{l_4}$. The methylene groups were assigned to ortho–ortho groups at ~30 ppm, ortho–para groups at ~35 ppm, and para-para groups at ~40 ppm, and also the integration values can give their ratio values that are characteristic of the catalyst kinds and reaction procedures used in synthesis of novolac resins.

28.1.5 Summary

A shortened version of a symposium presentation titled "^{13}C NMR study of phenol-formaldehyde resins" is discussed as an introduction to the chemical structures of liquid, uncured PF resole resins. Various monomers and dimers of PF resin molecules were analyzed by the ^{13}C NMR, and all aliphatic and phenolic ring carbons are identified by their chemical shift values, which were also found to be changing to small extents depending on the level of sodium hydroxide and water present in the resin sample. Using the chemical shift values of monomers and dimers a green PF resole resin, an advanced PF resole resin, and an acid-cooked PF novolac resin was analyzed and their structures identified. Thus, ortho–ortho methylene groups between phenolic rings occur at about 30 ppm, ortho–para methylene groups at about 35 ppm, and para-para methylene groups at about 40 ppm. Further, phenolic ring carbons occur between about 116 and 166 ppm depending on their positions on the ring and also on whether the ortho and para carbons are substituted or not, or whether they are substituted by hydroxymethyl or methylene groups. Hydroxymethyl groups occur at 60–64 ppm The chemical shift values of phenolic ring carbons, especially of advanced PF resins, overlap severely, but the methylene group/phenolic carbon ratios, for example, can be determined to give the degree of polymerization values of resins. Also, indications are that the phenolic polymer molecular chains are branched in most resole-type PF resins. Overall, ^{13}C NMR analysis results of various PF resins are in accord with polymer structures advanced by early pioneering researchers with more details.

28.2 Reaction Rates and Structures of a PF Resole Resin Synthesized at 70 °C

(Adapted with permission from the publisher and minor revisions from Kim, M. G., L. W. Amos, and E. E. Barnes. 1990. Study of the reaction rates and structures of a phenol-formaldehyde resole resin by ^{13}C NMR and gel permeation chromatography. Industrial and Engineering Chemistry Research. 29(10): 2032–2037.)

28.2.1 Introduction

Synthesis procedure of a PF resole resin.

- 22.6 kg 90.6% phenol; 32.8 kg 49.2% Formaldehyde; and 2.04 kg 100% NaOH.
  1. Charged all and mixed at 45 °C.
  2. Heated 1 °C/min to 70 °C and held for 360 minutes.
     (Samples #1–#6 were taken at 60 minutes intervals.)
  3. Added 0.73 kg NaOH at 360 minutes.
  4. Added 10.92 kg H_2O/3.40 kg NaOH at 490 minutes.
  5. Cooled at 510 minutes reaction time.
     (Samples #7–#10 were taken at ~35 minutes intervals.)
- F/P ratio 2.51.
- Total reaction time 510 minutes.
- End pH 10.2
- Viscosity 350 cps.
- Total batch 72.49 kg.
- NaOH content 8.51%.
- PF resin-solids, calculated, 38.63%.

A PF resole resin was prepared as described above and the ten time-samples taken during the synthesis are analyzed with ^{13}C NMR and GPC. The synthesis characteristics of the resin are low cook temperature (70 °C), long cook time, high caustic, low resin solids content, and final medium viscosity, which is similar to Resin #3 or #4 type discussed

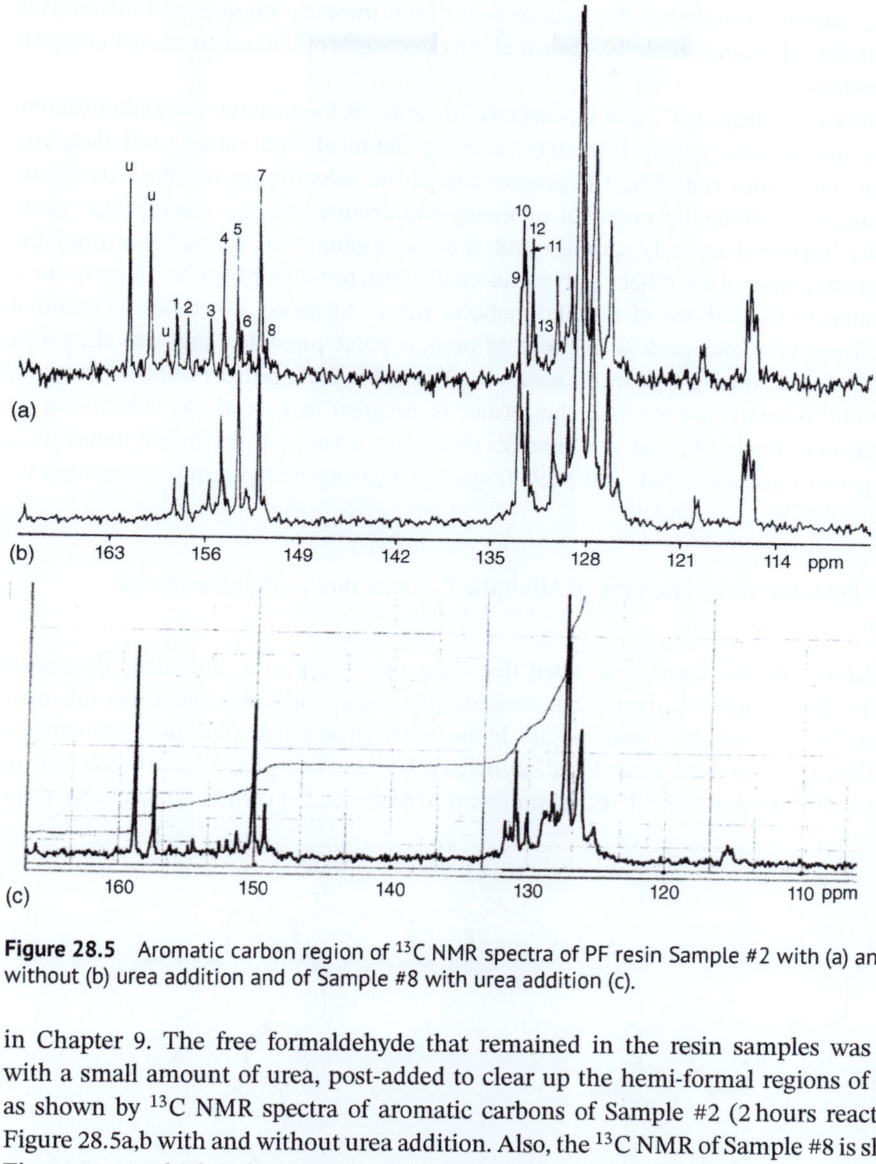

Figure 28.5 Aromatic carbon region of ^{13}C NMR spectra of PF resin Sample #2 with (a) and without (b) urea addition and of Sample #8 with urea addition (c).

in Chapter 9. The free formaldehyde that remained in the resin samples was reacted with a small amount of urea, post-added to clear up the hemi-formal regions of spectra, as shown by ^{13}C NMR spectra of aromatic carbons of Sample #2 (2 hours reaction), in Figure 28.5a,b with and without urea addition. Also, the ^{13}C NMR of Sample #8 is shown in Figure 28.5c, which indicates the changing aromatic carbon intensities from advancement of polymerization.

28.2.2 Peak Intensity Changes of Aromatic Carbons Due to Polymerization During Resin Synthesis

Intensity changes of C_2, C_6, and C_4 carbons: The peak intensity changes of C_2, C_6, and C_4 aromatic carbons for all 10 samples are listed in Table 28.2. First of all, most of the chemical shift values can be related to those of model compounds shown in Table 28.2 in Section 28.1 and in a recent report [9]. As the resin synthesis reaction progresses from Sample 1 to Sample 10, it is noted in Table 28.2 that the substitution of C_4 carbons goes faster than

C_2 and C_6 carbons, and these substitutions occur from the early samples and increases to the maximum values in Sample 10, shown also by decreases of the unsubstituted aromatic carbons intensities.

Peak intensity changes of C_1 and C_4 carbons: In Table 28.3 are shown the eight different C_1 carbon peaks, 1–8, arising from their varying chemical shift values, and their rise and fall in intensities reflecting the progression of the substitution reactions occurring on $C_{2,6}$ and C_4 carbons by methylol or methylene groups. Among these peaks, *peaks 6, 7,* and *8* appeared in early samples and kept increasing their prominence until the end, indicating that they reflect polymeric chain structures of phenolic rings; peak 6 is assignable to C_1 carbons of middle phenolic rings (A), peak 7 to those of terminal phenolic rings (A), and peak 8 to those of branch point phenolic rings (B) shown in Figure 28.6. Also, among C_4 carbon peaks *9–13* in Table 28.3, *peaks 11* and *13* grew up steadily until the end sample, and the former is assigned to C_4 carbons substituted by p–p methylene bonds (A) and the latter to those bonded to o–p methylene bonds (C). *Peak 10* goes to unsubstituted, and *peak 12* goes to hydroxymethyl group substituted C_4 carbons.

28.2.3 Peak Intensity Changes of Aliphatic Carbons due to Polymerization Reaction

For aliphatic carbons, it was first noted that their intensities were uniformly depressed relative to the aromatic carbon intensities, despite the quantitative measures taken in obtaining ^{13}C NMR spectra. Therefore, the hydroxymethyl and methylene group intensities were scaled with one fixed value for all samples relative to one phenolic carbon intensity value set at 20.0, so that the F/P value comes out to be about 2.5, the materials charge ratio

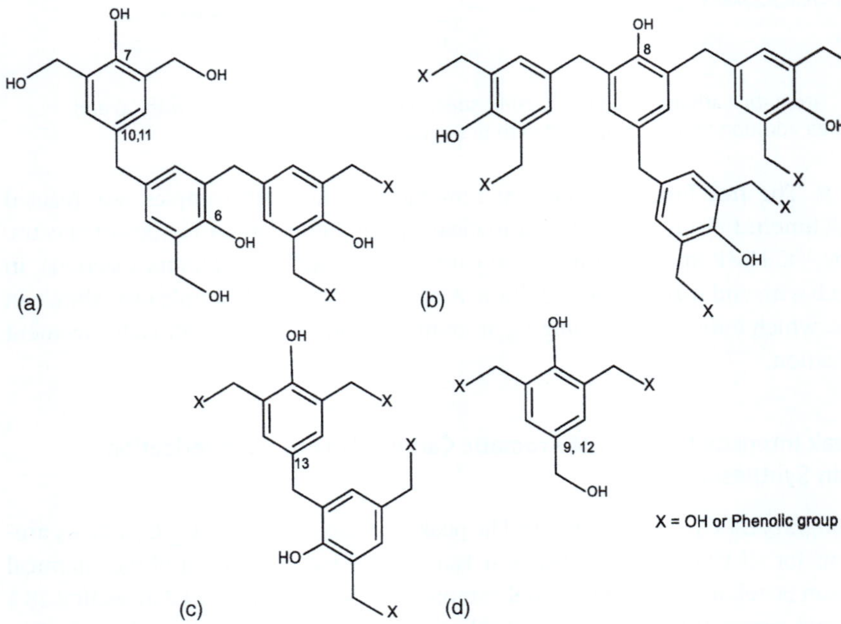

Figure 28.6 Assignment of some aromatic carbon chemical shifts.

Table 28.2 Changes of aromatic carbon intensities normalized for each carbon type to 20.0.

Sample	C_{4s}	C_{2356s}	C_{26u}	C_{4u}
1	15.44	65.5	14.5	4.56
2	17.03	67.5	12.5	2.97
3	18.00	71.9	8.1	2.00
4	19.00	72.0	8.0	1.00
5	19.46	73.5	6.5	0.54
6	20.00	74.8	5.2	0.00
7	20.00	75.2	4.8	0.00
8	20.00	76.7	3.3	0.00
9			2.0	
10			1.0	

s: substituted; *u*: unsubstituted.

Table 28.3 Chemical shift and intensity changes of C_1 (peaks 1–8) and C_4 carbon (peaks 9–13).

		Sample							
Peak	Chem shift, ppm	1	2	3	4	5	6	7	8
1	158.25 ± 0.06	11	7	5	4	2			
2	157.34 ± 0.07	10	7	6	5	4	3	2	2
3	155.73 ± 0.06	10	7	5	3	2			
4	154.77 ± 0.09	17	14	10	9	10	5	8	7
5	153.71 ± 0.02	18	14	14	2				
6	153.54 ± 0.10	5	6	8	17	15	14	10	10
7	152.10 ± 0.15	12	21	33	35	35	37	40	38
8	151.60 ± 0.04	2	10	6	7	10	14	17	20
9	133.07 ± 0.17	10	7	7	5	3	2		
10	132.90 ± 0.08	10	17						
11	132.81 ± 0.11	18	20	47	55	50	65	60	60
12	132.32 ± 0.13	33	20	12	12	2	1		
13	132.02 ± 0.06	0	5	15	15	22	22	25	30

The intensity is the integrated percentage of each peak based on each carbon.

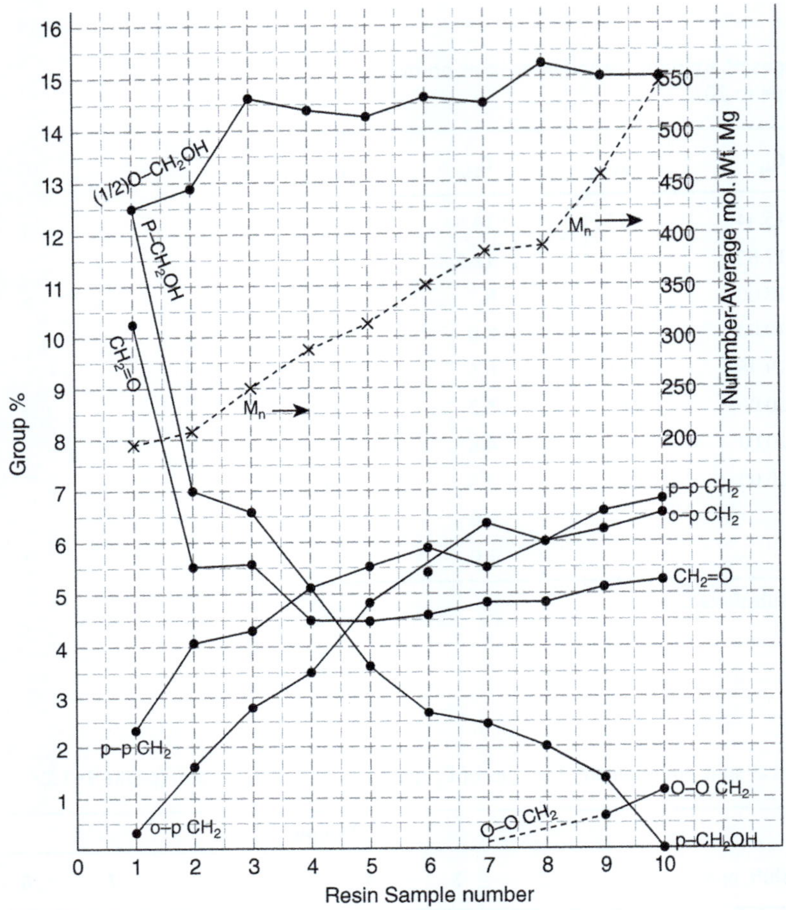

Figure 28.7 Aliphatic carbon intensities normalized with corresponding aromatic carbons changing with polymerization extents. Intensities are calculated with each type of aromatic carbons set to 20.0. M_n values are calculated using Eq. (28.1) and (28.2) and combined methylene group values, divided by 20.0. CH_2O % values were obtained by titration using the hydroxylamine method.

value used in the resin synthesis. The resultant aliphatic group changes in percentages are summarized in graphic form in Figure 28.7, including the free formaldehyde contents determined by titration using the hydroxylamine method [8].

$$n = 1/\{1 - (CH_2)\} \tag{28.1}$$

$$M_n = n\,\{136.1\text{–}30(Us)\} + 48 \tag{28.2}$$

(CH_2): Total methylene bonds formed per mole of phenol; Us: unsubstituted *o*- and *p*-carbons, combined.

o-Hydroxymethyl groups started out at a high value of about 25%, stayed high, and ended at about 30%, while *p-hydroxymethyl group* started out at 12.5% and continuously decreased to zero value. The hydroxymethylation reactions in the first hour progressed to 76% of available aromatic carbons. This reaction extent corresponds to an approximate second-order reaction rate constant (k_2) value shown in Eq. (28.3), which is roughly comparable with the combined rates for the formation of *p*- and *o*-(hydroxymethyl) phenols, which can be separately obtained from the respective group values (Eqs. 28.4 and 28.5).

$$k_2 = 9.10 \times 10^{-5}\ \text{l/mol s}\ (70\ °\text{C}) \tag{28.3}$$

$$k_2p = 5.95 \times 10^{-5}\ \text{l/mol s}\ (70\ °\text{C}) \tag{28.4}$$

$$k_2o = 4.85 \times 10^{-5}\ \text{l/mol s}\ (70\ °\text{C}) \tag{28.5}$$

$$k_{2pp} = 5.94 \times 10^{-5}\ \text{l/mol s}\ (70\ °\text{C}) \tag{28.6}$$

$$k_{2op} = 5.78 \times 10^{-6}\ \text{l/mol s}\ (70\ °\text{C}) \tag{28.7}$$

$$k_{2oo} = 6.00 \times 10^{-7}\ \text{l/mol s}\ (70\ °\text{C}) \tag{28.8}$$

p–p Methylene groups started out and maintained at a higher percentage value than the *o–p methylene bond* until Sample #6 and then both increased slowly to 6.8%. *o–o Methylene bond* started out only late in Sample #9. The calculated total methylene bond contents allowed the calculation of the degree of polymerization (n) and number-average molecular weight (M_n) using Eqs. (28.1) and (28.2). M_n values showed a gradual increase from 192 for Sample#1 to 556 for Sample #10. Assuming two hydroxymethyl groups are needed for methylene bond-forming reactions, the bi-molecular reaction rates are obtained from the NMR results as shown in Eqs. (28.6)–(28.8). Thus, *p–p* methylene bonds form about 10 times faster than *o–p* bonds, which are in turn about 10 times faster than the *o–o* bond formation.

Distribution of methylene bonds: Overall, in Sample #10, as shown in Figure 28.7, the methylene bond contents and orientations are as follows: 1.26 *o–o* bonds (8.6%), 6.60 *o–p* bonds (45%), and 6.80 *p–p* bonds (46.4%) per 20 phenolic rings. The result appears to be a characteristic of PF resole resin cooked at relatively low temperature of 70 °C and with high F/P mole ratio values.

Formaldehyde species (Figure 28.7), determined by the hydroxylamine hydrochloride titration method [8] indicate first decreases to about 4.5% levels at Samples #4–#6 and then increased slightly to 5.3% levels in Sample #10, indicating that the polymerization reaction in the later stages results in generating some free formaldehyde due to the reaction occurring between two hydroxymethyl groups to form one methylene group.

28.2.4 Polymerization of PF Resole Resins Observed by Gel Permeation Chromatography (GPC)

GPC analyses were done using 0.1 NaOH solution as eluent on Sephacryl 200S gel column with UV detection, and the calibration curve was created using polyethylene glycol molecular weight standards, which were, as shown in Figure 28.8a, slightly modified with phenol (MW 94.1) and a dimer phenolic compound of known MW value of 320 [10]. The GPC result of Sample #7 is shown in Figure 28.8b as an example. A convenient method used in

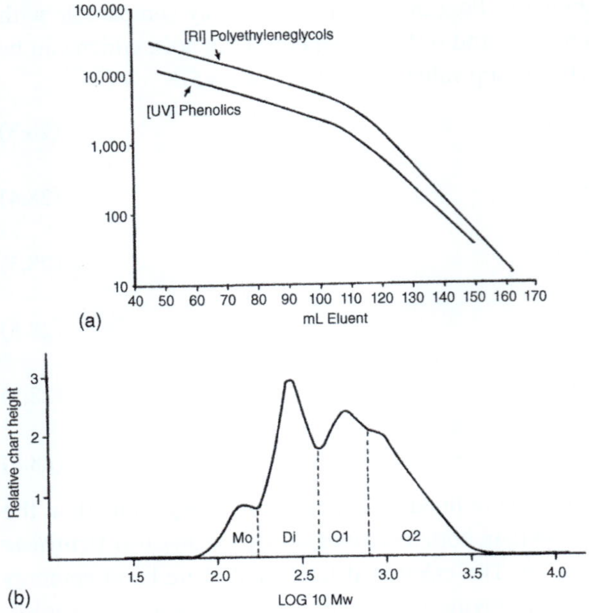

Figure 28.8 (a) GPC calibration and (b) GPC chromatogram (*b*) of PF resin sample #7.

this study for comparing the GPC chromatograms of ten (10) resin samples is to divide them into monomer (Mo), dimer (Di), trimer (O_1), and the rest (O_2) sections, and each section's changing percentage values are plotted as shown in Figure 28.9. As expected, with the sample number increased from Sample #1 to Sample #10, the Mo section decreased drastically, Di and O_1 sections increased and decreased, while O_2 section increased continuously as with the number-average average molecular weight values calculated. The disappearance rate of Mo section observed is calculated to give the bi-molecular rate constant shown in Eq. (28.9). These results generally agree with the ^{13} C NMR result for pp-methylene bond formation rate constant discussed above. A schematic PF resole resin structure derived is presented in Figure 28.10.

$$k_{2gpc} = (4.5\text{–}6.5) \times 10^{-5}\ 1/(\text{mol} \cdot \text{s}) \tag{28.9}$$

28.2.5 Summary

One typical PF resole resin was synthesized: F/P = 2.51, 3-step NaOH addition to 8.51%, reacted at 70 °C, resin solids level of 38.63%, reaction time 510 minutes. Samples were taken at 60- and 35-minutes intervals and analyzed by ^{13}C NMR and GPC. The analysis results showed that phenolic *p*-carbons are more reactive than *o*-carbons in the hydroxymethylation as well as in the methylene bond-forming reactions. Therefore, the polymer chain structures of resin are increasingly formed where *p*–*p* methylene bonds are firstly appeared followed by *o*–*p* methylene bonds to form dimers and trimers, and then higher polymers appeared by forming *o*–*o* methylene bonds, resulting in branched polymer structures. The terminal groups of polymer molecules are mostly phenolic rings that have two *o*-hydroxymethyl groups, which generally agreed with suggestions made by early pioneering investigators. Thus, further reactions for this type of resins in the curing stages would be mostly between *o*-hydroxymethyl groups to form *o*–*o* methylene groups.

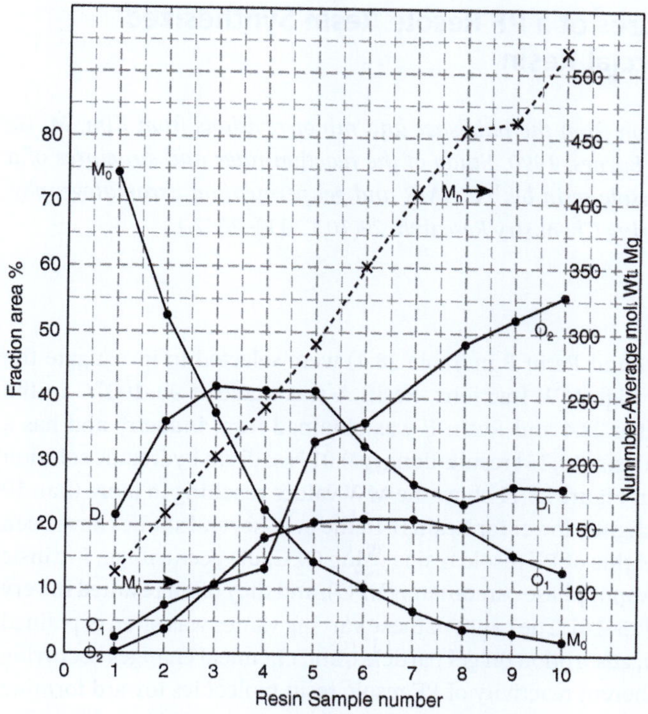

Figure 28.9 Gel permeation chromatography data of resin Samples #1–#10 showing the progression of polymer formation (O_2 fraction) from monomers, dimers, and trimers. % Values of Mo, Di, O_1, and O_2 sections are obtained as shown in Figure 28.8 (b).

Figure 28.10 Schematic chemical structure of PF resole resins synthesized at low temperature (70 °C), high F/P ratio (2.50), and high NaOH content.

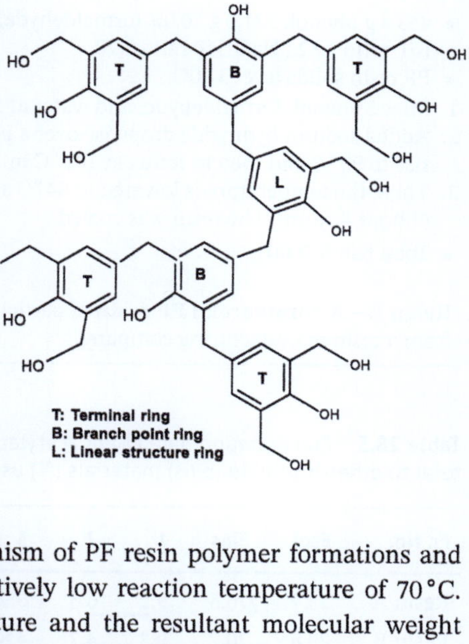

Overall, the study showed the detailed mechanism of PF resin polymer formations and resultant polymer structures attained at a relatively low reaction temperature of 70 °C. Resin syntheses at a higher cooking temperature and the resultant molecular weight structures of high molecular weight fractions are presented in the next section.

28.3 Polymer Structures of a PF Resole Resin Synthesized at 102 °C vs. a Commercial Resin

(Adapted with permission from the publisher and minor revisions from Kim, M. G., L. W. Amos, and E. E. Barnes. 1990. Study of the reaction rates and structures of a phenol-formaldehyde resole resin by ^{13}C NMR and gel permeation chromatography. Industrial and Engineering Chemistry Research. 29(10): 2032–2037.)

28.3.1 Introduction

Two PF resole resins (*Resin A* and *Resin B* in Table 28.4) were selected to investigate the polymeric structures of their high MW fractions. Resin A was made at 90–102 °C in the laboratory as described in Table 28.4 and Resin B was obtained from industry and has a high caustic content. GPC analysis [10, 13] results done in 0.01 N sodium hydroxide solution on a Sephacryl 200S column in Figure 28.11 show the MW values reaching to more than 10 thousand Da for Resin A and slightly lower for Resin B. In the past, several studies have been reported on the molecular weights of PF resole resins using GPC, ultracentrifuge, intrinsic viscosity measurements, and vapor phase osmometry (VPO), and they often resulted in very high and varying MW values [13–16]. The cause of these varying values was often explained to be of molecular association, adsorption on gel particles, and chemical changes occurring during analyses due to the inherent reactivity of PF resole resin molecules toward forming higher MW materials.

Table 28.4 A PF resole resin synthesized with a high temperature and high caustic regime and a commercial PF resole resin.

Resin A – Synthesis procedure of a wet process hardboard binder type PF resin.

- 453.4 g phenol; 721,0 g 50.0% formaldehyde, 299.0 g 50.0% NaOH. 526.0 g Water.
- F/P ratio 1.82.; NaOH/P ratio 0.78.
- PF resin solids level 31.0%.

1. Mixed phenol, formaldehyde, and water at 30 °C.
2. Added sodium hydroxide dropwise over a period of 25 min, while allowing the temperature to rise to 90 °C and then to reflux at 102 °C in 10 min, which is maintained for 15 min.
3. Then, the temperature is lowered to 84 °C and held till the resin reaches a viscosity of 300 cps (1 hour 45 min). The resin was cooled.

- Total batch 2000 g.

Resin B – A commercial PF resin of particleboard or oriented strand board core binder type from a resin manufacturing company.

Table 28.5 Fractionation data (%) of acetylated resins A and B: fractions 1–10 in percentages that total to ether-insoluble (Eins) materials (%) used in the fractionations.

Fr. No.	Esol	Eins	1	2	3	4	5	6	7	8	9	10
Resin A	23.7	76.3	9.2	6.6	6.9	7.5	4.8	4.2	6.3	10.7	10.4	10.2
Resin B	29.8	70.2	9.3	4.7	4.8	6.7	5.2	4.3	13.6	11.6	3.8	6.2

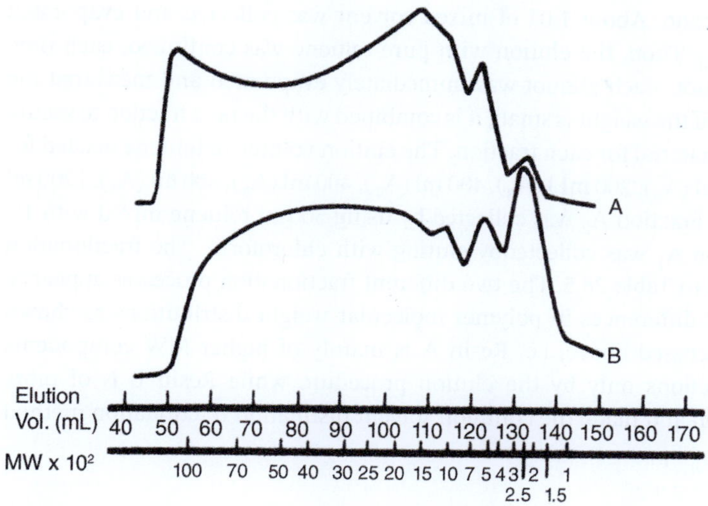

Figure 28.11 Gel permeation chromatograms of Resin A and B on Sephacryl 200S and 0.01 N NaOH as eluent.

28.3.2 Acetylation of PF Resin Samples and Fractionation

Acetylation and separation: In this study, the resin molecules are stabilized by acetylation of all hydroxyl groups using the reported acetic anhydride/pyridine procedure [16]. Three hundred (300) g of liquid resin resulted in 162 g of acetylated product for Resin A and 112 g for Resin B.

Fractionation of acetylated Resin B by precipitation: The acetylated resin material was dissolved in 500 ml of ethyl acetate and 500 ml of ether was slowly added while stirring vigorously to separate into ether-soluble (low MW components) and ether-insoluble (high MW components) fractions. Then, the ether-insoluble material of Resin B was fractionated by the precipitation method [17] as follows: 40 g of this material was dispersed/dissolved in a mixture of 1500 ml benzene (bp 80.1 °C) and 3500 ml toluene (bp 110.6 °C) and equilibrated at 23 °C overnight and the insoluble precipitates were collected as Fraction B_1. The residual solution was subjected to a slow distillation of benzene/toluene components to residual volumes of 4100, 3100, 2300, 1300, and 600 ml, and each time equilibrated at 23 °C and collected the precipitated materials as Fractions B_2, B_3, B_4, B_5, and B_6, respectively. The residual solution was diluted to 1900 ml toluene, and fractions B_{7-9} were obtained by adding hexane, heating the mixture to 60 °C, and equilibrating overnight at 23 °C, followed by filtration. The amounts of hexane added in succession to obtain each fraction were 100 ml (B_7), 210 ml (B_8), and 400 ml (B_9). Fraction B_{10} was obtained by evaporating the whole solution.

Fractionation of acetylated Resin A by elution: For the ether-insoluble material of acetylated Resin A, the distillation/precipitation method was found to be inapplicable. The following fractionation procedure [17] was used: 36.0 g of the acetylated material was dissolved in 400 g chloroform and added to it 1600 g of cleaned sand, followed by evaporating off the chloroform until dry while the sand is continuously stirred with a spatula. The resin-coated sand was then packed in a 5.0 cm diameter glass column and eluted with 400 ml hexane, followed by eluting with hexane mixed with increasing amounts of toluene up to 4 : 96 ratio. About 1.01 of mixed solvent was collected and evaporated to result in Fraction A_{10}. Then, the elution with pure toluene was continued, each time collecting a 200 ml aliquot. Each aliquot was immediately evaporated and measured the resin solids weight and, if the weight is small, it is combined with the next fraction to secure a workable amount of material for each fraction. The elution volume of toluene needed for each fraction was: 400 ml (A_9), 200 mLl (A_8), 400 ml (A_7), 400 ml (A_6), 800 ml (A_5), 1200 ml (A_4), and 1200 ml (A_3). Fraction A_2 was collected by using 500 ml toluene mixed with 1% chloroform, and fraction A_1 was collectedby eluting with chloroform. The fractionation results are summarized in Table 28.5. The two different fractionation processes appear to indicate the two resins' differences in polymer molecular weight distributions as shown by the GPC results discussed above, i.e. Resin A is mainly of higher MW components and separated into fractions only by the elution procedure while Resin B is of more gradual MW distribution and made the common solvent distillation fractionation method applicable.

28.3.3 ^{13}C NMR Chemical Shift Values of Model Compounds and PF Resin Fractions

^{13}C NMR results of acetylated PF resin model compounds, isolated by column chromatography or synthesized in this work, are reported in Table 28.6. The introduction of acetyl groups causes changes in the phenolic carbon chemical shift values slightly up-field on C_1 and down-field on C_2 and C_4, and minor changes on C_3. The chemical shift values changed only a little for *o*-hydroxymethyl carbons (61.6 and 61.3 ppm) and *o–o*, *o–p*, and *p–p* methylene carbons' (40.4, 35.9, and 30.9 ppm). ^{13}C NMR spectra of Fractions A_4 and B_4 and proton NMR spectra of fraction B_4 are shown as examples in Figures 28.12a/b and 28.13, respectively. They are very similar to those of unacetylated PF resins shown in Section 28.2.1, except the presence of high acetyl group peaks at about 20 ppm (CH_3) and about 170 ppm (CO).

The acetyl carbonyl carbon peaks showed a useful resolution between those bonded to phenolic oxygen (169.1, 168.9, 168.7 ppm) and those bonded to hydroxymethyl groups (170.4 ppm) groups, allowing an independent estimation possible for the ratios of total phenolic ring carbons vs. hydroxymethyl carbons.

Proton NMR was used extensively before ^{13}C NMR came on board for acetylated PF resins [9]. Hydrogens on phenolic rings, methylol groups, methylene groups, and acetyl groups occur at ~7; ~5; ~3.8; and ~2 ppm, respectively, as shown in Figure 28.13 for fraction B_4. The integration values allow to calculate molecular properties of PF resins and the results are reported in Table 28.7. It is noted that proton NMR is ineffective for analysis of un-acetylated liquid PF resins due to presence of water in resins.

28.3.4 ^{13}C NMR Results of PF Resin Model Compounds and Resin Fractions

$$[\boldsymbol{d}] = 3.0–2(CH_2/Ar) - (CH_2OH/Ar) - Us/Ar \qquad (28.10)$$

Results of ^{13}C and proton NMR of PF resin fractions are summarized in Table 28.7. First of all, the internal consistency of peak intensity values of ^{13}C NMR results was checked by calculating the ratio of aromatic carbon (151–122 ppm) vs. phenolic acetyl carbonyl carbon

Table 28.6 ^{13}C NMR chemical shift assignments of model compounds.

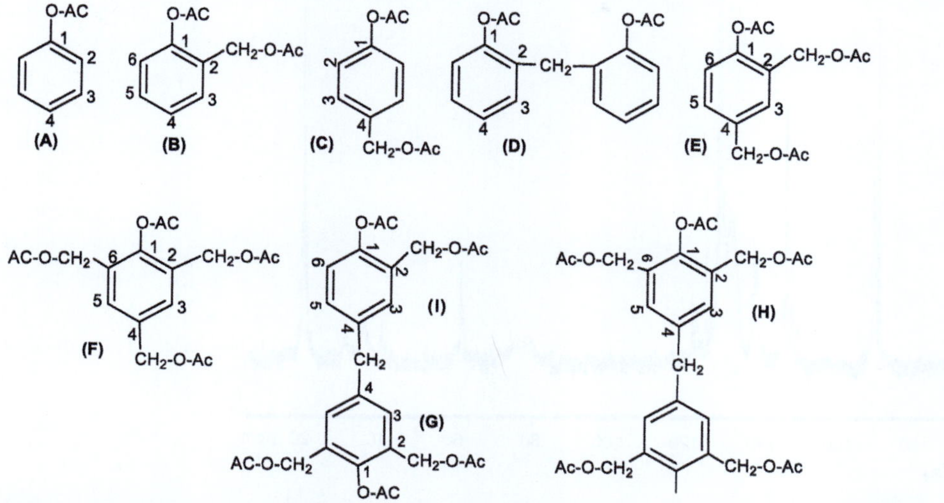

Ar	C_1	C_2	C_3	C_4	C_5	C_6
A	151.10	121.70	129.40	125.60		
B	149.50	128.70	129.60	126.17	130.36	122.93
C	150.77	121.86	129.47	134.07		
D	148.97	127.57	130.70	126.11	131.60	122.37
E	148.90	128.48	130.20	134.14	129.50	122.96
F	147.54	129.44	130.48	134.30		
G	147.65	128.17	130.04	138.13	130.81	122.83
H	146.20	129.17	130.07	138.43		
I	146.22	129.14	130.04	138.75	131.04	129.14

o-CH_2^*OAc, 61.1–61.8 ppm; —CO^*OAr, 169.2–168.6 ppm; *P*-CH_2^*OAc, 65.2–65.5 ppm; —CH_3, 20.2–20.6 ppm; —CO^*OCH_2, 170.5–170.7 ppm.

(168–169 ppm) ratio. The ($1/6Ar/CO_{AR}$) values were found to be higher than the theoretical value of 1.0 for some higher MW fractions, indicating the carbonyl values are depressed against the aromatic carbon values. This deviation was higher for Resin B fractions than Resin A fractions. Resin B fractions showed some unsubstituted *o*- and *p*-phenolic carbons, which might have shown extra residual ***nOe*** effects to increase the aromatic carbon values. Another way of checking internal consistency of the ^{13}C NMR results is to calculate the deviation factor [*d*] values as defined in Eq. (28.10), where CH_2/Ar, CH_2OH/Ar, and Us/Ar ratio values are, respectively, the number of methylene, hydroxymethyl, and unsubstituted *o*- and *p*- carbons per phenolic ring. The [d] values calculated also showed similar deviations to indicate that either methylene or hydroxymethyl or both values are depressed vs. aromatic carbon values. This effect appears to be also related to molecular associations for high MW factions in ^{13}C NMR solutions, to be further discussed later in this Chapter. Therefore,

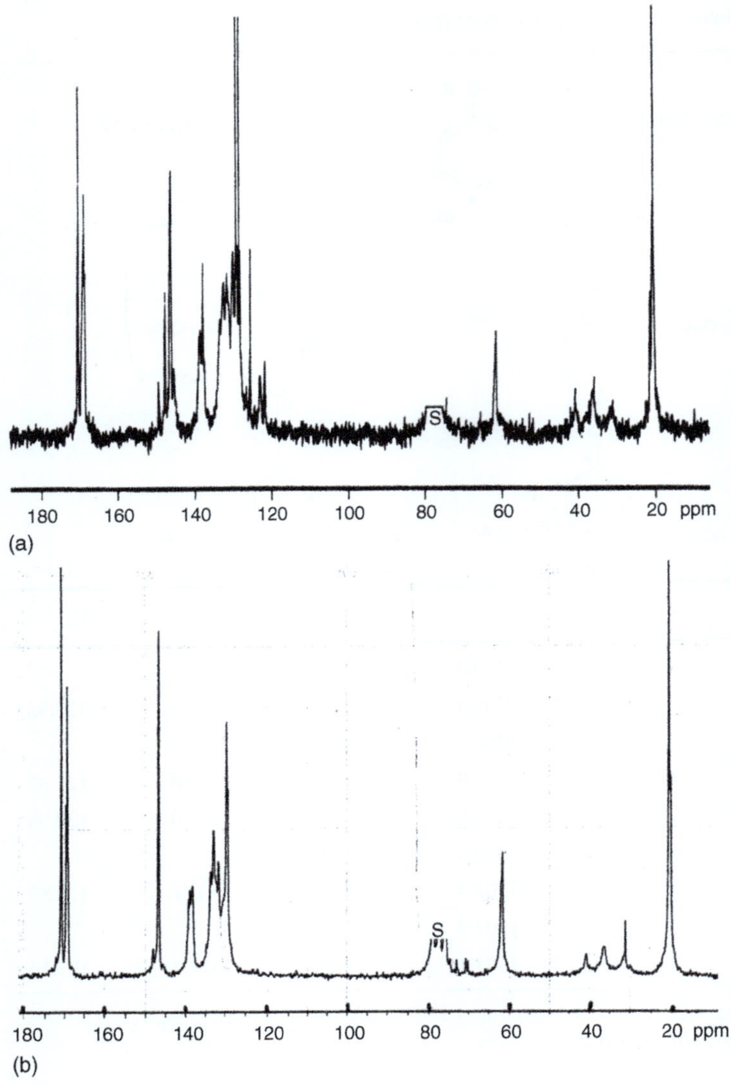

Figure 28.12 (a) ^{13}C NMR spectrum of acetylated Resin fractions B_4 (top) and (b) of fraction A_4 (bottom) in chloroform-d (~78 ppm). Peaks at ~20 and 170 ppm are methyl and carbonyl carbons of acetyl groups introduced in the resin structure.

the CH_2/Ar values obtained from ^{13}C NMR appeared to be somewhat unreliable for calculating MW values of PF resole resins. Proton NMR results (Figure 28.13) also showed somewhat broad methylene group peaks and the CH_2/Ar ratio values were more consistent but exceeded unity for some higher MW fractions, indicating the presence of intramolecular (cross-linked) methylene bonds [18] or highly branched polymer structures [19], despite the samples' complete dissolution in the NMR solvent chloroform. The intramolecular or branched methylene bonds would restrict the molecular motions in solution to explain the broad peaks of methylene groups in the proton NMR spectra. Therefore, molecular weight values were not calculated using the CH_2/Ar values in Table 28.7.

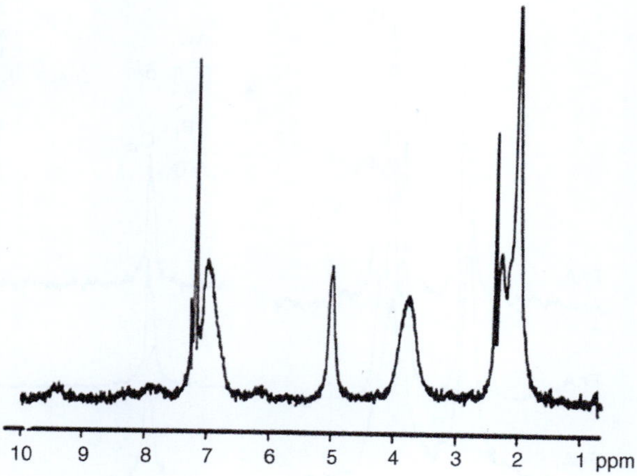

Figure 28.13 Proton NMR spectrum of acetylated fraction B_4 in chloroform-d (residual peaks at ~7.3 ppm).

Table 28.7 NMR data of acetylated PF resin fractions with [d] defined as in Eq. (28.10).

	CH_2O/Ar		CH_2O/Ar [a)]					CH_2 group ratio (%)		
	H_{NMR}	C_{NMR}	H_{NMR}	C_{NMR}	$[U_s]$	d	$(^1/_6Ar)/Ar_{co}$ [b)]	pp	po	oo
A_1	0.94	0.95	1.26	0.89	0	0.26	1.26	33	39	29
A_4	0.97	0.88	1.24	1.01	0	0.10	1.04	26	39	36
A_8	0.88	0.84	1.01	1.05	0	0.06	1.05	27	41	32
$A_{E_{is}}$	0.88	0.94	0.78	0.87	9	0.33	1.22	31	42	27
$A_{E_{sol}}$	1.06	1.27	0.64	0.77	0	0.19	1.06	36	43	21
B_1	0.60	0.47	1.20	0.66	0.35	0.88	1.38	30	47	23
B_4	0.58	0.52	1.05	0.72	0.50	0.54	1.29	27	42	31
B_8	0.50	0.58	0.89	0.72	0.57	0.41	1.35	27	42	31
$B_{E_{is}}$	0.65	0.64	1.05	0.83	0.52	0.18	0.91	30	46	24
$B_{E_{sol}}$	0.86	0.84	0.59	0.57	1.09	−0.07	1.33	35	40	25

Abbreviations: A_1 resin A fraction 1, etc.; E_{is}, ether-insoluble fraction; E_{sol}, ether-soluble fraction; Ar, phenol ring; $[U_s]$, unsubstituted $C_{2,4,6}$ per phenol ring; *d*, term defined in Eq. 1; pp:po:oo, methylene group orientation ratio.

a) Methylene peaks were broard.

b) The ratio obtained from ⅙ of the total aromatic carbon divided by the phenolic carbonyl carbon intensity.

Another set of data from ^{13}*C NMR results* in Table 28.7 is the *o–o*, *o–p*, and *p–p* methylene group ratios of fractions. Fractions of Resin A, synthesized by increasing the reaction temperature fast to 102 °C in the beginning, the methylene group values ranged in the order of *o–p* (39%–43%) > *p–p* (27%–36%) > *o–o* (21–36%) bonds, which is in contrast to the values observed for the resin synthesized at 70 °C, *o–p* (44.5%) ≤ *p–p* (46.5%) > *o–o* (8.9%) for Resin sample #10 discussed in the previous section (S. 28.2.3.) For the commercial Resin B fractions, the ratios are *o–p* (40%–47%) > *p–p* (27%–35%) > *o–o* (23%–31%), indicating that the resin's cooking temperature history in resin synthesis would be similar to Resin A. One conclusion is that the use of higher temperatures in the polymerization step results in more

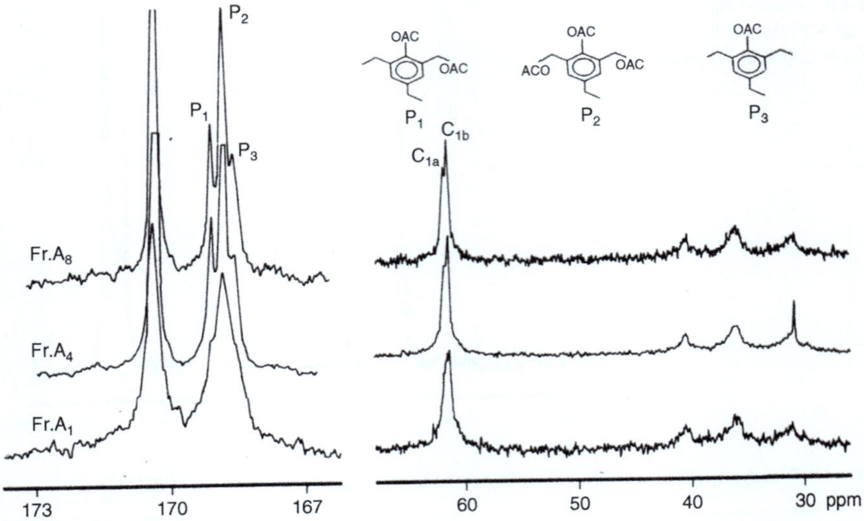

Figure 28.14 ^{13}C NMR spectra of Resin A fractions showing aliphatic (170.4 ppm) and aromatic (169.1, 168.9, 168.7 ppm) carbonyl carbons, *o*-hydroxymethyl (61.6, 61.3 ppm) and *o*−*o*, *o*−*p*, and *p*−*p* methylene (40.4, 35.9, 30.85 ppm) carbons. In certain fractions a small peak at 65.2 ppm is observed assignable to *p*-hydroxymethyl carbons.

o–*p* methylene groups. In theory, *o*–*p* methylene bonds mean more linear, less branched PF resin structures, while the presence of *p*–*p* methylene indicates that the polymer chain is more likely to be branched and extended not only through *o*–*p* but also *o*–*o* methylene bonds. These varying methylene bond orientations appear to be one of the key parameters to use in synthesizing PF resole resins for faster curing speeds.

Another interesting observation is that the phenolic acetyl carbonyl peaks are resolved into three distinct peaks for resin fractions A_4 and A_8, which have low [d] and low Us values: P_1 (169.1 ppm), P_2 (168.9 ppm), P_3 (168.7 ppm) as shown in Figure 28.14. The three peaks were assignable to the phenolic ring structures that are substituted with methyleneic groups in different ways: P_1 in the middle of linear phenolic polymer chain, P_2 in the ends of phenolic polymer chain, and P_3 in the middle of branched phenolic polymer chain, similar to the structures discussed in previous section (Section 28.2.3) with respect to C_1 carbon peaks of unacetylated PF resins. Also, hydroxymethyl groups are often resolved into two peaks, C_{1a} for P_1 and C_{1b} for P_2 structures as shown. Overall, phenolic ring structures in the polymer chain can be represented as P_1:P_2:P_3 peak area ratios, roughly in a 1 : 2 : 1 ratio for Resin A fraction. Resin B fractions did not show such resolutions for phenolic acetyl carbonyl peaks, because of the presence of unreacted ortho and para carbons. Thus, ^{13}C NMR study of Resin A fractions results in some details of PF resin polymer structures because most ortho and para carbons are substituted with methylene or hydroxymethyl groups.

28.3.5 Intrinsic Viscosity ([η]), Vapor Phase Osmometric (VPO) Analysis, and M–H Equation

The intrinsic viscosity and VPO molecular weight (M_n) measurement results of selected fractions are reported in Table 28.8. Since the fractionation was carried out with the

Table 28.8 Intrinsic viscosity (IV, mL/g), molecular weight (M_n), specific volume (SpV, mL/g), and the Huggins' constant (k') for PF resin fractions in benzene (B) and chloroform (C); IV ratio in C and in B (C/B).

	In benzene			In chloroform				
Fraction	IV	SpV	k'	IV	SpV	k'	$^C/_B$	M_n
$A_{E_{is}}$	3.85	0.793	0.79	6.97	0.762	1.88	1.81	4300
A_8	5.19	0.775	1.07	6.00	0.752	1.96	1.16	4610
A_4	6.85	0.776	0.74	8.76	0.769	1.06	1.28	4690
A_1	9.36	0.769	0.73	12.86	0.776	1.01	1.37	6070
$B_{E_{is}}$	4.23	0.793	2.60	5.76	0.803	1.46	1.36	1650
B_8	4.64	0.788	1.32	5.24	0.769	0.98	1.13	3060
B_4	4.58	0.779	3.02	6.44	0.801	1.23	1.41	3230
B_1	4.89	0.800	4.36	7.47	0.733	0.99	1.53	5600

SpV, values obtained as the average between the first and last dilutions in IV measurements that were in the range of $(10–46) \times 10^{-3}$ $^{mL}/_{g}$ in benzene and $(5–25) \times 10^{-3}$ $^{mL}/_{g}$ in chloroform; k', constant obtained in most cases with R^2values greater than 0.98 in least-squares regression plots; $^C/_B$, intrinsic viscosity in chloroform/in benzene; M_a, vapour pressure osmometric molecular weight with most values obtained with R^2values greater than 0.90 in least-squares plots. E_{is}, ether-insoluble fraction. (To convert the Mn of acetylated resin molecules to unacetylated forms, multiply the values with 0.614.)

ether-insoluble materials, the data reflect not the full resins but only the relatively higher MW materials.

Definition of intrinsic viscosity and measurement equation

$$\eta_{sp} = [\eta] + k'[\eta]^2 c \tag{28.11}$$

η_{sp}: specific viscosity
$[\eta]$: intrinsic viscosity
k': Huggins' constant
c: concentration.

Intrinsic viscosity (IV) is a measure of polymer molecules' individual molecular volume (mL/g) measurable in an infinitely dilute solution and is commonly measured using a dilution viscometer at several (four or more) dilute concentrations and the resultant specific viscosity (η_{sp}) vs concentration (c) curve is made and extrapolated to zero concentration, where the molecules are to be individually separated if they are in a good solvent. The intercept value is the intrinsic viscosity of the polymer, according to Eq. (28.11), and it is a measure of the molecular volume or molecular weight in the solution. Further, the intrinsic viscosity value, $[\eta]$, depends on how well the polymer molecule interacts with or whether it is soluble in the solvent used. In good solvents, polymer molecules swell more and therefore higher intrinsic viscosity values are obtained than in poor solvents. The IV values obtained for Resin A and Resin B fractions shown in Table 28.8 increased as the Mn values measured by VPO method increased, and their magnitudes are in a comparable range with those of PF novolac resins [20]. Furthermore, the IV values are greater in chloroform (C) than in benzene (B), i.e. the hydrodynamic volume expanded in chloroform, and the C/B ratio increased as the molecular weight increased to a similar extent as PF novolac resins, which are known to have no crosslinks or intramolecular methylene bonds [21].

Vapor pressure osmometry (VPO)[22] measures the temperature differences that arise from different rates of solvent evaporation from droplets of pure solvent vs. droplets of polymer solution in an atmosphere of solvent vapor. The temperature difference is proportional to the vapor pressure lowering of the polymer solution at equilibrium and thus results in the number-average molecular weight (M_n) of polymers. Four or more dilutions of polymer in a solvent, such as acetone are measured and extrapolated to zero concentration. The VPO values of fractions reported in Table 28.8 indicate that Resin A fractions are all of relatively high M_n materials, i.e. the ether-insoluble starting fraction did not contain gradually lower and lower molecular weight materials, which also agrees with the fact that the fractionation process was difficult to carry out using the normal precipitation method and done with the column elution method.

Mark–Houwink Equation

$$[\eta] = KM^{\alpha} \tag{28.12}$$

K: MH constant.
M: molecular weight.
α: expansion constant.

Mark–Houwink (M–H) equation [17]: IV values of a series of fractions of a polymer are commonly plotted against their MW values to result in the M–H equation (Eq. 28.12), in which the slope of the equation (α) is referred to as the expansion factor, a measure of polymer molecules' swelling (expansion) extent in the solvent, and the value in good solvent is normally 0.5 for linear polymers. The M–H correlation of Resin B fractions resulted in Eqs. (28.13a) and (28.13b). *Resin A fractions* behaved differently than resin B fractions with the M–H equation as shown in Figure 28.15 with dotted lines, i.e. the elution fractionation process used for this resin did not yield a good spread of molecular weights but resulted in fractionation of molecules more according to the IV values. The GPC results in Figure 28.11

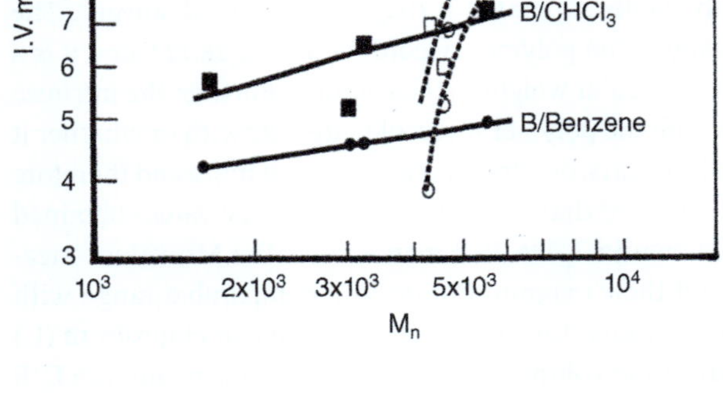

Figure 28.15 Mark–Houwink plot of ether-insoluble acetylated resin A and B fractions in chloroform and benzene.

indicate that the ether-insoluble components of Resin A used for fractionation were all of relatively

Mark–Houwink correlation equations of PF resin B fractions with number-average MW values in chloroform and benzene

$$[\eta] = 1.196\ \mathrm{Mn}^{0.213}\ \textit{Resin B}\ \text{in}\ \mathrm{CHCl_3} \tag{28.13a}$$

$$[\eta] = 1.735\mathrm{Mn}^{0.120}\ \textit{Resin B}\ \text{in}\ \mathrm{C_6H_6} \tag{28.13b}$$

$$1650 < M_n < 5600$$

narrow range of high molecular weight materials, and they have varying low and high IV values in the same solvents. This appears to indicate that the low IV value components might harbor more intramolecular methylene bonds than high IV value components present, a result expected of the high molecular weight components to be formed in the synthesis procedure of Resin A: low F/Phenol mole ratio of 1.82, high NaOH content, high cook temperature, and low solids content, etc. Resin A fractions overall showed similar IV increases as resin B fractions in the solvent change from benzene to chloroform, indicating a similarly branched structure.

28.3.6 Conclusion

One typical advanced PF resole resin (Resin A) synthesized (F/P = 1.82, NaOH/P = 0.78, reacted at 102 °C, resin solids level 31%, viscosity 310 cps) and one commercial wood binder-type PF resole resin (Resin B) were studied. GPC indicated that Resin A has a higher molecular weight than Resin B. The two resins were acetylated and fractionated into ether-soluble and ether-insoluble fractions and the latter were further fractionated into 10 fractions. Some acetylated PF resole resin model compounds were also isolated from the ether-soluble fractions, and their ^{13}C NMR chemical shift values were measured. Selected resin fractions were analyzed by ^{13}C NMR and proton NMR, and the methylene and hydroxymethyl group contents and the unsubstituted ortho and para carbons were quantified. The phenolic ring carbon/phenolic acetyl carbon ratio of resin fractions deviated variously from 1.0 and also the deviation factors (d) varied from 0.0 to indicate some inaccuracies involved in NMR measurements for high molecular weight fractions, but the polymer structures of PF resole resins can be invariably described as **polymeric (hydroxymethyl)-methylene-phenols.**

The methylene group/phenolic ring ratio values showed greater than 1.0, especially by proton NMR data, and molecular weights were indeterminate. However, the different NMR deviations observed between Resin A and B fractions were indicative of differences existing in their molecular structures. Further, fractions of Resin A, synthesized by increasing the reaction temperature up to 102 °C, the methylene group orientation value ranged in the order of *o–p* (39–43%) > *p–p* (27–36%) > *o–o* (21–36%) bonds, which were similar to those of resin B fractions but quite contrasting to the values observed for the resin synthesized at 70 °C, discussed in the previous Section (28.2). One conclusion is the use of higher temperatures in the polymerization step results in more *o–p* methylene groups. Molecular weight and intrinsic viscosity measurements of resin fractions and Mark–Houwink correlation results indicated that the high molecular weight Resin A fractions are composed of narrow range molecular weight components and harbor cross-linked structures in varying degrees, while those of resin B fractions have more uniform molecular weight distributions. Both resin fractions appear to be branched and compact in molecular structures but also have rooms to expand and shrink depending on the solvent. Overall, the differences observed between Resin A and B appear to be in line with synthetic procedures, and the results of this study appear to be useful in planning of PF resin synthesis procedures.

28.4 Polymer Structures of High Molecular Weight Fractions of a PF Resole Resin

(Adapted with permission of the American Chemical Society and minor revisions from Kim, M. G., W. L. Nieh, T. Sellers, Jr., W. W. Wilson, and J. W. Mays. 1992. Polymer solution properties of a phenol-formaldehyde resole resin by gel permeation chromatography, intrinsic viscosity, static light scattering, and VPO methods. Ind. & EC Research. 32(3): 973–979.)

28.4.1 Synthesis of an Oriented Strand Board Binder-type PF Resin and GPC Analysis

The polymer chain structures of high molecular weight fractions of a PF resole resin were investigated. An oriented strand board (OSB) binder-type PF resole resin similar to Resin A investigated in Section 28.3 was synthesized (Table 28.9) by using the synthesis temperature of 80 °C with the end viscosity of about 600 cps at 36.6% resin solids level. The NaOH/P mole ratio was 0.35, which is quite lower than Resin A. Aqueous GPC was done on Sephacryl 200S in 0.1 N NaOH solution with UV detection [22–24], resulting in a chromatogram shown in

Table 28.9 Synthesis of a softwood plywood binder-type PF resin at 80 °C and various analysis results.

- 1044.4 g 90% Phenol
- 120.0 g 50% NaOH
- 1500.0 g 50.0% Formaldehyde
- F/P ratio 2.50.
  1. Mixed phenol and sodium hydroxide and heated to 70 °C
  2. Added formaldehyde dropwise over 1.0-hour while keeping the temperature at 80 °C
  3. The reaction was continued at 80 °C until 275 cps measured by Gardener-Holdt viscometer
  4. Added 80 g 50% NaOH and 245 g water and continued the reaction until 375 cps
  5. Added 80 g of 50% NaOH and 460 g water and continued the reaction until 600 cps
  6. Cooled
- Total batch 3529.4 g;
- NaOH/P mole ratio 0.35.
- PF resin solids level 36.6%.

GPC in aqueous medium: Unacetylated resin analyzed on a Sephacryl column in 0.1 N NaOH solution.

Acetylation: carried out according to a known procedure using 200 g of liquid resin as described in the text.

Ether-soluble and insoluble fractions: the acetylated product was dissolved in 500 ml ethyl acetate and precipitated with 1870 ml ether to obtain 43.3 g ether-insolubles (54.6%) and the residual solution evaporated to give 36.0 g ether-solubles (45.4%).

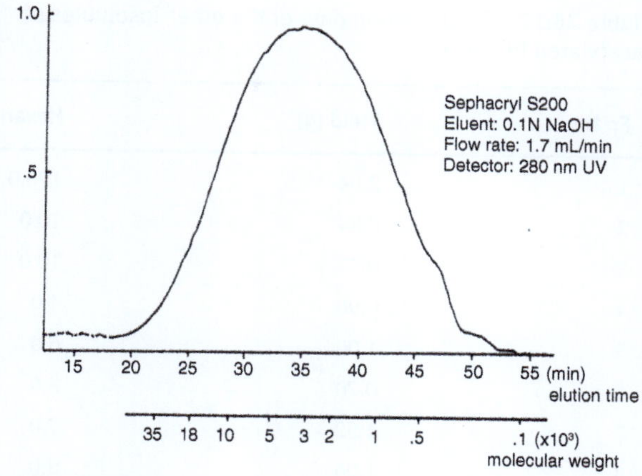

Figure 28.16 Aqueous GPC of the synthesized PF resin calibrated with polystyrene sulfonates.

Figure 28.16. The highest molecular weight fraction appears to show MW value of about 36 000 Da. The MW value is for the **unacetylated** resin and therefore, would be lower than acetylated resin samples, by the acetylation factor of 0.616 for conversion. In order to investigate the polymer structures of the high molecular weight fractions, the acetylation and fractionation procedures were followed.

28.4.2 Acetylation and Fractionation

The acetylation of 200 g of the liquid PF resin was conducted as before according to a known procedure [16]. The acetylated product was separated into ether-solubles (45.4%) and ether-insolubles (54.6%) as described in Table 28.9. Then, 40.0 g of ether-insolubles dissolved in 500 ml ethyl acetate (solvent) was fractionated into 21 fractions using the precipitation method of adding increasing amounts of hexane (non-solvent) similarly as in the previous Section 28.3, the first fraction being the highest MW fraction, as shown in Table 28.10. It is noted that fraction #1 and #2 amounted to 8.93% of total resin solids and the light scattering results in tetrahydrofuran (THF) as solvent showed the highest MW values but with negative A_2 (second virial coefficient) values, indicating that the association propensities could have affected or increased the molecular weight values.

28.4.3 Instrumental Analyses of Acetylated PF Resin Fractions

High-pressure GPC of acetylated fractions was done on Ultrastyragel column in THF as solvent with differential refractive index detection and weight average M_w values were obtained.

Proton and ^{13}C *NMR spectra,* vapor pressure osmometric MW, and intrinsic viscosity measurements of acetylated fractions were obtained similarly as in previous Section 28.3.

Light-scattering MW measurements of acetylated fractions were done on a DWAN Model F from Wyatt Technology [25] using THF or ethyl acetate as solvent: in this instrument, 15 photodiode detectors are arranged in a plane around the sample in tube at fixed distances and between 22.6 to 144.5° angles relative to the incident light. A vertically polarized laser light with a wavelength of 632.8 nm from helium–neon laser is used. As shown in Eq. (28.14), the quantity of scattered light or turbidity (R_Θ) of a polymer solution increases with polymer's concentrations (c) and also with polymers MW values. The turbidity also increases with decreasing angles (Θ) of detectors. Association of polymer molecules can

Table 28.10 Fractionation data of the ether-insolubles of acetylated PF resin.

Fr. No.	Yield (g)	Hexane (ml)
1	2.94	170.0
2	0.84	10.0
3	0.77	10.0
4	0.90	4.0
5	1.00	6.0
6	0.70	8.0
7	1.52	7.0
8	1.06	8.0
9	1.99	10.0
10	1.72	15.0
11	2.63	20.0
12	1.33	20.0
13	2.03	20.0
14	2.80	30.0
15	1.36	35.0
16	1.36	40.0
17	1.75	70.0
18	1.40	150.0
19	0.90	150.0
20	0.59	200.0
21	1.86	*b*
Total	31.41(78.5%)	

40.0 g of ether-insoluble fraction was dissolved in 500 ml of ethyl acetate, followed by the addition of hexane in indicated amounts.
b – Fraction 21 was obtained by evaporation of the entire solvent.

also increase the turbidity because small particles become larger particles by association, and therefore, an extrapolation to zero concentration and zero angle is needed for nullifying the polymer association and angle effects. The zero-angle value desired cannot be directly observed since a detector cannot be put at the position of the incident light. Thus, a measurement includes measuring the quantity of scattering light of a sample in solution at 15 different angles in a plane and repeats it at lower and lower concentrations and the results are extrapolated to zero concentration and zero angle (Zimm plot) [26], as shown in Figure 28.17 for fraction 1 and fraction 3. The intercept value gives weight-average MW (M_w) and coefficient A_2 values by Eq. (28.14)(1).

28.4.4 Results and Discussion

Proton and ^{13}C NMR results of all fractions reported in Table 28.11 showed the familiar structural features as acetylated PF resole resins, consisting of methylol groups, methylene

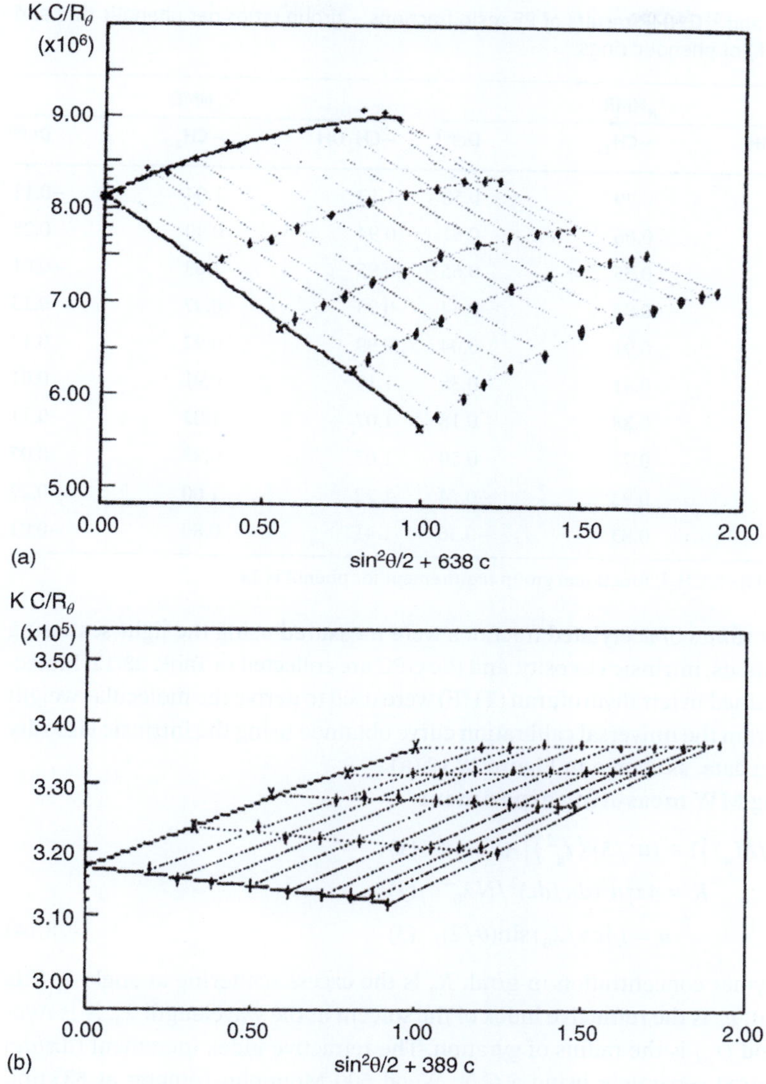

Figure 28.17 Light-scattering results (Zimm plots) of (a) fraction #1 and (b) fraction #3.

groups, and phenolic rings. The methylene group values per phenolic ring in ^{13}C NMR appear depressed and similarly adjusted as in the previous Sections 28.3.2–28.3.5. Small amounts of unreacted C_2 and C_4 carbons were present at 125–125 ppm. The adjusted methylene group values per phenolic ring are in a reasonable range (0.9–1.04) for high MW materials, but the unadjusted ^1H NMR values showed depressed values. In general, however, the values for the number of methylene and methylol groups per aromatic ring were not accurate enough to give reliable number-average M_n values when compared to the values obtained by light-scattering or VPO methods. Also, the [def] values calculated showed similar variations in both NMR results.

Table 28.11 Proton and ^{13}C NMR results of PF resin fractions – group ratios per phenolic ring and deficiency factors (def.) of phenolic rings.

Fr no.	$^{1}_{H}$NMR			$^{13}_{C}$NMR		
	$-CH_2OH$	$-CH_2-$	Def[a)]	$-CH_2OH$	$-CH_2-$	Def[a)]
1	0.84	0.79	0.58	1.13	1.04	−0.11
2	0.87	0.66	0.81	0.94	0.90	0.26
3	0.85	0.75	0.65	0.95	1.03	−0.01
4	1.02	0.82	0.34	0.93	0.97	0.13
7	1.14	0.91	0.04	0.99	0.92	0.17
10	1.00	0.81	0.38	1.17	0.91	0.01
13	1.06	0.88	0.18	1.07	1.02	−0.11
16	0.94	0.78	0.50	1.07	0.93	0.07
18	1.30	0.93	−0.24	1.22	1.00	−0.22
21	1.70	0.83	−0.36	1.41	0.80	−0.01

a) Def = 3.0 -[CH_2OH]−2[CH_2]; functional group requirement for phenol is 3.0.

Molecular weight values of acetylated fractions were measured using the light-scattering (M_w) and VPO methods, intrinsic viscosity, and the GPC are collected in Table 28.12. Molecular properties obtained in tetrahydrofuran (THF) were used to derive the molecular weight spreads (M_w/M_n) from the universal calibration curve obtained using the intrinsic viscosity and light-scattering data, as shown in Figure 28.19 (a).

Light scattering MW measuring experiment

$$Kc/R_\theta = (1/M_w)\left[1 + (u^2/3)\left\langle {r_g}^2\right\rangle\right] + 2A_2c \quad (1)$$
$$K = 4\pi^2 n^2 (dn/dc)^2/N{\lambda_0}^4 \quad (2)$$
$$u = (4\pi n/\lambda_0)\sin(\theta/2) \quad (3) \qquad (28.14)$$

where c is the polymer concentration in g/ml, R_θ is the excess scattering at angle θ, K is the optical constant, n_0 is the refractive index of the solvent at the wavelength λ_0, N is Avogadro's number, and $\langle r_g\rangle$ is the radius of gyration. The refractive index increment (dn/dc) values were Measured separately using a C.N. Wood 500 Monopho-Tometer at 633 nm. From the obtained R_θ values as a Function of poymer concentration and scattering angle, The Dawn-F "AURORA" software package provided the Zimm plot [26] in terms of Kc/R_θ versus $\sin^2(\theta/2)$

The light-scattering analysis procedure was calibrated using molecular weight standard materials and the results showed a good agreement with the reported M_w values. Since the quantity of scattered light depends on the volume of polymer molecules, the data obtained from light-scattering is related to weight average M_w values. Figure 28.17a shows the Zimm plot [26] of fraction #1 in ethyl acetate and the intercept gave a M_w of 114 000 Da. The concentration dependence of data extrapolated to negative slope, giving a negative second virial coefficient of $A_2 = -6.5 \times 10^{-4}$ mol·ml g^{-2}. In like manner, fraction #2 shows a similar negative value for A_2 with a M_w value of 69 300 Da. This negative A_2 values suggest that ethyl acetate is a thermodynamically poor solvent for fractions #1 and #2 [24], which were obtained with the least amounts of nonsolvent (hexane) added to the

Table 28.12 Molecular properties of acetylated PF resin fractions.

Fr. No.	M_w	$10^{-4}A_2$	$^{dn}/_{dc}$	M_n(VPO)	$^{M_w}/_{M_n}$(GPC)	$[\eta]_{THF}$(mL/g)	k'_{THF}	$[\eta]_{EA}$, $^{mL}/_g$	k'_{EA}
1	114 000	−6.5	0.175		7.65	8.98	0.60	5.82	1.90
2	69 300	−6.3	0.178		9.52			6.36	0.07
3	31 500	3.6	0.180		6.63			5.55	0.92
4	30 400	1.6	0.181		4.20			5.68	1.20
5	26 000	4.5	0.888		3.69	5.66	4.18	4.88	4.45
6	21 700	1.7	0.182		3.34	5.68	1.00	6.73	3.34
7	20 300	6.3	0.188		3.65	6.45	0.40	6.18	3.33
8	21 700	3.2	0.184		5.20	5.93	0.77	5.49	0.50
9	22 300	2.8	0.178		2.92	6.10	0.83	5.51	0.44
10	17 000	3.8	0.170		2.37	6.04	0.51	5.55	0.66
11	16 100	4.6	0.185		2.20	5.81	0.37	4.87	0.79
12	8 800	4.0	0.170		1.62	4.36	4.24	4.68	0.48
13	8 700	4.9	0.186	6710	1.64	5.12	0.57	4.67	0.34
14	6 900	2.9	0.161	2660	1.45	4.86	0.42	3.87	0.55
15	6 000	4.4	0.173	2250	1.42	4.78	0.30	4.09	2.19
16	4 900	2.1	0.163	3400	1.44	4.50	0.64	4.14	0.83
17	3 900	5.4	0.183	2350	1.25	4.17	0.98	4.03	0.94
18	3 100	3.3	0.161	3160	1.20			3.84	0.83
19	2 200	4.0	0.179	1420	1.24	3.79	0.18	3.97	0.14
20					1.19			3.71	0.46
21	1 200	−0.8	0.149	770	1.14				
dimer[a]	(660)				1.04	2.57	1.23		

M_w, molecular weight by light scattering in ethyl acetate; A_2, second viral coefficient from light scatteting in ml· mol g^{-2}; dn/dc, differential refractive index; M_n(VPO), molecular weight by vapor pressure osmometry in THF; $^{M_w}/_{M_n}$(GPC), polydispersity value obtained from GPC data using universal calibration results based on the light scattering M_w and intrinsic viscosity; $[\eta]_{THF}$, intrinsic viscosity $(^{mL}/_g)$ in THF; k', Huggins' constant in THF; $[\eta]_{EA}$, intrinsic visco viscosity $(^{mL}/_g)$ in ethyl acetate; k', Huggins constant in ethyl acetate.

a) Dimer was the acetate of 2,5,3′,5-tetrakis (hydroxymethyl)-4,4′-dihydroxydiphenylmethane.

ethyl acetate solution of PF polymer, i.e. fractions #1 and #2 molecules are self-associating even in ethyl acetate, the solvent used in the solvent/non-solvent fractionation procedure. Beginning with fraction #3, of which Zimm plot is shown in Figure 28.17b, M_w values drop consistently from about 31 500 Da to about 1200 Da for fraction #21. Overall, with the observed normal behavior of most Zimm plots and other experimental considerations, the authors believe the M_w values obtained are realistic values for these relatively low molecular weight resin fractions. Further, the fact that the third and subsequent resin fractions have appreciably lower molecular weight values appears to indicate that the advancing fractions in polyfunctional resin polymerization reactions, such as fractions #1 and #2 (combined, 8.93% of total resin solids), appear to be increasing in molecular weight rapidly during resin synthesis. Further, the obtained second virial coefficients, A_2, for low molecular weight fractions were somewhat erratic although most fractions show values in a reasonable range. Since A_2 is related to molecular interactions with solvent, the irregularity might have arisen from the structural heterogeneity among the fractions due to either the residual (unacetylated) hydroxyl groups, which was observed in ^{13}C NMR, or the different branching extents. The M_n values obtained by VPO were also somewhat erratic, especially for the high molecular weight fractions.

Correlation with the Mark–Houwink (M–H) equation [17]: Further, intrinsic viscosity data in THF and ethyl acetate are correlated with the light-scattering M_w values using the Mark–Houwink (M–H) equation as shown in Figure 28.18 and in Table 28.13 with exponent α values of 0.21 and 0.15, respectively, indicating the PF resin fractions swell slightly more in ethyl acetate than in THF. However, both α values are low to indicate that the polymers have compact solution structures agreeing with the previous [27] and other reported results [28, 29]. The exponent α value for linear polystyrene polymers in THF was measured in our laboratory to be 0.725. ($[\eta] = 1.02{M_w}^{0.725}$ for linear polystyrene in THF; $3.0 < [\eta] < 30$ ml/g.) Also, the Huggins constant (k') in Table 28.12 was obtained

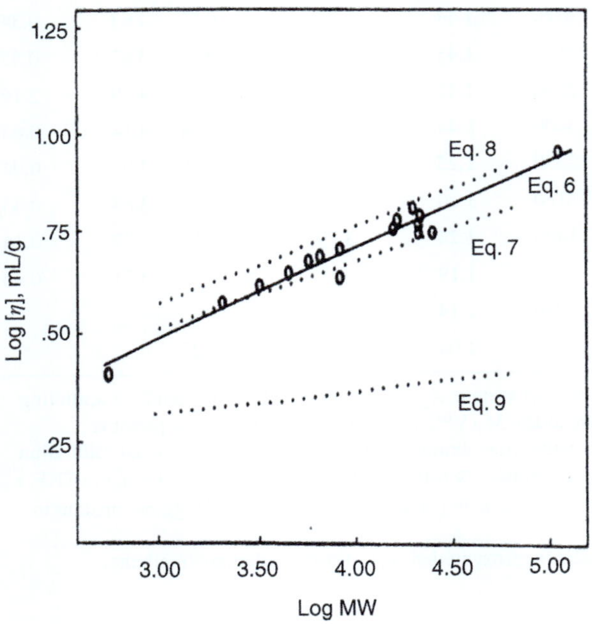

Figure 28.18 Mark–Houwink correlation of PF resin fractions in THF (solid line marked with "o" and Eq. 6). Eq. 7 is the correlation in ethyl acetate. Curves of Eq. 8 and 9 are of unacetylated PF resin fractions with their MW measured by diffusion method (M_d) reported in literature [28, 29].

Table 28.13 Mark–Houwink correlation equations for PF resin fractions using light scattering M_w values.

$[\eta] = 0.77M_w^{0.21}$ $(r^2 = 0.92; THF; 25\,°C)$
$[\eta] = 1.16M_w^{0.15}$ $(r^2 = 0.71; EA; 25\,°C)$
$3.8 < [\eta] < 9.0$ mL/g (*THF*)
$3.7 < [\eta] < 6.7$ mL/g (*EA*)
$2200 < M_w < 114000$ (acetylated)

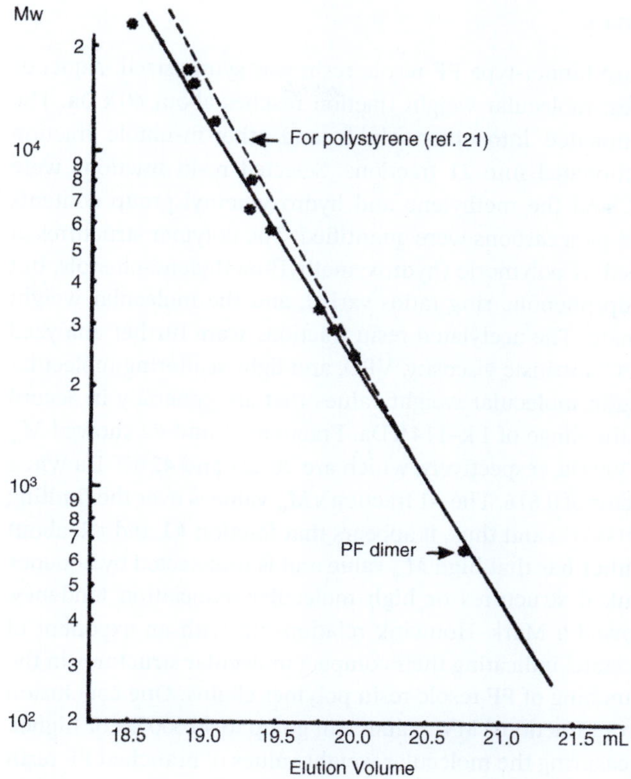

Figure 28.19 GPC calibration of acetylated PF resin fractions in ethyl acetate (●) by the universal calibration method (a). Similar curve (dotted line) for polystyrene (b).

from the intrinsic viscosity measurements (Eq. 28.13a in Section 28.3.5) of resin fractions were somewhat scattered but generally much larger than 1/3, the expected value for linear polymer chains in good solvents [30]. Large *k'* values have previously been reported for branched polymers. Overall, though, since the limiting α value of PF resins is not known, the extent of branching of PF resin fractions can be reckoned only in a relative term.

$$\log M_w[\eta] = -1.024\ V_e + 24.53\backslash(r^2 = 0.94;\ \mathrm{THF}) \tag{28.15}$$

GPC and universal calibration results: The high-pressure GPC analyses in THF of resin fractions initially resulted in elution volumes of fractions and corresponding M_w values based on polystyrene polymer molecular weight standards. The elution volumes (V_e) were then correlated with light-scattering M_w values and intrinsic viscosity values into the universal calibration equation, resulting in Eq. (28.15) and graphically in Figure 28.19(a). The equation is in relatively good agreement with a reported similar curve for linear polystyrenes obtained under the same chromatographic conditions [27] as shown in Figure 28.19(b). Based on these universal calibration results, the polymer dispersity values {(M_w/M_n(GPC)) for resin fractions were calculated as reported in Table 28.12. The later eluting resin fractions (#12–#21) have relatively narrow MW distributions with M_w/M_n ratios ranging below 1.62, but the early eluting fractions (#1–#11) showed broader MW distributions to indicate the less than adequate fractionation method used, as expected for branched polymers.

28.4.5 Conclusion and Summary

One typical oriented strand board binder-type PF resole resin was synthesized. Aqueous GPC indicated the resin's highest molecular weight fraction reaches about 40 k Da. The resin was acetylated and fractionated into ether-soluble and ether-insoluble fraction and the latter was further fractionated into 21 fractions. Selected resin fractions were analyzed by ^{13}C and ^{1}H NMR, and the methylene and hydroxymethyl group contents and the unsubstituted ortho and para carbons were quantified. The polymer structures of PF resole resins can be described as polymeric (hydroxymethyl)-methylene-phenols, but the accuracy of methylene group/phenolic ring ratios varied, and the molecular weight measurements were indeterminate. The acetylated resin fractions were further analyzed in ethyl acetate and THF by GPC, intrinsic viscosity, VPO, and light-scattering molecular weight measurements, resulting in molecular weight values that are generally in accord with the fractionation order in the range of 1 k–114 k Da. Fraction #1 and #2 showed M_w values of about 114 000 and 69 300 Da, respectively, which are 70 224 and 42 690 Da when multiplied by the acetylation factor of 0.616. The #1 fraction's M_w value is over the limiting aqueous GPC value of about 40 000 Da and thus, it appears that fraction #1 and #2, about 8.93% of total PF resin solids, either has that high M_w value and is undetected by aqueous GPC due to branched/cross-linked structures or high molecular association tendency. Overall, the resin fractions showed a Mark–Houwink relationship with an exponent of 0.21 in THF and 0.15 in ethyl acetate, indicating their compact molecular structures in the solution, likely the result of branching of PF resole resin polymer chains. One conclusion of this work is that PF resins' front fractions' MW values can go up to 40 000 Da or higher. The scope and limitations in measuring the molecular weight values of branched PF resin polymer structures using various current analytical methods are also shown.

28.5 Polymer Structures of Cured PF Resole Resins by Solid-state ^{13}C NMR

(Adapted with permission of the publisher and minor revisions from Kim, M. G., Y. Wu, and L. W. Amos. 1997. Polymer structures of cured alkaline phenol-formaldehyde resole resins with respect to resin synthesis mole ratio and oxidative side reactions. J. of Polymer Science: Part A: Polymer Chemistry, 35, 3275–3285.)

Solid-state ^{13}C NMR analysis method was developed based on the magic-angle spinning (MAS) and cross-polarization (CP) technique for decoupling the strong carbon-hydrogen spin interactions present in the solid-state organic materials. Application of this MAS/CP ^{13}C NMR method on cured PF resole resins was previously reported to provide useful information regarding the fully cured PF resole resin structures, [31, 35] however, the quantitative details as well as the resin synthesis and curing procedures were desired. Therefore, this research was carried out using the solid-state ^{13}C NMR analysis method with well-defined PF resin synthesis and curing procedures.

28.5.1 PF Resin Synthesis and Curing Procedures

Resin synthesis: As shown in Table 28.14, a PF resole resin having a resin-solids level of 39.13%, F/P ratio of 2.10, and NaOH/P ratio of 0.388 was synthesized with a temperature schedule of 90–80 °C.

Table 28.14 Synthesis of a PF resole resin.

308.6 g 90% phenol. 47.2 g 50% NaOH.

371.8 g 50% formaldehyde. F/P = 2.10.

(i) In a reactor under dry nitrogen, phenol, and sodium hydroxide is added and heated to 65 °C and formaldehyde was added over 20 min.
(ii) Heated to 95 °C and held for 40 min, vis. = GH.
(iii) Lowered to 85 °C and added 44.2 g 50% NaOH and 138.4 g H_2O, and held at 85 °C for 50 min.
(iv) Cooled to room temperature, vis. = K. NaOH/P = 0.388, Resin-solids = 39.13%.

Table 28.15 Curing methods of PF resin used for NMR Sample A and Sample B.

Sample A **:** 24.04 g of the PF resin was spread in a 250 ml Erlenmeyer flask and a slow dry steam of nitrogen was set to sweep over the spread resin mass into a scrubbing impinger placed under 200 ml of deionized water to capture the exiting volatiles. The flask was put in a sand bath and: (1) Heated slowly to 190 °C for 100 min, (2) Kept at 190 °C for 60 min, and (3) Cooled to room temperature in 60 min.

The curing resulted in black solid residues (10.9 g, **45.4%**); its sodium ion content was calculated to be 2.89%; Formaldehyde content in the scrubbing water was less than 0.05% based on the liquid resin weight. The cured resin product was finely ground, sealed, and kept in the refrigerator.

Sample B **:** 23.8 g of the liquid PF resin was cured with the same procedure, but the temperature was lower: heated to 125 °C for 45 min and held at the temperature for 270 min. The curing resulted in black solid residues 11.6 g (**48.7%**), similarly ground and kept. Formaldehyde content in the scrub water was less than 0.5% based on the liquid resin weight.

Curing of resin and washing of cured resin: As shown in Table 28.15, the cured resin *Sample A* was obtained by curing the PF resin under nitrogen stream at 190 °C for 60 minutes. The volatiles in the nitrogen stream were captured in water and analyzed. The cured resin residues were obtained at a solids yield of 45.4% with a calculated sodium hydroxide content of 2.89%. Also, the cured resin *Sample B* was obtained similarly but cured at 125 °C for 270 minutes, with a solids yield of 48.7%. Cured Sample A and B were ground into fine powders and sealed and kept in a refrigerator. Also, a portion of Sample A was washed with water, resulting in *Sample C* as shown in Table 28.16.

Table 28.16 Washing of Sample A for Sample C for NMR.

(1) A portion of Sample A was finely ground and a sample (2.511 g) was suspended in 70 g distilled water containing 0.84 g of acetic acid, for removal of sodium ions, (2) stirred for 3 days at room temperature and the particles were collected by filtration and drying at 105 °C for 10 hours, yielding 1.980 g (79.0%). (3) Sealed and kept in the refrigerator until analysis. Sodium ion content analyzed: 0.11%, a 96.1% sodium ion removal.

Isolation of water-soluble products

(1) 4.322 g of finely powdered Sample A was stirred in 50 g of distilled water for 3 hours at room temperature and filtered; (2) the particles were washed once more with 50 g water and filtered and dried to yield 3.70 g (97.7%). The combined filtrate water was analyzed for formate, bicarbonate, and carbonate species, which amounted to 0.33 mol equivalent of formaldehyde per mol of phenol.

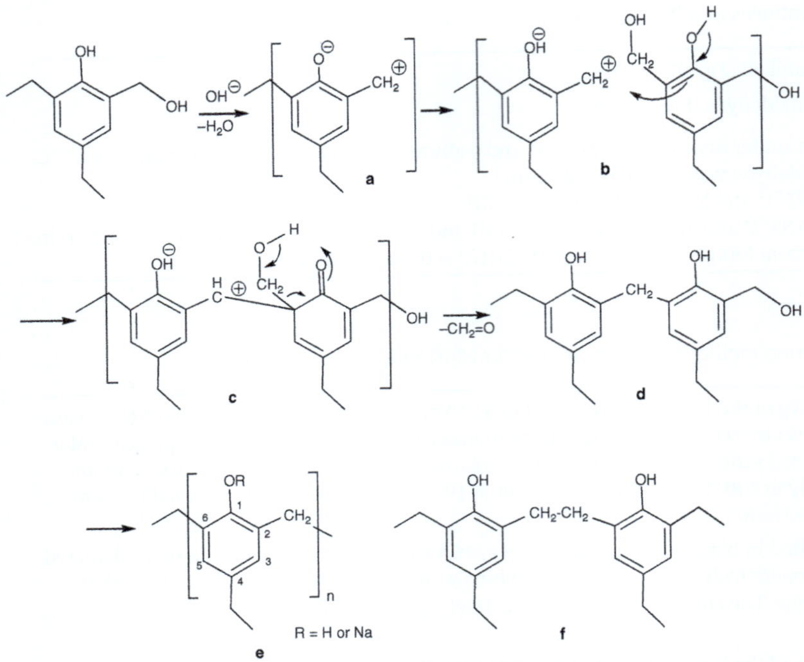

Figure 28.20 Chemical reactions occurring in synthesis and curing of PF resole resins.

28.5.2 Curing Mechanisms of PF Resole Resins and Loss of Some Formaldehyde? – a Review

$$F/P = (2.0 + 2.0/n)(1 - U_{o,p}/3P) - F_{un}/P \tag{28.16}$$

In theory, assuming no ring structure of phenol molecules is forming and no loss of formaldehyde occurred in uncured resins, as discussed in Chapter 9, Eq. (28.16) stands for PF resole resins, where F and P stand for moles of reactants used in resin synthesis, and $U_{o,p}$ and F_{un} are moles of unsubstituted ortho and para carbons and unreacted formaldehyde, respectively [36]. For most wood adhesive-type PF resole resins $U_{o,p}$ and F_{un} values are small. Then, the curing process of alkaline PF resole resins is a continuation of the substitution reactions that occurred in the resin synthesis stage, shown in Figure 28.20a-d [36]. One hydroxymethyl group of one phenolic molecule, activated as a benzylic cation (**a**), is attacked by a base-activated ortho- or para-carbon of another phenolic molecule that is substituted with a hydroxymethyl group or unsubstituted (**b, c**), to form a methylene bond (**d**), and expel a water or formaldehyde molecule. According to this reaction process, for PF resins made with an F/P ratio greater than 1.50, the completion of curing would lead to a fully methylene group-substituted phenolic polymer system shown in Figure 28.20e, for which the methylene group/phenolic ring ratio becomes 1.50, as shown in Eq. (28.17) [36]

The fact that most wood adhesive-type PF resole resins are synthesized with F/P mole ratio values of 1.7–2.5 for better curing speeds implies that about 0.2–1.0 mol of formaldehyde can become unaccounted materials or lost in the curing process, a technologically important question that needs to be ascertained. In the past, various decomposition and side reaction products including evolution of free formaldehyde were reported or proposed [37, 38]. The general feature of the curing process of PF resole resins reported using the solid-state ^{13}C NMR spectroscopy showed the transformation of hydroxymethyl groups into methylene groups [31–35], partly ascertaining the traditional theory. These studies also showed that no reactions occur on C_3 and C_5 carbons of the phenolic ring and no formation of ethylene-type bonds (Figure 28.20f). However, they showed qualitatively the presence of some hydroxymethyl or ether-type methylene bonds in their cured samples and also implied a possibility of forming ether-type methylene bonds on the phenolic C_1 carbons [32].

$$F/P = 3/2 = 1.50 \tag{28.17}$$

However, their extents or methods of curing of resins were not well defined and overall results were desired to be more quantitative. In the present study, one wood adhesive-type PF resole resin was synthesized and cured at two different temperatures: Sample A was cured at 190 °C and Sample B was cured at 125 °C, and Sample C was also obtained by washing the cured Sample A. Also, the curing experiments were carried out under a stream of dry nitrogen and the exiting gas was impinged into an ample amount of water and tested for formaldehyde content. Also, the washing water of cured resin Sample C was tested for oxidized formaldehyde-derived products, i.e. carbonate and bicarbonate, etc.

28.5.3 Solid-state CP/MAS ^{13}C NMR Spectra

Solid-state CP/MAS ^{13}C NMR spectra were obtained on GE S-100, observing ^{13}C at 25.2 MHz with decoupling of proton (^1H) done at 100.2 MHz. Various relaxation parameters were checked and optimized to result in using a CP contact time of 1.0 ms with an effective power of 12 gauss, followed by signal acquisition time of 25 ms as 512 data points and 2.5 s recovery (repetition) time. Four thousand transients were accumulated, and data were Fourier-transformed to give spectra shown in Figures 28.21a–c. Sample spinning rates were 2750 rotations per second, and phenolic carbons' (rotational) sidebands occurred at 20–40 ppm (up-field) and 236–260 ppm (down-field) with the up-field band partially overlapping the methylene carbons. Sidebands tend to occur for aromatic carbons and even C_1 carbons' down-field side band is separately observed at about 267 ppm and up-field side band is buried under the methylene group peaks. Sideband integrals were added to phenolic carbon integrals. Recently, this method was applied to cured PF resins and data reported on the morphology and molecular motions [33].

28.5.4 Solid-state CP/MAS ^{13}C NMR Chemical Shift Assignments of Aliphatic Carbons

Since aliphatic ≡CH and —CH_3 carbons are thought to be present in very low levels in cured PF resins as shown in Figure 28.22a–d, the methylene (—CH_2—) group integral was taken to be all that of 0–50 ppm region minus the integral of the (measurable) down-field sideband (236–268 ppm), which should be equal to the integral of the up-field sideband. As formaldehyde-derived groups, hydroxymethyl groups (60–65 ppm), ether groups (65–73 ppm), oxymethylene oligomers, or those bonded as hemi-formals (73–90 ppm), carbonate and bicarbonate (168–173 ppm) groups are identified. Further, aldehydic peaks (183–209 ppm) were present and assigned to structures of Figure 28.22e/f [39]. As a comparison, the solid-state ^{13}C NMR spectrum of *o*-hydroxymethyl-phenol (solid powder from Aldrich Chemical) resulted in the following chemical shift assignments: $\mathbf{C_1}$, 155.9 ppm; $\mathbf{C_2}$, 125.5 ppm; $\mathbf{C_3}$, 130.4 ppm; $\mathbf{C_4}$, 119.8 ppm; $\mathbf{C_5}$, 133.7 ppm; $\mathbf{C_6}$, 117.2 ppm; $\mathbf{C_7}$, 62.1 ppm

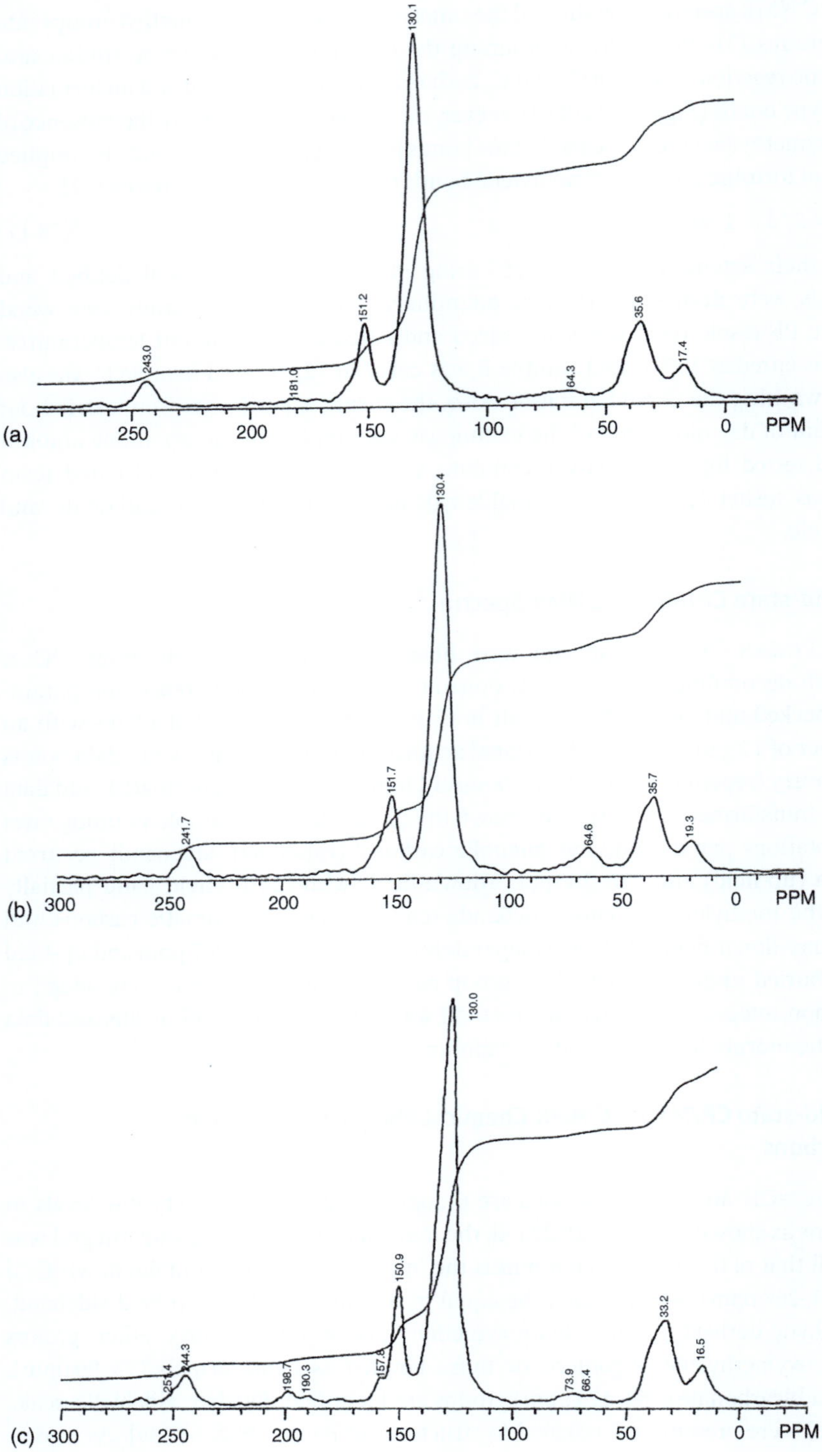

Figure 28.21 (a) Solid-state ^{13}C NMR spectrum of Sample A, (b) Solid-state ^{13}C NMR spectrum of Sample B (c) Solid-state ^{13}C NMR spectrum of Sample C.

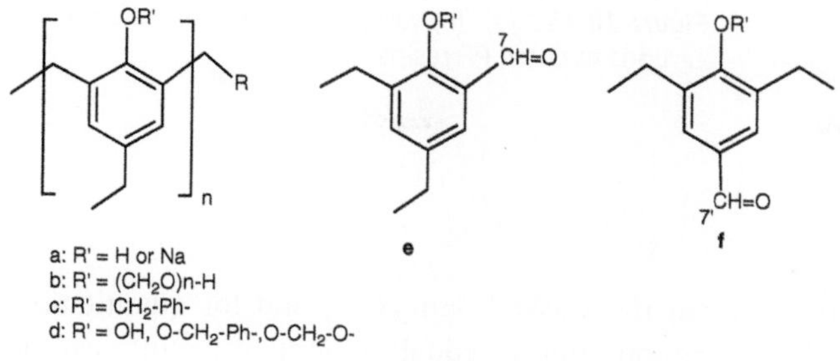

Figure 28.22 (a)–(f). Various structures of cured PF resins.

These chemical shift values are similar to those of liquid-state NMR data discussed in Section 28.1 [9, 10].

28.5.5 Solid-state CP/MAS ^{13}C NMR Chemical Shift Assignments of Phenolic Ring Carbons

Peaks in the 110–120 ppm region of unsubstituted C_2 and C_6 carbons were of very minor intensity. The major peaks of substituted $C_{2,6}$ carbons and unsubstituted $C_{3,5}$ carbons appeared in the 125–136 ppm region with the maximum centered at about 130 ppm, and the phenolic C_1 carbon peak appeared at about 145–168 ppm. These phenolic carbon peaks produced sidebands as discussed above (Section 28.5.4). Sidebands occur in solid-state ^{13}C NMR spectra due to the fast spinning of sample, and their chemical shift (ppm) values are varied depending on the sample spinning rates. Further, no noticeable peaks were observed in the 138–144 ppm region that would be of methylene or hydroxymethyl group-substituted phenolic $C_{3,5}$ carbons, once more showing that no reaction occurs on the meta-carbons of phenolic rings even under the high-temperature curing conditions [31–35]. The chemical shift value of C_3 carbon of (non-acetylated) model compound *3-hydroxymethyl phenol* in solvent is 144 ppm [9]. Further, the phenolic C_1 carbon (Figure 28.22a, R′ = H or Na) peaks showed the maximum at ~155 ppm and the weak shoulder peak noticeable at 156–163 ppm was suggested to be due to the forming of methylene-ether bonds in the curing process (Figure 28.22b,c) [32, 33]. This peak showed a poor resolution in Samples A and B and overlapped with a weak peak at 165 ppm assignable to the bicarbonate ion peaks. The shoulder peak became clearer and appeared separately from the main peak in Sample C (Figure 28.21c), probably due to the removal of sodium and bicarbonate ions by washing of the sample. Thus, although the hemi-formal bond formation constant of phenolic C_1 is only 1/70 of that with methanol [39], the curing process of PF resole resins made with high F/P ratios appears to force the formation of some methylene ether bonds on the phenolic C_1 carbons, which has never been suggested or observed in liquid PF resins. Further, the peaks at 183–209 ppm were assigned to aldehyde groups (Figure 28.22e/f), which were counted as aliphatic groups for quantitation.

28.5.6 Solid-state CP/MAS ^{13}C NMR Chemical Shift Assignment of Quinone Structures

Another small peak that appeared at 181 ppm was assigned to the quinonoid structures (Figure 28.23g/h) [36], which were suggested to occur in the curing process of PF resins

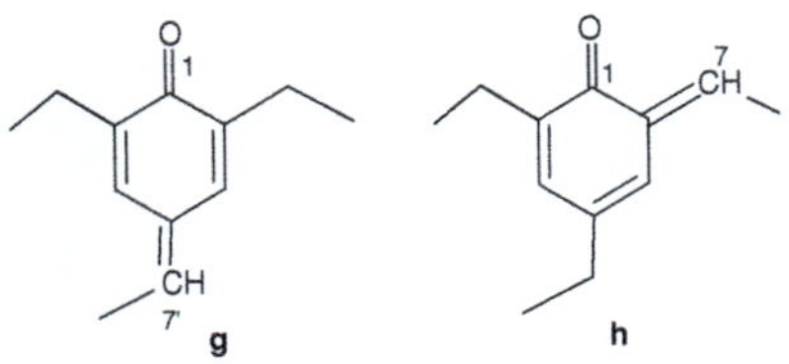

Figure 28.23 (g)–(h) Various C_1 quinone and olefinic carbons of cured PF resins.

[36, 37]. This peak did not disappear in the washed Sample C, and for quantitation, was counted as part of phenolic C_1 carbon, and an equal amount of olefinic carbon (Figure 28.23g and h, C_7 and C_7') was deducted from the phenolic $C_{2\text{–}6}$ carbons and added to the methylene group content.

28.5.7 Results and Discussion

All integration values are summed up for phenolic and aliphatic carbons, and appropriate group ratio values were calculated and presented in Table 28.17. Several observations are noted in the following.

Total substitution value of phenolic ring was slightly over 3.0 for Sample A, the functionality of phenol, and slightly lower than 3.0 for Sample B. Formaldehyde-derived materials present in the cured products were methylene groups, hydroxymethyl groups, freed oxymethylene oligomers or those bonded as hemi-formal, carbonate and bicarbonate, and aldehydic materials.

CH_2/P mole ratio values of 1.46 and 1.35 obtained, respectively, for Sample A and B, indicate that the cured resins reached the very highly cross-linked states approaching closely to the maximum value (1.50) of Eq. (28.17). Though, for Sample A in comparison with Sample B, the methylene groups content is significantly higher and the hydroxymethyl and methylene-ether group contents were lower, indicating that Sample A was cured more thoroughly due to the higher temperature used in the curing. The CH_2/P mol ratio values used in practice for curing of novolac-type PF resins is about 1.40 [41].

Total formaldehyde-derived materials amounted to 1.80 for Sample A and 1.86 for Sample B, which are less than 2.10, the amount used in the resin synthesis, by 0.30 mol and 0.24 mol, respectively. These amounts of formaldehyde thus appeared lost as some components not belonging to the organic cured resin structures. It may be the carbonate/bicarbonate components whose peaks (168–173 ppm) are excessively depressed in intensity in this ^{13}C NMR determination. Resin-solids determinations commonly done in industry for use in pricing are similar to Sample B (48.7%) obtained by curing at 125 °C and not like Sample A (45.4%) in Table 28.16. If the resin-solids value calculation is done by assuming the solids value of all formaldehyde fully as that of methylene groups (CH_2) and sodium hydroxide value as is (NaOH), the result comes out as 45.1%, very close to Sample A value. Thus, all formaldehyde used in PF resole resin synthesis appears to be counted as resin solids weights in industrial resin solids determinations with some weight gain (3.6%) attained, probably due to the oxidized formaldehyde species. Another note to make in this regard is that the missing formaldehyde in cured PF resins from ^{13}C NMR data at least does not cause emission problems of the bonded wood products to the extent of UF resin-bonded products. PF resin-bonded products are currently exempted from the current federal formaldehyde emission regulations [42].

Table 28.17 Solid-state ^{13}C NMR integrals and calculated group mole ratios (in parentheses) of cured PF resole resin Samples A, B, and C.

	Sample A	Sample B	Sample c	Chem. shift (ppm)
Phenolic carbons				
C_1, dfsb	0.41	0.23	0.76	256–275
C_1	11.52	11.10	11.80	145–168
C_1, ufsb	0.41	0.23	0.76	~34
$C_1 = O$, qn	0.73	0.52	0.73	177–184
C_1, total	13.07	12.08	14.05	
$C_{2\text{-}6}$, dfsb	3.64	3.46	4.00	235–256
$C_{2\text{-}6}$	57.48	57.34	56.56	102–144
$C_{2\text{-}6}$, ufsb	3.64	3.46	4.00	~17
$= CH-$, qn	−0.73	−0.52	−0.73	
$C_{2\text{-}6}(/C_1)$	64.03(4.90)	63.74(5.28)	63.83(4.54)	
$C_{1\text{-}6}(x1/6 = P)$	77.10(12.85)	75.82(12.64)	77.88(12.98)	
Aliphatic carbons				
$-CH- +$ ufsb	22.06	20.23	20.24	0–52
Ufsb	−4.05	−3.69	−4.76	
$= CH_2-$, qn	0.73	0.52	0.73	
$-CH_2-(/P)$	18.74(1.46)	17.06(1.35)	16.22(1.25)	
$-CH_2OH(/P)$	1.06(0.08)	2.71(0.21)	0.55(0.04)	60–65
$-CH_2-O-(/P)$	1.06(0.08)	1.25(0.10)	0.46(0.04)	65–73
(Total Subst'n on p)	(3.08)	(2.89)	(2.46)	
$(CH_2O)_x(/P)$	1.85(0.14)	2.00(0.16)	2.02(0.16)	73–90
$-CO_x(/P)$	0.55(0.04)	0.51(0.04)	0.05(0.000	168–173
$-CH = O(/P)$	0.05(0.00)	0.10(0.01)	2.07(0.16)	183–209
$C_{\text{all aliph}}/P$	(1.80)	(1.86)	(1.65)	

Samples: as defined in the text. ufsb: up field side band. dfsb: down field side band. $C_1 = O$, qn: C_1 in quinonoid form. $= CH-$, qn methylene groups in quinonoid structure, assumed to be equal to $C_1 = O$. P: mol of phenol taken as 1/6 of total phenolic carbons. $-CH_2OH$: phenolic hydroxymethyl groups. $-CH_2O-$: phenolic methyle ether groups. $(CH_2O)_x$: oxymethylene ether type species. CO_x: formate, bicarbonate, and carbonate species. $C_{\text{all aliph}}/P$ includes all formaldehyde-derived groups and ionic species.

Shoulder peak of C_1 carbon: In the washed Sample C, the C_1 peak centered at 151 ppm became sharpened considerably in comparison with Sample A, and the shoulder peak at 157.4 ppm became more distinct probably due to the transformation of sodium phenolate ions (Figure 28.22, R′ = Na) to phenolic hydroxyl groups (Figure 28.22a, R′ = H) and also due to removal of the oxidized products, i.e. bicarbonate, carbonate, and formate ions in the 168–173 ppm region. The shoulder peak was reported in the past to be assignable to two different ether-bond structures (Figure 28.22b/c, Ph = phenolic rings) implying two extra routes present in the curing of PF resole resins to more tightly crosslinked structures [32, 33]. This shoulder peak of Sample C was estimated by the cut-and-weighing method to be about 0.10 mol per phenolic ring, a relatively high value. However, since the phenolic methylene-ether group content of this Sample is only 0.04 mole per phenolic ring ($Ph-CH_2-O-$ group in Table 28.17), the methylene ether-bond structure of (Figure 28.22c) would be similarly 0.04 mole per phenolic ring at most. Also, since other ether-type bonds, such as phenoxyphenylether group (Ph-O-Ph), are unlikely to form in the basic resin synthesis or curing condition, much of the shoulder peak (0.06 mole per phenolc ring) appears to be due to the phenolic ether-bond structure of (Figure 28.22b), i.e. hemi-formal type oxymethylene groups (Figure 28.22b), forcibly formed by the excess F/P mole ratio present, of which contribution to actual cross-linking of polymer structure appears to be very limited at best.

CH_2/P mole ratio of the washed Sample C showed a significantly lower *CH_2/P mole ratio* value of 1.25 in comparison with the un-washed Sample A, indicating that the durable PF resin components of cured resins do not reach to the CH_2/P ratio of 1.50 of Eq. (28.17), which was suggested to be an impossibility for various steric reasons [37]. One factor for the lower *CH_2/P mole ratio* value may be traced to the two peaks at 190 and 198 ppm in the spectrum of Sample C of modestly high intensity of 0.16 mol per phenolic ring, which are assignable to carbaldehyde groups (Figure 28.22e and f) bonded to ortho- and para-carbons of phenolic ring, [40] which was suggested to be formed in the curing of PF resole resins by oxidation and resultant breakage of hydroxymethyl groups [36, 37]. Thus, it appears that the full substitution for all phenolic rings with methylene groups is precluded and oxidations of extra hydroxymethyl groups appeared to be forced to go in the curing. Sample C showed a weight loss of about 21% in the washing that included PF resin materials (6.9%), sodium ions (2.89%), and bicarbonate, carbonate, and formate anions (11.8%), as revealed from the analyses of the wash water. These relatively large amounts of water-dissolvable materials present in cured PF resins indicate a possible problem of water-resistant properties. Oriented strand boards bonded with about 5% PF resole resin solids loading levels of the type resin examined in this study lose about 50% of the dry IB strengths when boards are boiled in water for 2 hours. It is a harsh test, but some improvement appears necessary in the future.

Polymer molecular structure of cured PF resins of Sample C: On the other hand, the solid-state ^{13}C NMR of the washed Sample C was considered to reflect the more realistic cured polymer structures. First, the relatively high level of carbaldehyde groups found in Sample C is very interesting and explains the cured PF resole resin structures and why they could not come to close to Eq. (28.17) suggested by K. Hultzsch in 1950 [36]. To make it clear on this subject, a schematic polymer molecular structure of the PF resin of Sample C can be drawn using the ^{13}C NMR data as shown in Figure 28.24, where the CH_2/P mole ratio of 1.25 is translated into a unit cell formed by eight (8) phenolic rings and ten (10) methylene bonds. The eight (8) methylene groups connect to make a ring of 8 phenolic molecules wherein four of them are tri-substituted by the other two (2) methylene groups, so that they also connect to four other unit cells of 8 phenolic molecules, etc., leading to an infinite polymer network. The other four phenolic rings in the unit cell are bonded only with two methylene groups and therefore, there are four (4) un-bonded ortho or para positions in the unit cell, each position of which is bonded with 32% ($0.16 \times 8/4$) carbaldehydic groups (total 1.28 groups) and with 8% ($0.04 \times 8/4$) hydroxymethyl and methylene-ether groups (total 0.32 group), respectively, totaling about 48% positions, occupied by these formaldehyde-derived pendant groups, and the balance of 0.52 position

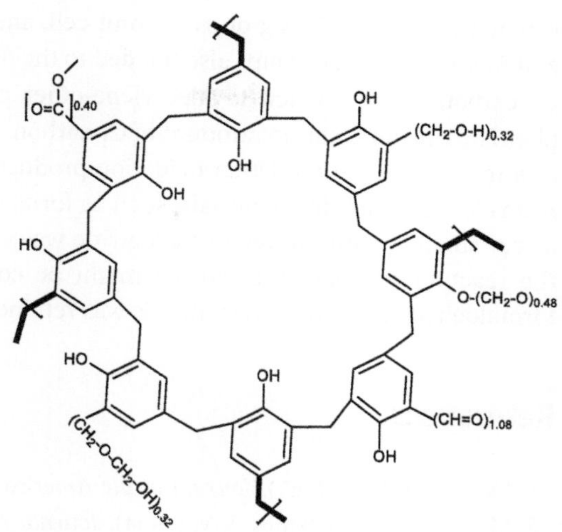

Figure 28.24 A schematic representation of PF resin polymer component of cured PF resins having CH_2 /P ratio of 1.25. (Unreported results.)

of each four (4) phenolic ring remains unoccupied. This result also indicates that the effective methylene-bond forming reaction is blocked at the level of 1.25 CH_2/P mole ratio probably due to steric restrictions. Further, all eight phenolic rings in a unit cell have about 15% of their C_1 carbons in the methylene-ether forms (total 1.5 groups). Also, about 4.0% of phenolic rings ($0.04 \times 8 = 0.32$) are in the quinone form but not drawn in the figure for simplicity. One additional thought on cured PF resin molecular structures is that the pendant carbaldehyde and hemi-formal ether groups can participate in wood bonding by reacting to form ketal and ether group bonds with hydroxyl groups on wood surface. Also, the present study was done in the absence of wood substrate and therefore a future curing study of PF resins in the presence of wood would give somewhat different results and an increased occurrence of -CH_2-O- bonds than in the current results (0.04 mol/phenolic ring) would indicate a direct participation of the abundant hydroxymethyl groups of PF resins in forming primary bonds with wood hydroxyl groups, proving the good durability properties of PF resin-bonded wood products.

28.5.8 Conclusion and Summary

A wood adhesive-type alkaline PF resole resin made with an F/P charge ratio of 2.10 and cured at 190 or 125 °C was analyzed by quantitative solid-state ^{13}C NMR spectroscopy. The results showed that a methylene/phenol mole ratio lies between 1.35 and 1.46, closely approaching the highly cross-linked polymer system suggested by early investigators. However, when the sample cured at 190 °C was washed with water, about 21.0% of the materials were dissolved out and the residue materials showed a methylene/phenol ratio of 1.25, or 10 methylene groups per 8 phenolic molecules. This methylene group ratio indicates a cured (durable) polymer network consisting of unit-cells made of 8 phenol molecules and 8 methylene groups, of which four phenolic molecules are tri-substituted and connected to other 8 phenolic ring unit-cells, and so on. A looser chemical structure with an average phenolic substitution value of 2.50 (1.25×2) rather than the full substitution value of 3.0 was indicated. There were also other pendant formaldehyde-derived groups, the prominent one being the carbaldehyde groups bonded to disubstituted phenol ring to the extent of 1.28 groups per unit-cell, and some minor amounts of hydroxymethyl and methylene-ether groups also bonded to the di-substituted phenol rings. Also, phenolic C_1 carbons were bonded to methylene-ether groups to the extent of 0.04 groups per phenolic ring and a minor amount of C_1 carbons was transformed into quinone structures as a result of oxidation. Other oxidation products of formaldehyde were present in cured resins as water-soluble materials, such as formate, bicarbonate, and carbonate ions. Only a very small amount of free formaldehyde was emitted during the resin curing. Although the reaction byproducts of curing might be considered as wastes, the weight of most formaldehyde used in resin synthesis was retained in the cured resins.

References

1 Lauterbur, P.C. (1961). *Journal of the American Chemical Society* 83: 1846.
2 Maciel, G.E. and James, R.V. (1964). *Journal of the American Chemical Society* 86: 3893.
3 Stothers, J.B. (1972). *Carbon-13 NMR Spectroscopy*. New York: Academic Press.
4 Levy, G.C. and Nelson, G.L. (1972). *Carbon-13 NMR for Organic Chemists*. New York: Interscience.
5 deBreet, A.J.J., Dankelman, W.D., Huysman, W.G.B., and deWit, J. (1977). *Angewandte Makromolekulare Chemie* 62: 7.
6 Dradi, E.G., Casiraghi, G., Satori, G., and Casnati, G. (1978). *Macromolecules* 11: 1295.
7 Pantke, O. (1933). *Low temperature phenol-formaldehyde resins and process of making same*. US Patent, 1,909,787.
8 Walker, J.F. (1964). *Formaldehyde*. New York: Reinhold.
9 Werstler, D.D. (1986). *Polymer* 27: 750.
10 Kim, M.G., Amos, L.W., and Tiedeman, G.T. (1979). *Weyerhaeuser Science Symposium II*, p. 263. Tacoma, WA: Weyerhaeuser Co.
11 Freeman, J.H. and Lewis, C.W. (1954). *Journal of the American Chemical Society* 76: 2080.
12 Reese, H. and Hultzsch, K. (1952). German Patent 960436. *Methoden der Org. Chem.* 14 (2): 273.
13 Wooten, A.L., Prewitt, M.I., Sellers, T., and Teller, D.C. (1988). *Journal of Chromatography* 345: 37.
14 Gollob, L., Krahmer, R.L., Wellons, J.D., and A. W. (1985). *Forest Products Journal* 35: 42.
15 Tobiason, F.L., Chandler, C., and Negstad, P. (1972). *Advances in Chemistry Series No. 125*, p. 194. Washington, DC: American Chemical Society.
16 Woodbrey, J.C., Higginbottom, H.P., and Culbertson, H.M. (1965). *Journal of Polymer Science Part A: Polymer Chemistry* 3: 1079.
17 Flory, P.J. (1953). *Principles of Polymer Chemistry*. Ithaca, N.Y: Cornel University Press.
18 Andreis, M. and Koenig, J.L. (1989). *Advances in Polymer Science No. 89*. New York: Springer-Verlag.
19 Ford, W.T. and Balakrishnan, T. (1983). ^{13}C NMR spectra of cross-linked poly(styrene-*co*-chloromethylstyrene) gels. In: *Polymer Characterization* (ed. C.P. Craver), p. 477. Washington, DC: American Chemical Society.
20 Tobiason, F.L., Chandler, C., and Schwarz, F.E. (1972). *Macromolecules* 5: 321.

21 Longi, P., Greco, F., and Rossi, U. (1969). *Die Makromolekulare Chemie* 129: 157.

22 Bersted, B.H. (1973). *Journal of Polymer Science* 17: 1415.

23 Aldersley, J.W., Bertram, V.M., Harper, G.R., and Stark, B.P. (1969). *British Polymer Journal* 1: 101.

24 Duval, M., Bloch, B., and Kohn, S. (1972). *Journal of Applied Polymer Science* 16: 1585.

25 Wyatt, P.J., Jackson, C., and G. K. Wyatt. (1988). *American Laboratory* 86–91 and 108–113.

26 Zimm, B.H. (1948). *The Journal of Chemical Physics* 16: 1093.

27 Kim, M.G. and Amos, L.W. (1991). *Industrial & Engineering Chemistry Research* 29: 1151.

28 Laurent, P. and Gallot, Z. (1982). *Journal of Chromatography* 236: 212.

29 Ishida, S., Chika, M., Nishikawa, K., and Kaneko, K. (1977). *Asahi Garasu Kogyo Gisuits Siourei Kenkyu Hogo* 30: 279.

30 Bodanecky, M. and Kovar, J. (1982). *Viscosity of Polymer Solutions*, 187. Amsterdam: Elsevier.

31 Fyfe, C.A., Rudin, A., and Tchir, W. (1980). *Macromolecules* 3: 1320.

32 Maciel, G.E., Chuang, I., and Gollob, L. (1984). *Macromolecules* 17: 1081.

33 Bryson, R.L., Hatfield, G.R., Early, T.A., Palmer, A.R., and Maciel, G.E. (1669). *Macromolecules 16.*

34 Fyfe, C.A., Mckinnon, M.S., and Rudin, A. (1983). *Macromolecules* 16: 1216.

35 Kelusky, E.C., Fyfe, C.A., and Mckinnon, M.S. (1986). *Macromolecules* 19: 329.

36 Hultzsch, K. (1950). *Chemie der Phenolharze*, p. 129. Berlin: Springer Verlag.

37 Megson, N.J.L. (1958). *Phenolic Resin Chemistry*. New York: p. 15 and 239.

38 Waters, G.D. (1993). *Proceedingsof 27th Washington State University Particleboard Symposium*, p. 201. Pullman, WA: WSU.

39 Kopf, P.W. and Wagner, E.R. (1973). *Journal of Polymer Science: Polymer Chemistry* 11: 939.

40 Kim, M.G.L., Amos, W., and Barnes, E.E. (1990). *Industrial & Engineering Chemistry Research* 29: 2032.

41 Knop, A. and Pilato, L.A. (1985). *Phenolic Resins*, p. 201. Berlin: Springer-Verlag.

42 U. S. Senate Bill. S 1660. (2001). The Formaldehyde Standards for Composite Wood Products Act.

Index

a

acid dyes 345
acid-curing PF and PMF resole resin and dispersion wood adhesives 162–163
acidic buffer capacity 19, 74–77
acidity 72, 74–77, 119, 162, 284
acrylic resins 4, 323–324, 350
acrylics 4, 321, 323, 335
additives to coatings 332
adhesion strength 20, 223, 270, 307
adhesive and interphase layers 285, 286
adhesive application rates in wood composites 26, 96, 106, 176–177, 198, 233, 237, 243, 246
adhesive bonding theory for wood 292–301
  adsorption (secondary bond forces) theory of adhesion 292–295
  diffusion theory of adhesion 297–298
  mechanical interlocking theory of adhesion 296–297
  primary chemical bond (covalent bond) theory 298–299
adhesive bonds, evaluation of 28, 302–316
adhesive cost 26, 166, 201, 278, 280
adhesive failure versus wood failure 269, 285–286
adhesive-layer or interphase failures in wood adhesive bonds 284–285
adhesive property of polymer solids 20
adhesive strength *vs.* substrate strength 283
adsorption theory in wood bonding 292–295
airless spray guns 339
alkyd resins 5, 320–321, 328, 337
alkyl resorcinol-formaldehyde resins 185
aluminum sulfate or chloride and free acids 74
American National Standard for Wood Products–Structural Glued Laminated Timber 181
ammonia 42, 46, 62–63, 72–73, 77, 88, 92, 129–130, 161, 469
anchoring in wood adhesive bonding 30, 296–297
animal blood-based wood adhesives 278
animal protein-based wood adhesive 276–277
antioxidant 266–267, 269
apparent viscosity 15–16, 200
Aspenites 239
attrition mill blenders 104–105
average molecular weight values 12–13, 57, 186, 488

b

backer materials of laminates 344
backstand sander 343
belt blending of OSB strands 105
birch plywood 313–314
biuret-type bonds formed in isocyanates resins 222
bleaching of wood in finishing 344–345
bonding adhesives of sander abrasives 344
bonding of fiber-reinforced plastics with PRF resin adhesives 182–183
bonding of wood preservative and fire-retardant treated lumber 182
Brookfield viscometer 17, 56, 175
brush back Vonnegut wheel sander 343

Adhesives and Finishes for Wood: For Practitioners and Students, First Edition. Moon G. Kim.

buffering of acidic catalysts 77–78
burner ash problem in the particle drying process 77

c

capillary contraction temperature measurement of PF novolac resins 132
carboxymethyl cellulose (ether) 8
carriers (solvents) of coating formulations 332
casein (protein from milk) 25
casein wood adhesives 275–277
 application and curing of 275
 production and formulation of 275–276
 uses of 276
catalytic curing agents of epoxy resins 271
catalyzation of UF resins by wood acids 74–77
cellulose, polymers 8
cellulose acetate (CA) 8, 323
cellulose acetate butyrates (CAB) 323, 346
cellulose nitrate (ester) 8
cellulosics 200–201, 251, 322–323
chroma values 331
class I hardboard coating finishes 350
class II hardboard coating finishes 350
coalescing agent of latices 36, 333–334
coalescing of PVAc polymer particles 252–253
coatings technology of wood 329–338
 additives to coatings 332
 carriers (solvents) of coating formulations 332
 color control methods 329–331
 color scales 331–332
 exterior coatings for wood 336–337
 film formation mechanisms 333–334
 manufacturing procedures of coatings 332–333
 pigment volume concentration (PVC) 317–319
 pigments and fillers 329
 three components of coatings 317
 vehicle polymers 319–328
 water-borne coatings and coatings for less VOC emissions 334–336
cohesive property of adhesive polymer solids 20
cohesive strength 20, 36, 156, 265–266, 283
 measurement and viscous and elastic responses 20–21
color control methods 329–332
color scales 331–332
colorimeter for checking plywood veneers 196
commercial PVAc emulsions 250–251
compression force 21, 307–308, 315
contact adhesives in wood bonding 270–271, 295
contact angles of adhesives on wood surfaces *vs.* good bonding 290
continuous meter-mixing technology 176
continuous presses or long hot-presses 27, 207
control filler 210–213
co-polymerization 90, 252, 265
critical pigment volume concentration (cPVC) 318
cross-laminated timber (CLT) 183
cup methods of viscosity measurements 17
cured epoxy resin structures 271–272
cured PF resin structures 514
cured UF resin structures 78–81, 84–85
Cure rate I of DMA tests of UF resins 434, 438
Cure rate II of DMA tests of resins 435, 438–439, 441–442
curing of PRF adhesive in a laminated beam 177
 curing of thermoplastic emulsion adhesives 36–37
 curing of thermosetting adhesive resins at elevated temperatures 29–33
 curing of thermosetting adhesives at room temperature 35–36
curing reaction mechanisms of UF resins 78–79
curing speeds of UF resins 81
curtain coating of adhesives and finishes 26, 198, 200, 203, 341, 347, 350

custom lumber laminators 180
cyclic delamination test procedure 181, 312

d

decorative (high pressure) laminates 161
defect theory of adhesive bond failures 300
degree of polymerization, defined 12
diisocyanate resins 219, 277, 326
diethylene tricarbamide-formaldehyde resins
- chemical and physical properties of 469
- co-polymer DMF resins syntheses 471
- co-polymer DUF resins synthesis 471
- efficient synthesis method of 470
- higher functionality urea analogue 469
- PE formaldehyde content values of particleboards 474
- physical strength and water soak test values 473–474
- pot-lives and curing characteristics of DF and copolymer resins 472–473
- properties of (aqueous) DF resins 471
- room temperature storage properties of DF and copolymer resins 472
- synthesis chemistry of DF resins 470–471
- synthesis formulations of DF resins *vs.* UF resins 471–472
- synthesis reaction characteristics of DF resins *vs.* UF and UMF resins 470

diffusion theory of adhesion 148, 223, 297–298
diol 5–6, 218–219, 223, 227, 272, 321–322, 326
dispersion forces in adhesive bonding of wood 292, 294–295
distribution zone of particleboard short retention time blende 96
DMA curing test results of UMF and UF resins 432–443
- 1.05, 1.15, and 1.25 by curing on DMA at 150 °C 441
- comparison of Resins $UFA_{1.15}$ and $UMF_{1.15}$ by curing on DMA at three different temperatures 439–440
- comparison of Resins UFA, UMF_{6D}, and UMF_{12D} made with F/(U+M) mole ratios of
- curing chemistry of Resin $UFA_{1.15}$ 435–438
- effects of catalyst mix temperatures and durations on DMA cure properties 438
- effects of different M addition points in resin synthesis on DMA resin curing properties 440
- first segment of curing, the gelation stage 435
- gel time, cure rates, and cure times determined cured on DMA at 120 °C 438
- second segment of curing, the rubbery stage 437
- third segment of curing, thermosetting stage 437
- three curing segments of DMA test of Resin $UFA_{1.15}$ 433–435

door skins (panels) 159
drawer sander 344
drum sander 342–344
dry internal bond strengths of particleboards 443–445
dryers of OSB strands 231
dryers of alkyd resins 319
durable adhesive bonds 297, 298
dynamic mechanical analysis (DMA) 82–85, 153–157, 178–179, 196, 242–243, 246, 432–445, 452, 462

e

edge sander 343–344
Einstein equation of viscosity 16
elastic response in mechanical testing of wood bonds 20–21
elasticity of solid polymer materials 21–22
electron-releasing tendency phenol molecule 123
electrostatic spray methods 33, 35, 339
emulsion coatings 332, 333
emulsion form of wax 102, 114
end-grain to end-grain joints 176–177
end-jointed lumber 27, 181, 305
engineered wood flooring (EWF) materials 273

epoxy resin adhesives 7, 271–273, 321, 325
aliphatic and aromatic primary and secondary amine hardeners 271–272
cold-set adhesives in wood bonding 273
commercial epoxy resin adhesives 272
cyclic acid anhydride curing catalysts 272
durable epoxy resin adhesives 273
tertiary amine catalysts 27
ethylene 3, 4, 10, 252, 264, 265, 469, 511
ethylene-vinyl acetate (EVA) co-polymer resin 4, 252, 264–268
excitation (resonance) of nuclear spins in ^{13}C NMR 351–354
extenders in softwood plywood adhesives 201–202
extractives of wood 195, 285, 290, 291, 313, 315

f

face-laminated lumber, testing of 302, 305
fast-drying oils 320
fatty oils 318–321, 333
fiber manufacture and blowline resin blending in MDF manufacture 102–104
fillers of wood adhesives 175, 252, 266
fillers in softwood plywood adhesives 200–201
film formation mechanisms of coatings 333–334
flake board 239
flat line finishing procedures 346–347
flexible and rigid urethane foams 219
flow coating 342
formaldehyde 41–42, 123, 137
emission in curing of UF resins 79–80
emission test results of particleboard (PB) 381–382
emission tests, small-chamber and perforator methods 443
emission values of particleboards 394–396
reactivity with lignins 279
species in PF resole resins 487
fracture mechanics test 306
free carboxylic acid and phenolic species in wood 76
free formaldehyde and hemi-formal groups in UF resins 358, 374
free formaldehyde species in UF resins 52, 363, 369–372, 379, 394, 407–408
free induction decays (FID) in ^{13}C NMR 354
free urea 45, 48, 63, 69, 358, 362, 369, 374, 377, 380, 391, 392, 402–405, 407, 408, 449, 458
F/U_1 mole ratios of UF resins, definition of 52–54
Furafil 100 200

g

gap-filling capability of softwood plywood adhesives 199
Gardener–Holdt bubble tube method of testing viscosity 17
gel permeation chromatography (GPC) 13, 57, 127, 171, 480, 482, 487–488, 490, 500
gel time measurement by DMA, defined 434
gelation of adhesive layer, definition of 32
gelation point of UF and PF resins- 32, 83, 84, 151, 155
gem-dihydroxymethyl group in UF resins 360
general acid catalysis mechanism 407, 418
glass-wool insulation 161
glazes in wood finishing 346
glueline's curing problems 284
good wetting and low contact angle 287

h

hand-held block sander 344
hardboard air-suspension process of manufacturing 160
hardboard and hardwood plywood wall panels 335
hardboard panels 159, 349
finishing procedures 349–350
hardboard wet process of manufacturing 159–160
hardeners of PRF resins 173–176, 191
hardeners of epoxy resins 272, 273
hardwood plywood 109–111
plywood grades 111

hardwood plywood industry 109–110
manufacturing procedure with UF resin adhesives 110
troubleshooting methods 111
hardwood plywood panel pre-finishing 348–349
heated airless spray guns of coatings 339
heavy oil in paraffin waxes 112, 113, 115
hot-pressing of wood composite mats 34–35
extents of resin cure in hot-pressing 35
platen pressure, mat thickness, and density profiles 33–34
temperature changes and moisture movements in board mat 34–35
high solids coatings 335
high-speed paddle blender of wood fiber 104
hot-melt adhesives 29, 263–268
advantages 266
disadvantages 266
materials used for 265–266
requirements for 263–265
uses of 267
hot-melt edge-banding operation 269–270
hot-plate cure time of PF resins 131
hot-stacking 33, 35, 99, 152, 206, 238
house log industry, PRF resin-bonded 181
hydrogen bonds 80, 277, 292–295, 298
hydrogen ion 18, 19, 69, 73, 74, 76, 78, 92, 124, 139, 406
hydroxyethyl cellulose (ether) 9
o-hydroxymethyl groups of PF resins 476, 480, 486–488
hydroxymethyl-methylene-phenols 499, 508
hydroxymethylphenols, formation of 125, 137, 138
hydroxymethyl resorcinol (HMR) 90, 168, 179, 182, 183, 221

i

inclined plate flow test of PF novolac resins 132
industrial coating application processes 339–350
airless spray guns 339
curtain coating 341
electrostatic spray methods 339
flat line finishing procedures 346–347
flow coating 342
hardboard panel finishing procedures 349–350
hardwood plywood panel pre-finishing 348–349
kitchen cabinet finishing procedures 347–348
knife coating 341
precision roll coaters 341
reverse roll coater 340–341
roll coating method 340
sanding abrasives, construction, and flexing 345
sanding processes 342–344
typical furniture finishing procedures and materials 344–346
industrial (high pressure) laminates 162
injection molding processes 132
inorganic polymers 2
in-situ adhesion testing of wood adhesive bonds 315
interfacial tensions between a liquid and solid 289
interfacial tension values 289, 290
interior-grade laminated lumber 276
interphase in wood bonding, defined 283
interphase failures 284–285
interphase in wood adhesive bonds 283–287
intrinsic viscosity (IV) 16, 17, 185, 189–191, 490, 496–501, 504–508
isocyanate groups (—N=C=O) 218
isocyanate (ISO) resins 6, 25, 217–227
bond properties and chemical reactions of pMDI resins as OSB binders 224–226
chemical structures of polymeric-MDI (pMDI) resins 217–227
general uses of 217
handling procedures of 218
mat moisture levels in using pMDI resins for OSB bonding 223–224
mix with an emulsion polyol as wood adhesives 220
pMDI resins used in bonding of other wood composite products 226–227

isocyanate (ISO) resins (*contd.*)
  reaction of isocyanate resins with water 218–223
  reaction with hydroxyl groups of polyols 219
  urethane resins 326
  wood adhesives and coatings made by pre-mixing pMDI resins and polyols 219–220

k

kitchen cabinet finishing procedures 347–348
knife coating 341
knife-planed lumber 176, 182
Kraft lignin 278–280

l

laminated veneer lumber (LVL) 27, 109, 110, 159, 207, 220
laminates 140, 149–151, 161, 259, 267, 268
lap shear test of plywood 303, 304
Large-Chamber test method formaldehyde emission from boards 447
Larmor frequency of nuclear spin in ^{13}C NMR 351, 352, 354
length of stirring in viscosity measurement 15
light scattering method of MW measurement 13, 144, 190, 500–508
lignin 1, 9, 10, 26, 74, 223, 239, 278, 281, 297, 298
lignin uses in alkaline PF resole resins 279–280
lignin uses with tannin materials 280
lignosulfonates, ammonium 279–280
calcium lignosulfonates, calcium 280
lignosulfonates, sodium 280
liquid adhesives 1, 15
liquid hardeners of PRF resin adhesives 175
liquid PF resin binders of OSB 234
loss modulus in mechanical testing of wood bonds 20, 82, 83, 155, 433–435, 437, 452
lower durability grade adhesive mixes of plywood bonds 203
low F/P mole ratio PF resins 147–148, 150
low pressure laminates 161

m

Mark–Houwink (M–H) equation 16, 498, 506
mastic adhesives 269–270
medium density fiberboard (MDF) bonded with UF resins 102–109
  blowline configuration 106
  dry-out and re-activation of fiber 105
  economics of MDF manufacturing process 106
  fiber manufacture and blowline resin blending 102–104
  fiber velocity 106
  fire hazard 105
  industry 109
  maintenance 106
  mat-forming and hot-pressing parameters 107–108
  paddle blending system of resins 104–105
  physical properties of 108
  refiner sizes 106
  resin spots in boards 106
  resin usage levels 105–106
melamine-formaldehyde (MF) resins 7, 323
melamine-formaldehyde (MF) resins, curing of 89–90
  melamine-urea formaldehyde (MUF) resin 23, 25, 29, 40, 63, 65
  curing of MUF resins 65, 90–91
  melamine, manufacture 63–64
  melamine-formaldehyde (MF) reaction products 64
  melamine syrup 64
  melamine-urea-formaldehyde (MUF) copolymer formation 65
  uses of MUF resins 65–66
meta-carbons of phenol molecule 123, 513
methyl methacrylic acid esters 323
methyl or ethyl cyanoacrylates 324
methylene groups of UF and MUF resins 44, 360
  bond formation in curing 44, 89–90
  groups' limited heat stability 81–82

methylene-di-urea 54, 62, 373, 380–382, 403–405, 409, 410, 414–416
methylene-ether groups in UF resins, formation of 44, 360
methylene-ether-cellulose bonds 80–81
methyleneic groups 356, 366, 373, 422, 449, 496
mineral abrading materials in wood sanding 344
modulus of elasticity (MOE) 21, 308, 393, 443, 474
modulus of rupture (MOR) 308, 393, 443, 474
moisture content (MC) of wood, effects in bonding 255–259
  differences between wood substrates 255
  effects of substrate condition 258
  gluing problems 259
  manufacturing plant and location of use 256
  quality control methods of wood bonds 259, 315
  radial and tangential differences in shrinkage of wood 255
  water in water-based adhesives 258
moisture-curing systems of isocyanate resin adhesives 221, 222, 227, 273
molded wood products 66, 160–161
mold sander 343
mono-hydroxymethylurea in UF resins 371, 380, 402–405, 417
mono-substituted urea in UF resins 362, 369, 371, 377, 380, 401–406, 450
multi-opening hot presses 26–27, 98

n

natural or reclaimed rubber-based mastics 269
natural pre-finishing of wood panels 348
neoprene (chloroprene) rubber-based mastics 270
Newtonian liquid/solution, definition of 14
nitrocelluloses 322, 323
no-formaldehyde-added resins 87
nondestructive testing of wood bonds 315
non-drying oil 320
non-Newtonian solution 14
novolac-type phenol-formaldehyde (PF) resins 122–133
number-average molecular weight (Mn), definition of 12, 13
nylon 2, 5, 10, 161, 343

o

oil-based paints 321, 332
oligomeric polymers in coatings 335
oligomeric or telechelic polymers of coatings 360
one-component, moisture-curing polyurethanes 221, 267, 273
one-step PF resins 5, 136
opacity of coatings 266, 329, 331, 337
open drum sander 343
orbital and reciprocating pad sander 344
orbital disc sander 344
organic/natural polymers 2
organic solvents 13, 14, 36, 130, 169, 218, 252, 269, 270, 321, 332–334, 345, 356
organosolv lignins 279
organic/synthetic/semi-synthetic polymers 2
oriented strand board (OSB) 5, 6, 27, 29, 132, 133, 140, 159, 194, 227, 229–240, 277, 500–501, 508
oriented strand board (OSB) manufacturing 229–246
  adhesive improvements needed 239
  binder resin types 231–234
  blending of binder resin and slack wax
    efficiency of resin binders *vs.* droplet sizes 233
    liquid PF or pMDI resin binders 233–234
    powder-type PF resin binders 233
    resin blender types and operation 232
    resinated wood strands 233
  drying and screening of wood strands 231
  durability of boards and span ratings 239–240
  hot-pressing of OSB 235–239
    industry hot pressing parameters 237–238

oriented strand board (OSB) manufacturing (*contd.*)
loading of mats 235
pre-cure or over-cure problems 236
press closing and effects of press closing rate 235
temperature and moisture content variation 235–236
under-cure problems 236
mat forming with strand orienters 234–235
mat moisture levels in using pMDI resins for OSB bonding 223–224
oriented strand lumber (OSL) 238–239
post-curing (hot stacking) and finishing of OSB 238
ripple effects of OSB roof decking 239
steam transfer and condensation mechanism of hot pressing of OSB 235
stranding (flaking) of wood boles 230–231
wafer board 239
wood species used 229–230
oriented strand board (OSB) made with PF resins synthesized and testing for 229–246
OSB binder PF resin synthesis and testing 241, 243–244
resin curing rates determined by DMA 242–243
test results of laboratory OSB 242
ortho- and *para*-carbons of phenol, reactivity and definitions of 123–126 136–139
over-curing of UF and PF resins 33, 61, 81, 82, 99, 102, 111, 306, 441, 444, 445

p

padding stains in wood finishing 346
paddle blending system of resins in MDF manufacturing 104–105
paints, coatings for wood 337
paper laminates 140, 149–151
paraffin wax 112–115
analyses of wax 113
application methods on wood substrates 114
melting point of commercial waxes 112
normality of paraffin wax 113
properties of 112–113
use levels based on wood weight 114
use of 112
paraformaldehyde 174–175
hardeners of PRF resin adhesives 174–175
pre-processing of 174
parallel (veneer) strand lumber (PSL) 207, 214
particleboard (PB), manufacturing of 94–102
Bison–Mende process 100
drying and tack development of binder UF resins 97–98
equilibration of boards and finishing 100
formaldehyde emissions of bonded boards 87–88
hot-pressing of particleboard, types 98
industry mat forming and hot-pressing parameters 107
mole ratio *vs.* board strength and formaldehyde emission levels
mat compaction and density profile of particleboards 98
mat formation and transport 97–98
mechanical forming machines of particleboard manufacture 97
methods proposed for reducing the formaldehyde emissions of boards 91
particleboard industry of U. S. 100–101
performance properties of industry particleboard 100
resin-solids loading levels 96–97
short-retention time blenders 95–96
softwood plywood trims 77
U. S. particleboard industry 100–101
ultra-low formooraldehyde-emitting resins 87
uses of 86
wood furnishes of 94–95
particleboards made in laboratory for tool wear testing 115–120
physical tests of manufactured boards 116–117

elemental analyses of boards 117
resin syntheses and additives 116
test methods of tool wear 117
test results of tool wear effects 117–119
tool wear effects of acidic curing catalysts of UF resins 119–120
perforator extraction (PE) tests of formaldehyde content in boards 443
PF resole wood adhesive resins, alkaline 136–165
acid-curing PF and PMF resole resin as dispersion wood adhesives 162–163
binder uses for various wood products 5, 24, 94, 122, 124–129, 159–162
curing of alkaline PF resole resin adhesives in wood bonding 151–153
decorative (high pressure) laminates 161
differential scanning calorimetric (DSC) analysis of 157, 158
durable matrix-forming material 202
dynamic mechanical analysis (DMA) test method of 153–157
F/P mole ratios of 147
formaldehyde emission problem 159
gel times and stroke cure times of 153
glass-wool insulation binders 161
good wood adhesion and exterior durability 158–159
industrial (high pressure) laminates 162
laminates 161
low pressure laminates 161
molded wood products 160–161
molecular weights 143–144
molecular weights of alkaline PF resole resins and wood cell wall penetration 148–149
oriented strandboard manufacturing, as binder 159, 231–232
paper laminates 150–151
pH value of 145
polymer molecular weight and structures 141, 195–196
polymer structures of 145–146
powder PF resole resin 149, 231, 233
reacted F/P mole ratio of PF resins 127, 129, 147
reaction chemistry of phenol with formaldehyde 123–124
resin #1 type PF resole resins 148
resin #2 and #3 type PF resole resins 147–148
resin #4 type PF resole resins 147
resole-type PF resins 122, 133, 136–165
resole resin chemistry 183
softwood plywood binders 150, 159–162
spray-drying of 149
synthesis chemistry and manufacturing practices of 136–139
synthesis procedures of resins for various uses of 140–143
tri-methylolphenol 137–138
viscosity values of 144–145
wafer board binders 161
pH of polymer solutions and buffers 18–19
phenol, manufacturing of 122–123
phenol-formaldehyde (PF) novolac resins 122–134
compounding of 129–130
finishing of resin manufacturing 127
gel permeation chromatography (GPC) of 127
initial reaction period resin manufacturing 125
novolac PF resin, ^{13}C NMR spectrum 477–478, 481
number-average molecular weight of 128–129
polymerization of hydroxymethylphenols 125–126
polymerization reaction mechanism 126
properties and structures of 127
raw materials of 122–123
reaction chemistry of phenol 123–124
reactions occurring in the final molding step 130
synthesis procedure of 124
wood adhesive uses of 132–133

phenol-formaldehyde (PF) resole resins (advanced reading materials) 476–517
 ^{13}C NMR analyses of model compounds and PF resole and novolac resins 477–480
 PF resole resin synthesized at 70 °C for reaction rate study 476
  changes of aromatic and aliphatic carbons from polymerization by ^{13}C NMR 484–487
  GPC and ^{13}C NMR analyses of resin samples taken during synthesis 487–488
  polymer structures of synthesized resins 489
  reaction rates of phenol-formaldehyde resin formation by ^{13}C NMR 487
  reaction rates of phenol-formaldehyde resin formation by GPC 488
 PF resole resins synthesized at 102 °C (A) *vs.* commercial resin (B) for comparison 490
  acetylation of the resins 491
  ^{13}C NMR analyses of the acetylated model compounds and fractions 492–496
  fractionation of acetylated resin A by elution 491–492
  fractionation of acetylated resin B by precipitation 491
  Mark–Houwink equation correlation results of resin fractions 498–499
  polymer structures determined by ^{13}C NMR, GPC, VPO, intrinsic viscosity 493–496
 PF resole resin synthesized at 80 °C (OSB binder) for high MW polymer structure study 501
  acetylation of resin and fractionation into 21 fractions 501–502
  Mark–Howink corelation and universal corelation equations of the fractions 506–507
  polymer structures by GPC, VPO, IV, ^{1}H and ^{13}C NMR and light-scattering MW determinations 505
  polymer structures of the highest MW fractions from light scattering results 506
 PF resole resin synthesized resin at 90 °C for study of cured PF resin structures 508
  chemical structures of sample C derived from solid-state ^{13}C NMR 515
  cured PF resole resin's polymer structure, Sample C 516–517
  curing of the resin at 125 (Sample B) and 190 °C (Sample A) 508–509
  polymer structures of cured PF resole resins by solid-state ^{13}C NMR 508–515
  solid-state ^{13}C NMR analysis methods, introduction of 508
  solid-state ^{13}C NMR analysis results of Samples A, B, C 512–515
  washing of Sample A with water, resulting in Sample C 509
phenol-resorcinol-formaldehyde (PRF) resins 166–193
 alkyl resorcinol-formaldehyde resins 185
 availability of resins in industry 173–174
 ^{13}C NMR analysis results of PRF resins 170–171
 co-polymer PRF resin manufacturing and chemistry of 167–173
 curing chemistry of resin with hardener 179
 determination of resorcinol/phenol mole ratios of resins by ^{13}C NMR 172
 GPC analysis results of a PRF resin 172
 hardeners for PRF resins 174–175
 homo-polymer RF resins 166
 mixing procedures of resin and hardener 175–176
 optimum MC in bonding of laminated beam 176
 powder hardeners of PRF resin adhesives 175, 176
 PRF resin-bonded laminated beams 36, 183
 resorcinol-formaldehyde (RF) resins 6

recent developments in 182–183
resole-type resorcinol-formaldehyde resins products 183
resorcinol-formaldehyde (RF) resin analysed for polymer structures 185–191
tannin-formaldehyde wood adhesive resins 184
use as adhesives in wood lamination 176–179
wood lamination industry 179–181
pigment volume concentration (PVC) 317–319, 332
pigment wiping stains 346
pigments and fillers 317, 329–332,
plasticizers 250–252, 265, 270, 273, 324, 325
plywood lap shear strength testing of 303
pneumatic drum sander 343
polyesters 5, 11, 265, 321–322, 335
polyethylene 2, 3, 5, 10, 12, 157, 265, 267, 290, 297, 480, 487
polymers, defined 2
polypropylene 3, 10, 265, 324
polystyrene 3, 10, 144, 501, 506, 507
polysulfide rubber-based mastics 270
polyvinyl acetate emulsion wood adhesive resins 249–262
additives 251–252
application methods 253
coatings vehicles 324
commercial 250–252,
curing mechanism of 253
manufacturing of 249–250
properties of 250, 252
polyvinyl acetate (PVAc) resins 3, 25–26, 29, 40
polyvinyl acetate (PVAc) resin adhesives, troubleshooting in gluing 259–262
assembly gluing by clamping process 262
cold-press lamination process 260
edge and face gluing by clamping process 261–262
hot-press laminating process 260–261
radiofrequency edge-gluing process 261
polyvinyl alcohol (PVA) 221, 250, 251, 324, 325
polyvinyl butyral 324, 325
polyvinyl chloride 4, 267, 325, 333
portable belt sander 344
post-curing of adhesive layers 33
powder coatings polymers 335
precision roll coaters 341
primary chemical bond (covalent bond) theory of wood adhesion 298–299
primer coatings 321, 337
printed wood panel finishing procedure 348
proteins 2, 10, 11, 276

r

radiation curing of coatings 335
radiofrequency (RF) heating processes of wood bonding 27
red cedar 77
reverse roll coater 340, 341, 347, 348
rigidity curve of DMA 82–83
roll coating method of wood adhesives and finishes 203, 340
rosins 321

s

sanding abrasives, construction, and flexing 344
sanding processes of wood 258, 342–344, 350
saturated polyesters 321–322, 335
SBR rubber-based mastics 270
scarfed end-jointed plywood, lap shear strength testing of 305
sealers in wood finishings 324, 325, 340, 341, 346
semi-drying oils 320
shear-thickening materials 15
shear-thinning materials 15
silicone resins 321, 327, 336
Six-Cycle Vacuum Soak Test 240
size coating and/or bleaching in wood finishings 344–345
small-chamber (SC) formaldehyde emission tests and results 395, 414

softwood plywood, manufacturing of 150, 194–216
  adhesive formulation and characteristics 198–203
  adhesive mixing procedure 197–198
  adhesive spread rate and method of application 202–203
  adhesive PF resole resin characteristics for 195–196
  bond performance characteristics of 197
  bonding with higher moisture content veneers 209
  dryout problems 208
  fillers and extenders 200–202
  hot-pressing parameters of 204–207
  hydrolysis residues of municipal newsprint waste as fillers 210–213
  laminated veneer lumber (LVL) 159, 207
  manufacturing industry 209–210
  manufacturing technology 194–214
  open and closed assembly times 204
  parallel (veneer) strand lumber (PSL) 207
  phenolic components in softwood plywood 209
  pre-pressing of softwood veneer assembly 204
  veneer, veneer drying, and adhesion problems 196–197
  wash water of PF resin plywood adhesives 203
solvent-borne coatings 332, 334, 335
solvent-borne or emulsion coatings 333
soybean meal wood adhesives
  production and formulation of 276–277
  recent research activities on 277
  uses of 277
stain or pigmented lacquer toner 345
stains in wood finishings 259, 317, 323, 336–338, 341, 345–347, 349
starch 2, 7, 9, 11, 196, 201, 202, 208, 251, 275, 276
steam-injection hot presses 27, 238
storage modulus 21, 452
stock laminating wood industry 180
stroke sander 343
styrene-butadiene resins 325
styrenics 321
substrate strength of wood in adhesive bonding 20
surface stability of wood in adhesive bonding 108
surface tensions of liquid and solid materials 287–288

t

tack property of UF resins, development of 57
tackifying agent of hot-melts 265
tan delta in DMA experiments 82, 83, 155–157, 433–435
tannin-formaldehyde wood adhesive resins 184
telechelic polymers in coatings 335
thermoplastic resin binder of wood 280
thermoplastic polymers 12, 20, 21, 143, 157, 250, 251, 264, 269, 435
thermosets, definition of 6, 29
thermosetting acrylic resins 323, 350
thermosetting coating vehicles 324
thermosetting hot-melt adhesives 267–268
thermosetting polymers 20, 21, 41, 217
thermosetting resins 28, 29, 37, 82, 85, 129, 142, 143, 153–157, 218, 411, 437
thixotropy 15, 182
toluene 122, 491–492
topcoats 337–338, 346–349
turning sander 343
TVA acid hydrolysis residue as fillers 210–214
two-component adhesives or coatings 220

u

UMF resins synthesized with 2.5% and 5.0% M added in the middle/end of the first step 457
  chemical structures of Resin UMF-2 and Resin UMF-3 determined by ^{13}C NMR 458
  ^{13}C NMR spectra of 5.0%$UMF_{1.05}$ 460
  poor storage stability of Resin UMF-2 and Resin UMF-3 459, 461

resin synthesis procedure for Resin UMF-2 456
resin synthesis procedure for Resin UMF-3 457
UMF resins synthesized with 2.5% and 5.0% M levels added in the first step 457–467
gel and cure times of catalyzed resins measured on rheometer 464
gel times of catalyzed resins 463
laboratory particleboard manufacturing parameters 464
particleboard test results of Resin UMF-1 *vs.* control UF resins 464, 465
pot-lives of catalyzed resins 461, 462
properties of catalyzed UMF and control UF resins
UMF resins synthesized with 2.5% and 5.0% M levels added in the 3rd step 447–456
chemical structures determined by ^{13}C NMR 450
^{13}C NMR spectra of 5.0%$UMF_{1.05}$ 449
gel times of catalyzed resins 452
gel and cure times of catalyzed resins measured with rheometer 452
particleboard test results of Resin UMF-4 *vs.* control UF resins 453–456
particleboard manufacturing parameters 455
pot-lives of catalyzed resins 451
storage stability of resin *vs.* control UF resin 451
synthesis procedure of resins 447
undercoats 337, 338
unsaturated polyesters 321, 322, 335
urea-formaldehyde (UF) and melamine-urea-formaldehyde (MUF) resins, curing of 72–93
aluminum sulfate or chloride and free acids as catalyst 74
ammonium salts catalysts 72–73
buffering capacities of various woods 75–76
buffering of acid catalysts 77–78
catalyzation by wood acids 74–75
composition of cured UF resins 80, 85–86
curing reaction mechanisms and cured resin structures 78–79
curing speed controls of UF resins 81
degradation of cured UF resins 83
formaldehyde emission in curing of UF resins 87–88
gelation, rubbery stage, vitrification, and full cure in DMA tests 82–83
latent/external catalysts based on ammonium salts 72–73
melamine-urea-formaldehyde (MUF) resins, definition of 64
methylene bond formation in curing 89–90
methylene groups' limited heat stability 81–82
performance properties of UF resin-bonded wood products 86
proposed methods for reducing the formaldehyde emission 89–92
reversibility of methylol groups to formaldehyde 82
role of methylene-ether bonds in UF resins 85
side reactions in curing of UF resins 81
stronger acid catalysts for curing of MUF and UMF resins 90, 422, 468
tertiary amine and other salt catalysts 73–74
urea-formaldehyde (UF) and melamine-urea-formaldehyde (MUF) wood adhesive resins 7, 24, 40–89
chemistry occurring in UF resin synthesis 46–50
effects of F/U_1 mole ratio in the polymerization step 53–54
hydroxymethyl group formation in the first and third steps 43, 45
methylene group formation and reaction rates in the second step 47–50
methylene-ether group formation in first step 43, 52–53
physical and chemical tests and properties of industrial UF resins. 55–58

urea-formaldehyde (UF) and melamine-urea-formaldehyde (MUF) wood adhesive resins (*contd.*)
polymer chain branch formation in UF resins 50–52
polymeric chemical structures of UF resin 63
reversible hydroxymethylation reactions of urea 46–47
side reactions occurring 52, 81
final F/U mole ratio and formaldehyde emission problem 55
formaldehyde emission mechanisms, F/U mole ratio, and board strengths 87–88
methylene-ether–cellulose bonds of binder UF resins 80
raw materials of UF resins, formaldehyde and urea 40–42
resin changes occurring after resin manufacture 58–60
resin manufacturing procedure, typical 42–46
urea-formaldehyde concentrate (UFC) 42, 448, 457, 458
uses of UF and MUF resins 86, 90
urea-formaldehyde (UF) wood adhesives resins (advanced reading materials) 351–420
^{13}C NMR analysis methods of UF resins, introduction 356–359
chemical shift values of various UF resin structures 357
example ^{13}C NMR spectra of UF resins of F/U ratio of 2.10 and 1.15 358
experimental procedure 359
peak integration values determine structural compositions of UF resins 358
^{13}C NMR analysis of reaction intermediates of UF resins with higher power NMR 366–374
bond group changes of resin samples observed by ^{13}C NMR 367
effect of the second urea addition and mild heat treatments 369
formation of UF polymers in the second step observed without and with the
rate of methylene bond formation at an F/U_1 mole ratio of 2.10 367
reaction paths to various intermediates in synthesis of UF resin polymers 368
second urea addition 370–373
^{13}C NMR analysis results of reaction intermediates taken in UF resin syntheses 360–366
bond group changes of resin samples observed by ^{13}C NMR 361
first step of resin synthesis 360–362
particleboards bonded with UF resins made with different F/U_1 mole ratios
rate of methylene bond formation depending on F/U_1 mole ratio 363
second step of resin synthesis 362–363
third step of resin synthesis 363
^{13}C NMR chemical changes occurring in UF resins on heat/stirring and room temperature storage treatments
effects of mild and intermediate level heat/stirring treatments 374–378
formation of methylene-di-urea type groups 380
formaldehyde emission test results of particleboards 381
irreversible methylene groups of UF resins 371
migration of types IIt/Iii hydroxymethyl groups 375
methylene group changes due to migration of IIt/IIi type hydroxymethyl groups 176
reaction of the second urea indicating some polymerization occurring 377
ripening times of UF resins 378
room temperature treatment effect of UF resins 378
treatment procedures 374–382
^{13}C NMR spectroscopic analysis method, introduction 351–356
applied magnetic field strength 351
CFT-20 NMR spectrometer 359, 360

chemical shifts 352
excitation (resonance) 352
magnetic spin population 351
nuclear Overhouser (nOe) effect 189, 355
precautions in reading ^{13}C NMR spectra 356
proton-decoupling 355
relaxation of excited nuclear spins 352
repeat scans and pulse-delays 354
T_1 relaxation time 355
T_2 relaxation time 354
Techmag 360 NMR spectrometer 359, 367, 449
tetramethylsilane (TMS) 354, 357
effects of mild heating/stirring treatments on UF resin synthesized with three different F/U_1 mole ratios 382–388
bond group changes occurring in the heating/stirring treatments by ^{13}C NMR 383
polymer chain branching increases with F/U_1 mole ratios 387
resin F/U_1 mole ratio effects on resin chemical structures 384–388
resin syntheses and heating/stirring treatment procedures 382–383
resin viscosity changes occurring in the heating/stirring treatments 384
effects of mild heat/stirring followed by room temperature storage treatments on UF resins made with three different F/U_1 mole ratios 397–418
bond group content changes by ^{13}C NMR analysis 399
bond group content changes occurring in graphic form 400
collection of treatment data of resin samples 398
formation of methylene-di-urea type insoluble compounds 403
formation of mono-substituted urea groups bonded to UF polymer 403
F/U_1 mole ratio and heating/storage treatment effects on PB properties 415–418
F/U_1 mole ratio and heating treatment effects on PB properties 409–411
F/U_1 mole ratio and posttreatment effects on resin polymeric structures 409–411
mechanisms proposed to formation of methylene-di-urea in alkaline pH 406–407
viscosity/turbidity changes of UF resins during storage 402
effects of room temperature storage treatments on UF resin synthesized with three different F/U_1 mole ratios 388–397
application of results to industry 396–397
bond group content changes by ^{13}C NMR analysis 390
bond group content changes occurring in graphic form 390–391
formaldehyde emissions of PBs, effects of F/U_1 ratio/storage times 394–396
methylene group increases indicating MW increases during treatment 392
optimum ripening times for UF resins 394
performances of PBs bonded with UF resins with four different storage days 393
room temperature treatment procedures (50 days) 389
viscosity/turbidity changes of UF resins during storage 389–392
urea-melamine-formaldehyde (UMF) resins 66–68
definition of 66
different addition points of melamine in resin synthesis 67–68
handling properties of UMF resins
resultant beneficial effects of melamine incorporation 67

urea-melamine-formaldehyde (UMF) resins modified with 6–12% M, syntheses of 420–432
  ^{13}C NMR analysis results of intermediates and UMF resins 425
  ^{13}C NMR spectra typical UMF resins with and without the addition of second urea 426
  ^{13}C NMR spectrum of an MF resin stabilized by urea addition 427
  different (four) M addition points in resin syntheses and resin properties 422, 424, 425
  methylene and methylene-ether group ratios of UF and UMF resins 432
  particleboard test results for UMF resins at an F/(U+M) mole ratio of 1.15 443–445
  reaction mechanisms of M and F of forming methylene or methylene-ether bonds 426
  reaction of melamine with formaldehyde, optimum pH 420, 428
  reaction of urea with formaldehyde, optimum pH 421
  reverse hydroxymethylation reactions of melamine 428
  syntheses of UF and MF intermediates of different extents polymerization 424
  syntheses of various resins of different extents polymerization of UF components 423
  viscosity increases of various UMF resins made with 2–12% melamine levels 430
urethane bonds 219, 223, 224, 326
urethane polymers 219, 321

V

van der Waals bonds 292, 294, 295
vapor pressure osmometry (VPO) 13, 490, 496–499, 503–505, 508
varnishes 100, 220, 270, 326, 332, 333, 336–338, 341, 347
vinyl resins 324–325
viscosity index, definition of 13–15
viscosity of polymer solutions 13–16, 141–142
viscous response 20
vitrification of adhesive layer, definition of 32–33
vitrification point of UF and PF resins on DMA tests 33, 83, 152, 156, 206, 237, 435
volatility index of paraffin wax 112

W

wafer boards 134, 161, 239
wash-coat of lacquer to smooth the wood surface 345
water-borne coatings and coatings for less VOC emissions 334–336
wax in hot-melt adhesive formulation 266
weight-average (M_w) molecular weights, definition of 12
white fir and sycamore 77
wide-belt sander 342
wind sifting-type former in particleboard manufacture 97
wood adhesive bond tests and evaluation methods 302–316
  automatic boil test (ABT) and exterior exposure test results 313–314
  bending strength tests 307–309
  certification of durability grades 310–311
  cleavage test 306–307
  comparison results of wood adhesives by accelerated aging an exterior exposure 313–315
  common laboratory aging test methods 312
  cyclic delamination test procedure 312
  durability evaluation results 309–310
  effects of aging of wood before gluing 313–314
  in-situ adhesion testing 315
  non-destructive testing methods 315
  objectives and scope of quality control processes 309
  peel tests 307
  shear strength test 302–303
  tensile strength tests 303–306

testing and certification organizations 313
torsion shear test 303
wood adhesives 1, 20–38
adhesive property of polymer solids 20
cohesive property of adhesive polymer solids 20
cohesive strength measurement and viscous and elastic responses 20–21
curing of thermoplastic emulsion adhesives 36–37
curing of thermosetting adhesives at room temperature 35–36
degree of polymerization and molecular weight 12–13
dependence of viscosity on the concentration and molecular weight of polymers 15–17
dependence of viscosity on the shear rate (stirring rate) 14–15
elasticity of solid polymer materials 21–22
flow and wetting of adhesive layers and molecular adsorption 31–32
gelation of adhesive layer 32
methods of viscosity measurement of liquid materials 17
pH of polymer solutions and buffers 18–19
polymer solids level and specific gravity of polymer solutions 18
post-curing of adhesive layer 33
selection criteria of 23–24
spreading and penetration of adhesive layers in hot-pressing 30–31
temperature dependence of viscosity of polymer solutions 15
thermal and moisture expansion/contraction of adhesive layers 38
viscosity changes of adhesive layer before hot-pressing 30
viscosity decreases by moisture content changes 31
viscosity increases and curing of adhesive layer by polymerization reaction 31
viscosity of polymer solutions 13–14
vitrification of adhesive layer 32–33
volume contractions of adhesive layers upon curing 37–38
wood phase failure 284
work of adhesion 291

y

Young's modulus 21, 33